T0215705

Lecture Notes in Computer Science 12036

More information about this series at http://www.springer.com/series/7409

Joemon M. Jose · Emine Yilmaz ·
João Magalhães · Pablo Castells ·
Nicola Ferro · Mário J. Silva ·
Flávio Martins (Eds.)

Advances in Information Retrieval

42nd European Conference on IR Research, ECIR 2020
Lisbon, Portugal, April 14–17, 2020
Proceedings, Part II

 Springer

Editors
Joemon M. Jose (ID)
University of Glasgow
Glasgow, UK

Emine Yilmaz (ID)
University College London
London, UK

João Magalhães (ID)
Universidade NOVA de Lisboa
Lisbon, Portugal

Pablo Castells (ID)
Universidad Autónoma de Madrid
Madrid, Spain

Nicola Ferro (ID)
University of Padua
Padua, Italy

Mário J. Silva (ID)
Universidade de Lisboa
Lisbon, Portugal

Flávio Martins (ID)
Universidade NOVA de Lisboa
Lisbon, Portugal

ISSN 0302-9743   ISSN 1611-3349   (electronic)
Lecture Notes in Computer Science
ISBN 978-3-030-45441-8   ISBN 978-3-030-45442-5   (eBook)
https://doi.org/10.1007/978-3-030-45442-5

LNCS Sublibrary: SL3 – Information Systems and Applications, incl. Internet/Web, and HCI

This Springer imprint is published by the registered company Springer Nature Switzerland AG
The registered company address is: Gewerbestrasse 11, 6330 Cham, Switzerland

Preface

The 42nd European Conference on Information Retrieval (ECIR) was held on April 14–17, 2020, and brought together hundreds of researchers from all over the world. ECIR 2020 was to be held in Lisbon, Portugal, but due to the COVID-19 lockdown and travel restrictions enforced worldwide, the conference was held online. The conference was organized by Universidade NOVA de Lisboa, Portugal, and Universidad Autónoma de Madrid, Spain, in cooperation with the British Computer Society's Information Retrieval Specialist Group (BCS-IRSG). It was supported by the ACM Special Interest Group on Information Retrieval (ACM SIGIR), Bloomberg, Amazon, Salesforce, TextKernel, NTENT, Google, and Levi Strauss.

These proceedings contain the papers presented and the summaries of workshops and tutorials given during the conference. This year the ECIR 2020 program boasted a variety of novel works from contributors located all around the world and also provided a platform for information retrieval-related (IR) activities from the CLEF Initiative. In addition, a new collaboration was instated between BCS-IRSG and the *Information Retrieval Journal*, whereby selected papers from the journal were presented at the conference, and a selection of ECIR 2020 papers were invited to submit an extended version for publication in a special issue of the journal.

In total, 457 submissions were fielded across all tracks from 57 different countries – adding the papers submitted to workshops, ECIR 2020 broke the 500 submissions barrier. The final program included 55 full papers (26% acceptance rate), 46 short papers (28% acceptance rate), 10 demonstration papers (30% acceptance rate), 8 reproducibility papers (38% acceptance rate), and 12 invited CLEF papers. All submissions were peer reviewed by at least three international Program Committee members to ensure that only submissions of the highest quality were included in the final program. The acceptance decisions were further informed by discussions among the reviewers for each submitted paper, led by a senior Program Committee member. A call for reviewers was set forth aiming to strengthen and update the Program Committee, integrating and catching up with both new and accomplished reviewing workforce in the field.

The accepted papers cover the state of the art in IR: deep learning based information retrieval techniques, use of entities and knowledge graphs, recommender systems, retrieval methods, information extraction, question answering, topic and prediction models, multimedia retrieval, etc. As with tradition, the ECIR 2020 program has seen a high proportion of papers with students as first authors, as well as papers from a variety of universities, research institutes, and commercial organizations.

In addition to the papers, the program also included three keynotes, three tutorials, four workshops, a doctoral consortium, and an industry day. The first keynote was presented by this year's BCS IRSG Karen Sparck Jones Award winner, Chirag Shah, the second keynote was presented by Jamie Callan, and the third keynote by Joana Gonçalves de Sá. The tutorials covered a range of topics including entity repositories, similar-question retrieval, and geographic IR, while the workshops brought together participants around such areas as narrative extraction, bibliometric IR, algorithmic bias, and health IR. ECIR 2020 also featured a CLEF session to enable CLEF organizers to report on and promote their upcoming tracks. The program introduced a new activity where recently published papers from the *Information Retrieval Journal* were presented at the conference – a selection of nine papers was included in this track. Such links between related forums added to the success and diversity of ECIR and helped build bridges between communities. The Industry Day was held on the last conference day, bringing together academic researchers and industry, offering a mix of talks by industry leaders (including Farfetch, Doctrine, Microsoft, TigerGraph, eBay) and presentations of novel and innovative ideas from industry research.

The success of ECIR 2020 would not have been possible without all the help from the team of volunteers and reviewers. We wish to thank all our track chairs for coordinating the different tracks, along with the teams of meta-reviewers and reviewers who helped ensure the high quality of the program. Thanks are due to the demo chairs: Nuno Correia and Ana Freire; reproducibility track chairs: Edleno Moura and Pável Calado; doctoral consortium chairs: Stefan Rueger and Suzan Verberne; workshop chairs: Suzane Little and Sérgio Nunes; tutorial chairs: Laura Dietz and Allan Hanbury; industry day chairs: Vanessa Murdock and Bruno Martins; publicity chair: Carla Teixeira Lopes and Ricardo Campos; and sponsorship chair: Dyaa Albakour. We would like to thank our webmaster, Flávio Martins, and our local chairs, Rui Nóbrega and Filipa Peleja, along with all the student volunteers who helped to create an excellent online and offline experience for participants and attendees.

ECIR 2020 was sponsored by: Bloomberg, Amazon, Salesforce, Google, Textkernel, Ntent, Levi Strauss, Signal AI, SIGIR, and Springer. We thank them all for their support and contributions to the conference. Finally, we wish to thank all the authors, contributors, and participants in the conference.

April 2020

Joemon M. Jose
Emine Yilmaz
João Magalhães
Pablo Castells
Nicola Ferro
Mário J. Silva
Flávio Martins

Organization

General Chairs

João Magalhães — Universidade NOVA de Lisboa, Portugal
Christina Lioma — University of Copenhagen, Denmark
Pablo Castells — Universidad Autónonoma de Madrid, Spain

Program Chairs

Joemon M. Jose — University of Glasgow, UK
Emine Yilmaz — University College London, UK

Short Paper Chairs

Nicola Ferro — University of Padua, Italy
Mário J. Silva — Universidade de Lisboa, Portugal

Workshop Chairs

Suzanne Little — Dublin City University, Ireland
Sérgio Nunes — Universidade do Porto, Portugal

Tutorial Chairs

Laura Dietz — University of New Hampshire, USA
Allan Hanbury — Technische Universität Wien, Austria

Demo Chairs

Nuno Correia — Universidade NOVA de Lisboa, Portugal
Ana Freire — Universitat Pompeu Fabra, Spain

Industry Day Chairs

Vanessa Murdock — Amazon, USA
Bruno Martins — Universidade de Lisboa, Portugal

Proceedings Chair

Flávio Martins — Universidade NOVA de Lisboa, Portugal

Reproducibility Track Chairs

Edleno Moura	Universidade Federal de Minas Gerais, Brazil
Pável Calado	Universidade de Lisboa, Portugal

Doctoral Consortium Chairs

Stefan Rueger	KMI, The Open University, UK
Suzan Verberne	Leiden University, Netherlands

Test of Time Award Chair

Maristella Agosti	University of Padua, Italy

Best Paper Award Chair

Ben Carterette	Spotify, USA

Publicity Chairs

Carla Teixeira Lopes	Universidade do Porto, Portugal
Ricardo Campos	INESC-TEC and Instituto Politécnico Tomar, Portugal

Sponsorship Chairs

Dyaa Albakour	Signal AI, UK
João Magalhães	Universidade NOVA de Lisboa, Portugal

Local Organization

Rui Nóbrega (Local Chair)	Universidade NOVA de Lisboa, Portugal
Filipa Peleja (Local Chair)	Levi Strauss & Co. Europe, Belgium
Flávio Martins (Web Chair)	Universidade NOVA de Lisboa, Portugal
João Magalhães (Support)	Universidade NOVA de Lisboa, Portugal
Nuno Correia (Support)	Universidade NOVA de Lisboa, Portugal
David Semedo (Support)	Universidade NOVA de Lisboa, Portugal
Pablo Castells (Support)	Universidad Autónonoma de Madrid, Spain

Program Committee

Full-Paper Meta-Reviewers

Ioannis Arapakis	Telefonica Research, Spain
Krisztian Balog	University of Stavanger, Norway
Ben Carterette	Spotify, USA
Fabio Crestani	University of Lugano (USI), Switzerland
Bruce Croft	University of Massachusetts Amherst, USA

Nicola Ferro	University of Padua, Italy
Norbert Fuhr	University of Duisburg-Essen, Germany
Lorraine Goeuriot	Grenoble Alpes University, France
Julio Gonzalo	UNED, Spain
Cathal Gurrin	Dublin City University, Ireland
Morgan Harvey	Northumbria University, UK
Claudia Hauff	Delft University of Technology, The Netherlands
Grace Huiyang	Georgetown University, USA
Gareth Jones	Dublin City University, Ireland
Diane Kelly	University of Tennessee, USA
Liadh Kelly	Maynooth University, Ireland
Udo Kruschwitz	University of Regensburg, Germany
Oren Kurland	Technion – Israel Institute of Technology, Israel
David Losada	University of Santiago de Compostela, Spain
Massimo Melucci	University of Padua, Italy
Boughanem Mohand	IRIT - Université Paul Sabatier Toulouse 3, France
Yashar Moshfeghi	University of Strathclyde, UK
Henning Müller	HES-SO, Switzerland
Iadh Ounis	University of Glasgow, UK
Gabriella Pasi	Università degli Studi di Milano-Bicocca, Italy
Benjamin Piwowarski	CNRS and University Pierre et Marie Curie, France

Full-Paper Program Committee

Mohamed Abdel Maksoud	Codoma.tech Advanced Technologies, Egypt
Ahmed Abdelali	Research Administration, Qatar
Karam Abdulahhad	GESIS, Germany
Dirk Ahlers	NTNU, Norway
Qingyao Ai	University of Utah, USA
Mohammad Akbari	University College London, UK
Ahmet Aker	University of Duisburg Essen, Germany
Navot Akiva	Bar Ilan University, Israel
Mehwish Alam	FIZ Karlsruhe, AIFB Institute, KIT, Germany
M-Dyaa Albakour	Signal AI, UK
Mohammad Aliannejadi	University of Amsterdam, The Netherlands
Pegah Alizadeh	University of Caen Normandy, France
Giambattista Amati	Fondazione Ugo Bordoni, Italy
Linda Andersson	Vienna University of Technology, Austria
Hassina Aouidad	CERIST, Algiers
Avi Arampatzis	Democritus University of Thrace, Greece
Ioannis Arapakis	Telefonica Research, Spain
Jaime Arguello	University of North Carolina at Chapel Hill, USA
Mozhdeh Ariannezhad	University of Amsterdam, The Netherlands
Nihal Yağmur Aydın	GREYC, France
Ebrahim Bagheri	Ryerson University, Canada
Seyed-Ali Bahrainian	Idsia, Swiss AI Lab, Switzerland

Krisztian Balog	University of Stavanger, Norway
Alvaro Barreiro	University of A Coruña, Spain
Alberto Barrn-Cede	University of Bologna, Italy
Alejandro Bellogin	Universidad Autónoma de Madrid, Spain
Patrice Bellot	Aix-Marseille Université, CNRS, LSIS, France
Anis Benammar	Regim, Tunisia
Klaus Berberich	Saarbruecken University of Applied Sciences, Germany
Pablo Bermejo	Universidad de Castilla-La Mancha, Spain
Catherine Berrut	LIG, Université Joseph Fourier Grenoble I, France
Sumit Bhatia	IBM, India
Pierre Bonnet	CIRAD, France
Gloria Bordogna	National Research Council of Italy (CNR), Italy
Larbi Boubchir	University of Paris 8, France
Pavel Braslavski	Ural Federal University, Russia
David Brazier	Northumbria University, UK
Paul Buitelaar	National University of Ireland Galway, Ireland
Guillaume Cabanac	IRIT, Université Paul Sabatier Toulouse 3, France
Luis Adrián Cabrera-Diego	Edge Hill University, UK
Fidel Cacheda	Universidade da Coruña, Spain
Sylvie Calabretto	LIRIS, CNRS, France
Pável Calado	INESC-ID, Universidade de Lisboa, Portugal
Rodrigo Calumby	University of Feira de Santana, Brazil
Ricardo Campos	INESC-TEC and Instituto Politécnico Tomar, Portugal
Fazli Can	Bilkent University, Turkey
Ivàn Cantador	Universidad Autónoma de Madrid, Spain
Annalina Caputo	University College Dublin, Ireland
Cornelia Caragea	University of Illinois at Chicago, USA
Ben Carterette	Spotify, USA
Pablo Castells	Universidad Autónoma de Madrid, Spain
Long Chen	University of Glasgow, UK
Max Chevalier	IRIT, France
Adrian-Gabriel Chifu	Aix-Marseille Université, Université de Toulon, France
Manoj Chinnakotla	Microsoft, India
Malcolm Clark	University of the Highlands and Islands, UK
Vincent Claveau	IRISA, CNRS, France
Jeremie Clos	University of Nottingham, UK
Fabio Crestani	University of Lugano (USI), Switzerland
Bruce Croft	University of Massachusetts Amherst, USA
Alfredo Cuzzocrea	ICAR-CNR and University of Calabria, Italy
Arthur Câmara	Delft University of Technology, The Netherlands
Zhuyun Dai	Carnegie Mellon University, USA
Jeffery Dalton	University of Glasgow, UK
Tirthankar Dasgupta	Tata Consultancy Services, India
Hélène De Ribaupierre	Cardiff University, UK
Martine DeCock	University of Washington, USA

Yashar Deldjoo	Polytechnic University of Bari, Italy
Kostantinos Demertzis	Democritus University of Thrace, Greece
José Devezas	University of Porto, Portugal
Kuntal Dey	IBM Research Lab, India
Emanuele DiBuccio	University of Padua, Italy
Giorgio DiNunzio	University of Padua, Italy
Laura Dietz	University of New Hampshire, USA
Inês Domingues	IPO Porto and Universidade de Coimbra, Portugal
Pan Du	University of Montreal, Canada
Mateusz Dubiel	University of Strathclyde, UK
Carsten Eickhoff	Brown University, USA
Mehdi Elahi	Free University of Bozen - Bolzano, Italy
Tamer Elsayed	Qatar University, Qatar
Liana Ermakova	Université de Bretagne Occidentale, France
Jose Esquivel	Signal AI, UK
Michael Faerber	University of Freiburg, Germany
Hui Fang	University of Delaware, USA
Hossein Fani	University of New Brunswick, Canada
Paulo Fernandes	Roberts College, USA
Nicola Ferro	University of Padua, Italy
Mustansar Fiaz	Kyungpook National University, South Korea
Sebastien Fournier	LSIS, France
Christoph M. Friedrich	University of Applied Science and Arts Dortmund, Germany
Ingo Frommholz	University of Bedfordshire, UK
Norbert Fuhr	University of Duisburg-Essen, Germany
Luke Gallagher	RMIT University, Australia
Patrick Gallinari	LIP6, University of Paris 6, France
Shreyansh Gandhi	WalmartLabs, USA
Debasis Ganguly	IBM Research Lab, Ireland
Wei Gao	Victoria University of Wellington, New Zealand
Dario Garigliotti	University of Stavanger, Norway
Anastasia Giachanou	Universitat Politècnica de València, Spain
Giorgos Giannopoulos	IMSI Institute, Athena Research Center, Greece
Alessandro Giuliani	University of Cagliari, Italy
Lorraine Goeuriot	Grenoble Alpes University, France
Julio Gonzalo	UNED, Spain
Pawan Goyal	IIT Kharagpur, India
Michael Granitzer	University of Passau, Germany
Guillaume Gravier	CNRS, IRISA, France
Adrien Guille	ERIC Lyon 2, EA 3083, Université de Lyon, France
Shashank Gupta	IIIT, India
Rajeev Gupta	Microsoft, India
Cathal Gurrin	Dublin City University, Ireland
Matthias Hagen	Martin-Luther-Universität Halle-Wittenberg, Germany
Lei Han	The University of Queensland, Australia

Shuguang Han	Google, USA
Preben Hansen	Stockholm University, Sweden
Donna Harman	NIST, USA
Morgan Harvey	Northumbria University, UK
Helia Hashemi	University of Massachusetts Amherst, USA
Claudia Hauff	Delft University of Technology, The Netherlands
Jer Hayes	Accenture, Ireland
Ben He	University of Chinese Academy of Sciences, China
Nathalie Hernandez	IRIT, France
Andreas Hotho	University of Würzburg, Germany
Gilles Hubert	IRIT, France
Grace Hui Yang	Georgetown University, USA
Ali Hürriyetoğlu	Koc University, Turkey
Adrian Iftene	Alexandru Ioan Cuza University of Iaşi, Romania
Dmitry Ignatov	National Research University Higher School of Economics, Russia
Bogdan Ionescu	University Politehnica of Bucharest, Romania
Radutudor Ionescu	University of Bucharest, Romania
Amir H. Jadidinejad	University of Glasgow, UK
Shoaib Jameel	Kent University, UK
Adam Jatowt	Kyoto University, Japan
Shen Jialie	Queen's University, Belfast, UK
Jiepu Jiang	Virginia Tech, USA
Alexis Joly	Inria, France
Gareth Jones	Dublin City University, Ireland
Jaap Kamps	University of Amsterdam, The Netherlands
Nattiya Kanhabua	Kasikorn Business - Technology Group, Thailand
Jaana Kekäläinen	University of Tampere, Finland
Diane Kelly	University of Tennessee, USA
Liadh Kelly	Maynooth University, Ireland
Roman Kern	Graz University of Technology, Austria
Daniel Kershaw	Elsevier, UK
Julia Kiseleva	Microsoft Research AI, USA
Prasanna Lakshmi Kompalli	GRIET, India
Yiannis Kompatsiaris	ITI-CERTH, Greece
Panos Kostakos	University of Oulu, Finland
Ralf Krestel	HPI, University of Potsdam, Germany
Kriste Krstovski	University of Massachusetts Amherst, USA
Udo Kruschwitz	University of Regensburg, Germany
Vaibhav Kumar	Carnegie Mellon University, USA
Oren Kurland	Technion – Israel Institute of Technology, Israel
Saar Kuzi	University of Illinois at Urbana-Champaign, USA
Wang-Chien Lee	The Pennsylvania State University, USA
Teerapong Leelanupab	King Mongkut's Institute of Technology Ladkrabang, Thailand
Mark Levene	Birkbeck, University of London, UK

Nut Limsopatham	Amazon, USA
Chunbin Lin	Amazon, USA
Aldo Lipani	University College London, UK
Nedim Lipka	Adobe Research, USA
Fernando Loizides	Cardiff University, UK
David Losada	University of Santiago de Compostela, Spain
Natalia Loukachevitch	Moscow State University, Russia
Bernd Ludwig	University Regensburg, Germany
Mihai Lupu	Research Studios Austria, Austria
Sean Macavaney	Georgetown University, USA
Craig Macdonald	University of Glasgow, UK
Andrew Macfarlane	City University of London, UK
João Magalhães	Universidade NOVA de Lisboa, Portugal
Walid Magdy	The University of Edinburgh, UK
Marco Maggini	University of Siena, Italy
Shikha Maheshwari	Chitkara University, India
Maria Maistro	University of Copenhagen, Denmark
Antonio Mallia	University of Pisa, Italy
Thomas Mandl	University of Hildesheim, UK
Behrooz Mansouri	Rochester Institute of Technology, USA
Jiaxin Mao	Tsinghua University, China
Stefania Marrara	Consorzio C2T, Italy
Miguel Martinez-Alvarez	Signal AI, UK
Bruno Martins	INESC-ID, Universidade de Lisboa, Portugal
Flávio Martins	Universidade NOVA de Lisboa, Portugal
Fernando Martínez-Santiago	Universidad de Jaén, Spain
Yosi Mass	IBM Haifa Research Lab, Israel
Sérgio Matos	IEETA, Universidade de Aveiro, Portugal
Philipp Mayr	GESIS, Germany
Parth Mehta	IRSI, India
Edgar Meij	Bloomberg L.P., UK
Massimo Melucci	University of Padua, Italy
Marcelo Mendoza	Universidad Técnica Federico Santa María, Chile
Zaiqiao Meng	University of Glasgow, UK
Alessandro Micarelli	Roma Tre University, Italy
Dmitrijs Milajevs	Queen Mary University of London, UK
Malik Muhammad Saad Missen	Research Lab L3I, Université de la Rochelle, France
Boughanem Mohand	IRIT, Université Paul Sabatier Toulouse 3, France
Ludovic Moncla	LIRIS, CNRS, France
Felipe Moraes	Delft University of Technology, The Netherlands
Ajinkya More	Netflix, USA
Jose Moreno	IRIT, UPS, France
Yashar Moshfeghi	University of Strathclyde, UK
Josiane Mothe	IRIT, Université de Toulouse, France
André Mourão	Universidade NOVA de Lisboa, Portugal

Henning Müller	HES-SO, Switzerland
Franco Maria Nardini	ISTI-CNR, Italy
Rekabsaz Navid	Johannes Kepler University Linz, Austria
Wolfgang Nejdl	L3S and University of Hannover, Germany
Massimo Nicosia	Google, Switzerland
Jian-Yun Nie	University of Montreal, Canada
Qiang Ning	University of Illinois at Urbana-Champaign, USA
Andreas Nuernberger	Otto von Guericke University Magdeburg, Germany
Neil O'Hare	Yahoo Research, USA
Anais Ollagnier	University of Exeter, UK
Teresa Onorati	Universidad Carlos III de Madrid, Spain
Salvatore Orlando	Università Ca' Foscari Venezia, Italy
Iadh Ounis	University of Glasgow, UK
Mourad Oussalah	University of Oulu, Finland
Deepak P.	Queen's University Belfast, UK
Jiaul Paik	IIT Kharagpur, India
Joao Palotti	Qatar Computing Research Institute, Qatar
Girish Palshikar	Tata Research Development and Design Centre, India
Panagiotis Papadakos	FORTH-ICS, Greece
Javier Parapar	University of A Coruña, Spain
Gabriella Pasi	Università degli Studi di Milano-Bicocca, Italy
Arian Pasquali	University of Porto, Portugal
Bidyut Kr. Patra	National Institute of Technology, India
Virgil Pavlu	Northeastern University, USA
Pavel Pecina	Charles University in Prague, Czech Republic
Gustavo Penha	Delft University of Technology, The Netherlands
Avar Pentel	Tallinn University, Estonia
Raffaele Perego	ISTI-CNR, Italy
Benjamin Piwowarski	CNRS, Pierre et Marie Curie University, France
Animesh Prasad	Amazon Alexa, UK
Chen Qu	University of Massachusetts Amherst, USA
Hossein A. Rahmani	University of Zanjan, Iran
Pengjie Ren	University of Amsterdam, The Netherlands
Kamal Sarkar	Jadavpur University, India
Ramit Sawhney	Netaji Subhas Institute of Technology, India
Philipp Schaer	TH Köln – University of Applied Sciences, Germany
Fabrizio Sebastiani	Italian National Council of Research, Italy
Florence Sedes	IRIT, Université Paul Sabatier Toulouse 3, France
Giovanni Semeraro	University of Bari, Italy
Procheta Sen	Indian Statistical Institute, India
Armin Seyeditabari	UNC Charlotte, USA
Gautam Kishore Shahi	University of Duisburg-Essen, Germany
Mahsa Shahshahani	University of Amsterdam, The Netherlands
Azadeh Shakery	University of Tehran, Iran
Ritvik Shrivastava	Columbia University, USA
Manish Shrivastava	IIIT Hyderabad, India

Gianmaria Silvello	University of Padua, Italy
Laure Soulier	Sorbonne Université, UPMC-LIP6, France
Marc Spaniol	Université de Caen Normandie, France
Günther Specht	University of Innsbruck, Austria
Rene Spijker	Cochran Netherlands, The Netherlands
Efstathios Stamatatos	University of the Aegean, Greece
L. Venkata Subramaniam	IBM Research Lab, India
Hanna Suominen	The ANU, Australia
Pascale Sébillot	IRISA, France
Lynda Tamine	IRIT, France
Thibaut Thonet	Naver Labs Europe, France
Antonela Tommasel	ISISTAN Research Institute, CONICET-UNCPBA, Argentina
Nicola Tonellotto	University of Pisa, Italy
Alina Trifan	University of Aveiro, Portugal
Theodora Tsikrika	ITI-CERTH, Greece
Ferhan Ture	Comcast Labs, USA
Yannis Tzitzikas	University of Crete and FORTH-ICS, Greece
Md Zia Ullah	CNRS, France
Julián Urbano	Delft University of Technology, The Netherlands
Daniel Valcarce	Google, Switzerland
Sumithra Velupillai	Institute of Psychiatry, Psychology and Neuroscience, UK
Nadimpalli Venkata Ganapathi Raju	GRIET, India
Suzan Verberne	Leiden University, The Netherlands
Manisha Verma	Verizon Media, USA
Vishwa Vinay	Adobe Research, India
Marco Viviani	Università degli Studi di Milano-Bicocca, Italy
Duc-Thuan Vo	Ryerson University, Canada
Stefanos Vrochidis	Information Technologies Institute, Greece
Shuohang Wang	Singapore Management University, Singapore
Christa Womser-Hacker	University of Hildesheim, UK
Peilin Yang	Twitter Inc., USA
Tao Yang	University of California at Santa Barbara, USA
Andrew Yates	Max Planck Institute for Informatics, Germany
Hai-Tao Yu	University of Tsukuba, Japan
Fattane Zarrinkalam	Ferdowsi University, Iran
Guido Zuccon	The University of Queensland, Australia
Arjen de Vries	Radboud University, The Netherlands

Short-Paper Program Committee

Dirk Ahlers	NTNU, Norway
Mehwish Alam	FIZ Karlsruhe, AIFB Institute, KIT, Germany
Mohammad Aliannejadi	University of Amsterdam, The Netherlands

Giambattista Amati	Fondazione Ugo Bordoni, Italy
Giuseppe Amato	ISTI-CNR, Italy
Maurizio Atzori	University of Cagliari, Italy
Ebrahim Bagheri	Ryerson University, Canada
Alvaro Barreiro	University of A Coruña, Spain
Alberto Barrón-Cedeño	University of Bologna, Italy
Alejandro Bellogin	Universidad Autónoma de Madrid, Spain
Patrice Bellot	Aix-Marseille Université, CNRS, LSIS, France
Anis Benammar	Regim, Tunisia
Klaus Berberich	Saarbruecken University of Applied Sciences, Germany
Catherine Berrut	LIG, Université Joseph Fourier Grenoble I, France
Pierre Bonnet	CIRAD, France
Gloria Bordogna	National Research Council of Italy (CNR), Italy
Luis Adrián Cabrera-Diego	Edge Hill University, UK
Fidel Cacheda	Universidade da Coruña, Spain
Pável Calado	INESC-ID, Universidade de Lisboa, Portugal
Iván Cantador	Universidad Autónoma de Madrid, Spain
Michelangelo Ceci	Università degli Studi di Bari, Italy
Adrian-Gabriel Chifu	Aix-Marseille Université, Université de Toulon, France
Malcolm Clark	University of the Highlands and Islands, UK
Arthur Câmara	Delft University of Technology, The Netherlands
Zhuyun Dai	Carnegie Mellon University, USA
Emanuele Di Buccio	University of Padua, Italy
Giorgio Maria Di Nunzio	University of Padua, Italy
Dennis Dosso	University of Padua, Italy
Tamer Elsayed	Qatar University, Qatar
Liana Ermakova	Université de Bretagne Occidentale, France
Andrea Esuli	Istituto di Scienza e Tecnologie dell'Informazione, Italy
Fabrizio Falchi	ISTI-CNR, Italy
Norbert Fuhr	University of Duisburg-Essen, Germany
Debasis Ganguly	IBM Research Lab, Ireland
Dario Garigliotti	University of Stavanger, Norway
Anastasia Giachanou	Universitat Politècnica de València, Spain
Giorgos Giannopoulos	IMSI Institute, Athena Research Center, Greece
Lorraine Goeuriot	Grenoble Alpes University, France
Adrien Guille	ERIC Lyon 2, Université de Lyon, France
Matthias Hagen	Martin-Luther-Universität Halle-Wittenberg, Germany
Donna Harman	NIST, USA
Helia Hashemi	University of Massachusetts Amherst, USA
Adrian Iftene	Alexandru Ioan Cuza University of Iaşi, Romania
Bogdan Ionescu	University Politehnica of Bucharest, Romania
Amir Jadidinejad	University of Glasgow, UK
Adam Jatowt	Kyoto University, Japan
Jaap Kamps	University of Amsterdam, The Netherlands

Mat Kelly	Drexel University, USA
Diane Kelly	University of Tennessee, USA
Daniel Kershaw	Elsevier, UK
Ioannis Kompatsiaris	ITI-CERTH, Greece
Kriste Krstovski	University of Massachusetts Amherst, USA
Vaibhav Kumar	Carnegie Mellon University, USA
Aldo Lipani	University College London, UK
Natalia Loukachevitch	Moscow State University, Russia
Claudio Lucchese	Ca' Foscari University of Venice, Italy
Mihai Lupu	Research Studios Austria, Austria
Andrew Macfarlane	City, University of London, UK
Marco Maggini	University of Siena, Italy
Maria Maistro	University of Copenhagen, Denmark
Antonio Mallia	University of Pisa, Italy
Jiaxin Mao	Tsinghua University, China
Stefano Marchesin	University of Padua, Italy
Bruno Martins	INESC-ID, Universidade de Lisboa, Portugal
Philipp Mayr	GESIS, Germany
Graham Mcdonald	University of Glasgow, UK
Edgar Meij	Bloomberg L.P., UK
Zaiqiao Meng	University of Glasgow, UK
Franco Maria Nardini	ISTI-CNR, Italy
Jian-Yun Nie	University of Montreal, Canada
Andreas Nuernberger	Otto von Guericke University Magdeburg, Germany
Sérgio Nunes	University of Porto, Portugal
Neil O'Hare	Yahoo Research, USA
Anais Ollagnier	University of Exeter, UK
Joao Palotti	Qatar Computing Research Institute, Qatar
Javier Parapar	University of A Coruña, Spain
Pavel Pecina	Charles University in Prague, Czech Republic
Raffaele Perego	ISTI-CNR, Italy
Martin Potthast	Leipzig University, Germany
Pengjie Ren	University of Amsterdam, The Netherlands
Paolo Rosso	Universitat Politècnica de València, Spain
Pasquale Savino	ISTI-CNR, Italy
Philipp Schaer	TH Köln University of Applied Sciences, Germany
Azadeh Shakery	University of Tehran, Iran
Gianmaria Silvello	University of Padua, Italy
Laure Soulier	Sorbonne Université, UPMC-LIP6, France
Marc Spaniol	Université de Caen Normandie, France
Hanna Suominen	The ANU, Australia
Lynda Tamine	IRIT, France
Thibaut Thonet	Naver Labs Europe, France
Nicola Tonellotto	University of Pisa, Italy
Theodora Tsikrika	ITI-CERTH, Greece
Yannis Tzitzikas	University of Crete and FORTH-ICS, Greece

Md Zia Ullah	CNRS, France
Julián Urbano	Delft University of Technology, The Netherlands
Gaurav Verma	Adobe Research, India
Marco Viviani	Università degli Studi di Milano-Bicocca, Italy
Christa Womser-Hacker	University of Hildesheim, UK
Yikun Xian	Rutgers University, USA
Andrew Yates	Max Planck Institute for Informatics, Germany
Justin Zobel	The University of Melbourne, Canada
Guido Zuccon	The University of Queensland, Australia
Arjen de Vries	Radboud University, The Netherlands

Reproducibility Track Reviewers

Jalal Alowibdi	University of Jeddah, Saudi Arabia
Leandro Balby Marinho	Federal University of Campina Grande, Brazil
José Borbinha	INESC-ID, Universidade de Lisboa, Portugal
Andre Carvalho	Universidade Federal do Amazonas, Brazil
André Carvalho	Universidade de Lisboa, Portugal
Edgar Chaves	CICESE, Mexico
Thierson Couto Rosa	Universidade Federal de Goiás, Brazil
Antonio Fariña	University of A Coruña, Spain
Juan M. Fernández-Luna	University of Granada, Spain
Marcos Goncalves	Federal University of Minas Gerais, Brazil
Xiaodong Liu	Microsoft, USA
Maria Maistro	University of Copenhagen, Denmark
David Matos	INESC-ID, Universidade de Lisboa, Portugal
Sérgio Matos	IEETA, Universidade de Aveiro, Portugal
Ajinkya More	Netflix, USA
Viviane P. Moreira	Universidade Federal do Rio Grande Do Sul, Brazil
Wolfgang Nejdl	L3S and University of Hannover, Germany
Özlem Özgöbek	NTNU, Norway
Altigran S. Da Silva	Universidade Federal do Amazonas, Brazil
Mahsa Shahshahani	University of Amsterdam, The Netherlands
Fei Sun	Alibaba Group, China
Ricardo Torres	NTNU, Norway
Alina Trifan	University of Aveiro, Portugal
Guido Zuccon	The University of Queensland, Australia

Demonstration Reviewers

Ahmed Abdelali	Research Administration, Qatar
Qingyao Ai	University of Utah, USA
M-Dyaa Albakour	Signal AI, UK
Diogo Cabral	ITI, LARSyS, Universidade de Lisboa, Portugal
Sylvie Calabretto	LIRIS, CNRS, France
Long Chen	University of Glasgow, UK

Manoj Chinnakotla	Microsoft, India
Alfredo Cuzzocrea	ICAR-CNR and University of Calabria, Italy
Yashar Deldjoo	Polytechnic University of Bari, Italy
Kuntal Dey	IBM India Research Lab, India
Jose Alberto Esquivel	Signal AI, UK
Hossein Fani	University of New Brunswick, Canada
Manuel Fonseca	Universidade de Lisboa, Portugal
Ingo Frommholz	University of Bedfordshire, UK
Michael Färber	University of Freiburg, Germany
Rui Nóbrega	Universidade NOVA de Lisboa, Portugal
时 冯	Northeastern University, China

Doctoral Consortium Reviewers

Carsten Eickhoff	Brown University, USA
Norbert Fuhr	University of Duisburg-Essen, Germany
Claudia Hauff	Delft University of Technology, The Netherlands
Gareth Jones	Dublin City University, Ireland
Udo Kruschwitz	University of Regensburg, Germany
Haiming Liu	University of Bedfordshire, UK
Philipp Mayr	GESIS, Germany
Josiane Mothe	IRIT, Université de Toulouse, France
Henning Müller	HES-SO, Switzerland

CLEF Track Reviewers

Giorgio Maria Di Nunzio	University of Padua, Italy
Nicola Ferro	University of Padua, Italy
Norbert Fuhr	University of Duisburg-Essen, Germany
Lorraine Goeuriot	University Grenoble Alpes, France
Donna Harman	NIST, USA
Bogdan Ionescu	University Politehnica of Bucharest, Romania
Mihai Lupu	Research Studios Austria, Austria
Maria Maistro	University of Copenhagen, Denmark
Jian-Yun Nie	University of Montreal, Canada
Aurélie Névéol	LIMSI, CNRS, Université Paris-Saclay, France
Raffaele Perego	ISTI-CNR, Italy
Martin Potthast	Leipzig University, Germany
Andreas Rauber	Vienna University of Technology, Austria
Paolo Rosso	Universitat Politècnica de València, Spain
Fabrizio Sebastiani	Italian National Council of Research, Italy
Laure Soulier	Sorbonne Université, UPMC-LIP6, France
Ellen Voorhees	NIST, USA

Additional Reviewers

Aditya Chandrasekar
Afraa Ahmad Alyosef
Alakananda Vempala
Alberto Purpura
Alessandra T. Cignarella
Alfonso Landin
Amir Jadidinejad
Amit Kumar Jaiswal
Ana Sabina Uban
Anastasia Moumtzidou
Andrea Iovine
Angelo Impedovo
Anna Nguyen
Behrooz Omidvar-Tehrani
Benjamin Murauer
Bilal Ghanem
Bishal Santra
Boteanu Bogdan Andrei
Cagri Toraman
Cataldo Musto
Charles Jochim
Christophe Rodrigues
Claudio Biancalana
Claudio Vairo
Daniel Campbell
Daniel Zoller
Dario Del Fante
David Otero
Debanjan Mahata
Despoina Chatzakou
Diana Nurbakova
Dilek Küçük
Dimitrios Effrosynidis
Disen Wang
Elisabeth Fischer
Emanuele Pio Barracchia
Eugene Yang
Fabian Hoppe
Fabio Carrara
Fang He

Fatima Haouari
Fedelucio Narducci
Felice Antonio Merra
Genet Asefa Gesese
Ghazal Fazelnia
Giulio Ermanno Pibiri
Giuseppe Sansonetti
Graziella De Martino
Gretel Sarracén
Guglielmo Faggioli
Himanshu Sharma
Hui-Ju Hung
Ilias Gialampoukidis
Janek Bevendorff
Jean-Michel Renders
Jinjin Shao
Johannes Jurgovsky
Johannes Kiesel
Johannes Schwerdt
Jun Ho Shin
Jussi Karlgren
Kalyani Roy
Konstantin Kobs
Kristian Noullet
Kuang Lu
Leopoldo Melo
Liviu-Daniel Ştefan
Lucia Vadicamo
Mahdi Dehghan
Maik Fröbe
Malte Bonart
Mandy Neumann
Manoj Kilaru
Maram Hasanain
Marco Polignano
Marco Ponza
Marcus Thiel
Matti Wiegmann
Michael Kotzyba
Mihai Dogariu

Mihai Gabriel Constantin
Mucahid Kutlu
Nirmal Roy
Paolo Mignone
Polina Panicheva
Reem Suwaileh
Reynier Ortega Bueno
Richard Mccreadie
Rob Koeling
Roberto Trani
Russa Biswas
Satarupa Guha
Sayantan Polley
Sevil Çalışkan
Shahbaz Syed
Shahrzad Naseri
Shikib Mehri
Shiyu Ji
Silvia Corbara
Silvio Moreira
Siwei Liu
Stefano Souza
Suresh Kumar Kaswan
Symeon Papadopoulos
Symeon Symeonidis
Tao-Yang Fu
Tarek Saier
Thanassis Mavropoulos
Thierson Couto-Rosa
Timo Breuer
Ting Su
Vaibhav Kasturia
Vikas Raunak
Wei-Fan Chen
Xiaoqi Ren
Yagmur Gizem Cinar
Yash Kumar Lal
Zuohui Fu

Test of Time Award Committee

Catherine Berrut	LIG, Université Joseph Fourier Grenoble I, France
Paul Clough	University of Sheffield, UK
Fabio Crestani	University of Lugano (USI), Switzerland
Gareth Jones	Dublin City University, Ireland
Josiane Mothe	IRIT, Université de Toulouse, France
Stefan Rueger	KMI, The Open University, UK

Best Paper Award Committee

Nicola Ferro	University of Padua, Italy
Udo Kruschwitz	University of Regensburg, Germany
Gabriella Pasi	Università degli Studi di Milano-Bicocca, Italy
Grace Hui Yang	Georgetown University, USA

Diamond Sponsors

Bloomberg Engineering

Gold Sponsors

Silver Sponsors

Bronze Sponsor

textkernel

Machine Intelligence for People and Jobs

Industry Impact Award Sponsor

SIGNAL

With Generous Support From

SIGIR
Special Interest Group
on Information Retrieval

Springer

Contents – Part II

CLEF Organizers Lab Track

Doctoral Consortium Papers

Workshops

Tutorials

Contents – Part I

Evaluation

Recommendation

Information Extraction

Deep Learning II

Retrieval

Multimedia

IR - General

Question Answering, Prediction, and Bias

Deep Learning IV

Abstracts of the IR Journal Papers

Reproducibility Papers

Knowledge Graph Entity Alignment with Graph Convolutional Networks: Lessons Learned

Max Berrendorf[1]([⊠]), Evgeniy Faerman[1], Valentyn Melnychuk[2],
Volker Tresp[1,3], and Thomas Seidl[1]

[1] Ludwig-Maximilians-Universität München, Munich, Germany
{berrendorf,faerman,seidl}@dbs.ifi.lmu.de
[2] Fraunhofer Institute for Integrated Circuits IIS, Erlangen, Germany
v.melnychuk@campus.lmu.de
[3] Siemens AG, Munich, Germany
volker.tresp@siemens.com

Abstract. In this work, we focus on the problem of entity alignment in Knowledge Graphs (KG) and we report on our experiences when applying a Graph Convolutional Network (GCN) based model for this task. Variants of GCN are used in multiple state-of-the-art approaches and therefore it is important to understand the specifics and limitations of GCN-based models. Despite serious efforts, we were not able to fully reproduce the results from the original paper and after a thorough audit of the code provided by authors, we concluded, that their implementation is different from the architecture described in the paper. In addition, several tricks are required to make the model work and some of them are not very intuitive. We provide an extensive ablation study to quantify the effects these tricks and changes of architecture have on final performance. Furthermore, we examine current evaluation approaches and systematize available benchmark datasets. We believe that people interested in KG matching might profit from our work, as well as novices entering the field. (Code: https://github.com/Valentyn1997/kg-alignment-lessons-learned).

1 Introduction

The success of information retrieval in a given task critically depends on the quality of the underlying data. Another issue is that in many domains knowledge bases are spread across various data sources [14] and it is crucial to be able to combine information from different sources. In this work, we focus on knowledge bases in the form of Knowledge Graphs (KGs), which are particularly suited for information retrieval [17]. Joining information from different KGs is non-trivial, as there is no unified schema or vocabulary. The goal of the entity alignment task is to overcome this problem by *learning* a matching between entities in different KGs. In the typical setting some of the alignments are known in advance (seed alignments) and the task is therefore supervised. More formally, we are given

© Springer Nature Switzerland AG 2020
J. M. Jose et al. (Eds.): ECIR 2020, LNCS 12036, pp. 3–11, 2020.
https://doi.org/10.1007/978-3-030-45442-5_1

graphs $G_L = (V_L, E_L)$ and $G_R = (V_R, E_R)$ with a seed alignment $A = (l_i, r_i)_i \subseteq V_L \times V_R$. It is commonly assumed that an entity $v \in V_L$ can match at most one entity $v' \in V_R$. Thus the goal is to infer alignments for the remaining nodes only.

Graph Convolutional Networks (GCN) [7,9], which have been recently become increasingly popular, are at the core of state-of-the-art methods for entity alignments in KGs [3,6,22,24,27]. In this paper, we thoroughly analyze one of the first GCN-based entity alignment methods, GCN-Align [22]. Since the other methods we are studying can be considered as extensions of this first paper and have a similar architecture, our goal is to understand the importance of its individual components and architecture choices.In summary, our contribution is as follows:

1. We investigate the reproducibility of the published results of a recent GCN-based method for entity alignment and uncover differences between the method's description in the paper and the authors' implementation.
2. We perform an ablation study to demonstrate the individual components' contribution.
3. We apply the method to numerous additional datasets of different sizes to investigate the consistency of results across datasets.

2   Related Work

In this section, we review previous work on entity alignment for Knowledge Graphs and revisit the current evaluation process. We believe that this is useful for practitioners, since we discover some pitfalls, especially when implementing evaluation scores and selecting datasets for comparison. An overview of methods, datasets and metrics is provided in Table 1.

Table 1. Overview of related work in the field of entity alignment for knowledge graphs with their used datasets and metrics.

Method	Datasets	Metrics	Code
MTransE [5]	WK3l-15K, WK3l-120K, CN3l	H@10(, MR)	yes
IPTransE [26]	DFB-{1,2,3}	H@{1,10}, MR	yes
JAPE [18]	DBP15K(JAPE)	H@{1,10,50}, MR	yes
KDCoE [4]	WK3l-60K	H@{1,10}, MR	yes
BootEA [19]	DBP15K(JAPE), DWY100K	H@{1,10}, MRR	yes
SEA [15]	WK3l-15K, WK3l-120K	H@{1,5,10}, MRR	yes
MultiKE [25]	DWY100K	H@{1,10}, MR, MRR	yes
AttrE [20]	DBP-LGD,DBP-GEO,DBP-YAGO	H@{1,10}, MR	yes
RSN [8]	custom DBP15K, DWY100K	H@{1,10}, MRR	yes
GCN-Align [22]	DBP15K(JAPE)	H@{1,10,50}	yes
CL-GNN [24]	DBP15K(JAPE)	H@{1,10}	yes
MuGNN [3]	DBP15K(JAPE), DWY100K	H@{1,10}, MRR	yes
NAEA [27]	DBP15K(JAPE), DWY100K	H@{1,10}, MRR	no

Methods. While the problem of entity alignments in Knowledge Graphs has been tackled historically by researching vocabularies which are as broad as possible, and establish them as a standard, recent approaches take a more data-driven view. Early methods use classical knowledge graph link prediction models such as TransE [2] to embed the entities of the individual knowledge graphs using an intra-KG link prediction loss, and differ in what they do with the aligned entities. For instance, MTransE [5] learns a linear transformation between the embedding spaces of the individual graphs using an L_2-loss. BootEA [19] adopts a bootstrapping approach and iteratively labels the most likely alignments to utilize them for further training. In addition to the alignment loss, embeddings of aligned entities are swapped regularly to calibrate embedding spaces against each other. SEA [15] learns a mapping between embedding spaces in both directions and additionally adds a cycle-consistency loss. Thereby, the distance between the original embedding of an entity, and the result of translating this embedding to the opposite space and back again, is penalized. IPTransE [26] embeds both KGs into the same embedding space and uses a margin-based loss to enforce the embeddings of aligned entities to become similar. RSN [8] generates sequences using different types of random walks which can move between graphs when visiting aligned entities. The generated sequences are feed to an adapted recurrent model. JAPE [18], KDCoE [4], MultiKE [25] and AttrE [20] utilize attributes available for some entities and additional information like the names of entities and relationships. Graph Convolutional Network (GCN) based models [3, 6, 22, 24, 27][1] have in common that they use GCN to create node representations by aggregating node representations together with representations of their neighbors. Most of GCN approaches do not distinguish between different relations and either consider all neighbors equally [6, 22, 24] or use attention [3] to weight the representations of the neighbors for the aggregation.

Datasets. The datasets used by entity alignments methods are generally based on large-scale open-source data sources such as DBPedia [1], YAGO [13], or Wikidata [23]. While there is the DWY-100K dataset, which comprises 100 K aligned entities across the three aforementioned individual knowledge graphs, most of the datasets, such as DBP15K, or WK3l are generated from a single multi-lingual database. There, subsets are formed according to a specific language, and entities which are linked across languages are used as alignments. A detailed description of most-used datasets can be found in Table 2.

As an interesting observation we found out that all papers which evaluate on DBP15, do not evaluate on the full DBP15K dataset[2] (which we refer to as *DBP15K (full)*), but rather use a smaller subset provided by the authors of JAPE [18] in their GitHub repository[3], which we call *DBP15K (JAPE)*. The smaller subsets were created by selecting a portion of entities (around 20K of 100K) which are popular,

---

[1] While [27] does not state explicitly that they use GCNs, their model is very similar to [21].

[2] Available at http://ws.nju.edu.cn/jape/.

[3] https://github.com/nju-websoft/JAPE/blob/master/data/dbp15k.tar.gz.

i.e. appear in many triples as head or tail. The number of aligned entities stays the same (15K). As [18] only reports the dataset statistics of the larger dataset, and does not mention the reduction of the dataset, subsequent papers also report the statistics of the larger dataset, although experiments use the smaller variant [3,18,19,22,26]. As the metrics rely on absolute ranks, the numbers are better than on the full dataset (cf. Table 3).

Scores. It is common practice to only consider the entities being part of the test alignment as potential matching candidates. Although we argue that ignoring entities exclusive to a single graph as potential candidates does not reflect well

Table 2. Overview of used datasets with their sizes in the number of triples (edges), entities (nodes), relations (different edge types) and alignments. For WK3l, the alignment is provided as a directed mapping on a entity level. However, there are additional triple alignments. Following a common practice as e.g. [15] we can assume that an alignment should be symmetric and that we can extract entity alignments from the triple alignments. Thereby, we obtain the number of alignments given in brackets.

Dataset	Subset	Graph	Triples	Entities	Relations	Alignments
DBP15K (full)	fr-en	fr	192, 191	66, 858	1, 379	15,000
		en	278, 590	105, 889	2, 209	
	ja-en	ja	164, 373	65, 744	2, 043	15,000
		en	233, 319	95, 680	2, 096	
	zh-en	zh	153, 929	66, 469	2, 830	15,000
		en	237, 674	98, 125	2, 317	
DBP15K (JAPE)	fr-en	fr	105, 998	19, 661	903	15,000
		en	115, 722	19, 993	1, 208	
	ja-en	ja	77, 214	19, 814	1, 299	15,000
		en	93, 484	19, 780	1, 153	
	zh-en	zh	70, 414	19, 388	1, 701	15,000
		en	95, 142	19, 572	1, 323	
WK3l-15K	en-de	en	209, 041	15, 127	1, 841	1,289 (10,383)
		de	144, 244	14, 603	596	1,140 (10,383)
	en-fr	en	203, 356	15, 170	2, 228	2,498 (8,024)
		fr	169, 329	15, 393	2, 422	3,812 (8,024)
WK3l-120K	en-de	en	624, 659	67, 650	2, 393	6,173 (50,280)
		de	389, 554	61, 942	861	4,820 (50,280)
	en-fr	en	1, 375, 406	119, 749	3, 109	36,749 (87,836)
		fr	760, 497	118, 592	2, 336	36,013 (87,836)
DWY-100K	dbp-wd	dbp	463, 294	100, 000	330	100,000
		wd	448, 774	100, 000	220	
	dbp-yg	dbp	428, 952	100, 000	302	100,000
		yg	502, 563	100, 000	31	

the use-case situation[4], we follow this evaluation scheme for our experiments to maintain comparability.

3   Method

GCN-Align [22] is a GCN-based approach to embed all entities from both graphs into a common embedding space. Each entity i is associated with *structural* features $h_i \in \mathbb{R}^d$, which are initialized randomly and updated during training. The features of all entities in a single graph are combined to the feature matrix H. Subsequently, a two-layer GCN is applied. A single GCN layer is described by $H^{(i+1)} = \sigma\left(\hat{D}^{-\frac{1}{2}}\hat{A}\hat{D}^{-\frac{1}{2}}H^{(i)}W^{(i)}\right)$ with $\hat{A} = A + I$, where A is the adjacency matrix, and $\hat{D}_{ii} = \sum_{j=1}^{n}\hat{A}_{ij}$ is the diagonal node degree matrix. The input of the first layer is set to $H^{(0)} = H$, and σ is non-linear activation function, chosen as ReLU. The output of the last layer is considered as the structural representation, denoted by $s_i = H_i^{(2)} \in \mathbb{R}^d$. Both graphs are equipped with their own node features, but the convolution weights $W^{(i)}$ are shared across the graphs.

The adjacency matrix is derived from the knowledge graph by first computing a score, called *functionality*, for each relation as the ratio between the number of different entities which occur as head, and the number of triples in which the relation occurs α_r. Analogously, the *inverse functionality* α_r' is obtained by replacing the nominator by the number of different tail entities. The final adjacency matrix is obtained as $A_{ij} = \sum_{(e_i,r,e_j)} \alpha_r' + \sum_{(e_j,r,e_j)} \alpha_r$. Note, that analogously to structural features GCN-Align is able to process the attributes and integrate them in final representation. However, since attributes have little effect on final score, and to be consistent with other GNN models, here we focus only on structural representations.

Implementation Specifics. The code[5] provided by the authors differs in a few aspects from the method described in the paper. First, when computing the adjacency matrix, $fun(r)$ and $ifun(f)$ are set to at least 0.3. S, the node embeddings are initialized with values drawn from a normal distribution with variance $n^{-1/2}$, where n is the number of nodes[6]. Additionally, the node features are always normalised to unit Euclidean length before passing them into the network. Finally, there are no convolution weights. This means that the whole GCN does not contain a single parameter, but is just a fixed function on the learned node embeddings.

---

[4] In the typical scenario it is not known in advance, which entities have matching and which not. Therefore the resulting score is too optimistic. We advocate to investigate this shortcoming further in future work.

[5] https://github.com/1049451037/GCN-Align.

[6] We could not find any justification for that in literature.

4   Experiments

In initial experiments we were able to reproduce the results reported in the paper using the implementation provided by the authors. Moreover, we are able to reproduce the results using our own implementation, and settings adjusted to the authors' code. In addition, we replaced the adjacency matrix based on functionality and inverse functionality by a simpler version, where $a_{ij} = \{(h, r, t) \in T \mid h = e_i, t = e_j\}$. We additionally use $\hat{D}^{-1}\hat{A}$ instead of the symmetric normalization. In total, we see no difference in performance between our simplified adjacency matrix, and the authors' one. We identified two aspects which affect the model's performance: Not using convolutional weights, and normalizing the variance when initializing node embeddings. We provide empirical evidence for this finding across numerous datasets. Our results regarding Hits@1 (H@1) are summarised in Table 3.

Table 3. Ablation study on using convolution weights and different embedding initialisation.We fix using convolution weights and the variance for the normal distribution from which the embedding vectors are initialized and optimize the other hyperparameters according to validation H@1 (80/20% train-validation split) on *DBP15K (JAPE)* *zh-en* in a large-scale hyperparameter search, comprising 1,440 experiments, with the following grid: optim. $\in$ {Adam, SGD}, lr $\in$ {0.1, 0.5, 1, 10, 20}, #layers $\in$ {1, 2, 3}, #neg. samples $\in$ {5, 50, 100}, #epochs $\in$ {10, 500, 2000, 3000}. Hence, we obtain four sets of hyperparameters. For each dataset, we perform a smaller hyperparameter search to fine-tune LR, #epochs & #layers for each dataset (again 80/20 split) and evaluate the best models on the official test set with standard deviation computed across 5 runs.

Weights		No		Yes	
Variance. Emb. Init.		1	$n^{-1/2}$	1	$n^{-1/2}$
DBP15K (full)	fr-en	**31.51 ± 0.16**	27.64 ± 0.22	21.82 ± 0.39	16.73 ± 0.59
	ja-en	**33.26 ± 0.10**	29.06 ± 0.23	26.21 ± 0.33	20.78 ± 0.16
	zh-en	**31.15 ± 0.15**	22.55 ± 0.27	24.96 ± 0.71	18.85 ± 0.99
DBP15K (JAPE)	fr-en	**45.37 ± 0.13**	41.03 ± 0.13	35.36 ± 0.33	30.50 ± 0.38
	ja-en	**45.53 ± 0.18**	40.29 ± 0.09	35.81 ± 0.53	31.46 ± 0.15
	zh-en	**43.30 ± 0.12**	39.37 ± 0.20	33.61 ± 0.49	29.94 ± 0.35
DWY100K	wd	**58.50 ± 0.05**	54.07 ± 0.05	50.13 ± 0.11	38.85 ± 0.31
	yg	**72.82 ± 0.06**	67.06 ± 0.03	67.36 ± 0.10	60.67 ± 0.30
WK3l-120K	en-de	**10.10 ± 0.03**	9.17 ± 0.05	9.02 ± 0.17	6.75 ± 0.12
	en-fr	**8.28 ± 0.03**	7.38 ± 0.03	7.26 ± 0.11	5.07 ± 0.16
WK3l-15K	en-de	16.57 ± 0.12	14.41 ± 0.23	**17.43 ± 0.38**	12.66 ± 0.30
	en-fr	**17.07 ± 0.15**	16.16 ± 0.16	15.98 ± 0.16	12.41 ± 0.18

Node Embedding Initialization. Comparing the columns of Table 3 we can observe the influence of the node embedding initialization. Using the settings from the authors' code, i.e. not using weights, a choosing a variance of $n^{-1/2}$ actually results in inferior performance in terms of H@1, as compared to use a standard normal distribution. These findings are consistent across datasets.

Convolution Weights. The first column of Table 3 corresponds to the weight usage and initialization settings used in the code for GCN-Align. We achieve slightly better results than published in [22], which we attribute to a more exhaustive parameter search. Interestingly, all best configurations use Adam optimizer instead of SGD. Adding convolution weights degrades the performance across all datasets and subsets thereof but one as witnessed by comparing the first two columns with the last two columns.

5   Conclusion

In this work, we reported our experiences when implementing the Knowledge Graph alignment method GCN-Align. We pointed at important differences between the model described in the paper and the actual implementation and quantified their effects in the ablation study. For future work, we plan to include other methods for entity alignments in our framework.

Acknowledgements. This work has been funded by the German Federal Ministry of Education and Research (BMBF) under Grant No. 01IS18036A and by the Bavarian Ministry for Economic Affairs, Infrastructure, Transport and Technology through the Center for Analytics-Data-Applications (ADA-Center) within the framework of "BAYERN DIGITAL II". The authors of this work take full responsibilities for its content.

References

1. Auer, S., Bizer, C., Kobilarov, G., Lehmann, J., Cyganiak, R., Ives, Z.: DBpedia: a nucleus for a web of open data. In: Aberer, K., et al. (eds.) ASWC/ISWC -2007. LNCS, vol. 4825, pp. 722–735. Springer, Heidelberg (2007). https://doi.org/10.1007/978-3-540-76298-0_52
2. Bordes, A., Usunier, N., García-Durán, A., Weston, J., Yakhnenko, O.: Translating embeddings for modeling multi-relational data. In: Burges, C.J.C., Bottou, L., Ghahramani, Z., Weinberger, K.Q. (eds.) Advances in Neural Information Processing Systems 26: 27th Annual Conference on Neural Information Processing Systems 2013. Proceedings of a Meeting, Lake Tahoe, Nevada, United States, 5–8 December 2013. pp. 2787–2795 (2013). http://papers.nips.cc/paper/5071-translating-embeddings-for-modeling-multi-relational-data
3. Cao, Y., Liu, Z., Li, C., Liu, Z., Li, J., Chua, T.: Multi-channel graph neural network for entity alignment. In: Korhonen et al. [10], pp. 1452–1461. https://www.aclweb.org/anthology/P19-1140/

4. Chen, M., Tian, Y., Chang, K., Skiena, S., Zaniolo, C.: Co-training embeddings of knowledge graphs and entity descriptions for cross-lingual entity alignment. In: Lang [12], pp. 3998–4004. https://doi.org/10.24963/ijcai.2018/556
5. Chen, M., Tian, Y., Yang, M., Zaniolo, C.: Multilingual knowledge graph embeddings for cross-lingual knowledge alignment. In: Sierra [16], pp. 1511–1517. https://doi.org/10.24963/ijcai.2017/209
6. Fey, M., Lenssen, J.E., Morris, C., Masci, J., Kriege, N.M.: Deep graph matching consensus. In: International Conference on Learning Representations (2020). https://openreview.net/forum?id=HyeJf1HKvS
7. Gilmer, J., Schoenholz, S.S., Riley, P.F., Vinyals, O., Dahl, G.E.: Neural message passing for quantum chemistry. In: Proceedings of the 34th International Conference on Machine Learning-Volume 70, pp. 1263–1272. JMLR. org (2017)
8. Guo, L., Sun, Z., Hu, W.: Learning to exploit long-term relational dependencies in knowledge graphs. In: Chaudhuri, K., Salakhutdinov, R. (eds.) Proceedings of the 36th International Conference on Machine Learning (ICML 2019), Long Beach, California, USA, 9–15 June 2019. Proceedings of Machine Learning Research, vol. 97, pp. 2505–2514. PMLR (2019). http://proceedings.mlr.press/v97/guo19c.html
9. Kipf, T.N., Welling, M.: Semi-supervised classification with graph convolutional networks. arXiv preprint arXiv:1609.02907 (2016)
10. Korhonen, A., Traum, D.R., Màrquez, L. (eds.): Proceedings of the 57th Conference of the Association for Computational Linguistics (ACL 2019), Florence, Italy, 28 July – 2 August 2019, Volume 1: Long Papers. Association for Computational Linguistics (2019). https://www.aclweb.org/anthology/volumes/P19-1/
11. Kraus, S. (ed.): Proceedings of the Twenty-Eighth International Joint Conference on Artificial Intelligence (IJCAI 2019), Macao, China, 10–16 August 2019. ijcai.org (2019). https://doi.org/10.24963/ijcai.2019
12. Lang, J. (ed.): Proceedings of the Twenty-Seventh International Joint Conference on Artificial Intelligence (IJCAI 2018), Stockholm, Sweden, 13–19 July 2018. ijcai.org (2018). http://www.ijcai.org/proceedings/2018/
13. Mahdisoltani, F., Biega, J., Suchanek, F.M.: YAGO3: a knowledge base from multilingual wikipedias. In: Seventh Biennial Conference on Innovative Data Systems Research (CIDR 2015), Asilomar, CA, USA, 4–7 January 2015, Online Proceedings. www.cidrdb.org (2015). http://cidrdb.org/cidr2015/Papers/CIDR15_Paper1.pdf
14. Nickel, M., Murphy, K., Tresp, V., Gabrilovich, E.: A review of relational machine learning for knowledge graphs. Proc. IEEE 104(1), 11–33 (2015)
15. Pei, S., Yu, L., Hoehndorf, R., Zhang, X.: Semi-supervised entity alignment via knowledge graph embedding with awareness of degree difference. In: Liu, L., et al. (eds.) The World Wide Web Conference (WWW 2019), San Francisco, CA, USA, 13–17 May 2019, pp. 3130–3136. ACM (2019). https://doi.org/10.1145/3308558.3313646
16. Sierra, C. (ed.): Proceedings of the Twenty-Sixth International Joint Conference on Artificial Intelligence (IJCAI 2017), Melbourne, Australia, 19–25 August 2017. ijcai.org (2017). http://www.ijcai.org/Proceedings/2017/
17. Singhal, A.: Introducing the knowledge graph: things, not strings. Official Google Blog 5, (2012)
18. Sun, Z., Hu, W., Li, C.: Cross-lingual entity alignment via joint attribute-preserving embedding. In: d'Amato, C., et al. (eds.) ISWC 2017. LNCS, vol. 10587, pp. 628–644. Springer, Cham (2017). https://doi.org/10.1007/978-3-319-68288-4_37

19. Sun, Z., Hu, W., Zhang, Q., Qu, Y.: Bootstrapping entity alignment with knowledge graph embedding. In: Lang [12], pp. 4396–4402. https://doi.org/10.24963/ijcai.2018/611
20. Trisedya, B.D., Qi, J., Zhang, R.: Entity alignment between knowledge graphs using attribute embeddings. In: The Thirty-Third AAAI Conference on Artificial Intelligence (AAAI 2019), The Thirty-First Innovative Applications of Artificial Intelligence Conference, IAAI 2019, The Ninth AAAI Symposium on Educational Advances in Artificial Intelligence (EAAI 2019), Honolulu, Hawaii, USA, 27 January – 1 February 2019, pp. 297–304. AAAI Press (2019). https://aaai.org/ojs/index.php/AAAI/article/view/3798
21. Veličković, P., Cucurull, G., Casanova, A., Romero, A., Lio, P., Bengio, Y.: Graph attention networks. arXiv preprint arXiv:1710.10903 (2017)
22. Wang, Z., Lv, Q., Lan, X., Zhang, Y.: Cross-lingual knowledge graph alignment via graph convolutional networks. In: Riloff, E., Chiang, D., Hockenmaier, J., Tsujii, J. (eds.) Proceedings of the 2018 Conference on Empirical Methods in Natural Language Processing, Brussels, Belgium, 31 October – 4 November 2018, pp. 349–357. Association for Computational Linguistics (2018). https://www.aclweb.org/anthology/D18-1032/
23. Wikidata. https://www.wikidata.org/
24. Xu, K., et al.: Cross-lingual knowledge graph alignment via graph matching neural network. In: Korhonen et al. [10], pp. 3156–3161 (2019). https://www.aclweb.org/anthology/P19-1304/. arXiv preprint arXiv:1905.11605
25. Zhang, Q., Sun, Z., Hu, W., Chen, M., Guo, L., Qu, Y.: Multi-view knowledge graph embedding for entity alignment. In: Kraus [11], pp. 5429–5435. https://doi.org/10.24963/ijcai.2019/754
26. Zhu, H., Xie, R., Liu, Z., Sun, M.: Iterative entity alignment via joint knowledge embeddings. In: Sierra [17], pp. 4258–4264. https://doi.org/10.24963/ijcai.2017/595
27. Zhu, Q., Zhou, X., Wu, J., Tan, J., Guo, L.: Neighborhood-aware attentional representation for multilingual knowledge graphs. In: Kraus [11], pp. 1943–1949. https://doi.org/10.24963/ijcai.2019/269

The Effect of Content-Equivalent Near-Duplicates on the Evaluation of Search Engines

Maik Fröbe[1]([⊠]), Jan Philipp Bittner[1], Martin Potthast[2], and Matthias Hagen[1]

[1] Martin-Luther-Universität Halle-Wittenberg, Halle, Germany
maik.froebe@informatik.unihalle.de
[2] Leipzig University, Leipzig, Germany

Abstract. Current best practices for the evaluation of search engines do not take into account duplicate documents. Dependent on their prevalence, not discounting duplicates during evaluation artificially inflates performance scores, and, it penalizes those whose search systems diligently filter them. Although these negative effects have already been demonstrated a long time ago by Bernstein and Zobel [4], we find that this has failed to move the community. In this paper, we reproduce the aforementioned study and extend it to incorporate all TREC Terabyte, Web, and Core tracks. The worst-case penalty of having filtered duplicates in any of these tracks were losses between 8 and 53 ranks.

1 Introduction

Web crawls contain pages that are duplicates or near-duplicate of other pages [15]. Although there can be legitimate reasons for a web publisher to host pages that duplicate other publishers' pages, the user of a web search engine gains nothing from viewing basically the same search result twice or more while browsing search results. Therefore, web search engines typically identify duplicates, either at crawl time, at indexing time, or at retrieval time, in order to remove all but one of them from their search results, showing only the "best" version of a piece of content according to some selection criteria.

The fact that duplicate results are not "useful" to the users of a web search engine has not been overlooked: Treating results as irrelevant when they are found to be *content-equivalent* to a document the user has already seen, Bernstein and Zobel [4] applied this so-called "novelty principle" during their analysis of content-equivalent documents within the GOV collections. With respect to the TREC 2004 Terabyte Track, their research shows (1) that 16.6% of all relevant documents in submitted runs are content-equivalent, and (2) that the application of the novelty principle causes MAP scores to decrease by 20% on average. This situation is an obstacle to progress, since doing the right thing and filtering duplicates is penalized. Fifteen years have passed since the report by Bernstein and Zobel, and we are curious as to whether anything has changed: Are there still duplicate documents in commonly used benchmarks, and if so, how do they affect the evaluation of retrieval systems?

J. M. Jose et al. (Eds.): ECIR 2020, LNCS 12036, pp. 12–19, 2020.
https://doi.org/10.1007/978-3-030-45442-5_2

Corpus	Size		Equivalent	
	Gzip	Docs	Docs	Classes
GOV1	4.6 GB	1.2 m	6.54%	24,394
GOV2	81 GB	25.1 m	23.39%	794,889
ClueWeb09	4.0 TB	1.04 b	7.74%	49.2 m
ClueWeb12	4.6 TB	0.73 b	14.71%	39.1 m
NYT-AC	8.3 GB	1.8 m	2.11%	28,493
WaPo	1.6 GB	0.59 m	12.52%	28,471

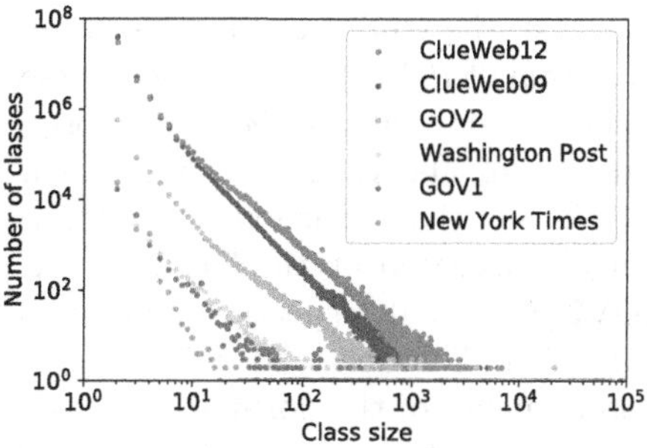

Fig. 1. Overview of the studied corpora (left), and the size of equivalence classes (right).

The two contributions of this paper are (1) a reproduction, and (2) an extension and generalization of the work of Bernstein and Zobel on the effects of content-equivalent documents on search engine evaluation. We independently confirm their findings on the TREC 2004 Terabyte Track, using our own reimplementation of their approach.[1] Thus validated, we go on to apply it to a selection of ad hoc retrieval tracks succeeding the one originally studied by Bernstein and Zobel: the Terabyte Track (2004–2006) [6,7,12], the Web Track (2009–2014) [8–11,13,14], and the recent Common Core Track (2017–2018) [1,2]. Applying the novelty principle causes changes in the evaluation scores of all shared tasks under consideration. These changes do not uniformly spread across participants. A participant who applies the novelty principle independently of others can loose up to 53 positions.

2 Corpus Analysis: Retrieval-Equivalent Documents

Following Bernstein and Zobel, we first analyzed the corpora employed in the shared tasks under consideration by grouping their documents into retrieval equivalence classes. Retrieval equivalence is determined using a fingerprint function based on selected elements of a standard indexing pipeline: the document string is lowercased, all HTML tags, punctuation, and stop words are replaced by a blank, all remaining words are stemmed, and all white space sequences are collapsed. The resulting string is fed to a cryptographic hash function, and the hash value is used as the document's fingerprint. We reimplement these steps using widespread open-source tools (e.g., Anserini [17]). Figure 1 shows the results of our analysis.

Terabyte Track. The Terabyte Track employed the GOV2 corpus, a crawl of web sites hosted under the .gov domain that took place in 2004 [7]. Bernstein and Zobel also analyzed its predecessor, the GOV1 corpus, which is why we do so as well. Although efforts were made to remove duplicates during crawling, this did not include retrieval-equivalent documents. As shown in the table in Fig. 1,

---

[1] Source code and resources: https://github.com/webis-de/trec-near-duplicates.

23.39% of the GOV2 documents are equivalent to at least one other document for a total of 794,889 equivalence classes. Our results for both GOV1 and GOV2 deviate by only about 1% from those reported by Bernstein and Zobel (22,870 and 865,362 classes, respectively), which, given the corpus sizes, is sufficiently close to say that their experiment can be successfully reproduced. This is further corroborated by the fact that the plot of the distribution of class sizes for GOV2, by visual inspection, has the same characteristics as that of Bernstein and Zobel. The observed differences are due to the lacking descriptions in the original paper of the normalization steps of the fingerprint function. Asking the authors for details was deemed unnecessary given the closeness of fit. Although the organizers of this track took note of the work of Bernstein and Zobel, they did not report on any activities to reduce the impact of near-duplicates [6,12].

Web Track. The Web Track employed the unrestricted web crawls ClueWeb09 and, as of 2013, ClueWeb12. Although the track organizers reported the removal of some duplicate documents up front, our analysis shows that both corpora still contain a large proportion of retrieval-equivalent documents (7.74% and 14.71%, respectively). The corpora are about 40 times larger than the GOV2 corpus, and so are the numbers of equivalence classes.

Core Track. The Core Track employed the New York Times Annotated Corpus (NYT-AC) in 2017, and the Washington Post (WaPo) corpus in 2018. One of this track's goals was to revisit the methodology of constructing evaluation corpora, implementing new ideas to avoid shortcomings of previous ones. The generation of relevance judgments for both corpora has been carefully carried out and documented. Unfortunately, we identify a proportion of retrieval-equivalent documents in the WaPo corpus similar to that of the ClueWeb12.

Altogether, the GOV2 corpus has the largest proportion of retrieval-equivalent documents, followed by ClueWeb12 and WaPo, ClueWeb09 and GOV1, and last the NYT-AC corpus with the least duplicates. Since each track has at least one corpus with significant amounts of duplication, this merits further investigation.

3   Assessment of Content-Equivalent Documents

Bernstein and Zobel consider two documents to be content-equivalent if they convey the same information. To quantify content equivalence, they employ a similarity measure that first fingerprints each document as a set of word 8-grams, and then divides the number of overlapping 8-grams between both fingerprints by the mean of their sizes (previously introduced as S_3 score in [3], and similar to the set-based resemblance measure of Broder [5]). An S_3 score of 0 indicates no syntactic overlap, and 1 retrieval-equivalence. From a user study, Bernstein and Zobel obtain an S_3 threshold of 0.58 above which content-equivalent pairs of GOV2 documents are identified with a precision of 0.95.

We repeated the user study for the GOV2 documents, as well as for the generic ClueWeb web pages and the news articles employed by the other tracks.

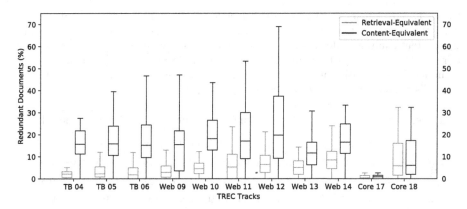

Fig. 2. Each box plots indicates the range of equivalent documents among the relevant documents across the topics of its respective track.

We sampled 100 document pairs per track at random and judged their content equivalence, ensuring that the sample uniformly covers the range of S_3 scores between 0.4 and 1. For the threshold reported by Bernstein and Zobel, we achieved only a precision of 87% on our sample under our interpretation of content equivalence. We hence chose the threshold 0.68 for GOV2 documents, 0.84 for generic ClueWeb pages, and 0.68 for news articles to obtain a precision of 0.95 for all three text genres. Given the threshold difference we obtained for the GOV2 documents, we compared all 50 topics of the Terabyte Track 2004 in detail with the results reported by Bernstein and Zobel and found only two with discrepancies; a reasonable result given the difficulty of repeating a user study.

Following Bernstein and Zobel, we implemented the SPEX algorithm [3] to identify all pairs of documents for which relevance judgments have been collected during one of the eleven editions of the three tracks, which exceed the aforementioned S_3 thresholds, respectively. Figure 2 shows box plots for each of the tracks, contrasting retrieval- and content-equivalent documents, when regarding only documents judged as relevant. Each box plot indicates the range of numbers of equivalent documents across the topics of its corresponding track. Except for the Core Track 2017, all tracks have topics with a high number of equivalent documents. Particularly striking is Topic 194 in the Web Track 2012 for the query `designer dog breeds`: among 47 relevant documents, there are 40 content-equivalent ones derived from the same Wikipedia article.

4   Impact of the Novelty Principle on Retrieval Evaluation

The novelty principle states that a document, though relevant in isolation, is irrelevant if it is content-equivalent to a document the user has already seen in the result list. We quantify the effect of the novelty principle on the shared tasks under consideration. Like Bernstein and Zobel, we removed poorly performing runs—keeping the best 75%—to discount the effect of those runs.

Table 1. The impact of the novelty principle on the ranking of retrieval-systems under the scenarios that: (1) content-equivalent documents are marked as irrelevant, (2) content-equivalent documents are removed by the search engine. We report the average nDCG ($\mathrm{avg_{nDCG}}$), the median (med_I) and maximum (max_I) ranking changes of the ideal participation model, changes in the average nDCG (Δ_{nDCG}), as well as Kendall's τ, and Kendall's τ of the top-5 systems ($\tau@5$).

Track		Runs		Equiv. irrelevant					Equiv. removed		
		#	$\mathrm{avg_{nDCG}}$	Δ_{nDCG}	τ	$\tau@5$	med_I	max_I	Δ_{nDCG}	τ	$\tau@5$
Terabyte	2004	70	0.425	−5.1%	0.96	0.80	−9.0	−19	+0.3%	0.98	1.00
	2005	58	0.586	−3.8%	0.95	0.20	−15.5	−27	+0.8%	0.98	0.80
	2006	80	0.654	−4.4%	0.94	1.00	−29.5	−53	−1.0%	0.94	0.86
Web	2009	71	0.323	−8.9%	0.89	0.80	-8.5	−24	−6.8%	0.91	0.80
	2010	56	0.302	−14.1%	0.49	0.42	−19.5	−39	−9.9%	0.57	0.33
	2011	37	0.341	−9.0%	0.85	0.40	−8.0	−13	−3.4%	0.92	0.80
	2012	28	0.295	−17.3%	0.72	0.61	−9.0	−16	−12.4%	0.81	0.73
	2013	34	0.324	−4.6%	0.86	0.80	−4.0	-8	−1.8%	0.90	0.80
	2014	30	0.380	−7.9%	0.87	1.00	−4.0	−11	−4.5%	0.94	1.00
Core	2017	75	0.560	−0.3%	0.99	1.00	−1.0	−9	+0.1%	1.00	1.00
	2018	72	0.541	−4.3%	0.92	1.00	−11.0	−26	−0.9%	0.93	0.73

We experiment with two strategies to model the novelty principle: local judgment manipulation as per Bernstein and Zobel, and a global judgment manipulation of our own design. In local judgment manipulation, judgments are manipulated for each run independently, so that a document that is content-equivalent to another document is judged irrelevant if the latter appears above in a search results list. Bernstein and Zobel employ only local judgment manipulation, which does not consistently implement the novelty principle.

Consider a ranking that does not contain any document from a given class of relevant, content-equivalent documents. In local judgment manipulation, all documents of that class are considered relevant for that ranking, whereas this is not the case for an alternative ranking that contains all documents of that class, where only one of them would be relevant. To resolve this contradiction, we propose a global judgment manipulation in which all documents of the same equivalence class—except for a representative document—are marked as irrelevant for a query. If a ranking contains documents of a relevant equivalence class, we apply the local judgment manipulation to choose the representative document. Otherwise, we chose a representative at random. In all our experiments, local manipulation amplified score changes compared to global manipulation.

Bernstein and Zobel find that many content-equivalent documents have inconsistent relevance judgments, i.e., one being judged relevant, but not an equivalent one. We confirm this observation for *all* considered tracks. The minimum of 33 inconsistent classes was found in the Core Track 2018, and the maximum of 604 in the Terabyte Track 2004. The inconsistencies are fixed by assigning the entire equivalence class the most frequently occurring judgment.

In their analysis, Bernstein and Zobel examine the impact of the novelty principle on Mean Average Precision (MAP) scores. Meanwhile, the use of MAP has been discouraged [16]. We hence also analyze the novelty principle's impact on nDCG scores, observing much greater changes in MAP scores than for nDCG. Due to space limitations, we only discuss the novelty principle's impact on nDCG scores under global manipulation; a reproduction of the MAP-based analysis is included in our accompanying repository (see link above).

Table 1 shows the impact of the novelty principle on all considered tracks. Regarding the Terabyte Track 2004 (column group "Equiv. irrelevant"), we observe a reduction Δ_{nDCG} of the avg$_{nDCG}$ by 5.1% from 0.425 to 0.403. This reduction may be ignored if it were rank-preserving with respect to the ranking of the participating systems. The original ranking of systems correlates to the cleansed one with 0.96 Kendall's τ. This seems acceptable, but we measure only 0.8 τ@5, regarding only the top five ranks; the best-performing systems are affected more strongly. We inspected how many ranks an "ideal system" I would drop, if it were the only one to remove content-equivalent documents from its rankings: In the median, it would drop 9 ranks (med$_I$) and in the worst case 19 (max$_I$). By contrast, if all systems had removed content-equivalent documents from their rankings/runs (column group "Equiv. removed"), the novelty principle's impact on the Terabyte Track 2004 would have been negligible. No ranking changes among the best-performing five systems occur, and the rank correlation of all systems increases to 0.98. Interestingly, even the average nDCG would have increased by 0.3%, caused by the fact that each run is only expected to return one document of a class of relevant content-equivalent documents under global judgment manipulation.

Similarly, Table 1 shows the novelty principle's impact on all other tracks under consideration. A complete discussion as exemplified for the Terabyte Track 2004 is beyond the space limitations; just a few more highlights: It turns out that the novelty principle has a strong impact on the Web Tracks of 2010 and 2012 in terms of Δ_{nDCG}. But we observe the maximum drop of ranks (53, max$_I$) in the Terabyte Track 2006. Unlike for the Terabyte Track 2004, for the Core Track 2018, if all participants were to remove duplicates ("Equiv. removed"), we still observe a large difference in Kendall's τ@5 of 0.73.

5   Conclusion

We successfully reproduced the work of Bernstein and Zobel [4], confirming their findings on the impact of near-duplicate and content-equivalent documents on the evaluation of the TREC Terabyte Track 2004. In addition, we extended their analysis to all Terabyte Tracks, the Web Tracks, and the two recent Core Tracks, and we improved upon their original implementation of the novelty principle. With the exception of the Core Track 2017, all of the tracks under consideration are (strongly) affected by the presence of content-equivalent duplicates.

Our findings are alarming. Not only are the evaluations carried out thus far invalidated to some extent, they also subdue newcomers: In practice, filtering duplicates from search results is done as a matter of course and without a second thought, and diligent participants may thus never learn that their retrieval systems would have actually outperformed the state of the art. One cannot expect anyone to realize that abstaining from filtering duplicates may result in better performance at TREC—a conclusion one can only draw from an in-depth run analysis. This is a call to action to all track organizers to henceforth take duplicates into account.

References

1. Allan, J., Harman, D., Kanoulas, E., Li, D., Gysel, C.V., Voorhees, E.M.: TREC 2017 common core track overview. In: Proceedings of The Twenty-Sixth Text REtrieval Conference, TREC 2017, Gaithersburg, Maryland, USA, 15–17 November 2017 (2017)
2. Allan, J., Harman, D., Kanoulas, E., Voorhees, E.M.: TREC 2018 common core track overview. In: Notebooks of The Twenty-Seventh Text REtrieval Conference (TREC 2018), Gaithersburg, Maryland, USA, 14–16 November 2018 (2018)
3. Bernstein, Y., Zobel, J.: A scalable system for identifying co-derivative documents. In: Apostolico, A., Melucci, M. (eds.) SPIRE 2004. LNCS, vol. 3246, pp. 55–67. Springer, Heidelberg (2004). https://doi.org/10.1007/978-3-540-30213-1_6
4. Bernstein, Y., Zobel, J.: Redundant documents and search effectiveness. In: Proceedings of the 2005 ACM CIKM International Conference on Information and Knowledge Management, Bremen, Germany, 31 October – 5 November 2005, pp. 736–743 (2005)
5. Broder, A.Z.: On the resemblance and containment of documents. In: Proceedings Compression and Complexity of SEQUENCES 1997, Positano, Amalfitan Coast, Salerno, Italy, 11–13 June 1997, pp. 21–29 (1997)
6. Büttcher, S., Clarke, C.L.A., Soboroff, I.: The TREC 2006 terabyte track. In: Proceedings of the Fifteenth Text REtrieval Conference (TREC 2006), Gaithersburg, Maryland, USA, 14–17 November 2006 (2006)
7. Clarke, C.L.A., Craswell, N., Soboroff, I.: Overview of the TREC 2004 terabyte track. In: Proceedings of the Thirteenth Text REtrieval Conference (TREC 2004), Gaithersburg, Maryland, USA, 16–19 November 2004 (2004)
8. Clarke, C.L.A., Craswell, N., Soboroff, I.: Overview of the TREC 2009 web track. In: Proceedings of The Eighteenth Text REtrieval Conference (TREC 2009), Gaithersburg, Maryland, USA, 17–20 November 2009 (2009)
9. Clarke, C.L.A., Craswell, N., Soboroff, I., Cormack, G.V.: Overview of the TREC 2010 web track. In: Proceedings of The Nineteenth Text REtrieval Conference (TREC 2010), Gaithersburg, Maryland, USA, 16–19 November 2010 (2010)
10. Clarke, C.L.A., Craswell, N., Soboroff, I., Voorhees, E.M.: Overview of the TREC 2011 web track. In: Proceedings of The Twentieth Text REtrieval Conference (TREC 2011), Gaithersburg, Maryland, USA, 15–18 November 2011 (2011)

11. Clarke, C.L.A., Craswell, N., Voorhees, E.M.: Overview of the TREC 2012 web track. In: Proceedings of The Twenty-First Text REtrieval Conference (TREC 2012), Gaithersburg, Maryland, USA, 6–9 November 2012 (2012)

12. Clarke, C.L.A., Scholer, F., Soboroff, I.: The TREC 2005 terabyte track. In: Proceedings of the Fourteenth Text REtrieval Conference (TREC 2005), Gaithersburg, Maryland, USA, 15–18 November 2005 (2005)

13. Collins-Thompson, K., Bennett, P.N., Diaz, F., Clarke, C., Voorhees, E.M.: TREC 2013 web track overview. In: Proceedings of The Twenty-Second Text REtrieval Conference (TREC 2013), Gaithersburg, Maryland, USA, 19–22 November 2013 (2013)

14. Collins-Thompson, K., Macdonald, C., Bennett, P.N., Diaz, F., Voorhees, E.M.: TREC 2014 web track overview. In: Proceedings of The Twenty-Third Text REtrieval Conference (TREC 2014), Gaithersburg, Maryland, USA, 19–21 November 2014 (2014)

15. Fetterly, D., Manasse, M.S., Najork, M.: On the evolution of clusters of near-duplicate web pages. In: 1st Latin American Web Congress (LA-WEB2003), Empowering Our Web, Sanitago, Chile, 10–12 November 2003, pp. 37–45 (2003)

16. Fuhr, N.: Some common mistakes in IR evaluation, and how they can be avoided. SIGIR Forum 51(3), 32–41 (2017)

17. Yang, P., Fang, H., Lin, J.: Anserini: enabling the use of Lucene for information retrieval research. In: Proceedings of the 40th International ACM SIGIR Conference on Research and Development in Information Retrieval, Shinjuku, Tokyo, Japan, 7–11 August 2017, pp. 1253–1256 (2017)

From MaxScore to Block-Max Wand: The Story of How Lucene Significantly Improved Query Evaluation Performance

Adrien Grand[1], Robert Muir[2], Jim Ferenczi[1], and Jimmy Lin[3(✉)]

[1] Elastic NV, Mountain View, USA
[2] Ntrepid Corporation, Herndon, USA
[3] David R. Cheriton School of Computer Science,
University of Waterloo, Waterloo, Canada
jimmylin@uwaterloo.ca

Abstract. The latest major release of Lucene (version 8) in March 2019 incorporates block-max indexes and exploits the block-max variant of Wand for query evaluation, which are innovations that originated from academia. This paper shares the story of how this came to be, which provides an interesting case study at the intersection of reproducibility and academic research achieving impact in the "real world". We offer additional thoughts on the often idiosyncratic processes by which academic research makes its way into deployed solutions.

Keywords: Open-source software · Technology adoption

1 Introduction

We share the story of how an innovation that originated from academia—block-max indexes and the corresponding block-max Wand query evaluation algorithm of Ding and Suel [6]—made its way into the open-source Lucene search library. This represents not only a case study in widespread reproducibility, since every recent deployment of Lucene has access to these features and thus their performance benefits can be easily measured, but also of academic research achieving significant impact. How did these innovations make their way from the "ivory tower" into the "real world"? We recount the sequence of events, including false starts, that finally led to the inclusion of block-max Wand in the latest major version of Lucene (version 8), released in March 2019.

We see this paper as having two main contributions beyond providing a narrative of events: First, we report results of experiments that attempt to match the original conditions of Ding and Suel [6] and present additional results on a number of standard academic IR test collections. These experiments characterize the performance of Lucene's implementation and show the extent to which performance improvements are retained when moving from a research prototype to a production codebase. Second, we offer a number of observations about the adoption of academic innovations, perhaps providing some insight into how academics might achieve greater real-world impact with their work.

© Springer Nature Switzerland AG 2020
J. M. Jose et al. (Eds.): ECIR 2020, LNCS 12036, pp. 20–27, 2020.
https://doi.org/10.1007/978-3-030-45442-5_3

2   Setting the Stage

From its very beginnings in 1999, Lucene has mostly existed in a "parallel universe" from academic IR researchers. Part of this can be attributed to its "target audience": developers who wish to build real-world search applications, as opposed to researchers who wish to write papers. Academic IR researchers have a long history of building and sharing search engines, dating back to the mid 1980s with Cornell's SMART system [4]. The tradition continues to this day, with Lemur/Indri [12,13] and Terrier [8,14] being the most successful examples of open-source academic search engines, still popular with many researchers today. Until recently, there has been little exchange between Lucene and these systems, other than a few academic workshops [16,21].

Lucene has, for the longest time, been somewhat maligned in the academic IR community. For much of its existence, its default ranking model was a variant of TF-IDF that was not only *ad hoc*, but demonstrably less effective than ranking models that were widely available in academic systems [18]. Okapi BM25 was not added to Lucene until 2011,[1] more than a decade after it gained widespread adoption in the research community; the consensus had long emerged that it was more effective than TF-IDF variants. This lag has contributed to the broad perception by researchers that Lucene produces poor search results and is ill-suited for information retrieval research.

This negative perception of Lucene, however, began to change a few years ago. In 2015, an evaluation exercise known as the "open-source reproducibility challenge" [7] benchmarked seven open-source search engines and demonstrated that Lucene was quite competitive in terms of both effectiveness and efficiency. It was the fourth fastest system (of seven) in terms of query evaluation, beating all the systems that were better than it in terms of effectiveness.

Since then, there has been a resurgence of interest in adopting Lucene for information retrieval research, including a number of workshops that brought together like-minded researchers over the past few years [1,2]. Anserini [19,20] is an open-source toolkit built on Lucene that was specifically designed to support replicable information retrieval research by providing many research-oriented features missing from Lucene, such as out-of-the-box support for a variety of common test collections. The project aims to better align IR researchers and practitioners, as Lucene has become the de facto platform used in industry to build production search solutions (typically via systems such as Elasticsearch and Solr). The experiments in this paper were conducted with Anserini.

3   From MAXSCORE to Block-Max WAND

At Berlin Buzzwords in 2012, Stefan Pohl gave a presentation about MAXSCORE [17] to raise awareness about efficient retrieval techniques in the Lucene community [15]. The presentation was accompanied by a working prototype.[2]

---

[1] https://issues.apache.org/jira/browse/LUCENE-2959.
[2] https://issues.apache.org/jira/browse/LUCENE-4100.

This contribution was exciting but also challenging to integrate as it conflicted with some of the flexibility that Lucene provides, requiring an index rewrite. There were ideas on how to address these issues, but they entailed a lot of effort, and so the issue remained stalled for about five years.

Five years is a long time and many changes occurred meanwhile. The switch from TF-IDF to BM25 as Lucene's default scoring function in 2015 created a natural upper bound on scores due to BM25's saturation effect, which made it possible to implement retrieval algorithms that reasoned about maximum scores without changes to Lucene's index format. This led to an effort to implement a general-purpose WAND [3], based on a previous implementation for BooleanQuery. Lucene received support for WAND at the end of 2017 (although it wasn't released until version 8.0 with block-max indexes).

Implementing WAND introduced two new issues. First, the total hit count would no longer be accurate, since not all matches are visited. Common analytics use cases depend on this count, and many search engines display this value in their interfaces (see additional discussion in Sect. 5). Second, the fact that some Lucene queries could produce negative scores became problematic, so Lucene now requires positive scores.[3]

Support for block-max indexes was the final feature that was implemented, based on the developers' reading of the paper by Ding and Suel [6], which required invasive changes to Lucene's index format. Note that the paper describes directly storing the maximum impact score per block, which fixes the scoring function at indexing time. To provide flexibility in being able to swap in different scoring functions, the Lucene implementation stores all tf (term frequency) and dl (document length) pairs that might yield the maximum score. If we have one such pair (tf_i, dl_i) then we can remove all other pairs (tf_j, dl_j) where $tf_j \leq tf_i \wedge dl_j \geq dl_i$, since they are guaranteed to yield lower (or equal) scores—based on the assumption that scores increase monotonically with increasing tf and decreasing dl. This is implemented by accumulating all such pairs in a tree-like structure during the indexing process. These pairs are stored in skip lists, so the information is available to groups of 8, 64, 512, 4096, ... blocks, allowing query evaluation to skip over more than one block at a time.

An interesting coda to this story is that academic researchers were exploring alternatives to per-block impact scores circa 2017, for exactly the same reason (to allow the scoring model to be defined at search time). For example, Macdonald and Tonellotto [10] showed how to derive tight approximate upper bounds for block-max WAND, based on work that dates back to 2011 [9]. Similarly, the recently-released PISA research system stores flexible block-level metadata [11]. Unfortunately, the Lucene developers were not aware of these developments during their implementation.

The journey from MAXSCORE to block-max WAND concluded in March 2019, with the rollout of all these features in the version 8.0 release of Lucene. They are now the out-of-the-box defaults in the world's most popular search library.

---

[3] https://issues.apache.org/jira/browse/LUCENE-7996.

Table 1. Per-query latency (ms), comparing Ding and Suel [6] with Lucene under similar experimental conditions, but on different hardware ($k = 10$).

	TREC 2005	TREC 2006
Ding and Suel [6]: exhaustive Or	369	226
Ding and Suel [6]: Wand	64	78
Ding and Suel [6]: Bmw	21	28
Lucene: exhaustive Or	98	188
Lucene: Bmw	32	55

4   Experimental Evaluation

During the implementation of block-max Wand, performance improvements were quantified in terms of Lucene's internal benchmark suite, which showed a 3× to 7× improvement in query evaluation performance. As part of a formal reproducibility effort, we present experiments that attempt to match, to the extent practical, the original conditions described by Ding and Suel [6].

According to the paper, experiments were conducted on the Gov2 web collection, on a randomly-selected subset of 1000 queries from the TREC 2005 and 2006 Efficiency Tracks, which we were able to obtain from the authors. For their experiments, the inverted index was completely loaded into main memory and query evaluation latency was measured to retrieval depth ten.

Our experiments were conducted with the Anserini IR toolkit,[4] comparing v0.5.1, which depends on Lucene 7.6 and uses an optimized exhaustive Or query evaluation strategy [5] with v0.6.0, which depends on Lucene 8.0 and uses block-max Wand. We used Anserini's standard regression test settings on the different collections, as described on its homepage. Results represent averages over three trials on a warm cache. While the indexes were not explicitly loaded into memory, Lucene benefits from caching at the OS level.

All experiments were conducted using a single thread on an otherwise idle server with dual Intel Xeon E5-2699 v4 processors and 1TB RAM running RHEL (release 7.7). Results are shown in Table 1, where figures in the top three rows are copied from Table 1 in the original paper. It is interesting that Ding and Suel report a much larger increase in performance comparing exhaustive Or to Bmw (18× on TREC 2005 and 8× on TREC 2006) than the comparable conditions in Lucene (a more modest improvement of around 3×). This is due to a more optimized implementation of exhaustive Or in Lucene, which, for example, implements block processing [5]. Interestingly, Ding and Suel report faster query evaluation in absolute terms, even on hardware that is much older: among the differences include C++ vs. Java, as well as the simplicity of a research prototype vs. the realities of a fully-featured search library. Beyond implementation differences, Lucene must additionally compute the upper bound scores per block from the stored (tf, dl) pairs on the fly.

---

[4] http://anserini.io/.

Table 2. Per-query latency (ms) for different queries, collections, and retrieval depths.

Collection	ClueWeb09b			ClueWeb12-B13		
Retrieval depth k	10	100	1000	10	100	1000
TREC 2005 queries						
Lucene: exhaustive OR	331	371	521	321	371	669
Lucene: BMW	96	137	370	94	148	489
Speedup	3.4×	2.7×	1.4×	3.4×	2.5×	1.4×
TREC 2006 queries						
Lucene: exhaustive OR	424	464	659	404	442	693
Lucene: BMW	123	191	431	127	186	526
Speedup	3.4×	2.4×	1.5×	3.2×	2.4×	1.3×

Table 3. Indexing time in seconds.

Collection	Lucene 7.6	Lucene 8.0	
Gov2	2528	2719	+7.6%
ClueWeb09b	6333	6817	+7.6%
ClueWeb12-B13	7514	7943	+5.7%

We also report performance evaluations on two other standard test collections frequently used in academic information retrieval: ClueWeb09b and ClueWeb12-B13, with the same sets of queries. These results are shown in Table 2, where we report figures for different values of retrieval depth k, also averaged over three trials. These numbers are consistent with Fig. 7 in Ding and Suel's paper: performance of exhaustive OR drops modestly as depth k increases, but BMW performance degrades much more quickly. This is exactly as expected.

Finally, we quantify the modest increase in indexing time due to the need to maintain (tf, dl) pairs in the inverted indexes, shown in Table 3 (averaged over three trials, using 44 threads in all cases). These experiments used Anserini's default regression settings on the respective collections, which builds full positional indexes and also stores the raw documents.

5 Discussion

The story of block-max WAND in Lucene provides a case study of how an innovation that originated in academia made its way into the world's most widely-used search library and achieved significant impact in the "real world" through hundreds of production deployments worldwide (if we consider the broader Lucene ecosystem, which includes systems such as Elasticsearch and Solr). As there are very few such successful case studies (the other prominent one being the incorporation of BM25 in Lucene), it is difficult to generalize these narratives into "lessons learned". However, here we attempt to offer a few observations about how academic research might achieve greater real-world impact.

In short, block-max Wand is in Lucene because the developers learned about Ding and Suel and decided to reimplement it. This is somewhat stating the obvious, but this fateful decision highlights the idiosyncratic nature of technology adoption. We could imagine alternatives where the Lucene developers had not come across the paper and developed a comparable solution in isolation, or they might have known about the paper and elected to take a different approach. In either case, the Lucene solution would likely differ from block-max Wand. This would be akin to convergent evolution in evolutionary biology, whereby different organisms independently evolve similar traits because they occupy similar environments. In such an "alternate reality", this paper would be comparing and contrasting different solutions to handling score outliers, not describing a reproducibility effort. To bring researchers and practitioners closer together, we recommend that the former be more proactive to "evangelize" their innovations, and the latter be more diligent in consulting the literature.

Eight years passed from the publication of the original paper (2011) until the release of Lucene that included block-max Wand (2019). The entire course of innovation was actually much longer if we trace the origins back to MaxScore (1995) and Wand (2003). One obvious question is: Why did it take so long?

There are many explanations, the most salient of which is the difference between a research prototype and a fully-featured search library that is already widely deployed. This decomposes into two related issues, the technical and the social. From a technical perspective, supporting BMW required invasive changes to Lucene's index format and a host of related changes in scoring functions— for example, scores could no longer be negative, and implementations could no longer access arbitrary fields (which was an API change). These had to be staged incrementally. Concomitant with technical changes and backwards-compatibility constraints were a host of "social" changes, which required changing users' expectations about the behavior of the software. In short, BMW was *not* simply a drop-in replacement. For example, as discussed in Sect. 3, the hit count was no longer accurate, which required workarounds for applications that depended on the value. Because such major changes can be somewhat painful, they need to be justified by the potential benefits. This means that only dramatic improvements really have any hope of adoption: multiple-fold, not marginal, performance gains. An interesting side effect is that entire generations of techniques might be skipped, in the case of Lucene, directly from exhaustive OR to BMW, leapfrogging intermediate innovations such as MaxScore and Wand.

6 Conclusions

Aiming to achieve real-world impact with academic research is a worthy goal, and we believe that this case study represents an endorsement of efforts to better align research prototypes with production systems, as exemplified by Lucene-based projects like Anserini. If academic researchers are able to look ahead "down the road" to see how their innovations might benefit end applications, the path from the "ivory tower" to the "real world" might become more smoothly paved.

Acknowledgments. This work was supported in part by the Natural Sciences and Engineering Research Council (NSERC) of Canada. We'd like to thank Craig Macdonald, Joel Mackenzie, Antonio Mallia, and Nicola Tonellotto for helpful discussions on the intricacies of computing flexible per-block score bounds, and Torsten Suel for providing us with the original queries used in their evaluations.

References

1. Azzopardi, L., et al.: The Lucene for information access and retrieval research (LIARR) workshop at SIGIR 2017. In: Proceedings of the 40th Annual International ACM SIGIR Conference on Research and Development in Information Retrieval (SIGIR 2017), pp. 1429–1430, Tokyo (2017)
2. Azzopardi, L., et al.: Lucene4IR: developing information retrieval evaluation resources using Lucene. In: SIGIR Forum, vol. 50, no. 2, pp. 58–75 (2017)
3. Broder, A.Z., Carmel, D., Herscovici, M., Soffer, A., Zien, J.: Efficient query evaluation using a two-level retrieval process. In: Proceedings of the Twelfth International Conference on Information and Knowledge Management (CIKM 2003), pp. 426–434, New Orleans (2003)
4. Buckley, C.: Implementation of the SMART information retrieval system. Department of Computer Science TR, pp. 85–686, Cornell University (1985)
5. Cutting, D.R., Pedersen, J.O.: Space optimizations for total ranking. In: Computer-Assisted Information Searching on Internet (RIAO 1997), pp. 401–412, Paris (1997)
6. Ding, S., Suel, T.: Faster top-k document retrieval using block-max indexes. In: Proceedings of the 34rd Annual International ACM SIGIR Conference on Research and Development in Information Retrieval (SIGIR 2011), pp. 993–1002, Beijing (2011)
7. Lin, J., et al.: Toward reproducible baselines: the open-source IR reproducibility challenge. In: Ferro, N., et al. (eds.) ECIR 2016. LNCS, vol. 9626, pp. 408–420. Springer, Cham (2016). https://doi.org/10.1007/978-3-319-30671-1_30
8. Macdonald, C., McCreadie, R., Santos, R.L., Ounis, I.: From puppy to maturity: experiences in developing Terrier. In: Proceedings of the SIGIR 2012 Workshop on Open Source Information Retrieval, pp. 60–63, Portland (2012)
9. Macdonald, C., Ounis, I., Tonellotto, N.: Upper-bound approximations for dynamic pruning. ACM Trans. Inf. Syst. **29**(4), 171–1728 (2011)
10. Macdonald, C., Tonellotto, N.: Upper bound approximation for BlockMaxWand. In: Proceedings of the ACM SIGIR International Conference on Theory of Information Retrieval (ICTIR 2017), pp. 273–276 (2017)
11. Mallia, A., Siedlaczek, M., Mackenzie, J., Suel, T.: PISA: performant indexes and search for academia. In: Proceedings of the Open-Source IR Replicability Challenge (OSIRRC 2019): CEUR Workshop Proceedings, vol. 2409, pp. 50–56, Paris (2019)
12. Metzler, D., Croft, W.B.: Combining the language model and inference network approaches to retrieval. Inf. Process. Manag. **40**(5), 735–750 (2004)
13. Metzler, D., Strohman, T., Turtle, H., Croft, W.B.: Indri at TREC 2004: Terabyte track. In: Proceedings of the Thirteenth Text Retrieval Conference (TREC 2004), Gaithersburg (2004)
14. Ounis, I., Amati, G., Plachouras, V., He, B., Macdonald, C., Lioma, C.: Terrier: a high performance and scalable information retrieval platform. In: Proceedings of the SIGIR 2006 Workshop on Open Source Information Retrieval, pp. 18–25 (2006)

15. Pohl, S.: Efficient scoring in Lucene. In: Berlin Buzzwords 2012 (2012)
16. Trotman, A., Clarke, C.L., Ounis, I., Culpepper, S., Cartright, M.A., Geva, S.: Open source information retrieval: a report on the SIGIR 2012 workshop. In: SIGIR Forum, vol. 46, no. 2, pp. 95–101 (2012)
17. Turtle, H., Flood, J.: Query evaluation: strategies and optimizations. Inf. Process. Manag. **31**(6), 831–850 (1995)
18. Turtle, H., Hegde, Y., Rowe, S.A.: Yet another comparison of Lucene and Indri performance. In: Proceedings of the SIGIR 2012 Workshop on Open Source Information Retrieval, pp. 64–67, Portland (2012)
19. Yang, P., Fang, H., Lin, J.: Anserini: enabling the use of Lucene for information retrieval research. In: Proceedings of the 40th Annual International ACM SIGIR Conference on Research and Development in Information Retrieval (SIGIR 2017), pp. 1253–1256, Tokyo (2017)
20. Yang, P., Fang, H., Lin, J.: Anserini: reproducible ranking baselines using Lucene. J. Data Inf. Qual. **10**(4), Article 16 (2018)
21. Yee, W.G., Beigbeder, M., Buntine, W.: SIGIR06 workshop report: open source information retrieval systems (OSIR06). In: SIGIR Forum, vol. 40, no. 2, pp. 61–65 (2006)

Which BM25 Do You Mean?
A Large-Scale Reproducibility Study
of Scoring Variants

Chris Kamphuis[1], Arjen P. de Vries[1], Leonid Boytsov[2], and Jimmy Lin[3(✉)]

[1] Radboud University, Nijmegen, The Netherlands
[2] Carnegie Mellon University, Pittsburgh, USA
[3] University of Waterloo, Waterloo, Canada
jimmylin@uwaterloo.ca

Abstract. When researchers speak of BM25, it is not entirely clear which variant they mean, since many tweaks to Robertson et al.'s original formulation have been proposed. When practitioners speak of BM25, they most likely refer to the implementation in the Lucene open-source search library. Does this ambiguity "matter"? We attempt to answer this question with a large-scale reproducibility study of BM25, considering eight variants. Experiments on three newswire collections show that there are no significant effectiveness differences between them, including Lucene's often maligned approximation of document length. As an added benefit, our empirical approach takes advantage of databases for rapid IR prototyping, which validates both the feasibility and methodological advantages claimed in previous work.

Keywords: Scoring functions · Relational databases

1 Introduction

BM25 [8] is perhaps the most well-known scoring function for "bag of words" document retrieval. It is derived from the binary independence relevance model to include within-document term frequency information and document length normalization in the probabilistic framework for IR [7]. Although learning-to-rank approaches and neural ranking models are widely used today, they are typically deployed as part of a multi-stage reranking architecture, over candidate documents supplied by a simple term-matching method using traditional inverted indexes [1]. Often, this is accomplished using BM25, and thus this decades-old scoring function remains a critical component of search applications today.

As many researchers have previously observed, e.g., Trotman et al. [11], the referent of BM25 is quite ambiguous. There are, in fact, many variants of the scoring function: beyond the original version proposed by Robertson et al. [8], many variants exist that include small tweaks by subsequent researchers. Also, researchers using different IR systems report (sometimes quite) different effectiveness measurements for their implementation of BM25, even on the same

© Springer Nature Switzerland AG 2020
J. M. Jose et al. (Eds.): ECIR 2020, LNCS 12036, pp. 28–34, 2020.
https://doi.org/10.1007/978-3-030-45442-5_4

test collections; consider for example the results reported in OSIRRC 2019, the open-source IR replicability challenge at SIGIR 2019 [2]. Furthermore, BM25 is parameterized in terms of k_1 and b (plus k_2, k_3 in the original formulation), and researchers often neglect to include the parameter settings in their papers.

Our goal is a large-scale reproducibility study to explore the nuances of different variants of BM25 and their impact on retrieval effectiveness. We include in our study the specifics of the implementation of BM25 in the Lucene open-source search library, a widely-deployed variant "in the real world". Outside of a small number of commercial search engine companies, Lucene—either stand-alone or via higher-level platforms such as Solr and Elasticsearch—has today become the *de facto* foundation for building search applications in industry.

Our approach enlists the aid of relational databases for rapid prototyping, an idea that goes back to the 1990s and was more recently revived by Mühleisen et al. [6]. Adding or revising scoring functions in any search engine requires custom code within some framework for postings traversal, making the exploration of many different scoring functions (as in our study) a tedious and error-prone process. As an alternative, it is possible to "export" the inverted index to a relational database and recast the document ranking problem into a database (specifically, SQL) query. Varying the scoring function, then, corresponds to varying the expression for calculating the score in the SQL query, allowing us to explore different BM25 variants by expressing them declaratively (instead of *programming* imperatively). We view our work as having two contributions:

- We conducted a large-scale reproducibility study of BM25 variants, focusing on the Lucene implementation and variants described by Trotman et al. [11]. Their findings are confirmed: effectiveness differences in IR experiments are unlikely to be the result of the choice of BM25 variant a system implemented.
- From the methodological perspective, our work can be viewed as reproducing and validating the work of Mühleisen et al. [6], the most recent advocate of using databases for rapid IR prototyping.

2   BM25 Variants

Table 1 summarizes the scoring functions of the BM25 variants we examined:

Robertson et al. [8] is the original formulation of BM25: N is the number of documents in the collection, df_t is the number of documents containing term t, tf_{td} is the term frequency of term t in document d. Document lengths L_d and L_{avg} are the number of tokens in document d and the average number of tokens in a document in the collection, respectively. Finally, k_1 and b are free parameters that can be optimized per collection.[1]

Lucene (default) is the variant implemented in Lucene (as of version 8), which introduces two main differences. First, since the IDF component of

---

[1] The original publication adds scoring components with constants k_2 and k_3 that are rarely used and thus not considered in our study.

Table 1. Scoring functions of the BM25 variants examined in this work.

Robertson et al.	$\sum_{t \in q} \log \left(\frac{N - df_t + 0.5}{df_t + 0.5} \right) \cdot \frac{tf_{td}}{k_1 \cdot \left(1 - b + b \cdot \left(\frac{L_d}{L_{avg}} \right) \right) + tf_{td}}$
Lucene (default)	$\sum_{t \in q} \log \left(1 + \frac{N - df_t + 0.5}{df_t + 0.5} \right) \cdot \frac{tf_{td}}{k_1 \cdot \left(1 - b + b \cdot \left(\frac{L_{dlossy}}{L_{avg}} \right) \right) + tf_{td}}$
Lucene (accurate)	$\sum_{t \in q} \log \left(1 + \frac{N - df_t + 0.5}{df_t + 0.5} \right) \cdot \frac{tf_{td}}{k_1 \cdot \left(1 - b + b \cdot \left(\frac{L_d}{L_{avg}} \right) \right) + tf_{td}}$
ATIRE	$\sum_{t \in q} \log \left(\frac{N}{df_t} \right) \cdot \frac{(k_1 + 1) \cdot tf_{td}}{k_1 \cdot \left(1 - b + b \cdot \left(\frac{L_d}{L_{avg}} \right) \right) + tf_{td}}$
BM25L	$\sum_{t \in q} \log \left(\frac{N + 1}{df_t + 0.5} \right) \cdot \frac{(k_1 + 1) \cdot (c_{td} + \delta)}{k_1 + c_{td} + \delta}$
BM25+	$\sum_{t \in q} \log \left(\frac{N + 1}{df_t} \right) \cdot \left(\frac{(k_1 + 1) \cdot tf_{td}}{k_1 \cdot \left(1 - b + b \cdot \left(\frac{L_d}{L_{avg}} \right) \right) + tf_{td}} + \delta \right)$
BM25-adpt	$\sum_{t \in q} G_q^1 \cdot \frac{(k_1' + 1) \cdot tf_{td}}{k_1' \cdot \left(1 - b + b \cdot \left(\frac{L_d}{L_{avg}} \right) \right) + tf_{td}}$
TF$_{lo\delta op} \times$IDF	$\sum_{t \in q} \log \left(\frac{N + 1}{df_t} \right) \cdot \left(1 + \log \left(1 + \log \left(\frac{tf_{td}}{1 - b + b \cdot \left(\frac{L_d}{L_{avg}} \right)} + \delta \right) \right) \right)$

Robertson et al. is negative when $df_t > N/2$, Lucene adds a constant one before calculating the log value. Second, the document length used in the scoring function is compressed (in a lossy manner) to a one byte value, denoted L_{dlossy}. With only 256 distinct document lengths, Lucene can pre-compute the value of $k_1 \cdot (1 - b + b \cdot (L_{dlossy}/L_{avg}))$ for each possible length, resulting in fewer computations at query time.

Lucene (accurate) represents our attempt to measure the impact of Lucene's lossy document length encoding. We implemented a variant that uses exact document lengths, but is otherwise identical to the Lucene default.

ATIRE [10] implements the IDF component of BM25 as $\log (N/df_t)$, which also avoids negative values. The TF component is multiplied by $k_1 + 1$ to make it look more like the classic RSJ weight; this has no effect on the resulting ranked list, as all scores are scaled linearly with this factor.

BM25L [5] builds on the observation that BM25 penalizes longer documents too much compared to shorter ones. The IDF component differs, to avoid negative values. The TF component is reformulated as $((k_1 + 1) \cdot c_{td}) / (k_1 + c_{td})$ with $c_{td} = tf_{td}/(1 - b + b \cdot (L_d/L_{avg}))$. The c_{td} component is further modified by adding a constant δ to it, boosting the score for longer documents. The authors report using $\delta = 0.5$ for highest effectiveness.

BM25+ [4] encodes a general approach for dealing with the issue that ranking functions unfairly prefer shorter documents over longer ones. The proposal is to add a lower-bound bonus when a term appears at least one time in a document. The difference with BM25L is a constant δ to the TF component. The IDF component is again changed to a variant that disallows negative values.

BM25-adpt [3] is an approach that varies k_1 per term (i.e., uses term specific k_1 values). In order to determine the optimal value for k_1, the method starts by identifying the probability of a term occurring at least once in a document as $(df_r + 0.5)/(N + 1)$. The probability of the term occurring one more time is then defined as $(df_{r+1} + 0.5)/(df_r + 1)$. The information gain of a term occurring $r + 1$ instead of r times is defined as $G_q^r = \log_2\left((df_{r+1} + 0.5)/(df_r + 1)\right) - \log_2\left((df_{tr} + 0.5)/(N + 1)\right)$, where df_r is defined as follows: $|D_{t|c_{td} \geq r-0.5}|$ if $r > 1$, df_t if $r = 1$, and N if $r = 0$ (c_{td} is the same as in BM25L). The information gain is calculated for $r \in \{0, \ldots, T\}$, until $G_q^r > G_q^{r+1}$. The optimal value for k_1 is then determined by finding the value for k_1 that minimizes the equation $k_1' = \mathrm{argmin}_{k_1} \sum_{r=0}^{T} \left(G_q^r/G_q^1 - ((k_1 + 1) \cdot r)/(k_1 + r)\right)^2$. Essentially, this gives a value for k_1 that maximizes information gain for that specific term; k_1' and G_q^1 are then plugged into the BM25-adpt formula.

We found that the optimal value of k_1' is actually not defined for about 90% of the terms. A unique optimal value for k_1' only exists when $r > 1$ while calculating G_q^r. For many terms, especially those with a low df, $G_q^r > G_q^{r+1}$ occurs before $r > 1$. In these cases, picking different values for k_1 has virtually no effect on retrieval effectiveness. For undefined values, we set k_1' to 0.001, the same as Trotman et al. [11].

TF$_{lo\delta op}$×**IDF** [9] models the non-linear gain of a term occurring multiple times in a document as $1 + \log\left(1 + \log\left(tf_{td}\right)\right)$. To ensure that terms occurring at least once in a document get boosted, the approach adds a fixed component δ, following BM25+. These parts are combined into the TF component using $tf_{td}/(1 - b + b \cdot (L_d/L_{avg}))$. The same IDF component as in BM25+ is used.

3   Experiments

Our experiments were conducted using Anserini (v0.6.0) on Java 11 to create an initial index, and subsequently using relational databases for rapid prototyping, which we dub "OldDog" after Mühleisen et al. [6]; following that work we use MonetDB as well. Evaluations with Lucene (default) and Lucene (accurate) were performed directly in Anserini; the latter was based on previously-released code that we updated and incorporated into Anserini.[2] The inverted index was exported from Lucene to OldDog, ensuring that all experiments share *exactly* the same document processing pipeline (tokenization, stemming, stopword removal, etc.). While exporting the inverted index, we precalculate all k_1' values for BM25-adpt as suggested by Lv and Zhai [3]. As an additional verification step, we implemented both Lucene (default) and Lucene (accurate) in OldDog and compared results to the output from Anserini. We are able to confirm that the results are the same, setting aside unavoidable differences related to floating point precision. All BM25 variants are then implemented in OldDog as minor variations upon the original SQL query provided in Mühleisen et al. [6]. The term-specific parameter optimization for the adpt variant was already calculated during the

---

[2] http://searchivarius.org/blog/accurate-bm25-similarity-lucene.

Table 2. Retrieval effectiveness.

	Robust04		Core17		Core18	
	AP	P@30	AP	P@30	AP	P@30
Robertson et al. [8]	.2526	.3086	.2094	.4327	.2465	**.3647**
Lucene (default)	.2531	.3102	.2087	.4293	.2495	.3567
Lucene (accurate)	.2533	.3104	.2094	.4327	.2495	.3593
ATIRE	.2533	.3104	.2094	.4327	.2495	.3593
BM25L	.2542	.3092	.1975	.4253	**.2501**	.3607
BM25+	.2526	.3071	.1931	.4260	.2447	.3513
BM25-adpt	**.2571**	**.3135**	**.2112**	.4133	.2480	.3533
$TF_{l \circ \delta \circ p} \times IDF$	.2516	.3084	.1932	**.4340**	.2465	**.3647**

index extraction stage, allowing us to upload the optimal (t, k) pairs and directly use the term-specific k values in the SQL query. The advantage of our experimental methodology is that we did not need to implement a single new ranking function from scratch. All the SQL variants implemented for this paper can be found on GitHub.[3]

The experiments use three TREC newswire test collections: TREC Disks 4 and 5, excluding Congressional Record, with topics and relevance judgments from the TREC 2004 Robust Track (Robust04); the New York Times Annotated Corpus, with topics and relevance judgments from the TREC 2017 Common Core Track (Core17); the TREC Washington Post Corpus, with topics and relevance judgments from the TREC 2018 Common Core Track (Core18). Following standard experimental practice, we assess ranked list output in terms of average precision (AP) and precision at rank 30 (P@30). The parameters shared by all models are set to $k_1 = 0.9$ and $b = 0.4$, Anserini's defaults. The parameter δ is set to the value reported as best in the corresponding source publication. Table 2 presents the effectiveness scores for the implemented retrieval functions on all three test collections.

All experiments were run on a Linux desktop (Fedora 30, Kernel 5.2.18, SELinux enabled) with 4 cores (Intel Xeon CPU E3-1226 v3 @ 3.30 GHz) and 16 GB of main memory; the MonetDB 11.33.11 server was compiled from source using the --enable-optimize flag. Table 3 presents the average retrieval time per query in milliseconds (without standard deviation for Anserini, which does not report time per query). MonetDB uses all cores for both inter- and intra-query parallelism, while Anserini is single-threaded.

The observed differences in effectiveness are very small and can be fully attributed to variations in the scoring function; our methodology fixes all other parts of the indexing pipeline (tag cleanup, tokenization, stopwords, etc.). Both an ANOVA and Tukey's HSD show no significant differences between any variant, on all test collections. This confirms the findings of Trotman et al. [11]:

---

[3] https://github.com/Chriskamphuis/olddog.

Table 3. Average retrieval time per query in ms: Anserini (top) and OldDog (bottom).

	Robust04	Core17	Core18
Lucene (default)	52	111	120
Lucene (accurate)	55	115	123
Robertson et al. [8]	158 ± 25	703 ± 162	331 ± 96
Lucene (default)	157 ± 24	699 ± 154	326 ± 90
Lucene (accurate)	157 ± 24	701 ± 156	324 ± 88
ATIRE	157 ± 24	698 ± 159	331 ± 94
BM25L	158 ± 25	697 ± 160	333 ± 96
BM25+	158 ± 25	700 ± 160	334 ± 96
BM25-adpt	158 ± 24	700 ± 157	330 ± 92
$TF_{l \circ \delta op} \times IDF$	158 ± 24	698 ± 158	331 ± 96

effectiveness differences are unlikely an effect of the choice of the BM25 variant. Across the IR literature, we find that differences due to more mundane settings (such as the choice of stopwords) are often larger than the differences we observe here. Although we find no significant improvements over the original Robertson et al. [8] formulation, it might still be worthwhile to use a variant of BM25 that avoids negative ranking scores.

Comparing Lucene (default) and Lucene (accurate), we find negligible differences in effectiveness. However, the differences in retrieval time are also negligible, which calls into question the motivation behind the original length approximation. Currently, the similarity function and thus the document length encoding are defined at index time. Storing exact document lengths would allow for different ranking functions to be swapped at query time more easily, as no information would be discarded at index time. Accurate document lengths might additionally benefit downstream modules that depend on Lucene. We therefore suggest that Lucene might benefit from storing exact document lengths.

4   Conclusions

In summary, this work describes a double reproducibility study—we methodologically validate the usefulness of databases for IR prototyping claimed by Mühleisen et al. [6] and performed a large-scale study of BM25 to confirm the findings of Trotman et al. [11]. Returning to our original motivating question regarding the multitude of BM25 variants: "Does it matter?", we conclude that the answer appears to be "no, it does not".

Acknowledgements. This work is part of the research program Commit2Data with project number 628.011.001, which is (partly) financed by the NWO. Additional support was provided by the Natural Sciences and Engineering Research Council (NSERC) of Canada.

References

1. Asadi, N., Lin, J.: Effectiveness/efficiency tradeoffs for candidate generation in multi-stage retrieval architectures. In: Proceedings of the 36th Annual International ACM SIGIR Conference on Research and Development in Information Retrieval (SIGIR 2013), pp. 997–1000, Dublin (2013)
2. Clancy, R., Ferro, N., Hauff, C., Lin, J., Sakai, T., Wu, Z.Z.: Overview of the 2019 Open-Source IR Replicability Challenge (OSIRRC 2019). In: CEUR Workshop Proceedings of the Open-Source IR Replicability Challenge (OSIRRC 2019) at SIGIR 2009, vol. 2409, Paris (2019)
3. Lv, Y., Zhai, C.: Adaptive term frequency normalization for BM25. In: Proceedings of the 20th ACM International Conference on Information and Knowledge Management (CIKM 2011), pp. 1985–1988, Glasgow (2011)
4. Lv, Y., Zhai, C.: Lower-bounding term frequency normalization. In: Proceedings of the 20th ACM International Conference on Information and Knowledge Management (CIKM 2011), pp. 7–16, Glasgow (2011)
5. Lv, Y., Zhai, C.: When documents are very long, BM25 fails! In: Proceeding of the 34th International ACM SIGIR Conference on Research and Development in Information Retrieval (SIGIR 2011), pp. 1103–1104, Beijing (2011)
6. Mühleisen, H., Samar, T., Lin, J., de Vries, A.: Old dogs are great at new tricks: column stores for IR prototyping. In: Proceedings of the 37th Annual International ACM SIGIR Conference on Research and Development in Information Retrieval (SIGIR 2014), pp. 863–866, Gold Coast (2014)
7. Robertson, S., Zaragoza, H.: The probabilistic relevance framework: BM25 and beyond. Found. Trends Inf. Retrieval **3**(4), 333–389 (2009)
8. Robertson, S.E., Walker, S., Jones, S., Hancock-Beaulieu, M., Gatford, M.: Okapi at TREC-3. In: Proceedings of the 3rd Text Retrieval Conference (TREC-3), pp. 109–126, Gaithersburg (1994)
9. Rousseau, F., Vazirgiannis, M.: Composition of TF normalizations: new insights on scoring functions for ad hoc IR. In: Proceedings of the 36th Annual International ACM SIGIR Conference on Research and Development in Information Retrieval (SIGIR 2013), pp. 917–920, Dublin (2013)
10. Trotman, A., Jia, X.F., Crane, M.: Towards an efficient and effective search engine. In: SIGIR 2012 Workshop on Open Source Information Retrieval, pp. 40–47, Portland (2012)
11. Trotman, A., Puurula, A., Burgess, B.: Improvements to BM25 and language models examined. In: Proceedings of the 2014 Australasian Document Computing Symposium (ADCS 2014), pp. 58–66, Melbourne (2014)

The Unfairness of Popularity Bias in Music Recommendation: A Reproducibility Study

Dominik Kowald[1]([⊠]), Markus Schedl[2], and Elisabeth Lex[3]

[1] Know-Center GmbH, Graz, Austria
dkowald@know-center.at
[2] Johannes Kepler University Linz, Linz, Austria
markus.schedl@jku.at
[3] Graz University of Technology, Graz, Austria
elisabeth.lex@tugraz.at

Abstract. Research has shown that recommender systems are typically biased towards popular items, which leads to less popular items being underrepresented in recommendations. The recent work of Abdollahpouri et al. in the context of movie recommendations has shown that this popularity bias leads to unfair treatment of both long-tail items as well as users with little interest in popular items. In this paper, we reproduce the analyses of Abdollahpouri et al. in the context of music recommendation. Specifically, we investigate three user groups from the Last.fm music platform that are categorized based on how much their listening preferences deviate from the most popular music among all Last.fm users in the dataset: (i) low-mainstream users, (ii) medium-mainstream users, and (iii) high-mainstream users. In line with Abdollahpouri et al., we find that state-of-the-art recommendation algorithms favor popular items also in the music domain. However, their proposed Group Average Popularity metric yields different results for Last.fm than for the movie domain, presumably due to the larger number of available items (i.e., music artists) in the Last.fm dataset we use. Finally, we compare the accuracy results of the recommendation algorithms for the three user groups and find that the low-mainstreaminess group significantly receives the worst recommendations.

Keywords: Algorithmic fairness · Recommender systems · Popularity bias · Item popularity · Music recommendation · Reproducibility

1 Introduction

Recommender systems are quintessential tools to support users in finding relevant information in large information spaces [10]. However, one limitation of typical recommender systems is the so-called popularity bias, which leads to the underrepresentation of less popular (i.e., long-tail) items in the recommendation

© Springer Nature Switzerland AG 2020
J. M. Jose et al. (Eds.): ECIR 2020, LNCS 12036, pp. 35–42, 2020.
https://doi.org/10.1007/978-3-030-45442-5_5

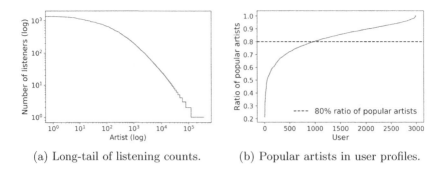

(a) Long-tail of listening counts.   (b) Popular artists in user profiles.

Fig. 1. Listening distribution of music artists. We find that around 1/3 (i.e., 1,000) of our users actually listen to at least 20% of unpopular artists.

lists [1,4,5]. The recent work of Abdollahpouri et al. [2] has investigated this popularity bias from the user perspective in the movie domain. The authors have shown that state-of-the-art recommendation algorithms tend to underserve users, who like unpopular items.

In this paper, we reproduce this study and conduct it in the music domain. As described in [16], there are several aspects of music recommendations that make them different to, e.g., movie recommendations such as the vast amount of available items. Therefore, we investigate music recommendations concerning popularity bias and, for reasons of comparability, raise the same two research questions as in [2]:

– *RQ1*: To what extent are users or groups of users interested in popular music artists?
– *RQ2*: To what extent does the popularity bias of recommendation algorithms affect users with different inclination to mainstream music?

For our experiments, we use a publicly available Last.fm dataset and address *RQ1* in Sect. 2 by analyzing the popularity of music artists in the user profiles. Next, we address *RQ2* in Sect. 3 by comparing six state-of-the-art music recommendation algorithms concerning their popularity bias propagation.

2   Popularity Bias in Music Data

For our reproducibility study, we use the freely available LFM-1b dataset [14]. Since this dataset contains 1.1 billion listening events of more than 120,000 Last.fm users and thus is much larger than the MovieLens dataset used in [2], we focus on a subset of it. Precisely, we extract 3,000 users that reflect the three user groups investigated in [2]. To this end, we use the mainstreaminess score, which is available for the users in the LFM-1b dataset and which is defined as the overlap between a user's listening history and the aggregated listening history of all Last.fm users in the dataset [3]. It thus represents a proxy for a user's inclination to popular music.

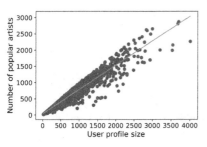
(a) Number of popular artists.

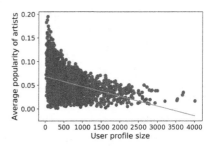
(b) Average popularity of artists.

Fig. 2. Correlation of user profile size and the popularity of artists in the user profile. While there is a positive correlation between profile size and number of popular artists, there is a negative correlation between profile size and the average artist popularity.

Our subset consists of the 1,000 users with lowest mainstreaminess scores (i.e., the LowMS group), the 1,000 users with a mainstreaminess score around the median (i.e., the MedMS group), and the 1,000 users with the highest mainstreaminess scores (i.e., the HighMS group). In total, we investigate 1,755,361 user-artist interactions between 3,000 users and 352,805 music artists. Compared to the MovieLens dataset with only 3,900 movies that Abdollahpouri et al. [2] have used in their study, our itemset is, consequently, much larger.

Listening Distribution of Music Artists. Fig. 1 depicts the listening distribution of music artists in our Last.fm dataset. As expected, in Fig. 1a, we observe a long-tail distribution of the listener counts of our items (i.e., artists). That is, only a few artists are listened to by many users, while most artists (i.e., the long-tail) are only listened to by a few users. Furthermore, in Fig. 1b, we plot the ratio of popular artists in the profiles of our 3,000 Last.fm users. As in [2], we define an artist as popular if the artist falls within the top 20% of artists with the highest number of listeners. We see that around 1,000 of our 3,000 users (i.e., around 1/3) have at least 20% of unpopular artists in their user profiles. This number also corresponds to the number of low-mainstream users we have in the LowMS user group.

User Profile Size and Popularity Bias in Music Data. Next, in Fig. 2, we investigate if there is a correlation between the user profile size (i.e., number of distinct items/artists) and the popularity of artists in the user profile. Therefore, in Fig. 2a, we plot the number of popular artists in the user profile over the profile size As expected, we find a positive correlation ($R = .965$) since the likelihood of having popular artists in the profile increases with the number of items in the profile. However, when plotting the average popularity of artists in the user profile over the profile size in Fig. 2b, we find a negative correlation ($R = -.372$), which means that users with a smaller profile size tend to listen to more popular artists. As in [2], we define the popularity of an artist as the ratio of users who have listened to this artist.

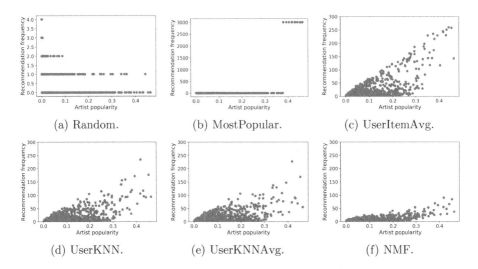

Fig. 3. Correlation of artist popularity and recommendation frequency. For all six algorithms, the recommendation frequency increases with the artist popularity.

Concerning *RQ1*, we find that one-third of our Last.fm users have at least 20% of unpopular artists in their profiles and thus, are also interested in low-mainstream music. Furthermore, we find that users with a small profile size tend to have more popular artists in their profiles than users with a more extensive profile size. These findings are in line with what Abdollahpouri et al. have found [2].

3   Popularity Bias in Music Recommendation

In this section, we study popularity bias in state-of-the-art music recommendation algorithms. To foster the reproducibility of our study, we calculate and evaluate all recommendations with the Python-based open-source recommendation toolkit Surprise[1]. Using Surprise, we formulate our music recommendations as a rating prediction problem, where we predict the preference of a target user u for a target artist a. We define the preference of a for u by scaling the listening count of a by u to a range of $[0, 1000]$ as also done in [15]. We then recommend the top-10 artists with the highest predicted preferences.

Recommendation of Popular Music Artists. We use the same evaluation protocol (i.e., 80/20 train/test split) and types of algorithms as in [2], which includes (i) baseline approaches, (ii) KNN-based approaches, and (iii) Matrix Factorization-based approaches. Specifically, we evaluate three baselines, i.e., Random, MostPopular, and UserItemAvg, which predicts the average listening count in the dataset by also accounting for deviations of u and a (e.g., if u tends

---

[1] http://surpriselib.com/.

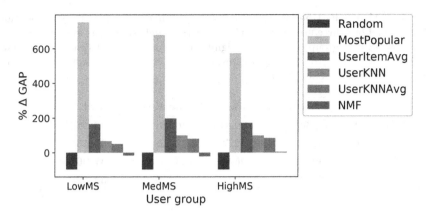

Fig. 4. Group Average Popularity (Δ GAP) of recommendation algorithms for LowMS, MedMS and HighMS. Except for the Random and NMF algorithms, all approaches provide too popular artist recommendations for all three user groups.

to have in general more listening events than the average Last.fm user) [6]. We also evaluate the two KNN-based approaches [13] UserKNN and UserKNNAvg, which is a hybrid combination of UserKNN and UserItemAvg. Finally, we include NMF (Non-Negative Matrix Factorization) into our study [9]. To reduce the computational effort of our study, in our evaluation, we exclude ItemKNN [12] as well as SVD++ [11] in contrast to [2]. In Fig. 3, we plot the correlation of artist popularity and how often the six algorithms recommend these artists. For all algorithms except for Random, we find a positive correlation, which means that popular items are recommended more often than unpopular items. As expected, this effect is most evident for the MostPopular algorithm and not present at all for the Random algorithm. It also seems that this popularity bias is not as strong in the case of NMF, which we will investigate further in the next section of this paper.

Popularity Bias for Different User Groups. To investigate the popularity bias of music recommendations for different user groups (i.e., LowMS, MedMS, and HighMS), we use the Group Average Popularity (GAP) metric proposed in [2]. Here, $GAP(g)_p$ measures the average popularity of the artists in the user profiles p of a specific user group g. We also define $GAP(g)_r$, which measures the average popularity of the artists recommended by a recommendation algorithm r to the users of group g. For each algorithm and user group, we are interested in the change in GAP (i.e., ΔGAP), which shows how the popularity of the recommended artists differs from the expected popularity of the artists in the user profiles. Hence, $\Delta GAP = 0$ would indicate fair recommendations in terms of item popularity, where fair means that the average artist popularity of the recommendations a user receives matches the average artist popularity in the user's profile. It is given by: $\Delta GAP = \frac{GAP(g)_r - GAP(g)_p}{GAP(g)_p}$.

Table 1. MAE results (the lower, the better) for four personalized recommendation algorithms and our three user groups. The worst (i.e., highest) results are always given for the LowMS user group (statistically significant according to a t-test with $p < .005$ as indicated by ***). Across the algorithms, the best (i.e., lowest) results are provided by NMF (indicated by bold numbers).

User group	UserItemAvg	UserKNN	UserKNNAvg	NMF
LowMS	42.991***	49.813***	46.631***	**38.515*****
MedMS	33.934	42.527	37.623	**30.555**
HighMS	40.727	46.036	43.284	**37.305**
All	38.599	45.678	41.927	**34.895**

In Fig. 4, we plot the ΔGAP for our six algorithms and three user groups. In contrast to the results presented in [2], where the LowMS group (i.e., the niche users) receives the highest values, we do not observe a clear difference between the groups except for MostPopular. We think that this is the case because of the large number of items we have in our Last.fm dataset (i.e., 352,805 artists compared to 3,900 movies in MovieLens). However, in line with Fig. 3, we again find that Random and NMF provide the fairest recommendations.

To further investigate $RQ2$, we analyze the Mean Average Error (MAE) [17] results of the four personalized algorithms for our user groups. As shown in Table 1, the LowMS group receives significantly worse (according to a t-test) recommendations than MedMS and HighMS for all algorithms. Interestingly, the MedMS group gets the best recommendations, probably since the users in this group have the largest profiles (i.e., on average 715 artists per user as compared to around 500 for the other two groups). Across the algorithms, NMF provides the best results. This is especially of interest since NMF also provided the fairest results in terms of artist popularity across the personalized algorithms.

4   Conclusion and Future Work

In this paper, we reproduced the study of [2] on the unfairness of popularity bias in movie recommender systems, which we adopted to the music domain. Similar to the original paper, we find (i) that users only have a limited interest in popular items ($RQ1$) and (ii) that users interested in unpopular items (i.e., LowMS) receive worse recommendations than users interested in popular items (i.e., HighMS). However, we also find that the proposed GAP metric does not provide the same results for Last.fm as it does for MovieLens, probably due to the high number of available items.

For future work, we plan to adapt this GAP metric in order to make it more robust for various domains. Furthermore, we want to study the characteristics of the LowMS users in order to better understand why they receive the worst recommendations and to potentially overcome this with novel algorithms (e.g., [7]).

Reproducibility and Acknowledgements. We provide our Last.fm dataset samples via Zenodo[2] [8] and our source code with all used parameter settings via Github[3]. This work was funded by the Know-Center GmbH (FFG COMET program) and the H2020 projects TRIPLE (GA: 863420) and AI4EU (GA: 825619).

References

1. Abdollahpouri, H., Burke, R., Mobasher, B.: Controlling popularity bias in learning-to-rank recommendation. In: Proceedings of the Eleventh ACM Conference on Recommender Systems, pp. 42–46. ACM (2017)
2. Abdollahpouri, H., Mansoury, M., Burke, R., Mobasher, B.: The unfairness of popularity bias in recommendation. In: Workshop on Recommendation in Multistakeholder Environments (RMSE 2019), in conjunction with the 13th ACM Conference on Recommender Systems, RecSys 2019 (2019)
3. Bauer, C., Schedl, M.: Global and country-specific mainstreaminess measures: definitions, analysis, and usage for improving personalized music recommendation systems. PloS one **14**(6), e0217389 (2019)
4. Brynjolfsson, E., Hu, Y.J., Smith, M.D.: From niches to riches: anatomy of the long tail. Sloan Manag. Rev. **47**(4), 67–71 (2006)
5. Jannach, D., Lerche, L., Kamehkhosh, I., Jugovac, M.: What recommenders recommend: an analysis of recommendation biases and possible countermeasures. User Model. User-Adap. Interact. **25**(5), 427–491 (2015). https://doi.org/10.1007/s11257-015-9165-3
6. Koren, Y.: Factor in the neighbors: scalable and accurate collaborative filtering. ACM Trans. Knowl. Discov. Data (TKDD) **4**(1), 1 (2010)
7. Kowald, D., Lex, E., Schedl, M.: Modeling artist preferences for personalized music recommendations. In: Proceedings of the Late-Breaking-Results Track of the 20th Annual Conference of the International Society for Music Information Retrieval (ISMIR 2019) (2019)
8. Kowald, D., Schedl, M., Lex, E.: LFM user groups (2019). https://doi.org/10.5281/zenodo.3475975
9. Luo, X., Zhou, M., Xia, Y., Zhu, Q.: An efficient non-negative matrix-factorization-based approach to collaborative filtering for recommender systems. IEEE Trans. Ind. Inform. **10**(2), 1273–1284 (2014)
10. Ricci, F., Rokach, L., Shapira, B.: Introduction to recommender systems handbook. In: Ricci, F., Rokach, L., Shapira, B., Kantor, P.B. (eds.) Recommender Systems Handbook, pp. 1–35. Springer, Boston (2011). https://doi.org/10.1007/978-0-387-85820-3_1
11. Sarwar, B., Karypis, G., Konstan, J., Riedl, J.: Incremental singular value decomposition algorithms for highly scalable recommender systems. In: Proceedings of the Fifth International Conference on Computer and Information Science, vol. 27, p. 28 (2002)
12. Sarwar, B.M., Karypis, G., Konstan, J.A., Riedl, J., et al.: Item-based collaborative filtering recommendation algorithms. In: WWW 1, 285–295 (2001)

---

[2] https://doi.org/10.5281/zenodo.3475975.
[3] https://github.com/domkowald/LFM1b-analyses.

13. Schafer, J.B., Frankowski, D., Herlocker, J., Sen, S.: Collaborative filtering recommender systems. In: Brusilovsky, P., Kobsa, A., Nejdl, W. (eds.) The Adaptive Web. LNCS, vol. 4321, pp. 291–324. Springer, Heidelberg (2007). https://doi.org/10.1007/978-3-540-72079-9_9
14. Schedl, M.: The LFM-1B dataset for music retrieval and recommendation. In: Proceedings of the 2016 ACM on International Conference on Multimedia Retrieval (ICMR 2016), pp. 103–110. ACM, New York (2016)
15. Schedl, M., Bauer, C.: Distance-and rank-based music mainstreaminess measurement. In: Adjunct Publication of the 25th Conference on User Modeling, Adaptation and Personalization, pp. 364–367. ACM (2017)
16. Schedl, M., Zamani, H., Chen, C.-W., Deldjoo, Y., Elahi, M.: Current challenges and visions in music recommender systems research. Int. J. Multimedia Inf. Retr. 7(2), 95–116 (2018). https://doi.org/10.1007/s13735-018-0154-2
17. Willmott, C.J., Matsuura, K.: Advantages of the mean absolute error (MAE) over the root mean square error (RMSE) in assessing average model performance. Clim. Res. 30(1), 79–82 (2005)

Reproducibility is a Process, Not an Achievement: The Replicability of IR Reproducibility Experiments

Jimmy Lin[✉] and Qian Zhang

David R. Cheriton School of Computer Science, University of Waterloo,
Waterloo, ON, Canada
jimmylin@uwaterloo.ca

Abstract. This paper espouses a view of reproducibility in the computational sciences as a *process* and not just a point-in-time "achievement". As a concrete case study, we revisit the Open-Source IR Reproducibility Challenge from 2015 and attempt to replicate those experiments: four years later, are those computational artifacts still functional? Perhaps not surprisingly, we are not able to replicate most of the retrieval runs encapsulated by those artifacts in a modern computational environment. We outline the various idiosyncratic reasons why, distilled into a series of "lessons learned" to help form an emerging set of best practices for the long-term sustainability of reproducibility efforts.

Keywords: Artifact evaluation · Community benchmarks

1 Introduction

In the broad discussion on reproducibility in the computational sciences, there has not been, at least in our view, much discussion of the fact that reproducibility is an *ongoing process*, not just an achievement—a "checkbox" we "tick off" and then continue going about our business.

There are many reasons why reproducibility is a worthwhile goal, ranging from demonstrating good stewardship of public resources (who fund a large portion of research worldwide) to the fact that reproducibility is intrinsic to the scientific process itself. Reproducibility enhances the veracity of findings and supports the iterative process of knowledge accumulation through which researchers build on each other's results.

The general discussion about reproducibility implicitly treats it as an achievement: a particular finding was reproduced successfully. And as a result, we (the scientific community) gain greater confidence in the veracity of the claims. This perspective is entirely appropriate in, for example, the physical or life sciences. Once we unravel the mystery of a particular physical phenomena, it is unlikely to change. Furthermore, reproduction efforts, for the most part, need to start "from scratch", with a completely independent set of experiments. Although further

© Springer Nature Switzerland AG 2020
J. M. Jose et al. (Eds.): ECIR 2020, LNCS 12036, pp. 43–49, 2020.
https://doi.org/10.1007/978-3-030-45442-5_6

studies refine our understanding over time (e.g., Newton's vs. Einstein's perspective on gravity), subsequent experiments are usually driven by new research questions, and cannot be considered reproduction efforts per se.

Reproducibility in the computational sciences, on the other hand, has many fundamentally different characteristics. A successful effort to reproduce a result typically yields a computational artifact that, ideally, should be executable by other researchers and remain functional over time. We explore exactly these desiderata and espouse the viewpoint that reproducibility in the computational sciences should not be viewed as an achievement (i.e., "we reproduced the technique of Yang et al. and confirm their findings") but rather a process.

Before proceeding, it is helpful to more precisely define our terminology. Here, we adopt recent ACM guidelines pertaining to artifact review and badging,[1] where reproducibility refers to artifacts created by an independent group to recreate a result, and replicability refers to using an existing artifact to recreate a result. Thus, we are concerned with the replicability of previous reproducibility (or replicability) experiments. That is, can we still rerun old computational artifacts to replicate their results?

As a case study, we attempt to replicate results from the Open-Source IR Reproducibility Challenge from 2015 [8] (henceforth, OSIRRC 2015). To quote: "the product is a repository that contains all code necessary to generate competitive *ad hoc* retrieval baselines, such that with a single script, anyone with a copy of the collection can reproduce *(sic)* the submitted runs." Four years later, can we still replicate those results? The answer, perhaps unsurprisingly, is *no* for most of the artifacts in a modern computational environment.

One might wonder, why would anyone want to run old code, and who cares? We present two compelling scenarios: First, findings in the computational sciences are always couched in some computational context. Take a simple example: algorithm A is faster than algorithm B, but the experiments were conducted on magnetic disks. With memory-resident data structures, does the finding still hold? Verification of results should be a continual process as computational contexts change: the shift from magnetic disks to SSDs, the advent of multi-core computing, etc. Although we do not undertake such an examination in this paper due to limited space, being able to replicate results using old computational artifacts is a pre-requisite. Second, another compelling scenario is long-term software preservation [11], especially for artifacts that have historical value, in the same way that there is active interest in preserving old video games today. For example, the SMART system [14] might fall into this category.

The contribution of this work is a discussion as well as a specific case study examining the long-term sustainability of reproducibility efforts. We grapple with issues that are under-explored in the community—for example, our questions do not naturally fit into the Platform, Research goal, Implementation, Method, Actor, and Data (PRIMAD) model [6] that attempts to capture the multi-faceted aspects of reproducibility. Of course, platforms change over time, but this change is an inevitable consequence of the passage of time; in other words, "platform" is

---

[1] https://www.acm.org/publications/policies/artifact-review-badging.

a dependent variable, not an independent variable, in our conception. PRIMAD implicitly assumes a static view of reproducibility, as opposed to the *process*-oriented view we advocate. More concretely, from the successes as well as failures in replicating OSIRRC 2015, we are able to extract a number of "lessons learned" that can be further distilled into best practices, especially to inform ongoing efforts such as the latest iteration of OSIRRC [3].

2 Replication Study

OSIRRC 2015 billed itself as providing scripts that allow any researcher to recreate, "out of the box", a number of baseline runs on the Gov2 web collection on topics 701–850 from the TREC Terabyte Tracks (2004–2006) [4]. Our experiments put these claims to the test. We cloned the git repository[2] and proceeded to run the dotgov2.sh script associated with each system after changing the location of the document collection in a common settings script. As designed, the single script should handle all aspects of producing an *ad hoc* experimental run from scratch, including building the inverted index, performing retrieval, and using trec_eval to generate effectiveness figures. In a few cases, the script worked exactly "as is", but more often than not, we encountered failures, which we then spent some time debugging (details below).

We ran experiments on two different servers, which we refer to as "old" and "new" for convenience. The "old server" was purchased in Spring 2017 and runs Ubuntu 16.04.2, with Oracle Java version 1.8.0_121 and gcc 5.4.0. The "new server" was purchased in Summer 2019 and runs Ubuntu 18.04.2, with Oracle Java version 11.0.2 and gcc 7.4.0.

The results of our replication study are summarized in Table 1; we refer the reader to the caption and our revised repository[3] for details. In summary, we were able to replicate results from five of the seven systems on the old server, and for two we obtained *exactly* the same AP scores; the remaining systems saw minor score differences. On the new server, we were only able to replicate results from one of the systems. Details for each system are presented below:

ATIRE [15] and **JASS** [9] are closely related in that the latter depends on the former. On the old server, the dotgov2.sh script worked without modification for both systems, and for ATIRE we obtained exactly the same AP scores as in the OSIRRC 2015 paper. Unfortunately, ATIRE fails to compile on the new server, and thus we were unable to replicate results there (and JASS by extension). Both systems are implemented in C++, so the critical difference here is the version of gcc (5.4.0 vs. 7.4.0); the compilation error arises from namespace clashes and different definitions of std::unordered_map between C++11 and C++17.

Indri [12] results were replicable on the old server without any modifications to the dotgov2.sh script, but the AP scores differed slightly from those reported

---

[2] https://github.com/lintool/IR-Reproducibility.

[3] https://github.com/lintool/IR-Reproducibility2.

Table 1. Summary of our replicability experiments, reporting average precision (AP) across different topics for different system combinations. Bolded headings represent rows copied directly from the OSIRRC 2015 paper; (old) and (new) refer to replication attempts on our "old" and "new" servers, respectively. AP scores that differ from OSIRRC 2015 are enclosed in parentheses.

System	Model	Index	Topics		
			701–750	751–800	801–850
ATIRE	BM25	Count	0.2616	0.3106	0.2978
ATIRE (old)	BM25	Count	0.2616	0.3106	0.2978
ATIRE (new)	BM25	Count	— Failed: compile error —		
ATIRE	Quantized BM25	Count + Quantized	0.2603	0.3108	0.2974
ATIRE (old)	Quantized BM25	Count + Quantized	0.2603	0.3108	0.2974
ATIRE (new)	Quantized BM25	Count + Quantized	— Failed: compile error —		
Galago	QL	Count	0.2776	0.2937	0.2845
Galago (old)	QL	Count	— Failed: exception —		
Galago (new)	QL	Count	— Failed: exception —		
Galago	SDM	Positions	0.2726	0.2911	0.3161
Galago (old)	SDM	Positions	— Failed: exception —		
Galago (new)	SDM	Positions	— Failed: exception —		
Indri	QL	Positions	0.2597	0.3179	0.2830
Indri (old)	QL	Positions	(0.2746)	(0.3182)	(0.2893)
Indri (new)	QL	Positions	— Failed: segmentation fault —		
Indri	SDM	Positions	0.2621	0.3086	0.3165
Indri (old)	SDM	Positions	(0.2624)	(0.3079)	(0.3244)
Indri (new)	SDM	Positions	— Failed: segmentation fault —		
JASS	1B Postings	Count	0.2603	0.3109	0.2972
JASS (repl. old)	1B Postings	Count	0.2603	(0.3108)	0.2972
JASS (repl. new)	1B Postings	Count	— Failed: compile error —		
JASS	2.5M Postings	Count	0.2579	0.3053	0.2959
JASS (repl. old)	2.5M Postings	Count	0.2579	0.3053	0.2959
JASS (repl. old)	2.5M Postings	Count	— Failed: compile error —		
Lucene	BM25	Count	0.2684	0.3347	0.3050
Lucene (old)	BM25	Count	0.2684	(0.3346)	0.3050
Lucene (new)	BM25	Count	— Failed: exceptions —		
Lucene	BM25	Positions	0.2684	0.3347	0.3050
Lucene (old)	BM25	Positions	0.2684	(0.3346)	0.3050
Lucene (new)	BM25	Positions	— Failed: exceptions —		
MG4J	BM25	Count	0.2640	0.3336	0.2999
MG4J	Model B	Count	0.2469	0.3207	0.3003
MG4J	Model B+	Positions	0.2322	0.3179	0.3257
Terrier	BM25	Count	0.2432	0.3039	0.2614
Terrier (old)	BM25	Count	0.2432	0.3039	0.2614
Terrier (new)	BM25	Count	0.2432	0.3039	0.2614
Terrier	DPH	Count	0.2768	0.3311	0.2899
Terrier (old)	DPH	Count	0.2768	0.3311	0.2899
Terrier (new)	DPH	Count	0.2768	0.3311	0.2899
Terrier	DPH + Bo1 QE	Count (inc direct)	0.3037	0.3742	0.3480
Terrier	DPH + Prox SD	Positions	0.2750	0.3297	0.2897

in the OSIRRC 2015 paper. On the new server, Indri compiles successfully but immediately encounters a segmentation fault when starting to build the index.

Galago [2] retrieval runs aborted due to a thrown exception after the indexing process finished without any obvious signs of error; the same issue was encountered on both servers. The exception appears to come from TupleFlow, a custom MapReduce-like framework used by Galago, which never gained widespread adoption and thus likely suffers from robustness issues.

Lucene runs successfully without any modification to the `dotgov2.sh` script on the old server; we encountered two tiny differences in AP scores compared to the OSIRRC 2015 paper, we suspect due to tie-breaking effects [10]. Lucene, however, failed to run on the new server, throwing exceptions during indexing. The critical difference is the version of Java (8 vs. 11). In the Lucene version used in these experiments (5.2.1), a bug was noted in Oracle's JRE implementation of `MMapDirectory`, which is unable to properly close the underlying OS file handle. The workaround deployed at that time used an undocumented internal cleanup functionality, which no longer works on Java 11.

MG4J [1] failed for the simple reason that the `dotgov2.sh` script attempts to fetch a tarball that was no longer available at the specified URL.

Terrier [13] was the only system where the `dotgov2.sh` script worked without modification on both the old server and the new server. Furthermore, we obtained exactly the same AP scores as the OSIRRC 2015 paper. However, the paper references two additional retrieval models, "DPH + Bo1 QE" and "DPH + Prox SD"; these runs were not included in the same script (unlike for most of the other systems). They were, in fact, provided in separate scripts (also available in the repository), but we did not discover this fact until much later; as a result, we were not able to replicate those two runs in time.

After encountering the initial errors with each of the systems, we did undertake efforts to debug the various issues. Ultimately, we did get two additional systems (ATIRE and Lucene) working on the new server. However, we consider it unreasonable to expect such debugging efforts from an envisioned user of the OSIRRC repository. Anything other than scripts "just working" should be consider a replicability failure, and a failure of OSIRRC to deliver on its promise.

3 Lessons Learned

For OSIRRC 2015, the passage of time has been ruthless in destroying the functionality of the computational artifacts. Although each artifact broke in its own idiosyncratic way, we do notice common themes. Given continued interest in reproducibility, most notably a new Docker-based iteration of OSIRRC in 2019 [3],[4] these "lessons learned" might form the basis of future best practices.

---

[4] https://osirrc.github.io/osirrc2019/.

Where Do These Results Come From? In many cases, the correspondence between figures in the paper and execution traces of the computational artifacts (even when successful) was not clear. As a simple example, Table 1 originally contained an "All" column, which, inexplicably, does not simply appear to be the average of each set of topics. The execution trace of each `dotgov2.sh` script was different: some printed out AP scores directly to stdout; others didn't, and silently completed execution. In the latter cases, it was necessary to dig through system outputs to find the actual scores; the fact that each system used completely different naming conventions for output runs, evaluation results, and log files made this task non-trivial.

Due to space restrictions, we focused on replicating effectiveness results in this paper, but OSIRRC 2015 also evaluated index size and query latency. However, the repository offered few details on how to extract figures corresponding to those reported in the paper. As one example, it wasn't clear where each system stored its index: some were contained in directories, some were single files, and others were groups of files (and in many cases, not intuitively named).

The high-level lesson here is that researchers should strive to make the correspondence between documentation, repository organization, execution traces, and reported figures in papers as explicit as possible—all these facets have to "line up" for successful replication. This can be viewed as an endorsement of "executable papers" [5,7]. Although this idea dates back decades (cf. literate programming), the popularity of notebooks (e.g., Jupyter) makes this much easier to realize today.

External Dependencies. Six of the seven systems in OSIRRC 2015 began by pulling in an external dependency (namely, the search engine itself) over the internet; the only exception, Lucene, had jars directly checked into the repository. "Link rot" is a well-known phenomenon, and the replication effort for MG4J failed at the outset because the artifact was no longer available. To ensure the long-term availability of resources at stable URLs, we advocate the use of dedicated archiving services such as Zenodo (or other domain-specific data repositories), which also provide citeable DOIs.

Platform Dependencies. As the saying goes, "change is the only constant", and computing platforms inevitably evolve, often in non-compatible ways: the two poignant examples here being JVM differences and compiler differences. Unlike external dependencies discussed above, it would be impractical to directly store and manage platform dependencies inside the OSIRRC 2015 repository, but here the use of Docker in OSIRRC 2019 represents an improvement. Docker allows platforms to be isolated in "layers" that can be composed to form images in a lightweight manner (compared to VMs); images themselves can be stored at stable locations (for example, in Docker Hub). Admittedly, this design does not guard against the obsolescence of Docker itself; despite the popularity of Docker today, one day it too will be superseded by new practices.

4   Conclusions

The 2019 edition of OSIRRC represents the next iteration of community efforts to advance the cause of reproducibility in information retrieval. It seems that many of the lessons discussed in the previous section have already been incorporated into its design. However, only time will tell if those efforts are successful: it would be interesting to try and rerun those experiments in 2023, the same distance we are from the initial OSIRRC efforts in 2015.

Acknowledgments. This research was supported by the Natural Sciences and Engineering Research Council (NSERC) of Canada and a Postdoctoral Fellowship in Software Curation grant from the Council on Library and Information Resources (CLIR), made possible by funding from the Andrew W. Mellon Foundation. We'd like to thank Craig Macdonald for catching a stupid mistake on our end.

References

1. Boldi, P., Vigna, S.: MG4J at TREC 2006. In: TREC (2006)
2. Cartright, M.A., Huston, S., Field, H.: Galago: a modular distributed processing and retrieval system. In: SIGIR 2012 Workshop on Open Source IR (2012)
3. Clancy, R., Ferro, N., Hauff, C., Lin, J., Sakai, T., Wu, Z.Z.: Overview of the 2019 open-source IR replicability challenge (OSIRRC 2019). In: CEUR Workshop Proceedings, Paris, France, vol. 2409, pp. 1–7 (2019)
4. Clarke, C., Craswell, N., Soboroff, I.: Overview of the TREC 2004 terabyte track. In: TREC (2004)
5. Dittrich, J., Bender, P.: Janiform intra-document analytics for reproducible research. Proc. VLDB Endow. **8**(12), 1972–1975 (2015)
6. Ferro, N., Fuhr, N., Järvelin, K., Kando, N., Lippold, M., Zobel, J.: Increasing reproducibility in IR: findings from the Dagstuhl seminar on "reproducibility of data-oriented experiments in e-science". SIGIR Forum **50**(1), 68–82 (2016)
7. Gorp, P.V., Mazanek, S.: SHARE: a web portal for creating and sharing executable research papers. Procedia Comput. Sci. **4**, 589–597 (2011)
8. Lin, J., et al.: Toward reproducible baselines: the open-source IR reproducibility challenge. In: ECIR, Padua, Italy, pp. 408–420 (2016)
9. Lin, J., Trotman, A.: Anytime ranking for impact-ordered indexes. In: ICTIR, pp. 301–304 (2015)
10. Lin, J., Yang, P.: The impact of score ties on repeatability in document ranking. In: SIGIR, Paris, France, pp. 1125–1128 (2019)
11. Matthews, B., Shaon, A., Bicarregui, J., Jones, C.: A framework for software preservation. Int. J. Digit. Curation **5**(1), 91–105 (2010)
12. Metzler, D., Croft, W.B.: Combining the language model and inference network approaches to retrieval. Inf. Process. Manag. **40**(5), 735–750 (2004)
13. Ounis, I., Amati, G., Plachouras, V., He, B., Macdonald, C., Lioma, C.: Terrier: a high performance and scalable information retrieval platform. In: SIGIR 2006 Workshop on Open Source IR (2006)
14. Salton, G.: The SMART Retrieval System-Experiments in Automatic Document Processing. Prentice-Hall, Englewood Cliffs (1971)
15. Trotman, A., Jia, X.F., Crane, M.: Towards an efficient and effective search engine. In: SIGIR 2012 Workshop on Open Source IR (2012)

On the Replicability of Combining Word Embeddings and Retrieval Models

Luca Papariello$^{(\boxtimes)}$, Alexandros Bampoulidis, and Mihai Lupu[ID]

Research Studio Data Science, RSA FG, Vienna, Austria
{luca.papariello,alexandros.bampoulidis,mihai.lupu}@researchstudio.at

Abstract. We replicate recent experiments attempting to demonstrate an attractive hypothesis about the use of the Fisher kernel framework and mixture models for aggregating word embeddings towards document representations and the use of these representations in document classification, clustering, and retrieval. Specifically, the hypothesis was that the use of a mixture model of von Mises-Fisher (VMF) distributions instead of Gaussian distributions would be beneficial because of the focus on cosine distances of both VMF and the vector space model traditionally used in information retrieval. Previous experiments had validated this hypothesis. Our replication was not able to validate it, despite a large parameter scan space.

1 Introduction

The last 5 years have seen proof that neural network-based word embedding models provide term representations that are a useful information source for a variety of tasks in natural language processing. In information retrieval (IR), "traditional" models remain a high baseline to beat, particularly when considering efficiency in addition to effectiveness [6]. Combining the word embedding models with the traditional IR models is therefore very attractive and several papers have attempted to improve the baseline by adding in, in a more or less ad-hoc fashion, word-embedding information. Onal et al. [10] summarized the various developments of the last half-decade in the field of neural IR and group the methods in two categories: *aggregate* and *learn*. The first one, also known as *compositional distributional semantics*, starts from term representations and uses some function to combine them into a document representation (a simple example is a weighted sum). The second method uses the word embedding as a first layer of another neural network to output a document representation.

The advantage of the first type of methods is that they often distill down to a linear combination (perhaps via a kernel), from which an explanation about the representation of the document is easier to induce than from the neural network layers built on top of a word embedding. Recently, the issue of explainability in IR and recommendation is generating a renewed interest [15].

In this sense, Zhang et al. [14] introduced a new model for combining high-dimensional vectors, using a mixture model of von Mises-Fisher (VMF) instead

© Springer Nature Switzerland AG 2020
J. M. Jose et al. (Eds.): ECIR 2020, LNCS 12036, pp. 50–57, 2020.
https://doi.org/10.1007/978-3-030-45442-5_7

of Gaussian distributions previously suggested by Clinchant and Perronnin [3]. This is an attractive hypothesis because the Gaussian Mixture Model (GMM) works on Euclidean distance, while the mixture of von Mises-Fisher (moVMF) model works on cosine distances—the typical distance function in IR.

In the following sections, we set up to replicate the experiments described by Zhang et al. [14]. They are grouped in three sets: classification, clustering, and information retrieval, and compare "standard" embedding methods with the novel moVMF representation.

2   Experimental Setup

In general, we follow the experimental setup of the original paper and, for lack of space, we do not repeat here many details, if they are clearly explained there.

2.1   Datasets

All experiments are conducted on publicly available datasets and are briefly described here below.

Classification. Two subsets of the movie review dataset: (i) the subjectivity dataset (subj) [11]; and (ii) the sentence polarity dataset (sent) [12].

Clustering. The 20 Newsgroups dataset[1] was used in the original paper, but the concrete version was not specified. We selected the "bydate" version, because it is, according to its creators, the most commonly used in the literature. It is also the version directly load-able in scikit-learn[2], making it therefore more likely that the authors had used this version.

Retrieval. The TREC Robust04 collection [13].

2.2   Models

The methods used to generate vectors for terms and documents are:

TF-IDF. The basic term frequency - inverse document frequency method [5]. *Implemented in the scikit-learn library*[3].

LSI. Latent Semantic Indexing [4].

LDA. Latent Dirichlet Allocation [2].

cBoW. Word2vec [9] in the Continuous Bag-of-Word (cBow) architecture.

PV-DBOW/DM. Paragraph vector (PV) is a document embedding algorithm that builds on Word2vec. We use here both its implementations: Distributed Bag-of-Words (PV-DBOW) and Distributed Memory (PV-DM) [7].

---

[1] http://qwone.com/~jason/20Newsgroups/.

[2] https://scikit-learn.org/stable/modules/generated/sklearn.datasets. fetch_20newsgroups.html.

[3] https://scikit-learn.org/stable/.

The LSI, LDA, cBoW, and PV implementations are available in the gensim library[4].

Fisher Kernel (FK). The FK framework offers the option to aggregate word embeddings to obtain fixed-length representations of documents. We use Fisher vectors (FV) based on (i) a Gaussian mixture model (FV-GMM) and (ii) a mixture of von Mises-Fisher distributions (FV-moVMF) [1].

We first fit (i) a GMM and (ii) a moVMF model on previously learnt continuous word embeddings. The fixed-length representation of a document X containing T words w_i—expressed as $X = \{E_{w_1}, \ldots, E_{w_T}\}$, where E_{w_i} is the word vector representation of word w_i—is then given by $\mathcal{G}^X = [\mathcal{G}_1^X, \ldots, \mathcal{G}_K^X]$, where K is the number of mixture components. The vectors $\mathcal{G}_i^X$, having the dimension (d) of the word vectors E_{w_i}, are explicitly given by [3,14]:

$$\text{(i)} \quad \mathcal{G}_i^X = \frac{1}{\sqrt{\omega_i}} \sum_{t=1}^{T} \gamma_t(i) \frac{x_t - \mu_i}{\sigma_i}, \quad \text{and} \quad \text{(ii)} \quad \mathcal{G}_i^X = \sum_{t=1}^{T} \frac{\gamma_t(i) x_t d}{\omega_i \kappa_i}, \quad (1)$$

where ω_i are the mixture weights, $\gamma_t(i) = p(i|x_t)$ is the soft assignment of x_t to (i) Gaussian and (ii) VMF distribution i, and $\sigma_i^2 = \text{diag}(\Sigma_i)$, with Σ_i the covariance matrix of Gaussian i. In (i), σ_i refers to the mean vector; in (ii) it indicates the mean direction and κ_i is the concentration parameter.

We implement the FK-based algorithms by ourselves, with the help of the scikit-learn library for fitting a mixture of Gaussian models and of the Spherecluster package[5] for fitting a mixture of von Mises-Fisher distributions to our data. The implementation details of each algorithm are described in what follows.

3   Experimental Results

Each of the following experiments is conceptually divided in three phases. First, text processing (e.g. tokenisation); second, creating a fixed-length vector representation for every document; finally, the third phase is determined by the goal to be achieved, i.e. classification, clustering, and retrieval.

For the first phase the same pre-processing is applied to all datasets. In the original paper, this phase was only briefly described as tokenisation and stopword removal. It is not given what tokeniser, linguistic filters (stemming, lemmatisation, etc.), or stop word list were used. Knowing that the gensim library was used, we took all standard parameters (see provided code[6]). Gensim however does not come with a pre-defined stopword list, and therefore, based on our own experience, we used the one provided in the NLTK library[7] for English.

---

[4] https://radimrehurek.com/gensim/.
[5] https://github.com/jasonlaska/spherecluster.
[6] https://rsagit.researchstudio.at/lpapariello/ecir_2020.git.
[7] https://www.nltk.org.

For the second phase, transforming terms and documents to vectors, Zhang et al. [14] specify that all trained models are 50 dimensional. We have additionally experimented with dimensionality 20 (used by Clinchant and Perronnin [3] for clustering) and 100, as we hypothesized that 50 might be too low. The TF-IDF model is 5000 dimensional (i.e. only the top 5000 terms based on their tf-idf value are used), while the Fischer-Kernel models are $15 \times d$ dimensional, where $d = \{20, 50, 100\}$, as just explained. In what follows, d refers to the dimensionality of LSI, LDA, cBow, and PV models.

The cBoW and PV models are trained using a default window size of 5, keeping both low and high-frequency terms, again following the setup of the original experiment. The LDA model is trained using a chunk size of 1000 documents and for a number of iterations over the corpus ranging from 20 to 100. For the FK methods, both fitting procedures (GMM and moVMF) are independently initialised 10 times and the best fitting model is kept.

For the third phase, parameters are explained in the following sections.

3.1   Classification

Logistic regression is used for classification in Zhang et al., and therefore also used here. The results of our experiments, for $d = 50$ and 100-dimensional feature vectors, are summarised in Table 1. For all the methods, we perform a parameter scan of the (inverse) regularisation strength of the logistic regression classifier, as shown in Fig. 1(a) and (b). Additionally, the learning algorithms are trained for a different number of epochs and the resulting classification accuracy assessed, cf. Fig. 1(c) and (d).

Table 1. Results of classification experiments on the *subj* and *sent* datasets. Shown are the mean accuracy and standard deviation, under 10-fold cross-validation, for optimally chosen hyperparameters (i.e. top values in Fig. 1).

Model	50-dim.		100-dim.	
	Subj	Sent	Subj	Sent
TF-IDF	89.3 ± 0.7	75.9 ± 1.3	89.3 ± 0.7	75.9 ± 1.3
LSI	82.5 ± 1.0	63.2 ± 0.9	84.3 ± 1.0	65.4 ± 1.6
LDA	62.2 ± 1.7	56.5 ± 1.6	62.1 ± 1.0	57.1 ± 1.5
cBow	89.0 ± 1.1	70.1 ± 1.3	89.2 ± 1.6	71.4 ± 1.1
PV-DBOW	88.7 ± 1.1	65.8 ± 1.4	89.4 ± 0.8	68.3 ± 1.3
PV-DM	84.0 ± 1.5	68.6 ± 1.3	88.0 ± 0.9	70.3 ± 1.1
FV-GMM	89.1 ± 0.9	68.5 ± 1.5	88.7 ± 0.9	72.4 ± 1.1
FV-moVMF	89.5 ± 0.9	70.1 ± 0.9	88.7 ± 1.4	71.2 ± 1.5

Figure 1(a) indicates that cBow, FV-GMM, FV-moVMF, and the simple TF-IDF, when properly tuned, exhibit a very similar accuracy on *subj*—the given confidence intervals do not indeed allow us to identify a single, best model. Surprisingly, TF-IDF outperforms all the others on the *sent* dataset (Fig. 1(b)). Increasing the dimensionality of the feature vectors, from $d = 50$ to 100, has the effect of reducing the gap between TF-IDF and the rest of the models on the *sent* dataset (see Table 1).

3.2   Clustering

For clustering experiments, the obtained feature vectors are passed to the k-means algorithm. The results of our experiments, measured in terms of Adjusted Rand Index (ARI) and Normalized Mutual Information (NMI), are summarised in Table 2. We used both $d = 20$ and 50-dimensional feature vectors. Note that the evaluation of the clustering algorithms is based on the knowledge of the ground truth class assignments, available in the 20 Newsgroups dataset.

As opposed to classification, clustering experiments show a generous imbalance in performance and firmly speak in favour of PV-DBOW. Interestingly, TF-IDF, FV-GMM, and FV-moVMF, all providing high-dimensional document representations, have a low clustering effectiveness.

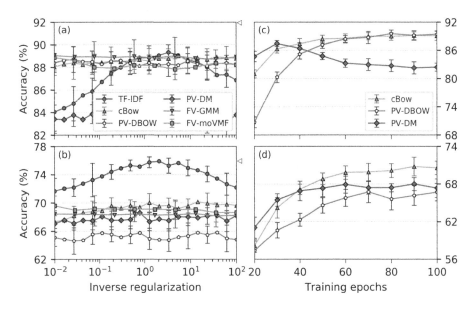

Fig. 1. Results of classification experiments, for 50-dimensional feature vectors, on the *subj* dataset [top panels, (a) and (c)] and *sent* dataset [bottom panels, (b) and (d)]. LSI and LDA achieve low accuracy (see Table 1) and are omitted here for visibility. The left panels [(a) and (b)] show the effect of (inverse) regularisation of the logistic regression classifier on the accuracy, while the right panels [(c) and (d)] display the effect of training for the learning algorithms. The two symbols on the right axis in panels (a) and (b) indicate the best (FV-moVMF) results reported in [14].

Table 2. Results of clustering experiments (mean performance and standard deviation over 10 runs) in terms of Adjusted Rand Index (ARI) and Normalised Mutual Information (NMI).

Model	20-dim.		50-dim.	
	ARI	NMI	ARI	NMI
TF-IDF	0.4 ± 0.1	5.6 ± 0.3	0.4 ± 0.1	5.6 ± 0.3
LSI	0.6 ± 0.1	5.3 ± 0.8	0.5 ± 0.2	5.8 ± 0.3
LDA	23.0 ± 0.7	39.5 ± 0.2	20.8 ± 1.1	40.2 ± 0.8
cBow	31.2 ± 0.4	50.4 ± 0.4	31.5 ± 0.3	51.2 ± 0.3
PV-DBOW	47.6 ± 1.2	63.4 ± 0.7	53.3 ± 1.3	66.1 ± 0.6
PV-DM	16.8 ± 0.6	44.8 ± 0.3	30.1 ± 0.9	53.2 ± 0.5
FV-GMM	1.4 ± 0.1	9.2 ± 0.8	1.0 ± 0.1	9.8 ± 1.3
FV-moVMF	2.1 ± 0.2	13.3 ± 1.3	1.6 ± 0.2	14.9 ± 1.6

3.3 Document Retrieval

For these experiments, we extracted from every document of the test collection all the raw text, and preprocessed it as described in the beginning of this section. The documents were indexed and retrieved for BM25 with the Lucene 8.2 search engine. We experimented with three topic processing ways: (1) title only, (2) description only, and (3) title and description. The third way produces the best results and closest to the ones reported by Zhang et al. [14], and hence are the only ones reported here.

An important aspect of BM25 is the fact that the variation of its parameters k_1 and b could bring significant improvement in performance, as reported by Lipani et al. [8]. Therefore, we performed a parameter scan for $k_1 \in [0,3]$ and $b \in [0,1]$ with a 0.05 step size for both parameters. For every TREC topic, the scores of the top 1000 documents retrieved from BM25 were normalised to [0,1]

Table 3. Results of retrieval experiments with 95% confidence intervals.

Model	Zhang et al.		Replicated		Replicated Best BM25	
	MAP	P@20	MAP	P@20	MAP	P@20
BM25	24.10	33.70	22.80 ± 2.51	33.21 ± 2.95	22.94 ± 2.50	33.63 ± 2.99
LSI	3.40	3.90	0.39 ± 0.33	1.20 ± 0.74	–	–
LDA	4.70	5.60	0.37 ± 0.13	0.96 ± 0.37	–	–
cBow	7.20	11.10	3.84 ± 0.78	8.84 ± 1.33	–	–
FV-GMM	9.80	12.40	5.72 ± 1.14	11.24 ± 1.79	–	–
FV-moVMF	11.20	13.90	3.16 ± 0.60	7.71 ± 1.23	–	–
BM25+LSI	25.30	36.60	23.11 ± 2.54	33.29 ± 2.95	23.22 ± 2.53	33.73 ± 3.00
BM25+LDA	25.30	36.30	22.88 ± 2.52	33.37 ± 2.96	22.99 ± 2.53	33.96 ± 2.98
BM25+cBow	25.30	36.50	23.78 ± 2.55	34.22 ± 2.95	23.88 ± 2.55	34.72 ± 2.98
BM25+FV-GMM	25.40	36.30	23.51 ± 2.55	34.08 ± 2.93	23.68 ± 2.54	34.54 ± 2.96
BM25+FV-moVMF	25.60	36.70	23.45 ± 2.53	33.88 ± 2.95	23.58 ± 2.54	34.34 ± 2.96

with the min-max normalisation method, and were used in calculating the scores of the documents for the combined models [14].

The original results, those of our replication experiments with standard ($k_1 = 1.2$ and $b = 0.75$) and best BM25 parameter values—measured in terms of Mean Average Precision (MAP) and Precision at 20 (P@20)—are outlined in Table 3.

4 Conclusions

We replicated previously reported experiments that presented evidence that a new mixture model, based on von Mises-Fisher distributions, outperformed a series of other models in three tasks (classification, clustering, and retrieval—when combined with standard retrieval models).

Since the source code was not released in the original paper, important implementation and formulation details were omitted, and the authors never replied to our request for information, a significant effort has been devoted to reverse engineer the experiments. In general, for none of the tasks were we able to confirm the conclusions of the previous experiments: we do not have enough evidence to conclude that FV-moVMF outperforms the other methods. The situation is rather different when considering the effectiveness of these document representations for clustering purposes: we find indeed that the FV-moVMF significantly underperforms, contradicting previous conclusions. In the case of retrieval, although Zhang et al.'s proposed method (FV-moVMF) indeed boosts BM25, it does not outperform most of the other models it was compared to.

Acknowledgments. Authors are partially supported by the H2020 Safe-DEED project (GA 825225).

References

1. Banerjee, A., Dhillon, I.S., Ghosh, J., Sra, S.: Clustering on the unit hypersphere using von Mises-Fisher distributions. J. Mach. Learn. Res. **6**, 1345–1382 (2005)
2. Blei, D.M., Ng, A.Y., Jordan, M.I.: Latent Dirichlet allocation. J. Mach. Learn. Res. **3**, 993–1022 (2003)
3. Clinchant, S., Perronnin, F.: Aggregating continuous word embeddings for information retrieval. In: Proceedings of the Workshop on Continuous Vector Space Models and Their Compositionality, pp. 100–109 (2013)
4. Deerwester, S., Dumais, S.T., Furnas, G.W., Landauer, T.K., Harshman, R.: Indexing by latent semantic analysis. J. Am. Soc. Inf. Sci. **41**(6), 391–407 (1990)
5. Harris, Z.S.: Distributional structure. Word **10**(2–3), 146–162 (1954)
6. Hofstätter, S., Hanbury, A.: Let's measure run time! Extending the IR replicability infrastructure to include performance aspects. In: Proceedings of the Open-Source IR Replicability Challenge Co-Located with 42nd International ACM SIGIR Conference on Research and Development in Information Retrieval, OSIRRC@SIGIR 2019, Paris, France, 25 July 2019, pp. 12–16 (2019)
7. Le, Q., Mikolov, T.: Distributed representations of sentences and documents. In: International Conference on Machine Learning, pp. 1188–1196 (2014)

8. Lipani, A., Lupu, M., Hanbury, A., Aizawa, A.: Verboseness fission for BM25 document length normalization. In: Proceedings of the 2015 International Conference on the Theory of Information Retrieval, pp. 385–388. ACM (2015)
9. Mikolov, T., Chen, K., Corrado, G., Dean, J.: Efficient estimation of word representations in vector space (2013)
10. Onal, K.D., et al.: Neural information retrieval: at the end of the early years. Inf. Retrieval J. **21**(2), 111–182 (2018). https://doi.org/10.1007/s10791-017-9321-y
11. Pang, B., Lee, L.: A sentimental education: sentiment analysis using subjectivity summarization based on minimum cuts. In: Proceedings of the 42nd Annual Meeting of the Association for Computational Linguistics, ACL 2004, pp. 271–278 (2004)
12. Pang, B., Lee, L.: Seeing stars: exploiting class relationships for sentiment categorization with respect to rating scales. In: Proceedings of the 43rd Annual Meeting of the Association for Computational Linguistics, ACL 200, pp. 115–124. Association for Computational Linguistics (2005)
13. Voorhees, E.M.: The TREC robust retrieval track. SIGIR Forum **39**(1), 11–20 (2005)
14. Zhang, R., Guo, J., Lan, Y., Xu, J., Cheng, X.: Aggregating neural word embeddings for document representation. In: Pasi, G., Piwowarski, B., Azzopardi, L., Hanbury, A. (eds.) ECIR 2018. LNCS, vol. 10772, pp. 303–315. Springer, Cham (2018). https://doi.org/10.1007/978-3-319-76941-7_23
15. Zhang, Y., Zhang, Y., Zhang, M., Shah, C.: EARS 2019: the 2nd international workshop on explainable recommendation and search. In Proceedings of the 42nd International ACM SIGIR Conference on Research and Development in Information Retrieval, SIGIR 2019, Paris, France, 21–25 July 2019, pp. 1438–1440 (2019)

Influence of Random Walk Parametrization on Graph Embeddings

Fabian Schliski$^{(\boxtimes)}$, Jörg Schlötterer , and Michael Granitzer

University of Passau, 94032 Passau, Germany
schliski@fim.uni-passau.de,
{joerg.schloetterer,michael.granitzer}@uni-passau.de

Abstract. Network or graph embedding has gained increasing attention in the research community during the last years. In particular, many methods to create graph embeddings using random walk based approaches have been developed. node2vec [10] introduced means to control the random walk behavior, guiding the walks. We aim to reproduce parts of their work and introduce two additional modifications (jump probabilities and attention to hubs), in order to investigate how guiding and modifying the walks influences the learned embeddings. The reproduction includes the case study illustrating homophily and structural equivalence subject to the chosen strategy and a node classification task. We were not able to illustrate structural equivalence and further results show that modifications of the walks only slightly improve node classification, if at all.

Keywords: Feature learning · Graph embedding · Random walk

1 Introduction

Network analysis involves methods to predict over nodes and edges, such as node classification [5], link prediction [12], clustering [8], and visualization [13].

Node classification aims at predicting the labels of unlabeled nodes based on a set of different labeled nodes and the network topology. An example is to predict the interests of a user in a social network based on other users with overlapping characteristics. Link prediction is used to predict missing or future links between nodes in the network. In a social network, it can be used to recommend new friends based on the current ones. Clustering attempts to identify similarities between nodes in the network and groups them into same-labeled clusters. This can be used to detect communities with similar interests in a social network. Visualization helps to gain quick insights about the structure of the network.

Graph embeddings are feature vector representations of the nodes and edges of a network and are used as input to the methods above. node2vec [10] uses a random-walk based approach to create those embeddings and introduces two additional parameters to guide the random walk, aiming at preserving both community structure and structural roles. Community structure in a graph is

© Springer Nature Switzerland AG 2020
J. M. Jose et al. (Eds.): ECIR 2020, LNCS 12036, pp. 58–65, 2020.
https://doi.org/10.1007/978-3-030-45442-5_8

based on proximity, i.e., nodes that are close together belong to a community. Structural roles can be nodes that act as bridges between sub-networks, or hubs, which are the main exchange point between many nodes. The two parameters guiding the random walk in node2vec [10] are:

- Return parameter p, controlling the likeliness of immediately revisiting a node.
- In-out parameter q, controlling how far outward the random walk should progress from the starting node.

These parameters allow to resemble depth-first (DFS) or breadth-first (BFS) search-like bevahiour in the most extreme setting, as well as a smooth interpolation between DFS and BFS. Grover and Leskovec suggest "that BFS and DFS strategies represent extreme ends on the spectrum of embedding nodes based on the principles of homophily (i.e., network communities) and structural equivalence (i.e., structural roles of nodes)" [10]. In a case study, they demonstrate that subject to the parameter settings, the resulting embeddings can capture homophily or structural equivalence. In this paper[1], we

- try to reproduce the case study illustrating homophily and structural equivalence.
- try to reproduce node2vec's node classification result.
- introduce two additional modifications to the random walk strategy and evaluate and compare them on the node classification task. The additional strategies comprise hub attention and jump probalities, where the latter can be seen as noise.

2 Related Work

Algorithms to create graph embeddings can be divided into three categories: Factorization based, deep learning based, and random walk based [9].

Factorization based algorithms represent the graph as a matrix and apply methods such as eigenvalue decomposition or gradient descent to obtain node embeddings [9]. Examples are LLE [19], Laplacian Eigenmaps [4], GraRep [6], and HOPE [16].

The deep learning based methods try to improve the performance of the factorization algorithms by computing non-linear functions on the graph. Examples are SDNE [20] (auto-encoder to reduce dimensionality), DNGR [7] (deep neural networks), and GCN [11] (graph convolutional networks).

Random walk based approaches create embeddings by processing sets of random walks through the graph. First in this line was DeepWalk [18], which samples purely random walks. These walks are then treated as sentence equivalents (where every node in the sequence corresponds to a word) and fed to a skip-gram model, a model variant of the word embeddings introduced by

---

[1] Sourcecode and datasets are available at https://doi.org/10.5281/zenodo.3514305.

Mikolov et al. [14], which became famous under the term word2vec. node2vec [10] follows this approach, but provides means to control the random walk behavior.

For further methods and additional details of the methods mentioned above consult a survey by Goyal and Ferrara [9].

3   Additional Random Walk Modifications

In the next sections, we will introduce modifications to the random walk strategy, similar to the one implemented by node2vec [10]. The modifications can take place in two sections of the random walk algorithm: During *sampling*, we can modify the transition probabilities between nodes to draw attention to specific ones, while during *walking*, we can directly influence how the random walk traverses.

3.1   Jump Probability

We introduce the parameter j to modify the random walk during walking. It controls the probability of jumping to a random node in the graph at any given time. Intuitively, j ranges from 0 to 1, where with 0 no jumps to a random node occur, and with 1 every walking step is a random jump. The latter allows to create a truly random "walk" through the graph, without drawing any attention to the structure of the graph and edges with their respective weights. We are sampling truncated random walks, i.e., we start walks with a fixed length from every node in the graph. Therefore, the jump probability can be seen as noise in the truncated walks, opposed to jump probability in a single (huge) walk (as used by PageRank [17] for example).

3.2   Hub Attention

Hubs are nodes in a graph with a degree that greatly exceeds the average [2]. The threshold ε is the sum of the mean node degree and the standard deviation of all degrees:

$$\varepsilon = \overline{deg} + \sigma_{deg}$$

Based on that, we define the subset of hubs H of the nodes N as every node with a higher degree than ε:

$$H = \{n \in N \mid deg(n) > \varepsilon\} \subset N$$

We are now modifying the random walk during sampling. Similar to node2vec [10], we change the unnormalized transition probability π_{vx} of a node v to its neighbors x on edges (v, x) to $\pi_{vx} = \alpha_h(x) * w_{vx}$, where w_{vx} is the weight of the edge, and

$$\alpha_h(x) = \begin{cases} \frac{1}{h} & if \ x \in H \\ 1 & else \end{cases}$$

The parameter h introduced here controls the random walk tendency towards and away from hubs. If this parameter is set to a high value (>1), the probability of directly revisiting hubs is reduced, and the less frequented nodes are explored. On the other hand, if h is small (<1), it increases the likelihood of traversing to hubs, so the walk is more focused on the areas around hubs in the graph.

4   Evaluation

We begin this section by introducing the datasets and parameters used in our experiments, followed by the different evaluation tasks performed.

Les Misérables [1] is a network which contains the characters and their co-appearances in the novel "Les Misérables" by Victor Hugo. Every node represents a character, and an edge between two characters indicates that they appeared in the same book chapter. The graph consists of 77 nodes, connected via 254 edges. BlogCatalog [21] is a social network where every node is a blogger, and an edge between two of them represents friendship. The graph consists of 10,312 nodes, connected via 333,983 edges and assigned to one or more of 39 classes (multi-label). The classes are the topics the blogger is interested in.

We define a set of parameters for all learning algorithms to create a basis for a fair comparison. These are the embedding dimension d, the walk length l, the number of walks n and the skip-gram model [15] window size w. Furthermore the algorithm-specific parameters p, q (node2vec [10]), h (Hub Attention, c.f. Sect. 3.2) and j (Jump Probability, c.f. Sect. 3.1).

4.1   Reproduction of Les Misérables Case Study

To create the visualizations, we first learn the embeddings of the Les Misérables dataset using the respective random-walk algorithm. We then cluster these embeddings using k-means and assign the nodes of the graph colors based on their cluster. These embeddings are then visualized as a graph with Gephi [3]. For the common parameters, we used the values reported by Grover and Leskovec: embedding dimension $d = 16$, walk length $l = 80$, number of walks $n = 10$ and context window size $w = 10$.

The upper graph in the original paper [10, figure 3] reflects community structure and the values $p = 1$ and $q = 0.5$ are reported for walk parametrization. Our result in Fig. 1a also represents the community structure of the network, comparable to that in the original paper.

For the lower graph in the original paper [10, figure 3], which as per description resembles structural equivalence, $p = 1$ and $q = 2$ are specified. Even with grid-search over these and further parameters (l, n and w), no result close to the original could be produced. The graph never represented structural equivalence, but community structure as well (with 3 instead of 6 clusters), as shown in Fig. 1b.

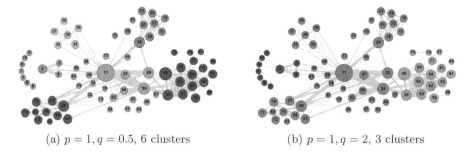

(a) $p = 1, q = 0.5$, 6 clusters (b) $p = 1, q = 2$, 3 clusters

Fig. 1. Les Misérables network with nodes colored corresponding to their cluster in embeddings created by different walk strategies. (Color figure online)

Table 1. Macro f1 scores and standard deviation ($\pm$) of the node classification task on the BlogCatalog dataset using different parameters for each algorithm.

Learner	Parameters	Score
node2vec	$p = 0.25, q = 0.25$	$\underline{26.72} \pm 0.72$
	$p = 1, q = 1$	25.85 ± 0.59
Jump probability	$j = 1$	03.64 ± 0.14
	$j = 0.25$	25.27 ± 0.68
	$j = 0.1$	$\underline{25.76} \pm 0.55$
Hub attention	$h = 0.5$	23.37 ± 0.59
	$h = 0.75$	25.21 ± 0.64
	$h = 4$	$\mathbf{\underline{27.44}} \pm 0.64$
	$h = 8$	27.17 ± 0.52
	$h = 10$	27.39 ± 0.55

4.2   Node Classification

We are running a multi-label node classification task on the BlogCatalog dataset. Same as in the original paper [10], we use a one-vs-rest logistic regression classifier with L2 regularization. The dataset is split into training and test data, with a training fraction of 50%, and the scores are averaged over 10 random splits. Again, we use the common parameters as reported by Grover and Leskovec [10]: embedding dimension $d = 128$, walk length $l = 80$, number of walks $n = 10$ and window size $w = 10$. All results of the following sections are reported in Table 1. The underlined values are the best per walk strategy, and the bold one is the best value overall.

Reproduction of Results. We used the values $p = q = 0.25$ as reported by node2vec [10] and also included results with parameters $p = q = 1$, eliminating the influence of the parameters on the random walk, resembling a "pure" random walk like DeepWalk [18]. We were not able to reproduce the results within a single iteration of the skip-gram model, but used five iterations instead. We suspect the difference to be due to an unrecognized change of the default hyper-parameters in gensim[2], the word2vec implementation used by node2vec. Our reproduced score is close to the reported one and even slightly better, which can be explained by the random factor of the experiment. From this, we can conclude that our values are consistent with those in the original paper. Furthermore, node2vec [10] performed better than non-parameterized random walks like DeepWalk, at least on this dataset and parameter settings.

Jump Probability. We set j to each of $\{1, 0.25, 0.1\}$. Expectably, the higher the jump probability is, the consistently worse the results get. However, at $j = 0.1$, i.e. 10% noise, the performance is close to DeepWalk. This indicates, that a small amount of noise in the walks does not drastically harm the performance of the resulting embeddings on the node classification task.

Hub Attention. We set h to each of $\{0.5, 0.75, 4, 8, 10\}$. As shown in Table 1, a value of $h = 4$ allows us to achieve the best value across the different strategies. When focussing on hubs ($h = 0.5, h = 0.75$), performance even drops below the score of a jump probability of 25% ($j = 0.25$), i.e. jumping to a random node in every 4th step on average. Conversely, we gain performance if we put more attention to otherwise less-frequently visited nodes (i.e. if we increase h), at least up to a certain point where the results stabilize.

5   Discussion and Conclusion

We attribute our inability to reproduce the case study in terms of structural equivalence to the skip-gram model. Its objective is to predict neighboring nodes and hence, nodes with similar neighbors are represented closeby in embedding space. Another factor is the context window size of the skip-gram model: No matter how far out a walk traverses, only nodes within this window will be considered as context. This also means, that the walks which start at a particular node are not that relevant to this particular node, but its embedding is determined by all the walks traversing through this node. With optimal parameter settings and taking the standard deviation into account, the performance difference between the walk strategies on the node classification task is negligible. In addition, Perozzi et al. report a macro-F1 score of 27.3 for DeepWalk [18], doing shorter, but more walks. They report performance to increase constantly with the number of walks until it finally stabilizes.

We conclude, that adapting the walk strategy can improve the embedding performance, if the number of sampled walks is insufficient. However, instead of tuning hyper-parameters of particular walk strategies, we can also increase the number of sampled walks per node instead. The nature of the skip-gram model

---

[2] https://radimrehurek.com/gensim/.

and our inability to reproduce the structural equivalence case study point to the embeddings always representing homophily.

Acknowledgments. The presented work was developed within the East-Bavarian Centre of Internet Competence, Big and Open Data Analytics for Small and Medium-sized Enterprises (BODA), funded by the Bavarian Ministry of Economic Affairs and Media, Energy and Technology.

References

1. Les misérables network dataset - KONECT, April 2017. http://konect.uni-koblenz.de/networks/moreno_lesmis
2. Barabási, A.L., et al.: Network Science. Cambridge University Press, Cambridge (2016)
3. Bastian, M., Heymann, S., Jacomy, M.: Gephi: an open source software for exploring and manipulating networks. In: Third International AAAI Conference on Weblogs and Social Media (2009)
4. Belkin, M., Niyogi, P.: Laplacian eigenmaps and spectral techniques for embedding and clustering. In: Advances in Neural Information Processing Systems, pp. 585–591 (2002)
5. Bhagat, S., Cormode, G., Muthukrishnan, S.: Node classification in social networks. Soc. Netw. Data Anal. 115-148 (2011). https://doi.org/10.1007/978-1-4419-8462-3_5
6. Cao, S., Lu, W., Xu, Q.: Grarep: learning graph representations with global structural information. In: Proceedings of the 24th ACM International on Conference on Information and Knowledge Management, pp. 891–900. ACM (2015)
7. Cao, S., Lu, W., Xu, Q.: Deep neural networks for learning graph representations. In: Thirtieth AAAI Conference on Artificial Intelligence (2016)
8. Ding, C.H.Q., He, X., Zha, H., Gu, M., Simon, H.D.: A min-max cut algorithm for graph partitioning and data clustering. In: Proceedings 2001 IEEE International Conference on Data Mining, pp. 107–114, November 2001. https://doi.org/10.1109/ICDM.2001.989507
9. Goyal, P., Ferrara, E.: Graph embedding techniques, applications, and performance: a survey. Knowl.-Based Syst. **151**, 78–94 (2018)
10. Grover, A., Leskovec, J.: node2vec: scalable feature learning for networks. In: Proceedings of the 22nd ACM SIGKDD International Conference on Knowledge Discovery and Data Mining (2016)
11. Kipf, T.N., Welling, M.: Semi-supervised classification with graph convolutional networks (2016). arXiv preprint arXiv:1609.02907
12. Liben-Nowell, D., Kleinberg, J.: The link-prediction problem for social networks. J. Am. Soc. Inf. Sci. Technol. **58**(7), 1019–1031 (2007). https://doi.org/10.1002/asi.20591
13. Maaten, L.V.D., Hinton, G.: Visualizing data using t-SNE. Mach. Learn. Res. **9**(Nov), 2579–2605 (2008)
14. Mikolov, T., Sutskever, I., Chen, K., Corrado, G., Dean, J.: Distributed representations of words and phrases and their compositionality. In: Proceedings of the 26th International Conference on Neural Information Processing Systems, NIPS 2013, pp. 3111–3119. Curran Associates Inc., USA (2013). http://dl.acm.org/citation.cfm?id=2999792.2999959

15. Mikolov, T., Sutskever, I., Chen, K., Corrado, G.S., Dean, J.: Distributed representations of words and phrases and their compositionality. In: Advances in Neural Information Processing Systems, pp. 3111–3119 (2013)

16. Ou, M., Cui, P., Pei, J., Zhang, Z., Zhu, W.: Asymmetric transitivity preserving graph embedding. In: Proceedings of the 22nd ACM SIGKDD International Conference on Knowledge Discovery and Data Mining, pp. 1105–1114. ACM (2016)

17. Page, L., Brin, S., Motwani, R., Winograd, T.: The pagerank citation ranking: bringing order to the web. Technical report Stanford InfoLab (1999)

18. Perozzi, B., Al-Rfou, R., Skiena, S.: Deepwalk: online learning of social representations. In: Proceedings of the 20th ACM SIGKDD International Conference on Knowledge Discovery and Data Mining. pp. 701–710. ACM (2014)

19. Roweis, S.T., Saul, L.K.: Nonlinear dimensionality reduction by locally linear embedding. Science 290(5500), 2323–2326 (2000)

20. Wang, D., Cui, P., Zhu, W.: Structural deep network embedding. In: Proceedings of the 22nd ACM SIGKDD International Conference on Knowledge Discovery and Data Mining, pp. 1225–1234. ACM (2016)

21. Zafarani, R., Liu, H.: Social computing data repository at ASU (2009). http://socialcomputing.asu.edu

Short Papers

Calling Attention to Passages
for Biomedical Question Answering

Tiago Almeida and Sérgio Matos[(⊠)]

DETI/IEETA, University of Aveiro, 3810-193 Aveiro, Portugal
{tiagomeloalmeida,aleixomatos}@ua.pt

Abstract. Question answering can be described as retrieving relevant information for questions expressed in natural language, possibly also generating a natural language answer. This paper presents a pipeline for document and passage retrieval for biomedical question answering built around a new variant of the DeepRank network model in which the recursive layer is replaced by a self-attention layer combined with a weighting mechanism. This adaptation halves the total number of parameters and makes the network more suited for identifying the relevant passages in each document. The overall retrieval system was evaluated on the BioASQ tasks 6 and 7, achieving similar retrieval performance when compared to more complex network architectures.

Keywords: Biomedical question answering · Neural networks · Attention mechanism · Snippet extraction

1 Introduction

Question Answering (QA) is a subfield of Information Retrieval (IR) that specializes in producing or retrieving a single answer for a natural language question. QA has received growing interest since users often look for a precise answer to a question instead of having to inspect full documents [4]. Similarly, biomedical question answering has also gained importance given the amount of information scattered over large specialized repositories such as MEDLINE. Research on biomedical QA has been pushed forward by community efforts such as the BioASQ challenge [13], originating a range of different approaches and systems.

Recent studies on the application of deep learning methods to IR have shown very good results. These neural models are commonly subdivided into two categories based on their architecture. **Representation-based** models, such as the Deep Structured Semantic Model (DSSM) [5] or the Convolutional Latent Semantic Model (CLSM) [12], learn semantic representations of texts and score each query-document pair based on the similarity of their representations. On the other hand, models such as the Deep Relevance Matching Model (DRMM) [3] or DeepRank [10] follow a **interaction-based** approach, in which matching signals between query and document are captured and used by the neural network to produces a ranking score.

© Springer Nature Switzerland AG 2020
J. M. Jose et al. (Eds.): ECIR 2020, LNCS 12036, pp. 69–77, 2020.
https://doi.org/10.1007/978-3-030-45442-5_9

The impact of neural IR approaches is also noticeable in biomedical question answering, as shown by the results on the most recent BioASQ challenges [9]. The top performing team in the document and snippet retrieval sub-tasks in 2017 [1], for example, used a variation of the DRMM [8] to rank the documents recovered by the traditional BM25 [11]. For the 2018 task, the same team extended their system with the inclusion of models based on BERT [2] and with joint training for document and snippet retrieval.

The main contribution of this work is a new variant of the DeepRank neural network architecture in which the recursive layer originally included in the final aggregation step is replaced by a self-attention layer followed by a weighting mechanism similar to the term gating layer of the DRMM. This adaptation not only halves the total number of network parameters, therefore speeding up training, but it is also more suited for identifying the relevant snippets in each document. The proposed model was evaluated on the BioASQ dataset, as part of a document and passage (snippet) retrieval pipeline for biomedical question answering, achieving similar retrieval performance when compared to more complex network architectures. The full network configuration is publicly available at https://github.com/bioinformatics-ua/BioASQ, together with code for replicating the results presented in this paper.

2   System Description

This section presents the overall retrieval pipeline and describes the neural network architecture proposed in this work for the document ranking step.

The retrieval system follows the pipeline presented in Fig. 1, encompassing three major modules, **Fast Retrieval**, **Neural Ranking** and **Snippet extraction**. The fast retrieval step is focused on minimizing the number of documents passed on to the computationally more demanding neural ranking module, while maintaining the highest possible recall. As in previous studies [1,7], we adopted Elasticsearch (ES) with the BM25 ranking function as the retrieval mechanism.

The documents returned by the first module are ranked by the neural network which also directly provides to the following module the information for extracting relevant snippets. These modules are detailed in Sects. 2.1 and 2.2.

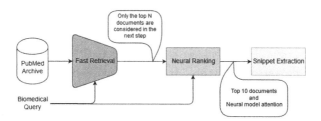

Fig. 1. Overview of the main modules of the proposed system. The number N of documents returned by the first module is considered an hyper-parameter.

2.1   Neural Ranking Model

The network follows a similar architecture to the original version of DeepRank [10], as illustrated in Fig. 2. Particularly, we build upon the best reported configuration, which uses a CNN in the measurement network and the reciprocal function as the position indicator. The inputs to the network are the **query**, a **set of document passages** aggregated by each query term, and the **absolute position** of each passage. For the remaining explanation, let us first define a query as a sequence of terms $q = \{u_0, u_1, ..., u_Q\}$, where u_i is the i-th term of the query; a set of document passages aggregated by each query term as $D(u_i) = \{p_0, p_1, ..., p_P\}$, where p_j corresponds to the j-th passage with respect to the query term u_i; and a document passage as $p = \{v_0, v_1, ..., v_S\}$, where v_k is the k-th term of the passage. We chose to aggregate the passages by their respective query term at the input level, since it simplifies the neural network flow and implementation.

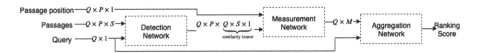

Fig. 2. High-level structure and data flow of the proposed version of DeepRank.

The **detection network** receives as input the **query** and the **set of document passages** and creates a similarity tensor (interaction matrix) $S \in [-1, 1]^{Q \times S}$ for each passage, where each entry S_{ij} corresponds to the cosine similarity between the embeddings of the i-th query term and j-th passage term, $S_{ij} = \frac{\vec{u_i}^T \cdot \vec{v_j}}{\|\vec{u_i}\| \times \|\vec{v_j}\|}$.

The **measurement network** step is the same used in the original DeepRank model. It takes as inputs the previously computed tensors S and the **absolute position** of each passage and applies a 2D convolution followed by a global max polling operation, to capture the local relevance present in each tensor S, as defined in Eq. 1:

$$h_{i,j}^m = \sum_{s=0}^{x-1} \sum_{t=0}^{y-1} w_{s,t}^m \times S_{i+s,j+t} + b^m \,,$$
$$h^m = \max_{i,j}(h_{i,j}^m), \ m = 1, ..., M \,. \tag{1}$$

At this point, the set of document passages for each query term is represented by their respective vectors $\vec{h}$, i.e, $D(u_i) = \{\vec{h}_{p_0}, \vec{h}_{p_1}, ..., \vec{h}_{p_P}\}$, where $\underset{M \times 1}{\vec{h}}$ encodes the local relevance captured by the M convolution kernels of size $x \times y$, plus an additional feature corresponding to the position of the passage.[1]

---

[1] For simplicity, we consider that the dimension M already accounts for the concatenated feature, i.e, $\underset{M \times 1}{\vec{h}} \leftarrow \underset{(M+1) \times 1}{\vec{h}}$.

The next step uses a self-attention layer [6] to obtain an aggregation $\underset{M \times 1}{\vec{c_{u_i}}}$ over the passages h_{p_j} for each query term u_i, as defined in Eq. 2. The weights a_{p_j}, which are computed by a feed forward network and converted to a probabilistic distribution using the softmax operation, represent the importance of each passage vector from the set $D(u_i)$. The addition of this self-attention layer, instead of the recurrent layer present in the original architecture, allows using the attention weights, that are directly correlated with the local relevance of each passage, to identify important passages within documents. Moreover, this layer has around $A \times M$ parameters, compared to up to three times more in the GRU layer (approximately $3 \times A \times (A + M)$), which in practice means reducing the overall number of network parameters to half.

$$
\begin{aligned}
s_{p_j} &= \underset{1 \times A}{w^T} \cdot \tanh \left(\underset{A \times M}{W} \cdot \underset{M \times 1}{\vec{h}_{p_j}} \right) , \\
a_{p_j} &= \frac{e^{s_{p_j}}}{\sum_{p_k \in D(u_i)} e^{s_{p_k}}} , \\
\underset{M \times 1}{\vec{c_{u_i}}} &= \sum_{p_j \in D(u_i)} \left(\underset{1 \times 1}{a_{p_j}} \times \underset{M \times 1}{\vec{h}_{p_j}} \right) .
\end{aligned}
\tag{2}
$$

Finally, the **aggregation network** combines the vectors $\underset{M \times 1}{\vec{c_{u_i}}}$ according to weights that reflect the importance of each individual query term u_i. We chose to employ a similar weighting mechanism to the term gating layer in DRMM [3], which uses the query term embedding to compute its importance, as defined in Eq. 3. This option replaces the use of a trainable parameter for each vocabulary term, as in the original work, which is less suited for modelling a rich vocabulary as in the case of biomedical documents.

The final aggregated vector $\vec{c}$ is then fed to a dense layer for computing the final ranking score.

$$
\begin{aligned}
s_{u_i} &= \underset{1 \times E}{\vec{w}} \cdot \underset{E \times 1}{\vec{x_{u_i}}} , \\
a_{u_i} &= \frac{e^{s_{u_i}}}{\sum_{u_k \in q} e^{s_{u_k}}} , \\
\underset{M \times 1}{\vec{c}} &= \sum_{u_i \in q} \left(\underset{1 \times 1}{a_{u_i}} \times \underset{M \times 1}{\vec{c_{u_i}}} \right) .
\end{aligned}
\tag{3}
$$

Optimization. We used the pairwise *hinge loss* as the objective function to be minimized by the *AdaDelta* optimizer. In this perspective, the training data is viewed as a set of triples, (q, d^+, d^-), composed of a query q, a positive document d^+ and a negative document d^-. Additionally, inspired by [14] and as successfully demonstrated by [16], we adopted a similar negative sampling strategy, where a negative document can be drawn from the following sets:

– **Partially irrelevant set**: Irrelevant documents that share some matching signals with the query. More precisely, this corresponds to documents

retrieved by the fast retrieval module but which do not appear in the training data as positive examples;
- **Completely irrelevant set**: Documents not in the positive training instances and not sharing any matching signal with the query.

2.2 Passage Extraction Details

Passage extraction is accomplished by looking at the attention weights of the neural ranking model. As described, the proposed neural ranking model includes two attention mechanisms. The first one computes a local passage attention with respect to each query term, a_{p_i}. The second is used to compute the importance of each query term, a_{u_k}. Therefore, a global attention weight for each passage can be obtained from the product of these two terms, $a_{g_{(k,i)}} = a_{u_k} \times a_{p_i}$, as shown in Eq. 4:

$$
\begin{aligned}
\vec{c}_{M \times 1} &= \sum_{u_k \in q} \left(a_{u_k}_{1 \times 1} \times \sum_{p_i \in D(u_k)} \left(a_{p_i}_{1 \times 1} \times \vec{h}_{p_i}_{M \times 1} \right) \right) \\
&= \sum_{u_k \in q} \left(\sum_{p_i \in D(u_k)} \left(\underbrace{a_{u_k}_{1 \times 1} \times a_{p_i}_{1 \times 1}}_{global\ attention} \times \vec{h}_{p_i}_{M \times 1} \right) \right).
\end{aligned}
\tag{4}
$$

3 Results and Discussion

This section presents the system evaluation results. We used the training data from the BioASQ 6b and 7b phase A challenges [13], containing 2251 and 2747 biomedical questions with the corresponding relevant documents, taken from the MEDLINE repository. The objective for a system is to retrieve the ten most relevant documents for each query, with the performance evaluated in terms of **Map@10** on five test sets containing 100 queries each.

3.1 Experiments

At first, a study was conducted to investigate the performance of the proposed neural ranking model. After that, the full system was compared against the results of systems submitted to the BioASQ 6 and 7 editions for the document retrieval task. Finally, we investigate if the attention given to each passage is indeed relevant.

In the results, we compare two variants of DeepRank: **BioDeepRank** refers to the model with the modified aggregation network and weighting mechanism, and using word embeddings for the biomedical domain [15]; **Attn-BioDeepRank** refers to the final model that additionally replaces the recurrent layer by a self-attention layer.[2]

---

[2] Configuration details of both variants, including all the hyperparameters used, are available in the code repository.

Neural Ranking Models. We compared both neural ranking versions against BM25 in terms of MAP@10 and Recall@10, on a 5-fold cross validation over the BioASQ training data. Table 1 summarizes the results.

Both models successfully improved the BM25 ranking order, achieving an increase of around 0.14 in MAP and 0.31 in recall. Results of Attn-BioDeepRank, although lower, suggest that this version is at least nearly as effective at ranking the documents as the model that uses the recursive layer.

Table 1. Evaluation of the retrieval models on 5-fold cross validation on the BioASQ 7b dataset. Results are presented as the average ± standard deviation over the 5 validation folds.

	BioASQ 7b	
	MAP	RECALL
BM25	0.153 ± 0.006	0.329 ± 0.013
BioDeepRank	0.298 ± 0.008	0.643 ± 0.035
Attn-BioDeepRank	0.289 ± 0.009	0.639 ± 0.038

Biomedical Document Retrieval. We report results on the BioASQ 6b and BioASQ 7b document ranking tasks (Table 2). Regarding BioASQ 6b, it should be noted that the retrieved documents were evaluated against the final gold-standard of the task, revised after reevaluating the documents submitted by the participating systems. Since we expect that some of the retrieved documents would have been revised as true positives, the results presented can be considered a lower bound of the system's performance. For BioASQ 7b, the results shown are against the gold-standard before the reevaluation, since the final annotations were not available at the time of writing. In this dataset both systems achieved performance nearer to the best result, including a top result on Batch 1.

Table 2. Evaluation of the retrieval models on BioASQ 6b and 7b test sets

6B Systems	Batch 1		Batch 2		Batch 3		Batch 4		Batch 5	
	MAP	RANK	MAP	RANK	MAP	RANK	MAP	RANK	MAP	RANK
Best result	0.2327	–	0.2512	–	0.2622	–	0.1843	–	0.1464	–
BioDeepRank	0.2051	(5/15)	0.2065	(11/22)	0.1857	(19/24)	0.1554	(11/21)	0.1116	(17/23)
Attn-DeepRank	0.1944	(5/15)	0.2080	(10/22)	0.2071	(15/24)	0.1556	(11/21)	0.1210	(12/23)
7B Systems	Batch 1		Batch 2		Batch 3		Batch 4		Batch 5	
	MAP	RANK	MAP	RANK	MAP	RANK	MAP	RANK	MAP	RANK
Best result	0.0809	–	0.0849	–	0.1199	–	0.1034	–	0.0425	–
BioDeepRank	0.0874	(1/12)	0.0760	(7/23)	0.1006	(6/21)	0.0922	(5/17)	0.0344	(9/18)
Attn-BioDeepRank	0.0865	(1/12)	0.0764	(7/23)	0.0995	(6/21)	0.0882	(6/17)	0.0373	(3/18)

Passage Evaluation. Finally, we analysed whether the information used by the model for ranking the documents, as given by the attention weights, corresponded to relevant passages in the gold-standard. For this, we calculated the precision of the passages, considering overlap with the gold-standard, and evaluated how it related to the confidence assigned by the model. Interestingly, although the model is not trained with this information, the attention weights seem to focus on these relevant passages, as indicated by the results in Fig. 3.

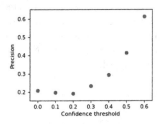

Fig. 3. Quality of retrieved passages as a function of the confidence attributed by the model.

4   Conclusion

This paper describes a new neural ranking model based on the DeepRank architecture. Evaluated on a biomedical question answering task, the proposed model achieved similar performance to a range of others strong systems.

We intend to further explore the proposed approach by considering semantic matching signals in the fast retrieval module, and by introducing joint learning for document and passage retrieval.

The network implementation and code for reproducing these results are available at https://github.com/bioinformatics-ua/BioASQ.

Acknowledgments. This work was partially supported by the European Regional Development Fund (ERDF) through the COMPETE 2020 operational programme, and by National Funds through FCT – Foundation for Science and Technology, projects PTDC/EEI-ESS/6815/2014 and UID/CEC/00127/2019.

References

1. Brokos, G.I., Liosis, P., McDonald, R., Pappas, D., Androutsopoulos, I.: AUEB at BioASQ 6: Document and Snippet Retrieval, September 2018. http://arxiv.org/abs/1809.06366
2. Devlin, J., Chang, M., Lee, K., Toutanova, K.: BERT: pre-training of deep bidirectional transformers for language understanding. CoRR abs/1810.04805 (2018). http://arxiv.org/abs/1810.04805

3. Guo, J., Fan, Y., Ai, Q., Croft, W.B.: A deep relevance matching model for ad-hoc retrieval. In: Proceedings of the 25th ACM International on Conference on Information and Knowledge Management - CIKM 2016, pp. 55–64. ACM Press, New York(2016). https://doi.org/10.1145/2983323.2983769, http://dl.acm. org/citation.cfm?doid=2983323.2983769

4. Hirschman, L., Gaizauskas, R.: Natural language question answering: the view from here. Nat. Lang. Eng. 7(04), 275–300 (2001). https://doi.org/10.1017/ S1351324901002807, http://www.journals.cambridge.org/abstract_S135132490100 2807

5. Huang, P.S., He, X., Gao, J., Deng, L., Acero, A., Heck, L.: Learning deep structured semantic models for web search using clickthrough data. In: Proceedings of the 22nd ACM International Conference on Information & Knowledge Management - CIKM 2013, pp. 2333–2338. ACM Press, New York (2013). https://doi.org/ 10.1145/2505515.2505665, http://dl.acm.org/citation.cfm?doid=2505515.2505665

6. Lin, Z., Feng, M., dos Santos, C.N., Yu, M., Xiang, B., Zhou, B., Bengio, Y.: A structured self-attentive sentence embedding. CoRR abs/1703.03130 (2017). http://arxiv.org/abs/1703.03130

7. Mateus, A., González, F., Montes, M.: Mindlab neural network approach at bioasq 6b, November 2018. 10.18653/v1/W18-5305

8. McDonald, R., Brokos, G.I., Androutsopoulos, I.: Deep Relevance Ranking Using Enhanced Document-Query Interactions, September 2018. http://arxiv.org/abs/ 1809.01682

9. Nentidis, A., Krithara, A., Bougiatiotis, K., Paliouras, G., Kakadiaris, I.: Results of the sixth edition of the BioASQ challenge. In: Proceedings of the 6th BioASQ Workshop A Challenge on Large-scale Biomedical Semantic Indexing and Question Answering, pp. 1–10. Association for Computational Linguistics, Brussels, November 2018. https://doi.org/10.18653/v1/W18-5301, https://www. aclweb.org/anthology/W18-5301

10. Pang, L., Lan, Y., Guo, J., Xu, J., Xu, J., Cheng, X.: DeepRank. In: Proceedings of the 2017 ACM on Conference on Information and Knowledge Management - CIKM 2017, pp. 257–266. ACM Press, New York (2017). https://doi.org/10.1145/ 3132847.3132914, http://dl.acm.org/citation.cfm?doid=3132847.3132914

11. Robertson, S., Zaragoza, H.: The probabilistic relevance framework: Bm25 and beyond. Found. Trends Inf. Retr. 3(4), 333–389, April 2009. https://doi.org/10. 1561/1500000019, http://dx.doi.org/10.1561/1500000019

12. Shen, Y., He, X., Gao, J., Deng, L., Mesnil, G.: A latent semantic model with convolutional-pooling structure for information retrieval. In: Proceedings of the 23rd ACM International Conference on Conference on Information and Knowledge Management - CIKM 2014, pp. 101–110. ACM Press, New York (2014). https:// doi.org/10.1145/2661829.2661935, http://dl.acm.org/citation.cfm?doid=2661829. 2661935

13. Tsatsaronis, G., et al.: An overview of the BioASQ large-scale biomedical semantic indexing and question answering competition. BMC Bioinform. **16**, 138 (2015). https://doi.org/10.1186/s12859-015-0564-6

14. Wang, J., Song, Y., Leung, T., Rosenberg, C., Wang, J., Philbin, J., Chen, B., Wu, Y.: Learning fine-grained image similarity with deep ranking. CoRR abs/1404.4661 (2014). http://arxiv.org/abs/1404.4661

15. Zhang, Y., Chen, Q., Yang, Z., Lin, H., Lu, Z.: BioWordVec, improving biomedical word embeddings with subword information and MeSH. Sci. Data **6**(1), 52 (2019). https://doi.org/10.1038/s41597-019-0055-0
16. Zhu, M., Ahuja, A., Wei, W., Reddy, C.K.: A hierarchical attention retrieval model for healthcare question answering. In: The World Wide Web Conference, pp. 2472–2482. WWW 2019. ACM, New York (2019). https://doi.org/10.1145/3308558.3313699, http://doi.acm.org/10.1145/3308558.3313699

Neural Embedding-Based Metrics for Pre-retrieval Query Performance Prediction

Negar Arabzadeh[1]([⊠]), Fattane Zarrinkalam[1], Jelena Jovanovic[2], and Ebrahim Bagheri[1]

[1] Ryerson University, Toronto, ON, Canada
{narabzad,fzarrinkalam,bagheri}@ryerson.ca
[2] University of Belgrade, Belgrade, Serbia
jelena.jovanovic@fon.bg.ac.rs

Abstract. Query Performance Prediction (QPP) is concerned with estimating the effectiveness of a query within the context of a retrieval model. It allows for operations such as query routing and segmentation, leading to improved retrieval performance. *Pre-retrieval* QPP methods are oblivious to the performance of the retrieval model as they predict query difficulty prior to observing the set of documents retrieved for the query. Since neural embedding-based models are showing wider adoption in the Information Retrieval (IR) community, we propose a set of pre-retrieval QPP metrics based on the properties of *pre-trained* neural embeddings and show that such metrics are more effective for query performance prediction compared to the widely known QPP metrics such as SCQ, PMI and SCS. We report our findings based on Robust04, ClueWeb09 and Gov2 corpora and their associated TREC topics.

Keywords: Query Performance Prediction · Neural embeddings · Specificity

1 Introduction

It is understood that the performance of retrieval models is not always consistent over different queries and corpora and there are some queries that have lower performance, often referred to as *hard* or *difficult* queries [1]. As such, the area of *Query Performance Prediction* is concerned with estimating the performance of a retrieval system for a given query. There is already a well-established body of work that explores query performance prediction through either a *post-retrieval* or a *pre-retrieval* strategy [2]. Methods in post-retrieval measure query difficulty, by analyzing the results obtained from the retrieval system as a response to the query. In contrast, pre-retrieval methods, which are the focus of this work as well, are based on linguistic and statistical features of the query and documents.

While existing work in pre-retrieval query performance has been predominantly focused on defining various statistical measures based on term and corpus-level frequency, the IR community has recently embarked on exploring the impact

© Springer Nature Switzerland AG 2020
J. M. Jose et al. (Eds.): ECIR 2020, LNCS 12036, pp. 78–85, 2020.
https://doi.org/10.1007/978-3-030-45442-5_10

and importance of neural IR techniques [5–7]. There are some recent work that propose to use neural networks for QPP based on a host of signals [8] but to the best of our knowledge, there is only one recent work that specifically utilizes *neural embeddings* of query terms for performing QPP [9]. Neural embeddings maintain interesting *geometric properties* between embedded terms [10] which are manifested by how term vectors are distributed in the embedding space. We explore exploiting the geometric properties of embeddings to define beyond-frequency QPP metrics. Our work *distinguishes* itself from the recent work [9], which proposes to cluster neural embeddings based on their vector similarity to perform QPP, by proposing to not only consider *term similarity* but also take term neighborhood and association into account through a *network representation* of neural embeddings. More specifically, we benefit from term vector associations in the neural embedding space for formalizing *term specificity*, which is correlated with query difficulty [3,4,11].

We base our work on the intuition that a term that has been closely surrounded by several other terms in the embedding space is more likely to be *specific* while a term with a *fewer number* of closely surrounded terms is more likely to be *generic*. We conceptualize the space surrounding a term by using an *ego network* representation where the term of interest serves as the *ego* and is contextualized by a set of *alter* nodes, which are other terms that are similar to it in the embedding space. We apply various measures of node centrality on the ego node to determine the specificity of the term that is being represented by the ego, which would then indicate query difficulty [16]. We have performed experiments based on three widely used TREC corpora, namely Robust04, ClueWeb09 and Gov2 and their corresponding topic sets. Our experiments show that the proposed metrics are effective in QPP using pre-trained neural embeddings.

2   Proposed Approach

This paper is concerned with the design of effective metrics for pre-retrieval QPP based on pre-trained neural embeddings. We focus on distribution of neural embedding vectors in the embedding space to define specificity metrics for QPP. Existing work in the literature [3,12] have already shown that measures of *term specificity* are suitable indicators of query difficulty, i.e., more specific terms are more discriminative and are hence easier to handle when used as queries.

Our work is driven by the *intuition* that more specific terms have a higher likelihood of being surrounded by a larger number of terms compared to generic terms. For instance, as shown in Fig. 1, the set of terms related to the specific term 'Arsenal', with an association degree (computed based on cosine similarity of terms' vector representation) above 0.75, includes terms such as 'Wenger', 'Tottenham', 'Everton', among others, which are also themselves very specific; whereas, the generic term 'soccer' has only one closely associated term (association degree above 0.75) and that is 'football', which is quite generic itself. While it is not possible to measure frequency information from neural embeddings, it is convenient to identify the set of highly similar terms to a term based on vector similarity. We benefit from this to formalize the notion of an *ego network*

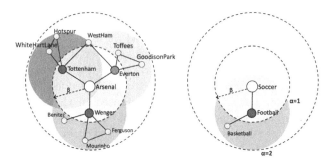

Fig. 1. Schematic of two α-depth β-cut ego networks.

that is based on vector similarities within the embedding space. We benefit from this to formalize our recursive definition of specificity, i.e., the extent to which a term is specific can be determined from the context created by the surrounding highly similar terms within the neural embedding space. In order to formalize *specificity*, we define an *ego network*, as follows:

Definition 1. *Let $\mathcal{P}(t_i, t_j)$ be the degree of similarity between vectors of terms t_i and t_j, V be the complete vocabulary set, and $\mathcal{P}_\mathcal{M}(t_i)$ be the highest degree of similarity to t_i from any term in V. We define an $\alpha - depth$ ego network for an ego node t_i in the form of a fully connected graph with a maximum depth α around the ego where the edge weights are $\mathcal{P}(t_k, t_l)$ between any two nodes t_k and t_l. We further refine the $\alpha - depth$ ego network into an $\alpha - depth$ $\beta - cut$ ego network where any edge with a weight less than $\beta \times \mathcal{P}_\mathcal{M}(t_i)$ is pruned.*

In simple terms, we propose to build an ego network for a term t_i such that t_i is the ego node and is connected directly to other adjacent terms only if the degree of similarity between the ego and the neighbor is above a discounted rate (β) of the most similar term to the ego. For instance, assuming 'Arsenal' is the ego and $\beta = 0.8$, given that 'Gunners' is the most similar term to the ego with a similarity of 0.854, the immediate neighbors of the ego will consist of all the terms in V that have a similarity above 0.6832 to 'Arsenal'. Furthermore, we allow the ego network to have a depth of α from the ego. For a depth of one ($\alpha = 1$), the ego network will only consist of the ego and its immediate neighbors. For a depth of two ($\alpha = 2$), each node in layer one will become the ego for another sub-ego network with a $\beta - cut$, as explained earlier. Figure 1 shows a schematic of the $\alpha - depth$ $\beta - cut$ ego network for the specific term 'Arsenal' and generic term 'soccer'. As seen, in Arsenal's case, the graph is populated with many terms closely related to the ego. In the second layer, the nodes immediately connected to the ego, e.g., 'Wenger', become an ego node for a second layer subgraph, which are in turn connected to their own alters, e.g., 'Mourinho', 'Benitez' and 'Ferguson'. In contrast, the network associated with the generic term 'soccer' is quite sparse with only two additional nodes present when $\alpha = 2$.

Based on the developed ego network, we propose to measure the *specificity* of the ego through the use of *node centrality* metrics [13,16]. Given queries can be

Table 1. Node centrality metrics on the ego network.

Metric	Description
Edge Count (EC)	This metric counts the number of edges in the ego network
Edge Weight Sum (EWS)	This metric calculates the sum of edge weights in the ego network where edge weights are degrees of term association
Inverse Edge Frequency (IEF)	This metric measures the log of the ratio of the number of edges in the network over the number of edges connected to the ego
Degree Centrality (DC)	This metric is the number of links incident upon the ego
Closeness Centrality (CC)	This metric calculates the average length of the shortest path between the ego and all other alters in the network
Betweenness Centrality (BC)	This metric measures the proportion of the shortest paths in the network that go through the ego
Page Rank (PR)	It is based on reciprocity of node importance

composed of more than one term, we adopt the integration approach that uses aggregation functions [14] over the specificity of individual query terms. Table 1 provides an overview of the metrics used in this paper.

3 Experiments

Corpora and Topics: We employed three widely used corpora, namely, Robust04, ClueWeb09, and Gov2. For Robust04, TREC topics 301–450 and 601–650, for Gov2, topics 701–850 and for ClueWeb09, topics 1–200 were used. Topic difficulty was based on Average Precision of each topic computed using QL [15].

Baselines: We adopt the widely used pre-retrieval metrics reported in [2]. The formulation of these metrics is provided in Table 2. As another baseline, we adopt the recent approach by Roy et al. [9] that utilizes embedded word vectors to predict query performance. Their *specificity metric*, known as $P_{clarity}$, is based on the idea that the number of clusters around the neighbourhood of a query term is a potential indicator of its specificity. To apply their approach on our embedding vectors, we have used the implementation provided by the authors.

Neural Embeddings: We used a pre-trained word2vec model based on the Google News corpus (https://goo.gl/wQ8eQ1).

Evaluation: A common approach for measuring the performance of a QPP metric is to use rank correlation metrics to measure the correlation between the list of queries (1) ordered by their difficulty for the retrieval method (ascending order of average precision), and (2) ordered by the QPP metric. Kendall's τ and Pearson's ρ co-efficient are common correlation metrics in this space.

Empirical studies on pre-retrieval QPP metrics have shown that there is no single or set of metrics that outperforms the others on all topics and corpora [2]. Our experiments confirm this. Therefore, to be able to rank the different metrics over a range of topics, we compute the rank of each metric in each topic set and report the rank of the median of each metric over all topics of each document collection. This is specified as *rank* and is reported separately for Kendall's τ and Pearson's ρ. These ranks show how a metric has performed over the different topic sets. Given our metrics are dependent on the α and β parameters, we set them using 5-fold cross validation optimized for Pearson correlation.

Table 2. Baseline metrics. t is a term in query q. d is a document in collection D. D_t is the set of documents with t. $tf(t, D)$ is term frequency of term t in D. $Pr(t|D) = tf(t, D)/|D|$. π_m is the prior probability of the most dominating sense of term t and $P(t|N(\mu_m, \Sigma_m))$ is the posterior probability of term t for the selected cluster.

Metric	Formulation	Ref
IDF	$idf(t) = log(\dfrac{|D|}{|D_t|})$	[2]
VAR	Variance of query term weights $w(t, d)$ in D	[2]
SCQ	$SCQ(t) = (1 + log(tf(t, D))).idf(t)$	[17]
SCS	$SCS(q) = \sum_{t \in q} Pr(t|q)log(\dfrac{Pr(t|q)}{Pr(t|D)})$	[3]
PMI	$PMI(t_1, t_2) = log\dfrac{Pr(t_1, t_2|D)}{Pr(t_1|D)Pr(t_2|D)}$	[18]
$P_{clarity}$	$P_{clarity}(t) = \pi_m P(t|N(\mu_m, \Sigma_m))$	[9]

Table 3. Results on Robust04. Gray rows are baselines. Bold metrics are the top-3 on Kendall τ (left) and Pearson ρ (right). † indicates statistical significance at alpha $= 0.05$.

Metric		301-350	351-400	401-450	601-650	Rank	Metric		301-350	351-400	401-450	601-650	Rank
VAR	Avg	0.17	$0.36^\dagger$	$0.35^\dagger$	$0.34^\dagger$	1	VAR	Avg	0.08	$0.46^\dagger$	0.23	$0.55\dagger$	12
SCQ	Max	0.13	$0.42^\dagger$	$0.5^\dagger$	$0.26^\dagger$	4	SCQ	Max	0.01	$0.5^\dagger$	$0.66^\dagger$	$0.35^\dagger$	4
IDF	Avg	$0.22^\dagger$	$0.29^\dagger$	$0.28^\dagger$	$0.32^\dagger$	5	**IDF**	Avg	$0.52^\dagger$	$0.42^\dagger$	$0.43^\dagger$	$0.36^\dagger$	1
SCS		$0.21^\dagger$	$0.24^\dagger$	$0.23^\dagger$	$0.3^\dagger$	9	SCS		$0.43^\dagger$	$0.35^\dagger$	$0.34^\dagger$	$0.42^\dagger$	7
PMI	Max	0.04	0.18	$0.18^\dagger$	$0.2^\dagger$	12	PMI	Max	0.06	0.12	$0.28^\dagger$	0.08	13
$P_{clarity}$		$0.28^\dagger$	$0.2^\dagger$	$0.25^\dagger$	$0.26^\dagger$	7	$P_{clarity}$		$0.34^\dagger$	$0.24^\dagger$	$0.36^\dagger$	$0.38^\dagger$	9
BC	Max	$0.31^\dagger$	$0.2^\dagger$	$0.3^\dagger$	$0.42^\dagger$	3	BC	Max	$0.35^\dagger$	0.17	$0.39^\dagger$	$0.4^\dagger$	5
IEF	Avg	$0.47^\dagger$	$0.36^\dagger$	$0.3^\dagger$	$0.33^\dagger$	1	**IEF**	Avg	$0.41^\dagger$	$0.44^\dagger$	$0.4^\dagger$	$0.43^\dagger$	1
DC	Max	$0.2^\dagger$	$0.29^\dagger$	$0.3^\dagger$	$0.33^\dagger$	6	DC	Max	$0.28^\dagger$	$0.42^\dagger$	$0.44^\dagger$	$0.44^\dagger$	10
CC	Avg	$0.24^\dagger$	$0.29^\dagger$	$0.28^\dagger$	$0.37^\dagger$	5	CC	Avg	$0.26^\dagger$	$0.41^\dagger$	$0.42^\dagger$	$0.45^\dagger$	5
PR	Max	$0.25^\dagger$	$0.4^\dagger$	$0.24^\dagger$	0.16	8	PR	Max	$0.33^\dagger$	$0.47^\dagger$	$0.31^\dagger$	$0.31^\dagger$	11
EWS	Min	0.17	$0.24^\dagger$	$0.29^\dagger$	$0.22^\dagger$	10	**EWS**	Min	$0.34^\dagger$	$0.49^\dagger$	$0.26^\dagger$	$0.46^\dagger$	3
EC	Min	0.17	0.15	$0.27^\dagger$	0.19	11	EC	Min	$0.37^\dagger$	$0.41^\dagger$	$0.26^\dagger$	$0.4^\dagger$	8

Findings: The results of our experiments are shown in Tables 3, 4 and 5. As shown, our metrics are among the top-3 on both measures on all corpora. On Robust04, two of our metrics, i.e., BC and IEF, are among the top-3 metrics based on Kendall τ. Based on Pearson ρ, IEF and EWS are among the top-3 along with IDF. On Robust04, there is little metric performance consistency on Kendall τ and Pearson ρ. When looking for those metrics that perform well on both measures, IEF and BC are consistent metrics where IEF ranks first on both Kendall τ and Pearson ρ whereas BC ranks third and fifth on these measures, respectively. The other metrics, both baseline metrics and the ones we proposed, have a high performance difference on the two measures. For instance, while the baseline VAR metric ranks first on τ, it ranks twelfth on ρ. On ClueWeb09 and Gov2, unlike Robust04, the top metrics are consistent for Kendall and Pearson where the top-3 metrics include the proposed DC and CC metrics for both measures. On ClueWeb09, these two metrics are accompanied by the BC and

PR metrics for τ and ρ, respectively. However, on Gov2, these metrics are followed by the baseline SCQ metric on τ and our IEF and EWS metrics on ρ. In summary, balancing between the evaluation measures and performance on all topics and corpora, we find our *CC metric* to perform well across the board. It is among the best metrics on Gov2 ahd ClueWeb09 and has a balanced performance on Robust04. However, CC has a high time complexity of $O(V^3)$. On the other hand, the DC metric performs well on both ClueWeb09 and Gov2 (in the top-3) but less effectiveness on Robust04. The benefit of DC is its low complexity: $O(1)$. Overall, CC is the preferred metric given QPP computations are performed offline. DC can serve as an alternative if computation limitations exist.

Table 4. Results on ClueWeb09. Table format is similar to Table 3.

Metric		1-50	51-100	101-150	150-200	Rank	Metric		1-50	51-100	101-150	150-200	Rank
VAR	Max	0.28†	0.23†	0.27†	0.01	5	VAR	Max	0.14	0.04	0.42†	0.08	8
SCQ	Max	0.19†	0.25†	0.3†	0.03	4	SCQ	Max	0.22	0.24	0.33†	0.09	7
IDF	Avg	0.24†	0.25†	0.16	0.05	7	IDF	Avg	0.18	0.21	0.27	0.01	10
SCS		0.23†	0.24†	0.1	0.07	9	SCS		0.18	0.19	0.16	0.05	12
PMI	Max	0.15	0.16	0.07	0.03	13	PMI	Max	0.2	0.12	0.04	0.04	12
P_clarity		0.3†	0.15	0.1	0.24†	9	P_clarity		0.29†	0.27	0.18	0.24†	6
BC	Avg	0.22†	0.28†	0.26†	0.22†	3	BC	Avg	0.28†	0.29†	0.37†	0.28†	4
IEF	Avg	0.17	0.2†	0.17	0.39†	8	IEF	Avg	0.29†	0.26	0.17	0.36†	5
DC	Avg	0.22†	0.27†	0.15	0.38†	1	DC	Avg	0.25†	0.31†	0.39†	0.34†	3
CC	Min	0.22†	0.2†	0.28†	0.36†	2	**CC**	Min	0.31†	0.17	0.47†	0.33†	1
PR	Max	0.13	0.29†	0.18	0.25†	6	**PR**	Max	0.36†	0.31†	0.37†	0.31†	1
EWS	Max	0.13	0.18	0.26†	0.06	12	EWS	Max	0.07	0.27	0.27	0.14	11
EC	Max	0.14	0.16	0.27†	0.05	11	EC	Max	0.06	0.27	0.29	0.14	9

Table 5. Results on Gov2. Table format is similar to Table 3.

Metric		701-750	751-800	801-850	Rank	Metric		701-750	751-800	801-850	Rank
VAR	Max	0.2†	0.02	0.05	13	VAR	Max	0.27	0.06	0.1	13
SCQ	Max	0.36†	0.29†	0.23†	3	SCQ	Max	0.53†	0.32†	0.35†	6
IDF	Avg	0.27†	0.22†	0.14	6	IDF	Avg	0.4†	0.25	0.17	10
SCS		0.23†	0.19†	0.11	9	SCS		0.34†	0.19	0.12	12
PMI	Max	0.28†	0.22†	0.16	6	PMI	Max	0.44†	0.26	0.22	9
P_clarity		0.25†	0.24†	0.25†	6	P_clarity		0.33†	0.33†	0.38†	5
BC	Min	0.18	0.11	0.3†	9	BC	Min	0.18	0.28†	0.5†	8
IEF	Max	0.09	0.25†	0.34†	5	**IEF**	Max	0.18	0.33†	0.47†	3
DC	Max	0.12	0.36†	0.41†	1	**DC**	Max	0.15	0.41†	0.52†	1
CC	Max	0.11	0.36†	0.41†	1	**CC**	Max	0.14	0.42†	0.51†	1
PR	Min	0.28†	0.19†	0.15	9	PR	Min	0.38†	0.32†	0.32†	6
EWS	Min	0.2†	0.26†	0.31†	4	**EWS**	Min	0.12	0.33†	0.44†	3
EC	Min	0.12	0.17	0.3†	12	EC	Min	0.11	0.24	0.38†	11

4   Concluding Remarks

We have shown that it is possible to devise metrics based on the neural embedding-based representation of terms to perform pre-retrieval QPP. Specifically, we have shown that specificity of a query term, estimated based on an *ego network* representation, can lead to better performance on QPP compared to several baselines such as the one that considers term clusters based on neural embeddings [9].

References

1. Mizzaro, S., Mothe, J.: Why do you think this query is difficult?: A user study on human query prediction. In: Proceedings of the 39th International ACM SIGIR Conference on Research and Development in Information Retrieval, pp. 1073–1076. ACM (2016)
2. Carmel, D., Yom-Tov, E.: Estimating the query difficulty for information retrieval. Synthesis Lectures Inf. Concepts Retrieval Serv. **2**(1), 1–89 (2010)
3. He, B., Ounis, I.: Inferring query performance using pre-retrieval predictors. In: Apostolico, A., Melucci, M. (eds.) SPIRE 2004. LNCS, vol. 3246, pp. 43–54. Springer, Heidelberg (2004). https://doi.org/10.1007/978-3-540-30213-1_5
4. He, J., Larson, M., de Rijke, M.: Using coherence-based measures to predict query difficulty. In: Macdonald, C., Ounis, I., Plachouras, V., Ruthven, I., White, R.W. (eds.) ECIR 2008. LNCS, vol. 4956, pp. 689–694. Springer, Heidelberg (2008). https://doi.org/10.1007/978-3-540-78646-7_80
5. Zuccon, G., Koopman, B., Bruza, P., Azzopardi, L.: Integrating and evaluating neural word embeddings in information retrieval. In: Proceedings of the 20th Australasian Document Computing Symposium, p. 12. ACM (2015)
6. Zhang, L., Zhang, S., Balog, K.: Table2Vec: neural word and entity embeddings for table population and retrieval. In: Proceedings of the 42nd International ACM SIGIR Conference on Research and Development in Information Retrieval, pp. 1029–1032. ACM (2019)
7. Mitra, B., Craswell, N.: An introduction to neural information retrieval. Found. Trends Inf. Retrieval **13**(1), 1–126 (2018)
8. Zamani, H., Croft, W.B., Culpepper, J.S.: Neural query performance prediction using weak supervision from multiple signals. In The 41st International ACM SIGIR Conference on Research & Development in Information Retrieval, pp. 105–114. ACM (2018)
9. Roy, D., Ganguly, D., Mitra, M., Jones, G.J.: Estimating Gaussian mixture models in the local neighbourhood of embedded word vectors for query performance prediction. Inf. Process. Manage. **56**(3), 1026–1045 (2019)
10. Mimno, D., Thompson, L.: The strange geometry of skip-gram with negative sampling. In: Empirical Methods in Natural Language Processing (2017)
11. Hauff, C., Hiemstra, D., de Jong, F.: A survey of pre-retrieval query performance predictors. In Proceedings of the 17th ACM Conference on Information and Knowledge Management, pp. 1419–1420. ACM (2008)
12. Thomas, P., Scholer, F., Bailey, P., Moffat, A.: Tasks, queries, and rankers in pre-retrieval performance prediction. In: Proceedings of the 22nd Australasian Document Computing Symposium, p. 11. ACM (2017)

13. Segarra, S., Ribeiro, A.: Stability and continuity of centrality measures in weighted graphs. IEEE Trans. Signal Process. **64**(3), 543–555 (2015)
14. Hauff, C., Kelly, D., Azzopardi, L.: A comparison of user and system query performance predictions. In: Proceedings of the 19th ACM International Conference on Information and Knowledge Management, pp. 979–988. ACM (2010)
15. Song, F., Croft, W.B.: A general language model for information retrieval. In: Proceedings of the Eighth International Conference on Information and Knowledge Management, pp. 316–321. ACM (1999)
16. Arabzadeh, N., Zarrinkalam, F., Jovanovic, J., Bagheri, E.: Geometric estimation of specificity within embedding spaces. In: Proceedings of the 28th ACM International Conference on Information and Knowledge Management, pp. 2109–2112. ACM (2019)
17. Zhao, Y., Scholer, F., Tsegay, Y.: Effective pre-retrieval query performance prediction using similarity and variability evidence. In: Macdonald, C., Ounis, I., Plachouras, V., Ruthven, I., White, R.W. (eds.) ECIR 2008. LNCS, vol. 4956, pp. 52–64. Springer, Heidelberg (2008). https://doi.org/10.1007/978-3-540-78646-7_8
18. Hauff, C.: Predicting the effectiveness of queries and retrieval systems. In: SIGIR Forum, vol. 44, no. 1, p. 88. ACM (2010)

A Latent Model for Ad Hoc Table Retrieval

Ebrahim Bagheri[1(✉)] and Feras Al-Obeidat[2]

[1] Laboratory for Systems, Software and Semantics (LS3),
Ryerson University, Toronto, Canada
`bagheri@ryerson.ca`
[2] College of Technological Innovation,
Zayed University, Abu Dhabi, United Arab Emirates
`Feras.Al-Obeidat@zu.ac.ae`

Abstract. The ad hoc table retrieval task is concerned with satisfying a query with a ranked list of tables. While there are strong baselines in the literature that exploit learning to rank and semantic matching techniques, there are still a set of *hard queries* that are difficult for these baseline methods to address. We find that such hard queries are those whose constituting tokens (i.e., terms or entities) are not fully or partially observed in the relevant tables. We focus on proposing a latent factor model to address such hard queries. Our proposed model factorizes the token-table co-occurrence matrix into two low dimensional latent factor matrices that can be used for measuring table and query similarity even if no shared tokens exist between them. We find that the variation of our proposed model that considers keywords provides statistically significant improvement over three strong baselines in terms of NDCG and ERR.

1 Introduction

Tables provide a structured representation of data that can be quickly interpreted by the users. There have been important work that focus on the automated retrieval and interpretation of data in tabular format. These works range from mining tables from documents [3,8] to extracting information from within tables [11] and semantic analysis of table content [2,14], to name a few. Recently, Zhang and Balog have systematically introduced the task of *ad hoc table retrieval* from an information retrieval perspective [16]. The idea is to retrieve a ranked list of relevant tables for an input query given a corpus of existing tables. The task shows close resemblance to the traditional ad hoc document retrieval problem while distinguishing itself in that data in tables are quite short, often including a few terms or just numbers, and as such making it challenging to extract the context that is needed to draw *relevance* conclusions.

Methods for ad hoc table retrieval show strong performance on $nDCG@20$ metric where supervised methods based on the learning to rank approach report between 0.5206 [1] to 0.6031 [16] while semantic relevance methods report up to 0.6825 on a corpus with 1.6M tables. Despite this strong performance, we note

© Springer Nature Switzerland AG 2020
J. M. Jose et al. (Eds.): ECIR 2020, LNCS 12036, pp. 86–93, 2020.
https://doi.org/10.1007/978-3-030-45442-5_11

that their performance is not satisfactory over *hard* (difficult) queries. In other words, while these methods perform very well on a subset of queries, they are not equally effective on another subset. This is inline with findings within the ad hoc document retrieval literature that distinguishes the performance of a retrieval method on soft (easy) and hard (difficult) queries [7,15]. More specifically, when looking at the bottom 20% of the queries ranked based on AP, it is possible to see that in the top-10 results retrieved per query, there are 0 out of 12, 2 out of 12 and 7 out of 12 queries that had retrieved at least one relevant table by WikiTable [1], Learn To Rank (LTR) and Semantic Table Retrieval (STR) [16] methods, respectively. We observe that methods that are primarily based on keyword-based features for determining relevance do not perform well on hard queries. For instance, as we will show with more details later, for a query such as *'pain medication'*, the most relevant table based on relevance judgements is one that does not include any of the query terms. Therefore, methods such as STR [16] that leverage semantics perform better on hard queries.

However, while STR shows improved performance by considering semantic information, the employed semantics are derived from word and graph embeddings learnt on generic corpora such as Google News and DBpedia. In this paper, we will present a systematic approach for learning *low dimensional latent factor matrices* to represent queries and tables based on the co-occurrence of terms and entities within the table corpus. The learnt latent factor matrices allow us to efficiently compute query-table similarities for ranking. We show that the learnt latent representation allow us to, statistically speaking, significantly improve the performance of ad hoc table retrieval on *hard queries*, which would otherwise not receive appropriate treatment. Our method is specially suited for hard queries, as the learnt latent representations extract transitive relations through observed entity and term co-occurrences, which are appropriate for measuring relevance when query terms are not present in relevant tables.

2   Proposed Approach

The objective of our work is to learn low dimensional latent factor matrices to represent tables and queries, which can then be used to measure query-table relevance. We position our work within the context of factored item-item collaborative filtering [12] where item-item similarity is learnt as the product of two matrices $\mathbf{P}$ and $\mathbf{Q}$. Hence, similarity between items i and j can be simply computed as $\mathbf{p}_i \cdot \mathbf{q}_i$. Let us assume that the set of tokens observed in the table corpus is denoted as $\mathcal{V}$ and the set of observed tokens in Table t is $\mathcal{V}_t^o$. We denote the set of tokens not observed in t as $\mathcal{V}_t^u = \mathcal{V} - \mathcal{V}_t^o$. Let us assume that $\mathbf{P}$ and $\mathbf{Q}$ are already computed; on this basis, it is possible to estimate the relevance (R) of a token i to a table t as:

$$\hat{R}_{t,i} = b_t + b_i + \sum_{j \in \mathcal{V}_t^o} \mathbf{p}_j \cdot \mathbf{q}_i^\mathsf{T} \tag{1}$$

Here, b_t and b_i are table and token biases. Now, the objective will be to efficiently learn matrices $\mathbf{P}$ and $\mathbf{Q}$ through a regularized optimization problem. Since our problem focuses on the effective ranking of tables, we are interested in minimizing *ranking error*, and so, we adopt the ranking-based loss function proposed in [13], which minimizes overall rank loss:

$$\sum_{t \in T} \sum_{i \in \mathcal{V}_t^o, j \in \mathcal{V}_t^u} \left((R_{t,i} - R_{t,j}) - (\hat{R}_{t,i} - \hat{R}_{t,j}) \right)^2. \tag{2}$$

where T is the set of tables in the table corpus, $R_{t,i}$ is the relevance of token i to table t and $\hat{R}_{t,i}$ is the predicted estimation for $R_{t,i}$.

With this loss function, we learn $\mathbf{P}$ and $\mathbf{Q}$ by minimizing a regularized optimization function that considers three factors within its optimization function:

$$minimize_{\mathbf{P},\mathbf{Q}} \frac{1}{2} \underbrace{\sum_{t \in T} \sum_{i \in \mathcal{V}_t^o, j \in \mathcal{V}_t^u} \|(R_{t,i} - R_{t,j}) - (\hat{R}_{t,i} - \hat{R}_{t,j})\|_F^2}_{\text{loss function}}$$

$$+ \underbrace{\frac{\beta}{2}(\|\mathbf{P}\|_F^2 + \|\mathbf{Q}\|_F^2) + \frac{\beta}{2}(\|b_i\|_2^2)}_{\text{regularization terms}} \tag{3}$$

$$+ \sum_{t_i, t_j \in T} \underbrace{\frac{\gamma}{2}\|\mathbf{p}_i \cdot \mathbf{q}_j - Sim(t_i, t_j)\|_F^2}_{\text{regularizing with baseline similarity}}.$$

The first part minimizes the loss function, which is based on the Bayesian Parameterized Ranking loss function [13]. The second part includes regularizers based on norms of the two matrices and the biases for tokens. The third part is included to ensure that additional feature-based similarity information are considered when optimizing $\mathbf{P}$ and $\mathbf{Q}$. Here the estimated similarity between two tables t_i and t_j, which is to be estimated based on $\mathbf{p}_i \cdot \mathbf{q}_j$, is compared to the value of a similarity function, $sim(t_i, t_j)$. We consider $sim(t_i, t_j)$ to be a similarity function derived from the baseline methods. This minimization function can be optimized using Stochastic Gradient Descent.

As indicated earlier, the tokens in a table can either be terms or entities. The tables in the corpus [16] are from Wikipedia and hence have links to other Wikipedia pages. We use these links as a representation of entities and the terms used in the table as representation of terms. Hence, we propose three variations, namely (1) Keyword, (2) Entity, and (3) Keyword + Entity. Within the Keyword variation, the relevance of token i to table t, i.e., $R_{t,i}$, is defined as: $R_{t,i} = 1$ if $i \in t$ and 0 otherwise. In the Entity model, $R_{t,i}$ is defined slightly differently because it is possible to determine the relevance of an entity to a table even if the entity is not observed in the table. We define $R_{t,i}$ for the Entity variation as $R_{t,i} = 1$ if $i \in t$ and $wmd(i, t)$ otherwise. In this case, if the entity is not present in the table, we compute the word mover's distance [9] to compute the similarity of entity i to table t. The Keyword + Entity variation is a combination of these two relevance functions.

The intuition for our approach is that tables are modelled as a collection of tokens (i.e., terms or entities), which can be distributed across multiple tables.

Table 1. Comparison with STR. † indicates statistical sig with paired t-test at 0.05.

	STR	Entity	Keyword	Keyword + Entity
NDCG@10	0.1603	0.1541	**0.1782**	0.1491
Δ		−3.88%	+11.14%†	−7.02%
NDCG@20	0.1919	0.1928	**0.2343**	0.1758
Δ		+0.44%	+22.09%†	−8.39%
ERR@10	0.1507	0.148	**0.187**	0.1522
Δ		−1.76%	+24.14%†	+1.01%
ERR@20	0.1757	0.1862	**0.2474**	0.1547
Δ		+6.00%	+40.83%†	−11.95%

Table 2. Comparison with LTR. † indicates statistical sig with paired t-test at 0.05.

	LTR	Entity	Keyword	Keyword + Entity
NDCG@10	0.0197	0.0466	**0.0872**	0.0417
Δ		+136.90%†	+343.31%†	+111.99%†
NDCG@20	0.0894	0.0952	**0.1195**	0.097
Δ		+6.49%	+33.66%	+8.50%†
ERR@10	0.0486	0.0569	**0.0965**	0.066
Δ		+17.08†	+98.56%†	+35.80%†
ERR@20	0.0710	0.0723	**0.1117**	0.0835
Δ		+1.83%	+57.32%	+17.61%†

The occurrence of tokens in multiple tables delivers indirect semantics on the relationship between the different tables. The two derived low dimensional latent factor matrices capture the semantics and allow us to compute the similarity between any two tables. For ranking tables for a query, we model the query similar to tables and use the same similarity function as a score of relevance.

3   Experimental Setup

Corpora: We used the corpus from [16] that includes 1.6M tables from Wikipedia.

Topics: We employ the 60 topics that accompany the table corpus to perform our experiments. We annotate the queries using TagMe as done in [4–6].

Baselines: The state of the art include the methods in [16]. Zhang and Balog report that WikiTables [2] is also a strong baseline. We define hard queries for each baseline as the bottom 20% of queries based on Average Precision (AP).

Neural Embeddings: The neural embeddings used for $R_{t,i}$ in Eq. 5 were based on the Hierarchical Category Embedding (HCE) model proposed in [10].

Metrics: Retrieval effectiveness was evaluated with NDCG and ERR.

4   Findings

Given the set of hard queries is different for each baseline, we report the performance improvements obtained through our proposed model separately for each baseline. Furthermore, given the $Sim(.,.)$ function employed in the third regularization term of Eq. 3 is dependent on the baseline, it is necessary to report the findings separately for each baseline. The results are reported in Tables 1, 2 and 3 for STR, LTR and WikiTable methods, respectively.

Table 3. Comparison with WikiTables. † indicates statistical sig with paired t-test at 0.05. * not possible to calculate delta improvement given divide by zero.

	WT	Entity	Keyword	Keyword + Entity
NDCG@10	0	0.0684	**0.0921**	0.0885
Δ		*	*†	*†
NDCG@20	0.09074	0.1053	**0.1195**	0.1107
Δ		*+16.06%†*	*+31.69%†*	*+21.99%†*
ERR@10	0	0.0521	0.1287	**0.1334**
Δ		*	*†	*†
ERR@20	0.0616	0.0749	**0.147**	0.1414
Δ		*+21.55%†*	*+138.68%†*	*+129.58%†*

When considering the three baselines and based on the three variations of our proposed approach, it is possible to see that the Keyword variation shows stronger performance compared to both Entity and Keyword+Entity approaches. The reason for the lower performance of the Entity-based variations was related to the sparsity of entities in tables compared to terms. It is more difficult for the latent factor model to identify table similarities based on sparse entity occurrence information. Furthermore, we found that the use of a semantic similarity score in Eq. 5 for cases when the entity was not explicitly observed in the table leads to undesirable derived table similarities. Hence, the performance of the variations that included entity information in our proposed approach is weaker than the variation that only uses terms.

We further observe that regardless of the variation, our approach improves over the baselines except for STR, which was only significantly outperformed using the Keyword variation. We postulate that this is due to the fact that STR already benefits from entity information in their semantic formulation and as such the use of sparse entity information in our model does not lead to observable improvements, while our variation based on Keywords outperforms STR. The other noticeable improvement is observed over ERR and NDCG at 10 for WikiTable. As seen in Table 3, there are no relevant retrievals at rank 10 by this baseline and hence both ERR@10 and NDCG@10 are reporting zero. However, all three variations of our approach have been able to improve WikiTable by identifying relevant tables for the hard queries at rank 10.

Table 4. Top-2 queries with most improvement by the *Keyword* model over baselines. [†]denotes tables not containing query terms, [‡]indicates partial query presence in table.

STR			
Query 7: Prime Ministers of England			
Table ID	Table Caption	Rel	Base	Ours
0406-281	Labour prime ministers[‡]	2	7	3
Query 51: Cereals nutrition value			
1573-730	Sesame seed kernels, toasted[‡]	2	9	2
LTR			
Query 7: Prime ministers of England			
0406-281	Labour prime ministers[‡]	2	17	7
Query 57: Board games number of players			
1098-540	List of Japanese board games[‡]	1	13	3
WikiTable			
Query 30: Pain medication			
1444-126	Threshold of pain[‡]	1	17	1
0520-188	Diseases and conditions[†]	1	12	2
Query 59: Constellations closest constellation			
0177-367	Deep space rendezvous[†]	1	17	3
1437-680	Constellations	1	20	6
1264-76	Symbols for zodiac constellations[‡]	1	16	9
1113-680	Solar encounter[‡]	1	19	16

Now, in order to analyze the performance of our approach in contrast with the baselines, we select two queries with the highest improvement over NDCG@10 for each baseline and report them in Table 4. The third column of the table shows the relevance score given to the table for the query in the relevance judgements, the fourth and fifth columns are the table rank produced by the baseline and the Keyword variation of our approach, respectively. For all the relevant tables for each query, our approach has been able to improve the ranking of the relevant table in these queries. We further looked into the characteristics of these tables for possible explanation of the better performance of our approach. We classify the tables into two types, denoted by [†] and [‡], which represent those tables that either have only a subset of the query tokens mentioned in them, or none of the query tokens mentioned in them, respectively. As seen in Table 4, all tables, except for Table 1437–680, are classified as either [†] or [‡]. Given our latent factor model is able to identify implicit relations between tables based on transitively shared tokens, it is able to identify query and table relevance even if the same tokens do not appear in them. For instance, for Query 30: 'Pain Medication', our approach has been able to identify Table 0520-188: 'Diseases and conditions' despite the fact that none of the tokens in the query appear in the table. A similar pattern can be observed in the other examples as well.

We observe that hard queries for the baselines are those queries which have relevant tables that are classified as [†] or [‡]. However, soft queries are those whose tokens are explicitly observed in the relevant tables. We believe that for such soft queries employing baseline methods that check for explicit query term occurrence would be a better strategy compared to our latent factor model that derive relevance based on transitive token-table occurrence patterns.

5   Concluding Remarks

In summary, we find that:

1. The Keyword variation of the proposed latent factor model is able to improve the performance of all the baseline in a statistically significant way.
2. While the variations that consider Entity information are able to also improve the performance of the baselines, they fail to do so when tested against the STR model, which already incorporates the semantics of entity information in its retrieval model.
3. Our model is suitable for satisfying those queries whose tokens do not appear or only partially appear in the relevant tables. This is due to the factorization that captures the implicit relationship between tables and queries.

References

1. Bhagavatula, C.S., Noraset, T., Downey, D.: Methods for exploring and mining tables on wikipedia. In: Interactive Data Exploration and Analytics, IDEA 2013, pp. 18–26 (2013)
2. Bhagavatula, C.S., Noraset, T., Downey, D. Tabel: entity linking in web tables. In: International Semantic Web Conference, pp. 425–441 (2015)
3. Clark, C.A., Divvala, S.: Looking beyond text: extracting figures, tables and captions from computer science papers. In: Workshops at the Twenty-Ninth AAAI Conference on Artificial Intelligence (2015)
4. Dargahi Nobari, A., Askari, A., Hasibi, F., Neshati, M.: Query understanding via entity attribute identification. In: The 27th ACM International Conference on Information and Knowledge Management, pp. 1759–1762 (2018)
5. Ensan, F., Bagheri, E.: Document retrieval model through semantic linking. In: Proceedings of the Tenth ACM International Conference on Web Search and Data Mining, WSDM 2017, pp. 181–190 (2017)
6. Hasibi, F., Balog, K., Bratsberg, S.E.: Dynamic factual summaries for entity cards. In: The 40th International ACM SIGIR Conference on Research and Development in Information Retrieval, SIGIR 2017, pp. 773–782 (2017)
7. Hauff, C., Hiemstra, D., de Jong, F.: A survey of pre-retrieval query performance predictors. In: The 17th ACM Conference on Information and Knowledge Management, pp. 1419–1420 (2008)
8. Khusro, S., Latif, A., Ullah, I.: On methods and tools of table detection, extraction and annotation in pdf documents. J. Inform. Sci. **41**(1), 41–57 (2015)

9. Kusner, M., Sun, Y., Kolkin, N., Weinberger, K.: From word embeddings to document distances. In: International Conference on Machine Learning, pp. 957–966 (2015)
10. Li, Y., Zheng, R., Tian, T., Hu, Z., Iyer, R., Sycara, K.P.: Joint embedding of hierarchical categories and entities for concept categorization and dataless classification. In: COLING 2016, 26th International Conference on Computational Linguistics, Proceedings of the Conference: Technical Papers, Osaka, Japan, 11–16 December 2016, pp. 2678–2688 (2016)
11. Muñoz, E., Hogan, A., Mileo, A.: Using linked data to mine rdf from wikipedia's tables. In: The 7th ACM International Conference on Web Search and Data Mining, WSDM 2014, pp. 533–542 (2014)
12. Ning, X., Karypis, G.: Slim: sparse linear methods for top-n recommender systems. In: 2011 IEEE 11th International Conference on Data Mining, pp. 497–506. IEEE (2011)
13. Rendle, S., Freudenthaler, C., Gantner, Z., Schmidt-Thieme, L.: BPR: Bayesian personalized ranking from implicit feedback. In: The Twenty-Fifth Conference on Uncertainty in Artificial Intelligence, pp. 452–461 (2009)
14. Venetis, P., et al.: Recovering semantics of tables on the web. Proc. VLDB Endow. 4(9), 528–538 (2011)
15. Zamani, H., Croft, W.B., Culpepper, J.S.: Neural query performance prediction using weak supervision from multiple signals. In: The 41st International ACM SIGIR Conference on Research & Development in Information Retrieval, pp. 105–114 (2018)
16. Zhang, S., Balog, K.: Ad hoc table retrieval using semantic similarity. In: The 2018 World Wide Web Conference, WWW 2018, International World Wide Web Conferences Steering Committee, Republic and Canton of Geneva, Switzerland, pp. 1553–1562 (2018)

Hybrid Semantic Recommender System
for Chemical Compounds

Márcia Barros[1,2]($\boxtimes$) ⓘ, André Moitinho[2] ⓘ, and Francisco M. Couto[1] ⓘ

[1] LASIGE, Departamento de Informática, Faculdade de Ciências,
Universidade de Lisboa, 1749-016 Lisbon, Portugal
mcbarros@fc.ul.pt
[2] CENTRA, Departamento de Física, Faculdade de Ciências,
Universidade de Lisboa, 1749-016 Lisbon, Portugal

Abstract. Recommending Chemical Compounds of interest to a particular researcher is a poorly explored field. The few existent datasets with information about the preferences of the researchers use implicit feedback. The lack of Recommender Systems in this particular field presents a challenge for the development of new recommendations models. In this work, we propose a Hybrid recommender model for recommending Chemical Compounds. The model integrates collaborative-filtering algorithms for implicit feedback (Alternating Least Squares (ALS) and Bayesian Personalized Ranking (BPR)) and semantic similarity between the Chemical Compounds in the ChEBI ontology (ONTO). We evaluated the model in an implicit dataset of Chemical Compounds, CheRM. The Hybrid model was able to improve the results of state-of-the-art collaborative-filtering algorithms, especially for Mean Reciprocal Rank, with an increase of 6.7% when comparing the collaborative-filtering ALS and the Hybrid ALS_ONTO.

Keywords: Recommender System · Implicit feedback · Ontology · Collaborative-Filtering · Semantic similarity

1 Introduction

The recommendation of Chemical Compounds of interest for scientific researchers has not been widely explored [9,23]. However, Recommender Systems (RSs) may help in the discovery of compounds, for example, by suggesting items not yet studied by the researchers. One challenge in this field is the lack of available datasets with the preferences of the researchers about the Chemical Compounds for testing the RS. More recently, alternatives have emerged with the development of datasets consisting of data collected from implicit feedback. Unlike what happens with other datasets, for example, Movielens [6],

This work was supported by the Fundação para a Ciência e Tecnologia (FCT), under LASIGE Strategic Project - UID/CEC/00408/2019, UIDB/00408/2020, CENTRA Strategic Project UID/FIS/00099/2019, FCT funded project PTDC/CCI-BIO/28685/2017 and PhD Scholarship SFRH/BD/128840/2017.

J. M. Jose et al. (Eds.): ECIR 2020, LNCS 12036, pp. 94–101, 2020.
https://doi.org/10.1007/978-3-030-45442-5_12

these datasets do not contain the specific interests of the researchers. Instead, this information is extracted from the activities of the researchers, for example, through scientific literature [3,15].

Datasets of explicit or implicit feedback require different recommender algorithms, especially because implicit feedback has some significant downgrades, such as the lack of negative feedback, and unbalanced ratio of positive vs unobserved ratings [11,18]. When dealing with implicit feedback datasets, the solution involves applying learning to rank (LtR) approaches. LtR consists in, given a set of items, identify in which order they should be recommended [17].

The main approaches in RSs are Collaborative-Filtering (CF) and Content-Based (CB) [20]. CF uses the similarity between the ratings of the users, and CB uses the similarity between the features of the items. CF approaches cannot deal with new items or new users in the system, i.e., items and users without ratings (cold start problem). CB does not need to deal with this problem for new items, and that is the main reason Hybrid RSs (CF + CB) exist. One of the tools used by CB are ontologies [27], which are related vocabularies of terms and definitions for a specific field of study [2,28]. Some examples of well-known ontologies are the Chemical Entities of Biological Interest (ChEBI)[1] [7], the Gene Ontology (GO)[2] [4], and the Disease Ontology (DO)[3] [21].

In this paper, we propose a Hybrid recommender model for recommending Chemical Compounds, consisting of a CF module and a CB module. In the CF module we tested two algorithms for implicit feedback datasets, Alternating Least Squares (ALS) [8] and Bayesian Personalized Ranking (BPR) [18], separately. In the CB module we explored the semantic similarity between the compounds in the ChEBI ontology (ONTO algorithm). The Hybrid model combines ALS + ONTO, and BPR + ONTO. The framework developed for this work is available at https://github.com/lasigeBioTM/ChemRecSys.

2   Related Work

There are a few studies using RS for recommending Chemical Compounds. [9] describes the use of CF methods for creating a Free-Wilson-like fragment recommender system. [23] use RS techniques for the discovery of new inorganic compounds, by applying machine-learning to find the similarity between the proposed and the existent compounds.

Next, we describe studies using ontologies for improving the performance of CF algorithms. [12] created a RS for recommending English collections of books in a library. The authors developed PORE, a personal ontology Recommender System, which consists of a personal ontology for each user and then the application of a CF method. They used a standard normalized cosine similarity for finding the similarity between the users. [26] also used an ontology for creating users' profiles for the domain of books. They calculated the similarity, not

---

[1] https://www.ebi.ac.uk/chebi/.
[2] http://geneontology.org/.
[3] http://disease-ontology.org/.

between the ratings of the users, but based on the interest scores derived from the ontology. The CF method used was the k-nearest neighbours. [24] developed a Trust–Semantic Fusion approach, tested on movies and Yahoo! datasets. Their approach incorporates semantic knowledge to the items primary information, using knowledge from the ontologies. They used the user-based Constrained Pearson Correlation and the user-based Jaccard similarity.

[16] presented a solution for the top@k recommendations specifically for implicit feedback data. The authors developed the Spank - semantic path-based ranking. They extracted path-based features of the items from DBpedia and used LtR algorithms to get the rank of the most relevant items. They tested the method on music and movies domains. [1] developed a new semantic similarity measure, the Inferential Ontology-based Semantic Similarity. The new measure improved the results of a user-based CF approach, using Pearson Correlation for calculating the similarity between the users. The authors tested the approach on the tourism domain. Most recently, [14] developed a Hybrid RS tested on the movies domain. The method used Single Value Decomposition for dimensionality reduction for the item and user-based CF, and ontologies for item-based semantic similarity, improving the CF results. They do not deal with implicit data.

To the best of our knowledge, our study is the first to use semantic similarity for recommending Chemical Compounds, dealing with implicit data by using state-of-the-art methods (ALS and BPR) and improving the results for the top@k in several evaluation metrics.

3   The Proposed Model

The proposed model has two modules: CF and CB. Figure 1 shows the general workflow of the model. The input data used in this model has the format of <user,item,rating>. The unrated set represents the items we want to rank to provide the best recommendations in the first positions to a user. The rated set are the items the users already rated. Since we will split the data into train and test, lets call train set to the rated set and test set to the unrated set. Both train and test sets are the input for CF and CB modules. Using CF algorithms for implicit feedback datasets, the CF module gives a score for each item in the test set. The CB module uses semantic similarity for providing a score for the items in the test set. In the last step, the scores from CF and CB modules are combined and sorted in descending order.

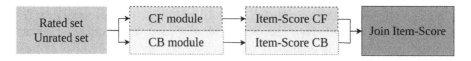

Fig. 1. Workflow of the Hybrid recommender model.

For the CF module, we selected state-of-the-art CF recommender algorithms for implicit data[4], ALS [8] and BPR [18]. ALS is a latent factor algorithm that addresses the confidence of a user-item pair rating. BPR is also a latent factor algorithm, but it is more appropriate for ranking a list of items. BPR does not just consider the unobserved user-item pairs as zeros, but instead, it takes into consideration the preference of a user between an observed and an unobserved rating.

The CB module (ONTO algorithm) is based on ChEBI ontology. This module assigns a score S to each item in the test set, calculating the semantic similarity between each item in the train and the test sets, as shown in Fig. 2. For calculating the similarity, we used DiShIn[5] [5], a tool for calculating semantic similarities between the entities represented by an ontology. Semantic similarity allows measuring how close two entities are in a semantic base. When using ontologies, the semantic similarity may be measured, for example, by calculating the shortest path connecting the nodes of two entities. DiShIn allows to calculate three similarity metrics: Resnik [19], Lin [13], and Jiang and Conrath [10]. For this work, we used the Lin metric. We intend to test the other metrics in the future.

Whereas the CF module uses all the ratings from the train set to train the model, CB module only takes into account the ratings of each user. Using DiShIn, we calculate the value of the similarity between each item in the train test and the items in test set.

Fig. 2. Example of ONTO algorithm. I_1 is a test item, I_2, I_3 and I_4 are train items. The semantic similarity is calculated for each pair of test-train items. The score for I_1 (SI_1) is the mean of the similarities of each test-train pair.

Lets I_1 be the item in test, and I_2, I_3 ... I_n the items in the train, with size m, for a user U. The score S for I_1 (S_{I1}) is calculated according to the Eq. 1. ONTO algorithm does not use any real rating of the test items when calculating the score for each item in the test set, thus we do not have the problem of introducing bias in the results.

$$S_{I1} = \frac{Sim_{1,2} + Sim_{1,3} + ... + Sim_{1,n}}{m} \tag{1}$$

For obtaining a final score (FS) for each item in the test, we combine the scores from CF module (S_{CF}) and CB module (S_{CB}), into a Hybrid recommendation approach, according to Eq. 2. Our goal is to prove that by combining both modules, we can improve the results of each module separately.

$$FS_{I1} = S_{CF} \times S_{CB} \tag{2}$$

---

[4] https://implicit.readthedocs.io/en/latest/index.html.
[5] https://github.com/lasigeBioTM/DiShIn.

4   Experiments and Results

Experiments. The data used in this work is a subset of a dataset of Chemical Compounds, CheRM, with the format of <user,item,rating> [3]. The users are authors from research articles, the items are Chemical Compounds present in ChEBI, and the ratings (implicit) are the number of articles the author wrote about the item[6]. The subset has 102 Chemical Compounds, 1184 authors, 5401 ratings, and a sparsity level of 95.5%. We used a subset of CheRM because it has more than 22,000 items and there is a bottleneck in the calculation of the similarity between all the items in real time.

The algorithms tested were ALS, BPR, ONTO, and the hybrids ALS_ONTO and BPR_ONTO. For ALS and BPR we tested different latent factors, achieving the best results for this data with 150 factors. We used offline methods [25] for evaluating the performance of the algorithms for the top@k, with k varying between 0 and 20, with steps of 1. From the vast range of metrics for evaluating recommender algorithms, we selected Classification Accuracy Metrics (CAMet) and Rank Accuracy Metrics (RAMet). CAMet measure the relevant and irrelevant items recommended in a ranked list. Examples of CAMet are Precision, Recall, and F-Measure. RAMet measure the ability of an algorithm for recommending the items in the correct order. Some well-known RAMet are Mean Reciprocal Rank (MRR), Normalized Discount Cumulative Gain (nDCG), and Limited Area Under the Curve (lAUC), a variation of AUC [22]. All the selected metrics range between 0 and 1, and values closest to 1 are better. For the segmentation of the dataset, we used a cross-validation approach, by splitting users and items in 5 folds. Each iteration had 1/5 of the users and the items as test and 4/5 as train data. All the positive ratings in the test set are considered as relevant items. We considered the unrated items as negative ratings, i.e., not relevant for the users.

Results. We present the results of this study in Fig. 3, for all the algorithms and all the metrics described previously. Analysing Fig. 3, the ONTO algorithm alone has the lowest results in all metrics. Nevertheless, in metrics such as Precision, Recall and F-measure, it follows the trend of the other algorithms, and when measuring these metric for the top@20, the results are similar. ONTO has the advantage of being a CB algorithm, therefore it does not have the problem of cold start for new items. ALS and BPR cannot be used if the item in the test set is not in the train set at least once (at least one author in the train set wrote about this Chemical Compound).

Between ALS and BPR, ALS achieved the best results. Since BPR is an algorithm for ranking, it was expected to obtain better results. We believe this is due the fact that the dataset has a large number of ratings equal to one, and many items have the same relevance (difficult to rank).

The approach with the best results in most of the metrics is the Hybrid ALS_ONTO. The use of ALS and ONTO algorithms together has a particularly

---

[6] https://github.com/lasigeBioTM/CheRM.

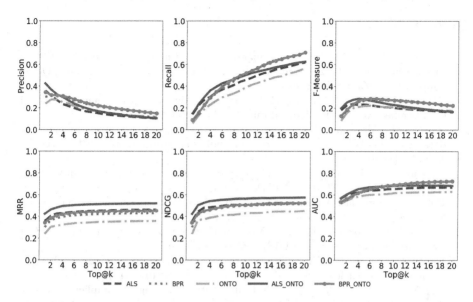

Fig. 3. Results comparing ALS, BPR, ONTO and the hybrids ALS_ONTO and BPR_ONTO, for Precision, Recall, F-measure, MRR, nDCG, and lAUC.

positive effect on the metrics measuring the ranking accuracy (MRR, nDCG and AUC), especially for MRR, with an increase of 6.7% when comparing the ALS algorithm and the Hybrid ALS_ONTO. This means that ONTO reorder ALS scores in a way that the first results in the top@k are more relevant.

These are preliminary results. The study needs to be replicated with the full CheRM dataset, and we need to perform more studies to see the real impact for the cold start problem. Nevertheless, the results seem promising, in the one hand for improving the relevant recommendations provided (CAMet), and on the other hand in enhancing the position of the most relevant items in a ranked list (RAMet). Our Hybrid algorithm may be applied to other areas, for example, for genes, phenotypes, and diseases, provided that exists an ontology for these items.

5   Conclusion

In this work, we presented a Hybrid recommendation model for recommending Chemical Compounds, based on CF algorithms for implicit data and a CB algorithm based on semantic similarity of the Chemical Compounds using the ChEBI ontology. The obtained results support our hypothesis that by using the semantic similarity between the Chemical Compounds, the results of state-of-the-art CF algorithms can be improved. For future work we intend to increase the length of the dataset, to test other similarity metrics, and to test other alternatives to calculate the final score of the Hybrid algorithm.

References

1. Al-Hassan, M., Lu, H., Lu, J.: A semantic enhanced hybrid recommendation approach: a case study of e-government tourism service recommendation system. Decis. Support Syst. **72**, 97–109 (2015)
2. Barros, M., Couto, F.M.: Knowledge representation and management: a linked data perspective. Yearb. Med. Inform. **25**(01), 178–183 (2016)
3. Barros, M., Moitinho, A., Couto, F.M.: Using research literature to generate datasets of implicit feedback for recommending scientific items. IEEE Access **7**, 176668–176680 (2019)
4. Consortium, G.O.: The gene ontology resource: 20 years and still going strong. Nucleic Acids Res. **47**(D1), D330–D338 (2018)
5. Couto, F., Lamurias, A.: Semantic similarity definition. In: Encyclopedia of Bioinformatics and Computational Biology, vol. 1 (2019)
6. Harper, F.M., Konstan, J.A.: The MovieLens datasets: history and context. ACM Trans. Interact. Intell. Syst. (TIIS) **5**(4), 1–19 (2015)
7. Hastings, J., et al.: ChEBI in 2016: improved services and an expanding collection of metabolites. Nucleic Acids Res. **44**(D1), D1214–D1219 (2015)
8. Hu, Y., Koren, Y., Volinsky, C.: Collaborative filtering for implicit feedback datasets. In: 2008 Eighth IEEE International Conference on Data Mining, pp. 263–272. IEEE (2008)
9. Ishihara, T., Koga, Y., Iwatsuki, Y., Hirayama, F.: Identification of potent orally active factor Xa inhibitors based on conjugation strategy and application of predictable fragment recommender system. Bioorg. Med. Chem. **23**(2), 277–289 (2015)
10. Jiang, J.J., Conrath, D.W.: Semantic similarity based on corpus statistics and lexical taxonomy. arXiv preprint cmp-lg/9709008 (1997)
11. Khawar, F., Zhang, N.L.: Conformative filtering for implicit feedback data. In: Azzopardi, L., Stein, B., Fuhr, N., Mayr, P., Hauff, C., Hiemstra, D. (eds.) ECIR 2019. LNCS, vol. 11437, pp. 164–178. Springer, Cham (2019). https://doi.org/10.1007/978-3-030-15712-8_11
12. Liao, I.E., Hsu, W.C., Cheng, M.S., Chen, L.P.: A library recommender system based on a personal ontology model and collaborative filtering technique for english collections. Electron. Libr. **28**(3), 386–400 (2010)
13. Lin, D., et al.: An information-theoretic definition of similarity. In: ICML, vol. 98, pp. 296–304. Citeseer (1998)
14. Nilashi, M., Ibrahim, O., Bagherifard, K.: A recommender system based on collaborative filtering using ontology and dimensionality reduction techniques. Expert Syst. Appl. **92**, 507–520 (2018)
15. Ortega, F., Bobadilla, J., Gutiérrez, A., Hurtado, R., Li, X.: Artificial intelligence scientific documentation dataset for recommender systems. IEEE Access **6**, 48543–48555 (2018)
16. Ostuni, V.C., Di Noia, T., Di Sciascio, E., Mirizzi, R.: Top-N recommendations from implicit feedback leveraging linked open data. In: Proceedings of the 7th ACM Conference on Recommender Systems, pp. 85–92. ACM (2013)
17. Rendle, S., Balby Marinho, L., Nanopoulos, A., Schmidt-Thieme, L.: Learning optimal ranking with tensor factorization for tag recommendation. In: Proceedings of the 15th ACM SIGKDD International Conference on Knowledge Discovery and Data Mining, pp. 727–736. ACM (2009)

18. Rendle, S., Freudenthaler, C., Gantner, Z., Schmidt-Thieme, L.: BPR: Bayesian personalized ranking from implicit feedback. In: Proceedings of the Twenty-Fifth Conference on Uncertainty in Artificial Intelligence, pp. 452–461. AUAI Press (2009)
19. Resnik, P.: Using information content to evaluate semantic similarity in a taxonomy. arXiv preprint cmp-lg/9511007 (1995)
20. Ricci, F., Rokach, L., Shapira, B.: Recommender systems: introduction and challenges. In: Ricci, F., Rokach, L., Shapira, B. (eds.) Recommender Systems Handbook, pp. 1–34. Springer, Boston (2015). https://doi.org/10.1007/978-1-4899-7637-6_1
21. Schriml, L.M., et al.: Human disease ontology 2018 update: classification, content and workflow expansion. Nucleic Acids Res. **47**(D1), D955–D962 (2018)
22. Schröder, G., Thiele, M., Lehner, W.: Setting goals and choosing metrics for recommender system evaluations. In: UCERSTI2 Workshop at the 5th ACM Conference on Recommender Systems, Chicago, USA, vol. 23, p. 53 (2011)
23. Seko, A., Hayashi, H., Tanaka, I.: Compositional descriptor-based recommender system for the materials discovery. J. Chem. Phys. **148**(24), 241719 (2018)
24. Shambour, Q., Lu, J.: A trust-semantic fusion-based recommendation approach for e-business applications. Decis. Support Syst. **54**(1), 768–780 (2012)
25. Shani, G., Gunawardana, A.: Evaluating recommendation systems. In: Ricci, F., Rokach, L., Shapira, B., Kantor, P.B. (eds.) Recommender Systems Handbook, pp. 257–297. Springer, Boston (2011). https://doi.org/10.1007/978-0-387-85820-3_8
26. Sieg, A., Mobasher, B., Burke, R.: Improving the effectiveness of collaborative recommendation with ontology-based user profiles. In: Proceedings of the 1st International Workshop on Information Heterogeneity and Fusion in Recommender Systems, pp. 39–46. ACM (2010)
27. Tarus, J.K., Niu, Z., Mustafa, G.: Knowledge-based recommendation: a review of ontology-based recommender systems for e-learning. Artif. Intell. Rev. **50**(1), 21–48 (2017). https://doi.org/10.1007/s10462-017-9539-5
28. Uschold, M., Gruninger, M.: Ontologies: principles, methods and applications. Knowl. Eng. Rev. **11**(2), 93–136 (1996)

Assessing the Impact of OCR Errors in Information Retrieval

Guilherme Torresan Bazzo, Gustavo Acauan Lorentz, Danny Suarez Vargas, and Viviane P. Moreira[✉] [iD]

Institute of Informatics, UFRGS, Porto Alegre, Brazil
{guilherme.bazzo,gustavo.lorentz,dsvargas,viviane}@inf.ufrgs.br

Abstract. A significant amount of the textual content available on the Web is stored in PDF files. These files are typically converted into plain text before they can be processed by information retrieval or text mining systems. Automatic conversion typically introduces various errors, especially if OCR is needed. In this empirical study, we simulate OCR errors and investigate the impact that misspelled words have on retrieval accuracy. In order to quantify such impact, errors were systematically inserted at varying rates in an initially clean IR collection. Our results showed that significant impacts are noticed starting at a 5% error rate. Furthermore, stemming has proven to make systems more robust to errors.

Keywords: OCR · Retrieval effectiveness · Noisy text

1 Introduction

Estimates say that most information useful for organizations is represented in an unstructured format, predominantly as free text [6]. A significant portion of this useful information is stored in PDF files – research articles, books, company reports, and presentations are typically disseminated in PDF format. PDF documents need to be converted into plain text before being processed by an Information Retrieval (IR) or a text mining system. These files can either be *digitally created* or created from *scanned* documents. While the former are generated from an original electronic version of a document (*i.e.,* contain the text characters), the later contain images of the original document and need to go through Optical Character Recognition (OCR) so that their contents can be extracted. Despite being addressed by researchers for decades, OCR is still imperfect. As a result, the extracted text contains errors that typically involve character exchanges. Although digitally created PDFs are cleaner, they are not problem-free since, for example, hyphenated terms (due to separation into syllables) may be identified as two tokens and indexed incorrectly.

Extraction errors can have a negative impact on the quality of IR systems and are found even in mainstream search engines. Figure 1 presents an excerpt of the result page generated by Google Scholar for the query "information retrieval techniques". In the small snippet from a matching document, we can see four

© Springer Nature Switzerland AG 2020
J. M. Jose et al. (Eds.): ECIR 2020, LNCS 12036, pp. 102–109, 2020.
https://doi.org/10.1007/978-3-030-45442-5_13

errors – three terms were erroneously segmented into two tokens, and two terms were concatenated into one token. The effect is that a query with the correct spelling for *e.g.,* "the barriers encountered in retrieving information" would be unable to retrieve that document. Approaches for treating misspelled queries cannot solve this problem as the issue is in the document, not in the query. Furthermore, there are important differences between the types of errors made by humans while typing and those made by OCR systems [9].

The fact that this is still an open issue is evidenced by two recent competitions organized in the scope of the International Conference on Document Analysis and Recognition (ICDAR) [2,12]. The best performing approaches employ state-of-the-art methods such as character-based Neural Machine Translation and recurrent networks (bidirectional LSTMs) taking BERT models as input. The best results for the error detection task were below 0.7 in terms of F1 in several languages [12], showing that there is still a lot of room for improvement.

Our goal is to revisit the problem of retrieving OCR-ed text and quantify the impact that these errors have on the accuracy of IR systems. Ideally, to quantify the impact, one needs an IR test collection with source PDF files, their extracted and corrected versions, a set of queries, and their corresponding relevance judgments. Unfortunately, to the best of our knowledge, such a collection does not exist and creating one would demand significant effort. In line with previous works on this topic [3,7], our approach was to systematically insert errors in a standard IR collection containing plain text documents, queries, and relevance judgments. Different error rates were tested so that we could gauge their effects. In order to simulate real errors, we collected, assessed, and manually corrected a sample of OCR-ed PDF documents. Statistics drawn from this sample were used to guide the error insertion approach. Our experiments were performed in an IR collection containing documents in Portuguese – a language that makes use of diacritics (*e.g.,* à, á, ã, â, é, í, ç *etc.*). These characters are typically among the ones with more extraction errors. The results showed that error rates starting at 5% can cause a significant impact in many system configurations and that stemming makes systems more robust to coping with errors.

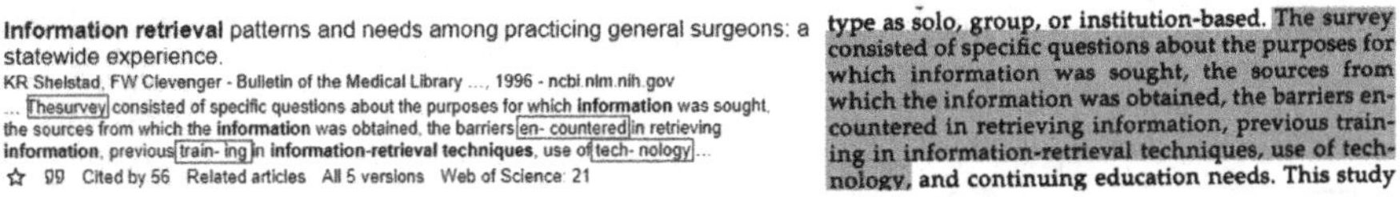
Information retrieval patterns and needs among practicing general surgeons: a statewide experience.
KR Shelstad, FW Clevenger - Bulletin of the Medical Library ..., 1996 - ncbi.nlm.nih.gov
... Thesurvey consisted of specific questions about the purposes for which **information** was sought, the sources from which the **information** was obtained, the barriers en- countered in retrieving **information**, previous train- ing in **information-retrieval techniques**, use of tech- nology ...
Cited by 56 Related articles All 5 versions Web of Science: 21

(a) Excerpt from GoogleScholar

type as solo, group, or institution-based. The survey consisted of specific questions about the purposes for which information was sought, the sources from which the information was obtained, the barriers en-countered in retrieving information, previous train-ing in information-retrieval techniques, use of tech-nology, and continuing education needs. This study

(b) Source PDF file

Fig. 1. Example of extraction errors identified in a mainstream search engine.

2 Related Work

Existing work on dealing with OCR-ed texts spans over a long period and focused on approaches for detecting and fixing errors [4,5,9,10]. Specifically on the topic of improving the retrieval of OCR text, Beitzel *et al.* [1] surveyed a number of solutions – most of which date to the late 1990s. TREC ran a confusion track to assess retrieval effectiveness on degraded collections. Their modified test collections had 5 and 20% character error rates. Five teams took part in the challenge. The organizers reported that counter-intuitive results had been found and that "there is still a great deal to be understood about the interaction of the diverse approaches" [7]. Croft *et al.* [3] share some similarities with our work. However, rather than injecting errors into a clean text collection, the authors opted to randomly select words to be discarded from the document and, as a consequence, they were not indexed. The limitation of such approach is that it does not account for issues with wrong segmentation (adding or suppressing the space character) or cases in which the error modifies the word into another valid word. The main finding was that performance degradation was more critical for very short documents. In a detailed investigation, Taghva *et al.* [13] observed that while the results seem to have insignificant degradation on average, individual queries can be greatly impacted. Furthermore, they report an impressive increase in the number of index terms in the presence of errors and that relevance feedback methods are unable to overcome OCR errors.

This paper differs from existing works in a number of aspects. The configurations we assess include the use of stemming, more recent ranking algorithms, and more levels of degradation. Finally, we experiment with a different test collection in a language that has not been extensively used for IR.

3   Simulating Errors

The methodology we propose to insert errors is shown in Fig. 2. Our goal is to replicate, as much as possible, the pattern of problems that actually happen in PDF conversions to plain text from both digitally created and scanned documents. In order to achieve that, one needs a sample of aligned pairs of extracted and expected contents (shown as input in Fig. 2). The expected contents need to be manually produced by correcting the extracted text. This is a laborious and time-consuming task. By comparing these <extracted, expected> pairs at character level, we generate a *character exchange list*.

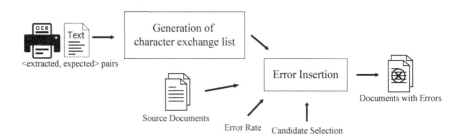

Fig. 2. Approach for error injection in documents.

To align the <extracted, expected> pairs, we used the Needleman-Wunsch [8] algorithm. This algorithm generates the best (global) alignment of two sequences, with the addition of gaps to account for mismatching characters. We found exchanges of one-to-one (*e.g.*, "inserted" → "insorted"), one-to-two (*e.g.*, "document" → "docurnent"), or two-to-one (*e.g.*, "light" → "hght") characters. The frequencies of the exchanges were computed and stored in the character exchange list. Then, they are used to bias the error insertion algorithm towards the most frequent exchanges.

By analyzing the pattern of errors found, we came up with a categorization of the types of issues. (*i*) *Exchange of characters.* This is the most common error found (90% of all errors) and it is caused by the low quality of the documents we are processing. Every exchange in our exchange list has assigned to itself the frequency of its appearance, which we use, in conjunction with the tournament selection, to elect one error to a given term. (*ii*) *Separated terms.* This error corresponds to 5% of the cases and it happens when a space character is erroneously inserted in the middle of a term. (*iii*) *Joined terms.* This error, which has a frequency of 4.9%, happens when the space between terms is omitted, resulting in the unexpected concatenation of terms. (*iv*) *Erroneous symbol.* This issue accounts for 0.1% of all errors, usually represents dirt or a printing error at the scanned document.

Issues (*i*) to (*iii*) can potentially affect recall as relevant documents containing terms with these problems will not be retrieved by the query. Issue (*iv*) can also lower precision since the fragment of a term can match a query for which the document is not relevant (*e.g.*, if the term "encounter" found in a document *d* is fragmented into the tokens "en" and "counter", then *d* can erroneously match a query with the term "counter").

Two alternatives for the selection of candidate terms were employed. In the first, any term from any document could be selected. In the second, a more targeted selection was made in which candidate terms were taken only from judged documents (*i.e.*, the documents in the qrels file).

Using the desired error rate, we iterate through every candidate term in the documents. The term is chosen to be modified with a probability equivalent to the given error rate. If the term is selected, then the choice of error is made taking the observed frequency. This was achieved by drawing a random float between 0 and 1 and matching it against the corresponding error frequency. The selection of which exchange to apply was made using tournament selection in ten rounds according to the frequency of the exchange.

4 Experimental Evaluation

This Section describes the experimental evaluation of the error insertion method to assess the impact of OCR errors in IR systems. The resources, tools, and configurations used in our experiments were as follows.

Data. To generate the character exchange list, we took a sample of 900 PDF documents containing abstracts from research articles published at the website of a Brazilian Oil Company[1]. The extracted text was manually checked and the extraction errors were fixed to create the list of <extracted, expected> pairs. The IR collection used was Folha de São Paulo, a Brazilian Newspaper. It has 103K documents, 100 queries, and it has been used in important evaluation campaigns such as CLEF [11].

Table 1. Number of index terms (in thousands) and the proportional increase in comparison to the baseline.

Setting	Baseline	1%	5%	10%	25%	50%
ALL-NS	273	355 (30%)	523 (91%)	659 (141%)	937 (243%)	1,243 (355%)
ALL_ST	203	253 (24%)	352 (73%)	434 (113%)	605 (197%)	801 (293%)
JD_NS	273	342 (25%)	473 (73%)	574 (110%)	770 (182%)	983 (260%)
JD_ST	203	245 (20%)	324 (59%)	386 (90%)	514 (153%)	660 (224%)

Tools. The OCR software used was Abbyy Finereader 14[2]. The choice was made after its good result compared to a number of other alternatives including Tesseract, a9t9, Omnipage, and Wondershare. The IR System was Apache Solr[3].

Experimental Procedure. In our experimental procedure, we varied the following parameters. The *Ranking function*, taking three possibilities: Cosine (COS) using TF-IDF weighting, BM25, and Divergence from Randomness (DFR). The *use of stemming*: applying a light stemmer (ST) and no stemming (NS). The *error rates* were 1%, 5%, 10%, 25%, and 50%. Baseline runs using the original documents were also created. The *candidate terms* for error insertion were either any term from any document (ALL) or any term from the judged documents (JD). These variations amounted to a total of 72 experimental runs, which were evaluated using standard IR metrics. Statistical significance was measured using T-tests. Queries were made by simply taking the title field from the topics. The goal was to simulate real queries that are typically short.

Table 1 shows the number of index terms for the combination of error rates, use of stemming, and candidate terms. As expected, the number of index entries grows remarkably with the error rates, reaching more than a four-fold increase for unstemmed runs with a 50% error rate.

The results for all experimental runs are in Table 2. The runs in which the mean average precision (MAP) decrease was found to be statistically significant (in relation to the baseline) at a 99% confidence interval are in a darker shade and the ones with a 95% significance are in a lighter shade.

The best ranking function in terms of absolute MAP values was DFR, followed by BM25. However, there were no differences on their robustness in

---

[1] http://publicacoes.petrobras.com.br/.

[2] https://www.abbyy.com/.

[3] https://lucene.apache.org/solr/.

Table 2. MAP Results for all configurations. The numbers in brackets indicate the proportional change.

Setting	Baseline	1%	5%	10%	25%	50%
ALL-BM25-NS	0.251	0.249 (−0.8%)	0.243 (−3.2%)	0.232 (−7.7%)	0.211 (−16.2%)	0.156 (−38.1%)
ALL-BM25-ST	0.294	0.291 (−0.8%)	0.288 (−1.9%)	0.281 (−4.4%)	0.263 (−10.4%)	0.223 (−23.9%)
ALL-COS-NS	0.242	0.243 (0.2%)	0.239 (−1.3%)	0.218 (−9.8%)	0.202 (−16.5%)	0.166 (−31.6%)
ALL-COS-ST	0.275	0.273 (−0.9%)	0.266 (−3.6%)	0.256 (−7.1%)	0.251 (−8.8%)	0.223 (−19.0%)
ALL-DFR-NS	0.263	0.262 (−0.2%)	0.258 (−1.8%)	0.240 (−8.8%)	0.222 (−15.6%)	0.180 (−31.5%)
ALL-DFR-ST	0.307	0.304 (−1.1%)	0.306 (−0.4%)	0.295 (−4.2%)	0.276 (−10.1%)	0.250 (−18.7%)
JD-BM25-NS	0.251	0.248 (−1.4%)	0.243 (−3.4%)	0.232 (−7.6%)	0.226 (−10.2%)	0.169 (−32.9%)
JD-BM25-ST	0.294	0.290 (−1.2%)	0.284 (−3.4%)	0.280 (−4.6%)	0.261 (−11.2%)	0.229 (−21.9%)
JD-COS-NS	0.242	0.240 (−0.6%)	0.237 (−2.0%)	0.218 (−10.0%)	0.200 (−17.3%)	0.152 (−37.1%)
JD-COS-ST	0.275	0.277 (0.7%)	0.271 (−1.7%)	0.263 (−4.5%)	0.247 (−10.4%)	0.209 (−24.2%)
JD-DFR-NS	0.263	0.263 (0.0%)	0.252 (−4.2%)	0.239 (−9.2%)	0.221 (−16.2%)	0.163 (−38.2%)
JD-DFR-ST	0.307	0.306 (−0.4%)	0.300 (−2.3%)	0.294 (−4.3%)	0.281 (−8.6%)	0.236 (−23.2%)

the presence of OCR errors as their pattern of MAP decrease was the same. Strangely, in two runs in which the cosine was used, the insertion of errors at a 1% rate improved the performance (ALL-COS-NS and JD-COS-ST). This can be explained by the fact that errors are inserted both relevant and non-relevant documents. In these cases, the errors were introduced in non-relevant documents which made relevant documents be ranked higher.

The use of stemming consistently improved the results – *i.e.,* all runs in which stemming was used had higher scores than their unstemmed counterparts. Stemming has made the runs more robust to the OCR errors. This can be seen comparing the loss in MAP of the runs with and without stemming. Nearly all runs in which stemming was used had smaller losses than their counterparts. Furthermore, the aid of stemming is more noticeable in the runs with higher error rates. The benefit of stemming can be explained by the fact that the OCR error can be in the suffix that is removed. Looking at the correlation between the number of index terms (Table 2) and MAP, we find a strong negative correlation of 0.86. When the correlation is measured for stemmed and unstemmed runs separately, the negative correlations are 0.81 and 0.90, respectively. This gives further support to the benefits of stemming.

Looking at our sample of aligned extracted and expected texts (assembled from real documents) we observed an error rate of 1.5%. Considering this rate and the results in Table 2, one can conclude that the errors do not have a severe impact on IR as significant impacts are observed starting at 5%. At a 10% rate, all runs are significantly affected. Recall that this small error rate was found using the software which provided the best results on relatively recent documents. Some studies that provide statistics of the proportion of errors found in OCR-ed documents report finding error rates of around 20% in historical documents [4,5,14]. At that error rate, the degradation is considered statistically significant.

Comparing the two choices of candidate terms for error insertion we find close scores. This means that the error injection targeting the judged documents did not have an influence on the results.

5   Conclusion

Despite having been investigated for decades, the issues associated with retrieving noisy text still remain unsolved in many IR systems. In this paper, we revisit this topic by assessing the impact that different error rates have on retrieval performance. We tested different setups, including ranking algorithms and the use of stemming. Our findings suggest that statistically significant degradation starts at a word error rate of 5% and that stemming is able to make systems more resilient to these errors. As future work, it would be useful to assess which type of error identified in (Sect. 3) has the greatest impact in retrieval quality.

Acknowledgments. This work was partially supported by Petrobras, CNPq/Brazil, and by CAPES Finance Code 001.

References

1. Beitzel, S.M., Jensen, E.C., Grossman, D.A.: A survey of retrieval strategies for OCR text collections. In: Proceedings of the Symposium on Document Image Understanding Technologies (2003)
2. Chiron, G., Doucet, A., Coustaty, M., Moreux, J.: ICDAR 2017 competition on post-OCR text correction. In: International Conference on Document Analysis and Recognition (ICDAR), vol. 01, pp. 1423–1428 (2017)
3. Croft, W.B., Harding, S., Taghva, K., Borsack, J.: An evaluation of information retrieval accuracy with simulated OCR output. In: Symposium of Document Analysis and Information Retrieval (1994)
4. Droettboom, M.: Correcting broken characters in the recognition of historical printed documents. In: Proceedings 2003 Joint Conference on Digital Libraries, pp. 364–366, May 2003
5. Evershed, J., Fitch, K.: Correcting noisy OCR: context beats confusion. In: Proceedings of the First International Conference on Digital Access to Textual Cultural Heritage (DATeCH 2014), pp. 45–51 (2014)
6. Grimes, S.: Unstructured data and the 80 percent rule, p. 10. Carabridge Bridgepoints (2008)
7. Kantor, P.B., Voorhees, E.M.: The TREC-5 confusion track: comparing retrieval methods for scanned text. Inf. Retrieval **2**(2), 165–176 (2000). https://doi.org/10.1023/A:1009902609570
8. Needleman, S., Wunsch, C.: A general method applicable to the search for similarities in the amino acid sequence of two proteins. J. Mol. Biol. **48**, 443–453 (1970)
9. Nguyen, T., Jatowt, A., Coustaty, M., Nguyen, N., Doucet, A.: Deep statistical analysis of OCR errors for effective post-OCR processing. In: ACM/IEEE Joint Conference on Digital Libraries (JCDL), pp. 29–38, June 2019

10. Parapar, J., Freire, A., Barreiro, Á.: Revisiting N-gram based models for retrieval in degraded large collections. In: Boughanem, M., Berrut, C., Mothe, J., Soule-Dupuy, C. (eds.) ECIR 2009. LNCS, vol. 5478, pp. 680–684. Springer, Heidelberg (2009). https://doi.org/10.1007/978-3-642-00958-7_66

11. Peters, C., Braschler, M.: European research letter: cross-language system evaluation: the CLEF campaigns. J. Am. Soc. Inf. Sci. Technol. **52**(12), 1067–1072 (2001)

12. Rigaud, C., Doucet, A., Coustaty, M., Moreux, J.P.: ICDAR 2019 competition on post-OCR text correction. In: International Conference on Document Analysis and Recognition (ICDAR) (2019)

13. Taghva, K., Borsack, J., Condit, A.: Evaluation of model-based retrieval effectiveness with OCR text. ACM Trans. Inf. Syst. **14**(1), 64–93 (1996)

14. Tanner, S., Muñoz, T., Ros, P.H.: Measuring mass text digitization quality and usefulness: lessons learned from assessing the OCR accuracy of the British library's 19th century online newspaper archive. D-Lib Mag. **15**(7/8), 1082–9873 (2009)

Towards Query Logs for Privacy Studies: On Deriving Search Queries from Questions

Asia J. Biega[1($\boxtimes$)], Jana Schmidt[2], and Rishiraj Saha Roy[2]

[1] Microsoft Research, Montreal, Canada
asia.biega@microsoft.com
[2] Max Planck Institute for Informatics,
Saarland Informatics Campus, Saarbrücken, Germany
{jschmidt,rishiraj}@mpi-inf.mpg.de

Abstract. Detailed query histories often contain a precise picture of a person's life, including sensitive and personally identifiable information. As sanitization of such logs is an unsolved research problem, commercial Web search engines that possess large datasets of this kind at their disposal refrain from disseminating them to the wider research community. Ironically, studies examining privacy in search often require detailed search logs with user profiles. This paper builds on an observation that information needs are also expressed in the form of questions in online Community Question Answering (CQA) communities. We take a step towards understanding the process of formulating queries from questions to form a basis for automatic derivation of search logs from CQA forums. Specifically, we sample natural language (NL) questions spanning diverse themes from the StackExchange platform, and conduct a large-scale conversion experiment where crowdworkers submit search queries they would use when looking for equivalent information. We also release a dataset of 7,000 question-query pairs from our study.

1 Introduction

Background. Commercial Web search engines refrain from disseminating detailed user search histories, as they may contain sensitive and personally identifiable information[1]. Studies examining privacy in search, however, require extensive search logs with user profiles to examine the sensitive semantics of queries or the topical distribution of user interests [1,5,6,8,16].

While there exist a number of public search query logs, none of them contain *detailed user histories*. Relevant among these, the TREC Sessions Track 2014 data [7] has 148 users, $4.5k$ queries, and about $17k$ relevance judgments. There are roughly ten sessions per user, where each session is usually a set of reformulations. Such collections with just a couple of queries per user are inadequate

---

[1] https://en.wikipedia.org/wiki/AOL_search_data_leak.

© Springer Nature Switzerland AG 2020
J. M. Jose et al. (Eds.): ECIR 2020, LNCS 12036, pp. 110–117, 2020.
https://doi.org/10.1007/978-3-030-45442-5_14

for driving research in privacy, especially research that focuses on topical profiling. The 2014 Yandex collection [15] is useful for evaluating personalization
algorithms. However, to protect the privacy of Yandex users, every query term
is replaced by a numeric ID. This anonymization strategy makes semantic interpretation impossible and may be a reason why this collection has not received
widespread adoption in privacy studies. Interpretability of log contents is vital
for understanding privacy threats [5,6,8].

Motivated by the lack of public query logs with rich user profiles, Biega
et al. [5] synthesized a query log from the StackExchange platform[2] – a collection of CQA subforums on a multitude of topics. Queries in the synthetic
log were derived from users' information needs posed as natural language questions. A collection like this has three advantages. First, it enables creation of rich
user profiles by stitching queries derived from questions asked by the same user
across different topical forums. Second, since it is derived from explicitly public
resources created by users under the StackExchange terms of service (allowing
reuse of data for research purposes), it escapes the ethical pitfalls intrinsic to
dissemination of private user data. Third, CQA forums contain questions and
assessments of relevance in the form of accepted answers *from the same user*,
which is vital for the correct interpretation of query intent [2,9]. Other signals
like similar queries and reformulations can also be simulated with related questions and duplicates, available on most CQA forums.

Contributions. We take a step towards better automatic question-query derivation methods to improve on the approach taken by Biega et al. [5] where queries
are constructed by choosing a random number of terms with highest TF-IDF
scores. An accurate approach like this would enable the creation of high-quality
search collections down the road. We make the following contributions: (1)
We conduct a large-scale user study where crowdworkers convert questions to
queries, controlling for several biases; (2) We provide insights from the collected
data that could drive strategies for automatic conversion at scale and be used
to derive synthetic search collections for privacy studies; (3) We release 7,000
question-query pairs collected from the study[3].

2   Setting up the User Study

Filtering Subforums. We used the StackExchange dump[4] from March 2018
with data for more than 150 different subforums. We are interested in textual
questions in English and thus exclude forums primarily dealing with programming, mathematics, and other languages. Moreover, we want to avoid highly-
specialized forums as an average AMT user may not have the background knowledge to generate queries for niche domains. We thus excluded all subforums with

---

[2] https://stackexchange.com/sites.

[3] https://www.mpi-inf.mpg.de/departments/databases-and-information-systems/
research/impact/mediator-accounts/.

[4] https://archive.org/details/stackexchange-snapshot-2018-03-14.

less than 100 questions, as a proxy for expression of a critical mass of interest, leaving us with a total of 75 subforums.

Sampling Questions. As a proxy for questions being understandable by users, we choose only those that have an answer accepted by the question author, and with at least five other answers provided. Under this constraint, we first sample 50 subforums from the 75 acceptable ones to have high diversity in question topics. Next, we draw 100 questions from each of these subforums, producing a sample of 5000 questions to be used as an input in the main study.

Setup. We recruited a total of 100 AMT Master workers[5] who had an approval rate of over 95%, to ensure quality of annotations. A unit task, i.e., an AMT HIT (Human Intelligence Task) consisted of converting *fifty NL questions* to Web queries to capture *user-specific* querying traits (thirty in our pilot study). Since this is significant effort expected to require more than an hour's work at a stretch, we paid \$9 per HIT (\$6 in our pilot, owing to fewer questions). The workers (Turkers) were given three hours to complete a HIT, while the actual average time taken turned out to be 1.6 h. This is about two minutes per question, which we deem as a reasonable time required for understanding the intent of a typical CQA question that often has a few hundred words.

Guidelines. Guidelines were kept to a minimum to avoid biasing participants towards certain query formulation behavior: they only stated the requirement of building a search query aimed at retrieving equivalent information as the source question. We provided five examples to better illustrate the task, that were meant to cover the various ways of arriving at a reasonable query. To build queries, we allowed workers to: (i) select exact words from the text of the question, (ii) modify question words (*'use'* ↦ *'using'*), or, (iii) use their own words to clarify the information need. These cases were not made explicit, but communicated by coloring words in the text and the query. Questions were presented as follows (some choices aimed at avoiding title bias, see Sect. 3.2):

[Subforum name] Title Body

Each question was a concatenation of the StackExchange post title and its body, prefixed with the subforum name of the post for context. The main task was accompanied by a *demographic survey* to help us understand if such features influence how people formulate queries.

Pilot Study. We tested the setup with a pilot containing five HITs with 30 questions each. The average query length came out to be 5.7 words with a standard deviation of 2.4 words. Out of the 150 questions in total, the forum name was included in the corresponding query 33 times. In nine of such cases, the subforum name was not present in the title or body of the question, which suggests that the presence of the subforum name is important in disambiguating the context. Most query words were chosen from the title, although title words are often repeated in the body of the question. Workers used their own words

---

[5] https://www.mturk.com/help#what_are_masters.

or words modified from the question 47 times. These results suggest that participants generally understood the instructions, and gave us the confidence that this setup can be used in the main study.

3 Conducting the User Study

Data Collection. In the main study, we asked 100 AMT users to convert 5000 questions to queries (50/Turker). Users who participated in control studies were not allowed to take part again, to *avoid familiarity biases* arising from such repetition. Guidelines were kept the same as in the pilot study. The mean query length was now 6.2 words: this reflects high complexity in the underlying information needs, and in turn, interesting research challenges for methods aiming at automated conversion strategies for query log derivation. Key features of the final dataset include: (i) question topics spanning 50 different subforums of StackExchange, and (ii) question-query pairs grouped by annotator IDs, making the testbed suitable for analyzing user-specific query formulation.

3.1 Analysis

We looked into three aspects of *question-query pairs* when trying to discriminate between words that are *selected* for querying, and those that are not.

Position. We measured relative positions of query and non-query words in the question, and found that a major chunk ($\simeq 60\%$) of the query words originate from the first 10% of the question. The next 10% of the question contributes an additional 17% of words to the query; the remaining 80% of the question, in a gently diminishing manner, produce the remaining 13% of the query. This is a typical top-heavy distribution, suggesting humans conceptualize the core *content* of the information need first and gradually add specifications or conditions of *intent* [13,14] towards the end. Notably, even the last 10% of the question contains 2.78% of the query, suggesting that we cannot disregard tail ends of questions. Finally, note that the title is positioned at the beginning of the question (Sect. 3.2), and alone accounts for 57% of the query. Title words, however, do repeat in the body. Further inspection reveals that only 12% of the query mass is comprised of words that appear exclusively in the title, signifying importance of the body. We also allowed users to use their *own words* in the queries. Our analysis reveals that a substantial 17% of query words fell into this category. Such aspects of this data pose interesting challenges for query generative models.

Part-of-Speech (POS). Words play various roles in NL, with a high-level distinction between *content words* (carrying the *core* information in a sentence) and *function words* (specifying *relationships* between content words). Web users have a mental model of what current search engines can handle: most people tend to drop function words (prepositions, conjunctions, etc.) when issuing queries [4], perhaps believing those are of little importance in query effectiveness. These intuitions are substantiated by our measurements: content words (nouns, verbs, adjectives, and adverbs) account for a total of 79% (47%, 15%, 13%, and 4%, respectively) of the query, while function words constitute only 21% of the query. For interpretability, we use the 12 Universal POS tags (UTS)[6]. Our findings partially concur with POS analysis of Yahoo! search queries from a decade back [4] where nouns and adjectives were observed to be the two most dominant tags; verbs featured in the seventh position with 2.4%. We believe that differences in our scenario can be attributed to more complex information needs that demand more content words to be present in queries. These insights from the POS analysis of queries can be applied to several tasks, like query segmentation [10].

Frequency. A verbose information need may be characterized by certain recurring units, which prompted us to measure the normalized frequency TF_{norm} of a term t in a question Q, as $TF_{norm}(t, Q) = TF(t, Q)/len(Q)$, where $len(Q)$ is the question length in words. Query terms were found to have a mean TF_{norm} of 0.032, significantly higher than that of non-query terms (0.018).

3.2   Control Studies

Title Position Bias. A vital component of any crowdsourced study is to check if participants are looking for quick workarounds for assigned tasks that would make it hard for requesters to reject payments, and to control for confounding biases. In the current study, a major source of bias stems from the fact that a question is not just a sequence of words but a semi-structured concept (subforum, title, body.) Web users might be aware that question titles often summarize questions. Thus, if the structure is apparent to the annotator, they might use words only from titles without examining the full question content.

To mitigate this concern, we present titles in the same font as the body, and do not separate them with newlines. Nevertheless, users may still be able to figure out that the first sentence is indeed the question title. To quantify such *position bias* of the title, we used ten HITs (500 questions) as a *control experiment* where, unknown to the Turkers, the title was appended as the *last sentence* in the question. These 500 questions were also annotated in the usual setup in the main study. We compare the main and the control studies by measuring how often users chose words from the first and the last sentences (Table 1). Values were normalized by the length of the question title, as raw counts could mislead the analysis (longer question titles contribute larger *numbers* of words to queries).

We make the following observations: (i) in both the main study and the control, users choose words from titles very often ($\simeq$97% and $\simeq$84%, respectively), showing similar task interpretation. Note that such high percentages are acceptable, as question titles typically do try to summarize intent. (ii) Relatively similar percentages of query words originate from titles in both cases (37.7% vs. 26.1%). (iii) If Turkers were trying to do the task just after skimming the first sentence (which they would perceive as the title), the percentage of words from

---

[6] https://github.com/slavpetrov/universal-pos-tags

the first sentence in the control would have been far higher than a paltry 12.2%, and the last sentence would contribute much lower than 26.1%. We also observed that in 4.1% of the cases, words were chosen *exclusively* from the last sentence.

Table 1. Measurements from the position bias control study.

Property	Main study	Control study
Times question title word chosen for query	96.6%	83.8%
Question title words in query	37.7%	26.1%
First sentence words in query	37.7%	12.2%
Last sentence words in query	9.0%	26.1%

User Agreement. While the main focus of the study was to construct a sizable collection of question-query pairs, we were also interested in observing the effect of individual differences on query formulation. To this end, we issued ten HITs (each with fifty questions) completed by three workers each. The validity of the comparison comes from the experimental design where query construction is conditioned on a specific information need. We computed the average Jaccard similarity coefficient between all pairs of queries (q_1, q_2) for the same question: $J(q_1, q_2) = \frac{|q_1 \cap q_2|}{|q_1 \cup q_2|}$, where q_1 and q_2 are the sets of words of the compared queries. We find the average overlap to be 0.33; the overlap was observed to typically arise from the most informative question words, again indicating generally correct task interpretation. Such query *variability* has been explored in [3].

3.3 Crowdworker Demographics

We asked about crowdworkers' gender, age, country of origin, highest educational degree earned, profession, income, and the frequency of using search engines in terms of the number of Web queries issued per day (such activity could be correlated with "search expertise", and this expertise may manifest itself subtly in the style of the generated queries). From the 100 subjects in our study, coincidentally, female and male participation was exactly 50 : 50. Nearly all workers lived in the USA except for three who lived in India. We found a weak positive correlation between the query length and age, and found that men formed slightly longer queries on average (6.56 words, versus 6.15 for women).

3.4 Dataset and Extended Analyses

The annotated dataset (with fields: study type, anonymous crowdworker ID, StackExchange user and post IDs, subforum name, post title, post body, and query) and an extended version of this paper with more analyses and details are available online[7]. The dataset contains 7,000 (question, query) pairs in total: 5,000 from the main study, 500 from the control experiment on title position bias, and 1,500 from the control on user agreement.

4   Conclusions

We conducted a user study to provide a better understanding of how humans formulate queries from information needs described by verbose questions, and released 7k crowdsourced question-query pairs from 50 domains. Gaining insights into this process forms an important foundation for automated conversion methods to create rich public search collections useful in privacy studies of profiling and beyond. In addition to such algorithmic conversion, potential future directions include an analysis of the quality of crowdsourced queries [12] for our setup (such as their potential for retrieval), as well as applying our general methodology to other CQA datasets [11].

References

1. Adar, E.: User 4xxxxx9: anonymizing query logs. In: Proceedings of Query Log Analysis Workshop, International Conference on World Wide Web (2007)
2. Bailey, P., Craswell, N., Soboroff, I., Thomas, P., de Vries, A.P., Yilmaz, E.: Relevance assessment: are judges exchangeable and does it matter. In: Proceedings of the 31st Annual International ACM SIGIR Conference on Research and Development in Information Retrieval, pp. 667–674. ACM (2008)
3. Bailey, P., Moffat, A., Scholer, F., Thomas, P.: UQV100: a test collection with query variability. In: Proceedings of the 39th International ACM SIGIR Conference on Research and Development in Information Retrieval, pp. 725–728. ACM (2016)
4. Barr, C., Jones, R., Regelson, M.: The linguistic structure of English web-search queries. In: Proceedings of the Conference on Empirical Methods in Natural Language Processing, pp. 1021–1030. Association for Computational Linguistics (2008)
5. Biega, A.J., Saha Roy, R., Weikum, G.: Privacy through solidarity: a user-utility-preserving framework to counter profiling. In: Proceedings of the 40th International ACM SIGIR Conference on Research and Development in Information Retrieval, pp. 675–684. ACM (2017)
6. Biega, J.A., Gummadi, K.P., Mele, I., Milchevski, D., Tryfonopoulos, C., Weikum, G.: R-susceptibility: an IR-centric approach to assessing privacy risks for users in online communities. In: Proceedings of the 39th International ACM SIGIR conference on Research and Development in Information Retrieval, pp. 365–374. ACM (2016)
7. Carterette, B., Kanoulas, E., Hall, M., Clough, P.: Overview of the TREC 2014 session track. Technical report, Delaware University Newark (2014)
8. Chen, G., Bai, H., Shou, L., Chen, K., Gao, Y.: UPS: efficient privacy protection in personalized web search. In: Proceedings of the 34th International ACM SIGIR Conference on Research and Development in Information Retrieval, pp. 615–624. ACM (2011)

[7] https://www.mpi-inf.mpg.de/departments/databases-and-information-systems/research/impact/mediator-accounts/.

9. Chouldechova, A., Mease, D.: Differences in search engine evaluations between query owners and non-owners. In: Proceedings of the sixth ACM International Conference on Web Search and Data Mining, pp. 103–112. ACM (2013)

10. Hagen, M., Potthast, M., Beyer, A., Stein, B.: Towards optimum query segmentation: in doubt without. In: Proceedings of the 21st ACM International Conference on Information and Knowledge Management, pp. 1015–1024. ACM (2012)

11. Hagen, M., Wägner, D., Stein, B.: A corpus of realistic known-item topics with associated web pages in the ClueWeb09. In: Hanbury, A., Kazai, G., Rauber, A., Fuhr, N. (eds.) ECIR 2015. LNCS, vol. 9022, pp. 513–525. Springer, Cham (2015). https://doi.org/10.1007/978-3-319-16354-3_57

12. Hauff, C., Hagen, M., Beyer, A., Stein, B.: Towards realistic known-item topics for the ClueWeb. In: Proceedings of the 4th Information Interaction in Context Symposium, pp. 274–277. ACM (2012)

13. Saha Roy, R., Katare, R., Ganguly, N., Laxman, S., Choudhury, M.: Discovering and understanding word level user intent in web search queries. J. Web Semant. **30**, 22–38 (2015)

14. Saha Roy, R., Suresh, A., Ganguly, N., Choudhury, M.: Place value: word position shifts vital to search dynamics. In: Proceedings of the 22nd International Conference on World Wide Web, pp. 153–154. ACM (2013)

15. Serdyukov, P., Dupret, G., Craswell, N.: Log-based personalization: the 4th web search click data (WSCD) workshop. In: Proceedings of the 7th ACM International Conference on Web Search and Data Mining, pp. 685–686. ACM (2014)

16. Zhang, S., Yang, G.H., Singh, L., Xiong, L.: Safelog: supporting web search and mining by differentially-private query logs. In: 2016 AAAI Fall Symposium Series (2016)

Machine-Actionable Data Management Plans: A Knowledge Retrieval Approach to Automate the Assessment of Funders' Requirements

João Cardoso[1,2(✉)] , Diogo Proença[1,2] , and José Borbinha[1,2]

[1] INESC-ID, Lisbon, Portugal
[2] Instituto Superior Técnico, Universidade de Lisboa, Lisbon, Portugal
{joao.m.f.cardoso,diogo.proenca,jlb}@tecnico.ulisboa.pt

Abstract. Funding bodies and other policy-makers are increasingly more concerned with Research Data Management (RDM). The Data Management Plan (DMP) is one of the tools available to perform RDM tasks, however it is not a perfect concept. The Machine-Actionable Data Management Plan (maDMP) is a concept that aims to make the DMP interoperable, automated and increasingly standardised. In this paper we showcase that through the usage of semantic technologies, it is possible to both express and exploit the features of the maDMP. In particular, we focus on showing how a maDMP formalised as an ontology can be used automate the assessment of a funder's requirements for a given organisation.

Keywords: Data Management Plan · Machine Actionable Data Management Plan · Semantic technologies

1 Introduction

Funding bodies and other policy-makers are increasingly more concerned with Research Data Management (RDM). One of the contributing factors is the general perception that research data should be a public good [16]. In order to guide researchers through the process of managing their data, many funding agencies (e.g. the National Science Foundation (NSF), the European Commission (EC), or the Fundação para a Ciência e Tecnologia (FCT) have created and published their own open access policies, as well as requiring that any grant proposals be accompanied by a Data Management Plan (DMP).

The DMP [7] is one of the tools available to researchers to aid in the management of their data. The DMP is a document describing the techniques, methods and policies on how data from a research project is to be created or collected, documented, accessed, preserved and disseminated. The DMP is not without issues, such as lack of standardisation, lack of continuous updates throught the

© Springer Nature Switzerland AG 2020
J. M. Jose et al. (Eds.): ECIR 2020, LNCS 12036, pp. 118–125, 2020.
https://doi.org/10.1007/978-3-030-45442-5_15

project life cycle, etc. As a result of these issues, the DMP is seen more as bureaucratic obligation, than as a valuable asset for data management.

The concept of Machine-Actionable DMP (maDMP) [13] (sometimes referred as "active", "dynamic", or "machine-readable" DMP), aims at addressing some of these issues by making the DMP machine-readable without compromising its human-readability. The adoption of an open, shared and interoperable concept of maDMP could bring multiple benefits, such as facilitating data discovery and reuse, and enabling automated evaluation and monitoring, etc.

It is clear that having funding agencies and universities push for DMP usage is not enough to achieve true adoption and standardisation. Researchers need to be convinced that the DMP can be a major tool to support RDM, especially if it is perceived as a living object, and there has to be a move towards standardising DMP documents. One of the attempts to tackle the standardisation issue, is being carried out by the Research Data Alliance (RDA) DMP Common Standards Working group [8]. Whose objective was to establish a common data model that would define a core set of elements for a DMP.

The objective of this paper is to explore the application of semantic technology to RDM. In particular, by resorting to semantic technologies to both express and exploit the features of the maDMP [4,14]. To that effect we set out to show a DMP formalised as an ontology can be used automate the assessment of a funder's requirements for a given organisation.

This paper is organised as follows. Section 2 offers definitions on the concepts of DMP, maDMP and Semantic Technologies. Section 3 describes our approach on how to establish a DMP creation service that allows for semantic based DMP representation. Section 4 closes the paper by presenting both a final appreciation on the proposed approach, and a description of possible future work on this topic.

2 Related Work

Ontologies. Semantic Technology has shifted from originally tackling syntax and structural issues, to focus on the exploitation of the semantics of information [11] to promote both system interoperability and enhance the web infrastructure [12].

Ontologies play a key role in Semantic Technology, for they enable knowledge representation through formal semantics. Studer describes ontologies as a "formal, explicit specification of a shared conceptualization" [14]. Ontology usage can be sorted into three categories [15]: Human communication by providing a common interpretation of knowledge, interoperability by enabling data exchange among heterogeneous sources, and systems engineering by providing a shared understating of problems.

Ontologies can be represented by formal languages, that have the dual purpose of encoding knowledge on specific domains, as well as including reasoning rules that aid in the processing of that knowledge. These formal languages are referred to as ontology languages (e.g. Resource Description Framework (RDF), Web Ontology Language (OWL), etc.).

Semantic Reasoning. According to Lenzerini et al. [5], "several reasoning tasks can be carried out to deduce implicit knowledge from the explicitly represented knowledge". Hence, reasoning mechanisms are used for carrying two tasks, ontology validation and ontology analysis. The validation of an ontology consists in checking if the ontology is correctly modelling the domain in focus, whereas the analysis of an ontology focuses on deducing facts about the modelled domain, processing and extracting new information from the original information. The same author also proposes a classification for different types of reasoning according to the results aimed for: deduction, induction and abduction.

Reasoners, sometimes referred as Discription Logics (DL) systems, can be organized into three generations [2]. The third and last generation of reasoners focus on optimized reasoners which are expressive, sound and complete. In this last generation are included FaCT++, RacerPro and HermiT.

Semantic Based Assessment. The relevance of ontology-based techniques is demonstrated by the growing use of ontologies in a diversity of domains.

Antunes proposed a "model for the representation and integration of ontologies" [1] which allows the analysis of an architecture. The model is based on meta-model and model integration consisting of an upper ontology and a collection of domain specific ontologies.

Bakhshandeh [3] proposed model for the representation, integration and analysis of EA models. The proposed model addresses the need for a representation that allows integration of different metamodels and models along with their analysis by computational needs. This proposal is based on the hypothesis that ontologies can represent, integrate and support the analysis of enterprise architecture models.

Proença [10] proposed a model and method to represent maturity models, namely its components, rules and assessment criteria using ontologies. It demonstrated how to use computational inference as an analysis technique to take advantage of the information already encoded in maturity models with the purpose of automating existing maturity assessment methods.

3 Proposed Approach and Validation

Method. This section details how the objective of this paper was approached, as well as a description of the methods used to provide a preliminary validation of said approach. Additionally, examples of the potential that semantic technologies can provide for maDMP exploitation are also given.

The overarching goal of this work to both express and exploit the features of the maDMP by resorting to semantic technologies. In particular, it Pfocuses on showing how a DMP formalised as an ontology can be used to automate the assessment of a funder's requirements for a given organisation.

Our approach for maDMP generation comprises of 8 tasks, that can be analysed in Fig. 1, can be interpreted into three main parts. The first part comprises of the execution of the first two tasks, which resulted into the creation of an

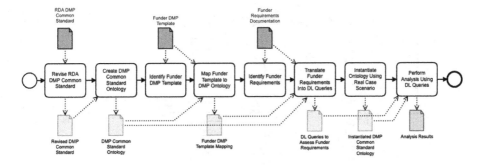

Fig. 1. The process for maDMP generation

ontology compliant with the DMP Common Standards model, that is detailed in this section. The second part comprises of the execution of the following four tasks and results in both the collection of the necessary mappings between the ontology and the identified DMP templates, and creation of DL queries based on the funders' requirements. The final part comprises of the last two tasks, and results in the creation of a poExpulated ontology that is subsequently analysed using the previously created DL queries. The second and third part of the approach are detailed in the last subsection.

With our approach researchers should be able to select a template and fill the necessary forms to generate a DMP. Our focus however is on having the generated DMP be machine-actionable, DMP Common Standards Model compliant, and expressed through the usage of semantic technologies.

DMP Common Standard Ontology. The DMP Common Standards Working group [8] was created with the objective of establishing a common data model that would define a core set of elements for a DMP. The resulting data model has a modular design allowing it to be extended by existing standards and vocabularies. The DMP Common Standards working group is also meant to provide reference implementations of the model in popular formats (e.g. JSON, XML, RDF).

Our approach called for the revision of the DMP Common Standards Model, and this resulted on a representation of the model using semantic technologies. The DMP Common Standard Ontology (DCSO)[1], was created with the objective of providing an implementation of the DMP Common Standards model expressed through the usage of semantic technology, which has been considered a possible solution in the data management and preservation domains [9]. The DCSO is represented using OWL [6], and its model can be analysed in Fig. 2.

The following step was to collect DMP templates from funding agencies. To that effect we collected the DMP template for both the EC Horizon 2020

---

[1] All the ontologies mentioned in this paper can be found in the DMP Common Standard Ontology Repository: https://github.com/RDA-DMP-Common/RDA-DMP-Common-Standard/tree/master/ontologies.

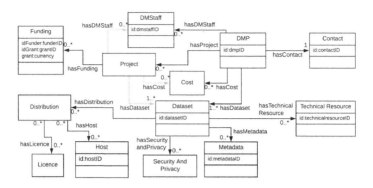

Fig. 2. The DMP common standards ontology with the proposed extension highlighted (Color figure online)

programme, and the FCT project funding. We then proceeded to attempt to validate that DCSO would cover the entirety of both the DMP templates, and establish the necessary mappings. However we came to the conclusion that we would have to extend the DCSO, in order to better address the mapping of the DMP templates to the DCSO. The extension was limited to the addition of three object properties that were added to the Project class, they can be identified in Fig. 2 highlighted in red.

Funders' Requirements Assessment Using DL Queries. Given the mappings between the DPM templates and the DCSO, the next step was to collect the funder's requirements for a DMP, and have them translated into DL Queries that were executable over the DCSO. It was however necessary to first instantiate the extended version of the DCSO. To that effect we resorted to areal case scenario, the Genomics Unit at the Instituto Gulbenkian de Ciência[2].

The Genomics unit provides Next Generation Sequencing services using state-of-the-art Illumina sequencers, and is currently engaged in two projects, namely *Oneida*[3] and *GenomePT*[4]. Our decision was to create a single DMP that would cover both projects. By using the instantiated ontology and the DL queries we were able to assert if the DMP complied with funder's requirements and compiled the analysis results in a document.

Figure 3 shows the verification of one of the FCT requirements, which state that every project must cover the cost to provide open access to their outputs. Some journals charge an additional fee in order to provide open access to a published paper, as a result researchers must include the costs of these fees in the DMP to be submitted in the project proposal, as well as, in the periodic reviews. As can be seen in Fig. 3 the costs for open access are detailed for the *GenomePT* project bot not for the *Oneida* project. Additionally, we can use these queries to get the list of costs covered by a project or the whole DMP.

---

[2] http://facilities.igc.gulbenkian.pt/genomics/genomics.php.

[3] https://www.itqb.unl.pt/oneida.

[4] https://www.genomept.pt/.

Query (class expression)

Cost and inverse(hasCost) some {genomept}

| Execute | Add to ontology |

Query results

Instances (4 of 4)
- ◆ equipment
- ◆ facilities
- ◆ open.access
- ◆ storage

Query (class expression)

{open.access} and inverse(hasCost) some {genomept}

| Execute | Add to ontology |

Query results

Instances (1 of 1)
- ◆ open.access

Query (class expression)

{open.access} and inverse(hasCost) some {oneida}

| Execute | Add to ontology |

Query results

Instances (0 of 0)

Fig. 3. Verifying if the *GenomePT* and *Oneida* projects cover costs related with open access.

Query (class expression)

DMStaff and inverse(hasDMStaff) some {oneida}

| Execute | Add to ontology |

Query results

Instances (3 of 3)
- ◆ joao.costa
- ◆ joao.sobral
- ◆ susana.ladeiro

Query (class expression)

DMStaff and inverse(hasDMStaff) some {genomept}

| Execute | Add to ontology |

Query results

Instances (3 of 3)
- ◆ carlos.penha-goncalves
- ◆ ricardo.leite.staff
- ◆ susana.ladeiro

Fig. 4. Verifying the researchers assigned to *GenomePT* and *Oneida* projects.

Another example is the researchers effort per project. Figure 4 shown the staff assigned to each project. The researcher "susana.ladeiro" is assigned to both projects which means that this researcher does not work in just one project. This means that the costs associated with this researcher must be covered in both projects.

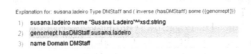

Explanation for: susana.ladeiro Type DMStaff and (inverse (hasDMStaff) some ({genomept}))

1) susana.ladeiro name "Susana Ladeiro"^^xsd:string
2) genomept hasDMStaff susana.ladeiro
3) name Domain DMStaff

Fig. 5. Explanation for the results provided by a DL query.

Another functionality provided by the use of ontologies is the provision of explanations for each of the results of a query. In this example, Fig. 5 depicts the explanation for assigning the researcher "susana.ladeiro" to the *GenomePT* project. This is a very simple example to show the potential of this feature, it can provide researchers an explanation on which funder's requirements are not compliant and the reasons for being non-compliant.

4   Conclusions and Future Work

The overall goal of this paper was to demonstrate the use of semantic technologies to both express an maDMP and exploit its features. With the approach described in Sect. 3, we focused on showcasing how a DMP expressed as an ontology can impact the assessment of funder's requirements for two organisations.

This paper is a report on an ongoing effort. As such, there are still action points that we consider in need of further development. The DCSO, due to the nature, is often too generic to adapt to specific DMP templates. That implies that more DCSO extensions will have to be created to cater for specific contexts. Another possible action point is use one of the many existing DMP creation frameworks (e.g. DMP Online[5], DMP Tool[6], Data Stewardship Wizzard[7] etc.) to automate the creation a instantiated DCSO. This would minimize the necessity for an ontology expert in the process.

Acknowledgements. This work was supported by national funds through FCT with reference UID/CEC/50021/2019, and by project PRECISE (LISBOA-01-0145-FEDER-016394).

References

1. Antunes, G.: Analysis of enterprise architecture models: an application of ontologies to the enterprise architecture domain. Ph.D. thesis, Instituto Superior Técnico, Universidade de Lisboa (2015)
2. Baader, F., Calvanese, D., McGuinness, D., Patel-Schneider, P., Nardi, D.: The Description Logic Handbook: Theory, Implementation and Applications. Cambridge University Press, Cambridge (2003)
3. Bakhshandeh, M.: Ontology-driven analysis of enterprise architecture models. Ph.D. thesis, Instituto Superior Técnico, Universidade de Lisboa (2016)
4. Breitman, K., Casanova, M.A., Truszkowski, W.: Semantic Web: Concepts, Technologies and Applications. Springer, Heidelberg (2007). https://doi.org/10.1007/978-1-84628-710-7
5. Lenzerini, M., Milano, D., Poggi, A.: Ontology representation & reasoning. Technical report, NoE InterOp (IST-508011) (2004)
6. McGuinness, D.L., Van Harmelen, F., et al.: Owl web ontology language overview. W3C Recommendation **10**(10), 2004 (2004)
7. Michener, W.K.: Ten simple rules for creating a good data management plan. PLoS Comput. Biol. **11**(10), e1004525 (2015)
8. Miksa, T., Neish, P., Walk, P.: WG DMP common standards case statement (2017)
9. Miksa, T., Vieira, R.J.C., Barateiro, J., Rauber, A.: VPlan-ontology for collection of process verification data (2014)
10. Proença, D.: Maturity assessment support with conceptual modelling methods and semantic techniques. Ph.D. thesis, Instituto Superior Técnico, Universidade de Lisboa (2018)

---

[5] http://dmponline.dcc.ac.uk.

[6] https://DMPTool.org.

[7] https://ds-wizard.org.

11. Sheth, A.P.: Changing focus on interoperability in information systems: from system, syntax, structure to semantics. In: Goodchild, M., Egenhofer, M., Fegeas, R., Kottman, C. (eds.) Interoperating Geographic Information Systems. The Springer International Series in Engineering and Computer Science, vol. 495, pp. 5–29. Springer, Boston (1999). https://doi.org/10.1007/978-1-4615-5189-8_2

12. Sheth, A.P., Ramakrishnan, C.: Semantic (web) technology in action: ontology driven information systems for search, integration, and analysis. IEEE Data Eng. Bull. **26**(4), 40 (2003)

13. Simms, S., Jones, S., Mietchen, D., Miksa, T.: Machine-actionable data management plans (maDMPs). Res. Ideas Outcomes **3**, e13086 (2017)

14. Studer, R., Benjamins, V.R., Fensel, D., et al.: Knowledge engineering: principles and methods. Data Knowl. Eng. **25**(1), 161–198 (1998)

15. Uschold, M., Gruninger, M.: Ontologies: principles, methods and applications. knowl. Eng. Rev. **11**(02), 93–136 (1996)

16. Whyte, A., Tedds, J.: Making the case for research data management. DCC Briefing Papers (2011)

Session-Based Path Prediction by Combining Local and Global Content Preferences

Kushal Chawla[1(✉)] and Niyati Chhaya[2]

[1] University of Southern California, Los Angeles, USA
kchawla@usc.edu
[2] Big Data Experience Lab, Adobe Research, Bengaluru, India
nchhaya@adobe.com

Abstract. Session-based future page prediction is important for online web experiences to understand user behavior, pre-fetching future content, and for creating future experiences for users. While webpages visited by the user in the current session capture the users' local preferences, in this work, we show how the global content preferences at the given instant can assist in this task. We present **DRS-LaG**, a Deep Reinforcement Learning System, based on Local and Global preferences. We capture these global content preferences by tracking a key analytics KPI, the number of views. The problem is formulated using an agent which predicts the next page to be visited by the user, based on the historic webpage content and analytics. In an offline setting, we show how the model can be used for predicting the next webpage that the user visits. The online evaluation shows how this framework can be deployed on a website for dynamic adaptation of web experiences, based on both local and global preferences.

1 Introduction

Users expect varied outcomes from their web experiences. Enterprises aim to create digital experience that not only cater to user intent but also help to improve their own business metrics. Given the variety of content, manual creation of customized and adaptive experiences is infeasible. Session-based future path prediction is necessary to understand user needs, pre-fetch future content, or even for adapting future experiences. Users' content creation and consumption patterns define their intent and needs. Their web tracks; i.e. the path that the user takes during their web journey is an essential ingredient for defining their interests and goals. In this work, we aim to create user intent models leveraging their consumption patterns combined with their website footprint to predict the potential user path and content needs.

Extensive studies have been conducted in the related space of recommender systems using traditional [2], deep-learning [6,7], and reinforcement learning [8]

K. Chawla—Work done when author was a full-time researcher at Adobe Research.

© Springer Nature Switzerland AG 2020
J. M. Jose et al. (Eds.): ECIR 2020, LNCS 12036, pp. 126–132, 2020.
https://doi.org/10.1007/978-3-030-45442-5_16

based techniques on both historic user-item interactions and session behavior. While the historic webpages visited in a session capture the users' local preferences, this work shows that the instantaneous global content preferences can further assist in understanding the future behavior of the users. We describe one such scenario in Fig. 1a. Specifically, we present a Deep Reinforcement Learning (RL) System, based on Local and Global preferences (**DRS-LaG**). Given the historic webpage content and analytics in a user session, our agent predicts the future preferences of the user. The model is trained on offline logs of a sports news website. Through offline evaluations, we show how the proposed model can be used to predict the next page user will go to. Our online evaluation shows how the predictions can be used to adapt future experiences of the users. RL allows our system to tackle the dynamic user preferences in news domain, while also incorporating expected future rewards when deployed in an online environment.

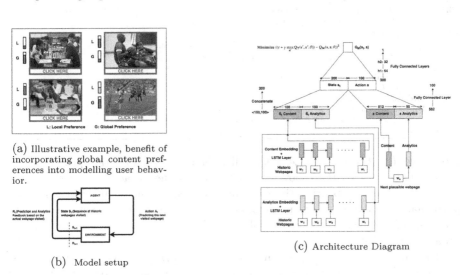

(a) Illustrative example, benefit of incorporating global content preferences into modelling user behavior.

(b) Model setup

(c) Architecture Diagram

Fig. 1. Internal workings of the proposed DRS-LaG framework.

2   DRS-LaG: Proposed Framework

Problem Formulation: We define an agent which models a user's session-level behavior to predict the next webpage user visits, based on the content and instantaneous analytics of the webpages visited in the current session. Since the predictions capture user preferences, they can then be recommended to the user or used to adapt future webpage experiences. At each timestep, the user (environment) provides feedback on the actions taken by the agent in the form of rewards. The agent is trained on offline session-level logs extracted from a sports news website. We illustrate this setup in Fig. 1b.

The task is modeled as a Markov Decision Process (MDP) with the tuple ($\mathcal{S}$, $\mathcal{A}$, $\mathcal{P}$, $\mathcal{R}$, γ): (1) State space $\mathcal{S}$: captures the current local and global content

Algorithm 1. Offline Training algorithm for our agent

1: Initialize replay memory M, Q-value model $Q_M(s,a)$ with random weights, target model $Q_T(s,a)$ with same wts as $Q_M(s,a)$.
2: Initialize webpage (action) pool P, webpage analytics A(action, kpi, time interval)
3: **for** e $= 1,E$ **do**   ▷ episodes or user web sessions in chronological order
4:     Reset environment state vector to a zero vector.
5:     **for** t$= 1,T$ **do**   ▷ timesteps in the current session
6:         Observe the current state s_t
7:         **for** n $= 1,N$ **do**   ▷ negative actions
8:             Sample a negative action a from pool P
9:             Observe the next state s_{t+1}, reward r
10:            Store transition (s_t, a, r, s_{t+1}, done=1) in M
11:   ▷ session ends after a negative sample
12:        **end for**
13:        Get the correct action a from the offline logs
14:        Observe the next state s_{t+1}, reward r
15:        Set done=1 if t==T, else 0
16:        Store transition (s_t, a, r, s_{t+1}, done=done) in M
17:        Update $s_t \leftarrow s_{t+1}$
18:        **if** M.length > batch-size **then**
19:            Sample a minibatch of transitions from M
20:            Set y =
$$\begin{cases} r, & \text{done=1} \\ r + \gamma \max_{a'} Q_T(s', a'; \theta), & \text{done=0} \end{cases}$$
21:            Minimize $(y - Q_M(s,a;\theta))^2$
22:        **end if**
23:    **end for**
24:    Update webpage pool P and webpage analytics A.
25:    Update target model after fixed number of iterations.
26: **end for**

---

preference, (2) Action space $\mathcal{A}$: set of all webpages, (3) Transition probabilities $\mathcal{P}$: probability $p(s'|s,a)$ to move to state s' by taking an action a in the state s, (4) Rewards $\mathcal{R}$: capturing the feedback received by the agent after taking a particular action and, (5) γ: the discount factor for future rewards in the current user session. The goal is to learn a policy $\pi : \mathcal{S} \to \mathcal{A}$ to maximize the cumulative reward of the system.

To deal with the dynamic action spaces, we use a Deep Q-Learning model-free approach. Figure 1c shows our architecture. Given a state-action pair, the network outputs the corresponding Q-value $Q(s,a)$. The optimal Q-value $Q^*(s,a)$ should follow the Bellman equation [1]: $Q^*(s,a) = E_{s'}[r + \gamma \max_{a'} Q^*(s', a')|s,a]$, where r is the corresponding reward for the given state-action pair.

Actions: Representing Webpages. The agent actions correspond to various webpages or URLs on the given website. Given the current state of a user, the agent returns a set of plausible webpages, using both the local and global content preferences. We hence represent webpages using both the content and the corresponding instantaneous analytics.

Webpage Content: The webpage text content is represented using Universal Sentence Encoder [3]. We leverage the pre-trained model using Tensorflow Hub[1] which returns a $d_C = 512$ dimensional representation for a given input.

Instantaneous Webpage Analytics: The incorporation of analytics allows the agent to better predict the future content preferences of the users, while

---

[1] https://tfhub.dev/google/universal-sentence-encoder/2.

also catering to business objectives. We divide the time scale into fixed-sized intervals. Let's consider a set of k analytics KPIs such as number of views and number of exits. While training and subsequent testing, we track the KPIs for all webpages seen until now. Analytics representation is obtained by combining the value for most recent d_A time intervals for each of the k KPIs hence resulting in a $d_A k$-dimensional vector. The final representation for an action is computed by concatenating both the content and analytics representations of the webpage, ending up with a $(d_C + d_A k)$ dimensional vector.

States: Historic Action Sequence. The current state must capture the session-level preference of the users. Hence, we aggregate the representations of all the historically visited webpages in the current session to define the state of the user. **DRS-LaG** uses two LSTM networks to combine the historic content and action analytics.

Defining Rewards. At each timestep, the agent receives a reward from the user, based on the action chosen in the given state. The complete reward for a given state-action pair $r(s, a)$ is a combination of prediction and instantaneous analytics: $r(s, a) = r_P(s, a) + (r_A^1(a) + r_A^2(a) + r_A^3(a)...r_A^k(a))$.

Where $r_P(s, a)$ refers to prediction reward, whether the corresponding webpage was visited by the user in the offline data logs, and r_A^i refers to the instantaneous analytics of the action a with respect to KPI i.

Learning Stage. The training algorithm is discussed here (see Algorithm 1). First, experience replay [4] and target network [5] are used to stabilize the training process. Second, at each timestep, apart from considering the actual action from the data, we also sample N negative actions from the webpage pool P. This is necessary as the offline logs only contain the positive samples for next-page prediction. Moreover, this allows the agent to explore the instantaneous analytics values of webpages, beyond those seen in the current session. Third, the model is trained using the Bellman Equation. Fourth, we skip the replay memory update for the first few webpages in every session, owing to the inadequacy of the initial webpages to capture the context. This detail is removed from Algorithm 1 for simplicity. Finally, since the model is trained on instantaneous analytics values, we update both the webpage pool and analytics values after each episode.

Test Stage - Offline: Given the state, the model is asked to predict the next webpage user will go to. Keeping $\gamma = 0$, the model is trained using Algorithm 1 to incorporate only the immediate reward, as appropriate for next-page prediction. The test data is parsed similar to the training procedure. At every timestep, the recall is observed based on the predictions from a trained model $Q_M(s, a)$ and the actual action from offline logs.

Online: We also evaluate our framework in an online simulated environment. Given the complexity of setting up an online evaluation, following prior work [8], we resort to a framework which effectively simulates the real-time environment with the capability to provide immediate feedback given state and action. We split our data into two and train this simulator on the first half, keeping the second for training. The simulator architecture is same as in Fig. 1c and is trained to only predict the immediate feedback. The performance of our model in the offline setting attests to the performance of the simulated environment.

3 Experiments

Dataset: The experiments are based on a snapshot of a sports website[2]. The clickstream is gathered using an enterprise analytics tool deployed on the website. The data consists of 37,667 user sessions. We maintain a temporal order in the paths based on timestamps associated with each session. Minimum path length is kept at 3 and maximum as 50. The data contains 1,599 unique urls. The first 33,900 paths are kept for training, next 1,883 paths for validation, and last 1,884 paths for testing.

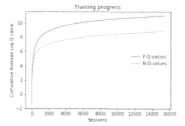 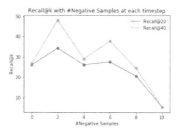

(a) Training progress of DRS-LaG. **P Q-values**: Q-values of the actual action from the data. **N Q-values**: Average Q-values for the negative actions, sampled uniformly from the action pool.

(b) Effect of varying the number of negative samples for DRS-LaG on Recall@20 and Recall@40 in the offline evaluation task. For this analysis, the models were trained on a 20% data.

Fig. 2. Training progress and the impact of the number of negative samples for DRS-LaG.

Hyperparameters: Content representations are 512 dimensions while the instantaneous analytics are 50-d. The batch size is 16, with learning rate for Adam optimizer as 0.01, number of negative samples as 2, interval size as 5 s and size of replay buffer as 5000 transitions. The weights are transferred to the target network after every 1000 replay iterations. The prediction reward is set to 3 for correct prediction and 0 otherwise, while the analytics reward is fixed to the total change in KPI value over the past 50 intervals. Number of views is considered as the KPI for all experiments. These parameters are tuned on the validation dataset. Once tuned, the models are trained on 'training+validation' data for evaluation on the test data.

Training Progress: Fig. 2a visualizes the training progress of **DRS-LaG**. We track two metrics: (1) **P Q-values**: Q-values of the actual action taken from the data and (2) **N Q-values**: Average Q-values for the negative webpages, sampled

---

[2] We cannot reveal the name of the website because of privacy constraints.

Table 1. Performance based on the offline logs for next-page prediction task.

Model	Offline	
	Recall@20	Recall@40
Random	2.00	3.18
Majority	27.40	38.75
W-Avg-c	27.16	41.19
LSTM-c	29.52	47.99
DRS-LaG: $r_A = 0$	35.55	50.46
DRS-LaG	**36.34**	**51.36**

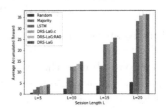

Fig. 3. Performance comparison on our online test based on average reward in a session.

uniformly from the action pool at each timestep. As expected, the two graphs for Cumulative Log values deviate as the training proceeds.

Offline Results: We use two metrics, Recall@20 and Recall@40: what percent of times the correct webpage visited by the user appears in top 20 and 40 webpages returned by the model respectively.

DRS-LaG is trained to predict only the immediate reward at every timestep by keeping $\gamma = 0$. Comparison against baseline models is provided. **Random** ignores the current state and returns a random set of webpages at every timestep. **Majority** returns a list of most-viewed webpages at every timestep. **Majority** can be a really strong baseline in hierarchical website environments. **W-Avg-c** combines only the content representations of the past webpages in the current session using a exponentially-decaying weighted average, to predict the future path. Given the dynamic nature of the websites, instead of predicting a softmax over all the webpages, given the historic webpages and a plausible next webpage, **W-Avg** is trained to predict a score that the plausible webpage will next be visited. At the time of testing, the model returns the webpages with the maximum scores. Similarly, **LSTM-c** uses a Long Short Term Memory recurrent network to capture the historic webpage content. **DRS-LaG:** $r_A = 0$ is trained with both local and global representations similar to **DRS-LaG** but without the analytics reward. Table 1 shows the results. **W-Avg-c** performs similar to the **Majority**, failing to capture the local context or preferences of the users. **LSTM-c** shows improvements, by using a recurrent network to combine historic content visited by the user. With the capability to incorporate both local and global content preferences, **DRS-LaG:** $r_A = 0$ outperforms the baseline methods. Using the analytics reward r_A, **DRS-LaG** shows further enhancement in the performance, attesting to the utility of our approach. We analyze the sensitivity of DRS-LaG in the offline evaluation task towards the number of negative examples sampled at each timestep in Fig. 2b. If the number is too low, the model may end up learning nothing, by learning to predict a high score for every webpage. If the number is too high, the model may consider some in-context webpages as negative, again countering its own learning mechanism. We empirically identify the value 2 for our experiments (see Fig. 2b).

Online Results: These experiments evaluate the model, if deployed to recommend webpages or adapt future experiences of the users, in a simulated environment. We observe the average rewards in a session to evaluate the models. Session length considered are 5, 10, 15 and 20. To incorporate cumulative future rewards, **DRS-LaG** is trained is $\gamma = 0.95$. **Random**, **Majority** and **LSTM** are implemented in the same manner as before. **DRS-LaG-c** and **DRS-LaG RA0** are trained similar to **DRS-LaG**. However, the former only considers the content (local preference) and the latter keeps $r_A = 0$. The results for our online experiments are plotted in Fig. 3. **DRS-LaG-c** outperforms **LSTM** which is only trained to predict the immediate feedback, attesting to the utility of reinforcement learning. This observation is more evident in longer sessions. **DRS-LaG** $R_A = 0$ and **DRS-LaG** further improve the performance.

4   Conclusion

We presented **DRS-LaG** framework, with the objective of improving user web experiences while simultaneously catering to analytics KPIs. Using Deep RL, our model incorporates both local and global content preferences. We show the proposed method effectively predicts user behavior in a dynamic web environment using both offline and online setups.

References

1. Bellman, R.: Dynamic programming. Science **153**(3731), 34–37 (1966)
2. Burke, R.: Hybrid recommender systems: survey and experiments. User Model. User-Adap. Inter. **12**(4), 331–370 (2002)
3. Cer, D., et al.: Universal sentence encoder. arXiv preprint arXiv:1803.11175 (2018)
4. Lin, L.J.: Reinforcement learning for robots using neural networks. Technical report, Carnegie-Mellon Univ Pittsburgh PA School of Computer Science (1993)
5. Mnih, V., et al.: Playing atari with deep reinforcement learning. arXiv preprint arXiv:1312.5602 (2013)
6. Wu, S., Ren, W., Yu, C., Chen, G., Zhang, D., Zhu, J.: Personal recommendation using deep recurrent neural networks in netease. In: 2016 IEEE 32nd International Conference on Data Engineering (ICDE), pp. 1218–1229. IEEE (2016)
7. Zhang, S., Yao, L., Sun, A., Tay, Y.: Deep learning based recommender system: a survey and new perspectives. ACM Comput. Surv. (CSUR) **52**(1), 5 (2019)
8. Zhao, X., Zhang, L., Ding, Z., Xia, L., Tang, J., Yin, D.: Recommendations with negative feedback via pairwise deep reinforcement learning. In: Proceedings of the 24th ACM SIGKDD International Conference on Knowledge Discovery and Data Mining, pp. 1040–1048. ACM (2018)

Unsupervised Ensemble of Ranking Models for News Comments Using Pseudo Answers

Soichiro Fujita[1]([⊠]), Hayato Kobayashi[2], and Manabu Okumura[1]

[1] Tokyo Institute of Technology, Kanagawa, Japan
fujiso@lr.pi.titech.ac.jp, oku@pi.titech.ac.jp
[2] Yahoo Japan Corporation/RIKEN AIP, Tokyo, Japan
hakobaya@yahoo-corp.jp

Abstract. Ranking comments on an online news service is a practically important task, and thus there have been many studies on this task. Although ensemble techniques are widely known to improve the performance of models, there is little types of research on ensemble neural-ranking models. In this paper, we investigate how to improve the performance on the comment-ranking task by using unsupervised ensemble methods. We propose a new hybrid method composed of an output selection method and a typical averaging method. Our method uses a pseudo answer represented by the average of multiple model outputs. The pseudo answer is used to evaluate multiple model outputs via ranking evaluation metrics, and the results are used to select and weight the models. Experimental results on the comment-ranking task show that our proposed method outperforms several ensemble baselines, including supervised one.

1 Introduction

User comments on online news services can be regarded as a useful content since users can read other users' opinions related to each news article. Many online news service sites rank comments in the order of the number of positive user-feedback for a comment, such as "Like"-button clicks, and preferentially display popular comments to readers. However, this type of user-feedback is not suitable to assess the comment quality, because this type of measurement is biased by where a comment appears [7]; Earlier comments tend to receive more feedback since they will be displayed at the top of the page. In attempt of solving this problem, several studies introduce some aspects of the comment quality to focus on, e.g., constructiveness [7,13] or persuasiveness [22]. In particular, Fujita et al. [7] proposed a new dataset to rank comments directly according to comment quality. This is a difficult task because we have various situations of judging whether a comment is good. For example, comments can indicate rare user experiences, provide new ideas, or cause discussions. Ranking models often fail to capture such information.

© Springer Nature Switzerland AG 2020
J. M. Jose et al. (Eds.): ECIR 2020, LNCS 12036, pp. 133–140, 2020.
https://doi.org/10.1007/978-3-030-45442-5_17

According to recent studies [2,12,15], ensemble techniques are widely known to improve the accuracy of machine learning models. These ensemble techniques can be roughly divided into two types: averaging and selecting. Averaging methods such as Naftaly et al. [17] simply average multiple model outputs. Selecting methods such as majority vote [15] select the most frequent label from the predicted labels of multiple classifiers in post-processing. These methods assist models to make up for other models' mistakes and to improve the results. Recently, Kobayashi [12] proposed an unsupervised ensemble method, post-ensemble, based on kernel density estimation, which was an extension of the majority vote to text generation models. He showed that this method outperformed averaging methods in a text summarization task.

In this paper, we propose a new unsupervised ensemble method, HPA, which is a hybrid of an output selection and a typical averaging method. In typical averaging methods, a lower accuracy model could merely be noise. A simple denoising method is to statically remove such lower accuracy models [19]. However, there is basically no model that fails for every inputs, particularly in neural models with the same architecture. In general, each model has its own strengths and weaknesses. Therefore, our method adopts dynamic denoising of outputs via a provisional averaging result. We use the provisional averaging result as a pseudo answer. Each predicted ranking is compared to the pseudo answer via a similarity function, and the similarity scores are used for selecting and weighting models. We adopt evaluation metrics as a kind of similarity to specialize in the ranking task. In experiments on a task of ranking constructive news comments, our proposed method HPA outperformed both previous unsupervised ensemble methods and a simple supervised ensemble method. Furthermore, we found that one of the evaluation metrics is useful as a similarity measure for the ensemble process.

2   Proposed Method

2.1   Problem Statement

Comment Ranking Task: Let an article be associated with comments $C = (c_1, ..., c_n)$. Each comment has a manually annotated score $S = (s_1, ..., s_n)$, such as the degree of comment quality. A ranking model m learns a scoring function $\tilde{s}_i = m(c_i)$. We consider a predicted score sequence as a ranking of the comments $r = (\tilde{s}_1, ..., \tilde{s}_n)$, because we can generate a ranked comment sequence using this score sequence.

Ensemble Problem: We prepare N rankings $R = (r_1, ..., r_N)$ from ranking models $M = (m_1, ..., m_N)$. The goal of the ensemble is to combine the ranking models to produce a better ranking than any of the individual ranking functions. A simple averaging method calculates the average of the comment scores, like $r^* = \sum_{r \in R} \frac{r}{|R|}$.

2.2 Post-ensemble

We introduce `PostNDCG` which applies the post-ensemble method [12] to the ranking task. Post-ensemble is an unsupervised ensemble method based on kernel density estimation for sequence generation. This method compares the similarity between model outputs and selects the majority-like output which is similar to the other outputs. This selection is equivalent to selecting the output whose estimated density is the highest in the outputs. `PostNDCG` calculates this scoring function: $f(r) = \frac{1}{|R|} \sum_{r' \in R} sim(r, r')$, where $sim(r, r')$ represents the similarity between r and r'. The final ranking of `PostNDCG` is defined as $r^* = \text{argmax}_{r \in R} f(r)$. We used the normalized discounted cumulative gain (NDCG@k) [1] as the similarity function $sim(\cdot)$ to compare each ranker.

2.3 HPA Ensemble

We propose a **H**ybrid method using the **P**seudo **A**nswer (HPA). Figure 1 illustrates an example of HPA. Here, HPA selects the top three rankings $\{r_2, r_3, r_5\}$ that are nearest to the pseudo answer. After that, it weights each selected ranking via a scoring function based on the pseudo answer. The concept of HPA is to denoise outputs via a pseudo answer $\bar{r}$, which is

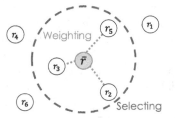

Fig. 1. Example of HPA.

represented by the average of each model output after the L2 normalization: $\bar{r} = \frac{1}{|R|} \sum_{r \in R} \frac{r}{||r||}$. The scoring function g is calculated as the similarity between the pseudo answer and the predicted ranking: $g(r) = sim(\bar{r}, r)$. Then, HPA selects the top k models with the highest scores. The final ranking r^* is represented as, $r^* = \sum_{r \in \bar{R}} g(r) \cdot r$, where $\bar{R}$ is the set of selected models (rankings).

3 Experiments

3.1 Experimental Settings

Dataset: We used a dataset for ranking constructive comments on Japanese articles in Yahoo! News[1], which was prepared in Fujita et al. [7]. The dataset consists of triplets of an article title, comment, and constructiveness score. The constructiveness score (C-score) is defined as the number of crowdsourced workers, out of 40, who have judged a comment to be constructive. Therefore, the C-score is an integer ranging from 0 to 40. In this research, 130,000 comments from 1,300 articles were used as training data, 11,300 comments from 113 articles were used as validation data, and 42,436 comments from 200 articles were

---

[1] https://research-lab.yahoo.co.jp/en/software/.

used as test data. In the training and validation data, 100 comments were randomly extracted in each article, whereas in the test data, all the comments were extracted assuming an actual service environment.

Preprocessing: We used a morphological analyzer MeCab[2] [14] with a neologism dictionary, NEologd[3] [20], for splitting Japanese texts into words. We replaced numbers with a special token and standardized the letter types by halfwidth to fullwidth[4]. We did not remove stop-words because function words will affect the performance in our task. We cutoff low-frequency words that appeared only three times or less in each dataset.

Model and Training: We used RankNet [1], a well-known pairwise ranking algorithm based on neural networks. Given a pair of two comments c_1 and c_2 on an article q, RankNet solves a binary classification problem of whether or not c_1 has a higher score than c_2. The score indicates the comment has high quality or not. We adopted the encoder-scorer structure for RankNet. The encoder consisted of two long short-term memory (LSTM) instances with 300 units to separately encode a comment and its title. The scorer predicted the ranking score of the comment via a fully-connected layer after concatenating the two encoded (comment and title) vectors. We used pre-trained word representations as the encoder input. They were obtained from a skip-gram model [16] trained with 1.5 million unlabeled news comments. We used the Adam optimizer ($\alpha = 0.0001$, $\beta_1 = 0.9$, $\beta_2 = 0.999$, $\epsilon = 1 \times 10^{-8}$) to train these models. Both the dimensions of the hidden states of the encoders of article titles and comments were 300. In the experiments, we trained 100 different models by random initialization for the ensemble methods.

Evaluation: We used normalized discounted cumulative gain (NDCG@k) [1]. The NDCG@k is typically calculated in the top-k comments ranked by the ranking model and denoted by NDCG@$k = Z_k \sum_{i=1}^{k} \frac{score_i}{\log_2{(i+1)}}$, where $score_i$ represents the true ranking score of the i-th comment ranked by the model, and Z_k is the normalization constant to scale the value between 0 and 1. In addition to NDCG@k, we use Precision@k as the second evaluation metrics. Precision@k is defined as the ratio of the correctly included comments in the inferred top-k comments to the true top-k comments. In the experiment, we evaluated the case of $k \in \{1, 5, 10\}$. Note that a well-known paper [10] in the information retrieval field determined NDCG to be more appropriate than Precision@k for graded-scores settings like ours.

3.2   Compared Methods

Ensemble Baselines: We prepared the following methods as baselines. `RankSVM` and `RankNet` are baselines of a single model. `ScoreAvg`, `RankAvg`, `TopkAvg`, and

---

[2] http://taku910.github.io/mecab/.

[3] https://github.com/neologd/mecab-ipadic-neologd.

[4] https://en.wikipedia.org/wiki/Halfwidth_and_fullwidth_forms.

`NormAvg` are commonly used ensemble methods that combine multiple models in post-processing without training. `SupWeight` is the popular supervised ensemble method based on weighting.

- `RankSVM`: The best single RankSVM model proposed in Fujita et al. [7].
- `RankNet`: The best single RankNet model in 100 models for ensemble.
- `ScoreAvg`: Average output scores of the models for each comment.
- `RankAvg`: Average rank orders of each comment.
- `TopkAvg`: Select comments with higher scores than a threshold from each ranking and average their scores [5].
- `NormAvg`: Average normalized output scores of the model outputs, as typified by [2]. We used L2 normalization to each ranking as $r' = r/\|r\|$.
- `SupWeight`: Average weighted scores of the model outputs [19]. Scores are weighted on the basis of NDCG@k on the validation dataset. Note that their weights are constant values per model.
- `PostNDCG`: Select the best single model per article introduced in Sect. 2.2.

Our Methods: We show proposed methods as following:

- `HPA`: Hybrid the output selection method and a typical averaging method proposed in Sect. 2.3. We set $k = 50$, which obtained the highest accuracy in $k = \{5 \times n, n = 1, ..., 20\}$ on the validation dataset.
- `SPA`: Select models using the pseudo answer and average them (equal to HPA without the weighting). We set $k = 50$ which is the same setting of HPA.
- `WPA`: Average weighted model outputs using the pseudo answer (equal to HPA without the selecting).

3.3   Experimental Results

Our experimental results are shown in Table 1. As a result of the ensemble, we confirmed that all ensemble methods perform better than when using a single model. In particular, the proposed method HPA has achieved the highest NDCG@k. `PostNDCG` achieved higher accuracy than `RankNet`. This implies that the method of calculating the similarity between models using evaluation metrics for each article is

Table 1. NDCG@k and Precision@k scores (%) on ranking comment task ($k \in \{1, 5, 10\}$).

	NDCG			Precision		
	@1	@5	@10	@1	@5	@10
RankSVM	73.38	74.59	76.01	15.5	30.20	38.95
RankNet	76.35	77.97	79.52	15.0	33.20	42.99
ScoreAvg	76.91	79.11	80.48	16.08	33.67	44.32
RankAvg	79.19	80.53	81.81	13.57	36.18	46.08
TopkAvg	78.38	80.52	81.57	14.07	35.38	46.08
NormAvg	79.83	80.77	82.16	**17.08**	37.18	46.48
SupWeight	78.64	80.33	81.94	16.28	35.47	46.58
PostNDCG	77.18	80.09	81.24	14.57	35.58	45.78
HPA	**79.87**	**81.43**	**82.33**	**17.08**	37.39	**47.34**
SPA	79.68	80.96	82.19	**17.08**	35.87	46.68
WPA	**79.87**	81.39	82.17	**17.08**	**37.88**	46.63

effective. However, it was less accurate than the common averaging ensemble method such as `NormAvg`. Since models were originally trained by a relative comparison of rankings, preserving the diversity of models is more effective for improving performance than selecting models with high confidence by using `PostNDCG`. The unsupervised method `HPA` outperformed the supervised method `SupWeight`. Therefore, we confirmed that it is better to determine the important model from the similarity between the predicted rankings rather than learning it in advance using the labeled data.

Furthermore, we verified the effectiveness of NDCG@k as a similarity function to calculate `HPA`, compared to other similarity functions. We selected Precision, cosine similarity, Kendall rank correlation coefficient [11], and Spearman rank correlation coefficient [21] as compared methods. Table 2 shows the results of `HPA` when the similarity function is changed. The NDCG@k functions outperformed other similarity functions. Furthermore, Precision@k performed better than cos. Note that Precision@k equals top-k cosine similarity. It indicates top-k focused measurement, evaluation metrics, is useful for the ensemble.

Table 2. Comparison of similarity functions for `HPA`.

	NDCG			Precision		
	@1	@5	@10	@1	@5	@10
NDCG@k	**79.87**	**81.43**	**82.33**	**17.08**	**37.39**	**47.34**
Precision@k	79.47	80.54	81.57	17.00	36.80	46.25
cos	77.80	80.21	81.82	14.07	35.90	46.93
kendall	78.10	80.44	81.61	16.28	36.88	46.85
spearman	78.70	80.52	81.62	15.50	37.18	46.58

4 Related Work

Analyzing comments on online forums, including news comments, has been widely studied in recent years. This line of research has included many studies on ranking comments according to user feedback [6,9,22]. On the other hand, there has also been much research on analyzing news comments in terms of "constructiveness" [7,13,18]. The most related research is Fujita et al. [7]. They ranked comments by using the C-score to evaluate the quality, instead of relying on user feedback. They created a news comment ranking dataset and improved the model performance from the viewpoint of the dataset structure. In our research, we further improve the the performance from the viewpoint of the model structure.

In the ensemble methods for ranking task, there are methods to average model outputs [2,5], as mentioned in Sect. 3.2. Our method expands those methods by denoising through the relationships between predicted rankings. There is also research on learning the query-dependent weights with semi-supervised ensemble learning in an information retrieval task [8]. This method focused on selecting documents that are highly relevant to a query (article). It is effective for information retrieval tasks but not for ranking news comments task, because almost all such comments would be associated with a news article.

There are also approaches that improve the ranking model according to evaluation metrics: NDCG@k, LambdaRank [3], and LambdaMART [4]. These methods handled model training by calculating NDCG@k between a gold ranking and a predicted one. It means NDCG@k was not used in inference. That fundamentally differs from our method which calculates NDCG@k between predicted rankings during inference.

5   Conclusion and Future Work

We proposed a hybrid unsupervised method of an output selection method and a typical averaging method. Our experiments showed that comparing predicted rankings using the evaluation metrics is effective for selecting and weighting models. For future work, we would like to compare the proposed method with the supervised ensemble method in terms of performance and speed. We also plan to combine various types of networks instead of using the same network structure.

References

1. Burges, C., et al.: Learning to rank using gradient descent. In: Proceedings of the 22nd International Conference on Machine Learning (ICML 2005), pp. 89–96. ACM (2005). https://dl.acm.org/doi/abs/10.1145/1102351.1102363
2. Burges, C., Svore, K., Bennett, P., Pastusiak, A., Wu, Q.: Learning to rank using an ensemble of lambda-gradient models. In: Proceedings of the Learning to Rank Challenge, pp. 25–35. PMLR (2011). http://proceedings.mlr.press/v14/burges11a
3. Burges, C.J., Ragno, R., Le, Q.V.: Learning to rank with nonsmooth cost functions. In: Advances in Neural Information Processing Systems 19 (NIPS 2007), pp. 193–200 (2007). https://papers.nips.cc/paper/2971-learning-to-rank-with-nonsmooth-cost-functions.pdf
4. Burges, C.J.: From RankNet to LambdaRank to LambdaMART: an overview. Learning 11(23–581), 81 (2010). https://www.microsoft.com/en-us/research/wp-content/uploads/2016/02/MSR-TR-2010-82.pdf
5. Cormack, G.V., Clarke, C.L., Buettcher, S.: Reciprocal rank fusion outperforms condorcet and individual rank learning methods. In: Proceedings of the 32nd International ACM SIGIR Conference on Research and Development in Information Retrieval (SIGIR 2009), pp. 758–759. ACM (2009). https://dl.acm.org/doi/10.1145/1571941.1572114
6. Das Sarma, A., Das Sarma, A., Gollapudi, S., Panigrahy, R.: Ranking mechanisms in Twitter-like forums. In: Proceedings of the Third ACM International Conference on Web Search and Data Mining (WSDM 2010), pp. 21–30. ACM (2010). https://doi.org/10.1145/1718487.1718491
7. Fujita, S., Kobayashi, H., Okumura, M.: Dataset creation for ranking constructive news comments. In: Proceedings of the 57th Annual Meeting of the Association for Computational Linguistics (ACL 2019), pp. 2619–2626. Association for Computational Linguistics (2019). https://www.aclweb.org/anthology/P19-1250
8. Hoi, S.C., Jin, R.: Semi-supervised ensemble ranking. In: Proceedings of the 23rd National Conference on Artificial Intelligence-Volume 2 (AAAI 2008), pp. 634–639. AAAI Press (2008). https://www.aaai.org/Papers/AAAI/2008/AAAI08-101.pdf
9. Hsu, C.F., Khabiri, E., Caverlee, J.: Ranking comments on the social web. In: Proceedings of the 2009 International Conference on Computational Science and Engineering (CSE 2009), vol. 4, pp. 90–97. IEEE (2009). https://doi.org/10.1109/CSE.2009.109
10. Järvelin, K., Kekäläinen, J.: Cumulated gain-based evaluation of IR techniques. ACM Trans. Inform. Syst. (TOIS) 20(4), 422–446 (2002). https://doi.org/10.1145/582415.582418

11. Kendall, M.G.: A new measure of rank correlation. Biometrika **30**(1/2), 81–93 (1938). https://www.jstor.org/stable/pdf/2332226.pdf
12. Kobayashi, H.: Frustratingly easy model ensemble for abstractive summarization. In: Proceedings of the 2018 Conference on Empirical Methods in Natural Language Processing (EMNLP 2018), pp. 4165–4176. Association for Computational Linguistics (2018). https://www.aclweb.org/anthology/D18-1449
13. Kolhatkar, V., Taboada, M.: Constructive language in news comments. In: Proceedings of the First Workshop on Abusive Language Online, pp. 11–17. Association for Computational Linguistics (2017). http://www.aclweb.org/anthology/W17-3002
14. Kudo, T., Yamamoto, K., Matsumoto, Y.: Applying conditional random fields to japanese morphological analysis. In: Proceedings of the 2004 Conference on Empirical Methods in Natural Language Processing (EMNLP 2004), pp. 230–237. Association for Computational Linguistics (2004). http://aclweb.org/anthology/W04-3230
15. Littlestone, N., Warmuth, M.K.: The weighted majority algorithm. Inform. Comput. **108**(2), 212–261 (1994). https://www.sciencedirect.com/science/article/pii/S0890540184710091
16. Mikolov, T., Sutskever, I., Chen, K., Corrado, G.S., Dean, J.: Distributed representations of words and phrases and their compositionality. In: Advances in Neural Information Processing Systems 26 (NIPS 2013), pp. 3111–3119 (2013). https://arxiv.org/abs/1310.4546
17. Naftaly, U., Intrator, N., Horn, D.: Optimal ensemble averaging of neural networks. Netw.: Comput. Neural Syst. **8**(3), 283–296 (1997). https://www.tandfonline.com/doi/abs/10.1088/0954-898X83004
18. Napoles, C., Pappu, A., Tetreault, J.R.: Automatically identifying good conversations online (yes, they do exist!). In: Proceedings of the Eleventh International AAAI Conference on Web and Social Media (ICWSM 2017), pp. 628–631. AAAI Press (2017). https://aaai.org/ocs/index.php/ICWSM/ICWSM17/paper/view/15673
19. Opitz, D.W., Shavlik, J.W.: Actively searching for an effective neural network ensemble. Conn. Sci. **8**(3–4), 337–354 (1996). https://research.cs.wisc.edu/machine-learning/shavlik-group/opitz.consci96.pdf
20. Sato, T., Hashimoto, T., Okumura, M.: Implementation of a word segmentation dictionary called mecab-ipadic-NEologd and study on how to use it effectively for information retrieval (in Japanese). In: Proceedings of the Twenty-three Annual Meeting of the Association for Natural Language Processing, pp. NLP2017-B6-1. The Association for Natural Language Processing (2017)
21. Spearman, C.: The proof and measurement of association between two things. Am. J. Psychol. **15**(1), 72–101 (1904). http://digamoo.free.fr/spearman1904a.pdf
22. Wei, Z., Liu, Y., Li, Y.: Is this post persuasive? Ranking argumentative comments in online forum. In: Proceedings of the 54th Annual Meeting of the Association for Computational Linguistics (ACL 2016), vol. 2, pp. 195–200. Association for Computational Linguistics (2016). https://www.aclweb.org/anthology/P16-2032

Irony Detection in a Multilingual Context

Bilal Ghanem[1(✉)], Jihen Karoui[2], Farah Benamara[3], Paolo Rosso[1],
and Véronique Moriceau[3]

[1] PRHLT Research Center, Universitat Politècnica de València, Valencia, Spain
bigha@doctor.upv.es, prosso@dsic.upv.es
[2] AUSY R&D, Paris, France
jkaroui@ausy.fr
[3] IRIT, CNRS, Université de Toulouse, Toulouse, France
{benamara,moriceau}@irit.fr

Abstract. This paper proposes the first multilingual (French, English and Arabic) and multicultural (Indo-European languages vs. less culturally close languages) irony detection system. We employ both feature-based models and neural architectures using monolingual word representation. We compare the performance of these systems with state-of-the-art systems to identify their capabilities. We show that these monolingual models trained separately on different languages using multilingual word representation or text-based features can open the door to irony detection in languages that lack of annotated data for irony.

Keywords: Irony detection · Social media · Multilingual embeddings

1 Motivations

Figurative language makes use of figures of speech to convey non-literal meaning [2,16]. It encompasses a variety of phenomena, including metaphor, humor, and irony. We focus here on irony and uses it as an umbrella term that covers satire, parody and sarcasm.

Irony detection (ID) has gained relevance recently, due to its importance to extract information from texts. For example, to go beyond the literal matches of user queries, Veale enriched information retrieval with new operators to enable the non-literal retrieval of creative expressions [40]. Also, the performances of sentiment analysis systems drastically decrease when applied to ironic texts [5,19]. Most related work concern English [17,21] with some efforts in French [23], Portuguese [7], Italian [14], Dutch [26], Hindi [37], Spanish variants [31] and Arabic [11,22]. Bilingual ID with one model per language has also been explored, like English-Czech [32] and English-Chinese [38], but not within a cross-lingual perspective.

In social media, such as Twitter, specific hashtags (#irony, #sarcasm) are often used as gold labels to detect irony in a supervised learning setting. Although recent studies pointed out the issue of false-alarm hashtags in self-labeled data [20], ID via hashtag filtering provides researchers positive examples

© Springer Nature Switzerland AG 2020
J. M. Jose et al. (Eds.): ECIR 2020, LNCS 12036, pp. 141–149, 2020.
https://doi.org/10.1007/978-3-030-45442-5_18

with high precision. On the other hand, systems are not able to detect irony in languages where such filtering is not always possible. Multilingual prediction (either relying on machine translation or multilingual embedding methods) is a common solution to tackle under-resourced languages [6,33]. While multilinguality has been widely investigated in information retrieval [27,34] and several NLP tasks (e.g., sentiment analysis [3,4] and named entity recognition [30]), no one explored it for irony.

We aim here to bridge the gap by tackling ID in tweets from both multilingual (French, English and Arabic) and multicultural perspectives (Indo-European languages whose speakers share quite the same cultural background vs. less culturally close languages). Our approach does not rely either on machine translation or parallel corpora (which are not always available), but rather builds on previous corpus-based studies that show that irony is a universal phenomenon and many languages share similar irony devices. For example, Karoui et al. [24] concluded that their multi-layer annotated schema, initially used to annotate French tweets, is portable to English and Italian, observing relatively the same tendencies in terms of irony categories and markers. Similarly, Chakhachiro [8] studies irony in English and Arabic, and shows that both languages share several similarities in the rhetorical (e.g., overstatement), grammatical (e.g., redundancy) and lexical (e.g., synonymy) usage of irony devices. The next step now is to show to what extent these observations are still valid from a computational point of view. Our contributions are:

I. *A new freely available corpus of Arabic tweets* manually annotated for irony detection[1].

II. *Monolingual ID*: We propose both feature-based models (relying on language-dependent and language-independent features) and neural models to measure to what extent ID is language dependent.

III. *Cross-lingual ID*: We experiment using cross-lingual word representation by training on one language and testing on another one to measure how the proposed models are culture-dependent. Our results are encouraging and open the door to ID in languages that lack of annotated data for irony.

2   Data

Arabic dataset ($AR = 11{,}225$ tweets). Our starting point was the corpus built by [22] that we extended to different political issues and events related to the Middle East and Maghreb that hold during the years 2011 to 2018. Tweets were collected using a set of predefined keywords (which targeted specific political figures or events) and containing or not Arabic ironic hashtags (#سخرية, #مسخرة, #تهكم, #استهزاء)[2]. The collection process resulted in a set of 6,809 ironic tweets (I) vs. 15,509 non ironic (NI) written using standard (formal) and different Arabic language varieties: Egypt, Gulf, Levantine, and Maghrebi dialects.

---

[1] The corpus is available at https://github.com/bilalghanem/multilingual_irony.

[2] All of these words are synonyms where they mean "Irony".

To investigate the validity of using the original tweets labels, a sample of 3,000 *I* and 3,000 *NI* was manually annotated by two Arabic native speakers which resulted in 2,636 *I* vs. 2,876 *NI*. The inter-annotator agreement using Cohen's Kappa was 0.76, while the agreement score between the annotators' labels and the original labels was 0.6. Agreements being relatively good knowing the difficulty of the task, we sampled 5,713 instances from the original unlabeled dataset to our manually labeled part. The added tweets have been manually checked to remove duplicates, very short tweets and tweets that depend on external links, images or videos to understand their meaning.

French dataset (FR = 7,307 tweets). We rely on the corpus used for the DEFT 2017 French shared task on irony [5] which consists of tweets relative to a set of topics discussed in the media between 2014 and 2016 and contains topic keywords and/or French irony hashtags (#ironie, #sarcasme). Tweets have been annotated by three annotators (after removing the original labels) with a reported Cohen's Kappa of 0.69.

English dataset (EN = 11,225 tweets). We use the corpus built by [32] which consists of 100,000 tweets collected using the hashtag #sarcasm. It was used as benchmark in several works [13,18]. We sliced a subset of approximately 11,200 tweets to match the sizes of the other languages' datasets.

Table 1 shows the tweet distribution in all corpora. Across the three languages, we keep a similar number of instances for train and test sets to have fair cross-lingual experiments as well (see Sect. 4). Also, for French, we use the original dataset without any modification, keeping the same number of records for train and test to better compare with state-of-the-art results. For the classes distribution (ironic vs. non ironic), we do not choose a specific ratio but we use the resulted distribution from the random shuffling process.

Table 1. Tweet distribution in all corpora.

	# Ironic	# Not-Ironic	Train	Test
AR	6,005	5,220	10,219	1,006
FR	2,425	4,882	5,843	1,464
EN	5,602	5,623	10,219	1,006

3   Monolingual Irony Detection

It is important to note that our aim is not to outperform state-of-the-art models in monolingual ID but to investigate which of the monolingual architectures (neural or feature-based) can achieve comparable results with existing systems. The result can show which kind of features works better in the monolingual settings and can be employed to detect irony in a multilingual setting. In addition, it can show us to what extend ID is language dependent by comparing their results to multilingual results. Two models have been built, as explained below. Prior to learning, basic preprocessing steps were performed for each language (e.g., removing foreign characters, ironic hashtags, mentions, and URLs).

Feature-Based Models. We used state-of-the-art features that have shown to be useful in ID: some of them are language-independent (e.g., punctuation marks, positive and negative emoticons, quotations, personal pronouns, tweet's length, named entities) while others are language-dependent relying on dedicated lexicons (e.g., negation, opinion lexicons, opposition words). Several classical machine learning classifiers were tested with several feature combinations, among them Random Forest (RF) achieved the best result with all features.

Neural Model with Monolingual Embeddings. We used Convolutional Neural Network (CNN) network whose structure is similar to the one proposed by [25]. For the embeddings, we relied on *AraVec* [36] for Arabic, FastText [15] for French, and Word2vec Google News [29] for English[3]. For the three languages, the size of the embeddings is 300 and the embeddings were fine-tuned during the training process. The CNN network was tuned with 20% of the training corpus using the *Hyperopt*[4] library.

Results. Table 2 shows the results obtained when using train-test configurations for each language. For English, our results, in terms of macro F-score (F), were not comparable to those of [32,39], as we used 11% of the original dataset. For French, our scores are in line with those reported in state of the art (cf. best system in the irony shared task achieved $F = 78.3$ [5]). They outperform those obtained for Arabic ($A = 71.7$) [22] and are comparable to those recently reported in the irony detection shared task in Arabic tweets [11,12] ($F = 84.4$). Overall, the results show that semantic-based information captured by the embedding space are more productive comparing to standard surface and lexicon-based features.

Table 2. Results of the monolingual experiments (in percentage) in terms of accuracy (A), precision (P), recall (R), and macro F-score (F).

	Arabic				French				English			
	A	P	R	F	A	P	R	F	A	P	R	F
RF	68.0	67.0	82.0	68.0	68.5	71.7	87.3	61.0	61.2	60.0	70.0	61.0
CNN	**80.5**	79.1	84.9	**80.4**	**77.6**	68.2	59.6	**73.5**	**77.9**	74.6	84.7	**77.8**

4   Cross-lingual Irony Detection

We use the previous CNN architecture with bilingual embedding and the RF model with surface features (e.g., use of personal pronoun, presence of

---

[3] Other available pretrained embeddings models have also been tested.

[4] https://github.com/hyperopt/hyperopt

interjections, emoticon or specific punctuation)[5] to verify which pair of the three languages: (a) has similar ironic pragmatic devices, and (b) uses similar text-based pattern in the narrative of the ironic tweets. As continuous word embedding spaces exhibit similar structures across (even distant) languages [28], we use a multilingual word representation which aims to learn a linear mapping from a source to a target embedding space. Many methods have been proposed to learn this mapping such as parallel data supervision and bilingual dictionaries [28] or unsupervised methods relying on monolingual corpora [1,10,41]. For our experiments, we use Conneau et al.'s approach as it showed superior results with respect to the literature [10]. We perform several experiments by training on one language ($lang_1$) and testing on another one ($lang_2$) (henceforth $lang_1 \rightarrow lang_2$). We get 6 configurations, plus two others to evaluate how irony devices are expressed cross-culturally, i.e. in European vs. non European languages. In each experiment, we took 20% from the training to validate the model before the testing process. Table 3 presents the results.

Table 3. Results of the cross-lingual experiments.

Train → Test	CNN				RF			
	A	P	R	F	A	P	R	F
Ar → Fr	60.1	37.2	26.6	**51.7**	47.03	29.9	43.9	46.0
Fr → Ar	57.8	62.9	45.7	**57.3**	51.11	61.1	24.0	54.0
Ar → En	48.5	26.5	17.9	34.1	49.67	49.7	66.2	**50.0**
En → Ar	56.7	57.7	62.3	**56.4**	52.5	58.6	38.5	53.0
Fr → En	53.0	67.9	11.0	42.9	52.38	52.0	63.6	**52.0**
En → Fr	56.7	33.5	29.5	50.0	56.44	74.6	52.7	**58.0**
(En/Fr) → Ar	62.4	66.1	56.8	**62.4**	55.08	56.7	68.5	62.0
Ar → (En/Fr)	56.3	33.9	09.5	42.7	59.84	60.0	98.7	**74.6**

From a semantic perspective, despite the language and cultural differences between Arabic and French languages, CNN results show a high performance comparing to the other languages pairs when we train on each of these two languages and test on the other one. Similarly, for the French and English pair, but when we train on French they are quite lower. We have a similar case when we train on Arabic and test on English. We can justify that by, the language presentation of the Arabic and French tweets are quite informal and have many dialect words that may not exist in the pretrained embeddings we used comparing to the English ones (lower embeddings coverage ratio), which become harder for the CNN to learn a clear semantic pattern. Another point is the presence of Arabic dialects, where some dialect words may not exist in the multilingual pretrained

---

[5] To avoid language dependencies, we rely on surface features only discarding those that require external semantic resources or morpho-syntactic parsing.

embedding model that we used. On the other hand, from the text-based per-spective, the results show that the text-based features can help in the case when the semantic aspect shows weak detection; this is the case for the $Ar \longrightarrow En$ configuration. It is worthy to mention that the highest result we get in this experiment is from the $En \rightarrow Fr$ pair, as both languages use Latin characters. Finally, when investigating the relatedness between European vs. non European languages (cf. $(En/Fr) \rightarrow Ar$), we obtain similar results than those obtained in the monolingual experiment (macro F-score 62.4 vs. 68.0) and best results are achieved by $Ar \rightarrow (En/Fr)$. This shows that there are pragmatic devices in common between both sides and, in a similar way, similar text-based patterns in the narrative way of the ironic tweets.

5   Discussions and Conclusion

This paper proposes the first multilingual ID in tweets. We show that simple monolingual architectures (either neural or feature-based) trained separately on each language can be successfully used in a multilingual setting providing a cross-lingual word representation or basic surface features. Our monolingual results are comparable to state of the art for the three languages. The CNN architecture trained on cross-lingual word representation shows that irony has a certain similarity between the languages we targeted despite the cultural differences which confirm that irony is a universal phenomena, as already shown in previous linguistic studies [9, 24, 35]. The manual analysis of the common misclassified tweets across the languages in the multilingual setup, shows that classification errors are due to three main factors. (1) First, the *absence of context* where writers did not provide sufficient information to capture the ironic sense even in the monolingual setting, as in مبارك حسني يسقط تاني نبدا !! (*Let's start again, get off get off Mubarak!!*) where the writer mocks the Egyptian revolution, as the actual president "Sisi" is viewed as Mubarak's fellows. (2) Second, the presence of *out of vocabulary (OOV) terms* because of the weak coverage of the multilingual embeddings which make the system fails to generalize when the OOV set of unseen words is large during the training process. We found tweets in all the three languages written in a very informal way, where some characters of the words were deleted, duplicated or written phonetically (e.g *phat* instead of *fat*). (3) Another important issue is the difficulty to *deal with the Arabic language*. Arabic tweets are often characterized by non-diacritised texts, a large variations of unstandardized dialectal Arabic (recall that our dataset has 4 main varieties, namely Egypt, Gulf, Levantine, and Maghrebi), presence of transliterated words (e.g. the word *table* becomes طابلة (*tabla*)), and finally linguistic code switching between Modern Standard Arabic and several dialects, and between Arabic and other languages like English and French. We found some tweets contain only words from one of the varieties and most of these words do not exist in the Arabic embeddings model. For example in مصر# ايه ولاه عيان هو .. مامتش يوم كام بقاله مبارك (*Since many days Mubarak didn't die .. is he sick or what? #Egypt*), only the words يَم (day), مبارك (Mubarak), and هو (he) exist in the embeddings. Clearly, considering only these three available words, we are not able to understand the context or the ironic meaning of the tweet.

To conclude, our multilingual experiments confirmed that the door is open towards multilingual approaches for ID. Furthermore, our results showed that ID can be applied to languages that lack of annotated data. Our next step is to experiment with other languages such as Hindi and Italian.

Acknowledgment. The work of Paolo Rosso was partially funded by the Spanish MICINN under the research project MISMIS-FAKEnHATE (PGC2018-096212-B-C31).

References

1. Artetxe, M., Labaka, G., Agirre, E., Cho, K.: Unsupervised neural machine translation. arXiv preprint (2017)
2. Attardo, S.: Irony as relevant inappropriateness. J. Pragmat. **32**(6), 793–826 (2000)
3. Balahur, A., Turchi, M.: Comparative experiments using supervised learning and machine translation for multilingual sentiment analysis. Comput. Speech Lang. **28**(1), 56–75 (2014)
4. Barnes, J., Klinger, R., Schulte im Walde, S.: Bilingual sentiment embeddings: joint projection of sentiment across languages. In: Proceedings of the 56th Annual Meeting of the Association for Computational Linguistics (Volume 1: Long Papers), pp. 2483–2493. Association for Computational Linguistics (2018)
5. Benamara, F., Grouin, C., Karoui, J., Moriceau, V., Robba, I.: Analyse d'opinion et langage figuratif dans des tweets présentation et résultats du Défi Fouille de Textes DEFT2017. In: Actes de DEFT@TALN2017, Orléans, France (2017)
6. Bikel, D., Zitouni, I.: Multilingual Natural Language Processing Applications: From Theory to Practice, 1st edn. IBM Press, Armonk (2012)
7. Carvalho, P., Sarmento, L., Silva, M.J., Oliveira, E.D.: Clues for detecting irony in user-generated contents: oh...!! it's "so easy";-). In: Proceedings of the 1st International CIKM Workshop on Topic-Sentiment Analysis for Mass Opinion, pp. 53–56. ACM (2009)
8. Chakhachiro, R.: Translating irony in political commentary texts from English into Arabic. Babel **53**(3), 216–240 (2007)
9. Colston, H.L.: Irony as indirectness cross-linguistically: on the scope of generic mechanisms. In: Capone, A., García-Carpintero, M., Falzone, A. (eds.) Indirect Reports and Pragmatics in the World Languages. PPPP, vol. 19, pp. 109–131. Springer, Cham (2019). https://doi.org/10.1007/978-3-319-78771-8_6
10. Conneau, A., Lample, G., Ranzato, M., Denoyer, L., Jégou, H.: Word translation without parallel data. arXiv preprint (2017)
11. Ghanem, B., Karoui, J., Benamara, F., Moriceau, V., Rosso, P.: IDAT@FIRE2019: overview of the track on irony detection in Arabic tweets. In: Proceedings of the 11th Forum for Information Retrieval Evaluation, pp. 10–13. ACM (2019)
12. Ghanem, B., Karoui, J., Benamara, F., Moriceau, V., Rosso, P.: IDAT@FIRE2019: overview of the track on irony detection in Arabic tweets. In: Working Notes of the Forum for Information Retrieval Evaluation (FIRE 2019). CEUR Workshop Proceedings, Kolkata, India, vol. 2517, pp. 380–390. CEUR-WS.org (2019)
13. Ghanem, B., Rangel, F., Rosso, P.: LDR at SemEval-2018 task 3: a low dimensional text representation for irony detection. In: Proceedings of the 12th International Workshop on Semantic Evaluation, pp. 531–536 (2018)

14. Gianti, A., Bosco, C., Patti, V., Bolioli, A., Caro, L.D.: Annotating irony in a novel Italian corpus for sentiment analysis. In: Proceedings of the 4th Workshop on Corpora for Research on Emotion Sentiment and Social Signals, Istanbul, Turkey, pp. 1–7 (2012)

15. Grave, E., Bojanowski, P., Gupta, P., Joulin, A., Mikolov, T.: Learning word vectors for 157 languages. CoRR abs/1802.06893 (2018)

16. Grice, H.P.: Logic and conversation. In: Cole, P., Morgan, J.L. (eds.) Speech Acts. Syntax and Semantics, vol. 3, pp. 41–58. Academic Press, New York (1975)

17. Hee, C.V., Lefever, E., Hoste, V.: SemEval-2018 task 3: irony detection in English tweets. In: Proceedings of The 12th International Workshop on Semantic Evaluation, SemEval@NAACL-HLT, New Orleans, Louisiana, 5–6 June 2018, pp. 39–50 (2018)

18. Hernández Farías, D.I., Bosco, C., Patti, V., Rosso, P.: Sentiment polarity classification of figurative language: exploring the role of irony-aware and multifaceted affect features. In: Gelbukh, A. (ed.) CICLing 2017. LNCS, vol. 10762, pp. 46–57. Springer, Cham (2018). https://doi.org/10.1007/978-3-319-77116-8_4

19. Hernández Farías, D.I., Patti, V., Rosso, P.: Irony detection in twitter: the role of affective content. ACM Trans. Technol. (TOIT) **16**(3), 19 (2016)

20. Huang, H.H., Chen, C.C., Chen, H.H.: Disambiguating false-alarm hashtag usages in tweets for irony detection. In: Proceedings of the 56th Annual Meeting of the Association for Computational Linguistics (Volume 2: Short Papers) (2018)

21. Huang, Y.-H., Huang, H.-H., Chen, H.-H.: Irony detection with attentive recurrent neural networks. In: Jose, J.M., et al. (eds.) ECIR 2017. LNCS, vol. 10193, pp. 534–540. Springer, Cham (2017). https://doi.org/10.1007/978-3-319-56608-5_45

22. Karoui, J., Benamara, F., Moriceau, V.: SOUKHRIA: towards an irony detection system for Arabic in social media. In: Third International Conference on Arabic Computational Linguistics, ACLING 2017, 5–6 November 2017, Dubai, United Arab Emirates, pp. 161–168 (2017)

23. Karoui, J., Benamara, F., Moriceau, V., Aussenac-Gilles, N., Belguith, L.H.: Towards a contextual pragmatic model to detect irony in tweets. In: Proceedings of the 53rd Annual Meeting of the Association for Computational Linguistics and the 7th International Joint Conference on Natural Language Processing of the Asian Federation of Natural Language Processing (Volume 2: Short Papers), ACL-IJCNLP 2015, pp. 644–650 (2015)

24. Karoui, J., Benamara, F., Moriceau, V., Patti, V., Bosco, C., Aussenac-Gilles, N.: Exploring the impact of pragmatic phenomena on irony detection in tweets: a multilingual corpus study. In: Proceedings of the 15th Conference of the European Chapter of the Association for Computational Linguistics (Volume 1: Long Papers), pp. 262–272. Association for Computational Linguistics (2017)

25. Kim, Y.: Convolutional neural networks for sentence classification. In: Proceedings of the 2014 Conference on Empirical Methods in Natural Language Processing (EMNLP), pp. 1746–1751. Association for Computational Linguistics (2014)

26. Liebrecht, C., Kunneman, F., van den Bosch, A.: The perfect solution for detecting sarcasm in tweets# not. In: Proceedings of the 4th Workshop on Computational Approaches to Subjectivity, Sentiment and Social Media Analysis, pp. 29–37. ACL, New Brunswick (2013)

27. Litschko, R., Glavaš, G., Ponzetto, S.P., Vulić, I.: Unsupervised cross-lingual information retrieval using monolingual data only. In: The 41st International ACM SIGIR Conference on Research and Development in Information Retrieval, SIGIR 2018, pp. 1253–1256 (2018)

28. Mikolov, T., Chen, K., Corrado, G., Dean, J.: Efficient estimation of word representations in vector space. CoRR abs/1301.3781 (2013)
29. Mikolov, T., Yih, W.T., Zweig, G.: Linguistic regularities in continuous space word representations. In: Proceedings of the 2013 Conference of the North American Chapter of the Association for Computational Linguistics: Human Language Technologies, pp. 746–751 (2013)
30. Ni, J., Florian, R.: Improving multilingual named entity recognition with Wikipedia entity type mapping. CoRR abs/1707.02459 (2017)
31. Ortega-Bueno, R., Rangel, F., Hernández Farıas, D., Rosso, P., Montes-y Gómez, M., Medina Pagola, J.E.: Overview of the task on irony detection in Spanish variants. In: Proceedings of the Iberian Languages Evaluation Forum (IberLEF 2019), Co-located with 34th Conference of the Spanish Society for Natural Language Processing (SEPLN 2019). CEUR-WS.org (2019)
32. Ptáček, T., Habernal, I., Hong, J.: Sarcasm detection on Czech and English twitter. In: Proceedings of the 25th International Conference on Computational Linguistics, COLING 2014: Technical Papers, pp. 213–223 (2014)
33. Ruder, S.: A survey of cross-lingual embedding models. CoRR abs/1706.04902 (2017)
34. Sasaki, S., Sun, S., Schamoni, S., Duh, K., Inui, K.: Cross-lingual learning-to-rank with shared representations. In: Proceedings of the 2018 Conference of the North American Chapter of the Association for Computational Linguistics: Human Language Technologies (Volume 2: Short Papers), pp. 458–463 (2018)
35. Sigar, A., Taha, Z.: A contrastive study of ironic expressions in English and Arabic. Coll. Basic Educ. Res. J. **12**(2), 795–817 (2012)
36. Soliman, A.B., Eissa, K., El-Beltagy, S.R.: AraVec: a set of Arabic word embedding models for use in Arabic NLP. In: Third International Conference on Arabic Computational Linguistics, ACLING 2017, 5–6 November 2017, Dubai, United Arab Emirates, pp. 256–265 (2017)
37. Swami, S., Khandelwal, A., Singh, V., Akhtar, S.S., Shrivastava, M.: A corpus of English-Hindi code-mixed tweets for sarcasm detection. In: 19th International Conference on Computational Linguistics and Intelligent Text Processing (CICLing) (2018)
38. Tang, Y., Chen, H.: Chinese irony corpus construction and ironic structure analysis. In: Proceedings of the 25th International Conference on Computational Linguistics, COLING 2014: Technical Papers, 23–29 August 2014, Dublin, Ireland, pp. 1269–1278 (2014)
39. Tay, Y., Luu, A.T., Hui, S.C., Su, J.: Reasoning with sarcasm by reading in-between. In: Proceedings of the 56th Annual Meeting of the Association for Computational Linguistics (Volume 1: Long Papers), pp. 1010–1020 (2018)
40. Veale, T.: Creative language retrieval: a robust hybrid of information retrieval and linguistic creativity. In: Proceedings of the 49th Annual Meeting of the Association for Computational Linguistics: Human Language Technologies, HLT 2011, vol. 1, pp. 278–287 (2011)
41. Wada, T., Iwata, T.: Unsupervised cross-lingual word embedding by multilingual neural language models. arXiv preprint (2018)

Document Network Projection
in Pretrained Word Embedding Space

Antoine Gourru$^{(\boxtimes)}$ (iD), Adrien Guille (iD), Julien Velcin (iD), and Julien Jacques (iD)

Université de Lyon, Lyon 2, ERIC EA3083, Lyon, France
{antoine.gourru,adrien.guille,julien.velcin,julien.jacques}@univ-lyon2.fr

Abstract. We present Regularized Linear Embedding (RLE), a novel method that projects a collection of linked documents (e.g., citation network) into a pretrained word embedding space. In addition to the textual content, we leverage a matrix of pairwise similarities providing complementary information (e.g., the network proximity of two documents in a citation graph). We first build a simple word vector average for each document, and we use the similarities to alter this average representation. The document representations can help to solve many information retrieval tasks, such as recommendation, classification and clustering. We demonstrate that our approach outperforms or matches existing document network embedding methods on node classification and link prediction tasks. Furthermore, we show that it helps identifying relevant keywords to describe document classes.

Keywords: Document network embedding · Representation Learning

1 Introduction

Information retrieval methods require relevant compact vector space representations of documents. The classical bag of words cannot capture all the useful semantic information. Representation Learning is a way to go beyond and boost the performances we can expect in many information retrieval tasks [6]. It aims at finding low dimensional and dense representations of high dimensional data such as words [12] and documents [2,10]. In this latent space, proximity reflects semantic closeness. Many recent methods use those representations for information retrieval tasks: capturing user interest [16], query expansion [9], link prediction and document classification [20].

In addition to the textual information, many corpora include links between documents, such as bibliographic networks (e.g., scientific articles linked with citations or co-authorship) and social networks (e.g., tweets with ReTweet relations). This information can be used to improve the accuracy of document representations. Several recent methods [11,20] study the embedding of networks with textual attributes associated to the nodes. Most of them learn continuous representations for nodes independently of a word-vector representation. That is to say, documents and words do not *lie* in the same space. It is interesting to find

© Springer Nature Switzerland AG 2020
J. M. Jose et al. (Eds.): ECIR 2020, LNCS 12036, pp. 150–157, 2020.
https://doi.org/10.1007/978-3-030-45442-5_19

a common space to represent documents and words when considering many tasks in information retrieval (query expansion) and document analysis (description of document clusters). Our approach allows to represent documents and words in the same semantic space. The method can be applied with word embedding learned on the data with any state-of-the art method [6,12], or with embeddings that were previously learned[1] to reduce the computation cost. Contrary to many existing methods that make use of deep and complex neural networks (see Sect. 2 for related works), our method is fast, and it has only one parameter to tune.

We propose to construct a weight vector for each document using both textual and network information. We can then project the documents into the prelearned word vector space using this vector (see Fig. 1). The method is straightforward to apply, as it only requires applying well studied word embedding methods and matrix multiplication. We show in Sect. 4 that it outperforms or matches existing methods in classification and link prediction tasks and we demonstrate that projecting the documents into the word embedding space can provide semantic insights.

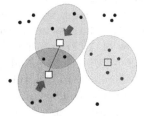

Fig. 1. Our method performs smoothing (represented as red arrows) on the documents' centroid representations (the square blocks). As the document in the blue circle (dots are words) is connected to the orange one, their representations get closer. The document in the green circle is isolated, thus it remains unchanged by the smoothing effect. (Color figure online)

2   Related Work

Several methods study the embedding of paragraph or short documents such as [10], generalizing the seminal word2vec models proposed by [12]. These approaches go beyond the simple method that consists in building a weighted average of representations of words that compose the document. For example in [2], authors propose to perturb weights for word average projection using Singular Value Decomposition (SVD). This last approach inspired our work as they show that word average is often a relevant baseline that can be improved in some cases using contextual smoothing.

As stated above, many corpora are structured in networks, providing additional information on documents semantics. TADW [20] is the first method that

---

[1] E.g., https://fasttext.cc/.

deals with this kind of data. It formulates network embedding [15] as a matrix tri-factorization problem to integrate textual information. Subsequent methods mainly adopt neural network based models: STNE [11] extends the seq2seq models, Graph2Gauss [3] learns both representations and variances via energy based learning, and VGAE [8] adopts a variational encoder. Even if these approaches yield good results, they require tuning a lot of hyperparameters. Two methods are based on factorization approaches: GVNR-t [4], that extends GloVe [14], and AANE [7]. None of these methods learn documents and words embedding in the same space. In [10] and [1], authors represent them in a comparable space. Yet, they do not consider network information, as opposed to LDE [19]. Nonetheless, this last method requires labels associated with nodes, making it a supervised approach. Our method projects the documents and the words into the same space in an unsupervised fashion, with only one hyperparameter to tune. We will now present the formulation of this approach.

3   RLE: Document Projection with Smoothing

In this section, we present our model to build vector representations for a collection of linked documents. From now on, we will refer to our method as Regularized Linear Embedding (RLE). Matrices are in capital letters, and if X is a matrix, we write x_i the i-th row of X. From a network of n nodes, we extract a pairwise similarity matrix $S \in \mathbb{R}^{n \times n}$, computed as $S = \frac{A+A^2}{2}$ with A the transition matrix of the graph. Similarly to [20], this matrix considers both first and second order similarities. v is the number of words in the vocabulary. The corpus is represented as a document-term matrix $T \in \mathbb{R}^{n \times v}$, with each entry of T being the relative frequency of a word in a given document.

With $U \in \mathbb{R}^{v \times k}$ a matrix of pretrained word embeddings in dimension k, our goal is to build a matrix $D \in \mathbb{R}^{n \times k}$ of document embeddings, in the same space as the word embeddings. We build, for each document, a weight vector $p_i \in \mathbb{R}^v$, stacked in a matrix P and define the embedding of a document as $d_i = p_i U$. We construct p_i as follows: we first compute a smoothing matrix $B \in \mathbb{R}^{n \times v}$ with:

$$b_i = \frac{1}{\sum_j S_{i,j}} \sum_j S_{i,j} t_j. \tag{1}$$

Each row b_i of this matrix is a centroid of the initial document-term frequency matrix T, weighted by the similarity between the document i and each of the other documents. Then, we compute the weight matrix P according to T and B, in matrix notation:

$$P = (1 - \lambda)T + \lambda B, \tag{2}$$

where $\lambda \in [0, 1]$ controls the smoothing intensity. Then, we compute $D = PU$. Our method implies matrix multiplication and normalization only, making it fast and easily scalable. When $\lambda = 0$, $P = T$, thus, we recover the word average method. When $\lambda = 1$, we obtain $P = B$ and thus embed the documents with respect to the contextual information only (i.e., the similar documents). We illustrate the effect of smoothing in Fig. 1.

4   Experiments

In this section, we present our experimental results on classification and link prediction tasks, followed by a qualitative analysis of document representations.

We use two citation networks: Cora [18] and DBLP [13,17]. We also use New York Times articles (https://www.nytimes.com/) from January 2007. We create a link between pairs of articles sharing a common tag. The class corresponds to the article section. Cora contains 2,211 labeled documents (7 classes) with 5,001 citation links. The dataset includes the abstract of each article. The New York Times dataset (Nyt) contains 5,135 documents, 3,050,513 edges and 4 classes. Dblp has 60,744 documents (4 classes) and 52,914 edges. It includes the title of the articles only. After pruning the vocabulary (removing stop words, filtering word occurring in more than 25% of the corpus and less than 10 times), we obtain vocabularies made of 2,421 features for the Cora dataset, 6,407 for the Nyt dataset, and 3,763 for Dblp.

All embeddings are in dimension 160. We use DeepWalk with 40 walks of length 40, and a window of size 10. We also experiment with Latent Semantic Analysis (LSA) [5] and a concatenation of LSA and DeepWalk representations in dimension 80 as done by [20], referred as "Concatenation". We also compare the performance of RLE with recent methods that embed attributed networks: STNE, Graph2Gauss, GVNR-t, VGAE, AANE and TADW. For STNE, we set the depth to 1 which leads to the best scores in our experiments. For Graph2Gauss, we set $K = 1$, depth $= 1$. We use default architecture for VGAE and determine optimal λ and ρ for AANE, and x_{min} for GVNR-t. For TADW, we use LSA in dimension 200 as a textual feature matrix and set regularization to 0.2, following authors' recommendation. For each method, we use the implementation provided by the authors. We discard LDE since it is semi supervised and will not lead to a fair comparison.

RLE needs prelearned word representations. Hence, we build word vectors using Skip-gram with negative sampling [12]. We use the implementation in gensim[2], with window size of 15 for Cora, 10 for Nyt and 5 for DBLP, and 5 negative examples for both. The procedure is fast (46 s for Cora, 84 on DBLP and 42 on Nyt). Similarly to baselines methods, we use the value of λ (0.7) that produces the optimal results on both datasets (see Fig. 2).

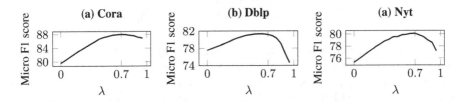

Fig. 2. Impact of λ on RLE in terms of document classification for $d = 160$. Optimum is achieved around 0.7 on each dataset (Cora, Nyt: 0.7, Dblp: 0.65).

---

[2] https://radimrehurek.com/gensim/.

4.1 Quantitative Results

We evaluate RLE in its ability to separate documents by classes in the embedding space and to predict links between documents. We perform SVM with L2 regularization on the vector representations of documents and report Micro F1 scores for different train/test ratios in Table 1. The regularisation strength is fixed through grid search. We also report computation times in second. For link prediction, we hide a percent of edges and compare the cosine similarity between hidden pairs and negative examples of unconnected documents. We report the Area Under the Roc Curve in Table 2.

Table 1. Comparison of Micro-F1 results on a classification task for different train/test ratios. The best score is in bold, second best is underlined. Execution time order is presented in seconds (Time).

	Cora				Dblp			
train/test ratio	10%	30%	50%	Time	10%	30%	50%	Time
DeepWalk	70.6 (2.0)	77.2 (0.9)	81.0 (0.7)	10^1	52.3 (0.4)	53.4 (0.1)	53.5 (0.2)	10^2
LSA	72.3 (1.9)	79.0 (0.7)	80.6 (0.7)	10^{-2}	73.5 (0.2)	74.1 (0.1)	74.2 (0.2)	10^1
Concatenation	71.4 (2.1)	80.5 (1.0)	84.0 (1.1)	10^1	77.5 (0.2)	78.0 (0.1)	78.2 (0.2)	10^2
TADW	81.9 (0.8)	86.3 (0.8)	87.4 (0.8)	10^{-1}	74.8 (0.1)	75.3 (0.2)	75.5 (0.1)	10^1
AANE	79.8 (0.9)	83.3 (1.1)	84.4 (0.7)	10^{-1}	73.3 (0.1)	73.9 (0.1)	74.2 (0.2)	10^2
GVNR-t	83.7 (1.2)	86.4 (0.7)	87.0 (0.8)	10^1	69.6 (0.1)	70.1 (0.1)	70.2 (0.2)	10^2
VGAE	72.3 (1.7)	79.2 (0.9)	81.1 (0.7)	10^1	Memory overflow			−
G2G	79.0 (1.5)	83.7 (0.8)	84.8 (0.7)	10^1	70.8 (0.1)	71.3 (0.2)	71.5 (0.2)	10^2
STNE	79.4 (1.0)	84.7 (0.7)	86.7 (0.8)	10^2	73.8 (0.2)	74.4 (0.1)	74.5 (0.1)	10^4
RLE	**84.0** (1.3)	**86.9** (0.5)	**87.7** (0.6)	10^1	**79.8** (0.2)	**80.9** (0.2)	**81.2** (0.1)	10^1

	Nyt			
train/test ratio	10%	30%	50%	Time
DeepWalk	66.9 (0.7)	68.2 (0.3)	68.7 (0.9)	10^2
LSA	71.6 (1.0)	75.7 (0.7)	76.7 (0.7)	10^{-2}
Concatenation	**77.9** (0.3)	**80.0** (0.5)	**81.1** (0.7)	10^2
TADW	75.8 (0.5)	78.4 (0.5)	79.4 (0.4)	10^1
AANE	71.7 (0.5)	75.6 (0.8)	76.9 (1.1)	10^1
GVNR-t	74.3 (0.4)	76.0 (0.6)	76.7 (0.6)	10^2
VGAE	68.1 (0.8)	69.3 (0.9)	70.1 (0.6)	10^2
G2G	69.0 (0.5)	70.5 (0.7)	71.5 (0.8)	10^2
STNE	75.1 (0.7)	77.3 (0.5)	78.1 (0.6)	10^2
RLE	77.7 (0.7)	79.3 (0.5)	80.0 (0.6)	10^1

In the classification task, RLE outperforms existing methods on Cora and Dblp, and is the second best method on Nyt. Interestingly, GVNR-t performs well with few training example, while TADW become second with 50% of training examples. Let us highlight that RLE runs fast, it is even faster than AANE on Dblp. Additionally, it is up to four orders of magnitude faster than STNE on Dblp. Additionally Fig. 2 shows that the optimal lambda values are similar for both datasets. Its tuning is not that crucial since RLE outperforms the baselines with $\lambda \in [0.6, 0.85]$ on Cora, $\lambda \in [0.15, 0.85]$ on DBLP, and every methods except Concatenation for $\lambda \in [0.45, 0.8]$ on Nyt.

Table 2. Comparison of AUC results on a link prediction task for different percents of edges hidden. The best score is in bold, second best is underlined.

% edges hidden	Cora		Dblp	
	50%	25%	50%	25%
DeepWalk	73.2 (0.6)	80.9 (1.0)	**89.7** (0.0)	**93.2** (0.2)
LSA	87.4 (0.6)	87.2 (0.8)	54.2 (0.1)	54.8 (0.0)
Concatenation	77.9 (0.3)	83.7 (0.8)	88.8 (0.0)	92.6 (0.3)
TADW	90.1 (0.4)	93.3 (0.4)	61.2 (0.1)	65.0 (0.5)
AANE	83.1 (0.8)	86.6 (0.8)	67.4 (0.1)	66.5 (0.1)
GVNR-t	83.9 (0.9)	91.5 (1.1)	88.1 (0.3)	91.4 (0.1)
VGAE	87.1 (0.4)	88.2 (0.7)	Does not scale	
Graph2Gauss	92.0 (0.3)	93.8 (1.0)	88.0 (0.1)	92.1 (0.5)
STNE	83.1 (0.5)	90.0 (1.0)	45.6 (0.0)	53.4 (0.1)
RLE	**94.3** (0.2)	**94.8** (0.2)	89.3 (0.1)	91.2 (0.2)

In link prediction, RLE outperforms existing methods on Cora, while Deep-Walk yields better results than baselines on Dblp. This might be due to the shortness of the documents (mean length is 6 while it is 49 for Cora): the textual information may not be as informative as the network information for link prediction.

4.2   Qualitative Insights

We compute a vector representation for a class by computing the centroid of the representations of the documents inside this class. We present the closest words to this representation in term of cosine similarity, which provides a general description of the class. In Table 3, we present a description using this method for the first four classes of the Cora Dataset. We also provide most weighted terms when computing the mean of documents $tf \cdot idf$ of the class. The $tf \cdot idf$ method produces too general words, such as "learning", "algorithm" and "model". RLE seems to provide specific words, which makes the descriptions more relevant.

Table 3. Classes description with our method as opposed to $tf \cdot idf$. Words that are repeated across classes are in bold. RLE produces more discriminative descriptions

Cora							
Class 1		Class 2		Class 3		Class 4	
RLE	$tf \cdot idf$	RLE	$tf \cdot idf$	RLE	$tf \cdot idf$	RLE	$tf \cdot idf$
hebbian	neural	reinforcement	**learning**	posterior	bayesian	pac	**learning**
network	network	discounted	reinforcement	gibbs	**model**	learnability	**algorithm**
layers	networks	qlearning	control	bayesian	models	polynomialtime	algorithms
multilayer	**learning**	rl	state	mcmc	**algorithm**	dnf	**model**
filters	**model**	multiagent	policy	sampler	belief	queries	decision

5 Conclusion

In this article, we presented the RLE method for embedding documents that are organized in a network. Despite its simplicity, RLE shows state-of-the art results for the three considered datasets. It is faster than most recent deep-learning methods. Furthermore, it provides informative qualitative insights. Future works will concentrate on automatically tuning λ, and exploring the effect of the similarity matrix S.

References

1. Ailem, M., Salah, A., Nadif, M.: Non-negative matrix factorization meets word embedding. In: Proceedings of the 40th International ACM SIGIR Conference on Research and Development in Information Retrieval, pp. 1081–1084. ACM (2017)
2. Arora, S., Liang, Y., Ma, T.: A simple but tough-to-beat baseline for sentence embeddings. In: International Conference on Learning Representations (2016)
3. Bojchevski, A., Günnemann, S.: Deep Gaussian embedding of graphs: unsupervised inductive learning via ranking. In: Proceeding of the International Conference on Learning Representations. ICLR (2018)
4. Brochier, R., Guille, A., Velcin, J.: Global vectors for node representations. In: Proceedings of the World Wide Web Conference, pp. 2587–2593. WWW (2019)
5. Deerwester, S., Dumais, S.T., Furnas, G.W., Landauer, T.K., Harshman, R.: Indexing by latent semantic analysis. J. Am. Soc. Inf. Sci. **41**(6), 391–407 (1990)
6. Devlin, J., Chang, M.W., Lee, K., Toutanova, K.: BERT: pre-training of deep bidirectional transformers for language understanding. arXiv preprint arXiv:1810.04805 (2018)
7. Huang, X., Li, J., Hu, X.: Accelerated attributed network embedding. In: Proceedings of the SIAM International Conference on Data Mining, pp. 633–641. SDM (2017)
8. Kipf, T.N., Welling, M.: Variational graph auto-encoders. In: Bayesian Deep Learning Workshop, BDL-NeurIPS (2016)
9. Kuzi, S., Shtok, A., Kurland, O.: Query expansion using word embeddings. In: Proceedings of the 25th ACM International Conference on Information and Knowledge Management, pp. 1929–1932. ACM (2016)
10. Le, Q., Mikolov, T.: Distributed representations of sentences and documents. In: International Conference on Machine Learning, pp. 1188–1196 (2014)
11. Liu, J., He, Z., Wei, L., Huang, Y.: Content to node: self-translation network embedding. In: Proceedings of the 24th ACM SIGKDD International Conference on Knowledge Discovery and Data Mining, pp. 1794–1802. ACM (2018)
12. Mikolov, T., Sutskever, I., Chen, K., Corrado, G.S., Dean, J.: Distributed representations of words and phrases and their compositionality. In: Advances in Neural Information Processing Systems, pp. 3111–3119 (2013)
13. Pan, S., Wu, J., Zhu, X., Zhang, C., Wang, Y.: Tri-party deep network representation. In: Proceedings of the International Joint Conference on Artificial Intelligence, pp. 1895–1901. IJCAI (2016)
14. Pennington, J., Socher, R., Manning, C.: GloVe: global vectors for word representation. In: Proceedings of the 2014 Conference on Empirical Methods in Natural Language Processing (EMNLP), pp. 1532–1543 (2014)

15. Perozzi, B., Al-Rfou, R., Skiena, S.: DeepWalk: online learning of social representations. In: Proceedings of the 20th ACM SIGKDD International Conference on
Knowledge Discovery and Data Mining, pp. 701–710. ACM (2014)
16. Seyler, D., Chandar, P., Davis, M.: An information retrieval framework for contextual suggestion based on heterogeneous information network embeddings. In:
The 41st International ACM SIGIR Conference on Research and Development in
Information Retrieval, pp. 953–956. ACM (2018)
17. Tang, J., Zhang, J., Yao, L., Li, J., Zhang, L., Su, Z.: ArnetMiner: extraction
and mining of academic social networks. In: Proceedings of the ACM SIGKDD
International Conference on Knowledge Discovery and Data Mining, pp. 990–998.
KDD (2008)
18. Tu, C., Liu, H., Liu, Z., Sun, M.: CANE: context-aware network embedding for
relation modeling. In: Proceedings of the 55th Annual Meeting of the Association
for Computational Linguistics (Volume 1: Long Papers), vol. 1, pp. 1722–1731
(2017)
19. Wang, S., Tang, J., Aggarwal, C., Liu, H.: Linked document embedding for classification. In: Proceedings of the 25th ACM International on Conference on Information and Knowledge Management, pp. 115–124. ACM (2016)
20. Yang, C., Liu, Z., Zhao, D., Sun, M., Chang, E.Y.: Network representation learning
with rich text information. In: International Joint Conference on Artificial Intelligence (2015)

Supervised Learning Methods for Diversification of Image Search Results

Burak Goynuk and Ismail Sengor Altingovde[(✉)] [ID]

Middle East Technical University, Ankara, Turkey
{burak.goynuk,altingovde}@ceng.metu.edu.tr

Abstract. We adopt a supervised learning framework, namely R-LTR [17], to diversify image search results, and extend it in various ways. Our experiments show that the adopted and proposed variants are superior to two well-known baselines, with relative gains up to 11.4%.

1 Introduction

Diversification of search results is a recent trend employed in various contexts (such as searching the web [11], social media [10], product reviews [8], structured databases [3], etc.), where the user query might be ambiguous/underspecified and/or user satisfaction can be increased by providing results related to the alternative aspects of a query. Image search is one such scenario that can benefit from diversification of results, as the diversification requirement is not only due to the different semantic intents of the queries, but may further stem from the visual properties of the images [9]. For instance, for the query "Hagia Sophia", there may be different photos of this landmark taken in daytime or nighttime, summer or winter, etc., and hence, diversification of results is still required.

In this work, we first apply a recently introduced supervised method to diversify web search results, namely, relational learning to rank (R-LTR) [17], for the diversification of image search results. To this end, we adopt the latter approach to capture both textual and visual diversity of images separately, which is referred to as R-LTR$_{IMG}$. To learn the feature weights, we employ a neural network (NN) framework with back-propagation, which also enables us to explore more general models, i,e., beyond the linear scoring function of R-LTR [17]. In particular, we train a fully connected two-layer NN using the same set of textual and visual features, called as R-LTR$_{IMG-NN}$.

Our second contribution is based on the following observation: R-LTR learns a ranking function based on an iterative selection process, where the diversity of a given document is computed wrt. the previously selected documents, i.e., following the paradigm of the well-known Maximal Marginal Relevance (MMR) diversification [2]. We extend R-LTR with an alternative approach, inspired by the Maximum Marginal Contribution (MMC) idea of [12]. While diversifying a result set, the MMC approach takes into account an upperbound for the *future*

© Springer Nature Switzerland AG 2020
J. M. Jose et al. (Eds.): ECIR 2020, LNCS 12036, pp. 158–165, 2020.
https://doi.org/10.1007/978-3-030-45442-5_20

diversity contribution that can be provided by the document being scored (details provided in Sect. 2). As far as we know, the earlier approaches for supervised diversification (such as [4,13,16,17]) essentially follow the MMR paradigm and hence, ours is the first attempt to *learn* an alternative ranking function.

Our experiments are conducted using the *Div150Cred* dataset employed in the 2014 Retrieving Diverse Social Images Task (of MediaEval Initiative) in a well-crafted framework. For the baseline strategies, MMR and MSD (described in Sec primer), we employed a dynamic feature weighting strategy for higher performance. For all the diversification methods, we used various pre-processing techniques, and employed a particular strategy based on representative images, to better capture the query-image relevance. We show that the adopted R-LTR$_{IMG}$ and its proposed variants outperform MMR and MSD in diversification effectiveness. Furthermore, according to the results reported in the Diversity task of MediaEval (in 2014), our best-performing variant, R-LTR$_{IMG-NN}$, is superior to all but one of the methods explored in this evaluation campaign.

2 Background and Preliminaries

The diversification of web search results based on textual evidence is well-explored in the literature [11]. We focus on so-called implicit diversification approaches that solely rely on features extracted from the documents in the ranking to be diversified. In this section, we review representative implicit strategies that are employed for textual diversification of search results, namely, MMR [2], MSD [5], MMC [12] and Relational-LTR (R-LTR) [17]. The former two approaches, MMR and MSD, have been widely employed in the literature (e.g., [11,14]), and hence, serve as the baselines in our setup. The last one, R-LTR, is a supervised strategy that we adopt and extend in this work.

Maximal Marginal Relevance (MMR) [2]. Given a query q and an initial result set D, MMR constructs a diversified ranking S of size k (typically, $k < |D|$) as follows. At first, the document with the highest relevance score is inserted into S. Then, in each iteration, the document that maximizes Eq. 1 is added to S. The score of a document $d \in D$ is computed as a weighted sum of its relevance to q, denoted as rel(q, d), and its average diversity from the documents that are already selected into the final result set S. Note that, the diversity part of MMR has different variants that employs minimum or maximum diversity wrt. the documents in S, and the version shown in Eq. 1 is based on [12]. λ is a trade-off parameter to balance the relevance and diversity in the final result set S.

$$MMR(d, q, S) = (1 - \lambda) * \text{rel}(q, d) + \frac{\lambda}{|S|} * \sum_{d_i \in S} \text{div}(d, d_i) \qquad (1)$$

Maximum Marginal Contribution (MMC) [12]. This approach is very similar to MMR, but in addition to taking into account the documents already selected in to S, MMC also considers an upperbound on the future diversity, i.e., computed as the contribution of the most diverse l documents (remaining in

D/S) to the current document d. In Eq. 2, the first two components are exactly same as MMR, while the third component captures the highest possible diversity that can be obtained based on d, in case that it is chosen into S.

$$MMC(d, q, S) = (1 - \lambda) * \mathrm{rel}(q, d) + \frac{\lambda}{|S|} * (\sum_{d_i \in S} \mathrm{div}(d, d_i) + \sum_{\substack{l=1 \\ d_j \in D - S - d}}^{k-|S|-1} \mathrm{div}(d, d_j))$$

(2)

Max-Sum Dispersion (MSD) [5]. At each iteration, MSD selects a pair of documents that are most relevant to the query and most diverse from each other. In particular, for all pairs of documents $(d_i, d_j) \in D$, Eq. 3 is calculated, and the pairs with the highest scores are selected until k results are obtained.

$$MSD(d_i, d_j, q) = (1 - \lambda) * (\mathrm{rel}(q, d_i) + \mathrm{rel}(q, d_j)) + 2 * \lambda * \mathrm{div}(d_i, d_j) \quad (3)$$

Relational-Learning to Rank (R-LTR) [17]. This is a supervised method that learns the weights for an MMR-like diversification approach using Stochastic Gradient Descent (SGD). Instead of relying on a single relevance and diversity score as in the aforementioned approaches, R-LTR computes multiple scores for each component and combines them using weight vectors, which are learnt over a training set.

$$\mathrm{R\text{-}LTR}(d_i, R_i, S) = \omega_r * \mathbf{x_i} + \omega_d * h_S(R_i) \quad (4)$$

Following the notation in [17], in Eq. 4, $\mathbf{x_i}$ denotes a relevance feature vector, i.e., a vector of $\mathrm{rel}(q, d)$ scores that are likely to be computed by different methods (e.g., tf-idf, BM25, etc.), while R_i is a matrix capturing the diversity scores of d_i to all other documents in D, again computed by various methods. Note that, R is a 3-way tensor that stores the relation, namely, diversity score, between each pair of documents in D computed using t different $\mathrm{div}(d_i, d_j)$ methods (e.g., body text diversity, title text diversity, anchor text diversity, etc.).

In this case, the function $h_S(R_i)$ is used to compute the aggregated diversity of d_i from the documents that are already in S, for each of these t diversity methods. As in the case of MMR, for a given diversity computation method, the aggregated diversity score between d and S can be computed using average, min or max function, and $h_S(R_i)$ will return the so-called diversity feature vector of t scores computed by using the selected aggregation function.

The ground truth ranking for a query is constructed in a greedy way, i.e., by choosing the document that maximizes a diversification metric, say SubTopic-recall, at each step, which is based on the document's relevance judgments (see [17] for the details). During training, in each iteration, the document chosen (i.e., maximizing Eq. 4) is compared to the document at the corresponding position in the ground truth so that the likelihood loss can be computed. Then, model parameters (i.e., ω_r and ω_d) are updated using SGD until the loss converges to a pre-defined value.

We are aware of a previous work [4] that has also exploited R-LTR for image diversification in a similar setup, i.e., MediaEval evaluation campaign. Our work differs from the latter in three ways: First, we implement R-LTR using a neural network framework with back-propagation, which allows us to train more general models, namely, a two-layer neural network, and hence, to go beyond the linear scoring function of R-LTR. Second, we extend R-LTR and propose a new variant that learns the MMC ranking function instead of the MMR. Third, we compare R-LTR variants to two baseline approaches with carefully tuned parameters (to optimize their performance), while the previous work reports only the results of a direct application of R-LTR.

3 Image Diversification Framework: R-LTR$_{\text{IMG}}$

In this paper, as in the Diversity task of MediaEval [6], we assume that a textual query is submitted to an image search engine, where each image is associated with textual metadata, and an initial result list D is retrieved. The goal is to obtain a ranking S that is both relevant to the query and including diverse images. The diversity of two images, again denoted as d_i and d_j, can be computed using textual and/or visual features. Therefore, we first adopt the R-LTR scoring function to separately capture these different types of diversity scores, as follows:

$$\text{R-LTR}_{\text{IMG}}(d_i, RT_i, RV_i, S) = \omega_r * \mathbf{x_i} + \omega_{textDiv} * h_S(RT_i) + \omega_{visDiv} * h_S(RV_i) \quad (5)$$

In Eq. 5, the 3-way tensors RT and RV store the pairwise image diversity scores based on the textual and visual diversity, respectively. This adopted version, referred to as R-LTR$_{\text{IMG}}$, also allows using different aggregation (h_S) functions for different types of diversity scores.

As a further extension, instead of considering an MMR-style approach in R-LTR$_{\text{IMG}}$, which only takes into account the diversity wrt. the images that are already in S, we apply the philosophy of aforementioned MMC approach. More specifically, for a given image d_i to be scored, we also compute the upperbound of the diversity that can be brought to S afterwards, i.e., if d_i is selected. To the best of our knowledge, earlier works on supervised approaches for implicit diversification are based on MMR, and ours is the first attempt to *learn* a framework that considers both the images selected into S and those *to be inserted* into S.

In Eq. 6, the last two components address the textual and visual diversity of d_i with respect to l images that are most dissimilar to it in the set of *remaining* images $U = D - S - d_i$. We refer to this version as R-LTR$_{\text{IMG-MMC}}$.

$$\text{R-LTR}_{\text{IMG-MMC}}(d_i, RT_i, RV_i, S, U) = \omega_r * \mathbf{x_i} + \omega_{textDiv} * h_S(RT_i) + \omega_{visDiv} * h_S(RV_i)$$
$$+ \omega_{textDivNext} * h_U(RT_i, l) + \omega_{visDivNext} * h_U(RV_i, l) \quad (6)$$

Finally, we implement these variants using PyTorch's neural network framework with back-propagation. This choice enables us to train more general models that can go beyond the linear scoring function of R-LTR. In particular, we implement a scoring function (taking the same input as Eq. 5) using a fully connected two-layer neural network and refer to this version as R-LTR$_{IMG-NN}$. The source codes are available at https://github.com/burakgoynuk/rltr-img.

4   Evaluation Setup and Results

Dataset. We employed *Div150Cred* dataset that has been validated in the 2014 Retrieving Diverse Social Images Task (of MediaEval Benchmarking Initiative) [6,7]. This dataset includes 45,375 images of around 150 landmark locations (e.g., Hagia Sophia) shared in Flickr. Each such location is considered as a query, for which the dataset provides the location's GPS coordinates, the link to its Wikipedia web page, some descriptive photos from Wikipedia and a ranked set of around 300 images retrieved from Flickr (each associated with various textual and visual features, described later). For the queries and retrieved images, relevance and diversity judgments (by human annotators) are also made available.

We use the default training and test splits as provided in the aforementioned evaluation campaign. Specifically, 30 locations (together with the retrieved images, all metadata and relevance judgments) are used for training, and 123 locations are held as the test set. Note that, while testing, we diversify the top-100 images retrieved for a query, as earlier works imply that going deeper in the ranking increases the likelihood of irrelevant results (e.g., [1]).

Following the common practice (e.g., [1]), we pre-processed all the images in training and test sets to reduce the noise. We first removed the images that may include people, using the Open-CV's built-in face detection algorithms. Secondly, the GPS coordinate of the queried location is compared to that of an image in the result set, and it is discarded if the distance is greater than 10 km.

Computing the Relevance Scores. To compute the $rel(q,d)$, we employ a strategy based on the *representative image* idea that is widely employed in the literature (e.g., [9]). This strategy aims to go beyond the textual relevance and make use of the visual features, even when the query is expressed textually. To train a model, we consider the top-ranked answer from Flickr as the representative image, i.e., the ground truth. Then, a neural network is trained with three basic features for each (query, image) pair, as follows: BM25 score (between query and image's textual metadata), the GPS distance of the query and image, and visual similarity between the Wikipedia image of the location and image in the result set. For all diversification methods employed here, we first automatically identify the representative image using the trained model, and obtain the $rel(q,d)$ scores based on visual features, described next.

Computing the Diversity Scores. We use both textual and visual features to compute various types of diversity scores between a pair of images, $div(d_i, d_j)$.

Using the textual features (i.e., image metadata), we compute two types of diversity scores, namely, the tf-idf weighted cosine similarity and Jaccard Coefficient, both of which are in the range $[0, 1]$. In terms of visual features, we consider four types of descriptors per image, namely, Global Histogram of Oriented Gradients, Global Color Structure Descriptor, Global Color Naming Histogram, and Global Color Moments on HSV Color Space (the other descriptors in the dataset are found to be unhelpful in our preliminary experiments on the training set and hence, discarded.). Thus, we compute four different types of diversity score between images, corresponding to each feature type. In particular, the diversity score based on the former two feature types are computed using Euclidean distance, while Cosine distance is found to work better for the latter two feature types. Note that, these scores (actually, their difference from 1) are also used to compute the relevance between a representative image and result image.

Baseline Diversification Methods. As the baselines, we employ MMR and MSD. For both methods, we again compute relevance scores based on the representative images, and diversity scores using the features described before. These scores are weighted using the dynamic feature weighting approach of [9] shown in Eq. 7. In particular, for each textual or visual feature $f \in F$, this approach weights the diversity score by θ_f, which denotes the variance of all diversity scores wrt. f. The trade-off parameter λ is learned over the training set.

$$div(d_i, d_j) = \frac{1}{|F|} * \sum_{f \in F} (\frac{1}{\theta_f^2} * \mathrm{div}_f(d_i, d_j)) \tag{7}$$

Parameters for R-LTR Variants. We implement all R-LTR variants using PyTorch's neural network framework with back-propagation and negative log likelihood loss (as in [13]). We set the number of epochs as 300 and learning rate 0.00001 based on the experiments on training data. For R-LTR$_{\text{IMG-MMC}}$, we set the parameter l in Eq. 6 as 1, i.e., we consider the diversity impact of the farthest image to the current one as an upperbound. For R-LTR$_{\text{IMG-NN}}$, the hidden layer has 3 nodes each with a sigmoid activation function, and a single output node combines the results of the hidden nodes. The ground truth ranking is obtained

Table 1. Diversification performance of baseline and proposed approaches. The symbol (*) denotes stat. significance wrt. MMR using paired t-test (at 0.05 confidence level).

Diversification method	$\alpha-$nDCG@20	ST-recall@20
Flickr (original ranking)	0.573	0.342
MSD	0.617	0.369
MMR	0.654	0.413
R-LTR$_{\text{IMG-MMC}}$	0.667	0.425
R-LTR$_{\text{IMG}}$	0.691*	0.455*
R-LTR$_{\text{IMG-NN}}$	**0.695***	**0.460***

by greedily selecting the image that maximizes the ST-recall metric (see [17] for details), which was among the official metrics for MediaEval task.

Results. We report the SubTopic-recall and α-nDCG scores at cut-off value of 20. Table 1 compares the diversification performance of Flickr's original ranking, MMR and MSD to the proposed R-LTR variants. We see that, as expected, non-diversified Flickr ranking has the lowest performance, and among the two traditional baselines, MMR is better. The proposed R-LTR$_{IMG-MMC}$ approach outperforms MMR, with the relative gains of 2% and 3% in terms of α-nDCG and ST-recall metrics, respectively. However, it is still inferior to R-LTR$_{IMG}$ and its variant, R-LTR$_{IMG-NN}$. Specifically, R-LTR$_{IMG}$ provides a relative improvement of 10.2% (5.7%) over the best baseline, MMR, in terms of the ST-recall (α-nDCG) metrics, respectively. R-LTR$_{IMG-NN}$ is the overall winner with a relative gain of 11.4% (1.1%) over MMR (R-LTR$_{IMG}$) in terms of ST-recall, respectively.

Comparison to Diversity 2014 Task Results at MediaEval. Among 14 participants of the Diversity task, 10 of them have submitted a run employing both textual and visual features, as we do here. Their median score for ST-recall is 0.4191 and 9 out of these 10 runs report a score less than 0.45, i.e., inferior to R-LTR$_{IMG}$ and R-LTR$_{IMG-NN}$. The run outperforming the R-LTR variants achieves a score of 0.473 [15], but they exploit additional features that are not provided in the dataset. Indeed, most submitted runs derive new features and/or employ different pre-processing techniques; so there is still a need for evaluating these methods within a common framework, which is left as a future work.

Concluding Summary. We adopted and extended a supervised learning framework to diversify image search results and showed that the proposed variants are superior to two well-known baselines. As our future work, we aim to train our models by directly optimizing the evaluation metrics, as suggested in [16].

Acknowledgements. This work is partially funded by The Scientific and Technological Research Council of Turkey (TÜBİTAK) grant 117E861 & TÜBA GEBIP award.

References

1. Boteanu, B., Mironică, I., Ionescu, B.: Pseudo-relevance feedback diversification of social image retrieval results. Multimed. Appl. **76**(9), 11889–11916 (2016). https://doi.org/10.1007/s11042-016-3678-6
2. Carbonell, J.G., Goldstein, J.: The use of MMR, diversity-based reranking for reordering documents and producing summaries. In: Proceedings of SIGIR, pp. 335–336 (1998)
3. Demidova, E., Fankhauser, P., Zhou, X., Nejdl, W.: DivQ: diversification for keyword search over structured databases. In: Proceedings of SIGIR, pp. 331–338 (2010)
4. Dudy, S., Bedrick, S.: OHSU @ mediaeval 2015: adapting textual techniques to multimedia search. In: Working Notes of the MediaEval 2015 Workshop (2015)

5. Gollapudi, S., Sharma, A.: An axiomatic approach for result diversification. In: Proceedings of WWW, pp. 381–390 (2009)
6. Ionescu, B., Gînsca, A.L., Boteanu, B., Popescu, A., Lupu, M., Müller, H.: Retrieving diverse social images at mediaeval 2014: challenge, dataset and evaluation. MediaEval **1263** (2014)
7. Ionescu, B., Popescu, A., Lupu, M., Gînscă, A.L., Boteanu, B., Müller, H.: Div150cred: a social image retrieval result diversification with user tagging credibility dataset. In: Proceedings of the 6th ACM Multimedia Systems Conference, pp. 207–212. ACM (2015)
8. Krestel, R., Dokoohaki, N.: Diversifying customer review rankings. Neural Netw. **66**, 36–45 (2015)
9. van Leuken, R.H., Pueyo, L.G., Olivares, X., van Zwol, R.: Visual diversification of image search results. In: Proceedings of WWW, pp. 341–350 (2009)
10. Onal, K.D., Altingovde, I.S., Karagoz, P.: Utilizing word embeddings for result diversification in tweet search. In: Proceedings of AIRS, pp. 366–378 (2015)
11. Santos, R.L.T., Macdonald, C., Ounis, I.: Search result diversification. Found. Trends Inf. Retr. **9**(1), 1–90 (2015)
12. Vieira, M.R., et al.: On query result diversification. In: Proceedings of ICDE, pp. 1163–1174 (2011)
13. Xia, L., Xu, J., Lan, Y., Guo, J., Cheng, X.: Modeling document novelty with neural tensor network for search result diversification. In: Proc. of SIGIR. pp. 395–404 (2016)
14. Xioufis, E.S., Papadopoulos, S., Gînsca, A., Popescu, A., Kompatsiaris, Y., Vlahavas, I.P.: Improving diversity in image search via supervised relevance scoring. In: Proceedings of International Conference on Multimedia Retrieval, pp. 323–330 (2015)
15. Xioufis, E.S., Papadopoulos, S., Kompatsiaris, Y., Vlahavas, I.P.: Socialsensor: finding diverse images at mediaeval 2014. In: Working Notes of the MediaEval 2014 Workshop (2014)
16. Xu, J., Xia, L., Lan, Y., Guo, J., Cheng, X.: Directly optimize diversity evaluation measures: a new approach to search result diversification. ACM TIST **8**(3), 41:1–41:26 (2017)
17. Zhu, Y., Lan, Y., Guo, J., Cheng, X., Niu, S.: Learning for search result diversification. In: Proceedings of SIGIR, pp. 293–302 (2014)

ANTIQUE: A Non-factoid Question Answering Benchmark

Helia Hashemi[1]([⊠]), Mohammad Aliannejadi[2], Hamed Zamani[1],
and W. Bruce Croft[1]

[1] Center for Intelligent Information Retrieval, University of Massachusetts Amherst,
Amherst, MA 01003, USA
{hhashemi,zamani,croft}@cs.umass.edu
[2] Information and Language Processing Systems (ILPS), University of Amsterdam,
Science Park 904, 1098 XH Amsterdam, The Netherlands
m.aliannejadi@uva.nl

Abstract. Considering the widespread use of mobile and voice search, answer passage retrieval for non-factoid questions plays a critical role in modern information retrieval systems. Despite the importance of the task, the community still feels the significant lack of large-scale non-factoid question answering collections with real questions and comprehensive relevance judgments. In this paper, we develop and release a collection of 2,626 open-domain non-factoid questions from a diverse set of categories. The dataset, called ANTIQUE, contains 34k manual relevance annotations. The questions were asked by real users in a community question answering service, i.e., Yahoo! Answers. Relevance judgments for all the answers to each question were collected through crowdsourcing. To facilitate further research, we also include a brief analysis of the data as well as baseline results on both classical and neural IR models.

1 Introduction

With the rising popularity of information access through devices with small screens, e.g., smartphones, and voice-only interfaces, e.g., Amazon's Alexa and Google Home, there is a growing need to develop retrieval models that satisfy user information needs with sentence-level and passage-level answers. This has motivated researchers to study answer sentence and passage retrieval, in particular in response to *non-factoid* questions [1,18]. Non-factoid questions are defined as open-ended questions that require complex answers, like descriptions, opinions, or explanations, which are mostly passage-level texts. Questions like "How to cook burgers?" are non-factoid. We believe this type of questions plays a pivotal role in the overall quality of question answering systems, since their

---

M. Aliannejadi—Work done while affiliated with Università della Svizzera italiana (USI), Switzerland.
Hamed Zamani is currently affiliated with Microsoft.

J. M. Jose et al. (Eds.): ECIR 2020, LNCS 12036, pp. 166–173, 2020.
https://doi.org/10.1007/978-3-030-45442-5_21

technologies are not as mature as those for factoid questions, which seek precise facts, such as "At what age did Rossini stop writing opera?".

Despite the widely-known importance of studying answer passage retrieval for non-factoid questions [1,2,8,18], the research progress for this task is limited by the availability of high-quality public data. Some existing collections, e.g., [8,13], consist of few queries, which are not sufficient to train sophisticated machine learning models for the task. Some others, e.g., [1], significantly suffer from incomplete judgments. Most recently, Cohen et al. [3] developed a publicly available collection for non-factoid question answering with a few thousands questions, which is called WikiPassageQA. Although WikiPassageQA is an invaluable contribution to the community, it does not cover all aspects of the non-factoid question answering task and has the following limitations: (i) it only contains an average of 1.7 relevant passages per question and does not cover many questions with multiple correct answers; (ii) it was created from the Wikipedia website, containing only formal text; (iii) more importantly, the questions in the WikiPassageQA dataset were generated by crowdworkers, which is different from the questions that users ask in real-world systems; (iv) the relevant passages in WikiPassageQA contain the answer to the question in addition to some surrounding text. Therefore, some parts of a relevant passage may not answer any aspects of the question; (v) it only provides binary relevance labels.

To address these shortcomings, in this paper, we create a novel dataset for non-factoid question answering research, called *ANTIQUE*, with a total of 2,626 questions. In more detail, we focus on the non-factoid questions that have been asked by users of Yahoo! Answers, a community question answering (CQA) service. Non-factoid CQA data without relevance annotation has been previously used in [1], however, as mentioned by the authors, it significantly suffers from incomplete judgments (see Sect. 2 for more information on existing collections). We collected four-level relevance labels through a careful crowdsourcing procedure involving multiple iterations and several automatic and manual quality checks. Note that we paid extra attention to collect reliable and comprehensive relevance judgments for the test set. Therefore, we annotated the answers after conducting result pooling among several term-matching and neural retrieval models. In summary, ANTIQUE provides annotations for 34,011 question-answer pairs, which is significantly larger than many comparable datasets.

We further provide brief analysis to uncover the characteristics of ANTIQUE. Moreover, we conduct extensive experiments with ANTIQUE to present benchmark results of various methods, including classical and neural IR models on the created dataset, demonstrating the unique challenges ANTIQUE introduces to the community. To foster research in this area, we release ANTIQUE.[1]

---

[1] https://ciir.cs.umass.edu/downloads/Antique/.

2   Existing Related Collections

Factoid QA Datasets. TREC QA [14] and WikiQA [17] are examples of factoid QA datasets whose answers are typically brief and concise facts, such as named entities and numbers. InsuranceQA [5] is another factoid dataset in the domain of insurance. ANTIQUE, on the other hand, consists of open-domain non-factoid questions that require explanatory answers. The answers to these questions are often passage level, which is contrary to the factoid QA datasets.

Non-factoid QA Datasets. There have been efforts for developing non-factoid question answering datasets [7,8,16]. Keikha et al. [8] introduced the WebAP dataset, which is a non-factoid QA dataset with 82 queries. The questions and answers in WebAP were not generated by real users. There exist a number of datasets that partially contain non-factoid questions and were collected from CQA websites, such as Yahoo! Webscope L6, Qatar Living [9], and StackExchange. These datasets are often restricted to a specific domain, suffer from incomplete judgments, and/or do not contain sufficient non-factoid questions for training sophisticated machine learning models. The nfL6 dataset [1] is a collection of non-factoid questions extracted from the Yahoo! Webscope L6. Its main drawback is the absence of complete relevance annotation. Previous work assumes that the only answer that the question writer has marked as correct is relevant, which is far from being realistic. That is why we aim to collect a complete set of relevance annotations. WikiPassageQA is another non-factoid QA dataset that has been recently created by Cohen et al. [3]. As mentioned in Sect. 1, despite its great potentials, it has a number of limitations. ANTIQUE addresses these limitations to provide a complementary benchmark for non-factoid question answering (see Sect. 1). More recently, Microsoft has released the MS MARCO V2.1 passage re-ranking dataset [10], containing a large number of queries sampled from the Bing search engine. In addition to not being specific to non-factoid QA, it significantly suffers from incomplete judgments. In contrast, ANTIQUE provides a reliable collection with complete relevance annotations for evaluating non-factoid QA models.

3   Data Collection

Following Cohen et al. [1], we used the publicly available dataset of non-factoid questions collected from the Yahoo! Webscope L6, called nfL6. We conducted the following steps for pre-processing and question sampling: (i) questions with less than 3 terms were omitted (excluding punctuation marks); (ii) questions with no best answer ($\hat{a}$) were removed; (iii) duplicate or near-duplicate questions were removed. We calculated term overlap between questions and from the questions with more than 90% term overlap, we only kept one, randomly; (iv) we omitted the questions under the categories of "Yahoo! Products" and "Computers & Internet" since they are beyond the expertise of most workers; (v) From the remaining data, we randomly sampled 2,626 questions (out of 66,634).

Each question q in nfL6 corresponds to a list of answers named 'nbest answers', which we denote with $\mathcal{A} = \{a_1, \ldots, a_n\}$. For every question, one answer is marked by the question author on the community web site as the best answer, denoted by $\hat{a}$. It is important to note that as different people have different information needs, this answer is not necessarily the best answer to the question. Also, many relevant answers have been added after the user has chosen the correct answer. Nevertheless, in this work, we respect the users' explicit feedback, assuming that the candidates selected by the actual user are relevant to the query. Therefore, we do not collect relevance assessments for those answers.

3.1   Relevance Assessment

We created a Human Intelligence Task (HIT) on Amazon Mechanical Turk, in which we presented workers with a question-answer pair, and instructed them to annotate the answer with a label between 1 to 4. The instructions started with a short introduction to the task and its motivations, followed by detailed annotation guidelines. Since workers needed background knowledge for answering the majority of the questions, we also included $\hat{a}$ in the instructions and called it a "possibly correct answer." In some cases, the question is very subjective and could have multiple correct answers. This is why it is called "possibly correct answer" to make it clear in the instructions that other answers could potentially be different from the provided answer, but still be correct.

Table 1. Statistics of ANTIQUE.

# training (test) questions: 2,426 (200)	# label 4: 13,067	# total workers: 577
# training (test) answers: 27,422 (6,589)	# label 3: 9,276	# total judgments: 148,252
average question length: 10.51	# label 2: 8,754	# rejected judgments: 17,460
average answer length: 47.75	# label 1: 2,914	% of rejections: 12%

Label Definitions. To facilitate the labeling procedure, we described labels in the form of a flowchart to users. Our aim was to preserve the notion of relevance in QA systems as we discriminate it with the typical topical relevance definition in ad-hoc retrieval tasks. The definition of each label is as follows: **Label 4:** It looks reasonable and convincing. Its quality is on par with or better than the "Possibly Correct Answer". Note that it does not have to provide the same answer as the "Possibly Correct Answer". **Label 3:** It can be an answer to the question, however, it is not sufficiently convincing. There should be an answer with much better quality for the question. **Label 2:** It does not answer the question or if it does, it provides an unreasonable answer, however, it is not out of context. Therefore, you cannot accept it as an answer to the question. **Label 1:** It is completely out of context or does not make any sense.

We included 15 diverse examples of annotated QA pairs with explanation of why and how the annotations were done. Overall, we launched 7 assignment batches, appointing 3 workers to each QA pair. In cases where the workers could agree on a label (i.e., majority vote), we considered the label as the ground truth. We then added all QA pairs with no agreement to a new batch and performed a second round of annotation. It is interesting to note that the ratio of pairs with no agreement was nearly the same among the 7 batches (∼13%). In the very rare cases of no agreement after two rounds of annotation (776 pairs), an expert annotator decided on the final label. To allow further analysis, we have added a flag in the dataset identifying the answers annotated by the expert annotator. In total, the annotation task costed 2,400 USD.

Quality Check. To ensure the quality of the data, we limited the HIT to the workers with over 98% approval rate, who have completed at least 5,000 assignments. 3% of QA pairs were selected from a set of quality check questions with obviously objective labels. It enabled us to identify workers who did not provide high-quality labels. Moreover, we recorded the click log of the workers to detect any abnormal behavior (e.g., employing automatic labeling scripts) that would affect the quality of the data. Finally, we constantly performed manual quality checks by reading the QA pairs and their respective labels. The manual inspection was done on the 20% of each worker's submission as well as the QA pairs with no agreement.

Training Set. In the training set, we annotate the list $\mathcal{A}$ (see Sect. 3) for each query, and assume that for each question, answers to the other questions are irrelevant. As we removed similar questions from the dataset, this assumption is fair. To test this assumption, we sampled 100 questions from the filtered version of nfL6 and annotated the top 10 results retrieved by BM25 using the same crowdsourcing procedure. The results showed that only 13.7% of the documents (excluding $\mathcal{A}$) were annotated as relevant (label 3 or 4). This error rate can be tolerated in the training process as it enables us to collect significantly larger amount of training labels. On the other hand, for the test set we performed pooling to label all possibly relevant answers. In total, the ANTIQUE's training set contains 27,422 answer annotations as it shown in Table 1, that is 11.3 annotated candidate answers per training question, which is significantly larger than its similar datasets, e.g., WikiPassageQA [3].

Test Set. The test set in ANTIQUE consists of 200 questions which were randomly sampled from nfL6 after pre-processing and filtering. Statistics of the test set can be found in Table 1. The set of candidate questions for annotation was selected by performing depth-k ($k = 10$) pooling. To do so, we considered the union of the top k results of various retrieval models, including term-matching and neural models (listed in Table 2). We took the union of this set and "nbest answers" (set $\mathcal{A}$) for annotation.

4   Data Analysis

Here, we present a brief analysis of ANTIQUE to highlight its characteristics.

Statistics of ANTIQUE. Table 1 lists general statistics of ANTIQUE. As we see, ANTIQUE consists of 2,426 non-factoid questions that can be used for training, followed by 200 questions as a test set. Furthermore, ANTIQUE contains 27.4k and 6.5k annotations (judged answers) for the train and test sets, respectively. We also report the total number of answers with specific labels.

Workers Performance. Overall, we launched 7 crowdsourcing batches to collect ANTIQUE. This allowed us to identify and ban less accurate workers. As reported in Table 1, a total number of 577 workers made over 148k annotations (257 per worker), out of which we rejected 12% because they failed to satisfy the quality criteria.

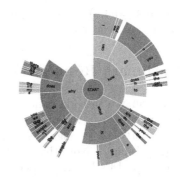

Fig. 1. Distribution of the top trigrams of ANTIQUE questions.

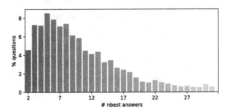

Fig. 2. Distribution of the length of $\mathcal{A}$ (i.e., nbest answers) per question.

Questions Distribution. Figure 1 shows how questions are distributed in ANTIQUE by reporting the top 40 starting trigrams of the questions. As shown in the figure, majority of the questions start with "how" and "why," constituting 38% and 36% of the questions, respectively. It is notable that, according to Fig. 1, a considerable number of questions start with "how do you," "how can you," "what do you," and "why do you," suggesting that their corresponding answers would be highly subjective and opinion based. Also, we can see a major fraction of questions start with "how can I" and "how do I," indicating the importance and dominance of personal questions.

Table 2. The benchmark results for a wide variety of retrieval models on ANTIQUE.

Method	MAP	MRR	P@1	P@3	P@10	nDCG@1	nDCG@3	nDCG@10
BM25	0.1977	0.4885	0.3333	0.2929	0.2485	0.4411	0.4237	0.4334
DRMM-TKS [6]	0.2315	0.5774	0.4337	0.3827	0.3005	0.4949	0.4626	0.4531
aNMM [15]	0.2563	0.6250	0.4847	0.4388	0.3306	0.5289	0.5127	0.4904
BERT [4]	**0.3771**	**0.7968**	**0.7092**	**0.6071**	**0.4791**	**0.7126**	**0.6570**	**0.6423**

Answers Distribution. Finally, in Fig. 2, we plot the distribution for the number of 'nbest answers' ($|\mathcal{A}|$). We see that the majority of questions have 9 or less nbest answers (=54%) and 82% of questions have 14 or less nbest answers. The distribution, however, has a long tail which is not shown in the figure.

5   Benchmark Results

In this section, we provide benchmark results on the ANTIQUE dataset. We report the results for a wide range of retrieval models in Table 2. In this experiment, we report a wide range of standard precision- and recall-oriented retrieval metrics (see Table 2). Note that for the metrics that require binary labels (i.e., MAP, MRR, and P@k), we assume that the labels 3 and 4 are relevant, while 1 and 2 are non-relevant. Due to the definition of our labels (see Sect. 3), we recommend this setting for future work. For nDCG, we use the four-level relevance annotations (we mapped our 1 to 4 labels to 0 to 3).

As shown in the table, the neural models significantly outperform BM25, an effective term-matching retrieval model. Among all, BERT [4] provides the best performance. Recent work on passage retrieval also made similar observations [11,12]. Since MAP is a recall-oriented metric, the results suggest that all the models still fail at retrieving all relevant answers. There is still a large room for improvement, in terms of both precision- and recall-oriented metrics.

6   Conclusions

This paper introduced ANTIQUE; a non-factoid question answering dataset. The questions in ANTIQUE were sampled from a wide range of categories on Yahoo! Answers, a community question answering service. We collected four-level relevance annotations through a multi-stage crowdsourcing as well as expert annotation. In summary, ANTIQUE consists of 34,011 QA-pair relevance annotations for 2,426 and 200 questions in the training and test sets, respectively. Additionally, we reported the benchmark results for a set of retrieval models, ranging from term-matching to recent neural ranking models, on ANTIQUE. Our data analysis and retrieval experiments demonstrated that ANTIQUE introduces unique challenges while fostering research for non-factoid question answering.

Acknowledgement. This work was supported in part by the Center for Intelligent Information Retrieval and in part by NSF IIS-1715095. Any opinions, findings and conclusions or recommendations expressed in this material are those of the authors and do not necessarily reflect those of the sponsor.

References

1. Cohen, D., Croft, W.B.: End to end long short term memory networks for non-factoid question answering. In: ICTIR 2016, pp. 143–146 (2016)

2. Cohen, D., Croft, W.B.: A hybrid embedding approach to noisy answer passage retrieval. In: Pasi, G., Piwowarski, B., Azzopardi, L., Hanbury, A. (eds.) ECIR 2018. LNCS, vol. 10772, pp. 127–140. Springer, Cham (2018). https://doi.org/10.1007/978-3-319-76941-7_10

3. Cohen, D., Yang, L., Croft, W.B.: WikiPassageQA: a benchmark collection for research on non-factoid answer passage retrieval. In: SIGIR 2018 (2018)

4. Devlin, J., Chang, M.W., Lee, K., Toutanova, K.: BERT: pre-training of deep bidirectional transformers for language understanding. CoRR (2018)

5. Feng, M., Xiang, B., Glass, M.R., Wang, L., Zhou, B.: Applying deep learning to answer selection: a study and an open task. CoRR (2015)

6. Guo, J., Fan, Y., Ai, Q., Croft, W.B.: A deep relevance matching model for ad-hoc retrieval. In: CIKM 2016 (2016)

7. Habernal, I., et al.: New collection announcement: Focused retrieval over the web. In: SIGIR 2016 (2016)

8. Keikha, M., Park, J., Croft, W.B.: Evaluating answer passages using summarization measures. In: SIGIR 2014, pp. 963–966 (2014)

9. Nakov, P., et al.: SemEval-2017 task 3: community question answering. In: SemEval 2017, pp. 27–48 (2017)

10. Nguyen, T., et al.: MS MARCO: a human generated machine reading comprehension dataset. CoRR abs/1611.09268 (2016)

11. Nogueira, R., Cho, K.: Passage re-ranking with BERT. CoRR abs/1901.04085 (2019)

12. Padigela, H., Zamani, H., Croft, W.B.: Investigating the successes and failures of BERT for passage re-ranking. CoRR abs/1903.06902 (2019)

13. Shah, C., Pomerantz, J.: Evaluating and predicting answer quality in community QA. In: SIGIR 2010 (2010)

14. Wang, M., Smith, N.A., Mitamura, T.: What is the Jeopardy model? A quasi-synchronous grammar for QA. In: EMNLP 2007 (2007)

15. Yang, L., Ai, Q., Guo, J., Croft, W.B.: aNMM: ranking short answer texts with attention-based neural matching model. In: CIKM 2016, pp. 287–296 (2016)

16. Yang, L., et al.: Beyond factoid QA: effective methods for non-factoid answer sentence retrieval. In: Ferro, N., et al. (eds.) ECIR 2016. LNCS, vol. 9626, pp. 115–128. Springer, Cham (2016). https://doi.org/10.1007/978-3-319-30671-1_9

17. Yang, Y., Yih, S.W., Meek, C.: WikiQA: a challenge dataset for open-domain question answering. In: ACL 2015 (2015)

18. Yulianti, E., Chen, R., Scholer, F., Croft, W.B., Sanderson, M.: Document summarization for answering non-factoid queries. In: TKDE (2018)

Neural Query-Biased Abstractive Summarization Using Copying Mechanism

Tatsuya Ishigaki[1(✉)], Hen-Hsen Huang[3], Hiroya Takamura[1,4], Hsin-Hsi Chen[2], and Manabu Okumura[1]

[1] Tokyo Institute of Technology, Tokyo, Japan
ishigaki@lr.pi.titech.ac.jp, {takamura,oku}@pi.titech.ac.jp
[2] National Taiwan University, Taipei, Taiwan
hhchen@ntu.edu.tw
[3] National Chengchi University, Taipei, Taiwan
hhhuang@nccu.edu.tw
[4] AIST, Tokyo, Japan

Abstract. This paper deals with the query-biased summarization task. Conventional non-neural network-based approaches have achieved better performance by primarily including the words overlapping between the source and the query in the summary. However, recurrent neural network (RNN)-based approaches do not explicitly model this phenomenon. Therefore, we model an RNN-based query-biased summarizer to primarily include the overlapping words in the summary, using a copying mechanism. Experimental results, in terms of both automatic evaluation with ROUGE and manual evaluation, show that the strategy to include the overlapping words also works well for neural query-biased summarizers.

Keywords: Abstractive summarization · Query-biased summarization

1 Introduction

A query-biased summarizer takes a query in addition to a source document as an input, and outputs a summary with respect to the query, as in Table 1. The generated summaries are intended to be used, for example, for snippets as the results of search engines. Query-biased summarization has been studied for decades [3,4,16,18]. Conventional approaches are mostly extractive, and often use the overlapping words as cues to calculate the salience score of a sentence [14,16,18].

On the other hand, recurrent neural network (RNN)-based approaches have enabled summarizers to generate fluent abstractive summaries [1,7,13], but do not explicitly model the strategy to primarily include the overlapping words. In this paper, therefore, we incorporate this strategy into RNN-based summarizers using copying mechanisms.

© Springer Nature Switzerland AG 2020
J. M. Jose et al. (Eds.): ECIR 2020, LNCS 12036, pp. 174–181, 2020.
https://doi.org/10.1007/978-3-030-45442-5_22

Table 1. Example of a source document, a query, the gold summary. The words overlapping between the source and query are shown in **bold**.

Source:	**Vigilanteism** simply causes more problems and will not fix the original problem. one should restore law and order rather than implementing disorder
Query:	Will **vigilanteism** restore law and order?
Gold Summary:	**Vigilanteism** merely instigates chaos

A copying mechanism is a network to primarily include the words in the source document in the summary [5,6,17]. To achieve this, the copying mechanism increases the probability of including the words in the source document. A copying mechanism can be seen as an extension of the *pointer-network* [19], which only copies words in the input and does not output words other than in the input. Gu et al. [5], Gulcehre et al. [6], and Miao et al. [12] extended the pointer-network to copying mechanisms by using a function to balance *copying* and *generation*. See et al. [17] and Chen and Lapata [2] applied the copying mechanism to single-document summarization tasks without a query. We came up with an idea of using copying mechanisms to include the overlapping words in the summary. However, the copying mechanisms were originally designed for the settings without the query information, and it is not necessarily clear how we can integrate the mechanisms into a query-biased summarizer.

Encoder-decoders for the query-biased setting have been proposed. Hasselqvist et al. [7] proposed an architecture being able to copy the words in the source document, while our copying mechanisms copy the overlapping words and their surroundings explicitly. Nema et al. [13] presented a dataset extracted from Debatepedia. They proposed a method to gain the diversity of the summary, while we focus on copying mechanisms.

We propose three copying mechanisms designed for query-biased summarizers: copying from the source, copying the overlapping words, and copying the overlapping words and their surroundings. We empirically show that the models copying the overlapping words perform better. These results support the fact that the strategy to include the overlapping words, which was shown useful for conventional query-biased summarizers, also works well for neural network-based query-biased summarizers.

2   Base Model

We first explain a base query-biased neural abstractive summarizer proposed by Nema et al. [13], into which we integrate our copying mechanisms in the next section.

Encoders: The base model has two bi-directional Long Short-term Memory (LSTM) [8]-based encoders; one is for the query $\mathbf{q} = \{q_1, ..., q_{|\mathbf{q}|}\}$ and another is for the source document $\mathbf{d} = \{w_1, ..., w_{|\mathbf{d}|}\}$. In each encoder, the outputs of the forward and the backward LSTM are concatenated into a vector. We refer

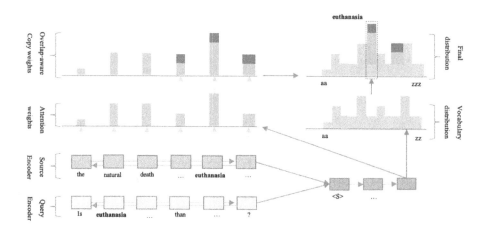

Fig. 1. Overview of a query-biased summarizer with a copying mechanism.

to the generated vector for the i-th word in the query as h_i^q, and the j-th word in the source document as h_j^d.

Decoder with Query- and Source Document- Attentions: The decoder outputs the summary. The final state $h_{|\mathbf{d}|}^d$ is used to initialize the first state of the LSTM in the decoder. In each time step t, the decoder calculates the attention weights $a_{t,i}^q = v_q \cdot \tanh(W_q s_t + U_q h_i^q)$ for every word in the query. $W_q, U_q \in \mathbb{R}^{l \times l}$ are weight matrices, and $v_q \in \mathbb{R}^l$ is a l−dimensional weight vector, where each element is automatically learned from the training data. $s_t = LSTM_d(s_{t-1}, [y_{t-1}; d_{t-1}])$ is the output of the LSTM in the decoder. Here, y_{t-1} is the embedding vector of the previously generated word and d_t is a document representation explained later. The weights are converted into probabilities $\alpha_{t,i}^q = \frac{\exp(a_{t,i}^q)}{\sum_{i=1}^{|\mathbf{q}|} \exp(a_{t,i}^q)}$. We now obtain a query vector: $q_t = \sum_{i=1}^{|\mathbf{q}|} \alpha_{t,i}^q h_i^q$.

The source document attention mechanism further calculates the attention weights $a_{t,j}^d$ for every word in the source document and converts them into probabilities $\alpha_{t,j}^d$:

$$a_{t,j}^d = v_d \cdot \tanh(W_d s_t + U_d h_j^d + Z q_t),$$
$$\alpha_{t,j}^d = \frac{\exp(a_{t,j}^d)}{\sum_{j=1}^{|\mathbf{w}|} \exp(a_{t,j}^d)}. \tag{1}$$

$W_d, U_d, Z \in \mathbb{R}^{l \times l}$ and $v_d \in \mathbb{R}^l$ are learnable parameters. Note that Eq. (1) contains q_t, which means that the weights are calculated by considering the query. We then take the weighted average to obtain a document vector: $d_t = \sum_{j=1}^{|\mathbf{w}|} \alpha_{t,j}^d h_j^d$.

Finally, the score of generating the word n in the pre-defined dictionary N is calculated as $a_{t,gen}(n) = \delta_n \cdot W_o(W_{dec} s_t + V_{dec} d_t)$. The scores are converted into a probability distribution:

$$p_{t,gen}(n) = \frac{\exp(a_{t,gen}(n))}{\sum_{m=1}^{|N|} \exp(a_{t,gen}(n_m))}, \tag{2}$$

where $W_o \in \mathbb{R}^{l \times N}$, $W_{dec} \in \mathbb{R}^{l \times l}$ and $V_{dec} \in \mathbb{R}^{l \times l}$ are learnable parameter matrices. $\delta_n \in \{0, 1\}^{|N|}$ is a one-hot vector where the element corresponding to the word n is 1, or 0 otherwise. Thus, the dot product of δ_n and $W_o(W_{dec}s_t + V_{dec}d_t)$ calculates the score of generating the word n. Equation (2) converts the score into a probability distribution.

Objective Function: All learnable matrices are tuned to minimize the negative likelihood for the reference summaries y in the training data D: $-\frac{1}{|D|} \sum_D log \ p(x|y)$.

3   Copying Mechanisms for Query-Biased Summarizers

We discuss the copying mechanisms for query-biased summarizers. Figure 1 shows the overview of a query-biased summarizer with a copying mechanism. In the following subsections, we present three mechanisms; SOURCE, OVERLAPand OVERLAP-WIND.

SOURCE: We explain SOURCE, which copies the words from the source document. The strategy is a straightforward extension of the existing copying mechanisms [5,6,17]. The neural query-biased summarizer proposed by Hasselqvist et al. [7] also adopted this strategy, but they did not report its impact. We further extend this mechanism in the following subsections. In this strategy, the output layer calculates the probability distribution over the set $N \cup M = \{n_1, ..., n_{|N|+|M|}\}$, where N is the set of words in the pre-defined dictionary and M is the set of words in the source document. Thus, $p_{t,gen}$ in Eq. (2) is modified to consider the extended vocabulary as follows:

$$p'_{t,gen}(n) = \begin{cases} p_{t,gen}(n) & (n \in N), \\ 0 & (n \notin N). \end{cases} \tag{3}$$

In the copying mechanism, we consider two different probabilities for the word n in the vocabulary; the generation probability $p'_{t,gen}$ and the copying probability $p_{t,copy}$. The switching probability $sw_t = \sigma(z_d \cdot d_t + z_s \cdot s_t + z_y \cdot y_{t-1})$ balances those probabilities as: $p_t(n) = sw_t p'_{t,gen}(n) + (1 - sw_t)p_{t,copy}(n)$. Here, $z_d, z_s \in \mathbb{R}^l$ and $z_y \in \mathbb{R}^{l_{emb}}$. σ represents a sigmoid function. l_{emb} is the dimension size of a word embedding. $p_{t,copy}$ is calculated as follows:

$$p_{t,copy}(n) = \begin{cases} \alpha_{t,idx\_s(n)}^d & (n \in M), \\ 0 & (n \notin M). \end{cases} \tag{4}$$

$idx\_s(n)$ is a function to return the position of the word n in the source document. The attention weight $\alpha_{t,idx\_s(n)}^d$ provided by the source document attention module is used as the score for outputting the word n in the source document.

OVERLAP: We propose OVERLAP, the model to copy the overlapping words. This model calculates the probability distribution over the set $N \cup M = \{n_1, ..., n_{|N|+|M|}\}$ in the same way as in SOURCE. This model increases the scores for the overlapping words as follows:

$$
a_{t,n}^o = \begin{cases} (1 + \lambda_d)a_{t,idx\_s(n)}^d & (n \in Q \cap M), \\ a_{t,idx\_s(n)}^d & (n \in M \backslash (Q \cap M)), \\ 0 & (n \notin M). \end{cases} \tag{5}
$$

The scores are converted into a probability distribution: $p_{t,copy}(n) = \frac{\exp(a_{t,idx(n)}^o)}{\sum_{j=1}^{|N|+|M|} \exp(a_{t,j}^o)}$.

In the equations above, Q refers to the set of content words[1] in the query. Thus, $Q \cap M$ represents the overlapping words. $\lambda_d \in \mathbb{R} > 0$ is a hyperparameter that controls the importance of the overlapping words. By using λ_d, this model can assign a relatively high probability for the overlapping words. Thus, the overlapping words are more likely to be included in the summary. λ_d is tuned on validation data.

OVERLAP-WINDOW: We finally explain OVERLAP-WIND, the model that copies the overlapping words and their surrounding words. We assume that the surroundings of overlapping words might also be important. This model calculates the scores for the overlapping words and their surroundings as follows:

$$
a_{t,idx\_s(n)}^{o\_w} = \begin{cases} (1 + \lambda_d)a_{t,idx\_s(n)}^d & (n \in M_{L_d}), \\ a_{t,idx\_s(n)}^d & (n \in M \backslash M_{L_d}), \\ 0 & (n \notin M), \end{cases} \tag{6}
$$

where M_{L_d} is the set that contains $L_d (\in \mathbb{N})$ words around the overlapping word in addition to the overlapping word itself. Then, the scores are converted into a probability distribution by using the softmax function: $p_{t,copy}(n) = \frac{\exp(a_{t,idx\_s(n)}^{o\_w})}{\sum_{j=1}^{|N|+|M|} \exp(a_{t,j}^{o\_w})}$.

4   Experiments

We used the publicly available dataset[2] provided by Nema et al. [13]. The data contains the tuples of a source document, a query and a summary, extracted from Debatepedia[3]. We used 80% of the data for training, and the remaining was equally split for parameter tuning and testing.

---

[1] We used the list of stop words defined in the nltk library for filtering to obtain content words.
[2] https://github.com/PrekshaNema25/DiverstiyBasedAttentionMechanism.
[3] http://www.debatepedia.org/en.

We used Adam [9] for the optimizer with $\beta_1 = 0.9$ and $\beta_2 = 0.999$. The initial learning rate was set to 0.0004. The word embeddings for both the query and the source document were initialized by GloVe [15] and further tuned during training the models. We selected the best-performing value from 200, 300 and 400 for the dimension size for LSTM by using the data for tuning. We used 32 for the batch size. We used all the vocabulary in the training data as the pre-defined dictionary.

We prepared three baselines without any copying mechanisms. The first, ENC-DEC, was a simple encoder-decoder based summarizer without the query encoder. This model uses Eq. (1) without q_t. The second, ENC-DEC QUERY, was the query-aware encoder-decoder explained in Sect. 2. The third, DIVERSE, was the state-of-the-art model proposed by Nema et al. [13]. We adopted full-length ROUGE [11] for the automatic evaluation metric. In addition, we conducted manual evaluation by human judges. 55 randomly selected document/query pairs and their summaries generated by DIVERSE, SOURCE, and OVERLAP-WINDwere shown to crowdworkers on Amazon Mechanical Turk. We assigned 10 workers for each set of document/query/summaries and asked them to rank the summaries. We adopted readability and responsiveness as the manual evaluation criteria, following the evaluation metric in DUC2007[4]. The workers were allowed to give the same rank to multiple summaries.

5 Results

We show the ROUGE scores[5] and the averaged rankings from human judges in Table 2.

Table 2. The full-length ROUGE-1, ROUGE-2, ROUGE-L (higher is better) and the averaged rankings (lower is better) from human judges. The best performing model is in bold.

	ROUGE-1	ROUGE-2	ROUGE-L	Readability	Responsiveness
Reference	–	–	–	1.50	1.55
ENC-DEC	13.73	2.06	12.84	–	–
ENC-DEC QUERY	29.28	10.24	28.21	–	–
DIVERSE [13]	41.02	26.44	40.78	3.36	3.39
SOURCE	43.32	29.12	42.96	1.99	1.93
OVERLAP	43.47	29.68	43.26	–	–
OVERLAP-WIND ($L_d = 1$)	44.41†	30.48†	44.20†	1.83†	1.85†
OVERLAP-WIND ($L_d = 2$)	43.16	29.15	42.90	–	–
OVERLAP-WIND ($L_d = 3$)	44.03	29.78	43.77	–	–

---

[4] https://duc.nist.gov/duc2007/tasks.html.
[5] The option for ROUGE is -a -n 2 -s.

ROUGE Scores: ENC-DEC, without query information, achieved very low performance. Adding the query encoder (ENC-DEC QUERY) improved the score. Furthermore, copying the words in the source document (SOURCE) achieved better scores than those of the best-performing model without a copying mechanism (DIVERSE). Thus, integrating the copying mechanism improved the performance even in the query-biased setting. Among our models, OVERLAP and OVERLAP-WIND $(L_d = 1)$ achieved the better performances than SOURCE. The dagger (†) indicates that the differences between the scores of SOURCE, OVERLAP and those of our best-performing model (OVERLAP-WIND($L_d = 1$)) are statistically significant with the paired bootstrap resampling test used in Koehn et al. [10] ($p < 0.05$). This supports our assumption that the strategy to copy the overlapping words is shown effective even for RNN-based summarizers.

Rankings by Human Judges: Our best model OVERLAP-WIND $(L_d = 1)$ is ranked higher than the state-of-the-art DIVERSE and SOURCE. The differences between OVERLAP-WIND $(L_d = 1)$ and SOURCE are statistically significant ($p < 0.05$) with the paired bootstrap resampling test [10]. The results of manual evaluation also support our assumption.

6   Conclusion

We proposed the copying mechanisms designed for query-biased summarizers to primarily include the words overlapping between the source document and the query. Our experimental results showed that the mechanisms to primarily include the overlapping words between the source document and the query achieved the better performances in terms of both ROUGE and rankings by human judges. The results suggested that the strategy to include the overlapping words, which has been shown useful for conventional non-neural summarizers, also works well for RNN-based summarizers.

References

1. Baumel, T., Eyal, M., Elhadad, M.: Query focused abstractive summarization: incorporating query relevance, multi-document coverage, and summary length constraints into seq2seq models. arXiv preprint arXiv:1801.07704 (2018)
2. Cheng, J., Lapata, M.: Neural summarization by extracting sentences and words. In: Proceedings of The 54th Annual Meeting of the Association for Computational Linguistics, ACL 2016, Berlin, Germany, pp. 484–49 (2016)
3. Dang, H.T.: Overview of DUC 2005. In: Proceedings of 2005 Document Understanding Conferences, DUC 2005, pp. 1–12. Citeseer (2005)
4. Daumé III, H., Marcu, D.: Bayesian query-focused summarization. In: Proceedings of the 54th Annual Meeting of the Association for Computational Linguistics, ACL 2006, pp. 305–312 (2006)
5. Gu, J., Lu, Z., Li, H., Li, V.O.: Incorporating copying mechanism in sequence-to-sequence learning. In: Proceedings of the 54th Annual Meeting of the Association for Computational Linguistics, ACL 2016. pp. 1631–1640 (2016)

6. Gulcehre, C., Ahn, S., Nallapati, R., Zhou, B., Bengio, Y.: Pointing the unknown words. In: Proceedings of the 54th Annual Meeting of the Association for Computational Linguistics, ACL 2016, vol. 1, pp. 140–149 (2016)
7. Hasselqvist, J., Helmertz, N., Kågebäck, M.: Query-based abstractive summarization using neural networks. arXiv preprint arXiv:1712.06100 (2017)
8. Hochreiter, S., Schmidhuber, J.: Long short-term memory. Neural Comput. **9**(8), 1735–1780 (1997)
9. Kingma, D.P., Ba, J.: Adam: a method for stochastic optimization. In: Proceedings of 3rd International Conference on Learning Representations, ICLR 2015 (2015)
10. Koehn, P.: Statistical significance tests for machine translation evaluation. In: Proceedings of the 2014 Conference on Empirical Methods on Natural Language Processing, EMNLP 2014, pp. 388–395 (2004)
11. Lin, C.Y.: Rouge: a package for automatic evaluation of summaries. In: Proceedings of ACL2004 Workshop, pp. 74–81 (2004)
12. Miao, Y., Blunsom, P.: Language as a latent variable: discrete generative models for sentence compression. In: Proceedings of the 2016 Conference on Empirical Methods in Natural Language Processing, EMNLP 2016, pp. 319–328 (2016)
13. Nema, P., Khapra, M.M., Laha, A., Ravindran, B.: Diversity driven attention model for query-based abstractive summarization. In: Proceedings of the 54th Annual Meeting of the Association for Computational Linguistics, ACL 2017, pp. 1063–1072, July 2017
14. Otterbacher, J., Erkan, G., Radev, D.R.: Biased LexRank: passage retrieval using random walks with question-based priors. Inf. Process. Manag. **45**(1), 42–54 (2009)
15. Pennington, J., Socher, R., Manning, C.: Glove: global vectors for word representation. In: Proceedings of the 2014 Conference on Empirical Methods in Natural Language Processing, EMNLP 2014, pp. 1532–1543 (2014)
16. Schilder, F., Kondadadi, R.: FastSum: fast and accurate query-based multi-document summarization. In: Proceedings of the 46th Annual Meeting of the Association for Computational Linguistics, ACL 2008, pp. 205–208 (2008)
17. See, A., Liu, P.J., Manning, C.D.: Get to the point: summarization with pointer-generator networks. In: Proceedings of the 55th Annual Meeting of the Association for Computational Linguistics, ACL 2017, vol. 1, pp. 1073–1083 (2017)
18. Tombros, A., Sanderson, M.: Advantages of query biased summaries in information retrieval. In: Proceedings of 21st Annual ACM/SIGIR International Conference on Research and Development in Information Retrieval, SIGIR 1998, pp. 2–10. ACM (1998)
19. Vinyals, O., Fortunato, M., Jaitly, N.: Pointer networks. In: Proceedings of Twenty-Ninth Conference on Neural Information Processing Systems, NIPS 2015, pp. 2692–2700 (2015)

Distant Supervision for Extractive Question Summarization

Tatsuya Ishigaki[1][(✉)], Kazuya Machida[1], Hayato Kobayashi[2],
Hiroya Takamura[1,3], and Manabu Okumura[1]

[1] Tokyo Institute of Technology, Tokyo, Japan
{ishigaki,machida}@lr.pi.titech.ac.jp, oku@pi.titech.ac.jp
[2] Yahoo Japan Corporation/RIKEN AIP, Tokyo, Japan
hakobaya@yahoo-corp.jp
[3] AIST, Tokyo, Japan
takamura.hiroya@aist.go.jp

Abstract. Questions are often lengthy and difficult to understand because they tend to contain peripheral information. Previous work relies on costly human-annotated data or question-title pairs. In this work, we propose a distant supervision framework that can train a question summarizer without annotation costs or question-title pairs, where sentences are automatically annotated by means of heuristic rules. The key idea is that a single-sentence question tends to have a summary-like property. We empirically show that our models trained on the framework perform competitively with respect to supervised models without the requirement of a costly human-annotated dataset.

Keywords: Question summarization · Extractive summarization

1 Introduction

People ask questions in various scenarios such as community question answering (CQA) sites, conference sessions, and e-mail conversations. However, questions tend to be lengthy and laborious to understand because they often contain peripheral information. Question summarization is the task of transforming such questions into shorter and concise ones. We focus on the setting used in existing studies that transforms an input question into a concise single-sentence summary [7,12,21]. This setting can be regarded as title or headline generation, and is particularly important for practical application on CQA sites where questions do not always have appropriate titles, unlike news articles. In fact, to reduce the burden on users who post questions, many CQA sites (including the biggest Japanese CQA site, Yahoo! Chiebukuro [23]), do not provide an input field for titles in the submission form to reduce the burden on users who post questions.On most CQA sites, the first sentence of a question is displayed as the headline, but this is not always appropriate. We show these examples in Table 1.

© Springer Nature Switzerland AG 2020
J. M. Jose et al. (Eds.): ECIR 2020, LNCS 12036, pp. 182–189, 2020.
https://doi.org/10.1007/978-3-030-45442-5_23

Table 1. Two translated examples of input question, gold labels (sentences labeled 1 should be included in the summary), and scores for the sentences given by our model `DistNet`. **Bold** and <u>underlined</u> sentences refer to ones selected by our model (`DistNet` +`Init`) and a baseline (`Lead` +`Q`), respectively.

Gold label	Salient score	Input question
0	0.17	I am an aged person
0	0.45	<u>Please kindly tell me the answer in detail</u>
...	...	...
1	0.99	**How can I write an email with a clickable url?**
0	0.05	I do stretches to get flexibility in the legs
1	0.95	**How am I able to do a front split?**
...	...	...
0	0.05	All of my family except me can do it
0	0.96	Is there any reason why only I cannot do a front split?

There are two approaches for summarization tasks: extractive and abstractive. The extractive approach selects salient units (e.g., sentences) in the input [3,8,13,15,16,20], while the abstractive approach generates a summary by possibly paraphrasing the information in the input [2,22]. The extractive approach has an advantage in that it does not generate errors when extracting one sentence for title generation and can be directly used for real services, as reported by Higurashi et al. [7]. Therefore, we take the extractive approach assuming that extracted titles will be used for a real service.

Previous studies on question summarization rely on costly human-annotated data or automatically collected question-title pairs. Tamura et al. [21] used an SVM-based classifier to extract the most important sentences to improve the performance of question answering systems. Higurashi et al. [7] proposed a 'learning to rank' approach for headline generation for CQA. Both of these methods require costly human-annotated data. Ishigaki et al. [12] trained extractive and abstractive summarizers by regarding a question-title pair posted on a CQA site as a question-summary pair. However, questions do not always have appropriate titles on other CQA sites, as mentioned above.

To overcome this problem, we take a distant supervision approach instead of creating costly human-annotated data. Distant supervision [17] involves training a model on pseudo data automatically labeled on the basis of heuristics, rules, and/or external resources. Various approaches for different tasks have been proposed, such as the use of FreeBase [1] for relation extraction, emoticons for sentiment classification [6], and topic labels on Wikipedia articles for blog topic classification [10]. In the summarization field, Nallapati et al. [18] and Chen and Lapata [3] automatically annotated labels for sentence extraction by using a dataset with human-generated abstractive summaries. However, we face

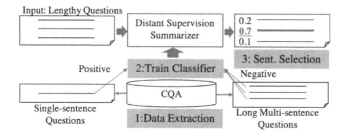

Fig. 1. Overview of our framework.

difficulties in directly applying their strategies due to the lack of human-generated abstractive summaries (or titles).

The key idea in this paper is to focus on the difference in characteristics between single-sentence questions and individual sentences in extremely long multi-sentence questions posted on CQA sites. We assume that the single-sentence questions are mostly self-contained questions, and can function as a summary (or title), while the individual sentences in long multi-sentence questions are not. Thus, we use a classifier to determine how likely it is that a sentence is similar to a single-sentence question and would therefore be appropriate as a title.

Our contributions are as follows:

- We propose a distant supervision framework for question summarization on CQA sites and verify the correctness of the assumption through manual analysis.
- We construct a large training dataset including 2.5M sentences with pseudo labels from Yahoo! Chiebukuro and will release it [11].
- We compare our models with several baselines on a human-annotated evaluation dataset and find that our models perform competitively with respect to the supervised baselines.

2   Framework

Correctness of Assumption: The ideal summaries for questions should be (1) a question sentence and (2) self-contained. We manually analyzed 300 randomly selected single-sentence questions on Yahoo! Chiebukuro and found that 98% of them were self-contained questions. In contrast, 91.2% of 315 sentences randomly selected from long multi-sentence posts consisting of more than ten sentences were not self-contained questions. In fact, they are often not even a question, which makes them more similar to sentences not to be included in a summary than to titles. Our main finding here is that single-sentence questions on CQA sites are mostly self-contained, which makes them similar to sentences used as a title. Note that small amounts of noise or incorrect labels can be ignored, as shown in later in the experimental results (Sect. 3).

Overview of Proposed Framework: Figure 1 shows the overview of our proposed framework, which includes the following three steps:

1. Extract single-sentence questions and extremely long multi-sentence questions from a CQA site, whose sentences can be automatically labeled by the assumption.
2. Train a classifier (that can predict a label with its confidence score) by using the automatically annotated sentences as pseudo training data.
3. Use confidence scores determined by the classifier to select a sentence from an input question document to form a summary.

In Step 1, we can obtain a large number of positive and negative instances without annotation costs. For positive instances, 28% of the questions on Yahoo! Chiebukuro are single-sentence questions. For negative instances, only a few long multi-sentence questions exist, but the number of sentences in the questions is large. In Step 2, we can use any classifier even with a large number of parameters thanks to the availability of large-scale data. In our experiments, we examine neural network and logistic regression classifiers. In Step 3, we prepare three selection strategies: Greedy, Init, and Q. Greedy selects the sentence with the highest confidence score, whereas Init selects the initial sentence among sentences with reasonably high confidence scores, which is expected to be effective for multi-focused questions. Q first extracts question sentences by means of a rule-based approach and then selects the sentence with the highest score among the extracted question sentences. We will explain these steps (i.e., how to create the pseudo data, the training settings of the classifiers, and the algorithms for the selection strategies) in more detail in the next section.

3   Experiments

Data: We created two datasets, Pseudo and Label, from the publicly available dataset [19] provided by Yahoo! Chiebukuro. Pseudo is a large dataset with pseudo labels for training classifiers. Label is a dataset with human-annotated correct answers.

Pseudo contains 800K single-sentence questions as positive instances and 1.7M sentences extracted from multi-sentence questions as negative instances. We randomly extracted single-sentence questions and multi-sentence questions consisting of more than ten sentences and assigned positive/negative labels on the basis of the assumption. Note that classifiers trained with Pseudo can accept any sentence even in a question with more than ten sentences since they do not care about the question length.

Label consists of 12,406 questions (including multiple sentences) separately sampled from Pseudo. Every sentence in each question has a binary label indicating whether the sentence is the most important, that is, only the best sentence has a label of 1, and the others have 0. We used crowdsourcing to annotate Label, where given a question, five workers were asked to select the best sentence that represented the main focus of the question. We included only questions for which at least four workers selected the same sentence.

Compared Models: We compared two models on our framework with several unsupervised and supervised baselines as follows. `DistNet` and `DistReg` are variants based on neural network and logistic regression models, respectively.

– `DistNet` selects the most important sentence on our framework with a neural network classifier trained on `Pseudo`, which uses an LSTM [9] to encode the input question into a fixed-length vector, and a softmax layer to convert the vector into a probability distribution expressing how likely it is that the sentence is similar to a single-sentence question.
– `DistReg` is a variant based on a logistic regression classifier [4] with n-gram, part-of-speech features.

The unsupervised baselines are as follows.

– `Lead` selects the first sentence, which is strong for generic summarization.
– `LexRank` selects the most important sentence using a graph-based extractive summarization method [5] trained on all the questions in the dataset.
– `SimEmb` selects the sentence that is the most similar to the input question in terms of the word mover's distance [14], where word embeddings were trained on all the questions. Taking account of the similarity between the input document and a sentence is common in unsupervised models for generic summarization.
– `TfIdf` selects the sentence with the highest averaged TF-IDF score, where the IDF was calculated by using all the questions.

The supervised baselines are variants of `DistNet` and `DistReg`.

– `SupNet` selects the most important sentence based on a neural network classifier trained on `Label` in the same way as `DistNet`.
– `SupReg` is a variant of `DistReg` based on a logistic regression classifier trained on `Label`.

Sentence Selection Strategies: We used three sentence selection strategies `Greedy`, `Init` and `Q`, as in Sect. 2.

– `Greedy` selects the sentence with the highest score.
– `Init` selects the initial sentence that is given a higher score than the specific threshold τ. τ was tuned on the validation data (as described later).
– `Q` selects the best one among the question sentences determined by the rule-based question extractor [21]. If there are no question sentences, `Q` is the same as `Greedy`.

Evaluation: For evaluating the performance, we used the accuracy measure calculated by dividing the number of questions in which the target method correctly selected their most important sentences by the number of questions used. Note that well-known metrics such as ROUGE and precision/recall are not appropriate since our task is to find only one sentence as a title (or snippet for a list item). We divided the labeled data Label into five sets (training:development:test=3:1:1) and performed 5-fold cross-validation for evaluating the supervised models and tuning τ.

Results: Table 2 shows the accuracy scores for the compared models with three strategies. The numbers in bold represent the best performing models that are trained without labeled data. There is no statistically significant difference between them and SupReg even though they do not use labeled data.

Looking at the column of Greedy, our model DistNet performed competitively with the best supervised model SupReg, although DistNet did not use the labeled dataset Label. Note that we didn't observe any statistically significant difference between DistNet and SupReg by the sign test. Comparing DistReg and SupReg, SupReg conversely outperformed DistReg because simple models like logistic regression can be trained on a small dataset. This implies that our distant supervision approach is suitable for complicated models like LSTM. Among unsupervised baselines, Lead performed the best, and the others did not perform well. This may be because LexRank, SimEmb, and TfIdf do not take into account whether a sentence is a question.

Table 2. Accuracy for compared models

	Greedy	Init	Q	Best
DistNet	**87.38**	**90.45**	**87.38**	**90.45**
DistReg	86.17	89.05	86.17	89.05
Lead	81.79	81.79	88.08	88.08
LexRank	78.49	81.79	84.95	84.95
SimEmb	59.46	81.79	71.17	81.79
TfIdf	52.03	81.79	69.68	81.79
SupNet	81.67	86.31	81.67	86.31
SupReg	87.89	91.21	87.89	91.21

The second column Init shows the results of selecting the initial sentence among the sentences with reasonably high scores. Comparing Greedy and Init for DistNet, we found that the simple strategy Init drastically improved the accuracy. This matches our intuition that former sentences tend to be important. Note that Lead with Init is the same as Lead with Greedy. LexRank, SimEmb, TfIdf with Init became equivalent to Lead as a result of tuning.

The third column Q shows the results of selecting the best sentence among the question sentences. This column is mainly for unsupervised baselines. The results of our models and supervised baselines are the same as Greedy since they were trained for basically extracting question sentences. The results in the column indicate that, overall, their accuracy scores were further enhanced, but as a result, none of the unsupervised baselines could overtake our best model DistNet with Init. We also examined the unsupervised baselines with both Init and Q, but the trends did not change.

The final column shows the best scores in the left three columns. Our model DistNet with Init outperformed the best all the baselines trained without Label. Furthermore, DistNet performed competitively with the best performing supervised model SupReg with Init, since we didn't observe any statistical significance between them.

Qualitative Analysis: Table 1 shows two examples of our model `DistNet` with `Init` (bold) and the strong baseline `Lead` with `Q` (underlined), where the scores were calculated by `DistNet`. The first example shows that our model successfully selected the self-contained question, while `Lead` with `Q` failed. The rule-based question extractor sometimes failed because it possibly selects non self-contained questions such as the second sentence in this example. The second example shows how effective `Init` was. We observed that `DistNet` with `Greedy` sometimes failed when the input was multi-focused, as shown in the last sentence (0.96), since the `DistNet` gave very high probabilities for question sentences. In contrast, `Init` mostly handled multi-focused questions well, thereby delivering better performances in terms of accuracy.

4   Conclusion

We proposed a distant supervision framework for question summarization based on the assumption that single sentence posts tend to have a summary-like property. For future research, we will examine if our assumption can be applied to other domains such as answers on CQA sites, user comments on news sites, and opinions on discussion forums.

References

1. Bollacker, K., Evans, C., Paritosh, P., Sturge, T., Taylor, J.: Freebase: a collaboratively created graph database for structuring human knowledge. In: Proceedings of the 2008 ACM SIGMOD International Conference on Management of Data (SIGMOD2008), pp. 1247–1250 (2008)
2. Chen, Q., Zhu, X., Ling, Z., Wei, S., Jiang, H.: Distraction-based neural networks for modeling documents. In: Proceedings of the 25th International Joint Conference on Artificial Intelligence (IJCAI2016), pp. 2754–2760 (2016)
3. Cheng, J., Lapata, M.: Neural summarization by extracting sentences and words. In: Proceedings of 2016 Annual Conference of the Association for Computational Linguistics (ACL2016), pp. 484–494 (2016)
4. Cox, D.R.: The regression analysis of binary sequences. J. R. Stat. Soc. Ser. B (Methodol.) **20**, 215–242 (1958)
5. Erkan, G., Radev, D.R.: LexRank: graph-based lexical centrality as salience in text summarization. J. Artif. Intell. Res. **22**, 457–479 (2004)
6. Go, A., Bhayani, R., Huang, L.: Twitter sentiment classification using distant supervision. In: Stanford Technical Report (2009)
7. Higurashi, T., Kobayashi, H., Masuyama, T., Murao, K.: Extractive headline generation based on learning to rank for community question answering. In: Proceedings of The 27th International Conference on Computational Linguistics (COLING2018), pp. 1742–1753 (2018)
8. Hirao, T., Isozaki, H., Maeda, E., Matsumoto, Y.: Extracting important sentences with support vector machines. In: Proceedings of The 11st International Conference on Computational Linguistics (COLING2002), pp. 1–7 (2002)
9. Hochreiter, S., Schmidhuber, J.: Long short-term memory. Neural Comput. **9**(8), 1735–1780 (1997)

10. Husby, S.D., Barbosa, D.: Topic classification of blog posts using distant supervision. In: Proceedings of the Workshop on Semantic Analysis in Social Media, pp. 28–36 (2012)
11. Ishigaki, T.: Scripts for Preprocessing Yahoo Chiebukuro dataset (2020). http:// lr-www.pi.titech.ac.jp/~ishigaki/chiebukuro/
12. Ishigaki, T., Takamura, H., Okumura, M.: Summarizing lengthy questions. In: Proceedings of The 8th International Joint Conference on Natural Language Processing (IJCNLP2017), vol. 1, pp. 792–800 (2017)
13. Kågebäck, M., Mogren, O., Tahmasebi, N., Dubhashi, D.: Extractive summarization using continuous vector space models. In: Proceedings of The 2nd Workshop on Continuous Vector Space Models and their Compositionality (CVSC2014), pp. 31–39 (2014)
14. Kusner, M., Sun, Y., Kolkin, N., Weinberger, K.: From word embeddings to document distances. In: Proceedings of the 32nd International Conference on Machine Learning (ICML2015), pp. 957–966 (2015)
15. Luhn, H.P.: The automatic creation of literature abstracts. IBM J. Res. Dev. $2(2)$, 159–165 (1958)
16. Mihalcea, R., Tarau, P.: TextRank: bridging order into texts. In: Proceedings of the 2004 Conference on Empirical Methods in Natural Language Processing (EMNLP2004), pp. 404–411 (2004)
17. Mintz, M., Bills, S., Snow, R., Jurafsky, D.: Distant supervision for relation extraction without labeled data. In: Proceedings of the Joint Conference of the 47th Annual Meeting of the ACL and the 4th International Joint Conference on Natural Language Processing (ACL-IJCNLP2009), pp. 1003–1011 (2009)
18. Nallapati, R., Zhai, F., Zhou, B.: SummaRuNNer: a recurrent neural network based sequence model for extractive summarization of documents. In: Proceedings of Thirty-First AAAI Conference on Artificial Intelligence (AAAI2017), pp. 3075–3081 (2017)
19. NII: Yahoo! Chiebukuro data, 2nd edn. (2018). https://www.nii.ac.jp/dsc/idr/en/ yahoo/
20. Takamura, H., Okumura, M.: Text summarization model based on maximum coverage problem and its variant. In: Proceedings of the 12th Conference of the European Chapter of the ACL (EACL 2009), pp. 781–789 (2009)
21. Tamura, A., Takamura, H., Okumura, M.: Classification of multiple-sentence questions. In: Dale, R., Wong, K.-F., Su, J., Kwong, O.Y. (eds.) IJCNLP 2005. LNCS (LNAI), vol. 3651, pp. 426–437. Springer, Heidelberg (2005). https://doi.org/10. 1007/11562214_38
22. Tan, J., Wan, X., Xiao, J.: Abstractive document summarization with a graph-based attentional neural model. In: Proceedings of The 55th Annual Meeting of the Association for Computational Linguistics (ACL2017), vol. 1, pp. 1171–1181 (2017)
23. Yahoo Japan Corp.: Yahoo! Chiebukuro (2019). https://chiebukuro.yahoo.co.jp/

Text-Image-Video Summary Generation Using Joint Integer Linear Programming

Anubhav Jangra[1], Adam Jatowt[2(✉)], Mohammad Hasanuzzaman[3], and Sriparna Saha[1]

[1] Department of Computer Science, Indian Institute of Technology Patna, Patna, India
anubhav0603@gmail.com, sriparna.saha@gmail.com
[2] Department of Social Informatics, Kyoto University, Kyoto, Japan
jatowt@gmail.com
[3] Department of Computer Science, Cork Institute of Technology, Cork, Ireland
hasanuzzaman.im@gmail.com

Abstract. Automatically generating a summary for asynchronous data can help users to keep up with the rapid growth of multi-modal information on the Internet. However, the current multi-modal systems usually generate summaries composed of text and images. In this paper, we propose a novel research problem of text-image-video summary generation (TIVS). We first develop a multi-modal dataset containing text documents, images and videos. We then propose a novel joint integer linear programming multi-modal summarization (JILP-MMS) framework. We report the performance of our model on the developed dataset.

Keywords: Multi-modal summarization · Integer Linear Programming

1 Introduction

Advancement in technology has led to rapid growth of multimedia data on the Internet, which prevent users from obtaining important information efficiently. Summarization can help tackle this problem by distilling the most significant information from the plethora of available content. Recent research in summarization [2,11,31] has proven that having multi-modal data can improve the quality of summary in comparison to uni-modal summaries. Multi-modal information can help users gain deeper insights. Including supportive representation of text can reach out to a larger set of people including those who have reading disabilities, users who have less proficiency in the language of text and skilled readers who are looking to skim the information quickly [26]. Although visual representation of information is more expressive and comprehensive in comparison to textual description of the same information, it is still not a thorough model of representation. Encoding abstract concepts like guilt or freedom [11], geographical locations or environmental features like temperature, humidity etc. via images is impractical. Also images are a static medium and cannot represent

J. M. Jose et al. (Eds.): ECIR 2020, LNCS 12036, pp. 190–198, 2020.
https://doi.org/10.1007/978-3-030-45442-5_24

dynamic and sequential information efficiently. Including videos could then help overcome these barriers since video contains both visual and verbal information. To the best of our knowledge, all the previous works have focused on creating text or text-image summaries, and the task of generating an extractive multi-modal output containing text, images and videos from a multi-modal input has not been done before. We thus focus on a novel research problem of text-image-video summary generation (TIVS). To tackle the TIVS task, we design a novel Integer Linear Programming (ILP) framework that extracts the most relevant information from the multimodal input. We set up three objectives for this task, (1) *salience within modality*, (2) *diversity within modality* and (3) *correspondence across modalities*. For preprocessing the input, we convert the audio into text using an Automatic Speech Recognition (ASR) system, and we extract the key-frames from video. The most relevant images and videos are then selected in accordance with the output generated by our ILP model.

To sum up, we make the following contributions: (1) We present a novel multi-modal summarization task which takes news with images and videos as input, and outputs text, images and video as summary. (2) We create an extension of the multi-modal summarization dataset [12] by constructing multi-modal references containing text, images and video for each topic. (3) We design a joint ILP framework to address the proposed multi-modal summarization task.

2   Related Work

Text summarization techniques are used to extract important information from textual data. A lot of research has been done in the area of extractive [10,21] and abstractive [3,4,19,23] summarization. Various techniques like graph-based methods [6,15,16], artificial neural networks [22] and deep learning based approaches [18,20,29] have been developed for text summarization. Integer linear programming (ILP) has also shown promising results in extractive document summarization [1,9]. Duan et al. [5] proposed a joint-ILP framework that produces summaries from temporally separate text documents.

Recent years have shown great promise in the emerging field of multi-modal summarization. Multi-modal summarization has various applications ranging from meeting recordings summarization [7], sports video summarization [25], movie summarization [8] to tutorial summarization [13]. Video summarization [17,28,30] is also a major sub-domain of multi-modal summarization. A few deep learning frameworks [2,11,31] show promising results, too. Li et al. [12] uses an asynchronous dataset containing text, images and videos to generate a textual summary. Although some work on document summarization has been done using ILP, to the best of our knowledge no one has ever used an ILP framework in the area of multi-modal summarization.

3   Problem Definition

Let M: $\{D_1, D_2, ..., D_{|D|}\} \bigcup \{I_1, I_2, ..., I_{|I|}\} \bigcup \{V_1, V_2, ..., V_{|V|}\}$ be a multi-modal dataset related to a topic T, where D_i is a text document, I_i is an image and V_i is a video. $|.|$ denotes the cardinality of a set. Each document D_i contains sentences x_j such that $D_i : \{x_{i,1}, x_{i,2}, ..., x_{i,|D_i|}\}$. Our objective is to generate a multi-modal summary $S = X_{sum} \bigcup I_{sum} \bigcup V_{sum}$ such that the final summary S covers up all the important information in the original data while minimizing the length of summary, where $X_{sum} \subseteq \{x_{i,j}|x_{i,j} \in D_i \wedge D_i \in M\}$, $I_{sum} \subseteq \{I_i, I_j, ..., I_k\}$, where $|I_{sum}| \leq |I|$ and $V_{sum} \subseteq \{V_l, V_m, ..., V_{|n|}\}$, where $|V_{sum}| \leq |V|^1$.

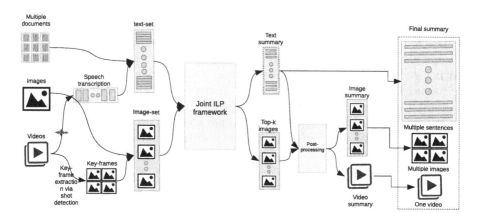

Fig. 1. The framework of our proposed model.

4   Proposed Method

4.1   Pre-processing

Each topic in our dataset comprises of text documents, images, audio and videos. As shown in Fig. 1, we firstlextract key-frames from the videos [32]. These key-frames together with images from the original data form the *image-set*. The audio is transcribed into text (IBM Watson Speech-to-Text Service: www.ibm.com/watson/developercloud/speech-to-text.html), which contributes to the *text-set* together with the sentences from text-documents. The images from then *image-set* are encoded by the VGG model [24] and the 4,096-dimensional vector from the pre-softmax layer is used as the image representation. Every sentence from the *text-set* is encoded using the Hybrid Gaussian-Laplacian Mixture Model (HGLMM) into a 6,000-dimensional vector. For text-image matching, these image and sentence vectors are fed into a two-branch neural network [27] to have a 512-dimensional vector for images and sentences in a shared space.

4.2   Joint-ILP Framework

ILP is a global optimization technique, used to maximize or minimize an objective function subject to some constraints. In this paper, we propose a joint-ILP

---

[1] We set $|V_{sum}| = 1$ for simplicity assuming that in many cases one video would be enough.

technique to optimize the output to have high salience, diversity and cross-modal correlation. The idea of joint-ILP is similar to the one applied in the field of across-time comparative summarization [5]. However, to the best of our knowledge, an ILP framework was not used to solve multi-modal summarization (Gurobi optimizer is used for ILP optimization: https://www.gurobi.com/).

Decision Variables. M_{txt} is a $n \times n$ binary matrix such that $m_{i,i}^{txt}$ indicates whether sentence s_i is selected as an exemplar or not and $m_{i,j\neq i}^{txt}$ indicates whether sentence s_i votes for s_j as its representative. Similarly, M_{img} is a $p \times p$ binary matrix that indicates the exemplars chosen in the image set. M_c is $n \times p$ binary matrix that indicates the cross-modal correlation. $m_{i,j}^c$ is true when there is some correlation between sentence s_i and image I_j.

Objective Function

$$\underset{max}{Arg}[\lambda \cdot m \cdot \{Sal(M_{txt}) + Sal(M_{img})\} + (1-\lambda) \cdot (k_{txt} + k_{img}) \cdot MCorr(M_c)] \quad (1)$$

$$Sal(M_{mod}) = \sum_{i=1}^{t} m_{i,i}^{mod} \cdot Imp(item_i, G(item_i)); i \in \{1,2,\ldots,t\} \quad (2)$$

$$Imp(item_i, G(item_i)) = \sum_{item_j \in G(item_i)} Sim_{cosine}(item_i, item_j); i,j \in \{1,2,\ldots,t\} \quad (3)$$

$$G(item_i) = \{item_j | m_{ji}^{mod} = 1; j \in \{1,2,\ldots,t\}\}; i \in \{1,2,\ldots,t\} \quad (4)$$

where $\langle mod, t, item \rangle \in \{\langle text, n, s \rangle, \langle img, p, I \rangle\}$ is used to represent multiple modalities together in a simple way.

$$MCorr = \sum_{i=1}^{n} \sum_{j=1}^{p} m_{i,j}^c \cdot Sim_{cosine}(s_i, I_j) \quad (5)$$

We need to maximize the objective function in Eq. 1, containing salience of text, images and cross-modal correlation. Similar to the joint-ILP formulation in [5] the diversity objective is implicit in this model. Equation 4 generates the set of entities that are a part of the cluster whose exemplar is $item_i$. The salience is calculated by Eqs. 2 and 3 by taking cosine similarity over all the exemplars with the items belonging to their representative clusters separately for each modality. The cross-modal correlation score is calculated in Eq. 5.

Constraints

$$m_{i,j}^{mod} \in \{0,1\}; mod \in \{txt, img, c\} \quad (6)$$

$$\sum_{i=1}^{n} m_{i,j}^{txt} = k_{txt} \text{ and } \sum_{i=1}^{p} m_{i,j}^{img} = k_{img} \quad (7)$$

$$\sum_{j=1}^{n} m_{i,j}^{txt} = 1; i \in \{1,2,\ldots,n\} \text{ and } \sum_{j=1}^{p} m_{i,j}^{img} = 1; i \in \{1,2,\ldots,p\} \quad (8)$$

$$m_{j,j}^{mod} - m_{i,j}^{mod} \geq 0; \ mod \in \{txt, img\} \tag{9}$$

Equation 7 ensures that exactly k_{txt} and k_{img} clusters are formed in their respective uni-modal vector space. Equation 8 guarantees that an entity can either be an exemplar or be part of a single cluster. According to Eq. 9, a sentence or image must be exemplar in their respective vector space to be included in the sentence-image summary pairs. Values of m, k_{txt} and k_{img} are set to be 10, same as in [5].

4.3 Post-processing

The Joint-ILP framework outputs the text summary (X_{sum}) and $top{-}m$ images from the $image{-}set$. This output is used to prepare the image and video summary.

Extracting Images

$$I_{sum} = I_{sum1} \cup I_{sum2} \tag{10}$$

$$I_{sum1} = \{I_z | I_z \in \ top \ m \wedge I_z \neq keyframes\} \tag{11}$$

$$I_{sum2} = \{I_z | \alpha \leq Sim_{cosine}(I_z, I_y) \leq \beta \wedge I_y \in I_{sum1}\} \tag{12}$$

Equation 11 selects all those images from $top10$ images that are not key-frames. Assuming that images which look similar would have similar annotation scores and would help users gain more insight, the images relevant to the images in I_{sum1} (at least with α cosine similarity) but not too similar (at max with β cosine similarity) to avoid redundancy are also selected to be a part of the final image summary I_{sum} (Eq. 12). α is set to 0.4 and β is 0.8 in our experiments.

Extracting Video. For each video, weighted sum of visual (Eq. 13) and verbal (Eq. 14) scores is computed. The video with the highest score is selected as our video summary.

$$visual - score = \sum_{I_k \in KF} \sum_{I_f \in I_{sum}} Sim_{cosine}(I_f, I_k) \tag{13}$$

$$verbal - score = \sum_{x_j \in ST} \sum_{x_i \in X_{sum}} Sim_{cosine}(x_i, x_j) \tag{14}$$

where KF is the set of all key-frames and ST is the set of speech transcriptions.

5   Dataset Preparation

There is no benchmark dataset for the TIVS task. Therefore, we created our own text-image-video dataset by extending and manually annotating the multi-modal summarization dataset introduced by Li et al. [12]. Their dataset comprised of 25 new topics. Each topic was composed of 20 text documents, 3 to 9 images, and 3 to 8 videos. The final summary however was unimodal, that is, in the form of only a textual summary containing around 300 words. We then extended it by selecting some images and a video for each topic that summarize the topic well. Three undergraduate students were employed to score the images and videos with respect to the benchmark text references. All annotators scored each image and video on a scale of 1 to 5, on the basis of similarity between the image/video and the text references (1 indicating no similarity and 5 denoting the highest level of similarity). Average annotation scores (AAS) were calculated for each image and video. The value of the minimum average annotation score for images was kept as a hyper-parameter to evaluate the performance of our model in various settings[2]. The video with the highest score is chosen to be the video component of the multi-modal summary[3].

Table 1. Overall performance comparison for textual summary using ROUGE.

System	ROUGE-1	ROUGE-2	ROUGE-l
Baseline-1	0.254	0.065	0.216
Baseline-2	**0.261**	0.068	0.225
Baseline-3	0.249	0.068	0.212
JILP-MMS	0.260	**0.074**	**0.226**

Table 2. Overall performance of image summary using precision and recall. AAS denotes here the threshold value of image summary generation.

AAS	Average precision	Average recall	Variance precision	Variance recall
5	0.016	0.060	0.003	0.048
4.5	0.099	0.084	0.067	0.041
4	0.258	0.313	0.105	**0.151**
3.5	0.335	0.332	0.111	0.125
3	**0.599**	**0.383**	**0.139**	0.076

6   Experimental Settings and Results

We evaluate the performance of our model using the dataset as described above. We use the ROUGE scores [14] to evaluate the textual summary, and based on them we compare our results with the ones of three baselines.

We use the multi-document summarization model proposed in [1]. For **Baseline-1** we feed the model with embedded sentences from all the original documents together. The central vector is calculated as the average of all the sentence vectors. The model is given vectors for sentences from the *text-set* and images from the *image-set* in the joint space for other baselines. For **Baseline-2**, the

---

[2] Every topic had at least one image when we set threshold for average annotation score to 3.

[3] In case two videos have the same score, the video with shorter length was chosen.

average of all the vectors is taken as the central vector. For **Baseline-3**, the central vector is calculated as the weighted average of all the sentence and image vectors. We give equal weights to text, speech and images for simplicity.

As shown in Table 1, our model produces better results than the prepared baselines in terms of ROUGE-2 and ROUGE-1 scores. Table 2 shows the average precision and recall scores as well as the variance. We set various threshold values for the annotation scores to generate multiple image test sets in order to evaluate the performance of our model. We get a higher precision score for low AAS value, because the number of images in the final solution increases on decreasing the threshold values. The proposed model gave **44%** accuracy in extracting the most appropriate video (whereas random selection of images for 10 different iterations gives an average 16% accuracy).

7   Conclusion

Unlike other problems that focus on text-image summarization, we propose to generate a truly multi-modal summary comprising of text, images and video. We also develop a dataset for this task, and propose a novel joint ILP framework to tackle this problem.

Acknowledgement. Dr. Sriparna Saha would like to acknowledge the support of Early Career Research Award of Science and Engineering Research Board (SERB) of Department of Science and Technology, India to carry out this research. Mohammed Hasanuzzaman would like to acknowledge ADAPT Centre for Digital Content Technology which is funded under the SFI Research Centres Programme (Grant 13/RC/2106).

References

1. Alguliev, R., Aliguliyev, R., Hajirahimova, M.: Multi-document summarization model based on integer linear programming. Intell. Control Autom. **1**(02), 105 (2010)
2. Chen, J., Zhuge, H.: Abstractive text-image summarization using multi-modal attentional hierarchical RNN. In: Proceedings of the 2018 Conference on Empirical Methods in Natural Language Processing, pp. 4046–4056 (2018)
3. Chen, Y.C., Bansal, M.: Fast abstractive summarization with reinforce-selected sentence rewriting. arXiv preprint arXiv:1805.11080 (2018)
4. Chopra, S., Auli, M., Rush, A.M.: Abstractive sentence summarization with attentive recurrent neural networks. In: Proceedings of the 2016 Conference of the North American Chapter of the Association for Computational Linguistics: Human Language Technologies, pp. 93–98 (2016)
5. Duan, Y., Jatowt, A.: Across-time comparative summarization of news articles. In: Proceedings of the Twelfth ACM International Conference on Web Search and Data Mining, pp. 735–743. ACM (2019)
6. Erkan, G., Radev, D.R.: LexRank: graph-based lexical centrality as salience in text summarization. J. Artif. Intell. Res. **22**, 457–479 (2004)

7. Erol, B., Lee, D.S., Hull, J.: Multimodal summarization of meeting recordings. In: Proceedings of the 2003 International Conference on Multimedia and Expo, ICME 2003, (Cat. No. 03TH8698), vol. 3, pp. III–25. IEEE (2003)

8. Evangelopoulos, G., et al.: Multimodal saliency and fusion for movie summarization based on aural, visual, and textual attention. IEEE Trans. Multimed. **15**(7), 1553–1568 (2013)

9. Galanis, D., Lampouras, G., Androutsopoulos, I.: Extractive multi-document summarization with integer linear programming and support vector regression. In: Proceedings of COLING 2012, pp. 911–926 (2012)

10. Kupiec, J., Pedersen, J., Chen, F.: A trainable document summarizer. In: Proceedings of the 18th Annual International ACM SIGIR Conference on Research and Development in Information Retrieval, pp. 68–73. ACM (1995)

11. Li, H., Zhu, J., Liu, T., Zhang, J., Zong, C., et al.: Multi-modal sentence summarization with modality attention and image filtering (2018)

12. Li, H., Zhu, J., Ma, C., Zhang, J., Zong, C., et al.: Multi-modal summarization for asynchronous collection of text, image, audio and video (2017)

13. Libovický, J., Palaskar, S., Gella, S., Metze, F.: Multimodal abstractive summarization for open-domain videos. In: Proceedings of the Workshop on Visually Grounded Interaction and Language (ViGIL), NIPS (2018)

14. Lin, C.Y.: ROUGE: a package for automatic evaluation of summaries. In: Text Summarization Branches Out, pp. 74–81. Association for Computational Linguistics, Barcelona, July 2004. https://www.aclweb.org/anthology/W04-1013

15. Mihalcea, R.: Graph-based ranking algorithms for sentence extraction, applied to text summarization. In: Proceedings of the ACL Interactive Poster and Demonstration Sessions, pp. 170–173 (2004)

16. Mihalcea, R., Tarau, P.: TextRank: bringing order into text. In: Proceedings of the 2004 Conference on Empirical Methods in Natural Language Processing, pp. 404–411 (2004)

17. Mirzasoleiman, B., Jegelka, S., Krause, A.: Streaming non-monotone submodular maximization: personalized video summarization on the fly. In: Thirty-Second AAAI Conference on Artificial Intelligence (2018)

18. Nallapati, R., Zhai, F., Zhou, B.: Summarunner: a recurrent neural network based sequence model for extractive summarization of documents. In: Thirty-First AAAI Conference on Artificial Intelligence (2017)

19. Nallapati, R., Zhou, B., Gulcehre, C., Xiang, B., et al.: Abstractive text summarization using sequence-to-sequence RNNs and beyond. arXiv preprint arXiv:1602.06023 (2016)

20. Nallapati, R., Zhou, B., Ma, M.: Classify or select: neural architectures for extractive document summarization. arXiv preprint arXiv:1611.04244 (2016)

21. Paice, C.D.: Constructing literature abstracts by computer: techniques and prospects. Inf. Process. Manag. **26**(1), 171–186 (1990)

22. Saini, N., Saha, S., Jangra, A., Bhattacharyya, P.: Extractive single document summarization using multi-objective optimization: exploring self-organized differential evolution, grey wolf optimizer and water cycle algorithm. Knowl.-Based Syst. **164**, 45–67 (2019)

23. See, A., Liu, P.J., Manning, C.D.: Get to the point: summarization with pointer-generator networks. CoRR abs/1704.04368 (2017). http://arxiv.org/abs/1704.04368

24. Simonyan, K., Zisserman, A.: Very deep convolutional networks for large-scale image recognition. arXiv preprint arXiv:1409.1556 (2014)

25. Tjondronegoro, D., Tao, X., Sasongko, J., Lau, C.H.: Multi-modal summarization of key events and top players in sports tournament videos. In: 2011 IEEE Workshop on Applications of Computer Vision (WACV), pp. 471–478. IEEE (2011)
26. UzZaman, N., Bigham, J.P., Allen, J.F.: Multimodal summarization of complex sentences. In: Proceedings of the 16th International Conference on Intelligent User Interfaces, pp. 43–52. ACM (2011)
27. Wang, L., Li, Y., Lazebnik, S.: Learning deep structure-preserving image-text embeddings. In: Proceedings of the IEEE Conference on Computer Vision and Pattern Recognition, pp. 5005–5013 (2016)
28. Wei, H., Ni, B., Yan, Y., Yu, H., Yang, X., Yao, C.: Video summarization via semantic attended networks. In: Thirty-Second AAAI Conference on Artificial Intelligence (2018)
29. Zhang, Y., Er, M.J., Zhao, R., Pratama, M.: Multiview convolutional neural networks for multidocument extractive summarization. IEEE Trans. Cybern. **47**(10), 3230–3242 (2016)
30. Zhou, K., Qiao, Y., Xiang, T.: Deep reinforcement learning for unsupervised video summarization with diversity-representativeness reward. In: Thirty-Second AAAI Conference on Artificial Intelligence (2018)
31. Zhu, J., Li, H., Liu, T., Zhou, Y., Zhang, J., Zong, C.: MSMO: multimodal summarization with multimodal output. In: Proceedings of the 2018 Conference on Empirical Methods in Natural Language Processing, pp. 4154–4164 (2018)
32. Zhuang, Y., Rui, Y., Huang, T.S., Mehrotra, S.: Adaptive key frame extraction using unsupervised clustering. In: Proceedings 1998 International Conference on Image Processing, ICIP98 (Cat. No. 98CB36269), vol. 1, pp. 866–870. IEEE (1998)

Domain Adaptation via Context Prediction for Engineering Diagram Search

Harsh Jhamtani[1(✉)] and Taylor Berg-Kirkpatrick[2]

[1] Carnegie Mellon University, Pittsburgh, USA
jharsh@cs.cmu.edu
[2] UC San Diego, San Diego, USA
tberg@eng.ucsd.edu

Abstract. Effective search for engineering diagram images in larger collections is challenging because most existing feature extraction models are pre-trained on natural image data rather than diagrams. Surprisingly, we observe through experiments that even in-domain training with standard unsupervised representation learning techniques leads to poor results. We argue that, because of their structured nature, diagram images require more specially-tailored learning objectives. We propose a new method for unsupervised adaptation of out-of-domain feature extractors that asks the model to reason about spatial context. Specifically, we fine-tune a pre-trained image encoder by requiring it to correctly predict the relative orientation between pairs of nearby image regions. Experiments on the recently released Ikea Diagram Dataset show that our proposed method leads to substantial improvements on a downstream search task, more than doubling recall for certain query categories in the dataset.

Keywords: Diagram search · Image retrieval · Domain adaptation

1 Introduction

Many engineering enterprises maintain technical know-how about the design and working of various parts of their products via engineering diagrams. Such engineering diagrams typically show different parts used for a product or a portion of a product, and how these parts fit together in assemblies. In Fig. 1, the image on the left is a sample from an Ikea drawings dataset [1]. A common way of distributing such diagrams is in the form of images rather than detailed 3D models due to privacy issues and ease of use [1].

An important use case for such enterprises is the ability to automatically search for similar parts in diagram images based on a drawing or diagram image of a query part. A typical goal is to search a collection of engineering diagram images to find diagrams that contain the query part or a similar part as a component in the larger diagram, possibly at a different scales and rotations relative to the query. The matching diagram will likely contain many other parts that

© Springer Nature Switzerland AG 2020
J. M. Jose et al. (Eds.): ECIR 2020, LNCS 12036, pp. 199–206, 2020.
https://doi.org/10.1007/978-3-030-45442-5_25

are irrelevant to the query, which makes this task different from more standard image search scenarios where matched images will have the query as their central focus. In many ways, the most analogous task is that of matching a keyword query to a full-text document: matched keywords are usually surrounded by irrelevant text.

There have been only a few efforts directed towards search in engineering diagrams [1,6]. Dai et al. [1] recently released a new Ikea dataset, which consists of 13,464 diagrams from Ikea manuals and 16,940 query images, and proposed a neural search engine for performing diagram search. The image features used by Dai et al. [1] are produced by a VGGNET model [16], pre-trained to predict object classes on ImageNet data [3]. Since ImageNet consists of natural images, rather than data like engineering diagrams, it is reasonable to expect that learning representations for diagrams using in-domain engineering data might lead to better results. Somewhat surprisingly, we show in experiments that tuning image representations using standard unsupervised techniques (e.g. using autoencoding objectives) leads to worse performance than the out-of-domain pretrained model. This is probably a result of structured nature of engineering diagrams that makes them visually distinct from natural images, on which the baseline unsupervised techniques were developed and validated [8].

In this paper we propose new unsupervised methods for tuning the pretrained model for engineering diagram data using spatial context as a learning signal. Specifically, a classifier trained to predict relative direction of two randomly sampled image patches provides the spatial signal. Use of spatial context alone to learn image representations *from scratch* has been explored in prior work [5], though not in the context of image search. However, we report that such an approach leads to poor results for engineering diagram search. We instead build on prior work to propose unsupervised *fine-tuning* of pre-trained neural image encoder models using the *spatial context*. Such use of spatial context to finetune the image encoder model has not been explored earlier.

In experiments, we find that the proposed approach leads to substantial gains in retrieval performance relative to past work as well as compared to a variety of unsupervised baselines, yielding state-of-the-art results on the Ikea dataset.

2   Related Work

Our work is related to Dai et al. [1] who introduce Ikea dataset and propose neural search methods for the task. However, they use pre-trained VGGNET model to extract image features and do not attempt learn or fine tune image feature extractors. Many earlier image retrieval methods are based on SIFT features and bag-of-word models [2,9,17]. Recently, convolutional and other neural models have been used more extensively as feature extractors for image retrieval [15,20]. Use of region based image features has been found useful in some prior image retrieval work [14]. Our proposed method is also related to recent success on learning contextualized word representations [4,10,13] and image representations from scratch [5,11] through self-supervised training by using context prediction as a training signal.

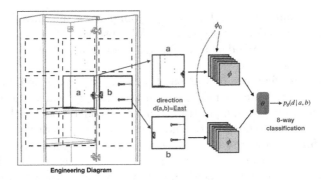

Fig. 1. Overview of the proposed method. For an image patch a, another patch b is sampled from one of the eight possible directions (shown with dotted borders). A classifier with parameters θ is trained to predict the relative direction d. The image encoder, with parameters ϕ, is biased to stay close to the original pre-trained parameters ϕ_0 via a L1-regularization term to encourage the model to retain useful features from the pre-trained model.

Most prior work on fine-tuning and domain adaptation for image representations requires in-domain supervised data, and either uses parameter fine-tuning [18,19] or feature selection [7,19] for domain adaptation. In contrast, our approach is a fully unsupervised method for domain adaptation of pre-trained models. Certain prior works on domain adaptation for image feature extractors require both the source and target domain data during training [19], which our approach does not require. Some prior works have explored using unsupervised methods like auto-encoders for domain adaptation [12].

3   Fine-Tuning by Context Prediction

We propose an unsupervised method that uses spatial context prediction as a learning signal to tune a pre-trained image feature extractor, which we refer to as an image encoder, for use in a neural engineering diagram search. Our approach extends prior work on context-based training objective [5] to perform domain adaptation and applies the method to a new domain and task. Specifically, our goal is to bias image encoders to capture more informative features of engineering diagrams by requiring them to predict spatial relationships between neighboring image regions. We shall refer to our fine-tuning method as **SPACES** (SPAtial Context for Engineering diagram Search).

Let $\mathcal{D}$ denote the set of images in the dataset. For an image $I \in \mathcal{D}$, we randomly pick a rectangular region in the image of size $M * M$ - an image patch (Fig. 1). Let us denote the identified patch by a. Thereafter, we choose one of the 8 cardinal directions (North, South, East, West, North-East, North-West, South-East, South-West), which we denote as d, uniformly at random. Then, a second rectangular patch b is identified close to the first patch in the sampled direction d such that b and a do not intersect. However, following [5], we identify candidates for b a minimum fixed distance x from patch a, and introduce random jitters in horizontal and vertical directions. We observed $M = 24$ and $x = 4$ to be a reasonable choice. We denote the distribution from which a, b, and d are sampled as the generator, G.

We extract the features of the patches using the image encoder model, denoted by f_ϕ, where ϕ are the model parameters. f_ϕ is typically a deep convolutional neural network. Given a pre-trained image encoder model with parameter weights ϕ_0, our task is to finetune the model using context prediction signal from a classifier defined as follows. A classifier with learnable parameters θ takes as input the extracted features of the two patches and makes a prediction about the relative direction of the patches. Specifically, we consider a two layer feed-forward neural network with a softmax function at the end to make a 8-way classification prediction (for 8 cardinal directions). Classification loss for a given patch pair a, b in relative direction d can be written as follows:

$$S(\phi, \theta) = \sum_{I \in \mathcal{D}} \mathbb{E}_{d,a,b \sim G(I)}[-\log(p_\theta(d|f_\phi(a), f_\phi(b)))] \tag{1}$$

Computing this exact loss is impractical due to an extremely large number of possible patch pairs. So we instead draw K random samples of pairs of patches for every image in the train set. Additionally, we regularize the image encoder model towards the pretrained model weights ϕ_0 by adding a L1 regularizer. This is done to encourage the model to retain many features from the pretrained model since abstract features like curves and shapes from have been shown to generalize well across tasks. We learn the image encoder and classifier jointly by optimizing for θ and ϕ to minimize the following loss function:

$$L\phi, \theta) = \sum_{I \in \mathcal{D}} \frac{1}{K} \sum_{k=1}^{K} [-\log(p_\theta(d^{(k)}|f_\phi(a^{(k)}), f_\phi(b^{(k)})))] + \lambda|\phi - \phi_0| \tag{2}$$

The regularization term biases the image encoder parameters to remain closer to the original pre-trained model values. In early experiments, we observe that using L1 for this term performs better than using a L2 version. The classification loss term is based on the work of Doersch et al. [5]. However, we use the spatial context loss to fine-tune a pre-trained image feature extractor for the target engineering diagrams domain. In contrast, Doersch et al. [5] learn image representations from scratch using a large dataset of natural images and focus on a different task. We demonstrate in experiments that the proposed tuning method substantially outperforms the training from scratch for our domain and task.

4   Experiments and Results

4.1   Dataset

We use the Ikea dataset [1] which consists of 13,464 furniture assembly diagrams. Each assembly diagram is a black-and-white image, and resembles a line drawing. Query images are generated automatically from a subset of documents using an iterative procedure proposed in past work [1]. The procedure begins with identifying a localized region of high density black pixels, and keeps on expanding it until the black pixel density is lower than a threshold. The Ikea dataset consists of 5 query types: psr, Psr, pSr, psR, PSR: Lowercase letters p, s, r signify that position, scale and rotation, respectively, are unchanged in the generated query relative to the original image from which the query was extracted. Capital letters denote the corresponding altered attribute. Thus, for **psr** queries set, the identified region is placed onto a white background image of size same as original image, and at same position as the identified region was in the original image. **psR** queries are constructed by rotating *psr* queries, **pSr** queries are constructed by scale transformations, and so on.

Table 1. Retrieval results on different query types in Ikea dataset.

Model	Invariant (psr)		Position (Psr)		Scale (pSr)		Rotation (psR)		ALL (PSR)	
	MRR	R@1	MRR	R@1	MRR	R@1	M–RR	R@1	MRR	R@1
VGG [1]	0.94	0.89	0.90	0.84	0.78	0.70	0.38	0.28	0.17	0.10
VGG-AE	0.93	0.89	0.89	0.84	0.76	0.69	0.41	0.30	0.16	0.08
CTXT [5]	0.83	0.78	0.54	0.45	0.02	0.01	0.15	0.08	0.03	0.0
SPACES	**0.98**	0.98	**0.96**	0.95	**0.88**	0.84	**0.65**	0.58	**0.22**	0.14
SPACES-L	0.95	0.90	0.91	0.86	0.87	0.81	0.61	0.52	0.21	0.14

4.2 Experiment Setup

We use our proposed method to fine-tune VGGNET [16] (a deep convolutional image encoder, pre-trained on ImageNet [3] data). We report recall and mean reciprocal rank in downstream search, using the DISHCONV [1], a neural retrieval method which utilizes pairwise training over features extracted from convolutional kernels over image representations. We perform early stopping during training based on *recall@1* for queries in the validation split. We consider following baselines: (1) **VGG** represents a fixed pre-trained VGGNET model, trained on ImageNet data, as used in prior work [1]. (2) **VGG-AE** fine-tunes a pre-trained VGGNET using an autoencoder (with a deconvolutional network as decoder) with reconstruction objective. (3) **CTXT** involves training image representations from scratch just using context prediction [5]. We consider VGGNET architecture (initialized randomly) to encode images, and trained to predict relative direction of pairs of image patches.

4.3 Results

Table 1 summarizes the results when evaluating downstream search performance. Overall, SPACES performs much better than the baselines VGG, VGG-AE and CTXT across different query types. Recall that *psr* are the most basic

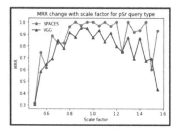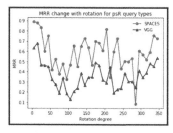

Fig. 2. MRR (Mean Reciprocal rank) plotted against (a) the scale transformation value of queries in pSr test set (b) rotation degrees in psR test set.

query types, *psR* are queries created by rotating basic *psr* queries by varying degrees, and so on. The largest improvement over baselines is observed for *pSr* and *psR* query types. SPACES performs better than VGG-AE model probably because it has a more suitable training signal given the structured nature of the engineering diagrams. The baseline CTXT model has to learn the image encoder model entirely from a relatively small number of images. In contrast, SPACES leverages the pre-trained model and is able to fine-tune on the Ikea dataset using only a few thousand images. This demonstrates the utility of SPACES in adapting large pre-trained image encoder models for engineering diagrams.

We report MRR for a range scale factor and rotations degrees (Fig. 2). SPACES performs better than the baseline almost all throughout different scale and rotation changes. We also report the results with SPACES-L which uses total loss on the validation split for early stopping instead of recall@1 scores. MRR and recall scores from these variants are observed to be very similar (Table 1) demonstrating that the proposed approach is robust to such changes in early stopping criteria.

A relative direction prediction classifier trained on the Ikea dataset images with features from pretrained VGGNET model, and then evaluated on 1000 patch pair samples achieves only 14.2% accuracy, which is close to performance of a random prediction classifier for a 8-way classification problem. The trained classifier within SPACES achieved 43% accuracy in the 8-way classification task, which demonstrates that features from our trained model encode more information about neighboring context.

5   Conclusions

In this paper we have proposed an unsupervised method to adapt a pre-trained neural image encoder on an engineering diagram dataset using spatial context prediction. We demonstrate that standard unsupervised representation learning methods such as autoencoder are not amenable to engineering diagrams, probably due to their structured nature. Our proposed method outperforms the original pre-trained feature extractor as well as other unsupervised baselines to achieve state-of-the-art results on Ikea dataset.

References

1. Dai, Z., Fan, Z., Rahman, H., Callan, J.: Local matching networks for engineering diagram search. In: The World Wide Web Conference, WWW 2019 (2019)
2. Datta, R., Joshi, D., Li, J., Wang, J.Z.: Image retrieval: ideas, influences, and trends of the new age. ACM Comput. Surv. (CSUR) **40**(2), 5 (2008)
3. Deng, J., Dong, W., Socher, R., Li, L.J., Li, K., Fei-Fei, L.: ImageNet: a large-scale hierarchical image database. In: 2009 IEEE Conference on Computer Vision and Pattern Recognition, pp. 248–255. IEEE (2009)
4. Devlin, J., Chang, M.W., Lee, K., Toutanova, K.: BERT: pre-training of deep bidirectional transformers for language understanding. arXiv preprint arXiv:1810.04805 (2018)
5. Doersch, C., Gupta, A., Efros, A.A.: Unsupervised visual representation learning by context prediction. In: Proceedings of the IEEE International Conference on Computer Vision, pp. 1422–1430 (2015)
6. Eitz, M., Hildebrand, K., Boubekeur, T., Alexa, M.: Sketch-based image retrieval: benchmark and bag-of-features descriptors. IEEE Trans. Vis. Comput. Graph. **17**(11), 1624–1636 (2010)
7. Krause, J., Jin, H., Yang, J., Fei-Fei, L.: Fine-grained recognition without part annotations. In: Proceedings of the IEEE Conference on Computer Vision and Pattern Recognition, pp. 5546–5555 (2015)
8. Krizhevsky, A., Hinton, G.E.: Using very deep autoencoders for content-based image retrieval. In: ESANN (2011)
9. Lowe, D.G., et al.: Object recognition from local scale-invariant features. In: ICCV 1999, pp. 1150–1157 (1999)
10. Mikolov, T., Sutskever, I., Chen, K., Corrado, G.S., Dean, J.: Distributed representations of words and phrases and their compositionality. In: Advances in Neural Information Processing Systems, pp. 3111–3119 (2013)
11. Noroozi, M., Favaro, P.: Unsupervised learning of visual representations by solving Jigsaw puzzles. In: Leibe, B., Matas, J., Sebe, N., Welling, M. (eds.) ECCV 2016. LNCS, vol. 9910, pp. 69–84. Springer, Cham (2016). https://doi.org/10.1007/978-3-319-46466-4_5
12. Parchami, M., Bashbaghi, S., Granger, E., Sayed, S.: Using deep autoencoders to learn robust domain-invariant representations for still-to-video face recognition. In: 2017 14th IEEE International Conference on Advanced Video and Signal Based Surveillance (AVSS), pp. 1–6. IEEE (2017)
13. Peters, M.E., et al.: Deep contextualized word representations. In: Proceedings of NAACL-HLT, pp. 2227–2237 (2018)
14. Pham, T.T., Maillot, N.E., Lim, J.H., Chevallet, J.P.: Latent semantic fusion model for image retrieval and annotation. In: Proceedings of the Sixteenth ACM Conference on Information and Knowledge Management, pp. 439–444. ACM (2007)
15. Razavian, A.S., Sullivan, J., Carlsson, S., Maki, A.: Visual instance retrieval with deep convolutional networks. ITE Trans. Media Technol. Appl. **4**(3), 251–258 (2016)
16. Simonyan, K., Zisserman, A.: Very deep convolutional networks for large-scale image recognition. arXiv preprint arXiv:1409.1556 (2014)

17. Sivic, J., Zisserman, A.: Video Google: a text retrieval approach to object matching in videos. In: Proceedings of the Ninth IEEE International Conference on Computer Vision, ICCV 2003, vol. 2, p. 1470. IEEE Computer Society, USA (2003)
18. Tajbakhsh, N., et al.: Convolutional neural networks for medical image analysis: full training or fine tuning? IEEE Trans. Med. Imaging **35**(5), 1299–1312 (2016)
19. Wang, M., Deng, W.: Deep visual domain adaptation: a survey. Neurocomputing **312**, 135–153 (2018)
20. Zhou, W., Li, H., Tian, Q.: Recent advance in content-based image retrieval: a literature survey. arXiv preprint arXiv:1706.06064 (2017)

Crowdsourcing Truthfulness: The Impact of Judgment Scale and Assessor Bias

David La Barbera[1], Kevin Roitero[1(✉)], Gianluca Demartini[2],
Stefano Mizzaro[1], and Damiano Spina[3]

[1] University of Udine, Udine, Italy
{labarbera.david,roitero.kevin}@spes.uniud.it, mizzaro@uniud.it
[2] University of Queensland, Brisbane, Australia
g.demartini@uq.edu.au
[3] RMIT University, Melbourne, Australia
damiano.spina@rmit.edu.au

Abstract. News content can sometimes be misleading and influence
users' decision making processes (e.g., voting decisions). Quantitatively
assessing the truthfulness of content becomes key, but it is often challenging and thus done by experts. In this work we look at how experts
and non-expert assess truthfulness of content by focusing on the effect of
the adopted judgment scale and of assessors' own bias on the judgments
they perform. Our results indicate a clear effect of the assessors' political
background on their judgments where they tend to trust content which is
aligned to their own belief, even if experts have marked it as false. Crowd
assessors also seem to have a preference towards coarse-grained scales,
as they tend to use a few extreme values rather than the full breadth of
fine-grained scales.

1 Introduction

The credibility of information available online may vary and the presence of
untrustworthy information has big implications on our safety online [5,12,15].
The recent increase of misinformation online is to be blamed on technologies
that have enabled the next level of strategic politic propaganda. Social media
platforms and their data allow for extreme personalization of content which
makes it possible to individually customise information. Given that the majority
of people access news from social media platforms [13] such strategies can be used
towards the goal of influencing decision making processes [1,14].

In this constantly evolving scenario, it is key to understand how people perceive the *truthfulness of information* presented to them. To this end, in this
paper we collect data from US-based crowd workers and compare it with expert
annotation data generated by fact-checkers such as PolitiFact. Our dataset contains multiple judgments of truthfulness of information collected from several
non-expert assessors to measure agreement levels and to identify controversial

© Springer Nature Switzerland AG 2020
J. M. Jose et al. (Eds.): ECIR 2020, LNCS 12036, pp. 207–214, 2020.
https://doi.org/10.1007/978-3-030-45442-5_26

content. We also collect judgments over two different judgment scales and collect information about assessors' background that allows us to analyse assessment bias. The dataset we created is publicly available at https://github.com/KevinRoitero/crowdsourcingTruthfulness.

The results of our analysis indicate that: (1) crowd judgments can be aggregated to approximate expert judgments, (2) there is a political bias in crowd-generated truthfulness labels where crowd assessors tend to believe more to statements coming from speakers off the same political party they have voted for in the last election; and (3) there seems to be a preference for coarse-grained scales where crowd assessors tend to use the extreme values in the scale more often than other values.

2   Related Work

Crowdsourcing has been previously used as a methodology in the context of information credibility research. For example, Zubiaga and Heng [17] looked at how tweet credibility can be assessed by means of Amazon MTurk workers in the context of disaster management. Their results show that it is difficult for crowd workers to properly assess the truthfulness of tweets in this context, but that the reliability of the source is a good indicator for trusted information. Kriplean et al. [6] analyse how volunteer crowdsourcing can be used for fact-checking by simulating the democratic process. The Fact-checking Lab at CLEF [3,9] looks at this problem by defining the task of ranking sentences according to their need to be fact-checked. Maddalena et al. [7] focus on the ability of the crowd to assess news quality along eight different quality dimensions. Roitero et al. [10] use crowdsourcing to study user perception of fake news statements. As compared to previous studies looking at crowdsourcing for information credibility tasks, we look at bias in the data due to the assessor and the rating scale used to collected labels in the context of the truthfulness of statements by US politicians.

3   Methodology

3.1   Dataset Description

In our study[1] we use the PolitiFact dataset constructed by Wang [16]. This dataset contains 12800 statements by politicians with truth labels produced by expert fact-checkers on a 6-level scale: i.e., **True**, **Mostly True**, **Half True**, **Barely True**, **False**, and **Lie**.[2] For this work, we selected a subset of 120 statements randomly sampled from the PolitiFact dataset to make sure that a balanced number of statements per class and per political party was included in the sample.

---

[1] The setup was reviewed and approved by the Human Research Ethics Committee at the University of Queensland.

[2] In the original dataset, **Pants on Fire** is used; we preferred **Lie** to facilitate workers.

Table 1. Most frequently used support URLs over both scales, with and without gold questions.

	PolitiFact	Wikipedia	WashingtonPost	Google	Youtube	CNN	NYTimes
S6 + Gold	0.55	0.16	0.01	0.02	0.01	0.00	0.1
S6 − Gold	0.73	0.06	0.01	0.01	0.00	0.01	0.01
S100 + Gold	0.54	0.13	0.02	0.02	0.03	0.00	0.1
S100 − Gold	0.72	0.03	0.01	0.01	0.00	0.01	0.1

3.2 Crowdsourcing Setup

We crowdsourced 120 statements each judged by 10 distinct crowd workers across 400 HITs on Amazon MTurk asking US-based workers to label the truthfulness of statements from the dataset. Each HIT, rewarded $1.20 (i.e., $0.15 for each statement), consisted of 8 statements for which we asked an assessment either using the original 6-level scale (S6) or a 100-level scale (from 0 to 100) using a slider set by default at 50 (S100). The 8 statements contained 2 gold questions used to quality check the workers' responses by means of providing judgments consistent with the expert ground truth. Other than gold questions, each HIT contained 3 statements by Republican party speakers and 3 by Democratic party speakers. More than the judgments, crowd workers where also asked to provide a justification for each of their judgments, and a URL pointing to the source of information supporting their judgment. At the beginning of the HIT each worker was asked to complete a demographics questionnaire; it also included questions about their political orientation, used to classify crowd assessors as aligned to the US Democratic party (Dem) or the US Republican party (Rep).

4 Results

Participants (400 in total) were well balanced across the political spectrum (108 Dem and 92 Rep for S6; 109 Dem and 91 Rep for S100) and most of them have a college degree. Many crowd assessors used the PolitiFact website as a source of evidence for their judgments (55.4% of judgments done with S6 and 54.9% of judgments done with S100) as shown in Table 1. The list of URLs in this table also shows how the majority of crowd assessors performed the task correctly, as they tried to refer to reliable sources such as PolitiFact, Wikipedia, Washington Post, CNN, and New York Times.

4.1 Judgment Distributions

Figure 1 shows the raw assessment score distributions given by crowd workers both for the S6 scale and the S100 scale. These results hint that workers tend to use fewer values than those available: the extremes of the scales are more used; for S100, the middle value (50) is also frequently used and some smaller

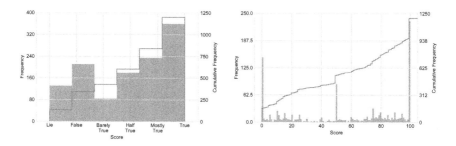

Fig. 1. Individual score distributions: S_6 (left, raw judgments), S_{100} (right, raw judgments). The red line shows the cumulative distribution of judgments. (Color figure online)

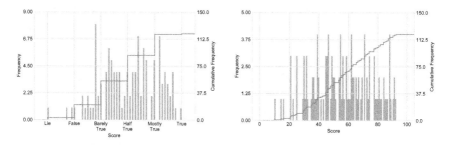

Fig. 2. Distribution of scores aggregated by mean: S_6 (left), S_{100} (right). The red line shows the cumulative distribution of judgments. (Color figure online)

peaks can be seen in correspondence of the multiples of 10. These outcomes are much less manifest in the aggregated scores (i.e., the arithmetic mean of the scores provided by the ten crowd workers judging the same statement), shown in Fig. 2: as usual, these are more evenly distributed. Also, values at the lower end of the scale are much less frequent. These outcomes suggest that perhaps a two- or three-level scale would be more appropriate for this task, although fine-grained scales have been successfully used in a crowdsourcing setting for relevance assessment [8, 11]. We intend to further address this issue in our future work also using the scale transformation techniques proposed by Han et al. [4].

4.2   Crowd vs. Experts

Figure 3 shows the crowd assessor labels as compared to expert judgment of truthfulness over both judgments scales. Crowd assessors seem able to distinguish among the different levels of the S6 scale as the median values are increasing following the expert assessments over the levels of the scale. A t-test comparing crowd assessor scores across expert judgment levels shows that crowd scores are significantly different ($p < 0.01$) across all levels except for the class combinations

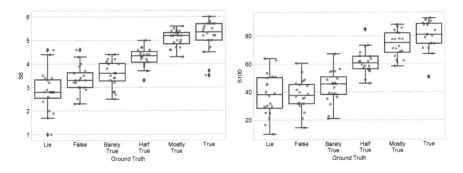

Fig. 3. Comparison of crowd labels with expert ground truth: S_6 (left), S_{100} (right).

Lie–False, False–Barely True, and Mostly True–True. The crowd appears to be more lenient than experts in assessing the truthfulness of statements as scores for the lowest categories tend not to reach the bottom end of the scale in both S6 and S100. This suggests the need to align scales when used by crowd assessors and experts in order to identify misleading content using crowdsourcing. Overall, S6 seems more adequate than S100, not only because (as noted above) workers tend to use coarse-grained scales but also because the agreement with experts seems higher for S6: S100 presents wider boxplots and less separable categories, especially for the first three classes (Lie, False, Barely True).

4.3   Crowd Assessor Bias

Figure 4 shows how crowd assessors labelled statements as compared to ground truth expert labels based on their political background. We can see that crowd assessors who voted for the Rep party tend to assign higher truthfulness scores, especially for the Lie and False ground truth labels, showing how, on average, they believe to content more than crowd assessors who voted for the Dem party.

When comparing how crowd workers assess statements differently based on who the speaker is, we can observe that True statements obtain higher scores from crowd assessors who voted for the speaker's party. That is, Dem workers assigned an average score of 84.54 on S100 and 5.48 on S6 to True statements by Dem speakers and only 81.83 on S100 and 5.00 on S6 to True statements by Rep speakers. Rep workers assigned an average score of 81.89 on S100 and 5.35 on S6 to True statements by Rep speakers and only 73.24 on S100 and 4.73 on S6 to True statements by Dem speakers. While this is an expected behaviour, we also notice that Dem crowd assessors appear to be more skeptical than Rep crowd assessors by showing a lower average judgment score for untrue statements (e.g., Fig. 4, top row).

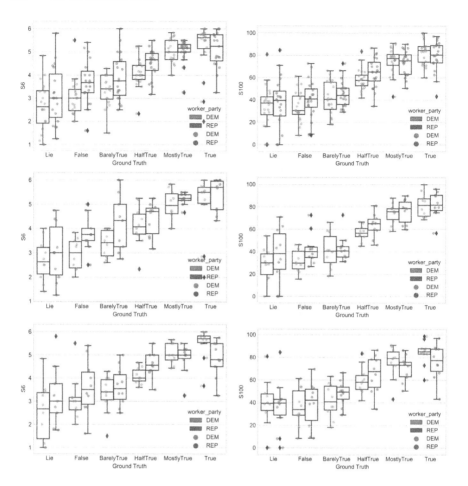

Fig. 4. Comparison with ground truth for Dem workers (blue) and Rep workers (red): S_6 (left), S_{100} (right). All statements (first row), Rep speaker statements (second row) and Dem speaker statements (third row). (Color figure online)

5   Conclusions

In this paper we presented a dataset of crowdsourced truthfulness judgments for political statements and compared the collected judgments across different crowd assessors, judgments scales, and with expert judgments.

Our results show that (1) crowd judgments, if properly aggregated, are comparable to expert ones (2) crowd assessors political background has an impact on how they label political statements: they show a tendency to be more lenient towards statements by politicians of the same political orientation as their own; and (3) crowd assessors seem to have a preference towards coarse-grained judgment scales for truthfulness judgements.

Acknowledgements. This work is partially supported by an Australian Research Council Discovery Project (DP190102141) and a Facebook Research award.

References

1. Bittman, L., Godson, R.: The KGB and Soviet Disinformation: An Insider's View. Pergamon-Brassey's (1985)
2. Cuzzocrea, A., Bonchi, F., Gunopulos, D. (eds.): Proceedings of the CIKM 2018 Workshops co-located with 27th ACM International Conference on Information and Knowledge Management (CIKM 2018), Torino, Italy, 22 October 2018, CEUR Workshop Proceedings, vol. 2482. CEUR-WS.org (2019). http://ceur-ws.org/Vol-2482
3. Elsayed, T., et al.: Overview of the CLEF-2019 CheckThat! Lab: automatic identification and verification of claims. In: Crestani, F., et al. (eds.) CLEF 2019. LNCS, vol. 11696, pp. 301–321. Springer, Cham (2019). https://doi.org/10.1007/978-3-030-28577-7_25
4. Han, L., Roitero, K., Maddalena, E., Mizzaro, S., Demartini, G.: On transforming relevance scales. In: Proceedings of the 28th ACM International Conference on Information and Knowledge Management (CIKM) (2019)
5. Jøsang, A., Ismail, R., Boyd, C.: A survey of trust and reputation systems for online service provision. Decis. Support Syst. **43**(2), 618–644 (2007)
6. Kriplean, T., Bonnar, C., Borning, A., Kinney, B., Gill, B.: Integrating on-demand fact-checking with public dialogue. In: Proceedings of CSCW, pp. 1188–1199 (2014)
7. Maddalena, E., Ceolin, D., Mizzaro, S.: Multidimensional news quality: a comparison of crowdsourcing and nichesourcing. In: Cuzzocrea et al. [2]. http://ceur-ws.org/Vol-2482/paper17.pdf
8. Maddalena, E., Mizzaro, S., Scholer, F., Turpin, A.: On crowdsourcing relevance magnitudes for information retrieval evaluation. ACM Trans. Inf. Syst. **35**(3), 19:1–19:32 (2017). https://doi.org/10.1145/3002172
9. Nakov, P., et al.: Overview of the CLEF-2018 CheckThat! lab on automatic identification and verification of political claims. In: Bellot, P., et al. (eds.) CLEF 2018. LNCS, vol. 11018, pp. 372–387. Springer, Cham (2018). https://doi.org/10.1007/978-3-319-98932-7_32
10. Roitero, K., Demartini, G., Mizzaro, S., Spina, D.: How many truth levels? Six? One hundred? Even more? Validating truthfulness of statements via crowdsourcing. In: Cuzzocrea et al. [2]. http://ceur-ws.org/Vol-2482/paper38.pdf
11. Roitero, K., Maddalena, E., Demartini, G., Mizzaro, S.: On fine-grained relevance scales. In: The 41st International ACM SIGIR Conference on Research & Development in Information Retrieval, SIGIR 2018, pp. 675–684. ACM, New York (2018). https://doi.org/10.1145/3209978.3210052
12. Self, C.C.: Credibility. In: An Integrated Approach to Communication Theory and Research, pp. 449–470. Routledge (2014)
13. Shearer, E., Gottfried, J.: News use across social media platforms 2017. Pew Research Center 7 (2017)
14. Starbird, K.: Disinformation's spread: bots, trolls and all of us. Nature **571**, 449 (2019)

15. Viviani, M., Pasi, G.: Credibility in social media: opinions, news, and health information—a survey. Wiley Interdis. Rev.: Data Min. Knowl. Discov. **7**(5), e1209 (2017)
16. Wang, W.Y.: "Liar, Liar Pants on Fire": a new benchmark dataset for fake news detection. In: Proceedings of the 55th Annual Meeting of the Association for Computational Linguistics (Volume 2: Short Papers), pp. 422–426 (2017)
17. Zubiaga, A., Ji, H.: Tweet, but verify: epistemic study of information verification on Twitter. Soc. Netw. Anal. Min. **4**(1), 163 (2014)

Novel and Diverse Recommendations by Leveraging Linear Models with User and Item Embeddings

Alfonso Landin$^{(\boxtimes)}$ (ID), Javier Parapar (ID), and Álvaro Barreiro (ID)

Information Retrieval Lab, Centro de Investigación TIC (CITIC),
Universidade da Coruña, A Coruña, Spain
{alfonso.landin,javierparapar,barreiro}@udc.es

Abstract. Nowadays, item recommendation is an increasing concern for many companies. Users tend to be more reactive than proactive for solving information needs. Recommendation accuracy became the most studied aspect of the quality of the suggestions. However, novel and diverse suggestions also contribute to user satisfaction. Unfortunately, it is common to harm those two aspects when optimizing recommendation accuracy. In this paper, we present EER, a linear model for the top-N recommendation task, which takes advantage of user and item embeddings for improving novelty and diversity without harming accuracy.

Keywords: Collaborative filtering · Novelty · Diversity · User and item embeddings

1 Introduction

In recent years, the way users access services has shifted from a proactive approach, where the user actively looks for the information, to one where the users take a more passive role, and content is suggested to them. Within this transformation, Recommender Systems have played a pivotal role, enabling an increase in user engagement and revenue.

Recommender Systems are usually classified into three families [1]. The first approach, content-based systems, use item metadata to produce recommendations [7]. The second family, collaborative filtering, is composed of systems that exploit the past interactions of the users with the items to compute the recommendations [10,17]. These interactions can take several forms, such as ratings, clicks, purchases. Finally, hybrid approaches combine both to generate suggestions. Collaborative Filtering (CF) systems can be divided into memory-based systems, that use the information about these interactions directly to compute the recommendations, and model-based systems, that build models from this information that are later used to make the recommendations.

In this paper, we will present a CF model to address the top-N recommendation task [4]. The objective of a top-N recommender is to produce a ranked list

© Springer Nature Switzerland AG 2020
J. M. Jose et al. (Eds.): ECIR 2020, LNCS 12036, pp. 215–222, 2020.
https://doi.org/10.1007/978-3-030-45442-5_27

of items for each user. These systems can be evaluated using traditional IR metrics over the rankings [2,4]. In that evaluation approach, accuracy is usually the most important metric and has been the focus of previous research and competitions [3]. Nevertheless, other properties are also important, such as diversity and novelty [8,13]. Diversity is the ability of the system to make recommendations that include items equitably from the whole catalog, which is usually desired by vendors [5,22]. On the other hand, novelty is the capacity of the system to produce unexpected recommendations. This characteristic is a proxy for serendipity, associated with higher user engagement and satisfaction [6]. All these properties, accuracy, diversity and novelty, are linked to the extent that raising accuracy usually lowers the best achievable results in the other properties [11].

In this paper, we propose a method to augment an existing recommendation linear model to make more diverse and novel recommendations, while maintaining similar accuracy results. We do so by making use of user and item embeddings that are able to capture non-linear relations thanks to the way they are obtained [21]. Experiments conducted on three datasets show that our proposal outperforms the original model in both novelty and diversity while maintaining similar levels of accuracy. With reproducibility in mind, we also make the software used for the experiments publicly available[1].

2   Background

In this section, we introduce FISM, the existing recommendation method we augment in our proposal. After that, we introduce `prefs2vec`, the user and item embedding model used to make this enhancement.

2.1   FISM

FISM is a state-of-the-art model-based recommender system proposed by Kabbur et al [9]. This method learns a low rank factorization of an item-item similarity matrix, which is later used to compute the scores to make the predictions. This method is an evolution of a previous method, SLIM [16], that learns this matrix without factorizing it. Factorizing the similarity matrix allows FISM to overcome SLIM's limitation of not being able to learn a similarity other than zero for items that have never been rated both by at least one user. As a side effect of this factorization, it lowers the space complexity from $\mathcal{O}(|\mathcal{I}|^2)$ to $\mathcal{O}(|\mathcal{I}| \times k)$, $k \ll |\mathcal{I}|$. It also drops the non-negativity constraint and the constraint that the diagonal of the similarity matrix has to contain zeroes. As a consequence of these changes, the optimization problem can be solved using regular gradient descent algorithms, instead of the coordinated gradient descent used by SLIM, leading to faster training times.

---

[1] https://gitlab.irlab.org/irlab/eer.

2.2 User and Item Embeddings

Embedding models allow transforming high-dimensional and sparse vector representations, such as classical one-hot and bag-of-words, into a space with much lower dimensionality. In particular, previous word embedding models, that produce fixed-length dense representations, have proven to be more effective in several NPL tasks [14,15,19].

Recently, prefs2vec [21], a new embedding model for obtaining dense user and item representations, an adaptation of the CBOW model [14], has shown that these embeddings can be useful for the top-N recommendation task. When used with a memory-based recommender, they are more efficient than the classical representation [21]. The results show that not only they can improve the accuracy of the results, but also their novelty and diversity. The versatility of this embedding model, in particular of the underlying neural model and the way it is trained, is also shown in [12]. Here the prediction capabilities of the neural model are used directly in a probabilistic recommender.

3 Proposal

In this section, we present our method to enhance diversity and novelty in recommendation, explaining how the model is trained and used to produce recommendations. Firstly, we introduce how the product of user and item embeddings (based on prefs2vec) can be used to make recommendations, which is later used as part of the proposal.

3.1 User and Item Embeddings Product

As representations of users and items in a space with much lower dimensionality, prefs2vec embeddings can be viewed as latent vectors. However, there is no sense in multiplying both item and user vectors as they have different basis even when they have the same dimensions. This is a consequence of learning the item and user representations independently, how prefs2vec initializes the parameters of the model and how the training is performed.

However, it is possible to make this product if we can compute a change of basis matrix $T \in \mathbb{R}^{d \times d}$ to transform the user embeddings into the item embeddings space. This way we can calculate an estimated ratings matrix $\hat{R}$ using the simple matrix multiplication:

$$\hat{R} = ETF^{\mathsf{T}} \tag{1}$$

where $E \in \mathbb{R}^{|\mathcal{U}| \times d}$ is the matrix of user embeddings, and $F \in \mathbb{R}^{|\mathcal{I}| \times d}$ is the matrix of item embeddings, one embedding in each row. The transformation matrix T is learned by solving the optimization problem with ℓ_2 regularization:

$$\underset{T}{\text{minimize}} \frac{1}{2} \|R - \hat{R}\|_F^2 + \frac{\beta_e}{2} \|T\|_F^2 \tag{2}$$

where R is the ratings matrix and β_e is the regularization hyperparameter. This problem can be solved using gradient descent algorithms.

Once the transformation matrix has been trained, recommendations can be produced by computing the estimated rating matrix $\hat{\boldsymbol{R}}$ as described in Eq. 1. Recommendations are made to each user by sorting the corresponding row and picking the top-N items not already rated by the user. We dubbed this recommender ELP, short for Embedding Linear Product, and we present its performance in Table 3 in the experiments section.

3.2   Embedding Enhanced Recommender

We have seen that linear methods, like FISM, can obtain good accuracy figures. On the other side, as results in Table 3 show, ELP is able to provide good figures in novelty and diversity, thanks to the embedding model capturing non-linear relations between users and items.

We propose to capture both properties by joining the models together in the EER model (Embedding Enhanced Recommender). We choose the RMSE variant of FISM as it matches the loss used in ELP. We also use a trainable scalar parameter α to joint the models, as the scores obtained from each recommender need not be on the same scale. This results in the following equation to calculate the estimated ratings matrix:

$$\hat{\boldsymbol{R}} = \boldsymbol{RPQ} + \alpha \boldsymbol{ETF}^{\mathsf{T}} \tag{3}$$

where $\boldsymbol{P} \in \mathbb{R}^{|\mathcal{I}| \times k}$ and $\boldsymbol{Q} \in \mathbb{R}^{k \times |\mathcal{I}|}$ are the low rank factorization of the item-item similarity matrix. The parameters of the model, $\boldsymbol{P}$, $\boldsymbol{Q}$, $\boldsymbol{T}$ and α, are learned by solving the joint ℓ_2 regularized optimization problem resulting from the previous joint equation, using standard gradient descent algorithms:

$$\underset{P,Q,T,\alpha}{\text{minimize}} \frac{1}{2} \|\boldsymbol{R} - \hat{\boldsymbol{R}}\|_F^2 + \frac{\beta}{2} \left(\|\boldsymbol{P}\|_F^2 + \|\boldsymbol{Q}\|_F^2 \right) + \frac{\beta_e}{2} \|\boldsymbol{T}\|_F^2 \tag{4}$$

Similar to the case of ELP, once the parameters are learned, we make the recommendations by calculating the estimated ratings matrix using Eq. 3, sorting each row and picking the top-N items not yet rated by the user corresponding to that row.

4   Experiments and Results

In this section, we introduce the datasets used to perform our experiments, the evaluation protocol followed and the metrics used. After that, we present the results of our experiments.

4.1   Datasets

To evaluate our proposal, we conducted a series of experiments on several datasets, from different domains: the MovieLens 20M dataset[2], a movie dataset,

---

[2] https://grouplens.org/datasets/movielens/20m/.

the book dataset LibraryThing, and the BeerAdvocate dataset[3], consisting of beer reviews. Table 1 shows statistics of each collection. In order to perform the experiments, the datasets were divided randomly into train and test sets. The training dataset consisted of 80% of the ratings of each user, with the remaining 20% forming the test dataset.

Table 1. Statistics of the collections.

Dataset	Users	Items	Ratings	Density
MovieLens 20M	138,493	26,744	20,000,263	0.540%
LibraryThing	7,279	37,232	749,401	0.277%
BeerAdvocate	33,388	66,055	1,571,808	0.071%

4.2 Evaluation Protocol

We follow the TestItems evaluation methodology [2] to evaluate the performance. To assess the accuracy of the rankings, we use Normalized Discounted Cumulative Gain (nDCG), using the *standard formulation* as described in [23], with the ratings in the test set as graded relevance judgments. We considered only items with a rating of 4 or more, on a 5 point scale, to be relevant for evaluation purposes. We also measured the diversity of the recommendations using the complement of the Gini index [5]. Finally, we use the mean self-information (MSI) [24] to assess the novelty of the recommendations. All the metrics are evaluated at cut-off 100 because it has shown to be more robust with respect to the sparsity and popularity biases than sallower cut-offs [20]. We perform a Wilcoxon test [18] to asses the statistical significance of the improvements regarding nDCG@100 and MSI@100, with $p < 0.01$. We cannot apply it to the Gini index because we are using a paired test and Gini is a global metric. Results in Table 3 are annotated with their statistical significance.

4.3 Results and Discussion

We performed a grid search over the hyperparameters of the original model and our proposal tuning them to maximize nDCG@100. Although we aim to increase diversity and novelty, we want the recommendations to be effective, which is why the tuning is done over accuracy. For the parameters of the `prefs2vec` model, we took those that performed better in [21]. For reproducibility's sake, values for the best hyperparameters for each collection can be consulted in Table 2.

Table 3 shows the values of nDCG@100, Gini@100 and MSI@100 for FISM, EER and ELP. The results show that EER outperforms the baseline (FISM) on both novelty and diversity. It also surpasses it on accuracy on the MovieLens 20M and LibraryThing datasets. In the case of diversity, we can see important

---

[3] https://snap.stanford.edu/data/web-BeerAdvocate.html.

Table 2. Best values of the hyperparameters for nDCG@100 for FISM and our proposals EER and ELP.

Model	MovieLens 20M	LibraryThing	BeerAdvocate
FISM	$\beta = 1, k = 1000$	$\beta = 1000, k = 1000$	$\beta = 50, k = 1000$
ELP	$\beta_e = 0.1$	$\beta_e = 10$	$\beta_e = 10$
EER	$\beta = 0.1, \beta_e = 1, k = 1000$	$\beta = 500, \beta_e = 10, k = 1000$	$\beta = 10, \beta_e = 1, k = 1000$

Table 3. Values of nDCG@100, Gini@100 and MSI@100 on MovieLens 20M, Library-Thing and BeerAdvocate datasets. Statistical significant improvements, according to Wilcoxon test with $p < 0.01$, in nDCG@100 and MSI@100 with respect to FISM and out proposals EER and ELP are superscripted with a, b and c respectively.

Model	Metric	MovieLens 20M	LibraryThing	BeerAdvocate
FISM	nDCG@100	$0{,}4641^{c}$	$0{,}2878^{c}$	$\mathbf{0{,}1502}^{bc}$
	Gini@100	0,0390	0,0896	0,0363
	MSI@100	230,5480	414,3157	324,4954
EER	nDCG@100	$\mathbf{0{,}4665}^{ac}$	$\mathbf{0{,}3017}^{ac}$	$0{,}1452^{c}$
	Gini@100	0,0412	0,1072	0,0521
	MSI@100	$234{,}0325^{a}$	$416{,}6850^{a}$	$328{,}2118^{a}$
ELP	nDCG@100	0,3322	0,1850	0,0855
	Gini@100	**0,0808**	**0,2901**	**0,3221**
	MSI@100	$\mathbf{307{,}9538}^{ab}$	$\mathbf{532{,}9078}^{ab}$	$\mathbf{519{,}5824}^{ab}$

improvements. ELP, on the other hand, obtains the best diversity and novelty values, but this comes with a big reduction in accuracy. It is common in the field of recommender systems for methods with lower accuracy to have higher values in diversity and novelty. We believe that the ability of the embeddings to find non-linear relationships contributes to the model novelty and diversity. This property of the model allows it, for example, to discover relationships between popular and not so popular items leading to better diversity. Moreover, the integration in the linear model allows to keep its advantage in terms on accuracy, clearly suparssing the use of embeddings in isolation (ELP).

5   Conclusions and Future Work

In this paper, we presented EER, a method to enhance an existing recommendation algorithm to produce recommendations that are both more diverse and novel, while maintaining similar levels on accuracy. This process is done by combining two models, a linear one that is able to obtain good levels of accuracy, with a model based in an embedding technique that extracts non-linear relationships, allowing it to produce more diverse and novel recommendations.

As future work, we plan to apply the same technique to other recommender systems, examining if it can be applied in general to enhance the recommendations, independently of the base algorithm chosen for the task. We also envision studying the effects that varying the value of α in Eq. 3 has on the recommendations.

Acknowledgements. This work was supported by project RTI2018-093336-B-C22 (MCIU & ERDF), project GPC ED431B 2019/03 (Xunta de Galicia & ERDF) and accreditation ED431G 2019/01 (Xunta de Galicia & ERDF). The first author also acknowledges the support of grant FPU17/03210 (MCIU).

References

1. Balabanović, M., Shoham, Y.: Fab: content-based, collaborative recommendation. Commun. ACM **40**(3), 66–72 (1997). https://doi.org/10.1145/245108.245124
2. Bellogín, A., Castells, P., Cantador, I.: Precision-oriented evaluation of recommender systems. In: Proceedings of the 5th ACM Conference on Recommender Systems, RecSys 2011, pp. 333–336. ACM, New York (2011). https://doi.org/10.1145/2043932.2043996
3. Bennett, J., Lanning, S., et al.: The netflix prize. In: Proceedings of KDD Cup and Workshop, vol. 2007, p. 35. ACM, New York (2007)
4. Cremonesi, P., Koren, Y., Turrin, R.: Performance of recommender algorithms on top-n recommendation tasks. In: Proceedings of the 4th ACM Conference on Recommender Systems, RecSys 2010, pp. 39–46. ACM, New York (2010). https://doi.org/10.1145/1864708.1864721
5. Fleder, D., Hosanagar, K.: Blockbuster culture's next rise or fall: the impact of recommender systems on sales diversity. Manag. Sci. **55**(5), 697–712 (2009). https://doi.org/10.1287/mnsc.1080.0974
6. Ge, M., Delgado-Battenfeld, C., Jannach, D.: Beyond accuracy: evaluating recommender systems by coverage and serendipity. In: RecSys 2010, pp. 257–260. ACM (2010). https://doi.org/10.1145/1864708.1864761
7. de Gemmis, M., Lops, P., Musto, C., Narducci, F., Semeraro, G.: Semantics-aware content-based recommender systems. In: Ricci, F., Rokach, L., Shapira, B. (eds.) Recommender Systems Handbook, pp. 119–159. Springer, Boston (2015). https://doi.org/10.1007/978-1-4899-7637-6_4
8. Herlocker, J.L., Konstan, J.A., Terveen, L.G., Riedl, J.T.: Evaluating collaborative filtering recommender systems. ACM Trans. Inf. Syst. **22**(1), 5–53 (2004). https://doi.org/10.1145/963770.963772
9. Kabbur, S., Ning, X., Karypis, G.: FISM: factored item similarity models for top-n recommender systems. In: Proceedings of the 19th ACM SIGKDD International Conference on Knowledge Discovery and Data Mining, KDD 2013, pp. 659–667. ACM, New York (2013). https://doi.org/10.1145/2487575.2487589
10. Koren, Y., Bell, R.: Advances in collaborative filtering. In: Ricci, F., Rokach, L., Shapira, B. (eds.) Recommender Systems Handbook, pp. 77–118. Springer, Boston (2015). https://doi.org/10.1007/978-1-4899-7637-6_3
11. Landin, A., Suárez-García, E., Valcarce, D.: When diversity met accuracy: a story of recommender systems. Proceedings **2**(18) (2018). https://doi.org/10.3390/proceedings2181178

12. Landin, A., Valcarce, D., Parapar, J., Barreiro, Á.: PRIN: a probabilistic recommender with item priors and neural models. In: Azzopardi, L., Stein, B., Fuhr, N., Mayr, P., Hauff, C., Hiemstra, D. (eds.) ECIR 2019. LNCS, vol. 11437, pp. 133–147. Springer, Cham (2019). https://doi.org/10.1007/978-3-030-15712-8_9
13. McNee, S.M., Riedl, J., Konstan, J.A.: Being accurate is not enough: how accuracy metrics have hurt recommender systems. In: Extended Abstracts on Human Factors in Computing Systems, CHI EA 2006. p. 1097. ACM Press, New York (2006). https://doi.org/10.1145/1125451.1125659
14. Mikolov, T., Chen, K., Corrado, G., Dean, J.: Efficient Estimation of Word Representations in Vector Space. CoRR abs/1301.3, pp. 1–12 (2013)
15. Mikolov, T., Sutskever, I., Chen, K., Corrado, G., Dean, J.: Distributed representations of words and phrases and their compositionality. In: Proceedings of the 26th International Conference on Neural Information Processing Systems, NIPS 2013, pp. 3111–3119. Curran Associates Inc., USA (2013)
16. Ning, X., Karypis, G.: Slim: sparse linear methods for top-n recommender systems. In: 2011 IEEE 11th International Conference on Data Mining, pp. 497–506 (2011). https://doi.org/10.1109/ICDM.2011.134
17. Ning, X., Desrosiers, C., Karypis, G.: A comprehensive survey of neighborhood-based recommendation methods. In: Ricci, F., Rokach, L., Shapira, B. (eds.) Recommender Systems Handbook, pp. 37–76. Springer, Boston (2015). https://doi.org/10.1007/978-1-4899-7637-6_2
18. Parapar, J., Losada, D.E., Presedo-Quindimil, M.A., Barreiro, A.: Using score distributions to compare statistical significance tests for information retrieval evaluation. J. Assoc. Inf. Sci. Technol. (2019). https://doi.org/10.1002/asi.24203
19. Pennington, J., Socher, R., Manning, C.: Glove: global vectors for word representation. In: Proceedings of the 2014 Conference on Empirical Methods in Natural Language Processing, EMNLP 2014, pp. 1532–1543. ACL, Stroudsburg (2014). https://doi.org/10.3115/v1/D14-1162
20. Valcarce, D., Bellogín, A., Parapar, J., Castells, P.: On the robustness and discriminative power of information retrieval metrics for top-n recommendation. In: Proceedings of the 12th ACM Conference on Recommender Systems, RecSys 2018, pp. 260–268. ACM, New York (2018). https://doi.org/10.1145/3240323.3240347
21. Valcarce, D., Landin, A., Parapar, J., Barreiro, Á.: Collaborative filtering embeddings for memory-based recommender systems. Eng. Appl. Artif. Intell. **85**, 347–356 (2019). https://doi.org/10.1016/j.engappai.2019.06.020
22. Valcarce, D., Parapar, J., Barreiro, Á.: Item-based relevance modelling of recommendations for getting rid of long tail products. Knowl.-Based Syst. **103**, 41–51 (2016). https://doi.org/10.1016/j.knosys.2016.03.021
23. Wang, Y., Wang, L., Li, Y., He, D., Chen, W., Liu, T.Y.: A theoretical analysis of NDCG ranking measures. In: Proceedings of the 26th Annual Conference on Learning Theory, COLT 2013. pp. 1–30. JMLR.org (2013)
24. Zhou, T., Kuscsik, Z., Liu, J.G., Medo, M., Wakeling, J.R., Zhang, Y.C.: Solving the apparent diversity-accuracy dilemma of recommender systems. Proc. Natl. Acad. Sci. **107**(10), 4511–4515 (2010). https://doi.org/10.1073/pnas.1000488107

A Multi-task Approach to Open Domain Suggestion Mining Using Language Model for Text Over-Sampling

Maitree Leekha$^{(\boxtimes)}$ ⓘ, Mononito Goswami ⓘ, and Minni Jain

Delhi Technological University, New Delhi, India
{maitreeleekha_bt2k16,mononito_bt2k16,minnijain}@dtu.ac.in

Abstract. Consumer reviews online may contain suggestions useful for improving commercial products and services. Mining suggestions is challenging due to the absence of large labeled and balanced datasets. Furthermore, most prior studies attempting to mine suggestions, have focused on a single domain such as `Hotel` or `Travel` only. In this work, we introduce a novel over-sampling technique to address the problem of class imbalance, and propose a multi-task deep learning approach for mining suggestions from multiple domains. Experimental results on a publicly available dataset show that our over-sampling technique, coupled with the multi-task framework outperforms state-of-the-art open domain suggestion mining models in terms of the F-1 measure and AUC.

Keywords: Open domain suggestion mining · Multi-task learning · Over-sampling techniques · Deep learning

1 Introduction

Consumers often express their opinions towards products and services through online reviews and discussion forums. These reviews may include useful suggestions that can help companies better understand consumer needs and improve their products and services. However, manually mining *suggestions* amid vast numbers of *non-suggestions* can be cumbersome, and equated to finding needles in a haystack. Therefore, designing systems that can automatically mine suggestions is essential. The recent *SemEval* [6] challenge on Suggestion Mining saw many researchers using different techniques to tackle the domain-specific task (*in-domain Suggestion Mining*). However, *open-domain suggestion mining*, which obviates the need for developing separate suggestion mining systems for different domains, is still an emerging research problem. We formally define the problem of open-domain suggestion mining as follows:

Definition 1 (Open-domain Suggestion Mining). *Given a set of reviews* $\mathcal{R} = \{r_1, r_2 \ldots r_n\}$ *from multiple domains in* $\mathcal{D} = d_1 \cup d_2 \cup \ldots d_m$, *train a*

M. Leekha, M. Goswami and M. Jain—Contributed equally and would like to be consider as joint first authors.

© Springer Nature Switzerland AG 2020
J. M. Jose et al. (Eds.): ECIR 2020, LNCS 12036, pp. 223–229, 2020.
https://doi.org/10.1007/978-3-030-45442-5_28

classifier C using $\mathcal{D}$ to predict the nature $n_i \in \{$ 'suggestion', 'non-suggestion'$\}$ of each review r_i.

Building on the work of [5], we design a framework to detect suggestions from multiple domains. We formulate a multitask classification problem to identify both the domain and nature (*suggestion* or *non-suggestion*) of reviews. Furthermore, we also propose a novel language model-based text over-sampling approach to address the class imbalance problem.

2   Methodology

2.1   Dataset and Pre-processing

We use the first publicly available and annotated dataset for suggestion mining from multiple domains created by [5]. It comprises of reviews from four domains namely, `hotel`, `electronics`, `travel` and `software`. During pre-processing, we remove all URLs (eg. *https://* ...) and punctuation marks, convert the reviews to lower case and lemmatize them. We also pad the text with start `S` and end `E` symbols for over-sampling.

2.2   Over-Sampling Using Language Model: LMOTE

One of the major challenges in mining suggestions is the imbalanced distribution of classes, *i.e.* the number of non-suggestions greatly outweigh the number of suggestions (refer Table 1). To this end, studies frequently utilize *Synthetic Minority Over-sampling Technique* (SMOTE) [1] to over-sample the minority class samples using the text embeddings as features. However, SMOTE works in

Table 1. Datasets and their sources used in our study [5]. The class ratio column highlights the extent of class imbalance in the datasets. The `travel` datasets have lower inter-annotator agreement than the rest, indicating that they may contain confusing reviews which are hard to confidently classify as suggestions or non-suggestions. This also reflects in our classification results.

Dataset	Source	Class ratio (suggestion: non-suggestion)	Inter-annotator agreement
Hotel train	Tripadvisor	$448/7086 \approx 6{:}100$	0.86
Hotel test	Tripadvisor	$404/3000 \approx 13{:}100$	0.86
Electronics train	Amazon	$324/3458 \approx 9{:}100$	0.83
Electronics test	Amazon	$101/1070 \approx 9{:}100$	0.83
Travel train	Insight vacations, Fodors	$1314/3869 \approx 34{:}100$	0.72
Travel test	Fodors	$229/871 \approx 26{:}100$	0.72
Software train	Uservoice suggestion forum	$1428/4296 \approx 33{:}100$	0.81
Software test	Uservoice suggestion forum	$296/742 \approx 39{:}100$	0.81

Table 2. Most frequent 5-grams and their corresponding suggestions sampled using LMOTE. While the suggestions as a whole may not be grammatically correct, their constituent phrases are nevertheless semantically sensible.

Domain	Most frequent 5-gram	A suggestion sampled using LMOTE
Hotel	I would definitely recommend hotel	I would definitely recommend hotel good value full ocean view great food worth
Electronics	Suggestion get lens protector help	Suggestion get lens protector help protect long lens coating uv 52 lens last long must try
Travel	Tipping remember shape luggage concerned	Tipping remember shape luggage concerned heavy luggage rough adviced wheeled duffle wont heavy
Software	It would be good if oversight	It would be good if oversight bixby developed bug feels wide back content zoom should be an option

the euclidean space and therefore does not allow an intuitive understanding and representation of the over-sampled data, which is essential for qualitative and error analysis of the classification models. We introduce a novel over-sampling technique, **L**anguage **M**odel-based **O**ver-sampling **T**echnique (LMOTE), exclusively for text data and note comparable (and even slightly better sometimes) performance to SMOTE. We use LMOTE to over-sample the number of suggestions before training our classification model. For each domain, LMOTE uses the following procedure to over-sample suggestions:

Find Top η n-Grams: From all reviews labelled as suggestions (positive samples), sample the top $\eta = 100$ most frequently occurring n-grams ($n = 5$). For example, the phrase *"nice to be able to"* occurred frequently in many domains.

Train Language Model on Positive Samples: Train a BiLSTM language model on the positive samples (suggestions). The BiLSTM model predicts the probability distribution of the next word (w_t) over the whole vocabulary ($V \cup E$) based on the last $n = 5$ words ($w_{t-5}, \ldots, w_{t-1}$), *i.e.*, the model learns to predict the probability distribution $P(w_i \mid w_{t-5} \ w_{t-4} \ w_{t-3} \ w_{t-2} \ w_{t-1}) \ \forall i \in (V \cup E)$, such that $w_t = \arg\max_{w_i} P(w_i \mid w_{t-5} \ w_{t-4} \ w_{t-3} \ w_{t-2} \ w_{t-1})$.

Generate Synthetic Text Using Language Model and Frequent n-Grams: Using the language model and a randomly chosen frequent 5-gram as the seed, we generate text by repeatedly predicting the most probable next word (w_t), until the end symbol E is predicted.

Table 2 comprises of the most frequent 5-grams and their corresponding suggestions 'sampled' using LMOTE. In our study, we generate synthetic positive reviews till the number of suggestion and non-suggestion class samples becomes equal in the training set.

Algorithm 1. Language Model-based Over-sampling Technique (LMOTE)

Input:
$\mathcal{D}_{sugg} = \{r_i \in \mathcal{D} \mid n_i = \{\text{'}suggestion\text{'}\}\}$ Suggestions from a particular domain $\mathcal{D}$;
η- Number n-grams to use in LMOTE;
n- type of n-grams $e.g.$ 2 for bi-grams, etc.;
$\mathcal{N}$- Number of suggestion samples required.
Output: $\mathcal{S} : \mathcal{N}$ over-sampled suggestions

1: $n\_grams \leftarrow$ **NGrams**$(\mathcal{D}_{sugg}, \eta, n)$
2: $language\_model \leftarrow$ **TrainLanguageModel**$(\mathcal{D}_{sugg}, n)$
3: **Initialize** $\mathcal{S} \leftarrow \mathcal{D}_{sugg}$
4: **while** $|\mathcal{S}| < \mathcal{N}$ **do**
5:      $seed \leftarrow$ **random**(n_grams)
6:      $sample \leftarrow$ **LMOTEGenerate**$(language\_model, seed)$
7:      $\mathcal{S} \leftarrow \mathcal{S} \cup sample$
8: **end while**
9: **return** $\mathcal{S}$

Algorithm 1 summarizes the LMOTE over-sampling methodology. Following is a brief description of the sub-procedures used in the algorithm:

- **NGrams**$(\mathcal{D}_{sugg}, \eta, n)$: It returns the top η $n\text{-}grams$ from the set of suggestions, $\mathcal{D}_{sugg}$.
- **TrainLanguageModel**$(\mathcal{D}_{sugg}, n)$: This procedure trains an $n\text{-}gram$ BiL-STM Language Model on $\mathcal{D}_{sugg}$.
- **random**(n_grams)- Randomly selects an $n\text{-}gram$ from the input set.
- **LMOTEGenerate**$(language\_model, seed)$: The procedure takes as input the trained language model and a randomly chosen $n\text{-}gram$ from the set of top η $n\text{-}grams$ as $seed$, and starts generating a review till the end tag, E is produced. The procedure is repeated until we have a total of $\mathcal{N}$ suggestion reviews.

2.3   Mining Suggestion Using Multi-task Learning

Multi-task learning (MTL) has been successful in many applications of machine learning since sharing representations between auxiliary tasks allows models to generalize better on the primary task. Figure 1B illustrates 3-dimensional UMAP [4] visualization of *text embeddings* of suggestions, coloured by their domain. These embeddings are outputs of the penultimate layer (dense layer before the final softmax layer) of the *Single task* (STL) ensemble baseline. It can be clearly seen that suggestions from different domains may have varying feature representations. Therefore, we hypothesize that we can identify suggestions better by leveraging domain-specific information using MTL. Therefore, in the MTL setting, given a review r_i in the dataset, D, we aim to identify both the domain of the review, as well as its nature.

2.4   Classification Model

We use an ensemble of three architectures namely, CNN [2] to mirror the spatial perspective and preserve the n-gram representations; Attention Network to learn the most important features automatically; and a BiLSTM-based text RCNN [3] model to capture the context of a text sequence (Fig. 2). In the MTL setting, the ensemble has two output softmax layers, to predict the domain and nature of a review. The STL baselines on the contrary, only have a singe softmax layer to predict the nature of the review. We use ELMo [7] word embeddings trained on the dataset, as input to the models.

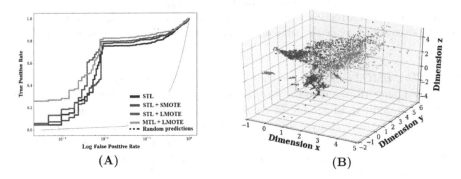

(A)   (B)

Fig. 1. (A) Receiver operating characteristics (TPR vs. Log FPR) curve pooled across all domains for all models used in this work demonstrates that LMOTE coupled with our multi-task model outperforms other considered alternatives across domains (B) 3-dimensional UMAP visualization of text embeddings of *suggestions* coloured by domain. Suggestions from different domains have distinct feature representations.

3   Results and Discussion

We conducted experiments to assess the impact of over-sampling, the performance of LMOTE and the multi-task model. We used the same train-test split as provided in the dataset for our experiments. All comparisons have been made in terms of the F-1 score of the suggestion class for a fair comparison with prior work on representational learning for open domain suggestion mining [5] (refer *Baseline* in Table 3). For a more insightful evaluation, we also compute the Area under Receiver Operating Characteristic (ROC) curves for all models used in this work. Tables 3, 4 and Figs. 3 and 1A summarize the results of our experiments, and there are several interesting findings:

Over-Sampling Improves Performance. To examine the impact of over-sampling, we compared the performance of our ensemble classifier with and without over-sampling *i.e.* we compared results under the *STL, STL + SMOTE* and *STL + LMOTE* columns. Our results confirm that in general, over-sampling suggestions to obtain a balanced dataset improves the performance (F-1 score & AUC) of our classifiers.

LMOTE Performs Comparably to SMOTE. We compared the performance of SMOTE and LMOTE in the single task settings (*STL + SMOTE*

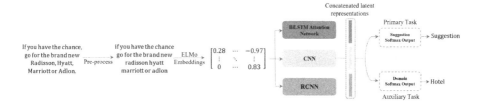

Fig. 2. Our multi-task classification model which consists of an ensemble of RCNN, CNN and BiLSTM attention network. The primary task is predicting the nature of a review (suggestion), while the auxiliary task involves predicting its domain (`hotel`).

and *STL + LMOTE*) and found that LMOTE performs comparably to SMOTE (and even outperforms it in the `electronics` and `software` domains). LMOTE also has the added advantage of resulting in intelligible samples which can be used to qualitatively analyze and troubleshoot deep learning based systems. For instance, consider suggestions *created* by LMOTE in Table 2. While the suggestions may not be grammatically correct, their constituent phrases are nevertheless semantically sensible.

Table 3. Performance evaluation using F-1 score. Multi-task learning with LMOTE outperforms other alternatives in open-domain suggestion mining. Furthermore, owing to potentially confusing reviews in the travel domain (Table 1), its F-1 scores are significantly lower than the other domains.

Domain	Baseline	STL	STL+SMOTE	STL+LMOTE	MTL+LMOTE
Hotel	0.77 (LSTM)	0.79	0.83	0.83	**0.86**
Electronics	0.78 (SVM)	0.80	0.80	0.83	**0.83**
Travel	0.66 (SVM)	0.65	0.68	0.69	**0.71**
Software	0.80 (LSTM)	0.79	0.81	0.84	**0.88**

Table 4. Performance evaluation using area under ROC with 95% confidence intervals. Multi-task learning with LMOTE outperforms other alternatives in open-domain suggestion mining. Multi-task learning leads to a significant improvement in AUC over its single task counterpart. (AUCs for baseline models proposed by [5] were unavailable.)

Domain	STL	STL+SMOTE	STL+LMOTE	MTL+LMOTE
Hotel	0.878 ± 0.022	**0.897 ± 0.021**	0.828 ± 0.025	0.894 ± 0.012
Electronics	0.897 ± 0.041	0.92 ± 0.037	0.912 ± 0.037	**0.944 ± 0.031**
Travel	0.828 ± 0.034	0.848 ± 0.033	0.835 ± 0.025	**0.852 ± 0.032**
Software	0.894 ± 0.025	0.893 ± 0.025	0.919 ± 0.022	**0.956 ± 0.015**
Pooled AUC	0.876 ± 0.014	0.883 ± 0.013	0.897 ± 0.012	**0.907 ± 0.012**

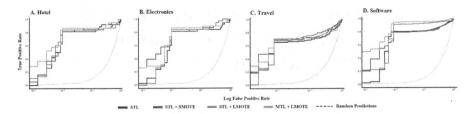

Fig. 3. Domain wise receiver operating characteristics (ROC) curves.

Multi-task Learning Outperforms Single-Task Learning. We compared the performance of our classifier in single and multi-task settings (*STL + LMOTE* and *MTL + LMOTE*) and found that by multi-task learning improves the performance of our classifier. We qualitatively analysed the single and multi task models, and found many instances where by leveraging domain-specific information the multi task model was able to accurately identify suggestions. For instance, consider the following review: *"Bring a Lan cable and charger for your laptop because house-keeping doesn't provide it."* While the review appears to be an assertion (*non-suggestion*), by predicting its domain (`hotel`), the multi-task model was able to accurately classify it as a suggestion.

4   Conclusion

In this work, we proposed a Multi-task learning framework for Open Domain Suggestion Mining along with a novel language model based over-sampling technique for text–LMOTE. Our experiments revealed that Multi-task learning combined with LMOTE over-sampling outperformed considered alternatives in terms of both the F1-score of the suggestion class and AUC.

References

1. Chawla, N.V., Bowyer, K.W., Hall, L.O., Kegelmeyer, W.P.: SMOTE: synthetic minority over-sampling technique. J. Artif. Intell. Res. **16**, 321–357 (2002)
2. Kim, Y.: Convolutional neural networks for sentence classification. In: Proceedings of the 2014 Conference on Empirical Methods in Natural Language Processing (EMNLP) (2014). https://doi.org/10.3115/v1/d14-1181
3. Lai, S., Xu, L., Liu, K., Zhao, J.: Recurrent convolutional neural networks for text classification. In: Twenty-Ninth AAAI Conference on Artificial Intelligence (2015)
4. McInnes, L., Healy, J., Melville, J.: UMAP: uniform manifold approximation and projection for dimension reduction. arXiv preprint arXiv:1802.03426 (2018)
5. Negi, S.: Suggestion mining from text. Ph.D. thesis, National University of Ireland Galway (NUIG) (2019)
6. Negi, S., Daudert, T., Buitelaar, P.: SemEval-2019 task 9: suggestion mining from online reviews and forums. In: SemEval@NAACL-HLT (2019)
7. Peters, M.E., et al.: Deep contextualized word representations. arXiv preprint arXiv:1802.05365 (2018)

MedLinker: Medical Entity Linking
with Neural Representations
and Dictionary Matching

Daniel Loureiro$^{(\boxtimes)}$ and Alípio Mário Jorge

LIAAD - INESCTEC, Porto, Portugal
{dloureiro,amjorge}@fc.up.pt

Abstract. Progress in the field of Natural Language Processing (NLP) has been closely followed by applications in the medical domain. Recent advancements in Neural Language Models (NLMs) have transformed the field and are currently motivating numerous works exploring their application in different domains. In this paper, we explore how NLMs can be used for Medical Entity Linking with the recently introduced MedMentions dataset, which presents two major challenges: (1) a large target ontology of over 2M concepts, and (2) low overlap between concepts in train, validation and test sets. We introduce a solution, MedLinker, that addresses these issues by leveraging specialized NLMs with Approximate Dictionary Matching, and show that it performs competitively on semantic type linking, while improving the state-of-the-art on the more fine-grained task of concept linking (+4 F1 on MedMentions main task).

Keywords: Entity Linking · Bioinformatics · Neural Language Models

1 Introduction

Medical Entity Recognition and Linking remain challenging tasks at the intersection between Natural Language Processing (NLP) and Information Retrieval (IR). The main difficulty arises from the fact that annotated datasets are scarce and particularly expensive to collect (require domain expertise), while the ontologies used in this domain are also especially large. From the standpoint of NLP, the relatively small and low-coverage datasets are hard to model using current neural approaches, whereas IR is limited by the subtle semantics underlying the different concepts that constitute the ontologies.

The recently introduced MedMentions [1] dataset provides the largest set of mention-level annotations targeting the UMLS (Unified Medical Language System) ontology. UMLS [6] is a compilation of several medical ontologies, making it the most comprehensive and broad, spanning a range of topics from viruses to

---

The research leading to these results has received funding from the European Union's Horizon 2020 - The EU Framework Programme for Research and Innovation 2014–2020, under grant agreement No. 733280.

J. M. Jose et al. (Eds.): ECIR 2020, LNCS 12036, pp. 230–237, 2020.
https://doi.org/10.1007/978-3-030-45442-5_29

biomedical occupations. Even though the MedMentions annotation effort was a substantial undertaking, it falls short of covering the full set of concepts comprising UMLS ($\sim$1% coverage), as well as displaying low overlap between the set of concepts occurring in its training splits and the set of concepts occurring in the development and test splits ($\sim$50% overlap). In order to overcome the challenges presented in this dataset, we propose a solution that's based on Neural Language Models (NLMs) but designed to fallback on Approximate Dictionary Matching (ADM) for zero-shot entity linking, taking advantage of the large lexicon provided with the UMLS Metathesaurus. Unlike previous approaches using NLMs in the medical domain [2–5], our solution decouples mention recognition and entity linking, leveraging NLMs for these subtasks in separate modules, and allowing for other methods, namely ADM, to take part in the linking process.

In this work we explore approaches using NLMs for the related task of Word Sense Disambiguation (WSD), particularly pooling methods for representing spans in NLM-space [9]. We show that the Semantic Type (STY) and Concept Unique Identifiers (CUI) embeddings learned in the process are useful for our linking tasks, and can be effectively combined with ADM for improved performance over previous solutions. While our solution, MedLinker[1], is designed for MedMentions, the breadth of the target ontology makes it useful elsewhere.

2   Related Work

In this section we focus on previous works using NLMs for biomedical NER, or addressing MedMentions directly. While there are already several works using the latest Transformer-based NLMs for biomedical tasks [2–4], these have, so far, focused only on the adaptation of the pretrained NLM for the medical domain, without considering how these can be leveraged with complementary approaches.

The authors of MedMentions reported results on a subset of their corpus (st21pv) using a popular method for biomedical NER called TaggerOne [8]. This method learns to jointly predict spans and link entities but relies mostly on discrete features. Also, MedMention's authors claimed that it took them several days to train their model with high-performance computing resources (e.g. 900 GB of RAM), raising tractability concerns about using TaggerOne.

The first results on applying NLMs to MedMentions have been recently reported by [5]. Their solution showed strong performance for semantic type linking, but followed the standard approach for NER tasks also used by [2–4].

3   Data

The MedMentions dataset is provided in two variants, one targeting the full ontology of UMLS, and another targeting a subset of that ontology[2] selected

---

[1] Package, code, and additional results: https://github.com/danlou/medlinker.
[2] Based on UMLS release 2017 AA Active - num. concepts: 3.4M (full), 2.3M (st21pv).

by domain experts as particularly interesting for medical document retrieval. MedLinker is trained and evaluated on this subset (st21pv) of MedMentions.

Regarding the concept aliases present in UMLS, we found it useful to introduce some additional restrictions that improve the processing speed of our string matching methods while maintaining task performance. We discard aliases longer than 5 tokens and aliases that include punctuation (except dashes). Aliases are lowercased, along with the query strings used with dictionary methods.

MedMentions uses the PubTator format, which annotates entities at the character-level. Since our methods require annotations at the token-level, we also preprocess the dataset with a tokenizer specialized for the medical domain. We use sciSpacy [7] for tokenization and sentence splitting, which is trained for the biomedical domain. Occasionally, sentence splitting errors incurred in misaligned mentions, and thus missing from training and evaluation. From a total of 203,282 annotations, this step produced 321 misalignments (0.16%).

4   Solution

In this section we describe our methods for mention recognition, entity matching using string-based methods and contextual embeddings, and finally how the different matchers are combined into our final predictions. In Fig. 1 we show how the major components of our solution interact with each other.

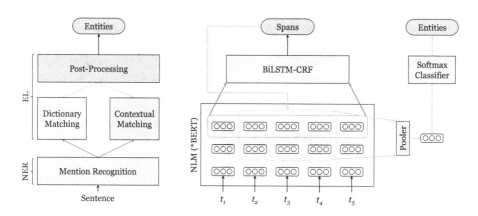

Fig. 1. Left: Overview of our solution, showing NER producing mentions that are matched to entities using independent approaches based on ngrams and contextual embeddings, which are combined in a post-processing step into the final entity predictions. Right: Detailed view into how we use NLMs to first derive spans from NER based on the last states of the NLM, then match entities based on a pooled representation of the predicted span (e.g. states for tokens t_2, t_3 and t_4 at layers $-1, -2$).

Mention Recognition Using NLMs. We follow the standard architecture for neural-based NER, using contextual embeddings from NLMs specialized for the medical domain [2–4]. This architecture is a BiLSTM that handles sequential encoding with long-term dependencies, together with a Conditional Random Field (CRF) which uses the BiLSTM's final states to improve dependencies between output labels. Similarly to [2], this model also employs character-level embeddings, learned during training, to capture morphological information.

Zero-Shot Linking with Approximate Dictionary Matching. Using SimString [10], we represent our restricted set of aliases from UMLS as character n-grams, similarly to previous works [11,12]. After experimenting with different sizes, we found that char n-grams of size 3 performed best. SimString matches strings (i.e. aliases) using the highly scalable CPMerge algorithm which is designed to find similar strings based on overlapping features (i.e. char n-grams).

Given aliases $a \in A_{\text{UMLS}}$, corresponding char n-gram features $\hat{a}$, and a function map for mapping entities (concepts) $e \in E_{\text{CUI/STY}}$ to aliases, we match query strings s, with char n-gram features $\hat{s}$, using the scoring function:

$$scoreSTR(s, e) = \max_{a \in map(e)} \cos(\hat{s}, \hat{a}) \quad (1) \qquad \cos(\hat{s}, \hat{a}) = \frac{|\hat{s} \cap \hat{a}|}{\sqrt{|\hat{s}| |\hat{a}|}} \quad (2)$$

Linking by Similarity to Entity Embeddings. Considering that MedMentions includes a large set of annotated spans, similarly to WSD corpora, we replicate the pooling method used in [9]. Essentially, we represent STYs and CUIs as the average of all their corresponding contextual embeddings, which are, in turn, represented by the sum of the embeddings from the last 4 layers of the NLM. This results in 21 STY embeddings, and 18,425 CUI embeddings that can be matched using Nearest Neighbors (1NN, only most similar) in NLM-space.

Given entities $e \in E_{\text{CUI/STY}}$, and corresponding precomputed embeddings $\overrightarrow{e}$, we match query strings s, with embedding $\overrightarrow{s}$, obtained by applying the same pooling procedure (with the full sentence), using the scoring function:

$$score1NN(s, e) = \cos(\overrightarrow{s}, \overrightarrow{e}) = \frac{\overrightarrow{s} \cdot \overrightarrow{e}}{\|\overrightarrow{s}\| \|\overrightarrow{e}\|} \quad (3)$$

Linking by Classifying Contextual Embeddings. Even though the 1NN approach is very successful for WSD, we also take this opportunity to experiment with training a classifier using contextual embeddings from NLMs, instead of averaging all contextual embeddings grouped by entity. In particular, we train a minimal softmax classifier, without hidden layers (# parameters = # embedding dimensions * # entities), on the annotations of the training set.

Using the same notation defined previously, we match query spans to entities using the following scoring function:

$$scoreCLF(s,e) = P(e|\overrightarrow{s}) = \frac{\exp^{f(\overrightarrow{s})_j}}{\sum_{i=1}^{|E|} \exp^{f(\overrightarrow{s})_i}} \quad (4) \qquad f : \mathbb{R}^{dim(\overrightarrow{s})} \to \mathbb{R}^{|E|} \quad (5)$$

The function f produces the output vector using weights learned during training with ADAM optimization and categorical cross-entropy loss. The output vector is processed by the softmax function to provide our class (entity) probabilities. We train for 100 epochs, with patience limit of 10, using a batch size of 64.

Combining String and Contextual Matching. We effectively combine our matchers using the following straightforward ensembling method (see Sect. 5):

$$scoreSTR\_CTX(s,e) = \max(scoreSTR(s,e), scoreCTX(s,e)) \quad (6)$$

where $scoreCTX$ can correspond either to $score1NN$ or $scoreCLF$, depending on the configuration being tested. Still, this method exhibits an undesirable bias towards higher recall and lower precision. Therefore, we introduce a post-processing (PP) step to minimize false positives. We train a Logistic Regression Binary Classifier, on the training set, using the following five features:

- $\max_{e \in E} scoreSTR(s,e)$ (max string).
- $\max_{e \in E} scoreCTX(s,e)$ (max context).
- $\max_{e \in E}(scoreSTR(s,e), scoreCTX(s,e))$ (max overall).
- $(\max_{e \in E} scoreSTR(s,e) + \max_{e \in E} scoreCTX(s,e))/2$ (average).
- $\arg\max_{e \in E} scoreSTR(s,e) == \arg\max_{e \in E} scoreCTX(s,e)$ (agreement).

Testing this classifier with different thresholds, we're able to determine the optimal thresholds for balanced Precision and Recall performance (see Fig. 2).

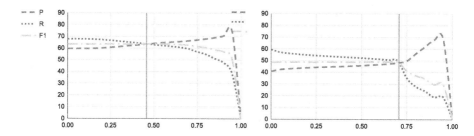

Fig. 2. Performance variation on validation set of MedMentions st21pv (Left: STY, Right: CUI) using different thresholds on the decision filter. Vertical line (green) marks the threshold which resulted in the best balance between Precision and Recall (Color figure online).

5   Evaluation

We evaluate our solution on the test split of MedMentions st21pv, following the metrics described in the MedMentions paper [1]. We require that $scoreSTR$, $score1NN$ and $scoreCLF$ have scores above 0.5 in order to reduce the incidence of false positives. In the event of ties, matches are sorted alphabetically.

Mention Recognition. As described in Sect. 4, our efforts on this task focus on experimenting with different NLMs specialized for the medical domain. On Table 1 we show that the various specialized NLMs we used to initialize the BiLSTM-CRF model produce comparable results on this task.

Table 1. Mention recognition performance using different specialized NLMs.

Model	P	R	F1
Exact match	51.32	32.96	40.14
NCBI BERT (uncased)	**69.44**	69.38	69.41
BioBERT 1.1 (cased)	70.00	70.43	70.21
SciBERT (SciVocab)			
- Uncased	69.42	**71.81**	**70.59**
- Cased	69.16	71.30	70.22

Mention Linking. On Table 2 we present our results for Entity Linking using the predicted spans from the SciBERT (uncased) based NER model that performed best for Mention Recognition. These results show that our solution performs competitively, achieving state-of-the-art on CUI Linking, and comparable results to the state-of-the-art on STY Linking. Additionally, we also report performance using the different scoring functions covered in this paper. $score1NN$ outperforms $scoreCLF$ on CUI Linking, by a small margin, but on STY Linking $scoreCLF$ substantially outperforms $score1NN$. We believe these differences can be explained from the fact that all STYs are represented in the training set, while the overlap between CUIs in the training and test sets is low.

Category Performance. Using gold spans to focus on linking performance, we notice[3] that types/categories such as 'T005-Virus' obtain 88 F1, while 'T022-Body System' obtains only 44 F1, on STY Linking. CUI Linking results show similar variations, although with a stronger tendency towards better performance on concepts belonging to narrower types (i.e. STYs encompassing fewer CUIs).

---

[3] Results for all categories: https://github.com/danlou/medlinker/categories.pdf.

Table 2. Semantic Type (STY) and Concept (CUI) Linking performance comparison.
[†] were produced using the same st21pv subset of UMLS release 2017 AA Active.

Model	STY Linking			CUI Linking		
	P	R	F1	P	R	F1
Exact match	49.04	31.97	38.71	47.12	31.11	37.48
QuickUMLS[†] (v1.3) [11]	14.51	16.87	15.60	17.98	26.11	21.30
ScispaCy[†] (v0.2.4) [7]	10.14	31.68	15.36	25.17	53.52	34.24
TaggerOne [1]	N/A	N/A	N/A	47.10	43.60	45.30
Nejadgholi et al. [5]						
- BioBERT	61	66	63	N/A	N/A	N/A
- BioBERT|BERT-base	**63**	65	**64**	N/A	N/A	N/A
MedLinker						
- $scoreSTR$	48.31	56.81	52.22	33.03	47.34	38.91
- $score1NN$	46.62	62.67	53.47	33.61	55.16	41.77
- $scoreCLF$	58.62	64.63	61.48	32.21	52.66	39.97
- $scoreSTR\_1NN$	53.06	65.94	58.80	40.46	**59.69**	48.23
- $scoreSTR\_CLF$	59.23	**67.81**	63.23	40.70	59.59	48.37
- $scoreSTR\_CLF$ (PP, bal. thresh.)	**63.13**	63.69	63.41	**48.43**	50.07	**49.24**

6 Conclusion

A major issue in biomedical NLP or IR research is the fact that annotated datasets are scarce and expensive to collect. While that problem is unlikely to improve in the near future, this work has shown that there's still significant room for improvement by simply adapting existing approaches, such as end-to-end neural models for NER or WSD, making them easier to integrate with previous works often unassociated to those approaches, such as ADM. From a more practical perspective, this work pushes the state-of-the-art on the new and challenging MedMentions dataset, while using a modular approach that can be further improved with the integration of additional IR methods in future work.

References

1. Mohan, S., Li, D.: MedMentions: a large biomedical corpus annotated with UMLS concepts. In: AKBC (2019)
2. Beltagy, I., Lo, K., Cohan, A.: SciBERT: a pretrained language model for scientific text. In: IJCNLP (2019)
3. Lee, J., et al.: BioBERT: a pre-trained biomedical language representation model for biomedical text mining. Bioinformatics **36**(4), 1234–1240 (2020)
4. Peng, Y., Yan, S., Lu, Z.: Transfer learning in biomedical natural language processing: an evaluation of BERT and ELMo on ten benchmarking datasets. In: BioNLP@ACL (2019)

5. Nejadgholi, I., Fraser, K.C., De Bruijn, B., Li, M., LaPlante, A., Abidine, K.Z.: Extracting UMLS concepts from medical text using general and domain-specific deep learning models. In: LOUHI@EMNLP (2019)

6. Bodenreider, O.: The unified medical language system (UMLS): integrating biomedical terminology. Nucleic Acids Res. **32**(suppl_1), 267–270 (2004)

7. Neumann, M., King, D., Beltagy, I., Ammar, W.: ScispaCy: fast and robust models for biomedical natural language processing. In: BioNLP@ACL (2019)

8. Leaman, R., Lu, Z.: TaggerOne: joint named entity recognition and normalization with semi-Markov models. Bioinformatics **32**(18), 2839–2846 (2016)

9. Loureiro, D., Jorge, A.M.: Language modelling makes sense: propagating representations through wordnet for full-coverage word sense disambiguation. In: ACL (2019)

10. Okazaki, N., Tsujii, J.: Simple and efficient algorithm for approximate dictionary matching. In: COLING (2010)

11. Soldaini, L., Goharian, N.: QuickUMLS: a fast, unsupervised approach for medical concept extraction. In: MedIR Workshop, SIGIR (2016)

12. Kaewphan, S., Hakala, K., Miekka, N., Salakoski, T., Ginter, F.: Wide-scope biomedical named entity recognition and normalization with CRFs, fuzzy matching and character level modeling. Database **2018**, (2018)

Ranking Significant Discrepancies
in Clinical Reports

Sean MacAvaney[1([⊠])], Arman Cohan[2], Nazli Goharian[1], and Ross Filice[3]

[1] IR Lab, Georgetown University, Washington DC, USA
{sean,nazli}@ir.cs.georgetown.edu
[2] Allen Institute for Artificial Intelligence, Seattle, WA, USA
armanc@allenai.org
[3] Department of Radiology, MedStar Georgetown University Hospital,
Washington DC, USA
Ross.W.Filice@medstar.net

Abstract. Medical errors are a major public health concern and a lead-ing cause of death worldwide. Many healthcare centers and hospitals use reporting systems where medical practitioners write a preliminary med-ical report and the report is later reviewed, revised, and finalized by a more experienced physician. The revisions range from stylistic to cor-rections of critical errors or misinterpretations of the case. Due to the large quantity of reports written daily, it is often difficult to manually and thoroughly review all the finalized reports to find such errors and learn from them. To address this challenge, we propose a novel ranking approach, consisting of textual and ontological overlaps between the pre-liminary and final versions of reports. The approach learns to rank the reports based on the degree of discrepancy between the versions. This allows medical practitioners to easily identify and learn from the reports in which their interpretation most substantially differed from that of the attending physician (who finalized the report). This is a crucial step towards uncovering potential errors and helping medical practitioners to learn from such errors, thus improving patient-care in the long run. We evaluate our model on a dataset of radiology reports and show that our approach outperforms both previously-proposed approaches and more recent language models by 4.5% to 15.4%.

1 Introduction

Medical errors are a pervasive problem in healthcare that can result in serious patient harm [11]. To identify and reduce the occurrence of preventable errors, many medical centers use reporting systems to document cases. Initial reports are often reviewed and revised by more experienced physicians. The revisions could be due to stylistic reasons or (more importantly) misinterpretations/errors in the initial report. In such cases, to prevent recurrence of the errors, is crucial to identify reports with substantive differences between the original and final report and discuss them with the clinician who wrote the initial report. It is

© Springer Nature Switzerland AG 2020
J. M. Jose et al. (Eds.): ECIR 2020, LNCS 12036, pp. 238–245, 2020.
https://doi.org/10.1007/978-3-030-45442-5_30

often challenging to manually identify such cases among the large number of daily written reports in a timely manner. In this work, we propose an approach for ranking revisions of medical reports by the degree of discrepancy between the different versions of the report. This allows medical practitioners to easily find the reports in which they made an error, which helps them learn from their mistakes and prevent future similar errors.

This is a challenging task to automate because the edits that an attending physician makes to a report can range from stylistic differences to significant discrepancies that may have a major effect on the patient (e.g., an unobserved mass). See Fig. 1 for an example of significant and non-significant discrepancies from radiology reports. As we can see, differences between the significant and non-significant discrepancies are often not trivial to identify and requires more than just comparing surface word changes in the versions of the reports. Furthermore, significant discrepancies can occur relatively frequently in practice; in our dataset collected from a large urban hospital, around 7% of reports contained significant errors. With hundreds of reports generated a week at some hospitals, this can amount to a considerable number of errors. We address this problem by proposing a supervised ranking approach for clinician's revised reports by the degree that there are significant discrepancies between their preliminary report and the final corrected report. I.e., our goal is to rank revisions that are more likely due to errors higher than revisions that are merely due to stylistic changes.

...with imaging features strongly suggestive of hepatocellular carcinoma (LI-RADS 4) ~~not well discernible~~ probably present but not conspicuous on prior examination...	... 3. Left renal artery: Single with ~~a~~ slightly early branching first branch point ~~averaging~~ averages 1.9 cm from the left lateral margin of the aorta. Left renal vein: Single without late confluence...
Significant discrepancy	**Non-Significant discrepancy**

Fig. 1. Example radiology impression revisions (strikeout removed, underline added). We aim to rank report revisions by the significance of the discrepancy.

Prior works have investigated significant discrepancies in medical reports through comparison of surface textual features [14,19], semantic similarity features [2], and word frequencies [7]. These works often treat the problem as classification and the most successful ones leverage a variety of textual similarity measures. Viewing this problem as ranking is a more suitable and practical form of evaluation; given a doctor's limited time, it is important for them to be presented with the reports that have the most significant discrepancies.

Document ranking in the broad medical domain have received extensive interest of researchers [8,13,15–17,21]. However, these efforts focus on conventional query-document retrieval. Our goal is to rank significant discrepancies by measuring the semantic overlap between the initial and final report. There have also been efforts to identify semantic similarity between two texts, e.g., for paraphrase identification [5,9,12,18], but these approaches operate on the sentence-level, making them unsuitable for documents (e.g., radiology reports).

To summarize, our contributions are: (i) we propose an end-to-end supervised ranking model for identifying significant discrepancies in medical reports. (ii) We demonstrate that our approach outperforms both previously proposed approaches and more recent language model approaches in a variety of metrics. (iii) We provide an analyses of the importance of different model components.

2   Model

We propose a supervised model that measures the overlap between the preliminary and final report for the purpose of ranking pairs of preliminary and final reports based on their significance over a given period of time. We observe that the central challenge of this task is being permissive of surface-level changes (which may be considerable), while emphasizing changes of substance, which may be subtle (see Fig. 1 for examples of such changes). To address this, we incorporate *importance* and *similarity* scores. The *importance score* weights each term/phrase and is learned during training. This score allows for terms that are not important to have less of an impact on the ranking score of the report (e.g., words like *well* and *but*). Note that this is a special application in which some function words that are often ignored actually have a big impact on the meaning of a report (e.g., *not* is often considered a stop word and removed). We let the model learn which terms are important during training. The *matching score* allows for the model to account for the replacement of similar terms using the cosine distance of word vectors (e.g., *averaging* and *averages* are similar) and synonym information from a domain-specific ontology (*chauffeur fracture* and *Hutchinson fracture* are synonymous). This allows the replacement of semantically-similar terms to have little impact on the ranking score. We calculate three *similarity scores* (addition, deletion, and overlap) using the importance and matching scores, and linearly combine them as a *ranking score*.

Notation and Task Definition. Let R be a set of clinical reports. Each report $r \in R$ consists of a preliminary and final version of the report (p and f, respectively), and a label $l \in \{0, 1, \ldots, L\}$ indicating the degree of discrepancy between p and f. Each version of the report consists of a sequence of tokens, denoted by p_i and f_i. The significant discrepancy ranking task produces a ranking score $s \in \mathbb{R}$ for each report $r \in R$ such that the reports with higher degrees of discrepancy are assigned a higher ranking score.

Similarity Scores. Our approach combines several similarity scores to produce a ranking score. Specifically, we measure the weighted soft additions, deletions, and overlap of unigrams, n-grams, and ontological entities. The addition score (S_a, Eq. 1) defines weighted soft similarity as the ratio between the similarity score (weighted by a learned importance score) and the total importance of all terms in the final report. Thus, terms from the final report that do not appear in the preliminary report (i.e., additions) yield a higher score. The deletion score (S_d, Eq. 2) is defined similarly, but in terms of the preliminary report; terms from the preliminary report that do not appear in the final report (deletions) yield a higher score. The overlap score (S_o, Eq. 3) combines the addition and deletion scores into one succinct measure. We use all three scores to measure term unigram, n-gram, and ontological differences (defined below). We define the similarity functions (where $M_X(y) \in [0,1]$ is a matching score of term y in X, and $I(y) \in [0,1]$ is the importance score of term y) as:

$$S_a(p,f) = -\frac{\sum_{f_i \in f} M_p(f_i)I(f_i)}{\sum_{f_i \in f} I(f_i)} \quad (1) \qquad S_d(p,f) = -\frac{\sum_{p_i \in p} M_f(p_i)I(p_i)}{\sum_{p_i \in p} I(p_i)} \quad (2)$$

$$S_o(p,f) = -\frac{\sum_{p_i \in p} M_f(p_i)I(p_i) + \sum_{f_i \in f} M_p(f_i)I(f_i)}{\sum_{p_i \in p} I(p_i) + \sum_{f_i \in f} I(f_i)} \quad (3)$$

Unigram and N-gram Matching. Unigram matching can provide valuable signals for significance in radiology reports. For instance, the addition *no* (e.g., *fracture* vs. *no fracture*) could change the meaning of the report considerably. We define the matching function for unigrams as the maximum cosine similarity between the word embeddings ($emb(\cdot)$) of the term and any term in the other report, and a unigram importance function using a simple feed-forward layer with sigmoid activation (W_{imp} and b_{imp} as model parameters):

$$M_X(y) = \max_{x \in X}(\cos(emb(x), emb(y))) \quad (4)$$

$$I(y) = \sigma(emb(y)W_{imp} + b_{imp}) \quad (5)$$

N-gram matching provides another important view of similarity, since there are many multi-word noun phrases in radiological notes. For instance, *right arm* and *left arm* represent completely different parts of the body, and should be treated differently. We handle n-grams by first taking the average of the embeddings over sliding windows. This is a simple and effective way to combine the representations. We use bi-grams and tri-grams in our experiments.

Ontological Matching. Since medical knowledge is broad and extensive, the model may never encounter certain medical entities during training. This knowledge may also not be captured effectively by embeddings. Thus, it is valuable to explicitly encode domain information into the model using an ontology. We use a mapping function that matches any exact ontological name to the corresponding concept, a constant similarity for exact entity matches, and constant weight for all ontology concepts. We use RadLex (v4.0, http://radlex.org/), an ontology of radiology concepts (e.g., procedures, diagnoses, etc.).

3   Experiments

Dataset. We train and evaluate using a dataset of 3,368 radiology reports from a large urban hospital. Each sample consists of a preliminary report written by a resident, and a final report revised by the attending radiologist, who labeled the edit by the degree of discrepancy between the two reports. The labels are 0 (attending doctor fully agrees with assessment of the resident, 81% of reports), 1 (errors exist, but they are insignificant to the overall impression, 12%), 2 (subtle, yet important, error exists, 6%), 3 (an obvious error exists, 1%). We split the dataset into 122 sets based on the combination of resident and week (*ranking sets*, average 27.6 reports per ranking set, min 5, max 148). Since residents often work weekly shifts, this is a valuable setting because it allows residents to review report discrepancies from the past week. We randomly split the ranking sets into 60-20-20 train-dev-test set splits. Each ranking set consists of at least 5 reports, each with at least one report discrepancy. Radiology reports contain several sections; we primarily concern ourselves with the summary section of the reports (called the *impression*) because it contains the main findings.

Baselines. To evaluate the effectiveness of the model, we compare with the variety of methods in the state-of-the-art, including ranking models, domain specific models and textual similarity models, briefly described below:

- **Vector space model (VSM).** We use the traditional TFIDF-weighted vector space similarity score between the preliminary and final report (from lucene).
- **BiPACRR.** We test the PACRR [6] neural IR model because it learns to identify n-gram similarity between two texts. We modify the architecture to learn two scores (one with the preliminary report as the query and the other with the final report as the query), and linearly combine them to produce a final ranking score. We call this variant BiPACRR. We also experimented with other neural rankers (e.g., KNRM [20]), but BiPACRR was the most effective.
- **Textual similarity regression (SimReg)** [2]. This approach uses logistic regression to combine several hand-crafted features (mostly consisting of textual similarity measures and lexical features) to identify significant discrepancies in radiology reports. Since this approach performs classification, we use the label score as the ranking score. Our experiments used the authors' implementation.
- **(Sci)BERT classification.** We use the standard fine-tuned BERT textual similarity method on both the pretrained BERT [4] (`base-uncased`) and SciBERT [1] (`scivocab-uncased`) models. Based on preliminary parameter tuning, we use a learning rate of 10^{-5} for fine-tuning these models.

Evaluation Metrics. Given the time constraints of doctors, we choose evaluation metrics that emphasize placing reports with higher discrepancies at the top. We evaluate using nDCG@1, nDCG@5, nDCG (without cutoff), P@1, P@5, and R-Prec (binary labels test any degree of discrepancy higher than label 0).

Parameters and Training. We train the neural models using pairwise cross-entropy loss [3]. Hyper-parameters are tuned using nCDG@5 on the dev set. We use SciBERT term embeddings [1] in our model and BiPACRR and tune for the optimal layer's embeddings akin to [10]. SciBERT is an adaptation of BERT to the biomedical and scientific domains, making it suitable for radiology notes.

Results. Test set performance of our best model configuration are shown in Table 1. Our optimal model consists of unigram, bi-gram, tri-gram, and RadLex scores. When compared to the best prior work (SimReg [2]), our model typically yields a considerable improvement in ranking performance. Our method improves R-Prec by 7.4%, nDCG@1 and nDCG@5 performance by 4.5%, and P@5 performance by 4.7%. In 54% of the test cases, our approach improves the nDCG@5 score over SimReg (decreases performance in only 27% of cases). Our model also outperforms leading language model classification approaches (BERT and SciBERT) and a leading neural ranking approach tuned for this task in most metrics (BiPACRR) by up to 15.4% in nDCG@1. We attribute this improved effectiveness of our approach to the explicit modeling of term importance and overlap, which are critical for the task.

Table 1. Ranking performance of our method and baselines.

Model	nDCG@1	nDCG@5	nDCG	P@1	P@5	R-Prec
VSM	48.1	54.0	70.9	65.4	42.3	49.4
BERT	59.0	69.8	78.7	69.2	53.8	53.9
SciBERT	62.2	68.2	79.3	76.9	51.5	58.3
BiPACRR	64.1	68.6	77.5	69.2	**56.2**	55.3
SimReg	69.9	70.7	81.1	**80.8**	51.5	51.8
Our method	**74.4**	**75.2**	**83.7**	**80.8**	**56.2**	**59.2**

Table 2. Ablation study of our method.

Model	nDCG@5
Full model	**75.2**
- Replace SciBERT with BERT	64.7
- Replace SciBERT with BioNLP (`pubmed-pmc`, bio.nlplab.org)	62.2
- Replace SciBERT with FastText (`wiki-news-300d-1M`, fasttext.cc)	59.1
- Without term importance	68.3
- Without ontology similarity	65.4
- Only overlap score (S_o)	70.8
- Only addition/deletion scores (S_a and S_d)	61.5

Ablations. Table 2 shows the ablation study examining the importance of different components in our system. We observe that both contextualization and domain-specificity of the word embeddings improve the performance of our approach. The term importance mechanism improves nDCG@5 by 6.9% and the ontology similarity improves performance by 9.8%. All three similarity measures appear to be important, however the overlap score alone can account for most of the performance (last row in table). This may be because it succinctly accounts for both additions and deletions.

Term Importance. To better understand the term importance mechanism of our approach, we present an example report in Fig. 2 (slightly altered for privacy). This report contains highly significant discrepancies and was ranked at position 3 by our approach and position 9 by SimReg (below several non-significant discrepancies). We observe that our model considers many radiological conditions as important, both when unmodified between the reports and when added/deleted (e.g., *fracture, dislocation, bankart*). Judging by the low textual similarity in this example, we conclude that the SimReg model may be relying too heavily on lexical features. We check the terms that are assigned high importance scores across all reports and find the most common are *no* (12% of reports), *cardiopulmonary* (3%), *process* (3%), and *abnormality* (3%).

anteroinferior dislocation of the left shoulder. mild hills sachs deformity without associated bankart lesion. no evidence of acute fracture or dislocation of the humerus.

Fig. 2. Example unigram importance scores (mean of preliminary and final report). Darker colors indicate higher scores. Underlines: additions. Strikeouts: deletions.

Conclusions. We presented a supervised ranking model based on lexical and ontological overlaps to rank medical reports by their discrepancy significance. On a real-world dataset of medical reports, we demonstrated that our approach outperforms existing approaches by large margins. This direction is a critical step towards addressing the problem of medical errors. By allowing medical practitioners to more easily find and learn from their previous errors, the chance of recurrent errors will be reduced, improving the well-being of patients.

Acknowledgements. This work was supported in part by ARCS Foundation.

References

1. Beltagy, I., Cohan, A., Lo, K.: SciBERT: pretrained contextualized embeddings for scientific text. In: EMNLP (2019)
2. Cohan, A., Soldaini, L., Goharian, N., Fong, A., Ross, F., Raj, R.: Identifying significance of discrepancies in radiology reports. In: SIAM International Conference on Data Mining (SDM) - Workshop on Data Mining for Medicine and Healthcare (DMMH) (2016)
3. Dehghani, M., Zamani, H., Severyn, A., Kamps, J., Croft, W.B.: Neural ranking models with weak supervision. In: SIGIR (2017)
4. Devlin, J., Chang, M.W., Lee, K., Toutanova, K.: BERT: pre-training of deep bidirectional transformers for language understanding. In: NAACL (2019)

5. Gan, Z., Pu, Y., Henao, R., Li, C., He, X., Carin, L.: Learning generic sentence representations using convolutional neural networks. In: EMNLP (2016)
6. Hui, K., Yates, A., Berberich, K., de Melo, G.: PACRR: a position-aware neural IR model for relevance matching. In: EMNLP (2017)
7. Kalaria, A.D., Filice, R.W.: Comparison-bot: an automated preliminary-final report comparison system. J. Digit. Imaging 29, 325–330 (2016). https://doi.org/ 10.1007/s10278-015-9840-2
8. Koopman, B., Cripwell, L., Zuccon, G.: Generating clinical queries from patient narratives: a comparison between machines and humans. In: SIGIR 2017 (2017)
9. Liu, L., Yang, W., Rao, J., Tang, R., Lin, J.: Incorporating contextual and syntactic structures improves semantic similarity modeling. In: EMNLP 2019 (2019)
10. MacAvaney, S., Yates, A., Cohan, A., Goharian, N.: CEDR: contextualized embeddings for document ranking. In: SIGIR (2019)
11. Makary, M.A., Daniel, M.: Medical error–the third leading cause of death in the US. BMJ 353, i2139 (2016)
12. Reimers, N., Gurevych, I.: Sentence-BERT: sentence embeddings using Siamese BERT-networks. In: EMNLP (2019)
13. Roberts, K., et al.: Overview of the TREC 2017 precision medicine track, pp. 500–324. NIST Special Publication (2017)
14. Ruutiainen, A.T., Scanlon, M.H., Itri, J.N.: Identifying benchmarks for discrepancy rates in preliminary interpretations provided by radiology trainees at an academic institution. J. Am. Coll. Radiol. 8(9), 644–648 (2011)
15. Saleh, S., Pecina, P.: Term selection for query expansion in medical cross-lingual information retrieval. In: Azzopardi, L., Stein, B., Fuhr, N., Mayr, P., Hauff, C., Hiemstra, D. (eds.) ECIR 2019. LNCS, vol. 11437, pp. 507–522. Springer, Cham (2019). https://doi.org/10.1007/978-3-030-15712-8_33
16. Sankhavara, J.: Biomedical document retrieval for clinical decision support system. In: ACL, pp. 84–90 (2018)
17. Soldaini, L., Yates, A., Goharian, N.: Denoising clinical notes for medical literature retrieval with convolutional neural model. In: CIKM (2017)
18. Tien, H.N., Le, M.N., Tomohiro, Y., Tatsuya, I.: Sentence modeling via multiple word embeddings and multi-level comparison for semantic textual similarity. Inf. Process. Manage. 56, 102090 (2018)
19. Walls, J., Hunter, N., Brasher, P.M., Ho, S.G.: The DePICTORS study: discrepancies in preliminary interpretation of CT scans between on-call residents and staff. Emerg. Radiol. 16(4), 303–308 (2009). https://doi.org/10.1007/s10140-009-0795-9
20. Xiong, C., Dai, Z., Callan, J.P., Liu, Z., Power, R.: End-to-end neural ad-hoc ranking with kernel pooling. In: SIGIR (2017)
21. Yates, A., Goharian, N., Frieder, O.: Relevance-ranked domain-specific synonym discovery. In: de Rijke, M., et al. (eds.) ECIR 2014. LNCS, vol. 8416, pp. 124–135. Springer, Cham (2014). https://doi.org/10.1007/978-3-319-06028-6_11

Teaching a New Dog Old Tricks: Resurrecting Multilingual Retrieval Using Zero-Shot Learning

Sean MacAvaney[1(✉)], Luca Soldaini[2], and Nazli Goharian[1]

[1] IR Lab, Georgetown University, Washington DC 20057, USA
{sean,nazli}@ir.cs.georgetown.edu
[2] Amazon Alexa Search, Manhattan Beach, CA 90266, USA
lssoldai@amazon.com

Abstract. While billions of non-English speaking users rely on search engines every day, the problem of ad-hoc information retrieval is rarely studied for non-English languages. This is primarily due to a lack of data set that are suitable to train ranking algorithms. In this paper, we tackle the lack of data by leveraging pre-trained multilingual language models to transfer a retrieval system trained on English collections to non-English queries and documents. Our model is evaluated in a zero-shot setting, meaning that we use them to predict relevance scores for query-document pairs in languages never seen during training. Our results show that the proposed approach can significantly outperform unsupervised retrieval techniques for Arabic, Chinese Mandarin, and Spanish. We also show that augmenting the English training collection with some examples from the target language can sometimes improve performance.

1 Introduction

Every day, billions of non-English speaking users [22] interact with search engines; however, commercial retrieval systems have been traditionally tailored to English queries, causing an information access divide between those who can and those who cannot speak this language [39]. Non-English search applications have been equally under-studied by most information retrieval researchers. Historically, ad-hoc retrieval systems have been primarily designed, trained, and evaluated on English corpora (e.g., [1,5,6,23]). More recently, a new wave of supervised state-of-the-art ranking models have been proposed by researchers [11,14,21,24,26,35,37]; these models rely on neural architectures to rerank the head of search results retrieved using a traditional unsupervised ranking algorithm, such as BM25. Like previous ad-hoc ranking algorithms, these methods are almost exclusively trained and evaluated on English queries and documents.

The absence of rankers designed to operate on languages other than English can largely be attributed to a lack of suitable publicly available data sets. This aspect particularly limits supervised ranking methods, as they require samples

© Springer Nature Switzerland AG 2020
J. M. Jose et al. (Eds.): ECIR 2020, LNCS 12036, pp. 246–254, 2020.
https://doi.org/10.1007/978-3-030-45442-5_31

for training and validation. For English, previous research relied on English collections such as TREC Robust 2004 [32], the 2009–2014 TREC Web Track [7], and MS MARCO [2]. No datasets of similar size exist for other languages.

While most of recent approaches have focused on ad hoc retrieval for English, some researchers have studied the problem of cross-lingual information retrieval. Under this setting, document collections are typically in English, while queries get translated to several languages; sometimes, the opposite setup is used. Throughout the years, several cross lingual tracks were included as part of TREC. TREC 6, 7, 8 [4] offer queries in English, German, Dutch, Spanish, French, and Italian. For all three years, the document collection was kept in English. CLEF also hosted multiple cross-lingual ad-hoc retrieval tasks from 2000 to 2009 [3]. Early systems for these tasks leveraged dictionary and statistical translation approaches, as well as other indexing optimizations [27]. More recently, approaches that rely on cross-lingual semantic representations (such as multilingual word embeddings) have been explored. For example, Vulić and Moens [34] proposed BWESG, an algorithm to learn word embeddings on aligned documents that can be used to calculate document-query similarity. Sasaki et al. [28] leveraged a data set of Wikipedia pages in 25 languages to train a learning to rank algorithm for Japanese-English and Swahili-English cross-language retrieval. Litschko et al. [20] proposed an unsupervised framework that relies on aligned word embeddings. Ultimately, while related, these approaches are only beneficial to users who can understand documents in two or more languages instead of directly tackling non-English document retrieval.

A few monolingual ad-hoc data sets exist, but most are too small to train a supervised ranking method. For example, TREC produced several non-English test collections: Spanish [12], Chinese Mandarin [31], and Arabic [25]. Other languages were explored, but the document collections are no longer available. The CLEF initiative includes some non-English monolingual datasets, though these are primarily focused on European languages [3]. Recently, Zheng et al. [40] introduced Sogou-QCL, a large query log dataset in Mandarin. Such datasets are only available for languages that already have large, established search engines.

Inspired by the success of neural retrieval methods, this work focuses on studying the problem of monolingual ad-hoc retrieval on non English languages using supervised neural approaches. In particular, to circumvent the lack of training data, we leverage transfer learning techniques to train Arabic, Mandarin, and Spanish retrieval models using English training data. In the past few years, transfer learning between languages has been proven to be a remarkably effective approach for low-resource multilingual tasks (e.g. [16,17,29,38]). Our model leverages a pre-trained multi-language transformer model to obtain an encoding for queries and documents in different languages; at train time, this encoding is used to predict relevance of query document pairs in English. We evaluate our models in a zero-shot setting; that is, we use them to predict relevance scores for query document pairs in languages never seen during training. By leveraging a pre-trained multilingual language model, which can be easily trained from abundant aligned [19] or unaligned [8] web text, we achieve competitive retrieval performance without having to rely on language specific relevance judgements. During the peer review of this article, a preprint [30] was published with similar observations as ours. In summary, our contributions are:

- We study zero shot transfer learning for IR in non-English languages.
- We propose a simple yet effective technique that leverages contextualized word embedding as multilingual encoder for query and document terms. Our approach outperforms several baselines on multiple non-English collections.
- We show that including additional in-language training samples may help further improve ranking performance.
- We release our code for pre-processing, initial retrieval, training, and evaluation of non-English datasets.[1] We hope that this encourages others to consider cross-lingual modeling implications in future work.

2   Methodology

Zero-Shot Multi-lingual Ranking. Because large-scale relevance judgments are largely absent in languages other than English, we propose a new setting to evaluate learning-to-rank approaches: zero-shot cross-lingual ranking. This setting makes use of relevance data from one language that has a considerable amount of training data (e.g., English) for model training and validation, and applies the trained model to a different language for testing.

More formally, let $\mathcal{S}$ be a collection of relevance tuples in the source language, and $\mathcal{T}$ be a collection of relevance judgments from another language. Each relevance tuple $\langle \mathbf{q}, \mathbf{d}, r \rangle$ consists of a query, document, and relevance score, respectively. In typical evaluation environments, $\mathcal{S}$ is segmented into multiple splits for training ($\mathcal{S}_{train}$) and testing ($\mathcal{S}_{test}$), such that there is no overlap of queries between the two splits. A ranking algorithm is tuned on $\mathcal{S}_{train}$ to define the ranking function $R_{\mathcal{S}_{train}}(\mathbf{q}, \mathbf{d}) \in \mathbb{R}$, which is subsequently tested on $\mathcal{S}_{test}$. We propose instead tuning a model on all data from the source language (i.e., training $R_{\mathcal{S}}(\cdot)$), and testing on a collection from the second language ($\mathcal{T}$).

Datasets. We evaluate on monolingual newswire datasets from three languages: Arabic, Mandarin, and Spanish. The Arabic document collection contains $384k$ documents (LDC2001T55), and we use topics/relevance information from the 2001–02 TREC Multilingual track (25 and 50 topics, respectively). For Mandarin, we use $130k$ news articles from LDC2000T52. Mandarin topics and relevance judgments are utilized from TREC 5 and 6 (26 and 28 topics, respectively). Finally, the Spanish collection contains $58k$ articles from LDC2000T51, and we use topics from TREC 3 and 4 (25 topics each). We use the topics, rather than the query descriptions, in all cases except TREC Spanish 4, in which only descriptions are provided. The topics more closely resemble real user queries than

---

[1] https://github.com/Georgetown-IR-Lab/multilingual-neural-ir

descriptions.[2] We test on these collections because they are the only document collections available from TREC at this time.[3]

We index the text content of each document using a modified version of Anserini with support for the languages we investigate [36]. Specifically, we add Anserini support for Lucene's Arabic and Spanish light stemming and stop word list (via `SpanishAnalyzer` and `ArabicAnalyzer`). We treat each character in Mandarin text as a single token.

Modeling. We explore the following ranking models:

- **Unsupervised baselines.** We use the Anserini [36] implementation of BM25, RM3 query expansion, and the Sequential Dependency Model (SDM) as unsupervised baselines. In the spirit of the zero-shot setting, we use the default parameters from Anserini (i.e., assuming no data of the target language).
- **PACRR** [14] models n-gram relationships in the text using learned 2D convolutions and max pooling atop a query-document similarity matrix.
- **KNRM** [35] uses learned Gaussian kernel pooling functions over the query-document similarity matrix to rank documents.
- **Vanilla BERT** [21] uses the BERT [10] transformer model, with a dense layer atop the classification token to compute a ranking score. To support multiple languages, we use the `base-multilingual-cased` pretrained weights. These weights were trained on Wikipedia text from 104 languages.

We use the embedding layer output from `base-multilingual-cased` model for PACRR and KNRM. In pilot studies, we investigated using cross-lingual MUSE vectors [8] and the output representations from BERT, but found the BERT embeddings to be more effective.

Experimental Setup. We train and validate models using TREC Robust 2004 collection [32]. TREC Robust 2004 contains 249 topics, $528k$ documents, and $311k$ relevance judgments in English (folds 1–4 from [15] for training, fold 5 for validation). Thus, the model is only exposed to English text in the training and validation stages (though the embedding and contextualized language models *are* trained on large amounts of unlabeled text in the languages). The validation dataset is used for parameter tuning and for the selection of the optimal training epoch (via nDCG@20). We train using pairwise softmax loss with Adam [18].

We evaluate the performance of the trained models by re-ranking the top 100 documents retrieved with BM25. We report MAP, Precision@20, and nDCG@20 to gauge the overall performance of our approach, and the percentage of judged documents in the top 20 ranked documents (judged@20) to evaluate how suitable the datasets are to approaches that did not contribute to the original judgments.

---

[2] Some have observed that the context provided by query descriptions are valuable for neural ranking, particularly when using contextualized language models [9].

[3] https://trec.nist.gov/data/docs_noneng.html.

Table 1. Zero-shot multi-lingual results for various baseline and neural methods. Significant improvements and reductions in performance compared with BM25 are indicated with ↑ and ↓, respectively (paired t-test by query, $p < 0.05$).

Ranker	P@20	nDCG@20	MAP	judged@20
Arabic (TREC 2002) [25]				
BM25	0.3470	0.3863	0.2804	99.0%
BM25 + RM3	0.3320	0.3705	↓0.2641	95.1%
SDM	0.3380	0.3775	↓0.2572	98.1%
PACRR multilingual	0.3270	0.3499	↓0.2517	96.4%
KNRM multilingual	0.3210	↓0.3415	↓0.2503	95.2%
Vanilla BERT multilingual	↑**0.3790**	**0.4205**	**0.2876**	97.4%
Arabic (TREC 2001) [25]				
BM25	**0.5420**	**0.5933**	**0.3462**	97.2%
BM25 + RM3	↓0.4700	0.5458	↓0.2903	85.6%
SDM	0.5140	0.5843	0.3213	96.2%
PACRR multilingual	↓0.3880	↓0.3933	↓0.2724	90.6%
KNRM multilingual	↓0.4140	↓0.4327	↓0.2742	91.0%
Vanilla BERT multilingual	0.5240	0.5628	0.3432	91.0%
Mandarin (TREC 6) [31]				
BM25	0.5962	0.6409	0.3316	89.6%
BM25 + RM3	↓0.5019	↓0.5571	0.2696	75.6%
SDM	0.5942	0.6320	0.3472	92.1%
PACRR multilingual	↓0.4923	↓0.5238	0.2856	79.0%
KNRM multilingual	↓0.5308	↓0.5497	↓0.3107	80.8%
Vanilla BERT multilingual	↑**0.6615**	↑**0.6959**	↑**0.3589**	92.7%
Mandarin (TREC 5) [33]				
BM25	0.3893	0.4113	0.2548	85.4%
BM25 + RM3	↓0.2768	↓0.3021	↓0.1698	64.6%
SDM	↑0.4536	↑0.4744	↑0.2855	94.1%
PACRR multilingual	0.3786	0.3998	0.2331	83.2%
KNRM multilingual	↓0.3232	↓0.3449	↓0.2223	77.5%
Vanilla BERT multilingual	↑**0.4589**	↑**0.5196**	↑**0.2906**	92.0%
Spanish (TREC 4) [12]				
BM25	0.3080	0.3314	0.1459	83.8%
BM25 + RM3	0.3360	0.3358	↑**0.2024**	85.2%
SDM	0.2780	0.3061	0.1377	78.6%
PACRR multilingual	0.2440	0.2494	0.1294	69.4%
KNRM multilingual	0.3120	0.3402	0.1444	79.2%
Vanilla BERT multilingual	↑**0.4400**	↑**0.4898**	↑0.1800	85.6%
Spanish (TREC 3) [13]				
BM25	0.5220	0.5536	0.2420	84.8%
BM25 + RM3	↑0.6100	0.6236	↑**0.3887**	93.0%
SDM	0.4920	0.5178	0.2258	83.8%
PACRR multilingual	↓0.4140	↓0.4092	0.2260	76.0%
KNRM multilingual	0.5560	0.5700	0.2449	85.2%
Vanilla BERT multilingual	↑**0.6400**	↑**0.6672**	↑0.2623	90.8%

3 Results

We present the ranking results in Table 1. We first point out that there is considerable variability in the performance of the unsupervised baselines; in some cases, RM3 and SDM outperform BM25, whereas in other cases they underperform. Similarly, the PACRR and KNRM neural models also vary in effectiveness, though more frequently perform much worse than BM25. This makes sense because these models capture matching characteristics that are specific to English. For instance, n-gram patterns captured by PACRR for English do not necessarily transfer well to languages with different constituent order, such as Arabic (VSO instead of SVO). An interesting observation is that the Vanilla BERT model (which recall is only tuned on English text) generally outperforms a variety of approaches across three test languages. This is particularly remarkable because it is a single trained model that is effective across all three languages, without any difference in parameters. The exceptions are the Arabic 2001 dataset, in which it performs only comparably to BM25 and the MAP results for Spanish. For Spanish, RM3 is able to substantially improve recall (as evidenced by MAP), and since Vanilla BERT acts as a re-ranker atop BM25, it is unable to take advantage of this improved recall, despite significantly improving the precision-focused metrics. In all cases, Vanilla BERT exhibits judged@20 above 85%, indicating that these test collections are still valuable for evaluation.

Table 2. Zero-Shot (ZS) and Few-Shot (FS) comparison for Vanilla BERT (multilingual) on each dataset. Within each metric and dataset, the top result is listed in bold. Significant increases from using FS are indicated with ↑ (paired t-test, $p < 0.05$).

Dataset	P@20		nDCG@20		MAP	
	ZS	FS	ZS	FS	ZS	FS
Arabic 2002	**0.3790**	0.3690	**0.4205**	0.3905	**0.2876**	0.2822
Arabic 2001	0.5240	↑**0.6020**	0.5628	↑**0.6405**	0.3432	**0.3529**
Mandarin 6	0.6615	**0.6808**	0.6959	**0.7099**	**0.3589**	0.3537
Mandarin 5	0.4589	**0.4643**	**0.5196**	0.5014	**0.2906**	0.2895
Spanish 4	0.4400	↑**0.5060**	0.4898	↑**0.5636**	0.1800	↑**0.2020**
Spanish 3	0.6400	**0.6560**	0.6672	**0.6825**	0.2623	**0.2684**

To test whether a small amount of in-language training data can further improve BERT ranking performance, we conduct an experiment that uses the other collection for each language as additional training data. The in-language samples are interleaved into the English training samples. Results for this few-shot setting are shown in Table 2. We find that the added topics for Arabic 2001 (+50) and Spanish 4 (+25) significantly improve the performance. This results in a model significantly better than BM25 for Arabic 2001, which suggests that there may be substantial distributional differences in the English TREC 2004 training and Arabic 2001 test collections. We further back this up by training an "oracle" BERT model (training on the test data) for Arabic 2001, which yields a model substantially better (P@20 = 0.7340, nDCG@20 = 0.8093, MAP = 0.4250).

4 Conclusion

We introduced a zero-shot multilingual setting for evaluation of neural ranking methods. This is an important setting due to the lack of training data available in many languages. We found that contextualized languages models (namely, BERT) have a big upper-hand, and are generally more suitable for cross-lingual performance than prior models (which may rely more heavily on phenomena exclusive to English). We also found that additional in-language training data may improve the performance, though not necessarily. By releasing our code and models, we hope that cross-lingual evaluation will become more commonplace.

Acknowledgements. This work was supported in part by ARCS Foundation.

References

1. Amati, G., Van Rijsbergen, C.J.: Probabilistic models of information retrieval based on measuring the divergence from randomness. ACM Trans. Inf. Syst. (TOIS) **20**(4), 357–389 (2002)
2. Bajaj, P., et al.: MS MARCO: a human generated machine reading comprehension dataset. arXiv preprint arXiv:1611.09268 (2016)
3. Braschler, M.: CLEF 2003 – overview of results. In: Peters, C., Gonzalo, J., Braschler, M., Kluck, M. (eds.) CLEF 2003. LNCS, vol. 3237, pp. 44–63. Springer, Heidelberg (2004). https://doi.org/10.1007/978-3-540-30222-3_5
4. Braschler, M., Schäuble, P., Peters, C.: Cross-language information retrieval (CLIR) track overview. In: TREC (2000)
5. Cao, Z., Qin, T., Liu, T.Y., Tsai, M.F., Li, H.: Learning to rank: from pairwise approach to listwise approach. In: Proceedings of the 24th International Conference on Machine Learning, pp. 129–136. ACM (2007)
6. Carpineto, C., Romano, G.: A survey of automatic query expansion in information retrieval. ACM Comput. Surv. (CSUR) **44**(1), 1 (2012)
7. Collins-Thompson, K., Macdonald, C., Bennett, P., Diaz, F., Voorhees, E.M.: TREC 2014 web track overview. Technical report, Michigan University, Ann Arbor (2015)
8. Conneau, A., Lample, G., Ranzato, M., Denoyer, L., Jégou, H.: Word translation without parallel data. arXiv preprint arXiv:1710.04087 (2017)
9. Dai, Z., Callan, J.: Deeper text understanding for IR with contextual neural language modeling. In: SIGIR (2019)
10. Devlin, J., Chang, M.W., Lee, K., Toutanova, K.: BERT: pre-training of deep bidirectional transformers for language understanding. In: Proceedings of the 2019 Conference of the North American Chapter of the Association for Computational Linguistics: Human Language Technologies, Volume 1 (Long and Short Papers), pp. 4171–4186. Association for Computational Linguistics, Minneapolis, June 2019

11. Guo, J., Fan, Y., Ai, Q., Croft, W.B.: A deep relevance matching model for ad-hoc retrieval. In: Proceedings of the 25th ACM International on Conference on Information and Knowledge Management, pp. 55–64. ACM (2016)
12. Harman, D.: Overview of the fourth text retrieval conference (TREC-4), pp. 1–24. NIST Special Publication (SP) (1996)
13. Harman, D.K.: Overview of the third text retrieval conference (TREC-3). DIANE Publishing (1995)
14. Hui, K., Yates, A., Berberich, K., de Melo, G.: PACRR: a position-aware neural IR model for relevance matching. arXiv preprint arXiv:1704.03940 (2017)
15. Huston, S., Croft, W.B.: Parameters learned in the comparison of retrieval models using term dependencies. Technical report (2014)
16. Johnson, M., et al.: Google's multilingual neural machine translation system: enabling zero-shot translation. Trans. Assoc. Comput. Linguis. 5, 339–351 (2017)
17. Kim, J.K., Kim, Y.B., Sarikaya, R., Fosler-Lussier, E.: Cross-lingual transfer learning for POS tagging without cross-lingual resources. In: Proceedings of the 2017 Conference on Empirical Methods in Natural Language Processing, pp. 2832–2838 (2017)
18. Kingma, D.P., Ba, J.: Adam: a method for stochastic optimization. In: ICLR (2015)
19. Lample, G., Conneau, A.: Cross-lingual language model pretraining. arXiv preprint arXiv:1901.07291 (2019)
20. Litschko, R., Glavaš, G., Ponzetto, S.P., Vulić, I.: Unsupervised cross-lingual information retrieval using monolingual data only. In: The 41st International ACM SIGIR Conference on Research & Development in Information Retrieval, pp. 1253–1256. ACM (2018)
21. MacAvaney, S., Yates, A., Cohan, A., Goharian, N.: CEDR: contextualized embeddings for document ranking. In: Proceedings of the 42nd International ACM SIGIR Conference on Research and Development in Information Retrieval, SIGIR 2019, pp. 1101–1104. ACM, New York (2019)
22. Roser, M., Ritchie, H., Ortiz-Ospina, E.: Internet (2019). https://ourworldindata.org/internet. Accessed 15 Sept 2019
23. Metzler, D., Croft, W.B.: A Markov random field model for term dependencies. In: Proceedings of the 28th Annual International ACM SIGIR Conference on Research and Development in Information Retrieval, SIGIR 2005, pp. 472–479. ACM, New York (2005)
24. Mitra, B., Craswell, N., et al.: An introduction to neural information retrieval. Found. Trends® Inf. Retrieval 13(1), 1–126 (2018)
25. Oard, D.W., Gey, F.C.: The TREC 2002 Arabic/English CLIR track. In: TREC (2002)
26. Onal, K.D., et al.: Neural information retrieval: at the end of the early years. Inf. Retrieval J. 21(2–3), 111–182 (2018). https://doi.org/10.1007/s10791-017-9321-y
27. Peters, C., Braschler, M., Clough, P.: Multilingual Information Retrieval: From Research to Practice. Springer, Heidelberg (2012). https://doi.org/10.1007/978-3-642-23008-0
28. Sasaki, S., Sun, S., Schamoni, S., Duh, K., Inui, K.: Cross-lingual learning-to-rank with shared representations. In: Proceedings of the 2018 Conference of the North American Chapter of the Association for Computational Linguistics: Human Language Technologies, Volume 2 (Short Papers), pp. 458–463. Association for Computational Linguistics, New Orleans, June 2018

29. Schuster, S., Gupta, S., Shah, R., Lewis, M.: Cross-lingual transfer learning for multilingual task oriented dialog. In: Proceedings of the 2019 Conference of the North American Chapter of the Association for Computational Linguistics: Human Language Technologies, Volume 1 (Long and Short Papers), pp. 3795–3805. Association for Computational Linguistics, Minneapolis June 2019

30. Shi, P., Lin, J.: Cross-lingual relevance transfer for document retrieval. ArXiv abs/1911.02989 (2019)

31. Voorhees, E., Harman, D., Wilkinson, R.: The sixth text retrieval conference (TREC-6). In: The Text REtrieval Conference (TREC), vol. 500, p. 240. NIST (1998)

32. Voorhees, E.M.: Overview of the TREC 2005 robust retrieval track. In: TREC (2005)

33. Voorhees, E.M., Harman, D.: Overview of the fifth text retrieval conference (TREC-5). In: TREC, vol. 97, pp. 1–28 (1996)

34. Vulić, I., Moens, M.F.: Monolingual and cross-lingual information retrieval models based on (bilingual) word embeddings. In: Proceedings of the 38th International ACM SIGIR Conference on Research and Development in Information Retrieval, SIGIR 2015, pp. 363–372. ACM, New York (2015)

35. Xiong, C., Dai, Z., Callan, J., Liu, Z., Power, R.: End-to-end neural ad-hoc ranking with kernel pooling. In: Proceedings of the 40th International ACM SIGIR Conference on Research and Development in Information Retrieval, pp. 55–64. ACM (2017)

36. Yang, P., Fang, H., Lin, J.: Anserini: reproducible ranking baselines using Lucene. J. Data Inf. Qual. **10**, 16:1–16:20 (2018)

37. Yang, W., Zhang, H., Lin, J.: Simple applications of BERT for ad hoc document retrieval. arXiv preprint arXiv:1903.10972 (2019)

38. Yang, Z., Salakhutdinov, R., Cohen, W.W.: Transfer learning for sequence tagging with hierarchical recurrent networks. arXiv preprint arXiv:1703.06345 (2017)

39. Young, H.: The digital language divide (2015). http://labs.theguardian.com/digital-language-divide/. Accessed 15 Sept 2019

40. Zheng, Y., Fan, Z., Liu, Y., Luo, C., Zhang, M., Ma, S.: Sogou-QCL: a new dataset with click relevance label. In: The 41st International ACM SIGIR Conference on Research & Development in Information Retrieval, pp. 1117–1120. ACM (2018)

Semi-supervised Extractive Question Summarization Using Question-Answer Pairs

Kazuya Machida[1], Tatsuya Ishigaki[1(✉)], Hayato Kobayashi[2],
Hiroya Takamura[1,3], and Manabu Okumura[1]

[1] Tokyo Institute of Technology, Yokohama, Japan
{machida,ishigaki}@lr.pi.titech.ac.jp, oku@pi.titech.ac.jp
[2] Yahoo Japan Corporation/RIKEN AIP, Tokyo, Japan
hakobaya@yahoo-corp.jp
[3] AIST, Tokyo, Japan
takamura.hiroya@aist.go.jp

Abstract. Neural extractive summarization methods often require much labeled training data, for which headlines or lead summaries of news articles can sometimes be used. Such directly useful summaries are not always available, however, especially for user-generated content, such as questions posted on community question answering services. In this paper, we address an extractive summarization (i.e., headline extraction) task for such questions as a case study and consider how to alleviate the problem by using question-answer pairs, instead of missing-headline pairs. To this end, we propose a framework to examine how to use such unlabeled paired data from the viewpoint of training methods. Experimental results show that multi-task training performs well with undersampling and distant supervision.

Keywords: Question summarization · Headline extraction

1 Introduction

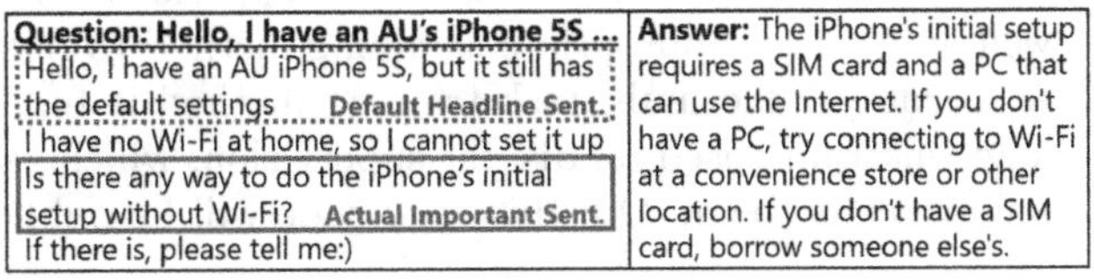

Questions are a means of acquiring knowledge, and since the advent of the Internet, many questions have been posted on community question-answering (CQA) sites.

Question: Hello, I have an AU's iPhone 5S ...	Answer: The iPhone's initial setup
Hello, I have an AU iPhone 5S, but it still has the default settings. **Default Headline Sent.** I have no Wi-Fi at home, so I cannot set it up	requires a SIM card and a PC that can use the Internet. If you don't have a PC, try connecting to Wi-Fi
Is there any way to do the iPhone's initial setup without Wi-Fi? **Actual Important Sent.** If there is, please tell me:)	at a convenience store or other location. If you don't have a SIM card, borrow someone else's.

Fig. 1. Example of a posted question and its answer.

Therefore, to find questions efficiently, we need a system by which the important parts of questions can be displayed in search results. On a CQA site, as

---

K. Machida and T. Ishigaki—Both equally contributed to this work.

© Springer Nature Switzerland AG 2020
J. M. Jose et al. (Eds.): ECIR 2020, LNCS 12036, pp. 255–264, 2020.
https://doi.org/10.1007/978-3-030-45442-5_32

represented by Yahoo! Chiebukuro [31], the first sentence of a question tends to be displayed as a headline (or list item) because of a restriction on the display area. Note that, to reduce the burden on users who post questions, many CQA sites do not provide an input field for headlines in the submission form. The most important sentence in a question, however, should be displayed instead of the first sentence, because sometimes the first sentence does not provide enough information, as shown in Fig. 1 (translated to English).

This task can be formalized as extractive summarization, which has long been addressed, e.g., by using a graph-based method [21], a topic-based method [10], or a features-based method [22]. The development of neural networks has led some studies [7,24] to report high-performance models that use large amounts of training data. Such large amounts of training data, however, incur a high cost to create and cannot always be prepared for practical use.

In this paper, we harness question-answer (QA) pairs to alleviate this problem. Many QA pairs on CQA sites can be easily obtained without annotation costs and are expected to be useful because, in general, each answer should be closely related to the most important sentence in the question. In fact, the answer in Fig. 1 includes keywords such as "initial setup" and "Wi-Fi" in the main question sentence. Our framework can be regarded as a semi-supervised approach with a small amount of labeled data and a large amount of unlabeled (paired) data. The main difference from classical semi-supervised settings is that unlabeled data has a paired structure. This allows us to formulate our problem as a multi-task problem of sentence extraction and answer generation. One of the difficulties of this formulation is "data imbalance", meaning that there is a small amount of data for sentence extraction and a large amount for answer generation. Therefore, we focus on this data imbalance problem and investigate how to use the unlabeled paired data from the viewpoint of training methods.

The contributions of our study are as follows.

- We address extractive question summarization with QA pairs as a case study of a semi-supervised setting with unlabeled paired data and we propose a simple framework to systematically examine different ways to use these pairs.
- We compare different training methods, namely, pretraining, separate training, and multi-task training, as well as normal training. Our experimental results show that (a) multi-task training performs the best but does not work well without an appropriate sampling method to reduce the data imbalance, and that (b) the multi-task training method is further enhanced with data augmentation based on distant supervision, which can simply solve the data imbalance problem. Our data and code will be publicly available [14].

2   Framework

Our framework consists of two models (Fig. 2); the **sentence extraction model (SEM)** based on a sequence labeling structure, and the **answer generation model (AGM)** based on a sequence-to-sequence structure. SEM directly solves our task, whereas AGM provides auxiliary information via attention weights.

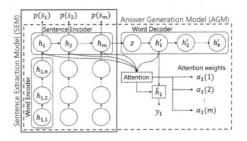

Fig. 2. Overview of our framework.

SEM first encodes a question with sentences $(s_1, \ldots, s_m)$ into sentence vectors $(h_1, \ldots, h_m)$ via a hierarchical encoder based on two LSTM units for words and sentences. Then, for each sentence s_i, the model calculates the extraction probability $p(s_i)$, which represents the importance score of s_i, by applying a binary softmax function with a linear transformation to h_i. In the training phase, we use the cross entropy loss L_{ext} based on $p(s_i)$ and the true label, similarly to classification tasks. We use SEM to define the importance score of s_i as $f_{\text{ext}}(s_i) = p(s_i)$, which is used for the evaluation phase, together with a score obtained by AGM as described below.

AGM encodes a question into sentence vectors in the same way as in SEM. The model uses these vectors to generate an answer (word sequence) by using an ordinary sequence-to-sequence model with an attention mechanism. We do not use a hierarchical decoder, because the main purpose of this study is not to improve the performance of answer generation. In the training phase, we use the negative log likelihood loss L_{gen} based on a predicted sequence and the correct sequence. In the evaluation phase, we calculate importance scores by using attention weights $\alpha_j(i)$, each of which represents the alignment level with respect to s_i at the j-th step in generation. Specifically, we define the importance score of s_i obtained by AGM as the average of the attention weights for s_i, i.e., $f_{\text{gen}}(s_i) = \frac{1}{k} \sum_{j=1}^{k} \alpha_j(i)$.

In our framework, we can thus simultaneously train two models in a multi-task setting (SEM and AGM are the respective main and auxiliary models) and combine their importance scores to estimate the most important sentence. We introduce two tuning parameters λ and κ for training and evaluation phases, respectively. The final loss function for the training phase is $\lambda L_{\text{ext}} + (1 - \lambda)L_{\text{gen}}$, and the score function for the evaluation phase is $\kappa f_{\text{ext}}(s_i) + (1 - \kappa)f_{\text{gen}}(s_i)$.

3   Experiment

Datasets: We prepared two datasets, `Pair` and `Label`, which were based on a publicly available CQA dataset [25] provided by Yahoo! Chiebukuro. These two datasets formed a semi-supervised setting with unlabeled paired data, in which `Pair` included many unlabeled QA pairs for training AGM, while `Label` included a few labeled questions for SEM.

`Pair` consisted of 100K QA pairs, each of which included a randomly sampled question and its best answer annotated in the CQA dataset. In the sampling procedure, we removed pairs including more than 10 sentences to reduce the computational cost, as these were less than 5% of the total. For the same reason, we removed pairs including sentences consisting of more than 50 words.

`Label` consisted of 775 questions sampled separately but in a similar way to `Pair`. Every sentence in each question had a binary label representing whether the sentence was the most important, meaning that only the best sentence had a label of 1, while the others had a label of 0. We used crowdsourcing to annotate `Label`. In the crowdsourcing, five workers were given a question and asked to select the best sentence representing the main focus of the question. We included only questions for which at least four workers selected the same sentence.

Unsupervised Baselines: We prepared the following unsupervised methods as simple baselines.

- `Lead`: Selects the initial sentence.
- `TfIdf`: Selects the sentence with the highest average tf-idf on the basis of the CQA dataset.
- `SimEmb`: Selects the sentence with the highest similarity on the basis of the word mover's distance [18] to the input question.
- `LexRank`: Uses a graph-based, unsupervised, extractive summarization model [8], which was trained with all the questions.

Compared Methods: We systematically compared the following methods to study how to effectively use `Pair` by changing the parameter settings of λ and κ in our framework.

- `Ext`: Trains and uses SEM only ($\lambda = 1, \kappa = 1$).
- `Gen`: Trains and uses AGM only ($\lambda = 0, \kappa = 0$).
- `Sep`: Trains SEM ($\lambda = 1$) and AGM ($\lambda = 0$) separately and combines them in the evaluation phase. Then, κ is tuned with the development set.
- `Pre`: Trains SEM ($\lambda = 1$) after initializing the encoder's parameters by using AGM ($\lambda = 0$). Prediction is done with SEM ($\kappa = 1$).
- `Multi`: Trains SEM and AGM simultaneously. Mini-batches are created for each dataset and shuffled, with the loss calculated per mini-batch. Then, λ and κ are tuned with the development set.

Oversampling/Undersampling: We additionally prepared two variants of `Multi` to reduce the data imbalance problem of `Label` and `Pair`, because the data size of the subtask is much larger than that of the main task. Specifically, we used oversampling and undersampling to reduce the imbalance as follows.

- `MultiOver`: Oversamples `Label` multiple times to be the same size as `Pair`.
- `MultiUnder`: Undersamples `Pair` to be the same size as `Label` in every epoch.

Distant Supervision: Furthermore, we prepared a pseudo labeled dataset Pseudo, which included pseudo (noisy) labels for all the questions in Pair. This pseudo labeling approach is often called distant supervision, in where unlabeled data is automatically annotated with some heuristic rules. Following Ishigaki et al. [15], we adopted their heuristic rule that single-sentence questions are basically self-contained and have summary-like characteristics. Because their labels for single-sentence questions could not be directly used for our questions with multiple sentences, we first trained a classifier with their labels and used it to make Pseudo. Thus, using Pseudo, we prepared the following variants of Multi, Ext, Sep, and Pre for comparison.

- MultiDist: Multi trained with Label, Pair, and Pseudo.
- ExtDist/SepDist/PreDist: Variants of Ext/Sep/Pre, similar to MultiDist.

Evaluation: For evaluating the performance, we used an accuracy measure calculated by dividing the number of questions for which the target method correctly selected the most important sentence by the number of questions used. Note that well-known metrics such as ROUGE and precision/recall were not appropriate, because our task was to find only one sentence as a (snippet) headline. We divided the labeled data Label into five sets (train:develop:test = 3:1:1) and performed five-fold cross-validation to evaluate the methods.

Results: Table 1 lists the results. The three row groups from top to bottom correspond to unsupervised, semi-supervised, and distantly supervised settings. In the first group, Lead performed the best, whereas the other methods (TfIdf, SimEmb, and LexRank) did not work well. This indicates the difficulty of our task and confirms that we need supervision to develop practical models.

In the second group, MultiUnder performed the best, although Multi (without sampling) performed worse than Ext did. This suggests that reducing the data imbalance is a key factor for our setting. MultiOver also worked well but did not reach the performance of MultiUnder. The reason seems to be that sampling the same data many times yields overfitting. Among other methods, Sep performed well because of an ensemble effect of

Table 1. Accuracy on the question summarization task. Each "✓" indicates that the corresponding dataset was used.

	Label	Pair	Pseudo	Acc
Lead	–	–	–	.690
TfIdf	–	–	–	.237
SimEmb	–	–	–	.472
LexRank	–	–	–	.587
Ext	✓	–	–	.813
Gen	–	✓	–	.649
Sep	✓	✓	–	.828
Pre	✓	✓	–	.788
Multi	✓	✓	–	.770
MultiOver	✓	✓	–	.833
MultiUnder	✓	✓	–	.857
ExtDist	✓	–	✓	.838
SepDist	✓	✓	✓	.855
PreDist	✓	✓	✓	.834
MultiDist	✓	✓	✓	**.875**

Ext and Gen, whereas Gen by itself performed the worst because it did not use any labels. Pre unexpectedly performed worse than Ext did, although Shimizu et al. [27] reported that sentiment classifiers were more enhanced by pretraining with tweet-reply pairs than by language model pretraining. This implies that the performance depends on the task settings, so our framework can be useful for other tasks.

In the third group, `MultiDist` (without sampling) performed the best. The differences from the other methods in this group were statistically significant according to the sign test ($p < 0.05$). Although distant supervision itself has positive effects as shown by the improvement for `ExtDist`, it has an extra bonus in that pseudo labels can simply solve the data imbalance. These results suggest that we have room to study the combinations of multi-task training and distant supervision for other NLP tasks. We also prepared a larger labeled dataset than `Label`. The experiments on this dataset showed similar tendencies. We will study how the data size of labeled data affects the performances in future work.

4   Related Work

Several studies have considered semi-supervised settings for summarization tasks [1,20,30], but in contrast to our main focus, none of them considered multi-task settings, especially using paired data. In the multi-task field, there have been several studies on summarization tasks. Guo et al. [11] improved an abstractive summarization model by using multi-task training with entailment and question generation tasks. Their work used human-annotated data from SQuAD dataset for these auxiliary tasks, whose sizes were much smaller than that of the main task, so their setting was completely different from ours. Angelidis et al. [2] addressed summarization of opinions from Amazon reviews by using multi-task training with aspect extraction and sentiment prediction tasks. Their work is related to ours in that they targeted user-generated content, but their auxiliary tasks were basic subtasks of opinion summarization with explicit aspect or sentiment labels. This implies that their task's usefulness was clearer than that of our task, in which we only assume a paired structure without any explicit labels. The study most related to ours is the work by Isonuma et al. [17], who proposed an extractive summarization method for news articles through multi-task training with a document classification task. Their strategy was similar to ours in that they used categories originally attached to news articles without costly annotation, but in many cases, we cannot access such categories or useful meta-information for documents, like CQA sites.

Several studies have used QA or similar structures for summarization tasks. Chen et al. [6] used a QA system to predict summarization quality in the evaluation phase, in contrast to our study, which uses QA paired data in the training phase. Arumae and Liu [3] used QA data to calculate a reward function for reinforcement learning in the training phase. They used Cloze-style (fill in the blank) questions, however, and we cannot directly apply their method to our task. Gao et al. [9] used an article-comments structure to personalize summaries in a multi-modal setting with multiple inputs, i.e., article and comments, rather than multi-task settings with multiple outputs, as in our study. Note that we did not consider such a multi-modal setting, as we assumed that answers would not always available for posted questions.

Many studies have used CQA data, but most have addressed different tasks, i.e., dealing with answering questions [4,5,23,28], retrieving similar questions [19,23,26], and generating questions [12]. Tamura et al. [29] focused on extracting a core sentence and identifying the question type as a classification task for answering multiple-sentence questions. Higurashi et al. [13] proposed a learning-to-rank approach for extracting an important substring from a question. Although their models are useful for retrieving important information, they considered methods that are trained with only labeled data. Finally, Ishigaki et al. [16] addressed neural abstractive and extractive approaches to summarize lengthy questions by using much paired data consisting of questions and headlines. Therefore, their method is not applicable to our task, in which we assume questions without headlines.

5   Conclusion

We have addressed an extractive question summarization task with QA pairs as a case study of a semi-supervised setting with unlabeled paired data. Our results suggest that multi-task training is effective especially with undersampling and distant supervision. For future work, we will apply our framework to other tasks with similar structures, such as news articles with comments.

References

1. Amini, M.R., Gallinari, P.: The use of unlabeled data to improve supervised learning for text summarization. In: Proceedings of the 25th Annual International ACM SIGIR Conference on Research and Development in Information Retrieval (SIGIR 2002), pp. 105–112. ACM (2002). https://doi.org/10.1145/564376.564397

2. Angelidis, S., Lapata, M.: Summarizing opinions: aspect extraction meets sentiment prediction and they are both weakly supervised. In: Proceedings of the 2018 Conference on Empirical Methods in Natural Language Processing (EMNLP 2018), pp. 3675–3686. Association for Computational Linguistics (2018). http://www.aclweb.org/anthology/D18-1403

3. Arumae, K., Liu, F.: Reinforced extractive summarization with question-focused rewards. In: Proceedings of ACL 2018, Student Research Workshop, pp. 105–111. Association for Computational Linguistics (2018). http://www.aclweb.org/anthology/P18-3015

4. Bhaskar, P.: Answering questions from multiple documents - the role of multi-document summarization. In: Proceedings of the Student Research Workshop Associated with RANLP 2013, pp. 14–21. INCOMA Ltd., Shoumen (2013). http://www.aclweb.org/anthology/R13-2003

5. Celikyilmaz, A., Thint, M., Huang, Z.: A graph-based semi-supervised learning for question-answering. In: Proceedings of the Joint Conference of the 47th Annual Meeting of the ACL and the 4th International Joint Conference on Natural Language Processing of the AFNLP, pp. 719–727. Association for Computational Linguistics (2009). http://www.aclweb.org/anthology/P/P09/P09-1081

6. Chen, P., Wu, F., Wang, T., Ding, W.: A semantic QA-based approach for text summarization evaluation. In: Proceedings of the Thirty-Second AAAI Conference on Artificial Intelligence, AAAI 2018, pp. 4800–4807. AAAI Press (2018). https://www.aaai.org/ocs/index.php/AAAI/AAAI18/paper/view/16115

7. Cheng, J., Lapata, M.: Neural summarization by extracting sentences and words. In: Proceedings of the 54th Annual Meeting of the Association for Computational Linguistics (ACL 2016), pp. 484–494. Association for Computational Linguistics (2016). http://www.aclweb.org/anthology/P16-1046

8. Erkan, G., Radev, D.R.: LexRank: graph-based lexical centrality as salience in text summarization. J. Artif. Intell. Res. (JAIR) 22(1), 457–479 (2004). https://doi.org/10.1613/jair.1523

9. Gao, S., et al.: Abstractive text summarization by incorporating reader comments. In: Proceedings of the Thirty-Third AAAI Conference on Artificial Intelligence, AAAI-2019. AAAI Press (2019). https://www.aaai.org/ojs/index.php/AAAI/article/view/4603/4481

10. Gong, Y., Liu, X.: Generic text summarization using relevance measure and latent semantic analysis. In: Proceedings of the 24th Annual International ACM SIGIR Conference on Research and Development in Information Retrieval (SIGIR2001), pp. 19–25 (2001). https://doi.org/10.1145/383952.383955

11. Guo, H., Pasunuru, R., Bansal, M.: Soft layer-specific multi-task summarization with entailment and question generation. In: Proceedings of the 56th Annual Meeting of the Association for Computational Linguistics (ACL 2018), pp. 687–697. Association for Computational Linguistics (2018). http://www.aclweb.org/anthology/P18-1064

12. Heilman, M., Smith, N.A.: Good question! Statistical ranking for question generation. In: Human Language Technologies: The 2010 Annual Conference of the North American Chapter of the Association for Computational Linguistics (NAACL-HLT 2010), pp. 609–617. Association for Computational Linguistics (2010). http://www.aclweb.org/anthology/N10-1086

13. Higurashi, T., Kobayashi, H., Masuyama, T., Murao, K.: Extractive headline generation based on learning to rank for community question answering. In: Proceedings of the 27th International Conference on Computational Linguistics (COLING 2018), pp. 1742–1753. Association for Computational Linguistics (2018). http://www.aclweb.org/anthology/C18-1148

14. Ishigaki, T.: Scripts for preprocessing Yahoo Chiebukuro dataset (2020). http://lr-www.pi.titech.ac.jp/~ishigaki/chiebukuro/

15. Ishigaki, T., Machida, K., Kobayashi, H., Takamura, H., Okumura, M.: Distant supervision for question summarization. In: Proceedings of the 42nd Annual European Conference on Information Retrieval (ECIR 2020) (2020)

16. Ishigaki, T., Takamura, H., Okumura, M.: Summarizing lengthy questions. In: Proceedings of the Eighth International Joint Conference on Natural Language Processing (IJCNLP 2017), pp. 792–800. Asian Federation of Natural Language Processing (2017). http://www.aclweb.org/anthology/I17-1080

17. Isonuma, M., Fujino, T., Mori, J., Matsuo, Y., Sakata, I.: Extractive summarization using multi-task learning with document classification. In: Proceedings of the 2017 Conference on Empirical Methods in Natural Language Processing (EMNLP 2017), pp. 2101–2110. Association for Computational Linguistics (2017). https://www.aclweb.org/anthology/D17-1223

18. Kusner, M., Sun, Y., Kolkin, N., Weinberger, K.: From word embeddings to document distances. In: Proceedings of the 32nd International Conference on Machine Learning (ICML 2015), vol. 37, pp. 957–966. PMLR (2015). http://proceedings.mlr.press/v37/kusnerb15.html

19. Lei, T., et al.: Semi-supervised question retrieval with gated convolutions. In: Proceedings of the 2016 Conference of the North American Chapter of the Association for Computational Linguistics: Human Language Technologies (NAACL-HLT 2016), pp. 1279–1289. Association for Computational Linguistics (2016). http://www.aclweb.org/anthology/N16-1153

20. Li, Y., Li, S.: Query-focused multi-document summarization: combining a topic model with graph-based semi-supervised learning. In: Proceedings of the 25th International Conference on Computational Linguistics (COLING 2014), pp. 1197–1207. Dublin City University and Association for Computational Linguistics (2014). http://www.aclweb.org/anthology/C14-1113

21. Mihalcea, R., Tarau, P.: TextRank: bringing order into text. In: Proceedings of the 2004 Conference on Empirical Methods in Natural Language Processing (2004). http://aclweb.org/anthology/W04-3252

22. Naik, S.S., Gaonkar, M.N.: Extractive text summarization by feature-based sentence extraction using rule-based concept. In: 2017 2nd IEEE International Conference on Recent Trends in Electronics, Information Communication Technology (RTEICT), pp. 1364–1368 (2017). https://ieeexplore.ieee.org/document/8256821

23. Nakov, P., et al.: SemEval-2017 task 3: community question answering. In: Proceedings of the 11th International Workshop on Semantic Evaluation (SemEval-2017), pp. 27–48. Association for Computational Linguistics (2017). http://www.aclweb.org/anthology/S17-2003

24. Nallapati, R., Zhai, F., Zhou, B.: SummaRuNNer: a recurrent neural network based sequence model for extractive summarization of documents. In: Proceedings of the Thirty-First AAAI Conference on Artificial Intelligence (AAAI 2017), pp. 3075–3081. AAAI Press (2017). http://aaai.org/ocs/index.php/AAAI/AAAI17/paper/view/14636

25. NII: Yahoo! Chiebukuro Data, 2nd edn (2018). https://www.nii.ac.jp/dsc/idr/en/yahoo/

26. Romeo, S., et al.: Neural attention for learning to rank questions in community question answering. In: Proceedings of the 26th International Conference on Computational Linguistics (COLING 2016), pp. 1734–1745. The COLING 2016 Organizing Committee (2016). http://aclweb.org/anthology/C16-1163

27. Shimizu, T., Shimizu, N., Kobayashi, H.: Pretraining sentiment classifiers with unlabeled dialog data. In: Proceedings of the 56th Annual Meeting of the Association for Computational Linguistics (ACL 2018), pp. 764–770. Association for Computational Linguistics (2018). http://www.aclweb.org/anthology/P18-2121

28. Surdeanu, M., Ciaramita, M., Zaragoza, H.: Learning to rank answers on large online QA collections. In: Proceedings of the 46th Annual Meeting of the Association for Computational Linguistics: Human Language Technologies (ACL-HLT 2008), pp. 719–727. Association for Computational Linguistics (2008). http://www.aclweb.org/anthology/P08-1082

29. Tamura, A., Takamura, H., Okumura, M.: Classification of multiple-sentence questions. In: Dale, R., Wong, K.-F., Su, J., Kwong, O.Y. (eds.) IJCNLP 2005. LNCS (LNAI), vol. 3651, pp. 426–437. Springer, Heidelberg (2005). https://doi.org/10.1007/11562214_38
30. Wong, K.F., Wu, M., Li, W.: Extractive summarization using supervised and semi-supervised learning. In: Proceedings of the 22nd International Conference on Computational Linguistics (COLING 2008), pp. 985–992. COLING 2008 Organizing Committee (2008). http://www.aclweb.org/anthology/C08-1124
31. Yahoo Japan Corp.: Yahoo! Chiebukuro (2019). https://chiebukuro.yahoo.co.jp/

Utilizing Temporal Psycholinguistic Cues for Suicidal Intent Estimation

Puneet Mathur[1(✉)], Ramit Sawhney[2], Shivang Chopra[3], Maitree Leekha[3], and Rajiv Ratn Shah[4]

[1] University of Maryland, College Park, USA
puneetm@cs.umd.edu
[2] Netaji Subhas Institute of Technology, Delhi, India
[3] Delhi Technological University, Delhi, India
[4] MIDAS Labs, IIIT Delhi, New Delhi, India

Abstract. Temporal psycholinguistics can play a crucial role in studying expressions of suicidal intent on social media. Current methods are limited in their approach in leveraging contextual psychological cues from online user communities. This work embarks in a novel direction to explore historical activities of users and homophily networks formed between Twitter users for extracting suicidality trends. Empirical evidence proves the advantages of incorporating historical user profiling and temporal graph convolutional modeling for automated detection of suicidal connotations on Twitter.

1 Introduction

Suicidal ideation detection is a well studied problem in social media analysis. Various works have tried to identify linguistic patterns correlated with suicidality intent. Despite the sustained efforts from the community, most approaches ignore the psychological relevance of temporal characteristics of suicidal behaviour. Moreover, there has been limited explorations in the space of homophily networks to identify collusive depressive users. We hypothesize that the contextual information embedded in social media engagement and historical activities of users can lead to substantial improvements in automated identification of suicidal ideation. We look beyond linguistic cues into temporal signals throughout this work, with the help of a publicly available dataset given by [14] of 34,306 tweets on suicidality detection.

2 Related Work

2.1 Challenges on Social Media

The growth of social media websites hosts a number of challenges such as cyberbullying, suicide pacts, and radicalism that motivate suicidal behavior and

---

P. Mathur and R. Sawhney—Equal contribution

© Springer Nature Switzerland AG 2020
J. M. Jose et al. (Eds.): ECIR 2020, LNCS 12036, pp. 265–271, 2020.
https://doi.org/10.1007/978-3-030-45442-5_33

impact the mental health of the users [10]. The associativity of suicide-related verbalizations on social media websites has been found to be strongly related to potential suicidal attempts. Prior studies show how suicidal intent declarations were significantly more assortative than chance, at times connected till 6 degrees of separation [5]. A patient's social media profile can help medical experts gain perspective into their mental health status and identify those at critical risk for suicide attempts [15]. The potential of technological interventions for suicidal risk assessment and mitigation needs to be explored in detail.

2.2   Text-Based Approaches

Various works have been recently proposed with an objective of automating the detection of social media posts expressing suicide ideation using textual information [3,7,17]. [4] performed a semi-automated content-based analysis on a small number of tweets related to depression in order to derive certain qualitative insights into the behavior of users displaying suicidal behavior. Self-disclosure helps to facilitate psychological well being in individuals with mental illness [2]. Textual descriptions of social media disclosures have been extensively studied in the past [7]. [19] explored deep learning based supervised classifiers for suicidal ideation detection.

2.3   Psycho-Linguistic Analysis

[13] used social graph based features and gained considerable improvement in the task of abuse detection. [16] performed a psycho-linguistic analysis of online users for a similar task. [1] tried to link users' psychological features such as personality traits including personalities, sentiment and emotion for cyberbulling and trolling. The contributions that we make in this work are different from previous efforts as there has been hardly any attempt to take a combined multi-faceted approach for solving the task of suicidal ideation in Twitter.

2.4   Signals from Temporal Data

Temporal graphs can capture the relationships in data with time so as to model new events and comparison to related entities and historical states [18]. [9] detected groups based on interesting features of the time-evolving networks. It studied several clustering frameworks for time-evolving networks for detecting group structure. [6] performed temporal sentiment analysis for early detection of cyberbulling and suicide ideation of a user through graph-based data mining approaches.

3   Methodology

The proposed methodology looks beyond text classifiers and leverages tweeting history of users as well as their social network communication patterns. User-based features were extracted from the historical tweeting activity and inter-user interactions was modeled as a social graph. The methodology is two-fold consisting of historical signal modeling and temporal graph convolutional modeling.

3.1 Classification Network

In order to learn from the textual information available in the raw tweets, we trained a **BLSTM + Attention** network [20]. We train a BLSTM model with 100 LSTM units, dropout rate of 0.25 and a recurrent dropout rate of 0.2. The attention layer was followed by another dropout layer of 0.2. This was followed by two dense layers having 256 units and 2 units, respectively.

3.2 Temporal Modeling of Suicidal Tendency

Motivation: The idea of temporal modeling of suicidal tendencies is inspired by [11] with additions. According to [11], a representation for the historical activity can be formulated as a temporal weighting scheme ϕ_i which is a sum of two independent time varying functions of suicidality - ideation build-up $\lambda_i(t)$ and sinusoidal episodes $\mu_i(t)$. Extrapolating from this, we add a third independent time-varying function - white Gaussian noise $z_i(t)$. Let Δt_i be the time offset from the original tweet and the temporal representation function z be given by Eq. 1.

$$z(u, H) = \sum_{h_i \in H} \phi_i(\Delta t) f(h_i) \qquad (1)$$

Suicidal Ideation Build-Up: Each user's historical tweets can be modeled as an exponential function in time given by Eq. 2 where α and β are hyper parameters tuned over training data.

$$\lambda_i(\Delta t) = \alpha e^{\beta \Delta t_i} \qquad (2)$$

Suicidal Episodes: Phased changes in suicidal intent are mathematically represented by Eq. 3. As per [12], the hyper parameters for the same are given by Table 1.

$$\mu_i(\Delta t) = \sum_1^Q (a_q cos(\frac{2\pi q \Delta t_i}{U}) + b_q sin(\frac{2\pi q \Delta t_i}{U})) \qquad (3)$$

Table 1. Hyper parameters for Eq. 3 [11]

Hyperparameters	Value
Q	3
U	{1, 2, 3, 4, 5, 6, 7}
a_q, b_q	$\approx \eta(0, \sigma^2)$

Temporal Surprise: Similar to any channel medium, social media platforms are prone to noise that adds randomness to the temporal suicidal patterns. The white Gaussian noise is modeled as being derived from a normal distribution with the expectation value of the noise term $\zeta_{(}t)$ equal to 1.

$$\phi_i(\Delta t) = \lambda_i(\Delta(t) + \mu_i(\Delta t) + \zeta_i(\Delta t) \tag{4}$$

For each of the tweet samples, the historical activity representation was an input to logistic regression model to learn temporal embeddings from these features which was used as an input to the final model.

3.3 Graph Convolutional Networks for User Profiling

Learning user representations can be significantly enriched by leveraging information derived from the inter-user interactions in social media channels. For this purpose, Graph Convolutional Networks (GCN) [8] can be effectively utilized that are capable of modeling social interactions in the form of features of nodes in the graph and allow contextual learning of information with respect to a node's neighbourhood.

Temporal GCN: We tried to incorporate the historical views into the extended graph by constructing time weighted TF-IDF vectors of the historical tweets. The author nodes were modified to consist of temporal weighting of TF-IDF representation of tweets. Let the TF-IDF vector f_k^t of tweet at timestamp t for k^{th} author be defined by Eq. 5, where C_k is the global noise parameter, $\hbar$ controls the margin of influence of a user on its neighbours social activity and ω is the rate of decay of the suicidal sentiment. The external parameters C_k, $\hbar_k$ and ω_k are learnt from the training portion of the dataset in an unsupervised fashion.

$$f^t = \hbar_k \exp^{\omega_k \Delta t} \tag{5}$$

4 Experiments

4.1 Data Description and Setup

To gauge the effectiveness of our proposed approach, we use the dataset from SNAP-BATNET [14] which consists of $34,306$ tweets with $3,984$ of them suicidal ideations. For each of these users, the tweet timelines were also collected to create the set of historical tweets. 10-fold stratified cross-validation was employed to evaluate models on each of the 10 train-val splits. The hyper parameters for the temporal weighted combination were tuned using a grid search over the grid $\alpha = \{0.1, 0.5, 1.0\}$, $\beta = \{0, 0.01, 0.1, 1\}$, $U = \{1, 2..., 7\}$ yielding $\alpha = 0.5$, $\beta = 1, U = 7$. t_0 was assigned to time series points with values equal to $argmax(|\mu|)_i$.

Table 2. Performance analysis

Model	F1	P	R
SNAP-BATNET [14]	92.60	72.20	93.52
BiLSTM + Attention (Text) [11]	91.26	70.02	91.23
Text + Temporal Modeling	92.75*	91.98*	93.70*
Temporal GCN	**93.89***	88.73	**94.54***

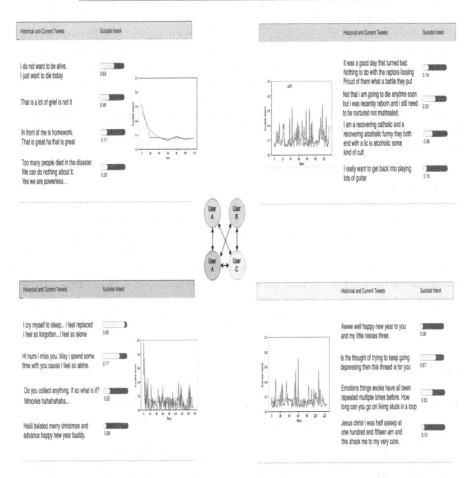

Fig. 1. Analysis of historical behaviour of users in a community over time

5   Results and Ablation Analysis

The ablation study of experimented features presented in Table 2 highlights the significance of temporal features extracted from social media in suicide ideation risk assessment. Temporal GCN provides a substantial gain over text in prediction confidence due to the user interactions. Additionally, it is interesting to observe the ability of the GCN model to better represent historical suicidal signals in comparison to naive historical and textual features to a sufficient degree. Empirically, temporal features help suppress false positives induced by text classifiers that try to overfit on the presence of anecdotal suicidal phrases such as *"kill me...hahaha !!"* that may be considered as noise in non-suicidal text. The most optimal weights for temporal signal modeling **Text + Builtup + Episodic + Surprise** were derived to be 0.52, 0.04, 0.04 and 0.32 through cross-validation experiments.

Figure 1 elucidates the impact of including psychological contextual cues on a small sample of connected users from the test dataset. It is evident from the historic trends of Users B and C that they follow a nearly episodic nature with scattered surprises. Analysing the trend plots for Users A and D reveals an inverse build-up thereby demonstrating that there can be either a positive or negative build-up in the suicidal intent of users. All these aspects when captured by our model has led to a statistically significant increase in the model's performance.

6   Conclusion

In spite of high importance of suicidal ideation identification on social media, little research has focused on looking beyond linguistic patterns. Through our work, we demonstrate that user interactions and past user behaviour are strong indicators of a potentially concerning mental state of online users. In this study, employing both qualitative and quantitative methods, we address this gap by investigating the impact of augmenting text based suicidal ideation detection models with contextual cues based on historical tweeting behavior and social media engagement.

References

1. Balakrishnan, V., Khan, S., Arabnia, H.R.: Improving cyberbullying detection using Twitter users' psychological features and machine learning. Comput. Secur. 101710 (2020)
2. Balani, S., De Choudhury, M.: Detecting and characterizing mental health related self-disclosure in social media. In: Proceedings of the 33rd Annual ACM Conference Extended Abstracts on Human Factors in Computing Systems, pp. 1373–1378. ACM (2015)
3. Benton, A., Mitchell, M., Hovy, D.: Multi-task learning for mental health using social media text. arXiv preprint arXiv:1712.03538 (2017)
4. Cavazos-Rehg, P.A., et al.: A content analysis of depression-related tweets. Comput. Hum. Behav. **54**, 351–357 (2016)
5. Cero, I., Witte, T.K.: Assortativity of suicide-related posting on social media. Am. Psychol. (2019)

6. Chatterjee, A., Das, A.: Temporal sentiment analysis of the data from social media to early detection of cyberbullicide ideation of a victim by using graph-based approach and data mining tools. In: Bhattacharyya, S., Mitra, S., Dutta, P. (eds.) Intelligence Enabled Research. AISC, vol. 1109, pp. 107–112. Springer, Singapore (2020). https://doi.org/10.1007/978-981-15-2021-1_12

7. De Choudhury, M., Gamon, M., Counts, S., Horvitz, E.: Predicting depression via social media. In: Seventh International AAAI Conference on Weblogs and Social Media (2013)

8. Kipf, T.N., Welling, M.: Semi-supervised classification with graph convolutional networks. arXiv preprint arXiv:1609.02907 (2016)

9. Lee, K.H., Xue, L., Hunter, D.R.: Model-based clustering of time-evolving networks through temporal exponential-family random graph models. J. Multivar. Anal. **175**, 104540 (2020)

10. Lopez-Castroman, J., et al.: Mining social networks to improve suicide prevention: a scoping review. J. Neurosci. Res. (2019)

11. Mathur, P., Sawhney, R., Shah, R.R.: Suicide risk assessment via temporal psycholinguistic modeling (student abstract). In: 2020 Proceedings of the 34th AAAI Conference on Artificial Intelligence. AAAI (2020)

12. Mathur, P., Shah, R., Sawhney, R., Mahata, D.: Detecting offensive tweets in Hindi-English code-switched language. In: Proceedings of the Sixth International Workshop on Natural Language Processing for Social Media, pp. 18–26 (2018)

13. Mishra, P., Del Tredici, M., Yannakoudakis, H., Shutova, E.: Author profiling for abuse detection. In: Proceedings of the 27th International Conference on Computational Linguistics, pp. 1088–1098 (2018)

14. Mishra, R., Sinha, P.P., Sawhney, R., Mahata, D., Mathur, P., Shah, R.R.: SNAP-BATNET: cascading author profiling and social network graphs for suicide ideation detection on social media. In: Proceedings of the 2019 NAACL Student Research Workshop, pp. 147–156 (2019)

15. Pourmand, A., Roberson, J., Caggiula, A., Monsalve, N., Rahimi, M., Torres-Llenza, V.: Social media and suicide: a review of technology-based epidemiology and risk assessment. Telemed. e-Health **25**(10), 880–888 (2019)

16. Qian, J., ElSherief, M., Belding, E.M., Wang, W.Y.: Leveraging intra-user and inter-user representation learning for automated hate speech detection. arXiv preprint arXiv:1804.03124 (2018)

17. Sawhney, R., Manchanda, P., Mathur, P., Shah, R., Singh, R.: Exploring and learning suicidal ideation connotations on social media with deep learning. In: Proceedings of the 9th Workshop on Computational Approaches to Subjectivity, Sentiment and Social Media Analysis, pp. 167–175 (2018)

18. Steer, B., Cuadrado, F., Clegg, R.: Raphtory: streaming analysis of distributed temporal graphs. Future Gener. Comput. Syst. **102**, 453–464 (2020)

19. Tadesse, M.M., Lin, H., Xu, B., Yang, L.: Detection of suicide ideation in social media forums using deep learning. Algorithms **13**(1), 7 (2020)

20. Zhou, P., Qi, Z., Zheng, S., Xu, J., Bao, H., Xu, B.: Text classification improved by integrating bidirectional lstm with two-dimensional max pooling. arXiv preprint arXiv:1611.06639 (2016)

PMD: An Optimal Transportation-Based User Distance for Recommender Systems

Yitong Meng[1]([✉]), Xinyan Dai[1], Xiao Yan[1], James Cheng[1], Weiwen Liu[1], Jun Guo[3], Benben Liao[2], and Guangyong Chen[2]

[1] The Chinese University of Hong Kong, Shatin, N.T., Hong Kong
{ytmeng,xydai,xyan,jcheng,wwliu}@cse.cuhk.edu.hk
[2] Tencent Quantum Lab, Shenzhen, Guangdong, China
{bliao,gycchen}@tencent.com
[3] Tsinghua University, Shenzhen, Guangdong, China
eeguojun@outlook.com

Abstract. Collaborative filtering predicts a user's preferences by aggregating ratings from similar users and thus the user similarity (or distance) measure is key to good performance. Existing similarity measures either consider only the co-rated items for a pair of users (but co-rated items are rare in real-world sparse datasets), or try to utilize the non-co-rated items via some heuristics. We propose a novel user distance measure, called Preference Mover's Distance (PMD), based on the optimal transportation theory. PMD exploits all ratings made by each user and works even if users do not share co-rated items at all. In addition, PMD is a metric and has favorable properties such as triangle inequality and zero self-distance. Experimental results show that PMD achieves superior recommendation accuracy compared with the state-of-the-art similarity measures, especially on highly sparse datasets.

Keywords: Recommendation · User similarity · Optimal transport

1 Introduction

Collaborative filtering (CF) is one of the most widely used recommendation techniques [14,47]. Given a user, CF recommends items by aggregating the preferences of similar users. Among CF recommendation approaches, methods based on nearest-neighbors (NN) are widely used, thanks to their simplicity, efficiency and ability to produce accurate and personalized recommendations [13,35,44]. Although deep learning (DL) methods [16,19,43] have attracted much attention in the recommendation community over the past few years, a very recent study [12] shows that NN-based CF is still a strong baseline and outperforms many DL methods. For NN-based methods, the user similarity measure plays an important role. It serves as the criterion to select a group of similar users whose ratings form the basis of recommendations, and is used to weigh users so that more similar users have greater impact on recommendations. Besides

© Springer Nature Switzerland AG 2020
J. M. Jose et al. (Eds.): ECIR 2020, LNCS 12036, pp. 272–280, 2020.
https://doi.org/10.1007/978-3-030-45442-5_34

CF, user similarity is also important for applications such as link prediction [4], community detection [34] and so on.

Related Work. Traditional similarity measures, such as cosine distance (COS) [9], Pearson's Correlation Coefficient (PCC) [9] and their variants [18, 29,38,39], have been widely used in CF [13,44]. However, such measures only consider *co-rated* items and ignore ratings on other items, and thus may only coarsely capture users' preferences as ratings are sparse and co-rated items are rare for many real-world datasets [35,40,44]. Some other similarity measures, such as Jaccard [22], MSD [39], JMSD [8], URP [27], NHSM [27], PIP [5] and BS [14] do not utilize all the rating information [6]. For example, Jaccard only uses the number of rated items and omits the specific rating values, while URP only uses the mean and the variance of the ratings. Critically, all these measures give zero similarity value when there are no co-rated items, which would harm recommendation performance. Recently, BCF [35] and HUSM [44] were proposed to alleviate the co-rating issue by modeling user similarity as a weighted sum of item similarities, where the weights are obtained using heuristics. As the weights are not derived in a principled manner, they do not satisfy important properties such as triangle inequality and zero self-distance, which are important for a high quality similarity measure.

The Earth Mover's Distance (EMD) is a distance metric on probabilistic space that originates from the optimal transportation theory [25,37]. EMD has been applied to many applications, such as computer vision [7], natural language processing [17,23] and signal processing [41]. EMD has also been applied to CF [48] but is used as a regularizer to force the latent variable to fit a Gaussian prior in auto-encoder training rather than a user similarity measure.

Our Solution. We propose the Preference Mover's Distance (PMD), which considers all ratings made by each user and is able to evaluate user similarity even in the absence of co-rated items. Similar to BCF and HUSM, PMD uses the item similarity as side information and assumes that if two users have similar opinions on similar items, then their tastes are similar. But the key difference is: PMD formulates the distance between a pair of users as an optimal transportation problem [26,36] such that the weights for item similarities can be derived in a principled manner. In fact, PMD can be viewed as a special case of EMD [33, 37,45], which is a metric that satisfies important properties such as triangle inequality and zero self-distance. We also make PMD practical for large datasets by employing the Sinkhorn algorithm [10] to speed up distance computation and using HNSW [30] to further accelerate the search for similar users. Experimental results show that PMD leads to superior recommendation accuracy over the state-of-the-art similarity measures, especially on sparse datasets.

2 Preference Mover's Distance

Problem Definition. Let $\mathcal{U}$ be a set of m users, and $\mathcal{I}$ a set of n items. The user-item interaction matrix is denoted by $\mathbf{R} \in \mathbb{R}^{m \times n}$, where $\mathbf{R}(u, i) \geq 0$ is the rating user u gives to item i. $\mathbf{R}$ is a partially observed matrix and usually highly sparse. For user $u \in \mathcal{U}$, her rated items are denoted by $\mathcal{I}_u \subset \mathcal{I}$. The item similarities are described by matrix $\mathbf{D}$ and $\mathbf{D}(i, j) \geq 0$ denotes the distance between items i and j. Item similarities can be derived from the ratings on items [35,44] or content information [46], such as item tags, comments, etc. In this paper, we assume $\mathbf{D}$ is given. We are interested in computing the distance between any pair (u, v) of users in $\mathcal{U}$ given $\mathbf{R}$ and $\mathbf{D}$. User similarity can be easily derived from the user distance as they are negatively correlated.

PMD. Let $\Sigma_k = \{\mathbf{p} \in [0,1]^k \mid \mathbf{p}^\top \mathbb{1} = 1\}$ denote a $(k-1)$-dimensional simplex and $\mathbb{1}$ is an all-1 column vector. We model a user's preferences as a probabilistic distribution $\mathbf{p}_u \in \Sigma_{|\mathcal{I}_u|}$ on $\mathcal{I}_u$, where $\mathbf{p}_u(i)$ indicates how much user u likes item i. In practice, the ground truth of $\mathbf{p}_u$ cannot be observed and we estimate it by normalizing user u's ratings on $\mathcal{I}_u$, i.e., $\mathbf{p}_u(i) \approx \frac{\mathbf{R}(u,i)}{\sum_{j \in \mathcal{I}_u} \mathbf{R}(u,j)}$ for $i \in \mathcal{I}_u$. We model the distance between users u and v, denoted by $d(\mathbf{p}_u, \mathbf{p}_v)$, as the weighted average of the distances among their rated items, i.e.,

$$\sum_{i \in \mathcal{I}_u} \sum_{j \in \mathcal{I}_v} \mathbf{W}_{u,v}(i, j) \mathbf{D}(i, j), \tag{1}$$

where $\mathbf{W}_{u,v}(i, j) \geq 0$ is the weight for an item pair (i, j) and we introduce the constraint $\sum_{i \in \mathcal{I}_u} \sum_{j \in \mathcal{I}_v} \mathbf{W}_{u,v}(i, j) = 1$ to control the scaling. $\sum_{j \in \mathcal{I}_v} \mathbf{W}_{u,v}(i, j)$ is the aggregate weight received by item i for user u and it should be large if $\mathbf{p}_u(i)$ is large such that $d(\mathbf{p}_u, \mathbf{p}_v)$ can focus on the items that user u likes. Similarly, $\sum_{i \in \mathcal{I}_u} \mathbf{W}_{u,v}(i, j)$ should also be large if $\mathbf{p}_v(j)$ is large. Thus, we constrain the marginal distributions of $\mathbf{W}_{u,v}$ follow $\mathbf{p}_u$ and $\mathbf{p}_v$, i.e., $\mathbf{W}_{u,v} \in U(\mathbf{p}_u, \mathbf{p}_v)$, where

$$U(\mathbf{p}_u, \mathbf{p}_v) := \left\{ \mathbf{W}_{u,v} \in [0,1]^{|\mathcal{I}_u| \times |\mathcal{I}_v|} \mid \mathbf{W}_{u,v} \mathbb{1} = \mathbf{p}_u, \mathbf{W}_{u,v}^T \mathbb{1} = \mathbf{p}_v \right\}. \tag{2}$$

However, $U(\mathbf{p}_u, \mathbf{p}_v)$ contains many different configurations of $\mathbf{W}_{u,v}$, which means that the user distance is indeterminate. Therefore, we define the user distance as the smallest among all possibilities:

$$d(\mathbf{p}_u, \mathbf{p}_v) := \min_{\mathbf{W}_{u,v} \in U(\mathbf{p}_u, \mathbf{p}_v)} \sum_{i \in \mathcal{I}_u} \sum_{j \in \mathcal{I}_v} \mathbf{W}_{u,v}(i, j) \mathbf{D}(i, j). \tag{3}$$

Equation (3) is a special case of the earth mover's distance (EMD) [11], when the moment parameter $p = 1$ and the probability space is discrete. Moreover, PMD is a metric as long as $\mathbf{D}$ is a metric [37]. We call $d(\mathbf{p}_u, \mathbf{p}_v)$ the preference mover's distance (PMD) to highlight its connection to EMD. Being a metric has some nice properties that make the user distance meaningful. For example, the triangle inequality indicates that if both user A and user B are similar to a third user C, then user A and user B are also similar. Moreover, a user should be most similar to himself among all users if $\mathbf{D}(i, i) = 0$. In contrast, it is unclear whether BCF and HUSM also have these properties as they determine weights using heuristics.

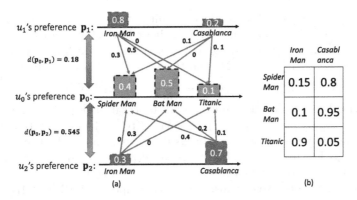

Fig. 1. An example of PMD. (a) shows the preference distributions of u_0, u_1 and u_2 using histogram and the arrows depict the optimal transportation plan (i.e., $\mathbf{W}_{u,v}$) between the preference distributions. (b) is the distance matrix for the 5 movies, in which movies with the same genre have smaller distance, i.e., are more similar.

Illustration. Intuitively, $d(\mathbf{p}_u, \mathbf{p}_v)$ can be viewed as the minimum cost of transforming the ratings of user u to the ratings of user v, which we show in Fig. 1. $\mathbf{p}_u$ and $\mathbf{p}_v$ define two distributions of mass, while $\mathbf{D}(i,j)$ models the cost of moving one unit of mass from $\mathbf{p}_u(i)$ to $\mathbf{p}_v(j)$. Therefore, PMD can model the similarity between u and v even if they have no co-rated items. If two users like similar items, $\mathbf{W}_{u,v}(i,j)$ takes a large value for item pairs with small $\mathbf{D}(i,j)$, which results in a small distance. This is the case for u_0 and u_1 in Fig. 1 as they both like science fiction movies. In contrast, if two users like dissimilar items, $\mathbf{W}_{u,v}(i,j)$ is large for item pairs with large $\mathbf{D}(i,j)$, which produces a large distance. In Fig. 1, u_0 likes science fiction movies while u_2 likes romantic movies, and thus $d(\mathbf{p}_{u_0}, \mathbf{p}_{u_2})$ is large. Even if u_0 has no co-rated movies with u_1 and u_2, PMD still gives $d(\mathbf{p}_{u_0}, \mathbf{p}_{u_1}) < d(\mathbf{p}_{u_0}, \mathbf{p}_{u_2})$, which implies that u_0 is more similar to u_1 than to u_2.

Computation Speedup. An exact solution to the optimization problem in Eq. (3) takes a time complexity of $O(q^3 \log q)$ [36], where $q = |\mathcal{I}_u \cup \mathcal{I}_v|$. To reduce the complexity, we use the Sinkhorn algorithm [10], which produces a high-quality approximate solution with a complexity of $O(q^2)$. To speed up the lookup for similar users in large datasets, we employ HNSW [30], which is the state-of-the-art algorithm for similarity search. HNSW builds a multi-layer k-nearest neighbour (KNN) graph for the dataset and returns high quality nearest neighbours for a query with $O(\log N)$ distance computations, in which N is the number of users. With these two techniques, looking up for the top 100 neighbours takes only 0.02 s on average for a user and achieves a high recall of 99.2% for the Epinions dataset in our experiments. We conduct the experiment on a machine with two 2.0 GHz E5-2620 Intel(R) Xeon(R) CPU (12 physical cores in total), 48 GB RAM, a 450 GB SATA disk (6 Gb/s, 10k rpm, 64 MB cache), and 64-bit CentOS release 7.2.

Positive/Negative Feedback. We can split the user ratings into positive ratings $\mathbf{R}^p$, e.g., 3, 4 and 5 if a score of 1–5 is allowed, which indicates that the user likes the item, and negative ratings $\mathbf{R}^n$, e.g., 1 and 2, which indicates that the user dislikes the item. Based on $\mathbf{R}^p$ and $\mathbf{R}^n$, we define positive preference $\mathbf{p}_u^p$ and negative preference $\mathbf{p}_u^n$, i.e., $\mathbf{p}_u^p(i) = \frac{\mathbf{R}^p(u,i)}{\sum_{j \in \mathbf{R}^p} \mathbf{R}^p(u,j)}$ and $\mathbf{p}_u^n(i) = \frac{\frac{1}{\mathbf{R}^n(u,i)}}{\sum_{j \in \mathbf{R}^n} \frac{1}{\mathbf{R}^n(u,j)}}$. Then we can define more fine-grained user distances using Eq. (3), e.g., $d(\mathbf{p}_u^p, \mathbf{p}_v^p)$, $d(\mathbf{p}_u^n, \mathbf{p}_v^n)$, $d(\mathbf{p}_u^n, \mathbf{p}_v^p)$ and $d(\mathbf{p}_u^p, \mathbf{p}_v^n)$. A small $d(\mathbf{p}_u^n, \mathbf{p}_v^n)$ indicates that the two users dislike similar items and can be used to avoid making bad recommendations that may lose users. A small $d(\mathbf{p}_u^p, \mathbf{p}_v^n)$ or $d(\mathbf{p}_u^n, \mathbf{p}_v^p)$ means that the interests of the two users complement each other and may be used for friend recommendation in social networks. We may also construct composite PMD (CPMD) such as:

$$\tilde{d}(\mathbf{p}_u, \mathbf{p}_v) := \mu d(\mathbf{p}_u^p, \mathbf{p}_v^p) + (1 - \mu) d(\mathbf{p}_u^n, \mathbf{p}_v^n), \tag{4}$$

where $\mu \in [0, 1]$ is a tuning parameter weighting the importance of the distances of positive and negative preferences.

3 Experiments

We evaluate PMD by comparing its performance for NN-based recommendation with various user similarity measures. Two well-known datasets, i.e., MovieLens-1M [2] and Epinions [1], are used and their statistics are reported in Table 1. The rating user u gives to item i is predicted as a weighted sum of its top-K neighbours in the training set, i.e., $\hat{\mathbf{R}}(u, i) = \bar{u} + \sum_{v \in \mathcal{N}_u} \frac{s(u,v) \times (\mathbf{R}(v,i) - \bar{v})}{\sum_{v \in \mathcal{N}_u} s(u,v)}$ [13], in which $\bar{u}$ is the average of the ratings given by user u, $\mathcal{N}_u$ contains the top-K neighbours of u and $s(u, v)$ is the similarity between a user pair u and v. We convert PMD into a similarity measure using $s(u, v) = 2 - d(\mathbf{p}_u, \mathbf{p}_v)$ and divide all ratings into train/validation/test sets, with an 8:1:1 ratio. Hyper-parameters are tuned to be optimal on the validation set for all methods. The mean absolute error (MAE) and the root mean square error (RMSE) [15,31,32] of the predicted ratings on the test set are used to evaluate the recommendation performance.

Table 1. Data statistics.

	MovieLens	Epinions
#user	6,040	116,260
#item	3,959	41,269
#rating	1,000,000	181,394
sparsity	4.14%	0.0038%
#rating/user	166	1.56
#rating/item	250	4.40

Table 2. CPMD under different K and μ.

	MovieLens ($K = 200$)		Epinions ($K = 50$)			MovieLens ($\mu = 0.6$)		Epinions ($\mu = 0.6$)	
μ	MAE	RMSE	MAE	RMSE	K	MAE	RMSE	MAE	RMSE
0.2	0.7126	0.9019	0.8542	1.1340	30	0.7148	0.9064	0.8518	1.1294
0.4	0.6970	0.8851	0.8506	1.1302	50	0.7084	0.9052	**0.8458**	**1.1260**
0.6	**0.6918**	**0.8817**	**0.8458**	**1.1260**	100	0.6972	0.8898	0.8550	1.1456
0.8	0.6955	0.8875	0.8550	1.1456	200	**0.6918**	**0.8817**	0.8592	1.1435
0.95	0.6989	0.8915	0.8596	1.1520	300	0.6938	0.8846	0.8667	1.1506

Table 3. Comparison with other user similarity measures.

Dataset	Metric	COS	PCC	MSD	Jaccard	JMSD	NHSM	BCF	HUSM	PMD	CPMD
Movie lens	MAE	0.7477	0.7234	0.7387	0.7109	0.7024	0.7079	0.7044	0.7034	0.7019	**0.6918**
	RMSE	0.9394	0.9182	0.9293	0.9125	0.8982	0.9080	0.9089	0.9067	0.8935	**0.8817**
Epin ions	MAE	1.0476	1.0468	1.0449	1.0340	1.0392	1.0213	0.9846	0.9734	0.8757	**0.8458**
	RMSE	1.4412	1.4384	1.4380	1.4226	1.4291	1.3969	1.3014	1.2846	1.1701	**1.1260**

Table 4. Comparison with latent factor models.

Dataset	Metric	NMF	SVD	SVD++	PMD	CPMD
Movie lens	MAE	0.7252	0.6864	**0.6739**	0.7019	0.6918
	RMSE	0.9177	0.8741	**0.8629**	0.8935	0.8817
Epin ions	MAE	0.9444	0.9482	0.9439	0.8757	**0.8458**
	RMSE	1.2096	1.2154	1.2091	1.1701	**1.1260**

Item Similarity. Both MovieLens and Epinions come with side information for computing item similarities. For MovieLens, we compute movie similarity using Tag-genomes [3,42]. For Epinions, we evaluate item similarity by applying Doc2Vec [24] on the comments. Since both Tag-genome and doc2vec derive item similarity by cosine, we convert item similarity into distance using $D(i,j) = \arccos(s(i,j))$, which is a metric on the item space. For fair comparison, the same item similarity matrix is used for PMD, BCF and HUSM[1].

Comparison Methods. COS, PCC and MSD are three classical user similarity measures. Jaccard, JMSD, NHSM, BCF, HUSM are five state-of-the-art measures. NMF [28], SVD [21] and SVD++ [20] are latent factor models for CF.

We report the performance of various similarity measures in Table 3, where PMD is based on Eq. (3) and CPMD is based on Eq. (4). The results show that PMD and CPMD consistently outperform other similarity measures and the improvement is more significant on the Epinions dataset which is much more sparse. We believe that our methods achieve good performance on sparse datasets mainly because it utilizes all rating information and derives the weights of the items using the optimal transportation theory, which works well when there are only few or no co-rated items. This is favorable as ratings are sparse in many real-world datasets [40]. CPMD achieves better performance than PMD, which suggests that it is beneficial to distinguish positive and negative feed-backs.

We also compare our methods with the latent factor models in Table 4. On the sparse Epinions dataset, both PMD and CPMD outperform the latent factor models. We report the performance of CPMD-based NN CF under different configurations of K and μ in Table 2. CPMD performs best when μ is around

---

[1] BCF and HUSM originally compute item similarity using the Bhattacharyya coefficient or the KL-divergence of ratings but we found that using the tag-genomes and doc2vec provides better performance.

0.6 on both datasets possibly because positive ratings can better represent the taste of a user than the negative ratings. In contrast, the optimal value of K is dataset dependent.

4   Conclusions

We proposed PMD, a novel user distance measure based on optimal transportation, which addresses the limitation of existing methods in dealing with datasets with few co-rated items. PMD also has the favorable properties of a metric. Experimental results show that PMD leads to better recommendation accuracy for NN-based CF than the state-of-the-art user similarity measures, especially when the ratings are highly sparse.

Acknowledgement. The authors thank Prof. Julian McAuley for his valuable suggestions on this paper, and Prof. Shengyu Zhang for his support. This work was supported by ITF 6904945, and GRF 14208318 & 14222816, and the National Natural Science Foundation of China (NSFC) (Grant No. 61672552).

References

1. https://cseweb.ucsd.edu/jmcauley/datasets.html
2. https://grouplens.org/datasets/movielens/
3. https://grouplens.org/datasets/movielens/tag-genome/
4. Farshad Aghabozorgi and Mohammad Reza Khayyambashi: A new similarity measure for link prediction based on local structures in social networks. Phys. A: Stat. Mech. Appl. **501**, 12–23 (2018)
5. Hyung Jun Ahn: A new similarity measure for collaborative filtering to alleviate the new user cold-starting problem. Inf. Sci. **178**(1), 37–51 (2008)
6. Al-bashiri, H., Abdulgabber, M.A., Romli, A., Hujainah, F.: Collaborative filtering similarity measures: revisiting. In: Proceedings of the International Conference on Advances in Image Processing, pp. 195–200. ACM (2017)
7. Arjovsky, M., Chintala, S., Bottou, L.: Wasserstein gan. arXiv preprint arXiv:1701.07875 (2017)
8. Bobadilla, J., Serradilla, F., Bernal, J.: A new collaborative filtering metric that improves the behavior of recommender systems. Knowl.-Based Syst. **23**(6), 520–528 (2010)
9. Breese, J.S., Heckerman, D., Kadie, C.: Empirical analysis of predictive algorithms for collaborative filtering. In: Proceedings of the Fourteenth Conference on Uncertainty in Artificial Intelligence, pp. 43–52. Morgan Kaufmann Publishers Inc. (1998)
10. Cuturi, M.: Sinkhorn distances: lightspeed computation of optimal transport. In: Advances in Neural Information Processing Systems, vol. 26, pp. 2292–2300 (2013)
11. Cuturi, M., Solomon, J.M.: A primer on optimal transport. In: Tutorial of 31st Conference on Neural Information Processing Systems (2017)
12. Dacrema, M.F., Cremonesi, P., Jannach, D.: Are we really making much progress? A worrying analysis of recent neural recommendation approaches. In: Proceedings of the 13th ACM Conference on Recommender Systems, pp. 101–109. ACM (2019)

13. Desrosiers, C., Karypis, G.: A comprehensive survey of neighborhood-based recommendation methods. In: Ricci, F., Rokach, L., Shapira, B., Kantor, P.B. (eds.) Recommender Systems Handbook, pp. 107–144. Springer, Boston (2011). https://doi.org/10.1007/978-0-387-85820-3_4

14. Guo, G., Zhang, J., Yorke-Smith, N.: A novel Bayesian similarity measure for recommender systems. In: Twenty-Third International Joint Conference on Artificial Intelligence (2013)

15. Guo, G., Zhang, J., Yorke-Smith, N.: TrustSVD: collaborative filtering with both the explicit and implicit influence of user trust and of item ratings. In: Twenty-Ninth AAAI Conference on Artificial Intelligence (2015)

16. He, X., Liao, L., Zhang, H., Nie, L., Hu, X., Chua, T.-S.: Neural collaborative filtering. In: Proceedings of the 26th International Conference on World Wide Web, pp. 173–182. International World Wide Web Conferences Steering Committee (2017)

17. Huang, G., et al.: Supervised word mover's distance. In: Proceedings of the 30th International Conference on Neural Information Processing Systems. NIPS 2016, pp. 4869–4877 (2016)

18. Jamali, M., Ester, M.: TrustWalker: a random walk model for combining trust-based and item-based recommendation. In: Proceedings of the 15th ACM SIGKDD International Conference on Knowledge Discovery and Data Mining, pp. 397–406. ACM (2009)

19. Karamanolakis, G., Cherian, K.R., Narayan, A.R., Yuan, J., Tang, D., Jebara, T.: Item recommendation with variational autoencoders and heterogeneous priors. In: Proceedings of the 3rd Workshop on Deep Learning for Recommender Systems, pp. 10–14. ACM (2018)

20. Koren, Y.: Factorization meets the neighborhood: a multifaceted collaborative filtering model. In: Proceedings of the 14th ACM SIGKDD International Conference on Knowledge Discovery and Data Mining, pp. 426–434. ACM (2008)

21. Koren, Y., Bell, R., Volinsky, C.: Matrix factorization techniques for recommender systems. Computer **8**, 30–37 (2009)

22. Koutrika, G., Bercovitz, B., Garcia-Molina, H.: FlexRecs: expressing and combining flexible recommendations. In: Proceedings of the 2009 ACM SIGMOD International Conference on Management of data, pp. 745–758. ACM (2009)

23. Kusner, M.J., Sun, Y., Kolkin, N.I., Weinberger, K.Q.: From word embeddings to document distances. In: Proceedings of The 32nd International Conference on Machine Learning, pp. 957–966 (2015)

24. Le, Q., Mikolov, T.: Distributed representations of sentences and documents. In: International Conference on Machine Learning, pp. 1188–1196 (2014)

25. Levina, E., Bickel, P.J.: The earth mover's distance is the mallows distance: some insights from statistics. In: Proceedings Eighth IEEE International Conference on Computer Vision. ICCV 2001, vol. 2, pp. 251–256 (2001)

26. Ling, H., Okada, K.: An efficient earth mover's distance algorithm for robust histogram comparison. IEEE Trans. Pattern Anal. Mach. Intell. **29**(5), 840–853 (2007)

27. Liu, H., Zheng, H., Mian, A., Tian, H., Zhu, X.: A new user similarity model to improve the accuracy of collaborative filtering. Knowl.-Based Syst. **56**, 156–166 (2014)

28. Luo, X., Zhou, M., Xia, Y., Zhu, Q.: An efficient non-negative matrix-factorization-based approach to collaborative filtering for recommender systems. IEEE Trans. Ind. Inform. **10**(2), 1273–1284 (2014)

29. Ma, H., King, I., Lyu, M.R.: Effective missing data prediction for collaborative filtering. In: Proceedings of the 30th Annual International ACM SIGIR Conference on Research and Development in Information Retrieval, pp. 39–46. ACM (2007)

30. Malkov, Y.A., Yashunin, D.A.: Efficient and robust approximate nearest neighbor search using hierarchical navigable small world graphs. IEEE Trans. Pattern Anal. Mach. Intell. **42**, 824–836 (2018)

31. Meng, Y., Chen, G., Li, J., Zhang, S.: Psrec: social recommendation with pseudo ratings. In: Proceedings of the 12th ACM Conference on Recommender Systems, pp. 397–401. ACM (2018)

32. Mnih, A., Salakhutdinov, R.R.: Probabilistic matrix factorization. In: Advances in Neural Information Processing Systems, pp. 1257–1264 (2008)

33. Monge, G.: Mémoire sur la théorie des déblais et des remblais. Histoire de l'Académie royale des sciences de Paris (1781)

34. Pan, Y., Li, D.-H., Liu, J.-G., Liang, J.-Z.: Detecting community structure in complex networks via node similarity. Phys. A: Stat. Mech. Appl. **389**(14), 2849–2857 (2010)

35. Patra, B.K., Launonen, R., Ollikainen, V., Nandi, S.: A new similarity measure using Bhattacharyya coefficient for collaborative filtering in sparse data. Knowl.-Based Syst. **82**, 163–177 (2015)

36. Pele, O., Werman, M.: Fast and robust earth mover's distances. In: 2009 IEEE 12th International Conference on Computer Vision, pp. 460–467. IEEE (2009)

37. Rubner, Y., Tomasi, C., Guibas, L.J.: A metric for distributions with applications to image databases. In: Sixth International Conference on Computer Vision (IEEE Cat. No.98CH36271), pp. 59–66 (1998)

38. Sarwar, B.M., Karypis, G., Konstan, J.A., Riedl, J., et al.: Item-based collaborative filtering recommendation algorithms. In: Www, vol. 1, pp. 285–295 (2001)

39. Shardanand, U., Maes, P.: Social information filtering: algorithms for automating "word of mouth". In: CHI, vol. 95, pp. 210–217. Citeseer (1995)

40. Symeonidis, P., Nanopoulos, A., Papadopoulos, A.N., Manolopoulos, Y.: Collaborative filtering: fallacies and insights in measuring similarity. Universitaet Kassel (2006)

41. Thorpe, M., Park, S., Kolouri, S., Rohde, G.K., Slepčev, D.: A transportation LP distance for signal analysis. J. Math. Imaging Vis. **59**(2), 187–210 (2017)

42. Vig, J., Sen, S., Riedl, J.: The tag genome: encoding community knowledge to support novel interaction. ACM Trans. Interact. Intell. Syst. (TiiS) **2**(3), 13 (2012)

43. Wang, H., Wang, N., Yeung, D.-Y.: Collaborative deep learning for recommender systems. In: Proceedings of the 21th ACM SIGKDD International Conference on Knowledge Discovery and Data Mining, pp. 1235–1244. ACM (2015)

44. Wang, Y., Deng, J., Gao, J., Zhang, P.: A hybrid user similarity model for collaborative filtering. Inf. Sci. **418**, 102–118 (2017)

45. Wolsey, L.A., Nemhauser, G.L.: Integer and combinatorial optimization. Wiley, Hoboken (2014)

46. Yao, Y., Harper, F.M.: Judging similarity: a user-centric study of related item recommendations. In: Proceedings of the 12th ACM Conference on Recommender Systems, pp. 288–296. ACM (2018)

47. Zheng, V.W., Cao, B., Zheng, Y., Xie, X., Yang, Q.: Collaborative filtering meets mobile recommendation: a user-centered approach. In: Twenty-Fourth AAAI Conference on Artificial Intelligence (2010)

48. Zhong, J., Zhang, X.: Wasserstein autoencoders for collaborative filtering. arXiv preprint arXiv:1809.05662 (2018)

On Biomedical Named Entity Recognition: Experiments in Interlingual Transfer for Clinical and Social Media Texts

Zulfat Miftahutdinov[1] (ID), Ilseyar Alimova[1] (ID), and Elena Tutubalina[1,2,3(✉)] (ID)

[1] Chemoinformatics and Molecular Modeling Laboratory, Kazan Federal University, Kazan, Russia
zulfatme@gmail.com, alimovailseyar@gmail.com, tutubalinaev@gmail.com
[2] Samsung-PDMI Joint AI Center, Steklov Mathematical Institute at St. Petersburg, Saint Petersburg, Russia
[3] Insilico Medicine Hong Kong Ltd., Pak Shek Kok, New Territories, Hong Kong

Abstract. Although deep neural networks yield state-of-the-art performance in biomedical named entity recognition (bioNER), much research shares one limitation: models are usually trained and evaluated on English texts from a single domain. In this work, we present a fine-grained evaluation intended to understand the efficiency of multilingual BERT-based models for bioNER of drug and disease mentions across two domains in two languages, namely clinical data and user-generated texts on drug therapy in English and Russian. We investigate the role of transfer learning (TL) strategies between four corpora to reduce the number of examples that have to be manually annotated. Evaluation results demonstrate that multi-BERT shows the best transfer capabilities in the zero-shot setting when training and test sets are either in the same language or in the same domain. TL reduces the amount of labeled data needed to achieve high performance on three out of four corpora: pretrained models reach 98–99% of the full dataset performance on both types of entities after training on 10–25% of sentences. We demonstrate that pretraining on data with one or both types of transfer can be effective.

Keywords: Biomedical entity recognition · BERT · Transfer learning

1 Introduction

Drugs and diseases play a central role in many areas of biomedical research and healthcare. A large part of the biomedical research has focused on scientific abstracts in English; see a good overview of the field in [9]. In contrast to the biomedical literature, research into the processing of electronic health records (EHRs) and user-generated texts (UGTs) about drug therapy has not reached the same level of maturity. The bottleneck of modern supervised models for named entity recognition (NER) is the human effort needed to annotate sufficient

© Springer Nature Switzerland AG 2020
J. M. Jose et al. (Eds.): ECIR 2020, LNCS 12036, pp. 281–288, 2020.
https://doi.org/10.1007/978-3-030-45442-5_35

training examples for each language or domain. Moreover, state of the art text processing models may perform extremely poorly under domain shift [14]. Recent advances in neural networks, especially deep contextualized word representations via language models [1,4,12,16] and Transformer-based architectures [19], offer new opportunities to improve NER models in the biomedical field.

In this work, we take the task a step further from existing monolingual research in a single domain [2,3,6,12,13,20,22] by exploring multilingual transfer between EHRs and UGTs in different languages. Our goal is not to outperform state of the art models on each dataset separately, but to ask whether we can transfer knowledge from a high-resource language, such as English, to a low-resource one, e.g., Russian, for NER of biomedical entities. Our transfer learning strategy involves pretraining the multilingual cased BERT [4] on one corpus and transferring the learned weights to initialize training on a gold-standard corpus in another language or domain. In this work, we seek to answer the following research questions: **RQ1:** How well does a BERT-based NER model trained on one corpus works for the detection of drugs and diseases from another language or domain in the zero-shot setting? **RQ2:** Given a small number of training examples, can the NER model perform as well as a model trained on much larger datasets? **RQ3:** Will transfer learning help achieve more stable performance on a varying size of training data?

All experiments are carried out on 4 datasets: English corpora CADEC [10] and n2c2 [7], a dataset of EHRs in Russian [17], and our novel dataset of UGTs in Russian. All three existing corpora share an entity of interest with our corpus. To our knowledge, this is the first work exploring the interlingual transfer ability of multilingual BERT on bioNER on two domains in English and Russian.

2   Data

Each corpus is characterized by two parameters: (i) language: English (EN) or Russian (RU); (ii) domain: electronic health records (EHRs) or user-generated texts (UGTs). A statistical summary of the datasets is presented in Table 1. Since all corpora have different annotation schemes for disease-related entities, and these subtypes are highly imbalanced in the corpus, we join them into a single primary type named *Disease*. Further, we unify the names of four datasets according to their characteristics.

CADEC (EN UGT). CSIRO Adverse Drug Event Corpus [10] contains medical forum posts taken from *AskaPatient.com* about 12 drugs of two categories: Diclofenac and Lipitor. Medical students and computer scientists annotated the dataset. The agreement between four annotators computed on a set of 55 user posts was approximately 78% for Diclofenac and 95% for Lipitor posts.

Our Dataset of UGTs (RU UGT). We utilized and annotated user posts in Russian from a publicly accessible source *Otzovik.com*; we note that we have obtained all reviews without accessing password-protected information. Four annotators from the I.M. Sechenov First Moscow State Medical University and

Table 1. Summary statistics of four datasets. Summary of each dataset includes the number of Drug and Disease entities, the number of documents and sentences, the average length of a document (in sentences), the average length of a sentence (in tokens), the average length of a Drug/Disease entity (in tokens).

Corpus	Disease subtypes	Drug	# doc.	# sent.	Avg. doc. len.	Avg. sen. len.	Avg. Drug len.	Avg. Dis. len.
CADEC (EN UGT) [10]	ADR, Symptom, Disease, Finding (6590)	Drug (1798)	1249	7670	6.14	8.27	1.11	2.48
n2c2 (EN EHR) [7]	ADE, Reason (7984)	Drug (26797)	503	70960	140.51	11.32	1.18	1.80
Our dataset (RU UGT)	ADR, Disease (2429)	Medication (1195)	400	4230	10.57	6.82	1.26	2.22
RU EHR [17]	Disease, Symptom (7874)	Drug (3479)	159	16835	105.86	6.14	1.27	2.91

the department of pharmacology of the Kazan Federal University were asked to read the review and highlight all spans of text including drug names and patient's health conditions experienced before/during/after the drug use. The agreement between two annotators computed on a set of 100 posts was 72%.

Russian EHRs (RU EHR). Shelmanov et al. [17] created a corpus of Russian clinical notes from a multi-disciplinary pediatric center. The authors extended an annotation scheme from the *CLEF eHealth 2014 Task 2*.

n2c2 (EN EHR). This corpus consists of de-identified EHRs [7]. Two independent annotators annotated each record in the dataset and a third annotator resolved conflicts. For both EHR corpora, the agreement rates were not provided.

3 Models

For NER, we utilize BERT with a softmax layer over all possible tags as the output. Word labels are encoded with the BIO tag scheme. The model was trained on a sentence level. Due to space constraints, we refer to [4,12] for more details. In particular, we use $BERT_{base}$, Multilingual Cased (Multi-BERT), which is pretrained on 104 languages and has 12 heads, 12 layers, 768 hidden units per layer, and a total of 110M parameters. All models were trained without fine-tuning or explicit selection of parameters. The loss function became stable (without significant decreases) after 35–40 epochs. We use Adam optimizer with polynomial decay to update the learning rate on each epoch with warm-up steps in the beginning. As baselines, we utilized LSTM-CRF with default settings from the Saber library [5] and BioBERT [12]. For LSTM-CRF, we adopted (i) 200-dim. *word2vec* embeddings trained on 2.5M of health-related posts in English [18] and (ii) 300-dim. *word2vec* embeddings trained on the Russian National Corpus [11].

4   Experiments and Evaluation

We randomly split each of the datasets into 70% training set and 30% test set. We trained a total of 720 models on one machine with 8 NVIDIA P40 GPUs. The training of all models took approximately 96 h. We compare all models in terms of precision (P), recall (R), and F1-score (F) on the test sets with exactly matching criteria via a CoNLL script. Our experiments are available at https://github.com/dartrevan/multilingual_experiments.

Comparison with Baselines. Table 2 shows the in-corpus (IC) performance of Multi-BERT with BioBERT and LSTM-CRF when trained and tested on the same corpus. On all datasets, BERT-based models achieve the best scores over LSTM-CRF based on word embeddings. The difference in the performance of BioBERT and Multi-BERT is not statistically significant; we measured significance with the two-tailed t-test ($p \leq 0.05$). All models achieve much higher performance for the detection of drugs rather than diseases; it can be explained by boundary problems in multi-word expressions (see the av. length in Table 1).

Zero-Shot Transfer. To answer **RQ1**, we trained Multi-BERT on one corpus and then applied it to another language/domain in a zero-shot fashion, i.e., without further training. Results of the out-of-corpus (OOC) performance of Multi-BERT are presented in Table 3. For drug recognition, the best generalizability is achieved when training on EHRs and evaluated on UGTs in English. For OOC performance on the EN UGT corpus, the model reaches F1-scores of 77.08% and 36.31% when trained on the EN EHR and RU UGT corpora, respectively, while IC reaches the F1-score of 84.88%. We note that the number of sentences in the EN EHR corpus is nine times higher than in the EN UGT corpus. 78% of Drug tokens in the EN UGT corpus are presented in the EN EHR set (see Table 4). For OOC performance on the RU UGT corpus, the model achieves F1-scores of 26.31% and 34.78% when trained on the EN UGT and EN EHR corpora, respectively, while the IC performance is F1-score of 60.45%.

For disease recognition, Multi-BERT generalizes much worse to corpora other than it was trained on. For OOC performance on the RU UGT corpus, the model achieves F1-scores of 24.12% and 30.86% when trained on the EN UGT and RU EHR corpora, respectively, while the IC performance is F1-score of 49.35%. For OOC performance on the EN UGT corpus, the model obtains F1-scores of 37.94% and 4.32% when trained on the RU UGT and EN EHR corpora,

Table 2. In-corpus (IC) performance of multi-BERT with comparison to BiLSTM-CRF and BioBERT, measured by Precision, Recall, and F1-score with an exact matching criteria.

Corpus	Models	Disease			Drug		
		P	R	F	P	R	F
EN EHR (n2c2)	Multi-BERT	55.05	63.91	59.15	92.21	92.58	92.39
	BioBERT	56.33	65.56	60.60	92.39	92.97	92.68
	LSTM-CRF	55.00	56.95	55.96	89.87	89.70	89.79
EN UGT (cadec)	Multi-BERT	65.62	68.96	67.25	79.40	91.18	84.88
	BioBERT	67.14	69.88	68.48	87.27	91.73	89.44
	LSTM-CRF	64.68	62.77	63.71	78.50	70.41	74.23
RU UGT	Multi-BERT	45.93	53.33	49.35	58.85	62.14	60.45
	LSTM-CRF	27.78	17.44	21.43	37.74	40.31	38.98
RU EHR	Multi-BERT	78.61	75.96	77.26	87.18	82.93	85.00
	LSTM-CRF	62.00	61.69	61.85	62.00	79.49	69.66

Table 3. Out-of-corpus (OOC) performance of Multi-BERT in the zero-shot setting. OOC performance is derived by training on one corpus (train) and testing on another (test).

Train	Test	Disease			Drug		
		P	R	F	P	R	F
EN UGT	EN EHR	43.05	7.47	12.73	58.23	71.9	64.35
	RU UGT	20.61	29.07	24.12	23.45	29.97	26.31
	RU EHR	7.73	44.33	13.17	5.81	91.18	10.92
EN EHR	EN UGT	2.25	51.58	4.32	78.09	76.09	77.08
	RU UGT	0.77	12.84	1.44	42.5	29.43	34.78
	RU EHR	3.33	2.85	3.07	5.35	72.65	9.97
RU UGT	EN EHR	11.9	3.23	5.08	14.63	75.3	24.50
	EN UGT	31.98	46.61	37.94	23.22	83.22	36.31
	RU EHR	10.22	41.37	16.4	28.75	44.27	34.86
RU EHR	EN EHR	0.5	20.00	0.97	46.15	37.5	41.38
	RU UGT	24.88	40.65	30.86	17.95	17.95	17.95
	EN UGT	43.78	28.12	34.24	35.90	23.73	28.57

Table 4. Summary statistics of Byte Pair Encoding (BPE) tokens of entities in four datasets. Summary includes the number of unique BPE tokens, intersection between un. tokens, percentage of shared tokens from unique set.

Dataset D_1	Dataset D_2	Entity Type	# un. BPE in D_1	# un. BPE in D_2	$D_1 \cap D_2$	% from D_1	% from D_2
EN UGT	EN EHR	Drug	528	2401	410	78%	17%
EN UGT	EN EHR	Disease	2338	2491	1172	50%	47%
RU UGT	RU EHR	Drug	696	896	381	55%	43%
RU UGT	RU EHR	Disease	1487	2427	1011	68%	42%

Table 5. Summary of the number of training sentences needed to achieve 99% of the full dataset performance with Multi-BERT with pretraining.

Pretrain	Entity type	EN UGT	RU UGT	RU EHR
Best pretrain	Drug	500	700	550
Worst pretrain	Drug	900	650	1200
No pretrain	Drug	1050	1000	1500
Best pretrain	Disease	1050	700	1850
Worst pretrain	Disease	1050	900	1850
No pretrain	Disease	1300	1100	1850

respectively, while the IC performance is F1-score of 67.25%. One possible explanation might be that there are well-known differences in layperson language and professional medical terms.

Few-Shot Transfer. Transfer learning aims to solve the problem on a "target" dataset using knowledge learned from a "source" dataset [5,15,21]. In the transfer learning setting, the BERT-based NER model was pretrained on one of three "source" datasets (see Table 2 for the IC performance of these models). To answer **RQ2** and **RQ3**, we begin with a random sampling of 50 sentences from a "target" training set, train the pretrained model on this subsampled dataset, and test it on the "target" test set. Next, we increase the sample size by 50 sentences of the "target" training set and repeat the described procedure, doing so up to 2000 sentences of the training set. In each round, we train from scratch to avoid overfitting, as suggested in [8].

For each pretraining setup, we record the size of the subset when the model achieves at least 99% of the F1-measure achieved on the full dataset. Results for the RU UGT, RU EHR, and EN UGT datasets are given in Table 5 and Fig. 1. Multi-BERT pretrained on the EN UGT set and trained on 2000 sentences from the EN EHR corpus (2.81% of the full corpus) obtains 92% F1 and 76% F1 of the full dataset performance on drugs and diseases respectively. As shown in Table 5 and Fig. 1, models with transfer knowledge outperform the models without the pretraining phase even in cases when both domain and language shifts between "source" and "target" sets. Using the transfer learning strategy could require up to 550 sentences less than training from scratch. In particular, models require only 10% and 23% of the EN UGT and RU URT corpora respectively to achieve results as good as full dataset performances. We believe that this observation is very crucial for low resource languages and new domains (e.g., social media, clinical trials). We observe that the performance of models with pretraining setup trained on the different numbers of sentences becomes more stable in terms of deviations between F1-scores (see Fig. 1).

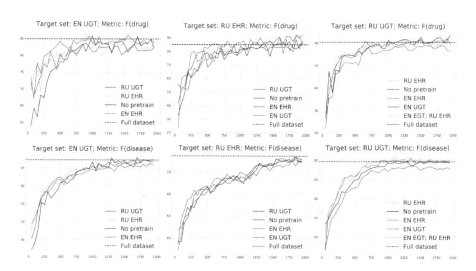

Fig. 1. Performance of Multi-BERT models with pre-training on the source dataset (a corpus's name in a legend) or without pre-training ("No pretrain" line) for the EN UGT, RU UGT, RU EHR datasets. Y-axis: F1-scores for detection of Drug or Disease mentions, X-axis: the number of sentences used for training.

5   Conclusion and Future Work

We studied the task of recognition of drug and disease mentions in English and a low-resource language in the biomedical area, using a newly collected Russian corpus of user reviews about drugs (RU UGT) with 3,624 manually annotated entities. We ask: can additional pretraining on an existing dataset be helpful for bioNER performance of multilingual BERT-based NER model on a new dataset with a small number of labeled examples if the domain, the language, or both shift between these datasets? Our study consisted of over 720 models trained on different subsets of two corpora in English and two corpora in Russian. For each language, we experimented with the clinical domain, i.e., electronic health records, and the social media domain, i.e., reviews about drug therapy. As expected, models with pretraining on data in the same language or the same domain obtain better results in zero-shot or few-shot settings. To our surprise, we found that pretraining on data with two shifts can be effective. The model with the best pretraining achieves 99% of the full dataset performance using only 23.56% of the training data on our RU URT corpus, while the model with pretraining on data with two shifts (the EN EHR set) used 26.1% of the training data. The model without pretraining achieves similar results on the RU URT corpus using 31.97% of the training set.

We foresee three directions for future work. First, transfer learning and multi-task strategies on three and more domains remain to be explored. Second, a promising research direction is the evaluation of multilingual BERT on a broad set of entities. Third, future research will focus on the creation of fine-grained entity types in our corpus of Russian reviews that can help in finding associations between drugs and adverse drug reactions.

Acknowledgments. We thank Sergey Nikolenko for helpful discussions. This research was supported by the Russian Science Foundation grant # 18-11-00284.

References

1. Alsentzer, E., et al.: Publicly available clinical BERT embeddings. In: Proceedings of the 2nd Clinical Natural Language Processing Workshop, pp. 72–78 (2019)
2. Crichton, G., Pyysalo, S., Chiu, B., Korhonen, A.: A neural network multi-task learning approach to biomedical named entity recognition. BMC Bioinform. **18**(1), 368 (2017)
3. Dang, T.H., Le, H.Q., Nguyen, T.M., Vu, S.T.: D3NER: biomedical named entity recognition using CRF-biLSTM improved with fine-tuned embeddings of various linguistic information. Bioinformatics **34**(20), 3539–3546 (2018)
4. Devlin, J., Chang, M.W., Lee, K., Toutanova, K.: BERT: pre-training of deep bidirectional transformers for language understanding. In: Proceedings of the 2019 Conference of the North American Chapter of the Association for Computational Linguistics: Human Language Technologies, Volume 1(Long and Short Papers), pp. 4171–4186 (2019)
5. Giorgi, J., Bader, G.: Towards reliable named entity recognition in the biomedical domain. Bioinformatics (2019)

6. Gupta, A., Goyal, P., Sarkar, S., Gattu, M.: Fully contextualized biomedical NER. In: Azzopardi, L., Stein, B., Fuhr, N., Mayr, P., Hauff, C., Hiemstra, D. (eds.) ECIR 2019. LNCS, vol. 11438, pp. 117–124. Springer, Cham (2019). https://doi.org/10.1007/978-3-030-15719-7_15

7. Henry, S., Buchan, K., Filannino, M., Stubbs, A., Uzuner, O.: 2018 n2c2 shared task on adverse drug events and medication extraction in electronic health records. J. Am. Med. Inform. Assoc. (2019)

8. Hu, P., Lipton, Z.C., Anandkumar, A., Ramanan, D.: Active learning with partial feedback. arXiv preprint arXiv:1802.07427 (2018)

9. Huang, C.C., Lu, Z.: Community challenges in biomedical text mining over 10 years: success, failure and the future. Brief. Bioinform. 17(1), 132–144 (2015)

10. Karimi, S., Metke-Jimenez, A., Kemp, M., Wang, C.: Cadec: a corpus of adverse drug event annotations. J. Biomed. Inform. 55, 73–81 (2015)

11. Kutuzov, A., Kunilovskaya, M.: Size vs. structure in training corpora for word embedding models: araneum russicum maximum and Russian National Corpus. In: van der Aalst, W.M.P., et al. (eds.) AIST 2017. LNCS, vol. 10716, pp. 47–58. Springer, Cham (2018). https://doi.org/10.1007/978-3-319-73013-4_5

12. Lee, J., et al.: BioBERT: pre-trained biomedical language representation model for biomedical text mining. Bioinformatics (2019)

13. Miftahutdinov, Z., Tutubalina, E.: Deep neural models for medical concept normalization in user-generated texts. In: Proceedings of the 57th Annual Meeting of the Association for Computational Linguistics: Student Research Workshop, pp. 393–399 (2019)

14. Neumann, M., King, D., Beltagy, I., Ammar, W.: ScispaCy: fast and robust models for biomedical natural language processing. arXiv preprint arXiv:1902.07669 (2019)

15. Pan, S., Yang, Q.: A survey on transfer learning. IEEE Trans. Knowl. Discov. Data Eng. 22(10) (2010)

16. Peters, M., et al.: Deep contextualized word representations. In: Proceedings of the 2018 Conference of the North American Chapter of the Association for Computational Linguistics: Human Language Technologies, Volume 1 (Long Papers), vol. 1, pp. 2227–2237 (2018)

17. Shelmanov, A., Smirnov, I., Vishneva, E.: Information extraction from clinical texts in Russian. In: Computational Linguistics and Intellectual Technologies: Papers from the Annual International Conference "Dialogue", vol. 14, pp. 537–549 (2015)

18. Tutubalina, E.V., et al.: Using semantic analysis of texts for the identification of drugs with similar therapeutic effects. Russ. Chem. Bull. 66(11), 2180–2189 (2017). https://doi.org/10.1007/s11172-017-2000-8

19. Vaswani, A., et al.: Attention is all you need. In: Advances in Neural Information Processing Systems, pp. 5998–6008 (2017)

20. Weber, L., Münchmeyer, J., Rocktäschel, T., Habibi, M., Leser, U.: HUNER: improving biomedical NER with pretraining. Bioinformatics (2019)

21. Weiss, K., Khoshgoftaar, T.M., Wang, D.D.: A survey of transfer learning. J. Big Data 3(1), 1–40 (2016). https://doi.org/10.1186/s40537-016-0043-6

22. Zhao, S., Liu, T., Zhao, S., Wang, F.: A neural multi-task learning framework to jointly model medical named entity recognition and normalization. In: Proceedings of the AAAI Conference on Artificial Intelligence, vol. 33, pp. 817–824 (2019)

SlideImages: A Dataset for Educational Image Classification

David Morris[1(✉)], Eric Müller-Budack[1], and Ralph Ewerth[1,2]

[1] TIB – Leibniz Information Centre for Science and Technology, Hannover, Germany
{David.Morris,Eric.Mueller,Ralph.Ewerth}@tib.eu
[2] L3S Research Center, Leibniz Universität Hannover, Hannover, Germany

Abstract. In the past few years, convolutional neural networks (CNNs) have achieved impressive results in computer vision tasks, which however mainly focus on photos with natural scene content. Besides, non-sensor derived images such as illustrations, data visualizations, figures, etc. are typically used to convey complex information or to explore large datasets. However, this kind of images has received little attention in computer vision. CNNs and similar techniques use large volumes of training data. Currently, many document analysis systems are trained in part on scene images due to the lack of large datasets of educational image data. In this paper, we address this issue and present SlideImages, a dataset for the task of classifying educational illustrations. SlideImages contains training data collected from various sources, e.g., Wikimedia Commons and the AI2D dataset, and test data collected from educational slides. We have reserved all the actual educational images as a test dataset in order to ensure that the approaches using this dataset generalize well to new educational images, and potentially other domains. Furthermore, we present a baseline system using a standard deep neural architecture and discuss dealing with the challenge of limited training data.

Keywords: Document figure classification · Educational documents · Classification dataset

1 Introduction

Convolutional neural networks (CNNs) are making great strides in computer vision, driven by large datasets of annotated photos, such as ImageNet [1]. Many images relevant for information retrieval, such as charts, tables, and diagrams, are created with software rather than through photography or scanning.

There are several applications in information retrieval for a robust classifier of educational illustrations. Search tools might directly expose filters by predicted label, natural language systems could choose images by type based on what information a user is seeking. Further analysis systems could be used to extract more information from an image to be indexed based on its class. In this case, we have classes such as pie charts and x-y graphs that indicate what type of

© Springer Nature Switzerland AG 2020
J. M. Jose et al. (Eds.): ECIR 2020, LNCS 12036, pp. 289–296, 2020.
https://doi.org/10.1007/978-3-030-45442-5_36

information is in the image (e.g., proportions, or the relationship of two numbers) and how it is symbolized (e.g., angular size, position along axes).

Most educational images are created with software and are qualitatively different from photos and scans. Neural networks designed and trained to make sense of the noise and spatial relationships in photos are sometimes suboptimal for born-digital images and educational images in general.

Educational images and illustrations are under-served in training datasets and challenges. Competitions such as the Contest on Robust Reading for Multi-Type Web Images [2] and ICDAR DeTEXT [3] have shown that these tasks are difficult and unsolved. Research on text extraction such as Morris et al. [4] and Nayef and Ogier [5] has shown that even noiseless born-digital images are sometimes better analyzed with neural nets than with handcrafted features and heuristics. Born-digital and educational images need further benchmarks on challenging information retrieval tasks in order to test generalization.

In this paper, we introduce SlideImages, a dataset which targets images from educational presentations. Most of these educational illustrations are created with diverse software, so the same symbols are drawn in different ways in different parts of the image. As a result, we expect that effective synthetic datasets will be hard to create, and methods effective on SlideImages will generalize well to other tasks with similar symbols. SlideImages contains eight classes of image types (e.g. bar charts and x-y plots) and a class for photos. The labels we have created were made with information extraction for image summarization in mind.

In the rest of this paper, we discuss related work in Sect. 2, details about our dataset and baseline method in Sect. 3, results of our baseline method in Sect. 4, and conclude with a discussion of potential future developments in Sect. 5.

2   Related Work

Prior information retrieval publications used or could use document figure classification. Charbonnier et al. [6] built a search engine with image type filters. Aletras and Mittal [7] automatically label topics in photos. Kembhavi et al.'s [8] diagram analysis assumes the input figure is a diagram. Hiippala and Orekhova extended that dataset by annotating it in terms of Relational Structure Theory, which implies that the same visual features communicate the same semantic relationships. De Herrera et al. [9] seek to classify image types to filter their search for medical professionals.

We intend to use document figure classification as a first step in automatic educational image summarization applications. A similar idea is followed by Morash et al. [10], who built one template for each type of image, then manually classified images and filled out the templates, and suggested automating the steps of that process. Moraes et al. [11] mentioned the same idea for their SIGHT (Summarizing Information GrapHics Textually) system.

A number of publications on document image classification such as Afzal et al. [12] and Harley et al. [13] use the RVL-CDIP (Ryerson Vision Lab Complex Document Information Processing) dataset, which covers scanned documents.

Table 1. Comparison of different datasets including the number of classes and images.

	Classes	Images		
		Train	Val	Test
SlideImages	9	2646	292	691
DocFigure	28	19795	0	13172
Head-to-head SlideImages	8	2331	257	575
Head-to-head DocFigure	8	11678	3886	3891

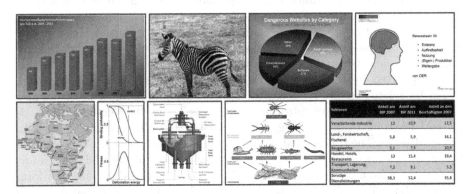

Fig. 1. Train set class examples clockwise from top left: bar charts, photos, pie charts, slide images, tables, structured diagrams, technical drawings, x-y plots, and maps.

While document scans and born-digital educational illustrations have materially different appearance, these papers show that the utility of deep neural networks is not limited to scene image tasks (Fig. 1).

A classification dataset of scientific illustrations was created for the NOA project [14]. However, their dataset is not publicly available, and does not draw as many distinctions between types of educational illustrations. Jobin et al.'s DocFigure [15] consists of 28 different categories of illustrations extracted from scientific publications totaling 33,000 images.

3   Dataset and Baseline System

Techniques that work well on DocFigure [15] do not generalize to the educational illustrations in our use case scenarios (as we also show in Sect. 4.2). Different intended uses or software cause sufficient differences in illustrations that a dataset of specifically educational illustrations is needed.

CNNs and related techniques are heavily data driven. An approach must consist of both an architecture and optimization technique, but also the data used for that optimization. In our case, we consider the dataset our main contribution.

3.1   SlideImages Dataset

When building our taxonomy, we have chosen classes such that one class would have the same types of salient features, and appropriate summaries would also be similar in structure. Our classes are also all common in educational materials. Beyond the requirements of our taxonomy, our datasets needed to be representative of common educational illustrations in order to fit real-world applications, and legally shareable to promote research on educational image classification. Educational illustrations are created by a variety of communities with varying expertise, techniques, and tools, so choosing a dataset from one source may eliminate certain variables in educational illustration. To identify these variables, we kept our training and test data sources separate.

We assembled training and validation datasets from various sources of open access illustrations. Bar charts, x-y plots, maps, photos, pie charts, slide images, table images, and technical drawings were manually selected by a student assistant (supported by the main author) using the Wikimedia Commons image search for related terms. We manually selected graph diagrams, which we also call node-edge diagrams or "structured diagrams," from the Kembhavi et al. [8] AllenAI Diagram Understanding (AI2D) dataset; not all AI2D images contain graph edges [8]. The training dataset of SlideImages consists of 2,938 images and is intended for fine-tuning CNNs, not training from scratch. The SlideImages test set is derived from a snapshot of SlideWiki open educational resource platform (https://slidewiki.org/) datastore obtained in 2018. From that snapshot, two annotators manually selected and labeled 691 images. Our data are available at our code repository: https://github.com/david-morris/SlideImages/.

3.2   Baseline Approach

The SlideImages training dataset is small compared to datasets like ImageNet [1], with over 14 million images, RVL-CDIP [13] with 400,000 images, or even DocFigure [15] with 33,000 images. Much of our methodology is shaped by needing to confront the challenges of a small dataset. In particular, we aim to avoid overfitting: the tendency of a classifier to identify individual images and patterns specific to the training set rather than the desired semantic concepts.

For our pre-training dataset, a large, diverse dataset is required that contains a large proportion of educational and scholarly images. We pre-trained on a dataset of almost 60,000 images labeled by Sohmen et al. [6] (NOA dataset), provided by the authors on request. The images are categorized as composite images, diagrams, medical imaging, photos, or visualizations/models.

To mitigate overfitting, we used data augmentation: distorting an image while keeping relevant traits. We used image stretching, brightness scaling, zooming, and color channel shifting as shown in our source code. We also added dropout with a rate of 0.1 on the extracted features before the fully connected and output layers. We used similar image augmentation for pre-training and training.

We use MobileNetV2 [16] as our network architecture. We chose MobileNetV2 as a compromise between a small number of parameters and performance on ImageNet. Intuitively, a smaller parameter space implies a model with more bias and

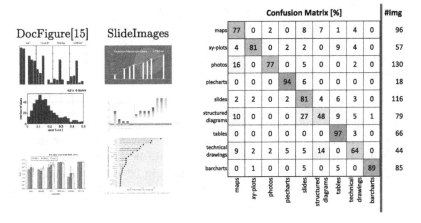

	Confusion Matrix [%]								#Img
maps	77	0	2	0	8	7	1	4	0	96
xy-plots	4	81	0	2	2	0	9	4	0	57
photos	16	0	77	0	5	0	0	2	0	130
piecharts	0	0	0	94	6	0	0	0	0	18
slides	2	2	0	2	81	4	6	3	0	116
structured diagrams	10	0	0	0	27	48	9	5	1	79
tables	0	0	0	0	0	0	97	3	0	66
technical drawings	9	2	2	5	5	14	0	64	0	44
barcharts	0	1	0	0	5	0	5	0	89	85
	maps	xy-plots	photos	piecharts	slides	structured diagrams	tables	technical drawings	barcharts	

Fig. 2. Left: examples of bar charts from the DocFigure [15] train set and our own test set. Right: confusion matrix of our baseline system on SlideImages. Entries show percent of true members of the class on the left margin labeled as on the bottom margin. Weighted accuracy average is 80% over all 691 images.

lower variance, which is better for smaller datasets. We initialized our weights from an ImageNet model and pre-trained for a further 40 epochs with early stopping on the NOA dataset using the Adam (adaptive moment estimation) [17] optimizer. This additional pre-training was intended to cause the lower levels of the network to extract more features specific to born-digital images. We then trained for 40 epochs with Adam and a learning rate schedule. Our schedule drops the learning rate by a factor of 10 at the 15th and 30th epoch. Our implementation is available at https://github.com/david-morris/SlideImages/.

4 Preliminary Results

We have performed two experiments, in order to show that this dataset represents a meaningful improvement over existing work, and to establish a baseline. Because our classes are unbalanced, we have reported summary statistics as accuracy averages of each class weighted by number of instances per class.

4.1 Baseline

We set a baseline for our dataset with the classifier described in Sect. 3.2. The confusion matrix in Fig. 2 shows that misclassifications do tend towards a few types of errors, but none of the classes have collapsed. While certain classes are likely to be misclassified as another specific class (such as structured diagrams as slides), those relationships do not happen in reverse, and a correct classification is more likely. Figure 2 shows that our baseline leaves room for improvement, and our test set helps to identify challenges in this task. Viewing individual classification errors highlighted a few problems with our training data. Our training

Table 2. Head-to-head comparison of accuracy (weighted averages).

	SlideImages train	DocFigure train	DocFigure baseline
SlideImages test	80%	78%	75%
DocFigure test	92%	99%	99%

data do not include sufficient structured diagrams with illustrated arrows, or edges which travel only at 90° increments, such as organigrams or some Unified Modeling Language diagrams. Our photos do not include examples with the background removed, but these are common in educational images. These problems should be remedied in future training datasets for this and similar problems.

4.2 Head-to-Head Comparison

The related DocFigure dataset covers similar images and has much more data than SlideImages. To justify SlideImages, we have created a head-to-head comparison of classifiers trained in the same way (as described in Sect. 3.2) on the SlideImages and DocFigure datasets. All the SlideImages classes except *slides* have an equivalent in DocFigure. We have shown the reduction in the data used, and the relative sizes of the datasets, in Table 1. The Head-to-head datasets contain only the matching classes, and in the case of the DocFigure dataset, the original test set has been split into validation and test sets.

After obtaining the two trained networks, we have tested each network on both the matching test set, and the other test set. Although we were unable to reproduce the VGG-V baseline used by Jobin et al., we used a linear SVM with VGG-16 features and achieved comparable results on the full DocFigure dataset (90% macro average compared to their 88.96% with a fully neural feature extractor). The results (Table 2) show that SlideImages is a more challenging and potentially more general task. The net trained on SlideImages did even better on the DocFigure test set than on the SlideImages test set. Despite having a different source and approximately a fifth of the size of the DocFigure dataset, the net trained on SlideImages training set was better on our test set.

5 Conclusions and Future Work

In this paper, we have presented the task of classifying educational illustrations and images in slides and introduced a novel dataset SlideImages. The classification remains an open problem despite our baseline and represents a useful task for information retrieval. We have provided a test set derived from actual educational illustrations, and a training set compiled from open access images. Finally, we have established a baseline system for the classification task. Other potential avenues for future research include experimenting with the DocFigure dataset in the pre-training and training phases, and experimenting with text extraction for multimodal classification.

Acknowledgement. This work is financially supported by the German Federal Ministry of Education and Research (BMBF) and European Social Fund (ESF) (InclusiveOCW project, no. 01PE17004).

References

1. Deng, J., Dong, W., Socher, R., Li, L.-J., Li, K., Fei-Fei, L.: ImageNet: a large-scale hierarchical image database. In: CVPR09 (2009)
2. He, M., et al.: ICPR2018 contest on robust reading for multi-type web images. In: 24th International Conference on Pattern Recognition. ICPR 2018, Beijing, China, 20–24 August 2018, pp. 7–12. IEEE Computer Society (2018)
3. Yang, C., Yin, X., Yu, H., Karatzas, D., Cao, Y.: ICDAR2017 robust reading challenge on text extraction from biomedical literature figures (detext). In: 14th IAPR International Conference on Document Analysis and Recognition. ICDAR 2017, Kyoto, Japan, 9–15 November 2017, pp. 1444–1447 (2017)
4. Morris, D.. Tang, P., Ewerth, R.: A neural approach for text extraction from scholarly figures. In: 15th International Conference on Document Analysis and Recognition. ICDAR 2019, Sydney, Australia, 20–25 September 2019, pp. 1438–1443 (2019, to appear)
5. Nayef, N., Ogier, J.: Semantic text detection in born-digital images via fully convolutional networks. In: 14th IAPR International Conference on Document Analysis and Recognition. ICDAR 2017, Kyoto, Japan, 9–15 November 2017, pp. 859–864 (2017)
6. Charbonnier, J., Sohmen, L., Rothman, J., Rohden, B., Wartena, C.: NOA: a search engine for reusable scientific images beyond the life sciences. In: Pasi, G., Piwowarski, B., Azzopardi, L., Hanbury, A. (eds.) ECIR 2018. LNCS, vol. 10772, pp. 797–800. Springer, Cham (2018). https://doi.org/10.1007/978-3-319-76941-7_78
7. Aletras, N., Mittal, A.: Labeling Topics with Images Using a Neural Network. In: Jose, J.M., et al. (eds.) ECIR 2017. LNCS, vol. 10193, pp. 500–505. Springer, Cham (2017). https://doi.org/10.1007/978-3-319-56608-5_40
8. Kembhavi, A., Salvato, M., Kolve, E., Seo, M., Hajishirzi, H., Farhadi, A.: A diagram is worth a dozen images. In: Leibe, B., Matas, J., Sebe, N., Welling, M. (eds.) ECCV 2016. LNCS, vol. 9908, pp. 235–251. Springer, Cham (2016). https://doi.org/10.1007/978-3-319-46493-0_15
9. García Seco de Herrera, A., Markonis, D., Joyseeree, R., Schaer, R., Foncubierta-Rodríguez, A., Müller, H.: Semi–supervised learning for image modality classification. In: Müller, H., Jimenez del Toro, O.A., Hanbury, A., Langs, G., Foncubierta Rodríguez, A. (eds.) Multimodal Retrieval in the Medical Domain. LNCS, vol. 9059, pp. 85–98. Springer, Cham (2015). https://doi.org/10.1007/978-3-319-24471-6_8
10. Morash, V.S., Siu, Y., Miele, J.A., Hasty, L., Landau, S.: Guiding novice web workers in making image descriptions using templates. TACCESS **7**(4), 12:1–12:21 (2015)
11. Moraes, P.S., Sina, G., McCoy, K.F., Carberry, S.: Evaluating the accessibility of line graphs through textual summaries for visually impaired users. In: Kurniawan, S., Richards, J. (eds.) Proceedings of the 16th International ACM SIGACCESS Conference on Computers & Accessibility. ASSETS 2014, Rochester, NY, USA, 20–22 October 2014, pp. 83–90. ACM (2014)

12. Afzal, M.Z., Kölsch, A., Ahmed, S., Liwicki, M.: Cutting the error by half: investigation of very deep CNN and advanced training strategies for document image classification. In: 14th IAPR International Conference on Document Analysis and Recognition. ICDAR 2017, Kyoto, Japan, 9–15 November 2017, pp. 883–888 (2017)
13. Harley, A.W., Ufkes, A., Derpanis, K.G.: Evaluation of deep convolutional nets for document image classification and retrieval. In: 13th International Conference on Document Analysis and Recognition. ICDAR 2015, Nancy, France, 23–26 August 2015, pp. 991–995. IEEE Computer Society (2015)
14. Sohmen, L., Charbonnier, J., Blümel, I., Wartena, C., Heller, L.: Figures in scientific open access publications. In: Méndez, E., Crestani, F., Ribeiro, C., David, G., Lopes, J.C. (eds.) TPDL 2018. LNCS, vol. 11057, pp. 220–226. Springer, Cham (2018). https://doi.org/10.1007/978-3-030-00066-0_19
15. Jobin, K.V., Mondal, A., Jawahar, C.V.: DocFigure: a dataset for scientific document figure classification. In: 13th IAPR International Workshop on Graphics Recognition. GREC 2019, Sydney, Australia, 20–22 September 2019 (2019, to appear)
16. Sandler, M., Howard, A.G., Zhu, M., Zhmoginov, A., Chen, L.: MobileNetV2: inverted residuals and linear bottlenecks. In: 2018 IEEE Conference on Computer Vision and Pattern Recognition. CVPR 2018, Salt Lake City, UT, USA, 18–22 June 2018, pp. 4510–4520. IEEE Computer Society (2018)
17. Kingma, D.P., Ba, J.: Adam: a method for stochastic optimization. In: Bengio, Y., LeCun, Y. (eds.) 3rd International Conference on Learning Representations. ICLR 2015, San Diego, CA, USA, 7–9 May 2015, Conference Track Proceedings (2015)

Rethinking Query Expansion for BERT Reranking

Ramith Padaki[(✉)], Zhuyun Dai, and Jamie Callan

Language Technologies Institute, Carnegie Mellon University, Pittsburgh, USA
{rpadaki,zhuyund,callan}@cs.cmu.edu

Abstract. Recent studies have shown promising results of using BERT for Information Retrieval with its advantages in understanding the text content of documents and queries. Compared to short, keywords queries, higher accuracy of BERT were observed on long, natural language queries, demonstrating BERT's ability in extracting rich information from complex queries. These results show the potential of using query expansion to generate *better* queries for BERT-based rankers. In this work, we explore BERT's sensitivity to the addition of structure and concepts. We find that traditional word-based query expansion is not entirely applicable, and provide insight into methods that produce better experimental results.

Keywords: Neural IR · BERT · Query expansion

1 Introduction

Prevalence of term-matching in several popular search engines has led users to be conditioned to produce short keyword-based queries when attempting to express their information need. A recent study from [2] shows that while applying BERT [4] for reranking, queries written in natural language enable better search results than keywords. The added structure between keywords, which is often ignored in classic retrieval systems, helps BERT better understand the query and leads to higher retrieval accuracy. These allow users to more effectively express their information need and help systems to better disambiguate documents with similar-looking content.

While this is true, there are several reasons why natural language queries may not be available from a user: due to lack of clarity in the information need, due to an inability to phrase the information need as a natural language statement, or quite simply because users are not aware of the capability of natural language understanding in search engines. In such situations, automatically expanding queries is a popularly accepted approach to add new terms and add additional context for the search engine [1]. We hypothesize that while BERT would benefit from expansion, traditional techniques may not be suitable as they do not contain inherent word order, and further lack natural language structure relating them.

© Springer Nature Switzerland AG 2020
J. M. Jose et al. (Eds.): ECIR 2020, LNCS 12036, pp. 297–304, 2020.
https://doi.org/10.1007/978-3-030-45442-5_37

This paper explores methods of expanding on an original query in a BERT-based reranker. We distinguish between two means of expanding the original query: first by adding structural words that help create a coherent natural language sentence, and second, by adding additional terms to add new concepts to the original query. We show that neither of the two are individually sufficient, and in-fact a combination of the two benefits reranking with BERT the most.

2   Related Work

Most retrieval models, including many previous state-of-the-art neural ranking models such as [3,8], assumed independence among words in a sentence or only consider short-term dependencies. This caused issues when attempting to dis-ambiguate word meanings and to understand word relations. Recently, large progress has been made on learning contextual word representations using deep neural language models [4,14]. In the domain of neural IR, BERT has been shown to be effective for passage retrieval [12] and document retrieval, specifically when provided long natural language queries for training and evaluation [2].

The addition of contextualization of query-text allows deep neural ranking models to capture several latent traits of languages that were previously difficult to capture. A recent study [2] shows that BERT-based neural rerankers achieved *better* performance on longer queries than on short keywords queries. This makes automatic query expansion desirable. Classic query expansion techniques expand the original query with terms selected from related documents; terms are usually added as a bag-of-words [1,9]. There is prior research that shows classic query expansion to be effective for a few neural ranking models where query terms and document terms are matched softly [10]. However, there is no prior work in exploration of new ways to add pseudo-relevance feedback to BERT-based rankers that rely on free-flowing natural language-text.

3   Discerning the Effect of Structure and Concepts

When people write a short query in the search engine, they often have a longer, more complex question in mind. An example is shown in Table 1: the query-title is the keyword query commonly used during search, and query-description is the real information need. It has been previously shown that applying a BERT reranker using descriptive queries written in natural language provide impressive performance gains over using classic keywords queries [2]. We hypothesize that this is due to a combination of two factors. First, the sentence *structure* in natural language that draws relations between different concepts, and second, the introduction of new *concepts* that are closely tied with the underlying information need. This work aims to understand the effects of these factors through various forms of expanding the original query, as follows.

Expansion with Structure. [2] has shown that non-concept words, such as stop-words, also contribute to BERT's effectiveness by building sentence structure. We therefore utilize two methods to gauge BERT's sensitivity to structure:

Table 1. Original title query, original description query, and variants of expansion for Query 607 in Robust04 dataset.

Title query	Human genetic code
Description query	What progress is being made in the effort to map and sequence the human genetic code
GeneratedStructure	What is the human genetic code
TemplateStructure	What is the human genetic code?
ClassicQEConcepts	Gene dna genome genes research
GoogleQuestions	How long is the human genetic code?
	How many genes are in the human genome 2018?
	Who broke the DNA genetic code?
	What is human code?

GeneratedStructure and TemplateStructure. GeneratedStructure uses a neural machine translation approach to generate new, synthetic questions from the original keyword question. This could have good practical use if proved an effective technique. We adopt the approach proposed in [5]. As shown in Table 1, GeneratedStructure tends to copy the keywords and adds a few question words, adding structure without new concepts.

TemplateStructure tests the maximum possible range of benefit from adding structure to queries. It uses a templating process by hand, manually converting the keyword queries to a question using one of several templates. Queries can be reformulated into "who, what, when, why, how" questions or a request to "describe xyz". These templated questions were generated by the authors, with care being taken to restrict addition of new words other than to relate keyword query words with each other. All original keyword terms are ensured to be included in the reformulation as well. The templated questions provides an upper bound of the effectiveness of expanding with structure.

Expansion with Concepts. This method expands the query with a set of related concepts, while grammar structures are not considered. Our method ClassicQEConcepts leverages RM3 [9], a classic pseudo-relevance feedback based query expansion model, to find related concepts. The expansion terms are concatenated to the original query, ordered by their scores estimated by RM3 as in Table 1.

Combining Structure and Concept Expansion. The last method expands the query with both sentence structure and new concepts. To do so, we rely on scraping Google's suggestions for reformulated queries to acquire additional related questions. If found during scraping, it is used in conjunction with the original title query, else just the title query is used. We refer to this approach as GoogleQuestions.

Qualitative analysis reveals that the suggested questions do not always align with the original query description. To fully verify the power of combined addition of structure and concepts, we again resort to an oracle to filter the suggested questions manually to ensure that there is description match. We achieve this by manually eliminating questions that do not align with the query description. The annotation was conducted by the authors, with an guideline that allows questions that re-formulate the original question in some other logical structure, thereby being redundant in semantics but not structure. Additionally, questions that were a direct consequence of the original question in concepts were allowed. The manual filtering leads to the FilteredGoogleQuestions.

4   Experimental Setup

Dataset. We use Robust04, a widely-used ad-hoc retrieval benchmark. It contains 249 queries and 0.5M documents. We use two types of queries: *Descriptions*, containing long natural language text describing the information need and *Titles*, the short keyword query text commonly used by search engine users.

Baselines and Experimental Methods. Baselines include three standard bag-of-words retrieval models using the Indri search engine: Indri-LM uses the query language model, Indri-SDM uses the sequential dependency model, and Indri-QE uses the classic query expansion algorithm RM3 [9]. Baselines also include two BERT rerankers from [2]. BERT-Title-Title was trained and evaluated on the query titles, and BERT-Desc-Desc was trained and evaluated on the query descriptions. The authors provided the rankings of all baselines except Indri-QE. For Indri-QE, the parameters were selected through a parameter sweep, including number of feedback documents, number of feedback terms, and weight of the original query. Our experimental methods replace the query titles in the BERT reranker with various types of expanded queries as described in Sect. 3, by concatenating the expansion to the original query title. Document-text remains unchanged.

BERT Reranker and Hyper Parameters. We adopt the model and data setup of the passage-based BERT reranker BERT-MaxP [2]. It splits documents into passages, estimates the relevance between the query and a passage using BERT's two-sentence classification model, and ranks documents using their max passage scores. We fine-tune the model as used in BERT-MaxP [2], using a batch-size of 16 and a learning rate of $1e^{-5}$ for 1000 iterations with 5-fold cross-validation. We train up to a depth of 1000 in the initial retrieved documents and only rerank the top 100 documents from the initial ranking at test time. We also sample 10% of all passages with overlap in addition to the first passage of every document during training to prevent over-fitting, as original proposed [2]. Other existing BERT rerankers use a similar architecture as BERT-MaxP, and apply more complex techniques to improve accuracy, e.g., domain adaptation [2,16], fusion with BM25 [16], and customized pooling layers [11]. This work focuses on using the simpler BERT-MaxP so that the effectiveness of queries is more clear.

Query Expansion Models and Hyperparameters. GeneratedStructure follows prior work [5] that uses CopyNet [7] to translate keywords to questions. We used the AllenNLP [6] implementation of CopyNet using an embedding dimension of 100 (initialized to GloVe [13] vectors) and an encoder/decoder of size 400 units. The model was trained over a processed Wiki-Answers dataset, containing 3M pairs of questions paired with a synthetically generated keyword question provided by authors of [5]. ClassicQEConcepts used the Indri implementation of RM3 [9]. The query expansions parameters were the same as used in Indri-QE.

5   Experimental Results

This section first studies whether the addition of new concepts has greater benefits during training or evaluation. Then experiments are conducted to find the contributions of different sources of expansion of structure and concepts.

5.1   Concepts and Structure During Training/Evaluation

It is not immediately apparent as to whether the usage of structure and concepts contribute more heavily at train or evaluation. This experiment compares

Table 2. Performance of the BERT reranker [2] on varying training and evaluation sources. Format for reference is BERT-<train>-<eval>.

Method	P@10	P@20	NDCG@10	NDCG@20	MAP
BERT-Desc-Desc	0.552	0.456	0.559	0.524	0.257
BERT-Title-Title	0.486	0.407	0.492	0.467	0.232
BERT-Title-Desc	0.474	0.386	0.479	0.439	0.196
BERT-Desc-Title	0.490	0.408	0.498	0.469	0.233

Table 3. Performance of the BERT reranker [2] when tested on various types of expansions of the original query title. The first 3 methods used Indri's bag-of-words retrieval. The other models used the BERT reranker trained on query descriptions.

Method	P@10	P@20	NDCG@10	NDCG@20	MAP
Indri-LM	0.425	0.358	0.437	0.417	0.211
Indri-SDM	0.432	0.367	0.448	0.427	0.222
Indri-QE	0.439	0.372	0.442	0.427	0.239
Query Title	0.490	0.408	0.498	0.469	0.233
ClassicQEConcepts	0.440	0.376	0.448	0.429	0.216
GeneratedSructure	0.472	0.399	0.486	0.463	0.232
TemplateStrucutre	0.484	0.402	0.489	0.460	0.224
GoogleQuestions	0.488	0.404	0.502	0.471	0.234
FilteredGoogleQuestions	**0.508**	**0.413**	**0.526**	**0.486**	**0.239**
Query description	0.552	0.456	0.559	0.524	0.257

performance when training on query titles and query descriptions, when evaluated on these criteria. Results are presented in Table 2. Format for reference is BERT-<train>-<eval>. For example, BERT-Desc-Title implies that the model was trained on query-descriptions and evaluated on query-titles.

The results confirm findings from [2] that using query-descriptions for training and evaluation benefit the BERT reranker over just using the query-title. More importantly, the gain of using query-descriptions when training (BERT-Desc-Desc) is better than when only using it for evaluation (BERT-Title-Desc). We adopt this approach going forward, training on descriptions while evaluating on a concatenation of the original query with extensions of the original query. Our experiments indicate that this performs better than training on expanded queries.

5.2 Query Expansion with Concepts and Structures for BERT

The next experiment tests the effectiveness of query expansion using structures/concepts in BERT. We expand the query titles with various approaches as described in Sect. 3. Results are in Table 3. Indri-LM/SDM/QE are classic bag-of-words retrieval baselines using Indri; Query Title/Description applies the BERT reranker with query titles/descriptions during evaluation. A good query expansion should be able to out-perform Query Title, and be close to Query Description.

Adding Structure. Results from GeneratedStructure and TemplateStructure reveal that neither of the two strategies out-perform the original query title. This leads us to conclude that structure alone does not provide much evidence for short queries, as short queries do not contain too many complex relations between them; structure is more important with addition of additional keywords, where they help to build the complex relations between the many concepts.

Adding New Concepts. Table 3 reveals that adding expansion terms to Indri produces results that are better than Indri-LM and Indri-SDM, but significantly worse than most BERT rerankers. On training ClassicQEConcepts, with the classical Indri-QE expansion terms, we find that the model under-performs all other BERT models. This indicates that classic query expansion using discrete terms is not suitable for BERT-based deep language models. From Table 1, classic QE expand the query with 'gene dna genome genes research', which is a bag of discrete words and lacks natural language structure that BERT is trained on [4]. This experiment helps establish that the new keywords and concepts must be related to each other in some coherent form, and that these relations actually benefit the ranking process.

Combining Structure and Concepts. FilteredGoogleQuestions provides significant lift in precision and NDCG when compared with the competing baseline Query Title. A win-loss analysis reveals that Google queries improve the original query by providing new concepts as well as several reformulations of the original query. Often, few but meaningful follow-up questions are more useful than several unrelated ones. Further, rephrasing the original sentence in multiple ways benefits the reranker. For instance for the query "opening adoption records", the questions "Are adoption records public?", and "Should adoption records be open?" give a boost of +0.3 in NDCG@5. This behaviour has been previously studied in previous work [15] wherein BERT favors text sequence pairs that are close in semantic meaning. Often, synonyms and multiple paraphrased versions of the original intent benefit the reranker. On the other hand, without filtering, the GoogleQuestions model does not produce any improvement in performance. This is mainly due to the off-topic questions, which take up about 60% of all retrieved queries. Our results reveal promising direction of query expansion for BERT using related questions that people often searched together. We show that these natural questions, when on-topic, provide valuable information to the original keyword queries, and are more effective than class query expansions or solely adding structures.

6 Conclusion and Future Work

BERT has shown to be good at long descriptive queries in document reranking tasks. With a new paradigm in which deep contextual representations of text show promise in the field of text retrieval, we provide insight into means of emulating descriptive queries after experimental analysis of the behaviour of BERT.

Our results reveal traditional word-based query expansion are not sufficient. A good query for BERT-based rerankers requires both a rich set of concepts and grammar structures that build word relations. However, a critical aspect is identifying extensions of the original query that are in-domain to the corpus, and align with the original intent. Further work in this field would involve automatic identification of questions that are in-domain to the source corpus and alternate means of generating the same.

References

1. Carpineto, C., Romano, G.: A survey of automatic query expansion in information retrieval. ACM Comput. Surv. (CSUR) 44, 1–50 (2012)
2. Dai, Z., Callan, J.: Deeper text understanding for IR with contextual neural language modeling. In: The 42nd International ACM SIGIR Conference on Research & Development in Information Retrieval (2019)
3. Dehghani, M., Zamani, H., Severyn, A., Kamps, J., Croft, W.B.: Neural ranking models with weak supervision. In: The 40th International ACM SIGIR Conference on Research & Development in Information Retrieval (2017)
4. Devlin, J., Chang, M.W., Lee, K., Toutanova, K.: Bert: pre-training of deep bidirectional transformers for language understanding. arXiv preprint arXiv:1810.04805 (2018)
5. Ding, H., Balog, K.: Generating synthetic data for neural keyword-to-question models. In: Proceedings of the 2018 ACM SIGIR International Conference on Theory of Information Retrieval. ACM (2018)

6. Gardner, M., et al.: AllenNLP: a deep semantic natural language processing platform (2017)
7. Gu, J., Lu, Z., Li, H., Li, V.O.K.: Incorporating copying mechanism in sequence-to-sequence learning. arXiv preprint arXiv:1603.06393 (2016)
8. Guo, J., Fan, Y., Ai, Q., Croft, W.B.: A deep relevance matching model for ad-hoc retrieval. In: Proceedings of the 25th ACM International on Conference on Information and Knowledge Management (2016)
9. Lavrenko, V., Croft, W.B.: Relevance based language models. In: Proceedings of the 24th Annual International ACM SIGIR Conference on Research & Development in Information Retrieval (2001)
10. Li, C., et al.: NPRF: a neural pseudo relevance feedback framework for ad-hoc information retrieval. In: Proceedings of the 2018 Conference on Empirical Methods in Natural Language Processing (EMNLP) (2018)
11. MacAvaney, S., Yates, A., Cohan, A., Goharian, N.: CEDR: contextualized embeddings for document ranking. In: Proceedings of the 42nd International ACM SIGIR Conference on Research & Development in Information Retrieval (2019)
12. Nogueira, R., Cho, K.: Passage re-ranking with BERT. arXiv preprint arXiv: 1901.04085 (2019)
13. Pennington, J., Socher, R., Manning, C.D.: Glove: global vectors for word representation. In: Proceedings of the 2014 Conference on Empirical Methods in Natural Language Processing (EMNLP) (2014)
14. Peters, M.E., et al.: Deep contextualized word representations. arXiv preprint arXiv:1802.05365 (2018)
15. Qiao, Y., Xiong, C., Liu, Z., Liu, Z.: Understanding the behaviors of BERT in ranking. arXiv preprint arXiv:1904.07531 (2019)
16. Yilmaz, Z.A., Wang, S., Yang, W., Zhang, H., Lin, J.: Applying BERT to document retrieval with birch. In: Proceedings of the 2019 Conference on Empirical Methods in Natural Language Processing and the 9th International Joint Conference on Natural Language Processing (EMNLP-IJCNLP): System Demonstrations (2019)

Personalized Video Summarization Based Exclusively on User Preferences

Costas Panagiotakis[1,2]([✉]) [iD], Harris Papadakis[3] [iD],
and Paraskevi Fragopoulou[2,3] [iD]

[1] Department of Management Science and Technology,
Hellenic Mediterranean University, 72100 Agios Nikolaos, Greece
cpanag@hmu.gr
[2] Institute of Computer Science, FORTH, Heraklion, Greece
fragopou@ics.forth.gr
[3] Department of Electrical and Computer Engineering,
Hellenic Mediterranean University, 71004 Heraklion, Greece
adanar@hmu.gr

Abstract. We propose a recommender system to detect personalized video summaries, that make visual content interesting for the subjective criteria of the user. In order to provide accurate video summarization, the video segmentation provided by the users and the features of the video segments' duration are combined using a Synthetic Coordinate based Recommendation system.

Keywords: Recommender system · Video summarization

1 Introduction

Video summarization is an application of recommender systems [9,13] that generally aims at providing users with targeted information about items that might interest them. Recommender systems are also used to provide users with suggestions for various entities such as e-shop items, web pages, news, articles, movies, music, hotels, television shows, books, restaurants, friends, etc.

In this work, we study the problem of personalized video summarization without an priori knowledge of the video categories. According to our knowledge, this is the first work that solves the personalized video summarization based exclusively on user preferences for a given dataset of videos. In order to solve this problem, we propose a video segmentation method that yields global video segments. The main contribution of this work is the proposed video segmentation method and the efficient combination of the video segments' duration attribute with the Synthetic Coordinate based Recommendation system (SCoR) [12] without the use complex audiovisual features.

2 Related Work

The problem of content recommendation can be described as follows. Given a set U of users, a set I of items and a set R of user ratings for items, we

© Springer Nature Switzerland AG 2020
J. M. Jose et al. (Eds.): ECIR 2020, LNCS 12036, pp. 305–311, 2020.
https://doi.org/10.1007/978-3-030-45442-5_38

need to predict ratings for user-item pairs which are not in R. One of the main recommender system techniques is similarity-based Collaborative Filtering [1]. Such algorithms are based on a similarity function which takes into account user preferences and outputs a similarity degree between pairs of users. Another important approach in recommender systems is Dimensionality Reduction. Each user or item in the system is represented by a vector. A user's vector is the set of his ratings for all items in the system (even those that have not been rated by the specific user). The Matrix Factorization method [5] that characterizes both items and users by vectors of latent factors inferred from item rating patterns, is also a Dimensionality Reduction technique. High correlation between item and user factors leads to a recommendation.

In [12], the SCoR recommender system has been proposed that assigns synthetic coordinates to users and items (nodes). SCoR assigns synthetic coordinates (vectors) to users and items as proposed in [2], but instead of using the dot product, SCoR uses the Euclidean distance between a user and an item in the Euclidean space, so that, when the system converges, the distance between a user-item pair provides an accurate prediction of that user's preference for the item. SCoR has been also successfully applied to the distributed community detection problem [11] and to the interactive image segmentation problem [10].

A video summary usually includes the most important scenes and events from a video, with the shortest possible description. Many traditional video summarization approaches, which are not personalized, [8,16] find a global optimal representation of a given video taking into account only its audiovisual features. As the given, video synopsis datasets and annotations increase, the computer vision community realized that the problem of video summarization can be also defined and solved separately for each user taking into account his preferences. Thus, the research on personalized video summarization is gaining increased attention recently [19].

There exist supervised methods based on complex audiovisual features that can become personalized by training on annotations coming from a single user [18]. Other personalized methods use text queries [17]. They suffer, however, from the cold start problem, not being able to provide recommendations for users that are not in the training set. In addition, only a small number of examples per user are often available. This limits the class of possible methods to simple models that can be trained from a handful of examples [6]. More recent methods use a ranking formulation, where the goal is to score interesting video segments higher than non-interesting ones [4,6,14,19] while combining audiovisual representation and user preferences. In [19], a novel pairwise deep ranking model is proposed that employs deep learning in order to learn the relationship between highlighted and non-highlighted video segments. A two-stream network structure is developed by representing video segments from complementary information on the appearance of video frames and temporal dynamics across frames for video highlight detection. Rather than training one model per user, the model proposed in [6] is personalized via its inputs, which allows to effectively adapt its predictions, given only a few user-specific examples. To train this model, a large-

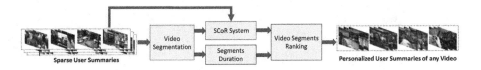

Fig. 1. The schema of the proposed system architecture.

scale dataset of users and GIFs is created, providing an accurate indication of their interests. In this work, we use the same dataset and a ranking formulation.

3 Personalized Video Summarization

In this Section, the proposed personalized video summarization method is described. Figure 1 depicts the two stages of the proposed framework. In the first stage, each video is segmented into non overlapping segments according to the preferences of the users. In the second stage, the personalized rankings of the video segments are provided.

3.1 Video Segmentation

The goal of video segmentation is to provide the candidate video segments that are included in the video summarization, significantly reducing the problem search space from the set of frames to the set of video segments. The simplest video segmentation is to use fixed segments (e.g. of 5 s duration) [6]. Several audiovisual based video summarization methods use shot detection [3] or other more complex temporal segmentation approaches [7,19] to provide accurate (non-overlapping) video segmentation. In this work, since the audiovisual data are not taken into account, we take advantage of the user preferences in the training set to derive the video segmentation.

Let F_v be the union of segment borders (frames) in ascending order, that the users provide in the training set according to the proposed video highlights of video v. As the number of users increases, the frames of F_v correspond to an over-segmentation of the given video v. So, in this work we simplify set F_v, so that there is a minimum duration for each video segment, e.g. at least 1 s. To do so, we repetitively remove the frame f from F_v according to Eq. 1, until the minimum segment length is at least 1 s.

$$f = \underset{i \in \{1,2,...,|F_v(i)|\}}{\arg\min} \quad min(\delta_v(i), \delta_v(i+1)) + \frac{1}{|v|} \cdot max(\delta_v(i), \delta_v(i+1)) \quad (1)$$

where $\delta_v(i) = |F_v(i) - F_v(i-1)|$ corresponds to the duration of video segment $[F_v(i), F_v(i-1)]$ and $|v|$ is the video length. This equation selects the frame that corresponds to the shortest segment. In order to decide which of the two border frames of a segment should be eliminated, we also take into account the size of the longest neighbor segment $(max(\delta_v(i), \delta_v(i+1)))$, so that the frame in between the two shorter in duration video segments is selected.

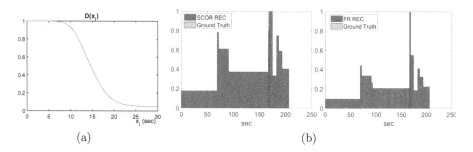

(a)   (b)

Fig. 2. (a) An example of the ranking function $D(x_i)$. (b) An example of $SCOR_u(i)$ (left) and $FR_u(i)$ (right) recommendations on a given video.

3.2   Video Segments Duration

Generally, it holds that the users select short video segments to be included in the proposed video synopsis (e.g. less than 20 s). In this work, we apply a statistical analysis approach with personalized components taking into account the average segment duration of a user (d_u), of a video (d_v), for dataset (d) and the standard deviation of the video segment duration in dataset (σ). So, for a user u and an unseen (for that user) video v, the ranking function $D(x_i)$ (see Fig. 2(a)) is computed, where x_i denotes the duration of segment $[F_v(i), F_v(i-1)]$.

$$D(x_i) = (1 - \lambda) \cdot (1 - CDF_{\mu,\sigma}(x_i)) + \lambda \qquad (2)$$

where $CDF_{\mu,\sigma}$ is the Cumulative Gamma distribution function with mean value $\mu = \frac{d_u + d_v + d}{3} + 3 \cdot \sigma$ and standard deviation σ. The popular two-parameter Gamma distribution is selected, since it is defined only for positive values, such as the duration attribute. The positive parameter λ (e.g. $\lambda = 0.05$) and the addition of $3 \cdot \sigma$ is used to relax the effect of the duration attribute to the whole ranking process, since it is a complementary feature in the final decision process.

3.3   Ranking Video Segments

In the final stage of the proposed method, the video segments are ranked by combining the segment duration based on the ranked function $D(x_i)$ and the ranking of video segments provided by the SCoR system.

Similarly to [12], in order to train SCoR, we get all video segments (see Sect. 3.1) of each video v that have been summarized by user u. Let $[F_v(i), F_v(i-1)]$ be the video segment i of video v, then the recommendation $R_u(i)$ of user u for this segment, that is used to train the SCoR, is given by the percentage of the video segment frames $[F_v(i), F_v(i-1)]$ that belong to the video summary that user u provides. This means that $R_u(i) \in [0, 1]$.

SCoR [12] assigns synthetic coordinates to users and items (video segments), so that the distance between a user and a video segment provides an accurate prediction of the user preference for that video segment. The lowest ranking value (recommendation) is assigned a distance of 1, whereas the highest ranking value is assigned a distance of 0. When the system converges, users and video segments have been placed in the same multi-dimensional Euclidean space. Let $p(u)$ and $p(i)$, be the position of user u and video segment i in this space. Then, for a pair of user u and video segment i, SCoR is able to provide a recommendation $SCOR_u(i) = max(0, 1 - ||p(u) - p(i)||_2)$. The final personalized recommendation $FR_u(i) \in [0, 1]$ is given by the product of SCoR and the duration based recommendations:

$$FR_u(i) = \frac{SCOR_u(i) \cdot D(x_i)}{max_j SCOR_u(j) \cdot D(x_j)} \qquad (3)$$

The denomination of Eq. 3, normalizes the final recommendation $FR_u(i)$ so that its maximum value is one. Figure 2(b) depicts an example of $SCOR_u(i)$ (left) and $FR_u(i)$ (right) recommendation for a given video.

4   Experimental Results

In our experimental results, we included the proposed method $(SCOR - D)$ and two methods from the literature $(PHD - CA + SVM - D$ [6] and $Video2GIF$ [4]) and the following three variants of the proposed method:

- $SCOR$: The variant of the proposed method that only uses the SCoR system.
- $SCOR - FIX$: The variant of the proposed method that combines SCoR with fixed length (5 s, as proposed in [6]) video segmentation.
- $RANDOM$: Random summaries based on the proposed video segmentation.

To obtain personalized video highlight data, we have used the large scale dataset proposed in [6], that contains *13,822 users and 222,015 annotations on 119,938 YouTube videos*. Due to the fact that our method is only based on user preferences, we keep users and videos with at least five annotations in order to be able to provide recommendations (cold start problem). The resulting dataset consists of *1822 users and 6347 annotations on 381 videos* with 129,890 candidate video segments under the proposed video segmentation with variable segment lengths, and 199,462 video segments with fixed, 5 s, segment length. The dataset was randomly separated into training and test sets, as proposed in [6]. In the test set, we included annotations from 191 users concerning their last (191) annotated videos (50% of the given videos).

To evaluate the performance of the video summarization methods, we report the mean Average Precision (mAP) [14] and the Normalized Meaningful Summary Duration $(NMSD)$ [6]. $NMSD$ rates how much of the video has to be watched before the majority of the ground truth selection is shown, given that the frames in the video are re-arranged in descending order of their predicted recommendation scores. In addition, we report the F_1 *score* that is computed by comparing the ground truth selection with the video summary of the same length $(recall = precision)$ that is created by adding frames in descending order of their predicted recommendation scores. Thus, the F_1 *score* measures the percentage of the video summary that belongs to the ground truth selection.

Table 1. Comparison with the state-of-the-art comparison

Criteria	PHD-CA + SVM-D	Video2GIF	SCOR-D	SCOR	SCOR-FIX	RANDOM
mAP	16.68	15.86	**21.65**	15.71	10.22	9.67
$nMSD$	40.26	42.06	**28.82**	42.48	44.52	55.96
$F_1\ score$	–	–	**18.32**	9.51	5.72	4.69

Table 1 presents the average mAP, $nMSD$ and $F_1\ score$. It holds that the proposed method $SCOR - D$ clearly outperforms all the remaining methods under any evaluation metric. The importance of the duration attribute and the proposed variable length video segmentation is verified by comparing the results of the proposed method against $SCOR$ and $SCOR - FIX$, respectively. The $F_1\ score$ of the proposed method is 9% and 13% higher than the $F_1\ score$ of $SCOR$ and $SCOR - FIX$, respectively. $SCOR$ is the second method in performance, while $SCOR - FIX$ is the third one, under any evaluation metric. Finally, it should be noted that the performances of $PHD - CA + SVM - D$ and $Video2GIF$ have been obtained in the whole dataset of [6], so they are not directly comparable with the other methods.

5   Conclusions

In this work, we presented a methodology to detect personalized video highlights without taking into account audiovisual features. The proposed method efficiently uses known user preferences to derive a video segmentation and it combines the segment duration attribute with the SCoR recommender system [12], yielding accurate personalized video summarization. According to our experimental results, the proposed system outperforms other variants and methods from literature. The proposed methodology can be extended to include rich audiovisual features [15], in order to be able to provide personalized user summaries even for unseen videos.

Acknowledgements. This research has been co-financed by the European Union and Greek national funds through the Operational Program Competitiveness, Entrepreneurship and Innovation, under the call RESEARCH - CREATE - INNO-VATE (project code: T1EDK-02147).

References

1. Adomavicius, G., Kwon, Y.: Improving aggregate recommendation diversity using ranking-based techniques. IEEE Trans. Knowl. Data Eng. **24**(5), 896–911 (2012)
2. Gorrell, G.: Generalized Hebbian algorithm for incremental singular value decomposition in natural language processing. In: EACL 2006, 11th Conference of the European Chapter of the Association for Computational Linguistics, Proceedings of the Conference, 3–7 April 2006, Trento, Italy (2006)

3. Gygli, M.: Ridiculously fast shot boundary detection with fully convolutional neural networks. In: 2018 International Conference on Content-Based Multimedia Indexing (CBMI), pp. 1–4. IEEE (2018)
4. Gygli, M., Song, Y., Cao, L.: Video2gif: automatic generation of animated gifs from video. In: Proceedings of the IEEE Conference on Computer Vision and Pattern Recognition, pp. 1001–1009 (2016)
5. Koren, Y., Bell, R., Volinsky, C.: Matrix factorization techniques for recommender systems. Computer 42(8), 30–37 (2009)
6. del Molino, A.G., Gygli, M.: Phd-gifs: personalized highlight detection for automatic gif creation. arXiv preprint arXiv:1804.06604 (2018)
7. Pal, G., Acharjee, S., Rudrapaul, D., Ashour, A.S., Dey, N.: Video segmentation using minimum ratio similarity measurement. Int. J. Image Min. 1(1), 87–110 (2015)
8. Panagiotakis, C., Doulamis, A., Tziritas, G.: Equivalent key frames selection based on iso-content principles. IEEE Trans. Circuits Syst. Video Technol. 19(3), 447–451 (2009)
9. Panagiotakis, C., Papadakis, H., Fragopoulou, P.: Detection of hurriedly created abnormal profiles in recommender systems. In: International Conference on Intelligent Systems (2018)
10. Panagiotakis, C., Papadakis, H., Grinias, E., Komodakis, N., Fragopoulou, P., Tziritas, G.: Interactive image segmentation based on synthetic graph coordinates. Pattern Recogn. 46(11), 2940–2952 (2013)
11. Papadakis, H., Panagiotakis, C., Fragopoulou, P.: Distributed detection of communities in complex networks using synthetic coordinates. J. Stat. Mech: Theory Exp. 2014(3), P03013 (2014)
12. Papadakis, H., Panagiotakis, C., Fragopoulou, P.: Scor: a synthetic coordinate based system for recommendations. Expert Syst. Appl. 79, 8–19 (2017)
13. Ricci, F., Rokach, L., Shapira, B.: Recommender systems: introduction and challenges. In: Ricci, F., Rokach, L., Shapira, B. (eds.) Recommender Systems Handbook, pp. 1–34. Springer, Boston (2015). https://doi.org/10.1007/978-1-4899-7637-6_1
14. Sun, M., Farhadi, A., Seitz, S.: Ranking domain-specific highlights by analyzing edited videos. In: Fleet, D., Pajdla, T., Schiele, B., Tuytelaars, T. (eds.) ECCV 2014. LNCS, vol. 8689, pp. 787–802. Springer, Cham (2014). https://doi.org/10.1007/978-3-319-10590-1_51
15. Tran, D., Bourdev, L., Fergus, R., Torresani, L., Paluri, M.: Learning spatiotemporal features with 3d convolutional networks. In: Proceedings of the IEEE International Conference on Computer Vision, pp. 4489–4497 (2015)
16. Truong, B.T., Venkatesh, S.: Video abstraction: a systematic review and classification. ACM Trans. Multimedia Comput. Commun. Appl. (TOMM) 3(1), 3 (2007)
17. Vasudevan, A.B., Gygli, M., Volokitin, A., Van Gool, L.: Query-adaptive video summarization via quality-aware relevance estimation. In: Proceedings of the 2017 ACM on Multimedia Conference, pp. 582–590. ACM (2017)
18. Xu, J., Mukherjee, L., Li, Y., Warner, J., Rehg, J.M., Singh, V.: Gaze-enabled egocentric video summarization via constrained submodular maximization. In: Proceedings of the IEEE Conference on Computer Vision and Pattern Recognition, pp. 2235–2244 (2015)
19. Yao, T., Mei, T., Rui, Y.: Highlight detection with pairwise deep ranking for first-person video summarization. In: Proceedings of the IEEE Conference on Computer Vision and Pattern Recognition, pp. 982–990 (2016)

SentiInc: Incorporating Sentiment Information into Sentiment Transfer Without Parallel Data

Kartikey Pant$^{(\boxtimes)}$, Yash Verma, and Radhika Mamidi

International Institute of Information Technology, Hyderabad, India
kartikey.pant@research.iiit.ac.in, yash.verma@students.iiit.ac.in,
radhika.mamidi@iiit.ac.in

Abstract. Sentiment-to-sentiment transfer involves changing the sentiment of the given text while preserving the underlying information. In this work, we present a model SentiInc for sentiment-to-sentiment transfer using unpaired mono-sentiment data. Existing sentiment-to-sentiment transfer models ignore the valuable sentiment-specific details already present in the text. We address this issue by providing a simple framework for encoding sentiment-specific information in the target sentence while preserving the content information. This is done by incorporating sentiment based loss in the back-translation based style transfer. Extensive experiments over the Yelp dataset show that the SentiInc outperforms state-of-the-art methods by a margin of as large as ∼11% in G-score. The results also demonstrate that our model produces sentiment-accurate and information-preserved sentences.

Keywords: Textual style transfer · Sentiment analysis

1 Introduction

Esoteric methods in sequence to sequence tasks use massive amounts of parallel data. However, in many style transfer tasks such as sentiment-to-sentiment transfer, such data is not readily available. Therefore, most of the recent work in language style transfer focuses on solving this task in an unsupervised setting [2,6]. Unsupervised learning involves learning a latent representation of data in a shared latent space, which provides fine control over the latent attributes in the data. Autoencoders have been used for generating sentences with controllable attributes by the disentangled latent representations [1,4]. Most of the work done previously on the task uses adversarial training for learning this latent representation [3,4,12,18].

The task of sentiment-to-sentiment transfer is equivalent to finding target sentence y that maximizes the conditional probability of y given a source sentence

---

K. Pant and Y. Verma—Contributed equally to the work.

© Springer Nature Switzerland AG 2020
J. M. Jose et al. (Eds.): ECIR 2020, LNCS 12036, pp. 312–319, 2020.
https://doi.org/10.1007/978-3-030-45442-5_39

x and sentiment s, i.e., $\max p(y|x, s)$. Sentiment-to-sentiment transfer can be seen as a special style transfer task. It involves changing the underlying sentiment of the source text while preserving the underlying non-semantic content.

In this work, we propose an architecture that extends the model proposed by [6], which performs machine translation relying on monolingual corpora in each language. We form language models for each sentiment and use iterative back-translation [10], thereby converting this unsupervised task into a semi-supervised task.

To summarize, the following are the contributions[1]:

- We propose a novel approach for sentiment-to-sentiment transfer incorporating sentiment-specific information in an unsupervised setting.
- The proposed method encodes sentiment-specific information in the target sentence by incorporating sentiment-based loss in the iterative back-translation algorithm.
- The proposed method also gives finer control over the trade-off between content-preservation and sentiment-transfer, thus making it adaptable for various applications.
- We extensively evaluate the performance of our model on a real-world dataset, and results reveal that it outperforms the state-of-the-art methods.

2   Related Work

The most closely related works are on the areas of neural unpaired sentiment-to-sentiment translation and encoding sentiment-specific information in training neural networks.

Generating sentences with controllable attributes has been addressed in [4] by learning disentangled latent representations [1]. Their model builds on variational auto-encoders (VAEs) and uses independency constraints to enforce reliable inference of attributes back from generated sentences. [12] focuses on separating the underlying content from style information, [19] employs an iterative back-translation algorithm by using a pseudo-parallel corpus created using a word-to-word transfer table built by cross-domain word embeddings and style specific language models. Use of back translation to facilitate unsupervised machine translation is addressed in [6]. This model was extended by [13] to address the limitations of attribute transfer by performing attribute conditioning and latent representation pooling.

On the other hand, sentiment-specific information is encoded with lexicon-based approaches [14,16,17], mostly sentiment-polarity pairs, incorporating negation and intensification to compute sentence polarity for each sentence. [15] proposed learning sentiment-specific word embeddings (SSWE) which encoded the sentiment information into a continuous representation of words by incorporating a sentiment-specific loss function in the training process.

---

[1] Made available at the following Github repository: https://github.com/kartikey pant/sentiinc-sentiment-transfer.

To the best of our knowledge, there is no previous work that facilitates sentiment-to-sentiment transfer while encoding sentiment-specific information, in an unsupervised setting.

3 Sentiment Transfer Using Sentiment-Specific Loss

In this section, we define the task and its preliminaries before presenting the details of learning sentiment-specific information for unsupervised sentiment transfer.

Let x be an input sentence (having a sentiment y), $\tilde{y}$ be the target sentiment, and $\tilde{x}$ be the target sentence. Our task is to map $(x, \tilde{y})$ to $\tilde{x}$, such that maximum amount of original content-based semantic information from x is preserved independent of the sentiment information. For the remainder of the paper, we denote the source space as S, and the target space as T. Let P_s and P_t denote the language models trained on datasets from each domain and $P_{s \rightarrow t}$ and $P_{t \rightarrow s}$ be the translation models from source to target and vice versa.

3.1 Style Transfer as Unsupervised Machine Translation

Instead of considering words, we use BPE tokens [11] as they reduce the vocabulary size and eliminate the presence of unknown words in the output. The data of both the sentiment domains (positive and negative) is jointly processed to form the BPE tokens which are shared across sentiment domains. Token embeddings [8] are now learned to initialize the lookup tables for encoder and decoder. We accomplish language modeling via denoising autoencoders [6], by minimizing the following loss:

$$L_{lm} = E_{x \sim S}[-log P_{s \rightarrow s}(x|C(x))] + E_{y \sim T}[-log P_{t \rightarrow t}(y|C(y))] \qquad (1)$$

where, C is a noise model with some words dropped and swapped. $P_{s \rightarrow s}$ and $P_{t \rightarrow t}$ are the composition of encoder and decoder both operating on the source and target sides, respectively.

For converting this unsupervised task to a semi-supervised setting, we use iterative back-translation. We consider $\forall x \in S$, let $v^\star(x) = \arg\max P_{s \rightarrow t}(v|x)$ and $\forall y \in T$, $u^\star(y) = \arg\max P_{t \rightarrow s}(x, v^\star(x))$ and $(u^\star(y), y)$ which forms automatically-generated parallel sentences. Using this, we train two transfer models by minimizing the following loss:

$$L_{back} = E_{y \sim T}[-log P_{s \rightarrow t}(y|u^\star(y))] + E_{x \sim S}[-log P_{t \rightarrow s}(x|v^\star(x))] \qquad (2)$$

3.2 Sentiment-Specific Loss

For incorporating the sentiment-specific loss into the training, we use pretrained fastText classifiers [5], which provide polarities from -2 to 2, -2 being the most negative. Let the score predicted by fastText classifier be denoted by $\delta(t)$. Now, let $f(t)$ be the indicator of the target sentiment, given a score $+1$ if the target sentiment is negative and -1 if it is positive. The sentiment-specific loss is modeled as:

$$L_{s(t)} = exp(n^2 \cdot k) \cdot max(0, f(t) \cdot \delta(t)) \tag{3}$$

Here, n denotes the number of epochs and k is a hyperparameter and $exp(n^2 \cdot k)$ is used to make the sentiment-specific information more dominant as the model learns to generate content-preserved sentences with an increase in n.

3.3   Shared Latent Representation

SentiInc uses latent representation for both language modeling and style transfer as it ensures inductive transfer across both tasks. To share the encoder representation, we share all the encoder as well as decoder parameters across the two sentiment domains.

While minimizing the loss function, backpropagation is not performed through the reverse model which generated the data since as observed by [6], no significant improvement were observed by doing so. The final loss function to be minimized is as follows:

$$L = L_{lm} + L_{back} + L_{s(t)} \tag{4}$$

4   Experiments

In this section, we introduce the datasets and briefly describe the evaluation metrics used by contemporary models. We also compare SentiInc's performance to the current state-of-the-art on these datasets.

4.1   Dataset

We conduct experiments on the publicly available Yelp food review dataset as previously used by [7,12,13]. Unlike most work in the area, we operate at the scale of complete reviews and do not assume that every sentence of the review inherits the sentiment of the complete review. We, therefore, relax constraints enforced in prior works [7,12] that discard reviews with more than 15 words and only consider the $10k$ most frequent words. Instead, we consider full reviews with up to 100 words and use byte-pair encodings (BPE) [11] eliminating the presence of unknown words.

Small Yelp Review Dataset (SYelp): It is used by many of the previous works conducted in this area [3,7,12], and contains sentences instead of complete reviews. It is based on the assumption that each sentence in a review inherits the sentiment of the complete review. Finally, the data is encoded in $10k$ BPE Codes.

Big Yelp Review Dataset (BYelp): It contains full reviews instead of individual sentences. Following previous work on reviews spanning multiple reviews [13], we preprocess this data to remove reviews that are not written in English according to a fastText classifier [5] which achieves competitive performance. Based on the rating associated with the review, we classify it as either positive or negative. Finally, the data is encoded in $60k$ BPE Codes.

4.2 Evaluation Metrics

In this work, we use a combination of multiple automatic evaluation criteria informed by our desiderata. We want our model to generate sentences that have the target sentiment while preserving the structure and content of the input.

Therefore, we evaluate samples from different models along the following two different criteria:

- **Transfer of Sentiment** (Accuracy): We measure the extent to which the sentiment is converted using the pretrained fastText classifier for the polarity of the reviews. The fastText classifier [5] achieves a high accuracy of 95.7% in determining the polarity of the review.
- **Content preservation** (BLEU): We measure the extent to which a model preserves the content present in a given input using n-gram statistics, by measuring the BLEU score [9] between generated text and the input itself.

It has been observed by [7,13] that as the BLEU score increases, the sentiment transfer accuracy decreases. As we want our model to produce sentences that have the target sentiment while preserving the content, we use the Geometric mean (G-score) of Accuracy and BLEU as an evaluation metric to evaluate the overall performance [18].

4.3 Baselines

We compare our proposed model with the following baselines:

1. **Style Embedding** [3]: In this method, the model learns a representation for the input sentence that only contains the content information after which it learns style embeddings in addition to the content representations.
2. **Multi-Decoder** [3]: This method uses a multi-decoder model with adversarial learning which uses different decoders, one for each style, to learn generation of sentences in each corresponding style.
3. **Cross-Alignment Auto-Encoder (CAE)** [12]: This model uses refined alignment of latent representations in hidden layers.
4. **Retrieval, DeleteOnly, and DeleteAndRetrieve** [7]: DeleteOnly works on extracting content words by deleting phrases associated with the sentence's original style. Retrieval works on retrieving new phrases associated with the target style. DeleteAndRetrieve is a neural model that smoothly combines the DeleteOnly and Retrieval method into a final output.

5   Results and Analysis

Table 1 compares the results of the baselines with our model on SYelp dataset where accuracy evaluates transfer of sentiment, BLEU evaluates semantic content preservation, and G-score is the geometric mean of accuracy and BLEU. Our models outperform the current state-of-the-art by a G-score of 11%.

In SYelp dataset, StyleEmbedding achieves a high BLEU score of 67.63; however, it is unable to transfer the sentiment significantly, showing a low sentiment transfer accuracy of 14.5%. MultiDecoder achieves a BLEU score of 40.22 showing sentiment transfer accuracy of 50.40%. CAE achieves a BLEU score of 20.28 and shows a sentiment transfer accuracy of 73.7%, obtaining a G-score of 38.66. Retrieval achieves a much lower BLEU score of 2.62; however, it shows a high sentiment transfer accuracy of 83.8%. DeleteOnly and DeleteAndRetrieve show competitive performances among the baselines, having BLEU scores of 37.45 and 35.55 respectively and sentiment transfer accuracy of 82.6% and 84% respectively.

Table 1. Results for the SYelp dataset.

Model	Accuracy	BLEU	G-score
StyleEmbedding [3]	14.5%	67.63	31.31
MultiDecoder [3]	50.4%	40.22	45.02
CAE [12]	73.7%	20.28	38.66
Retrieval [7]	83.8%	2.62	14.83
DeleteOnly [7]	82.6%	35.45	54.11
DeleteAndRetrieve [7]	84.0%	35.55	54.64
SentiInc w/o Sentiment Loss	74.1%	57.53	65.29
SentiInc	73.7%	59.56	**66.25**

We also compare SentiInc with and without sentiment loss. With sentiment loss, it shows the maximum G-score of 66.25, and obtains sentiment transfer accuracy of 73.7% and BLEU score of 59.56. This denotes that SentiInc produces better sentiment-accurate and content-preserved sentences than all the baselines. Without sentiment loss, we obtain a G-score of 65.29, observing a drop of 0.96 on the SYelp dataset.

The ablation study conducted on the BYelp dataset shows a 1.76 increase in G-score upon incorporation of sentiment loss. We also observe a decrease in the trade-off between BLEU and sentiment transfer accuracy with respect to the baseline without the sentiment loss. This shows a direct effect of the sentiment loss in reducing the limitations by the BLEU-accuracy trade-off, which makes the target sentences content-preserved while maintaining sentiment accuracy.

6   Conclusion and Future Work

In this paper, we focus on unpaired sentiment-to-sentiment translation and proposed our model SentiInc based on back-translation and sentiment analysis. SentiInc incorporates sentiment-based loss that enables training through only mono-sentiment data. Our experiments on review datasets (with varied maximum sentence length) show that our method substantially outperforms the state-of-the-art models in overall performance. We further show that by incorporating sentiment-loss into the back-translation based model, it is possible to decrease the limitations of the trade-off between content preservation and sentiment transfer accuracy. In the future, we would like to experiment with converting offensive text and hate-speech into polite forms and increasing the degrees of polarity in the sentiment transfer.

References

1. Chen, X., Duan, Y., Houthooft, R., Schulman, J., Sutskever, I., Abbeel, P.: Infogan: interpretable representation learning by information maximizing generative adversarial nets (2016). https://arxiv.org/abs/1606.03657
2. Conneau, A., Kiela, D., Schwenk, H., Barrault, L., Bordes, A.: Supervised learning of universal sentence representations from natural language inference data. arXiv preprint arXiv:1705.02364 (2017)
3. Fu, Z., Tan, X., Peng, N., Zhao, D., Yan, R.: Style transfer in text: exploration and evaluation. arXiv e-prints arXiv:1711.06861, November 2017
4. Hu, Z., Yang, Z., Liang, X., Salakhutdinov, R., Xing, E.P.: Toward controlled generation of text. arXiv e-prints arXiv:1703.00955, March 2017
5. Joulin, A., Grave, E., Bojanowski, P., Mikolov, T.: Bag of tricks for efficient text classification. arXiv preprint arXiv:1607.01759 (2016)
6. Lample, G., Ott, M., Conneau, A., Denoyer, L., Ranzato, M.: Phrase-based & neural unsupervised machine translation. CoRR abs/1804.07755 (2018). http://arxiv.org/abs/1804.07755
7. Li, J., Jia, R., He, H., Liang, P.: Delete, retrieve, generate: a simple approach to sentiment and style transfer. CoRR abs/1804.06437 (2018). http://arxiv.org/abs/1804.06437
8. Mikolov, T., Sutskever, I., Chen, K., Corrado, G., Dean, J.: Distributed representations of words and phrases and their compositionality. CoRR abs/1310.4546 (2013). http://arxiv.org/abs/1310.4546
9. Papineni, K., Roukos, S., Ward, T., Zhu, W.J.: Bleu: a method for automatic evaluation of machine translation. In: Proceedings of the 40th Annual Meeting on Association for Computational Linguistics, ACL 2002, pp. 311–318. Association for Computational Linguistics, Stroudsburg (2002). https://doi.org/10.3115/1073083.1073135
10. Sennrich, R., Haddow, B., Birch, A.: Improving neural machine translation models with monolingual data. CoRR abs/1511.06709 (2015). http://arxiv.org/abs/1511.06709
11. Sennrich, R., Haddow, B., Birch, A.: Neural machine translation of rare words with subword units. CoRR abs/1508.07909 (2015). http://arxiv.org/abs/1508.07909

12. Shen, T., Lei, T., Barzilay, R., Jaakkola, T.: Style transfer from non-parallel text by cross-alignment. arXiv e-prints arXiv:1705.09655, May 2017
13. Subramanian, S., Lample, G., Smith, E.M., Denoyer, L., Ranzato, M., Boureau, Y.L.: Multiple-attribute text style transfer. arXiv e-prints arXiv:1811.00552, November 2018
14. Taboada, M., Brooke, J., Tofiloski, M., Voll, K.D., Stede, M.: Lexicon-based methods for sentiment analysis. Comput. Linguist. **37**(2), 267–307 (2011). http://dblp.uni-trier.de/db/journals/coling/coling37.html#TaboadaBTVS11
15. Tang, D., Wei, F., Yang, N., Zhou, M., Liu, T., Qin, B.: Learning sentiment-specific word embedding for twitter sentiment classification, vol. 1, pp. 1555–1565 (2014). https://doi.org/10.3115/v1/P14-1146
16. Thelwall, M., Buckley, K., Paltoglou, G.: Sentiment strength detection for the social web. J. Am. Soc. Inf. Sci. Technol. **63**(1), 163–173 (2012). https://doi.org/10.1002/asi.21662
17. Turney, P.D.: Thumbs up or thumbs down? Semantic orientation applied to unsupervised classification of reviews. CoRR cs.LG/0212032 (2002). http://arxiv.org/abs/cs.LG/0212032
18. Xu, J., et al.: Unpaired sentiment-to-sentiment translation: a cycled reinforcement learning approach. CoRR abs/1805.05181 (2018). http://arxiv.org/abs/1805.05181
19. Zhang, Z., et al.: Style transfer as unsupervised machine translation. arXiv e-prints arXiv:1808.07894, August 2018

Dualism in Topical Relevance

Panagiotis Papadakos[1,2](✉) (iD) and Orfeas Kalipolitis[2]

[1] Institute of Computer Science, FORTH-ICS, Heraklion, Greece
papadako@ics.forth.gr
[2] Computer Science Department, University of Crete, Heraklion, Greece
csd3285@csd.uoc.gr

Abstract. There are several concepts whose interpretation and meaning is defined through their binary opposition with other opposite concepts. To this end, in this paper we elaborate on the idea of leveraging the available *antonyms* of the original query terms for eventually producing an answer which provides a better overview of the related conceptual and information space. Specifically, we sketch a method in which antonyms are used for producing dual queries, which can in turn be exploited for defining a multi-dimensional topical relevance based on the antonyms. We motivate this direction by providing examples and by conducting a preliminary evaluation that shows its importance to specific users.

Keywords: Information retrieval · Exploratory search · Dualism · Antonyms

1 Introduction: Why Dualism Is Important

Dualism denotes the state of two parts. The term was originally coined to denote co-eternal binary opposition and has been especially studied in philosophy. For example, there is duality in ethics (good - bad), in human beings (man - Nietzsche's übermensch or man - god) and in logic (true - false). In addition, dualism determines in a great extent our everyday lives (ugly - beautiful, happy - unhappy, etc.), and our relations with other people (rich - poor, black - white, love - hate, etc.). None of these concepts can be understood without their dual concepts, since this duality and opposition generates their meaning and interpretation. Dualism is also crucial in mathematics and physics (e.g., matter - antimatter), and is the power behind our whole information society and our binary data.

Moving from philosophy, sciences and everyday life to information retrieval, we find a very vague situation. Users of search engines are 'dictated' to provide a very concise and specific query that is extremely efficient for focalized search (e.g., looking for a specific hotel). On the other hand, studies show that 60% of user tasks are of exploratory nature [12]. In such tasks users do not accurately know their information need and can not be satisfied by a single 'hit' [5]. Consequently, users spend a lot of time reformulating queries and investigating

© Springer Nature Switzerland AG 2020
J. M. Jose et al. (Eds.): ECIR 2020, LNCS 12036, pp. 320–327, 2020.
https://doi.org/10.1007/978-3-030-45442-5_40

results, in order to construct a conceptual model regarding their information need. Information needs that include non-monosemous terms can be considered such exploratory tasks. However, the simplicity of inserting terms in an empty text box and 'magically' return the most relevant object(s), will always be a desired feature.

In this paper we elaborate on the idea of leveraging the available *antonyms* of the original query terms (if they exist), for eventually producing an answer which provides a better overview of the related information and conceptual space. We sketch a method in which antonyms are used for producing dual queries, which in turn can be exploited for defining a multi-dimensional topical relevance. This approach can be applied on demand, helping users to be aware of the various opposing dimensions and aspects of their topic of interest. A preliminary evaluation shows the value of the approach for some exploratory tasks and users.

To the best of our knowledge, the proposed direction is not covered by the existing literature. Antonyms have been studied in Fuzzy Logic [7] showing a relation with negates. In the IR domain, query expansion methods are based on synonyms and semantically related terms, but do not exploit antonyms explicitly, while in relevance and pseudo-relevance feedback techniques the antonyms are essentially penalized [1]. Results diversification can produce a kind of dual clusters, but this is neither guaranteed nor controlled [3].

2 Motivating Examples

"Capitalism and War". Consider a user exploring the relationship between *capitalism* and *war*. The user submits to a WSE (Web Search Engine) the query `"capitalism and war"` and starts inspecting the results. The left part of Fig. 1 shows the top-5 results for this query from a popular WSE. The results include articles about the connection of capitalism with war from research and academic domains, as well as from socialistic, communistic and theological sites. Considering a different direction, the user might also be interested about how capitalism can support *peace*, the dual of *war*. The top-5 results for the query `"capitalism and peace"` are shown at the right side of Fig. 1. They contain a wikipedia and a research article about the *Capitalist Peace Theory*, and articles about the importance of capitalism for the prosperity of modern societies and its association to peace from policy research organizations.

Analogously, since *socialism* is the economic system that opposes capitalism, the user could be interested about how socialism may promote war or support peace, by inspecting the results of the queries `"socialism and war"` and `"socialism and peace"` respectively. The top-5 results for each of the above queries are shown in Fig. 2. The results for the former query include the *Socialism and War* pamphlet written by Lenin, a collection of articles by the economist and philosopher Friedrich Hayek, a list of articles from two marxist domains, and a critical article for both left and right views from the Foundation for Economic Education. For the latter query, the results include articles connecting socialism with peace, like a chapter from the Encyclopedia of Anti-Revisionism, a wikipedia article about the theoretical magazine *Problems of Peace and Socialism*, and an article from a site supporting a far left U.S. Party.

The above hits indicate interesting directions to the original information need of the user. We argue that users should get aware of these directions for a better exploration of the domain at hand, since they can provide a more comprehensive view of the information and conceptual space. Furthermore, the exploration of these directions let available supportive or counter arguments of dual concepts to emerge, leading to better informed and responsible humans and citizens.

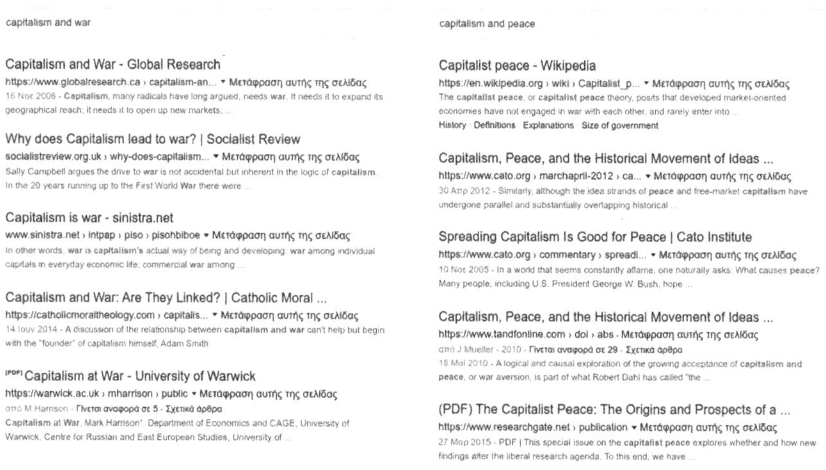

Fig. 1. Top-5 results for `"capitalism and war"` and `"capitalism and peace"`

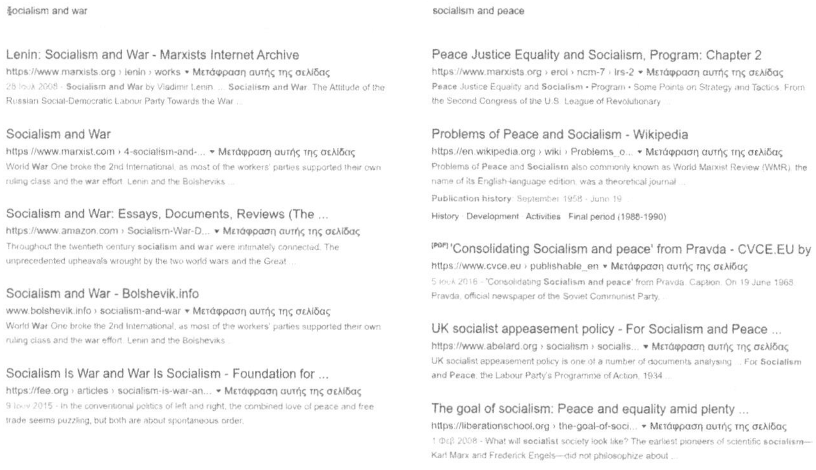

Fig. 2. Top-5 results for `"socialism and war"` and `"socialism and peace"`

"Aloe". A comprehensive view of the various different directions can be beneficial also for reducing *false-positive* results. For example, consider a pregnant woman that was advised to take *aloe vera* by mouth to relieve digestive discomfort. To check if this is true, she submits to a WSE the query "aloe vera indications". However, since aloe can stimulate uterine contractions, increasing the risk of miscarriage or premature birth, it is crucial to know also its contraindications. The proposed direction can alleviate this problem, because this information would be contained in the results of the query "aloe vera contraindications".

3   The Gordian Knot Solution

One can imagine various ways for leveraging antonyms. We shall hereafter use $t * t'$ to denote that the terms t and t' are antonyms. Building on the "capitalistic" example of the previous section, according to the online dictionary Wordnet[1], socialism $*$ capitalism, and war $*$ peace. Now, we can generate all possible queries, denoted by Q, where non-monosemous terms of the original query are substituted by their dual ones, as expressed by their antonyms. For example, the query "capitalism and war" will generate three extra queries: "socialism and peace", "capitalism and peace" and "socialism and war". Based on Q we can now define two vector spaces. In the first case, the space has $|Q|$ dimensions, where each query is a dimension of the space. Each document is placed in this space according to its relevenace to each query. In the second case we assume a space with only $\frac{|Q|}{2}$ dimensions. Each dimension represents a pair of *dual queries*, where each query in the pair contains the antonyms of the other. We denote with $q * q'$, that the queries q and q' are dual. For our running example, the first pair is ("capitalism and war","socialism and peace") and the second one is ("capitalism and peace","socialism and war"). Each pair defines an axis, therefore the two pairs define a 2D space against which we can evaluate the "value" of each document. For each axis we can consider policies for composing the relevance scores of each document to each member of a *dual* query.

Generally, there are various criteria that can be considered for assessing the value of each document or set of documents. Such criteria include the *bias* of documents to specific queries (e.g., the original user query), the *purity* to a specific query, the *overview* factor of a document regarding either a dual query or all queries, and the *diversity* of the returned set of documents with respect to these queries. In general, we need to define appropriate ranking methods, that will take into account the relevance of the documents to the available queries for different criteria. Therefore, we will explore whether the existing multiple-criteria approaches described in [4,6,9,13] are appropriate for the problem at hand.

---

[1] http://wordnet.princeton.edu/.

Regarding the process of finding the corresponding antonyms, we can use existing dictionaries like WordNet for nouns and adjectives or word-embedding antonym detection approaches like [8,11]. The case of verbs and adverbs is more complicated since they require a kind of grammatical and language analysis (i.e., exist ✳ not exist, lot ✳ total, a lot ✳ bit, etc). There are three categories of antonyms: (a) *gradable*, (b) *relational* and (c) *complementary*. We have *gradable* antonyms (e.g., hot ✳ cold) in cases where the definitions of the words lie on a continuous spectrum. We have *relational* antonyms (e.g., teacher ✳ student) in cases where the two meanings are opposite only within the context of their relationship. The rest are called *complementary* antonyms (e.g., day ✳ night). In general, the selection of the "right" antonyms raises various questions. In many cases more than one antonyms exist, so one should decide which one(s) to select. Sometimes this can depend on the context, e.g., the antonym of "action" is "apathy", but in terms of physics or sociology the dual of "action" is "reaction".

Notice that the proposed approach can be exploited in any context where the aim is to retrieve semantically opposing entities, information, etc. As an example consider the *Argument Web* [2], where the approach could be used for retrieving contradicting arguments and providing support for each one of them. From a system's perspective, the approach can be realized in various levels and settings. In the setting of an IR system, it can be implemented by changing accordingly the query processor and the ranking module, while in a meta-search setting, by changing the query rewriting, the query forwarding and the ranking components. It could also be exploited in the query autocompletion layer.

4   Preliminary Evaluation

To start with, we have conducted a preliminary evaluation. We have specified 15 information tasks which are shown in Table 1, that can exploit the proposed approach. The tasks are of exploratory nature and were created using the task refinement steps described in [10]. We have identified the following types of tasks: Explore Domain (*ED*), Medical Treatment (*MT*), Explore Product Reviews (*EPR*) and Person Qualities (*PQ*). For each task we provide a description of the information need, a representative query and the relevant antonyms, which were manually selected from the list of the respective WordNet antonyms.

We conducted our experiment over 9 female and 25 male users of various ages. For each task, they were given two lists of results. One contained the results of the query from a popular WSE, and the other one was constructed by interleaving the results of the same WSE for the dual queries of this task (i.e., first the top result of the original query, then the first result of its dual, etc.). The two kinds of lists were given in random order for each task. The users were asked to select the most preferred list and to provide a grade of preference taking values in {1, 2, 3, 4, 5}, where 5 means that the selected list was preferred much more than the other one. In the background, when users prefer the results of the dual approach, we change the sign of the score and make it negative. The users were not aware how the lists were constructed and were not guided in any way by the evaluator.

Table 1. Evaluation tasks

Id	Type	Information need	Query	Antonyms
Q_1	ED	is capitalism connected to war?	capitalism and war	socialism, peace
Q_2	ED	learn about the greek debt crisis	greek debt crisis	profit
Q_3	ED	learn about 9/11	the truth about 9/11	lie
Q_4	ED	learn about black holes	black holes	white
Q_5	ED	search about free will	free will	enslaved
Q_6	ED	what is open source?	open source	closed
Q_7	ED	learn about the syrian civil war	syrian civil war	peace
Q_8	MT	you are pregnant and you were suggested to drink Aloe Vera	aloe vera indications	contraindications
Q_9	MT	you were suggested that the drug accutane can help with acne	positive accutane experiences	negative
Q_{10}	EPR	which is the best smartphone?	best smartphone	worst
Q_{11}	EPR	android phones are better than iphone	android phones are better than iphone	worse
Q_{12}	EPR	where should I go for my summer?	best summer vacation	worst, winter
Q_{13}	EPR	which are the safest cars?	most safe cars	less, unsafe
Q_{14}	PQ	learn about Obama's peace nobel prize	Obama nobel peace prize	war
Q_{15}	PQ	is Greek politician XXX smart?	XXX is smart	stupid

In Fig. 3 we provide two graphs that describe the results of the evaluation. Figure 3 (a), shows the aggregated scores given by all users to each query, while Fig. 3 (b) shows the aggregated scores given by each participant to all queries. Regarding the first one the results are not the expected ones, although we hypothesize that the users mainly penalized the dual approach because of the 'irrelevant' results to the original query in terms of query tokens and not in terms of relevant information. For eleven of the queries there is a strong preference towards the non-dual approach. The *EPR* type of queries belong to this category, showing that users are probably not interested for reviews with the opposite direction of what they are looking for. This is especially true for Q_{12}, where the dual approach provided results about winter vacations and was the least preferred. For two of the tasks, the approaches are almost incomparable. Both of these tasks belong to the *MT* group. There are also two queries, Q_3 and Q_{15}, where the dual approach is better, especially in the last one. In their comments for these queries, users mention that the selected (i.e., dual) list "provides a more general picture" and "more relevant and interesting results, although contradicting". Regarding the second graph we have the interesting result that the proposed approach appeals to specific users. It seems that nine users (26% of the participants) have an exploratory nature and generally prefer the dual approach (six of them strongly), while for four of them the two approaches are incomparable. The rest are better served with the non-dual approach. This is an interesting outcome, and in the future we plan to identify the types of users that prefer the dual approach.

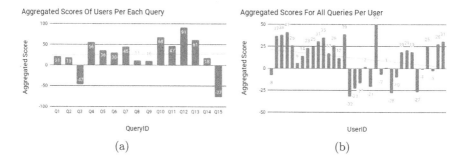

Fig. 3. Graph (a) shows the aggreggated users' scores for each query, while graph (b) the aggreggated scores of each user for all queries

5   Conclusion and Future Work

We have motivated with examples why it is worth investigating dualism for non-monosemous terms in the context of exploratory search and we have shown its importance at least for some types of users and tasks. For the future, we plan to define the appropriate antonyms selection algorithms and relevance metrics, implement the proposed functionality in a meta-search setting, and conduct a large scale evaluation with real users over exploratory tasks, to identify in which queries the dual approach is beneficial and to what types of users.

References

1. Azad, H.K., Deepak, A.: Query expansion techniques for information retrieval: a survey. Inf. Process. Manag. **56**(5), 1698–1735 (2019)
2. Bex, F., Lawrence, J., Snaith, M., Reed, C.: Implementing the argument web. Commun. ACM **56**(10), 66–73 (2013)
3. Carpineto, C., D'Amico, M., Romano, G.: Evaluating subtopic retrieval methods: clustering versus diversification of search results. Inf. Process. Manag. **48**(2), 358–373 (2012)
4. da Costa Pereira, C., Dragoni, M., Pasi, G.: Multidimensional relevance: a new aggregation criterion. In: Boughanem, M., Berrut, C., Mothe, J., Soule-Dupuy, C. (eds.) ECIR 2009. LNCS, vol. 5478, pp. 264–275. Springer, Heidelberg (2009). https://doi.org/10.1007/978-3-642-00958-7_25
5. Crawford, D. (ed.): Supporting Exploratory Search, vol. 49. ACM, New York (2006)
6. da Costa Pereira, C., Dragoni, M., Pasi, G.: Multidimensional relevance: prioritized aggregation in a personalized information retrieval setting. Inf. Process. Manag. **48**(2), 340–357 (2012)

7. De Soto, A.R., Trillas, E.: On antonym and negate in fuzzy logic. Int. J. Intell. Syst. **14**(3), 295–303 (1999)
8. Dou, Z., Wei, W., Wan, X.: Improving word embeddings for antonym detection using thesauri and sentiwordnet. In: Zhang, M., Ng, V., Zhao, D., Li, S., Zan, H. (eds.) NLPCC 2018. LNCS (LNAI), vol. 11109, pp. 67–79. Springer, Cham (2018). https://doi.org/10.1007/978-3-319-99501-4_6
9. Gabrielli, S., Mizzaro, S.: Negotiating a multidimensional framework for relevance space. In: Draper, S.W., Dunlop, M.D., Ruthven, I., van Rijsbergen, C.J. (eds.) MIRA, Workshops in Computing, BCS (1999)
10. Kules, B., Capra, R.: Creating exploratory tasks for a faceted search interface. In: Proceedings of the 2nd Workshop on Human-Computer Interaction (HCIR 2008) (2008)
11. Ono, M., Miwa, M., Sasaki, Y.: Word embedding-based antonym detection using thesauri and distributional information. In: Proceedings of the 2015 Conference of the North American Chapter of the Association for Computational Linguistics: Human Language Technologies, pp. 984–989 (2015)
12. Rose, D.E., Levinson, D.: Understanding user goals in web search. In: Proceedings of the 13th International Conference on World Wide Web, WWW 2004, pp. 13–19. ACM Press (2004)
13. Saracevic, T.: Relevance: a review of the literature and a framework for thinking on the notion in information science. Part ii: nature and manifestations of relevance. J. Am. Soc. Inf. Sci. Technol. **58**(13), 1915–1933 (2007)

Keyphrase Extraction as Sequence Labeling Using Contextualized Embeddings

Dhruva Sahrawat[1], Debanjan Mahata[2,3(✉)], Haimin Zhang[3],
Mayank Kulkarni[3], Agniv Sharma[2], Rakesh Gosangi[3], Amanda Stent[3],
Yaman Kumar[4], Rajiv Ratn Shah[2], and Roger Zimmermann[1]

[1] National University of Singapore, Singapore, Singapore
{dhruva,rogerz}@comp.nus.edu.sg
[2] Indraprastha Institute of Information Technology, New Delhi, India
rajivratn@iiitd.ac.in, agnivsharma96@gmail.com
[3] Bloomberg, New York City, USA
{dmahata,hzhang449,rgosangi,astent}@bloomberg.net
[4] Adobe, New Delhi, India
ykumar@adobe.com

Abstract. In this paper, we formulate keyphrase extraction from scholarly articles as a sequence labeling task solved using a BiLSTM-CRF, where the words in the input text are represented using deep contextualized embeddings. We evaluate the proposed architecture using both contextualized and fixed word embedding models on three different benchmark datasets, and compare with existing popular unsupervised and supervised techniques. Our results quantify the benefits of: (a) using contextualized embeddings over fixed word embeddings; (b) using a BiLSTM-CRF architecture with contextualized word embeddings over fine-tuning the contextualized embedding model directly; and (c) using domain-specific contextualized embeddings (SciBERT). Through error analysis, we also provide some insights into why particular models work better than the others. Lastly, we present a case study where we analyze different self-attention layers of the two best models (BERT and SciBERT) to better understand their predictions.

Keywords: Keyphrase extraction · Contextualized embeddings

1 Introduction

Keyphrase extraction is the process of selecting phrases that capture the most salient topics in a document [24]. They serve as an important piece of document metadata, often used in downstream tasks including information retrieval, document categorization, clustering and summarization. Classic techniques for keyphrase extraction involve a two stage approach [10]: (1) *candidate generation*, and (2) *pruning*. During the first stage, the document is processed to extract a set of candidate keyphrases. In the second stage, this candidate set is pruned to

© Springer Nature Switzerland AG 2020
J. M. Jose et al. (Eds.): ECIR 2020, LNCS 12036, pp. 328–335, 2020.
https://doi.org/10.1007/978-3-030-45442-5_41

select the most salient candidate keyphrases, using either supervised or unsupervised techniques. In the supervised setting, pruning is formulated as a binary classification problem: *determine if a given candidate is a keyphrase*. In the unsupervised setting, pruning is treated as a *ranking problem*, where the candidates are ranked based on some measure of *importance*.

Challenges. Researchers typically employ a combination of different techniques for candidate generation such as extracting named entities, finding noun phrases that adhere to pre-defined lexical patterns [3], or extracting n-grams that appear in an external knowledge base like Wikipedia [9]. The candidates are further cleaned up using stop word lists or gazetteers. Errors in any of these techniques reduces the quality of candidate keyphrases. For example, if a named entity is not identified, it misses out on being considered as a keyphrase; if there are errors in part of speech tagging, extracted noun phrases might be incomplete. Also, since candidate generation involves a combination of heuristics with specific parameters, thresholds, and external resources, it is hard to be reproduced or migrated to new domains.

Motivation. Recently, researchers have started to approach keyphrase extraction as a sequence labeling task [1,8]. This formulation completely bypasses the candidate generation stage and provides a unified approach to keyphrase extraction. Unlike binary classification where each keyphrase is classified independently, sequence labeling finds an optimal assignment of keyphrase labels for the entire document. Sequence labeling allows to capture long-term semantic dependencies in the document.

There have been significant advances in deep contextual language models [7,21]. These models can take an input text and provide contextual embeddings for each token for use in downstream architectures. They have been shown to achieve state-of-the-art results for many different NLP tasks. More recent works [4,16] have shown that contextual embedding models trained on domain-specific corpora can outperform general purpose models.

Contributions. Despite all the developments, to the best of our knowledge, there hasn't been any work on the use of contextual embeddings for keyphrase extraction. We expect that, as with other NLP tasks, keyphrase extraction can benefit from contextual embeddings. We also posit that domain-specific language models may further help improve performance. To explore these hypotheses, in this paper, we approach keyphrase extraction as a sequence labeling task solved using a BiLSTM-CRF [11], where the underlying words are represented using various contextual embedding architectures. Following are the main contributions of this paper:

- We quantify the benefits of using deep contextual embedding models for sequence-labeling-based keyphrase extraction over using fixed word embeddings.
- We demonstrate the benefits of using a BiLSTM-CRF architecture with contextualized word embeddings over fine-tuning the contextualized word embedding model for keyphrase extraction.
- We demonstrate improvements using contextualized embeddings trained on a large corpus of in-genre text (SciBERT) over ones trained on generic text.

- We perform a robust set of experiments on three benchmark datasets, achieving state-of-the-art results and provide insights into the working of the different self-attention layers of our top-performing models.

2   Methodology

We approach the problem of automated keyphrase extraction from scholarly articles as a sequence labeling task, which can be formally stated as: Let $d = \{w_1, w_2, ..., w_n\}$ be an input text, where w_t represents the tth token. Assign each w_t in the document one of three class labels $Y = \{k_B, k_I, k_O\}$, where k_B denotes that w_t marks the beginning of a keyphrase, k_I means that w_t is inside a keyphrase, and k_O indicates that w_t is not part of a keyphrase.

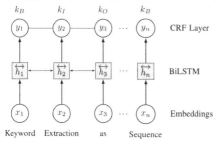

Fig. 1. BiLSTM-CRF architecture

In this paper, we employ a BiLSTM-CRF architecture (Fig. 1) to solve this sequence labeling problem. We map each token w_t in the input text to a fixed-size dense vector x_t. We then use a BiLSTM to encode sequential relations between the word representations. We then apply an affine transformation to map the output from the BiLSTM to the class space. The score outputs from the BiLSTM serve as input to a CRF [15] layer. In a CRF, the likelihood for a labeling sequence is generated by exponentiating the scores and normalizing over all possible output label sequences.

3   Experiments

Datasets. We ran our experiments on three different publicly available keyphrase datasets: Inspec [12], SemEval-2010 [14] (SE-2010), and SemEval-2017 [2] (SE-2017). Inspec consists of abstracts from 2000 scientific articles (train: 1000, validation: 500, and test: 500) where abstract is accompanied by both abstractive, i.e., not present in the documents, and extractive keyphrases. SE-2010 consists of 284 full length ACM articles (train: 144, validation: 40, and test: 100) containing both abstractive and extractive keyphrases. SE-2017 consists of 500 open access articles (train: 350, validation: 50, and test: 100) with location spans for keyphrases, i.e. all keyphrases are extractive.

Because we are modeling keyphrase extraction as a sequence labeling task, we only consider extractive keyphrases for our experiments. For Inspec and SE-2010, we identified the location spans for each extractive keyphrase. For SE-2010 and SE-2017, we only considered the abstract and discarded the remaining text due to memory constraints during inference with the contextual embedding models.

All the tokens in the datasets[1] were tagged using the B-I-O tagging scheme described in the previous section.

Experimental Settings. One of the main aims of this work is to study the effectiveness of contextual embeddings in keyphrase extraction. To this end, we use the BiLSTM-CRF with seven different pre-trained contextual embeddings: BERT [7] (small-cased, small-uncased, large-cased, large-uncased), SciBERT [4] (basevocab-cased, basevocab-uncased, scivocab-cased, scivocab-uncased), OpenAI GPT [22], ELMo [21], RoBERTa [17] (base, large), Transformer XL [5], and OpenAI GPT-2 [23] (small, medium). We also use 300 dimensional fixed embeddings from Glove [20], Word2Vec [19], and FastText [13] (common-crawl, wiki-news). We also compare the proposed architecture against four popular baselines: SGRank [6], SingleRank [26], Textrank [18], and KEA [27].

To train the BiLSTM-CRF models, we use stochastic gradient descent with Nesterov momentum in batched mode. The learning rate was set to 0.05 and the models were trained for a total of 100 epochs with patience value of 4 and annealing factor of 0.5. The hidden layers in the BiLSTM models were set to 128 units and word dropout set to 0.05. During inference, we run the model on a given abstract and identify keyphrases as all sequences of class labels that begin with the tag k_B followed by zero or more tokens tagged k_I. As in previous studies [14], we use F1-measure to compare different models. For each embedding model we report results for the best performing variant of that model (e.g. cased vs uncased) on each dataset.

4 Results

Table 1. Embedding models comparison (F1-score)

	Inspec	SE-2010	SE-2017
SciBERT	0.593	**0.357**	0.521
BERT	0.591	0.330	**0.522**
ELMo	0.568	0.225	0.504
Transformer-XL	0.521	0.222	0.445
OpenAI-GPT	0.523	0.235	0.439
OpenAI-GPT2	0.531	0.240	0.439
RoBERTa	**0.595**	0.278	0.508
Glove	0.457	0.111	0.345
FastText	0.524	0.225	0.426
Word2Vec	0.473	0.208	0.292
SGRank	0.271	0.229	0.211
SingleRank	0.123	0.142	0.155
TextRank	0.122	0.147	0.157
KEA	0.137	0.202	0.129

Of the ten embedding architectures, BERT or BERT-based models consistently obtained the best performance across all datasets (see Table 1). This was expected considering that BERT uses bidirectional pre-training which is more powerful. SciBERT was consistently one of the top performing models and was significantly better than any of the other models on SemEval-2010. Further analysis of the results shows that SciBERT was more accurate than other models in capturing keyphrases that contained scientific terms such as chemical names (e.g. Magnesium, Hydrozincite), software projects (e.g. HemeLB), and abbreviations (e.g. DSP, SIMLIB). SciBERT was also more accurate with keyphrases containing more than three tokens.

---

[1] https://github.com/midas-research/keyphrase-extraction-as-sequence-labeling-data.

Contextual embeddings outperformed their fixed counterparts for most of the experimental scenarios. The only exception was on SemEval-2010 where FastText outperformed Transformer-XL. Of the three fixed embedding models studied in this paper, FastText obtained the best performance across all datasets. Our model significantly outperforms all the four baseline methods for all three datasets irrespective of the embeddings used.

Contextualized embedding models can either serve as numerical representations of words for downstream architectures or they can be fine-tuned to be optimized for a specific task. Fine-tuning typically involves adding an untrained layer at the end and then optimizing the layer weights for the task-specific objective. We fine-tuned our best-performing contextualized embedding models (BERT and SciBERT) for each dataset and compared with the performance of the corresponding BiLSTM-CRF and BiLSTM models.

Both BiLSTM and BiLSTM-CRF models outperform fine-tuning across all datasets (see Table 2). We expect this might be due to the small sizes of the datasets on which the models were fine-tuned. The addition of the CRF layer improved the performance for all datasets. Analysis of results on the SemEval-2017 data shows that the CRF layer is more effective in capturing keyphrases that include prepositions (e.g. 'of'), conjunctions (e.g. 'and'), and articles (e.g. 'the'). We also observed that the CRF layer is more accurate with longer keyphrases (more than two tokens).

Case Study: Attention analysis is used to understand if self-attention patterns in the layers of BERT and SciBERT provide any insight into the linguistic properties learned by

Table 2. Fine-tuning vs Pretrained (F1-score)

	BERT			SciBERT		
	Inspec	SE-2010	SE-2017	Inspec	SE-2010	SE-2017
Fine-tuning	0.474	0.236	0.270	0.488	0.268	0.339
BiLSTM-CRF	**0.591**	**0.330**	**0.522**	**0.593**	**0.357**	**0.521**
BiLSTM	0.501	0.295	0.472	0.536	0.301	0.455

the models. We present a case study of attention analysis for keyphrase extraction on a randomly chosen abstract from SemEval2017.

Table 3. SciBERT vs BERT: keyphrase identification

SciBERT	BERT
An object-oriented version of SIMLIB -LRB- a simple simulation package -RRB- This paper introduces an object-oriented version of SIMLIB -LRB- an easy-to-understand discrete-event simulation package -RRB- . The object-oriented version is preferable to the original procedural language versions of SIMLIB in that it is easier to understand and teach simulation from an object point of view. A single-server queue simulation is demonstrated using the object-oriented SIMLIB	An object-oriented version of SIMLIB -LRB- a simple simulation package -RRB- This paper introduces an object-oriented version of SIMLIB -LRB- an easy-to-understand discrete-event simulation package -RRB- . The object-oriented version is preferable to the original procedural language versions of SIMLIB in that it is easier to understand and teach simulation from an object point of view. A single-server queue simulation is demonstrated using the object-oriented SIMLIB

Table 3 presents the classification results on this abstract from the BERT and SciBERT models; true positives in blue and false negatives in red. Using BertViz [25] we analyzed the aggregated attention of all 12 layers of both the models. We observed that keyphrase tokens (k_B and k_I) typically tend to pay most attention towards other keyphrase tokens. Contrarily, non-keyphrase tokens (k_O) usually pay uniform attention to their surrounding tokens. We found that both BERT and SciBERT exhibit similar attention patterns in the initial and final layers but they vary significantly in the middle layers. For example, the left figure in Table 4 compares the attention patterns in the fifth layer of both models. In SciBERT, the token 'object' is very strongly linked to other tokens from its keyphrase but the attentions are comparably weaker for BERT.

Table 4. Self-attention maps of layers in BERT and SciBERT

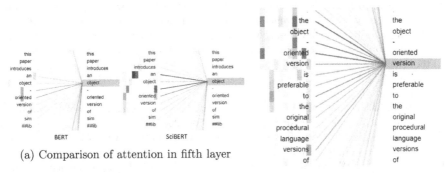

(a) Comparison of attention in fifth layer

(b) Attention towards similar tokens in SciBERT

We also observed that keyphrase tokens paid strong attention to similar tokens from other keyphrases. As shown in the right figure in Table 4, the token 'version' from 'object-oriented version' pays strong attention to 'versions' from 'procedural language versions'. This is a possible reason for both models failing to identify the third mention of 'object-oriented version' in the abstract as a keyphrase. We observed similar patterns in many other documents and we plan to quantify this analysis in future.

5   Conclusions

In this paper, we formulate keyphrase extraction as a sequence labeling task solved using BiLSTM-CRFs, where the underlying words are represented using various contextualized embeddings. We quantify the benefits of this architecture over direct fine tuning of the embedding models. We demonstrate how contextual embeddings significantly outperform their fixed counterparts in keyphrase extraction.

References

1. Alzaidy, R., Caragea, C., Giles, C.L.: Bi-LSTM-CRF sequence labeling for keyphrase extraction from scholarly documents. In: Proceedings of WWW (2019)
2. Augenstein, I., Das, M., Riedel, S., Vikraman, L., McCallum, A.: SemEval 2017 task 10: ScienceIE-extracting keyphrases and relations from scientific publications. In: Proceedings of SemEval (2017)
3. Barker, K., Cornacchia, N.: Using noun phrase heads to extract document keyphrases. In: Hamilton, H.J. (ed.) AI 2000. LNCS (LNAI), vol. 1822, pp. 40–52. Springer, Heidelberg (2000). https://doi.org/10.1007/3-540-45486-1_4
4. Beltagy, I., Cohan, A., Lo, K.: Scibert: pretrained contextualized embeddings for scientific text. arXiv:1903.10676 (2019)
5. Dai, Z., et al.: Transformer-XL: attentive language models beyond a fixed-length context. arXiv preprint arXiv:1901.02860 (2019)
6. Danesh, S., Sumner, T., Martin, J.H.: SGRank: combining statistical and graphical methods to improve the state of the art in unsupervised keyphrase extraction. In: Proceedings of Joint Conference on Lexical and Computational Semantics (2015)
7. Devlin, J., Chang, M.W., Lee, K., Toutanova, K.: BERT: pre-training of deep bidirectional transformers for language understanding. In: Proceedings of NAACL-HLT (2019)
8. Gollapalli, S.D., Li, X.l.: Keyphrase extraction using sequential labeling. arXiv preprint arXiv:1608.00329 (2016)
9. Grineva, M., Grinev, M., Lizorkin, D.: Extracting key terms from noisy and multitheme documents. In: Proceedings of WWW (2009)
10. Hasan, K.S., Ng, V.: Automatic keyphrase extraction: a survey of the state of the art. In: Proceedings of ACL (2014)
11. Huang, Z., Xu, W., Yu, K.: Bidirectional LSTM-CRF models for sequence tagging. CoRR abs/1508.01991 (2015). http://arxiv.org/abs/1508.01991
12. Hulth, A.: Improved automatic keyword extraction given more linguistic knowledge. In: Proceedings of EMNLP (2003)
13. Joulin, A., Grave, E., Bojanowski, P., Mikolov, T.: Bag of tricks for efficient text classification. In: Proceedings of EACL (2017)
14. Kim, S.N., Medelyan, O., Kan, M.Y., Baldwin, T.: SemEval-2010 task 5: automatic keyphrase extraction from scientific articles. In: Proceedings of SemEval (2010)
15. Lafferty, J., McCallum, A., Pereira, F.C.: Conditional random fields: probabilistic models for segmenting and labeling sequence data. In: Proceedings of ICML (2001)
16. Lee, J., et al.: BioBERT: pre-trained biomedical language representation model for biomedical text mining. arXiv preprint arXiv:1901.08746 (2019)
17. Liu, Y., et al.: RoBERTa: a robustly optimized BERT pretraining approach. arXiv preprint arXiv:1907.11692 (2019)
18. Mihalcea, R., Tarau, P.: TextRank: bringing order into text. In: Proceedings of EMNLP (2004)
19. Mikolov, T., Sutskever, I., Chen, K., Corrado, G.S., Dean, J.: Distributed representations of words and phrases and their compositionality. In: Advances in Neural Information Processing Systems, pp. 3111–3119 (2013)
20. Pennington, J., Socher, R., Manning, C.: Glove: global vectors for word representation. In: Proceedings of EMNLP (2014)
21. Peters, M.E., et al.: Deep contextualized word representations. arXiv preprint arXiv:1802.05365 (2018)

22. Radford, A., Narasimhan, K., Salimans, T., Sutskever, I.: Improving language understanding by generative pre-training (2018). https://s3-us-west-2. amazonaws.com/openai-assets/research-covers/language-unsupervised/language_ understanding_paper.pdf
23. Radford, A., Wu, J., Child, R., Luan, D., Amodei, D., Sutskever, I.: Language models are unsupervised multitask learners. OpenAI Blog, February 2019
24. Turney, P.D.: Learning to extract keyphrases from text. arXiv preprint cs/0212013 (2002)
25. Vig, J.: A multiscale visualization of attention in the transformer model. arXiv preprint arXiv:1906.05714 (2019)
26. Wan, X., Xiao, J.: Single document keyphrase extraction using neighborhood knowledge. In: Proceedings of AAAI (2008)
27. Witten, I.H., Paynter, G.W., Frank, E., Gutwin, C., Nevill-Manning, C.G.: Kea: practical automated keyphrase extraction. In: Design and Usability of Digital Libraries: Case Studies in the Asia Pacific, pp. 129–152. IGI Global (2005)

Easing Legal News Monitoring with Learning to Rank and BERT

Luis Sanchez[1,2], Jiyin He[2(✉)], Jarana Manotumruksa[1], Dyaa Albakour[2],
Miguel Martinez[2], and Aldo Lipani[1]

[1] University College of London, London, UK
{luis.izquierdo18,j.manotumruksa,aldo.lipani}@ucl.ac.uk
[2] Signal AI, London, UK
jiyin.he@signalmedia.co, research@signal-ai.com

Abstract. While ranking approaches have made rapid advances in the Web search, systems that cater to the complex information needs in professional search tasks are not widely developed, common issues and solutions typically rely on dedicated search strategies backed by ad-hoc retrieval models. In this paper we present a legal search problem where professionals monitor news articles with constant queries on a periodic basis. Firstly, we demonstrate the effectiveness of using traditional retrieval models against the Boolean search of documents in chronological order. In an attempt to capture the complex information needs of users, a learning to rank approach is adopted with user specified relevance criteria as features. This approach, however, only achieves mediocre results compared to the traditional models. However, we find that by fine-tuning a contextualised language model (e.g. BERT), significantly improved retrieval performance can be achieved, providing a flexible solution to satisfying complex information needs without explicit feature engineering.

Keywords: Professional search · Complex information needs · BERT

1 Introduction

In information retrieval (IR), there has been a long standing interest in professional search, as demonstrated by various TREC tracks dedicated to a diverse range of professional domains [4,8]. Unlike traditional Web search, an important characteristic of professional search is the complex information needs of the professional users. For instance, a professional user may ask for information within certain time range, written in a professional quality [24].

Although there have been ongoing discussions and studies calling for search systems addressing common issues faced by professional search, solutions typically rely on dedicated databases or specialised search strategies that are backed by ad-hoc retrieval models [20,23]. Meanwhile, although traditional retrieval models as well as learning to rank (L2R) approaches have made rapid advances

© Springer Nature Switzerland AG 2020
J. M. Jose et al. (Eds.): ECIR 2020, LNCS 12036, pp. 336–343, 2020.
https://doi.org/10.1007/978-3-030-45442-5_42

in Web search, retrieval models that cater to the diverse requirements in professional search tasks are not widely developed.

In this paper, we study a case in the context of *legal professional search* and investigate how different retrieval approaches can be employed to address the complex needs of professional users. The work task of our users is to monitor a number of legal topic of their interest in news and select articles to be included in a report periodically according to a set of clearly defined criteria, ranging from topical relevance to language quality. Like in many other professional search scenarios, our users setup their searches against a news stream with complex Boolean queries [20] where results are ranked in chronological order; and they deem recall an important metric as they do not want to miss relevant articles.

While Boolean queries are often preferred by professional searchers due to their needs of having results which are "efficient, trustable, explainable and accountable"[16,24], it falls short in addressing the complex relevance criteria beyond matching terms. Traditional retrieval models such as BM25 and Language Models (LM) capture topical relevance. As a first step going beyond the Boolean search practice, we answer the following research question:

RQ1. Do traditional IR models help our users in identifying relevant documents more effectively compared to the Boolean search practice?

Further, in order to satisfy users' complex information needs beyond topicality, it seems natural to encode indicators of different criteria as features, and combine them with a L2R approach. Therefore, our next research question is:

RQ2. Can we provide better results by adopting a L2R approach to satisfy users' complex information needs beyond topicality?

However, feature engineering for every criterion can be time-consuming and may not be convenient when switching from one use case to another. Recently, pre-trained contextualised language models (e.g. BERT [10]) have effectively addressed various NLP tasks [10,25] eliminating the need of feature engineering. This leads the investigation of the follow up research question:

RQ3. Can we further improve the quality of the search results by fine-tuning a pre-trained language model on our search task?

Our contributions can be summarised as follows:

1. Unlike simulation based studies such as TREC tasks where the information needs, relevance criteria and judgements are set by different parties rather than the actual users, the complex information needs in our study come from real users, who also define the relevance criteria and provide judgements. Our study not only reveals the practical challenges for professional search systems, but also demonstrates possible solutions to effectively address these challenges, and;
2. We also contribute to the generic solution to professional search (i.e. search with complex information needs). Our experiments show the potential of

employing pre-trained contextualised language models to learn relevance criteria without handcrafted features, which leads to a flexible solution that adapts to varying complex needs.

2   Methodology

In this case study, our users have three relevance criteria: *topical relevance, factual information*, and *language quality*. Specifically, the topicality of retrieved articles must be associated with a specific legal area[1]; only factual articles are considered relevant; and articles written in technical language and linguistically accurate are preferred.

We first explore the effectiveness of traditional models in satisfying the users' needs. We include four models: TF-IDF, BM25, unigram Language Model (LM) with Jelinek-Mercer and Dirichlet smoothing, applied to three fields of the news articles (title, summary, and content). As for query, we extract the keywords from the complex Boolean queries that our users created and concatenate them as a long query (typically ~100 words), where negation terms were ignored.

In order to estimate the relevance of a document with respect to the combined relevance criteria described above, we employ a L2R approach and encode these criteria as features. We devise 28 features (see Table 1) as follows.

Topical Relevance. We model topical relevance using the outputs of traditional retrieval models, as usually done in the literature [17].

Factual Information. We model factual information with three types of features: (1) Subjectivity: it measures the degree of subjectivity of an article, which is directly related to the "factuality" of the article. (2) Modality: it shows the degree of certainty of the statements of an article by looking at the verb tense in which the article is written. (3) Sentiment: it provides the degree of negativity or positivity of the language used–while not directly related to the factuality dimension, there can be entanglement between the subjective and opinionated dimensions [13]. We employ a lexicon based approach to compute these features [15,22]. We also include the number of lexicons assessed in an article as a normalisation factor for articles of different lexicon sizes.

Language Quality. Since the content our users request are technical and sometimes hard for non-expert to read, we employ readability scores as features, which measure the ease with which a reader can understand a written text.

Apart from devising task specific features, we exploit a pre-trained contextualised language model to automatically learn the complex relevance criteria. By fine tuning the model on our search task we expect to associate these language features with the relevance judgements. We use BERT [18], which shows the state-of-the-art performance on a wide range of NLP tasks [10,25]. Inspired by the work of MacAvaney et al. [18], we employ BERT in its regression form

---

[1] Information regarding the specific legal domain cannot be disclosed due to a non-disclosure agreements that we have with the legal professionals.

Table 1. Features for learning to rank

Type	Features	Description
Topicality	Retrieval model scores (12)	TF-IDF, BM25, LM (J-M), and LM (Dir) applied on an article's title, define summary, and content
Factual information	Subjectivity (1)	Degree of objectivity vs. subjectivity
	Modality (1)	Degree of certainty of the statements
	Sentiments (4)	Negative, positive, neutral, and compound scores [15]
	Lexicons (1)	# of the lexicon's vocabulary that appear in the article
Language quality	Readability (9)	Kincaid index [12], Readability Index [21], Coleman-Liau index [7], Flesch index [11], Gunning Fog index [14], Lasbarhets index [6], McLaughlin's SMOG index [19], John Aderson's index [5], and Dale-Chall index [9]

(known as Vanilla BERT in [18]). Specifically, the input consists of a query-document pair, and the output is a predicted relevance score. For document input, we use (i) a combined title and summary field (referred to as BERT on summary), and (ii) the content of the article (referred to as BERT on content).

3   Experimental Setup

Dataset. The dataset we use to evaluate our retrieval approaches comes from the interaction data of legal professionals with a news monitoring system over a one year period. The users monitor a specific legal topic by querying the news stream with that topic periodically and all the retrieved results are tagged with a relevance judgment for later usage. Given this context, we group the data into equal intervals corresponding to the report creation times and evaluate the retrieved results per interval. The initial ranked lists were generated by using the Boolean query created by the users, and ranked in a chronological order. We apply the alternative ranking approaches as a re-ranking task. In total the dataset consists 206 queries and 60,512 labelled news articles, among which 2,872 (21%) are marked as relevant. By grouping the searches into the equal interval and removing sessions with no relevant articles, we obtain 1,774 search sessions (i.e. query-results pairs). The average number of relevant articles per-session varies from 1.5 to 4.4 articles depending on the queries. We randomly split the dataset into training (80%), validation (10%), and test (10%) sets. The same setup holds for both traditional retrieval models as well as for L2R approaches. We use the validation set to tune the models' parameters.

Evaluation Measures. We use Mean Average Precision (MAP) to train and measure the performance of the retrieval models. Since recall is important in this user task, we use two recall oriented metrics: R@3 (given the small number of relevant documents per search); and average Search Length (SL) which measures the amount of effort a user needs to find *all* relevant documents.

Features. The topical features take the scores generated by the traditional IR models with their optimal parameter settings. For language usage features, we use an implementation of the CLiPS *pattern-en* module for subjectivity and modality [22]; and VADER [3] to compute sentiment scores. The 9 readability features are computed using the Python Readability package [2].

Models. As L2R approach we use the LambdaMART implementation from RankLib [1]. We apply a linear normalisation to our features as implemented by the library; each feature is normalised according to its minimum and maximum values. BERT is fine tuned using our labelled data as described by MacAvaney et al. [18]. The input of BERT is the concatenation of a [CLS] token, the query, a [SEP] token, and the document. A document is capped when longer than 512 tokens. The output of BERT is the vector representations for each input token. We use the BERT-base uncased version and each vector has a dimension of 764. For fine-tuning we stack a linear-layer on top of BERT, which takes as input the output vector for the [CLS] token. We rank documents according to the score output by the linear-layer.

Table 2. Traditional models vs. Boolean search. Models are run on Title (T), Summary (S), and Content (C). All differences are statistically significant compared to the Boolean search results (paired t-test with p-value < 0.01).

	MAP			R@3			SL		
Boolean	0.421			0.469			6.66		
Method	T	S	C	T	S	C	T	S	C
TF-IDF	0.518	0.509	0.501	0.580	0.579	0.591	5.632	5.800	5.859
BM25	0.517	0.520	0.546	0.585	0.576	0.594	5.643	5.622	**5.303**
J-M	0.534	0.522	0.551	0.599	0.589	0.588	5.616	5.638	5.660
Dir	0.531	0.521	**0.556**	**0.615**	0.586	0.594	5.465	5.627	5.595

4   Results and Discussion

Regarding RQ1, Table 2 lists the results of traditional retrieval models (TF-IDF, BM25 and Language Model (LM) with Jelinek-Mercer smoothing (J-M) and Dirichlet smoothing (Dir)) compared to that from the Boolean search with a chronological order, i.e. the working practice of our professional users. We see that all retrieval models significantly outperform the Boolean search results for all measures. This suggests that without further effort in terms of feature engineering and model fitting, traditional models already improve the ranking quality by capturing topical relevance. Further, we see that different fields may be best for one model but not for the other, suggesting that their combination in a L2R approach may be beneficial. Hereafter, we choose Dir on content, which has the best MAP score, as a baseline for the remaining experiments.

To address RQ2 and RQ3, Table 3 shows the results of LambdaMART with explicitly encoded features and BERT scores, compared to the Dir baseline.

Firstly, We see LambdaMART with explicitly encoded features has no significant improvements over the baseline. In particular, topical features (i.e. a linear combination of the traditional models and different fields) does not provide better performance compared to Dir. However, among the different type of features, LambdaMART with all features performs the best, suggesting that both types of features are somewhat useful in capturing the relevance criteria. In response to RQ2, these results imply that the L2R approach with explicit feature engineering does not achieve a competitive performance—perhaps, hand-crafted features were not able to match well with the user specified relevance criteria.

Table 3. Results of LambdaMART with feature variants. † indicates a statistical significant difference compared to the baseline Dir with p-value < 0.05 by paired t-test (p-value < 0.01 when ‡).

Method	MAP	R@3	SL
Dir	0.556	0.596	5.595
LambdaMART (all features)	0.543	0.616	5.205
LambdaMART (topicality)	0.550	0.607	5.400
LambdaMART (language)	0.514	0.572	5.503
BERT on summary	0.739‡	**0.831‡**	**1.746‡**
BERT on content	**0.763‡**	0.777†	2.059‡

Next, we observe that BERT based approaches significantly outperform Dir. In particular, in terms of SL, with the baseline a user would need to read on average 5.5 irrelevant documents before finding all relevant documents, while with BERT based models this is reduced to less than 2, providing potentially improved user experience. Moreover, compared to explicit feature engineering, fine tuning BERT seems to have captured the user information needs in an implicit manner. This is encouraging as it not only learns the complex relevance criteria more accurately, but also provides more flexibility as the model can be fine tuned for use cases with different criteria without dedicated feature engineering.

The above results show promising performance of different ranking approaches in terms of off-line IR evaluation, compared to the original Boolean setup. From a user perspective, this means users may be able to confidently stop reading results after seeing certain number of irrelevant results, which would be particularly useful when the result list is long and relevant articles are few. On the other hand, we should also be aware that as the model complexity increases, there is decreasing model explainability and user controllability—the properties of Boolean search appreciated by professional users [16,24]. Therefore for future work we find it crucial to investigate methods that explain and control complex models such as BERT.

5   Conclusion

We explored different retrieval approaches to address the complex information needs of professional users in a legal search context. We found that, compared to Boolean search, traditional retrieval models are effective in improving the ranking quality and reducing user effort in finding relevant information (e.g. measured by SL). Learning to rank with explicit feature encoding does not seem to be able to easily improve over traditional models. However, fine-tuning a pre-trained language model (BERT) shows strong improvements over both traditional models and L2R models, with the advantage of not requiring dedicated feature encoding. In particular, our study opens up a number of research questions in the context of professional search: (i) what kind of features allow pre-trained LMs to capture the implicit information needs from users' relevance judgements? (ii) what are the limitations of pre-trained LMs to capture fine-grained information needs? and; (iii) how does the above depend on the number and quality of the relevance judgements, particularly in the case of niche retrieval tasks?

References

1. Ranklib. https://sourceforge.net/p/lemur/wiki/RankLib/
2. Readability. https://pypi.org/project/readability/
3. Varder. https://github.com/cjhutto/vaderSentiment
4. Guidelines for the 2011 TREC medical records track (2011). https://www-nlpir.nist.gov/projects/trecmed/2011/. Accessed 26 Aug 2019
5. Anderson, J.: Lix and rix: variations on a little-known readability index. J. Read. **26**(6), 490–496 (1983)
6. Björnsson, C.H.: Läsbarhet. Liber (1968)
7. Coleman, M., Liau, T.L.: A computer readability formula designed for machine scoring. J. Appl. Psychol. **60**(2), 283 (1975)
8. Cormack, G.V., Grossman, M.R., Hedin, B., Oard, D.W.: Overview of the TREC 2010 legal track. In: Proceedings of TREC, vol. 1 (2010)
9. Dale, E., Chall, J.S.: A formula for predicting readability: instructions. Educ. Res. Bull. **27**, 37–54 (1948)
10. Devlin, J., Chang, M.W., Lee, K., Toutanova, K.: Bert: pre-training of deep bidirectional transformers for language understanding. In: Proceedings of NAACL (2019)
11. Farr, J.N., Jenkins, J.J., Paterson, D.G.: Simplification of flesch reading ease formula. J. Appl. Psychol. **35**(5), 333 (1951)
12. Flesch, R.: A new readability yardstick. J. Appl. Psychol. **32**(3), 221 (1948)

13. Fuhr, N., et al.: An information nutritional label for online documents. SIGIR Forum **51**, 46–66 (2017)
14. Gunning, R.: The Technique of Clear Writing. McGraw-Hill, New York (1952)
15. Hutto, C.J., Gilbert, E.: Vader: a parsimonious rule-based model for sentiment analysis of social media text. In: Proceedings of AAAI (2014)
16. Kim, Y., Seo, J., Croft, W.B.: Automatic Boolean query suggestion for professional search. In: Proceedings of SIGIR, pp. 825–834. ACM (2011)
17. Liu, T.Y., et al.: Learning to rank for information retrieval. Found. Trends Inf. Retr. **3**(3), 225–331 (2009)
18. MacAvaney, S., Yates, A., Cohan, A., Goharian, N.: Cedr: contextualized embeddings for document ranking. In: Proceedings of SIGIR (2019)
19. Mc Laughlin, G.H.: Smog grading-a new readability formula. J. Read. **12**(8), 639–646 (1969)
20. Russell-Rose, T., Chamberlain, J., Azzopardi, L.: Information retrieval in the workplace: a comparison of professional search practices. Inf. Process. Manag. **54**(6), 1042–1057 (2018)
21. Senter, R., Smith, E.A.: Automated readability index. Technical report, Cincinnati Univ. Ohio (1967)
22. Smedt, T.D., Daelemans, W.: Pattern for python. J. Mach. Learn. Res. **13**(Jun), 2063–2067 (2012)
23. Verberne, S., et al.: First international workshop on professional search. In: Proceedings of SIGIR, vol. 52, pp. 153–162. ACM (2019)
24. Verberne, S., He, J., Wiggers, G., Russell-Rose, T., Kruschwitz, U., de Vries, A.P.: Information search in a professional context-exploring a collection of professional search tasks. In: Proceedings of SIGIR, Paris, France, pp. 1–5 (2019)
25. Yang, Z., Dai, Z., Yang, Y., Carbonell, J., Salakhutdinov, R., Le, Q.V.: Xlnet: generalized autoregressive pretraining for language understanding (2019)

Generating Query Suggestions for Cross-language and Cross-terminology Health Information Retrieval

Paulo Miguel Santos[1] and Carla Teixeira Lopes[1,2]([⊠])

[1] Department of Informatics Engineering, Faculty of Engineering,
University of Porto, Porto, Portugal
{up201403745,ctl}@fe.up.pt
[2] INESC TEC, Porto, Portugal

Abstract. Medico-scientific concepts are not easily understood by laypeople that frequently use lay synonyms. For this reason, strategies that help users formulate health queries are essential. Health Suggestions is an existing extension for Google Chrome that provides suggestions in lay and medico-scientific terminologies, both in English and Portuguese. This work proposes, evaluates, and compares further strategies for generating suggestions based on the initial consumer query, using multi-concept recognition and the Unified Medical Language System (UMLS). The evaluation was done with an English and a Portuguese test collection, considering as baseline the suggestions initially provided by Health Suggestions. Given the importance of understandability, we used measures that combine relevance and understandability, namely, uRBP and uRBPgr. Our best method merges the Consumer Health Vocabulary (CHV)-preferred expression for each concept identified in the initial query for lay suggestions and the UMLS-preferred expressions for medico-scientific suggestions. Multi-concept recognition was critical for this improvement.

Keywords: Health information retrieval · Cross-language information retrieval · Cross-terminology information retrieval · Query suggestion

1 Introduction

Search engines are commonly used to seek health information, an activity that is considered the third most popular activity on the Internet [1]. Despite the increasing use of the Web to search for health-related information, there may exist inequalities in access to health information [6]. Users with low levels of health literacy can struggle to satisfy their information needs because health-related information usually contains medico-scientific expressions that are not easily understandable [13]. The gap between lay and medico-scientific terminologies limits this access and can be assisted through query modification techniques [10].

© Springer Nature Switzerland AG 2020
J. M. Jose et al. (Eds.): ECIR 2020, LNCS 12036, pp. 344–351, 2020.
https://doi.org/10.1007/978-3-030-45442-5_43

There is evidence that multilingual query suggestions in lay and medico-scientific terminologies improve health information retrieval by laypeople [9].

Taking this into account, Health Suggestions was developed as an extension for Google Chrome, suggesting queries in lay and medico-scientific terminologies, both in English and Portuguese, based on the Consumer Health Vocabulary (CHV) [8]. To improve the system, we propose and evaluate strategies for query suggestion that involve multi-concept recognition and information from the Unified Medical Language System (UMLS). For evaluation, the new generated query is used to retrieve documents from an English and a Portuguese test collection. The strategies are evaluated, taking into account the relevance of the documents and its understandability by lay users, comparing them with the results of queries initially suggested by Health Suggestions.

2   Related Work

When users are trying to express their information need, they might use keywords that are too general or different from the ones included in documents, as well as an insufficient number of terms, making the query difficult to "be understood" by the system [5]. Techniques such as query expansion, query refinement, and query suggestion have been proposed to solve this problem, improving the relevance and comprehension of the retrieved documents.

Zeng et al. [12] developed a system that suggests alternative or additional terms to the query using logs and the co-occurrence of concepts in medical documents, as well as the semantic relationships existing in medical vocabularies. Liu and Wesley [7] proposed a query expansion method that exploited the UMLS, appending additional relevant terms to the original query.

A query suggestion system was developed by Lopes and Ribeiro [9], combining multilingual alternatives (in Portuguese and English) with the use of lay and medico-scientific terminology. Authors used the CHV that maps technical terms to consumer-friendly language. For each query, they identify the associated concept and then return its CHV and UMLS-preferred names in English and Portuguese. Lopes and Fernandes [8] created HealthSuggestions, an extension for Google Chrome to assist users in obtaining high-quality search results in the health domain using the CHV.

3   Proposed Methods for Suggesting Queries

To generate the query suggestions, we implemented several methods that use multi-concept recognition to detect the medical concepts included in the initial query and use the information from UMLS as a knowledge source. All methods follow the approach described in Fig. 1. Briefly, the initial query is translated into English, and its medical concepts are identified. For each of these concepts, we select lay and medico-scientific expressions, concatenate them to compose the corresponding suggestions in English and, in the end, we translate them to the original language. All translations are done with Google Translator.

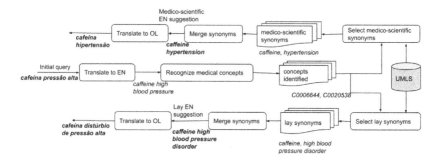

Fig. 1. Process followed for the generation of query suggestions.

Several strategies were analyzed for multi-concept recognition, and we decided to use MetaMap, a rule-based system of concept recognition, to discover UMLS concepts referred to in free text [2], which is interesting because we use UMLS as our knowledge source. MetaMap provides a list of mappings for each identified concept. In each query suggestion method, we used two approaches to select the best mapping. In the first approach, we choose the first mapping, that is, the one with the highest score. In the second, we used the Word-Sense Disambiguation (WSD) feature that favors those that are semantically consistent with the surrounding text [3]. For each approach, we used the UMLS Concept Unique Identifier (CUI) and the name of the concepts as input.

The selection of lay and medico-scientific synonyms is what differentiates the suggestion methods. All the methods use the UMLS, a knowledge base that aggregates multiple thesauri of the medical domain [4], each composed of concepts related to health, their various names, and the relationships that exist between them. One of the UMLS vocabularies is the CHV[1], a vocabulary that connects simple, everyday health words to technical terms used by health care professionals. For each concept, it stores the best way to express it for a lay audience (CHV-preferred) and the same for a professional audience (UMLS-preferred).

Differences between the methods are summarized in Table 1. In the *CHV-preferred/UMLS-preferred* method, the selected synonyms correspond to the

Table 1. Proposed methods.

Name	Vocabulary	Atoms	Relationships	Type of relationships
CHV-preferred/UMLS-preferred	Only CHV	All	–	–
Preferred Atoms	UMLS ones	Preferred atoms	–	–
All preferred/synonym atoms	UMLS ones	All English preferred/synonym atoms	–	–
All Atoms	UMLS ones	All English atoms	–	–
All Atoms + Child/Parent/Same Relations	UMLS ones	All English atoms	With atoms	Child/Parent/Same
Broader/Narrower Concepts	UMLS ones	All English atoms	With concepts	Broader/Narrower

---

[1] https://www.nlm.nih.gov/research/umls/sourcereleasedocs/current/CHV/.

CHV-preferred and UMLS-preferred expressions for each concept. This is the only method using exclusively one vocabulary.

The other methods use the overall UMLS to obtain an expression or a subset of expressions, from which we select the lay and medico-scientific synonymous. The lay synonymous is the expression with the highest value of similarity with the lay terminology, and the medico-scientific one is the expression closest to the medico-scientific terminology. To determine the closeness of the expressions to these terminologies, we used a previously created algorithm [11].

The *Preferred Atoms* method uses the default preferred atom associated with the CUI. The *All preferred/synonym atoms* method retrieves a list of all English atoms that are the preferred names or a synonym in the various vocabularies of the UMLS. The *All Atoms* method retrieves all the English atoms, instead of extracting only the preferred and synonym ones. To explore other atoms associated with a concept, the method *All Atoms + Child/Parent/Same Relations* identifies all English atoms associated with a concept and then retrieves atoms related to the first one through parent/child/same relationships. Finally, the *Broader/Narrower Concepts* recovers broader and narrower atoms that are directly connected with the initial identified concept, instead of looking for atoms associated with the concept.

4   Evaluation

To assess and compare the effectiveness of the developed methods, we used two test collections, one in English and the other in Portuguese. The English collection is provided by the Consumer Health Search Task in the 2018 edition of the CLEF eHealth Lab[2]. This task uses a set of 50 English queries and a document corpus with 5,535,120 web pages acquired from a CommonCrawl dump. It also provides 26,025 judgments of relevance and understandability.

The Portuguese collection was explicitly built for this work. We used the English queries provided by the User-Centred Health Information Retrieval[3] and Patient-Centred Information Retrieval[4] Tasks of the 2015 and 2016 editions of the CLEF eHealth Lab. We translated the 208 queries to the Portuguese language with the collaboration of a medical doctor. Although the dataset of the 2015 edition had Portuguese translations of the queries, they were in some cases in PT-BR, and for this reason, we decided to translate them to PT-PT manually.

The queries were used in a user study with 104 participants. These participants were students, and as part of one work assignment, they were assigned two tasks regarding two different queries. In each task, they were asked to judge the relevance and understandability of the 30-top documents retrieved by four search engines: Google, Bing, Yahoo!, and HONSearch. The 16,505 assessed documents and the judgments of the participants complete this collection[5]. The number

---

[2] https://sites.google.com/view/clef-ehealth-2018/task-3-consumer-health-search.

[3] https://sites.google.com/site/clefehealth2015/task-2.

[4] https://sites.google.com/site/clefehealth2016/task-3.

[5] Available at https://rdm.inesctec.pt/dataset/cs-2020-004.

of documents is different from 24,960 (208*4*30) because there was an overlap between documents retrieved by the four search engines and because the number of retrieved results may be inferior to 30.

We have indexed the document corpora in Elastic Search. For each query, we compute four types of suggestions, in lay and medico-scientific suggestions, both in English and Portuguese. Using the judgments of each test collection as ground truth, we assessed the performance of each suggestion through the top-10 documents retrieved by Elastic Search for that query. For this evaluation, our baseline is the performance of the suggestions provided by Health Suggestions.

The performance was assessed through the Understandability-based RBP (uRBP) and uRBP graded (uRBPgr). uRBP is a measure that increases when the user chooses a document that is considered both relevant and understandable, based on binary assessments. The uRBPgr allows graded assessment values [14]. For each method, we conduct one evaluation considering word-sense disambiguation and one without it.

5   Results

The best methods select the CHV-preferred expressions for lay suggestions and the UMLS-preferred expression for the medico-scientific suggestions (Table 2). Both methods outperform the baseline.

Table 2. Evaluation of the methods using the English and Portuguese test collections.

| Terminology | Method | English | | | | Portuguese | | | |
| | | Without WSD | | With WSD | | Without WSD | | With WSD | |
		uRBP	uRBPgr	uRBP	uRBPgr	uRBP	uRBPgr	uRBP	uRBPgr
Lay	HealthSuggestions (Baseline)	0.2869	0.1257	0.2869	0.1257	0.0404	0.0567	0.0404	0.0567
	CHV-preferred	**0.4961**	**0.2372**	**0.4846**	**0.2298**	**0.1237**	**0.1496**	**0.1258**	**0.1500**
	All preferred/ synonym atoms	0.3618	0.1750	0.3221	0.1558	0.0878	0.1070	0.0879	0.1102
	All Atoms	0.3189	0.1634	0.3530	0.1763	0.0813	0.1006	0.0888	0.1091
	All Atoms + Child Relations	0.1460	0.0665	0.2341	0.1001	0.0400	0.0507	0.0466	0.0578
	All Atoms + Parent Relations	0.1972	0.0948	0.2685	0.1226	0.0705	0.0918	0.0675	0.0870
	All Atoms + Same Relations	0.3477	0.1645	0.3693	0.1731	0.0919	0.1126	0.0910	0.1096
	Broader Concepts	0.2852	0.1307	0.3321	0.1525	0.0839	0.1056	0.0928	0.1147
	Narrower Concepts	0.1617	0.0775	0.2397	0.1121	0.0590	0.0772	0.0801	0.0999

(continued)

Table 2. (*continued*)

Terminology	Method	English				Portuguese			
		Without WSD		With WSD		Without WSD		With WSD	
		uRBP	uRBPgr	uRBP	uRBPgr	uRBP	uRBPgr	uRBP	uRBPgr
Medico-scientific	HealthSuggestions (Baseline)	0.2610	0.1167	0.2610	0.1167	0.0385	0.0537	0.0385	0.0537
	UMLS-preferred	**0.4155**	**0.2073**	**0.4280**	**0.2122**	**0.0969**	**0.1214**	**0.1081**	**0.1334**
	All preferred/synonym atoms	0.3164	0.1610	0.2510	0.1279	0.0821	0.1022	0.0823	0.1020
	All Atoms	0.3381	0.1690	0.3269	0.1634	0.0628	0.0805	0.0716	0.0894
	All Atoms + Child Relations	0.1192	0.0531	0.1772	0.0716	0.0496	0.0632	0.0573	0.0744
	All Atoms + Parent Relations	0.2031	0.0999	0.2655	0.1231	0.0715	0.0933	0.0715	0.0922
	All Atoms + Same Relations	0.3480	0.1753	0.3719	0.1873	0.0721	0.0932	0.0673	0.0860
	Broader Concepts	0.2273	0.1042	0.2798	0.1243	0.0848	0.1031	0.1017	0.1228
	Narrower Concepts	0.1774	0.0828	0.2547	0.1188	0.0597	0.0781	0.0803	0.1012
Both	Preferred Atoms	0.3662	0.1905	0.3469	0.1754	0.1011	0.1295	0.1084	0.1369

Globally, the methods with better performance are the ones that consider the preferred atoms of the different vocabularies from the UMLS, mainly the CHV. Using child relations does not help, probably due to the specificity of the suggestion. Using broader terms (parent and broader relations) proved to be more useful since other designations for the same concept are being explored.

In the English test collection, the use of WSD does not improve the performance of the methods that use UMLS-preferred terms but is useful when exploring relations. In the Portuguese collection, in general, there are slightly better results when using WSD. Nevertheless, this difference is so small that we conclude that it is better to disambiguate in methods that explore relations and the other way around in methods that pick the preferred terms. Note that context is essential in methods that use relations that may justify the importance of disambiguation.

The average number of seconds to formulate a suggestion is presented, for each method, in Table 3. As can be seen, methods that consider the relationships of atoms take a longer time compared to the others. The use of the relations from concepts should be preferred since it takes less time to process them, and the performance is similar. In English, the use of WSD helps to reduce the processing time because fewer atoms are retrieved and, therefore, less processing is needed afterward. The CHV/UMLS-preferred are the fastest methods since they only need to identify the concept and retrieve the corresponding CHV/UMLS-preferred expression.

Table 3. Average number of seconds to generate a suggestion.

Method	English		Portuguese	
	Without WSD	With WSD	Without WSD	With WSD
CHV/UMLS-preferred	**0.42**	**0.34**	**0.22**	**0.26**
Preferred Atoms	1.34	1.08	1.52	1.45
All preferred/synonym atoms	30.54	25.72	38.59	32.55
All Atoms	50.02	41.60	64.85	46.98
All Atoms + Child Relations	111.5	99.30	200.98	207.29
All Atoms + Parent Relations	175.44	134.72	321.52	250.51
All Atoms + Same Relations	127.34	100.72	257.44	205.61
Broader Concepts	17.42	10.80	18.29	21.04
Narrower Concepts	3.98	3.66	5.82	5.42

6 Conclusions

The majority of the developed methods proved to be better than the baseline, helping the user to retrieve more relevant and understandable documents. Using UMLS-preferred terms resulted in a better performance. Others explored broader terms, more specific terms, and similar terms but did not retrieve as good results. The best method to suggest lay queries is the one that uses the CHV-preferred expressions (the most familiar ones) to substitute the identified concepts. The best method to suggest medico-scientific suggestions uses UMLS-preferred expressions. These methods are better in the relevance and understandability but are also better in generation time. Since the word-sense disambiguation reduces the time that is necessary to generate new suggestions, and slightly improves or does not affect the overall performance, we conclude it should be used.

Acknowledgments. This work was financed by the Portuguese funding agency, FCT - Fundação para a Ciência e a Tecnologia, through national funds, and co-funded by the FEDER, where applicable.

References

1. Akerkar, S., Bichile, L.: Health information on the internet: patient empowerment or patient deceit? Indian J. Med. Sci. **58**(8), 321–6 (2004)
2. Aronson, A.R.: Effective mapping of biomedical text to the UMLS Metathesaurus: the MetaMap program. In: Proceedings of AMIA Symposium, pp. 17–21 (2001). https://www.ncbi.nlm.nih.gov/pmc/articles/PMC2243666/
3. Aronson, A.R., Lang, F.M.: An overview of MetaMap: historical perspective and recent advances. J. Am. Med. Inform. Assoc. **17**(3), 229–236 (2010). https://doi.org/10.1136/jamia.2009.002733
4. Bodenreider, O.: The Unified Medical Language System (UMLS): integrating biomedical terminology. Nucleic Acids Res. **32**(Database Issue), 267–70 (2004)
5. Ermakova, L., Mothe, J., Nikitina, E.: Proximity relevance model for query expansion. In: Proceedings of the 31st Annual ACM Symposium on Applied Computing, SAC 2016, pp. 1054–1059. ACM, New York (2016). https://doi.org/10.1145/2851613.2851696

6. Jacobs, W., Amuta, A.O., Jeon, K.C.: Health information seeking in the digital age: an analysis of health information seeking behavior among US adults. Cogent Soc. Sci. **3**(1), 1–11 (2017). https://doi.org/10.1080/23311886.2017.1302785

7. Liu, Z., Chu, W.W.: Knowledge-based query expansion to support scenario-specific retrieval of medical free text. Inf. Retrieval **10**(2), 173–202 (2007). https://doi.org/10.1007/s10791-006-9020-6

8. Lopes, C.T., Fernandes, T.A.: Health suggestions: a chrome extension to help laypersons search for health information. In: Fuhr, N., et al. (eds.) CLEF 2016. LNCS, vol. 9822, pp. 241–246. Springer, Cham (2016). https://doi.org/10.1007/978-3-319-44564-9_22

9. Lopes, C.T., Ribeiro, C.: Effects of language and terminology on the usage of health query suggestions. In: Fuhr, N., et al. (eds.) CLEF 2016. LNCS, vol. 9822, pp. 83–95. Springer, Cham (2016). https://doi.org/10.1007/978-3-319-44564-9_7

10. Ooi, J., Ma, X., Qin, H., Liew, S.C.: A survey of query expansion, query suggestion and query refinement techniques. In: 2015 4th International Conference on Software Engineering and Computer Systems (ICSECS), pp. 112–117 (2015). https://doi.org/10.1109/ICSECS.2015.7333094

11. Santos, P., Lopes, C.T.: Is it a lay or medico-scientific concept? Automatic classification in two languages. In: 2019 14th Iberian Conference on Information Systems and Technologies (CISTI), pp. 1–4 (2019). https://doi.org/10.23919/CISTI.2019.8760745

12. Zeng, Q.T., Crowell, J., Plovnick, R.M., Kim, E., Ngo, L., Dibble, E.: Assisting consumer health information retrieval with query recommendations. J. Am. Med. Inform. Assoc. **13**(1), 80–90 (2006). https://doi.org/10.1197/jamia.M1820

13. Zeng, Q.T., Kogan, S., Plovnick, R.M., Crowell, J., Lacroix, E.M., Greenes, R.A.: positive attitudes and failed queries: an exploration of the conundrums of consumer health information retrieval. Int. J. Med. Inform. **73**(1), 45–55 (2004). https://doi.org/10.1016/j.ijmedinf.2003.12.015

14. Zuccon, G.: Understandability biased evaluation for information retrieval. In: Ferro, N., et al. (eds.) ECIR 2016. LNCS, vol. 9626, pp. 280–292. Springer, Cham (2016). https://doi.org/10.1007/978-3-319-30671-1_21

Identifying Notable News Stories

Antonia Saravanou[1(✉)], Giorgio Stefanoni[2], and Edgar Meij[2]

[1] National and Kapodistrian University of Athens, Athens, Greece
antoniasar@di.uoa.gr
[2] Bloomberg, London, UK
giorgio.stefanoni@gmail.com, emeij@bloomberg.net

Abstract. The volume of news content has increased significantly in recent years and systems to process and deliver this information in an automated fashion at scale are becoming increasingly prevalent. One critical component that is required in such systems is a method to automatically determine how notable a certain news story is, in order to prioritize these stories during delivery. One way to do so is to compare each story in a stream of news stories to a notable event. In other words, the problem of detecting notable news can be defined as a ranking task; given a trusted source of notable events and a stream of candidate news stories, we aim to answer the question: "Which of the candidate news stories is most similar to the notable one?". We employ different combinations of features and learning to rank (LTR) models and gather relevance labels using crowdsourcing. In our approach, we use structured representations of candidate news stories (triples) and we link them to corresponding entities. Our evaluation shows that the features in our proposed method outperform standard ranking methods, and that the trained model generalizes well to unseen news stories.

1 Introduction

With the rise in popularity of social media and the increase in citizen journalism, news is increasing in volume and coverage all around the world. As a result, news consumers run the risk of either being overwhelmed due to the sheer amount of news being produced, or missing out on news stories due to heavy filtering. To deal with the information overload, it is crucial to develop systems that can filter the noise in an intelligent fashion. Due to the highly condensed language used in news, automated systems have been developed to process them and generate well-defined structured representations from their content [9]. Each structure is a so-called triple that represents an event in the form of *who* did *what* to *whom*, with additional metadata information about *when* and *where* this happened. Such representations (triples) form a *knowledge graph* (KG). There are multiple computational benefits when searching, labeling, and processing KGs due to their clean and simple structure [11,14].

A. Saravanou—This work was done whilst interning at Bloomberg.

J. M. Jose et al. (Eds.): ECIR 2020, LNCS 12036, pp. 352–358, 2020.
https://doi.org/10.1007/978-3-030-45442-5_44

Table 1. Example of a query q_0 and two candidate events c_0 and c_1.

Query q_0	Tagged Query
A suicide bomber detonates a vehicle full of explosives at a military camp in Gao, Mali, killing at least 76 people and wounding scores more in Mali's deadliest terrorist attack in history. *Date*: 17 January 2017	A [WIKI: Suicide_attack] detonates a [WIKI: Vehicle] full of [WIKI: Explosive] at a military camp in [WIKI: Gao], [WIKI: Mali], [WIKI: Murder] at least 76 people and wounding scores more in Mali's deadliest [WIKI: Terrorism] in history. *Date*: 17 January 2017

	Subject	Predicate	Object	Date	Location
c_0	Armed Gang	Carry out suicide bombing	Armed rebel	17 Jan. 2017	Gao, Mali
c_1	Armed Gang	Carry out suicide bombing	Military	17 Jan. 2017	Bamako, Mali

A common approach to measure notability of a news event is to track it through a proxy metric. For example, Naseri *et al.* [7] decide whether an article describes a notable event by counting the user interactions, while Setty *et al.* [10] cluster together similar news articles and then use the cluster size to decide if the common theme is notable. Wang *et al.* [12] propose a recommendation framework that takes as input a stream of news and predicts the user's click-through rate.

In this paper, we approach the problem of identifying notable news stories as a ranking task, i.e., we rank structured news stories represented as triples against notable events. We use *Wikipedia's Current Events Portal* (WCEP) [2] as curated notable events and, using a combination of textual and semantic features, we build a learning to rank (LTR) model to solve the ranking problem.

2   Problem Statement

Let $\mathcal{Q} = [q_0, \ldots, q_k]$ denote a stream of events, where each *query event* $q_i \in \mathcal{Q}$ is a notable event composed of a textual description and of a publication date. Let $\mathcal{C} = [c_0, \ldots, c_l]$ denote a stream of *candidate* events. Each $c_j \in \mathcal{C}$ is a structured representation of a news story that consists of a *triple* of the form (s, p, o), where s is the subject, p is the predicate, and o is the object, together with information about the *location (city, country)* and the *date* of the news story.

Given a query $q_i \in \mathcal{Q}$ and a stream of candidates $\mathcal{C}$, we aim to rank each candidate $c_j \in \mathcal{C}$ by its relevance to the query q_i. A pair (q_i, c_j) is considered as *very relevant* when the information from q_i and c_j matches completely; it is considered as *relevant* when some of the information matches; otherwise, it is considered as *not relevant*. Table 1 shows a query q_0 and two candidates c_0 and c_1. The pair (q_0, c_0) is very relevant because c_0 matches q_0 completely; in contrast, the pair (q_0, c_1) is relevant because q_0 and c_1 disagree only on the location of the event.

3   Method

In this section we present our method to identify notable news stories which consists of three steps: (1) creating (query, candidate) pairs, (2) extracting textual and semantic features, and (3) training a learning to rank (LTR) model.

(1) Creating Pairs. We create the set of all possible (query, candidate) pairs where (i) the query and the candidate have the same publication date, and (ii) the query and the candidate have at least one word in common as a pair is unlikely to be relevant if they share no words.

(2) Extracting Features. We extract a set of features for each constructed pair. Our features can be classified into three groups as follows.

(i) *Features related to a component.* We compute the size of the query or the candidate (i.e., the number of terms in the query/candidate).

(ii) *Features related to the pair.* We calculate the Okapi *BM25* score, the term frequency (*TF*) and the term frequency–inverse document frequency (*TF–IDF*) for the query/candidate in the pair. We calculate these scores using the stemmed versions of the query/candidate (using the Porter Stemmer [8]). We further define a similarity score, *element match*, $EM(q_i, ele_{c_j}) = |q_i \cap ele_{c_j}|/|ele_{c_j}|$ where an element ele_{c_j} is one of the: subject, predicate, object, description of the predicate, location, and the date in the candidate c_j. For each of those, we calculate the fraction of the number of common terms between the element ele_{c_j} and the query q_i to the total number of terms in ele_{c_j}. In addition, we compute all *EM* scores using the stemmed versions of the pair components. We also extract similarity scores for combinations of elements, as for example $EM(q_i, subject \cap predicate \cap object)$ and $EM(q_i, city \cap country)$.

(iii) *Semantic features.* An entity is a well-defined, meaningful and unique way to characterize a word/phrase. We therefore apply entity linking using the TagMe API [5] to identify entities (an example is shown in Table 1). Given the tagged query and the tagged candidate, we calculate the number of common entities using the Jaccard similarity.

(3) Ranking Pairs. We then use our features to train a learning to rank model in order to obtain a ranking of pairs. More details on the training and the tuning can be found in Sect. 4.

4   Experimental Setup

For the candidate news stories, we use the Integrated Crisis Early Warning System (ICEWS) [1] dataset which contains events that are automatically extracted from news articles using TABARI [3,9]. This system uses grammatical parsing to identify events (*who* did *what* to *whom*, *when* and *where*) using human-generated rules. The events are triples consisting of coded actions between socio-political actors. The actors refer to individuals, groups, sectors and nation states. The actions are coded into 312 categories. Geographical and temporal metadata are also associated with each triple (examples are shown in Table 1).

In our experiments, we use the same two weeks of data from ICEWS and WCEP. We remove triples with the generic action type "Make statement" as they do not convey any meaningful information. We then create pairs as described in Sect. 3. We build a crowdsourcing task (see below) to get golden truth labels. From the resulting annotated dataset, we only keep queries with at least one

Table 2. Distribution of the relevance labels in the dataset.

	Train	Validate	Test	Total
Very Relevant	220 (4%)	73 (4%)	47 (3%)	340 (3%)
Relevant	106 (2%)	20 (1%)	9 (1%)	135 (1%)
Not Relevant	5219 (94%)	1959 (95%)	1475 (96%)	8653 (96%)

relevant ICEWS triple as there are, e.g., sports events in the WCEP dataset but not in the ICEWS dataset. In total, the resulting dataset contains 9.1K pairs; 74 queries and 123 candidates per query on average. To evaluate our method in a real-world setting we split the dataset by date and use the first ten days for training, the next two days for validation, and the last two for testing.

Golden Truth. We employ crowdsourcing on the Figure-eight platform and ask annotators to judge the relevance of each pair on a 3-point scale (very relevant, relevant, not relevant).[1] Each pair (q_i, c_j) is annotated by at least three annotators and we use majority voting to obtain the gold labels. Our task obtains a inter-annotator agreement of 96.57%. Table 2 shows the distribution of relevance labels among pairs. The resulting dataset is highly skewed; with 3% annotated as *very relevant*, 1% as *relevant*, and 96% as *not relevant*.

Models. We explore various LTR algorithms and include results from Rank-Boost (RB) [6], lambdaMART (LM) [13], and Random Forest (RF) [4]. We experiment using different sets of features: *all* features (ALL), *all except entity-related* features (ALL⁻), *selected* features (SEL) and *baseline* features (B). SEL features include BM25 and TF–IDF scores calculated from the original/stemmed versions, *EM* scores for subject, predicate, object and location, and the number of entities in common and Jaccard similarity between the query and the candidate. For B features, we only consider BM25 and TF–IDF scores calculated from the original/stemmed versions of the query and the candidate. We evaluate using MAP, Precision@k, NDCG@k and MRR.

5   Results and Discussion

In this section we discuss our experimental results and answer the following research questions. How does our method compare against the baselines? Does the performance vary with different parameter settings? Does the number of *relevant* pairs affect performance? Do we benefit from tagging entities?

5.1   Overall Performance

We compare the three LTR models on the ALL and B feature sets and show the results in Table 3. Our method (using ALL features) achieves better results than using just the baseline B features. For each model and feature set, we only show the best tuned model on the validation set. Our method consistently

---

[1] https://www.figure-eight.com/.

Table 3. Main results of the LTR models on our dataset.

	MAP	P@5	P@10	NDCG@5	NDCG@10	MRR
RB$_{All}$	0.53	0.42	0.3	0.6	**0.62**	**0.75**
LM$_{All}$	0.44	0.38	0.31	0.51	0.56	0.65
RF$_{All}$	**0.56**	**0.47**	0.32	**0.64**	0.61	**0.75**
RB$_B$	0.37	0.33	0.29	0.37	0.45	0.6
LM$_B$	0.34	0.31	0.29	0.36	0.44	0.6
RF$_B$	0.44	0.4	**0.33**	0.42	0.57	0.62

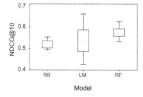 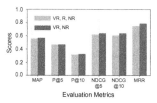

Fig. 1. (Left) Results for each model on the validation set. Each box shows the median and upper/lower quartiles. (Right) Performance using RB with selected features on two datasets.

outperforms all baselines, achieving 5–12% improvements on NDCG@10. These improvements are statistically significant with $p \leq 0.01$ using paired t-test.

We tune the parameters for each model on the validation set using NDCG@10. Figure 1 (left) shows the performance quartile plot using different parameter settings. RB and RF models show less sensitivity in the parameters tuning compared to LM. We evaluate the models when ranking pairs using all annotations (*VR, R,* and *NR*). We perform the same experiment using only the *VR* and *NR* labeled pairs. Figure 1 (right) shows that the model achieves better results when excluding the *R* labeled pairs. This is expected as the relevant label is very rare (only 1%, see Table 2) and the models tend to consider it as noise.

Our next step is to evaluate different combinations of features (ALL, ALL⁻, SEL, B). We show our findings in Table 4. First, we compare our method using ALL⁻ and B feature sets. We show that using the proposed features (Sect. 3) we achieve better performance for all LTR models. Second, we evaluate the performance of the models when we add the entity features by comparing ALL and ALL⁻. In Table 4, we show that there is a statistically significant improvement ($p \leq 0.01$) on MRR (+7%) when we add the entity–related features.

5.2 Analysis

In this section, we show examples of the output from our best performing setting, i.e., RF with ALL features using the *VR* and *NR* labeled pairs. We show our best and worst per–query NDCG@10 performance. The best one achieves a score of 1, which indicates that our method was able to rank all pairs in the right order. The top–1 ranked pair is the query *"At least 15 children are killed and 45 more are injured after a school bus collides with a truck in Etah, India.*

Table 4. Results using binary relevance labels.

	MAP	P@5	P@10	NDCG@5	NDCG@10	MRR
RB$_{All}$	**0.57**	0.42	0.3	0.61	**0.65**	0.69
LM$_{All}$	0.53	0.4	0.3	0.56	0.61	0.71
RF$_{All}$	**0.57**	**0.47**	**0.33**	0.64	0.64	**0.79**
RB$_{All-}$	0.52	**0.47**	0.3	0.62	0.62	0.68
LM$_{All-}$	0.44	0.33	0.28	0.47	0.54	0.65
RF$_{All-}$	**0.53**	0.44	**0.3**	**0.64**	**0.65**	**0.72**
RB$_{Sel}$	**0.61**	0.44	0.28	**0.67**	**0.67**	**0.81**
LM$_{Sel}$	0.53	0.4	0.27	0.56	0.6	0.75
RF$_{Sel}$	0.55	**0.47**	**0.31**	0.62	0.65	0.62
RB$_B$	**0.44**	**0.38**	0.28	**0.47**	0.51	0.6
LM$_B$	0.38	0.33	0.28	0.39	0.47	0.54
RF$_B$	0.42	0.31	**0.3**	0.42	**0.58**	**0.63**

Date: 20 Jan. 2017" and the candidate *<Attacker (from India), Kill by physical assault, Children (from India)>* with metadata *<Etah, India, 20 Jan. 2017>*. The item with the worst per–query NDCG@10 performance is *"Mexican drug lord Joaquin Guzman is extradited to the USA, where he will face charges for his role as leader of the Sinaloa Cartel. Date: 20 Jan. 2017"* paired with the candidate *<USA, Host a visit, Narendra Modi>* with metadata *<-, USA, 20 Jan. 2017>*. This query is about the extradition of a drug lord, while the candidate is about a visit of the Prime Minister of India. However, among the top–10 ranked candidates, the most relevant one is the triple *<USA, Arrest, detain, or charge with legal action, Men (from Mexico)>* with metadata *<Kansas City, USA, 20 Jan. 2017>*, ranked 9th. This shows that even in the worst ranking per–query, our method ranks a relevant candidate in the top–10.

In summary, we provide quantitative and qualitative performance analyses of our proposed method and we conclude that learning to rank is a viable method to determine notability of news stories. Among the key steps of our method are: (i) the extraction of textual and semantic features, and (ii) the exclusion of the pairs that do not convey strong signal, i.e., the ones labeled as *'relevant'*. The RF model outperforms all baselines and it is also more robust with respect to hyperparameter settings. These findings show that our approach to detect notable news through ranking is a promising one. Although our method obtains high performance (MRR = 81%), we believe we can attain further improvements by leveraging relations of the identified entities to discover implicitly relevant ones, such as *<Narendra_Modi, isPrimeMinisterOf, India>*.

6   Conclusion and Future Work

In this paper, we present a method to rank notable news representations which leverages textual and semantic features. Our evaluation on labeled pairs from WCEP and the ICEWS shows that our method is effective. In the future, we intend to include features based on the relations of the tagged entities from external KGs, such as DBPedia and Freebase.

References

1. Integrated Crisis Early Warning System (ICEWS). https://dataverse.harvard.edu/dataverse/icews. Accessed 17 Jan 2020
2. Wikipedia's Current Events Portal (WCEP). https://en.wikipedia.org/wiki/Portal:Current_events. Accessed 17 Jan 2020
3. Boschee, E., Lautenschlager, J., O'Brien, S., Shellman, S., Starz, J.: ICEWS automated daily event data. Harvard Dataverse (2018)
4. Breiman, L.: Random forests. Mach. Learn. **45**(1), 5–32 (2001). https://doi.org/10.1023/A:1010933404324
5. Ferragina, P., Scaiella, U.: TAGME: on-the-fly annotation of short text fragments (by Wikipedia entities). In: Proceedings of the 19th ACM International Conference on Information and Knowledge Management, CIKM 2010. ACM (2010)
6. Freund, Y., Iyer, R., Schapire, R.E., Singer, Y.: An efficient boosting algorithm for combining preferences. J. Mach. Learn. Res. **4**, 933–969 (2003)
7. Naseri, M., Zamani, H.: Analyzing and predicting news popularity in an instant messaging service. In: ACM SIGIR Conference on Research and Development in Information Retrieval. ACM (2019)
8. Porter, M.F.: An algorithm for suffix stripping. In: Sparck Jones, K., Willett, P. (eds.) Readings in Information Retrieval. Morgan Kaufmann Publishers Inc., Burlington (1997)
9. Schrodt, P.: Automated coding of international event data using sparse parsing techniques. In: Annual Meeting of the International Studies Association (2001)
10. Setty, V., Anand, A., Mishra, A., Anand, A.: Modeling event importance for ranking daily news events. In: ACM International Conference on Web Search and Data Mining, WSDM 2017. ACM (2017)
11. Voskarides, N., Meij, E., Tsagkias, M., de Rijke, M., Weerkamp, W.: Learning to explain entity relationships in knowledge graphs. In: ACL-IJCNLP. Association for Computational Linguistics (2015)
12. Wang, H., Zhang, F., Xie, X., Guo, M.: DKN: deep knowledge-aware network for news recommendation. In: World Wide Web Conference, WWW 2018 (2018)
13. Wu, Q., Burges, C.J., Svore, K.M., Gao, J.: Adapting boosting for information retrieval measures. Inf. Retrieval **13**, 254–270 (2010). https://doi.org/10.1007/s10791-009-9112-1
14. Yang, S., Han, F., Wu, Y., Yan, X.: Fast top-k search in knowledge graphs. In: 2016 IEEE 32nd International Conference on Data Engineering (ICDE). IEEE (2016)

BERT for Evidence Retrieval and Claim Verification

Amir Soleimani[✉], Christof Monz, and Marcel Worring

Informatics Institute, University of Amsterdam, Amsterdam, The Netherlands
{a.soleimani,c.monz,m.worring}@uva.nl

Abstract. We investigate BERT in an evidence retrieval and claim verification pipeline for the task of evidence-based claim verification. To this end, we propose to use two BERT models, one for retrieving evidence sentences supporting or rejecting claims, and another for verifying claims based on the retrieved evidence sentences. To train the BERT retrieval system, we use pointwise and pairwise loss functions and examine the effect of hard negative mining. Our system achieves a new state of the art recall of 87.1 for retrieving evidence sentences out of the FEVER dataset 50K Wikipedia pages, and scores second in the leaderboard with the FEVER score of 69.7.

Keywords: Evidence retrieval · Claim verification · BERT

1 Introduction

The constantly growing online textual information has been accompanied by an increasing spread of false claims. Therefore, there is a need for automatic claim verification and fact-checking. The Fact Extraction and VERification (FEVER) shared task [14] introduces a benchmark for evidence-based claim verification, making it possible to integrate information retrieval and natural language inference components. FEVER consists of 185 K claims labelled as 'Supported', 'Refuted' or 'NotEnoughInfo' ('NEI') based on a 50K Wikipedia pages dump. The task is to classify the claims and extract the corresponding evidence sentences (see Fig. 1). To evaluate the retrieval and verification performance together, FEVER score is defined as label accuracy conditioned on providing evidence sentence(s) unless the label is 'NEI'.

Verifying a claim based on 50K pages is a computational challenge and can be alleviated by a multi-step pipeline. Most work [7–9,16] on FEVER has adopted a three-step pipeline. (1) Document Retrieval: a set of documents, which possibly contain relevant information to support or reject a claim, are retrieved; (2) Sentence Retrieval: five sentences are extracted out of the retrieved documents; (3) Claim Verification: the claim is verified against the retrieved sentences.

Pre-trained language models, particularly Bidirectional Encoder Representations from Transformers (BERT) [6] has significantly advanced the performance

This research was partly supported by VIVAT.

J. M. Jose et al. (Eds.): ECIR 2020, LNCS 12036, pp. 359–366, 2020.
https://doi.org/10.1007/978-3-030-45442-5_45

Claim: Roman Atwood is a content creator. (**Supported**)   **Evidence:** [`wiki/Roman_Atwood`] He is best known for his vlogs, where he posts updates about his life on a daily basis.
Claim: Furia is adapted from a short story by Anna Politkovskaya. (**Refuted**)   **Evidence:** [`wiki/Furia_(film)`] Furia is a 1999 French romantic drama film directed by Alexandre Aja, ..., adapted from the science fiction short story Graffiti by Julio Cortázar.
Claim: Afghanistan is the source of the Kushan dynasty. (**NotEnoughInfo**)

Fig. 1. Three examples from the FEVER dataset [14].

in a wide variety of information retrieval and natural language processing tasks including passage re-ranking [2], question answering [1,6], and question retrieval [12].

In this paper, we examine BERT for evidence-based claim verification in a three-step pipeline. A first BERT model is trained to retrieve evidence sentences. We compare pointwise cross entropy loss and pairwise Hinge loss and Ranknet loss [3] for the BERT evidence retrieval. We also investigate the effect of Hard Negative Mining (HNM), which means training on harder negative samples. Next, we train another BERT model to verify claims against the retrieved evidence. The code is available online[1].

In summary, our contributions are as follows: (1) To the best of our knowledge, we are the first to use BERT for evidence retrieval and claim verification; (2) We compare pointwise and pairwise loss functions for training the BERT sentence retrieval; (3) We investigate and employ HNM to improve the retrieval performance; (4) We achieve second rank in the FEVER leaderboard.

2   Related Work

Thorne et al. [14] shortlists the k-nearest documents based on TF-IDF features similar to DrQA [4]. UCL [16] detects the pages titles in the claims and rank pages by logistic regression. UKP-Athene [7], the highest document retrieval scoring system, uses MediaWiki API[2] to search Wikipedia for the claims noun phrases.

To extract evidence sentences, Thorne et al. [14] use a TF-IDF approach similar to their document retrieval. UCL [16] trains a logistic regression model on a heuristic set of features. Enhanced Sequential Inference Model (ESIM) [5] has been used in [7,9]. ESIM uses two BiLSTMs and the co-attention mechanism to classify a hypothesis based on a premise.

Decomposable attention [10] is used by Thorne et al. [14] for claim verification, which compares and aggregates soft-aligned words in sentences. In [8], transformer networks pre-trained on language generation [11] are employed.

---

[1] http://github.com/asoleimanib/BERT_FEVER.

[2] mediawiki.org/wiki/api:main_page.

ESIM has been widely used for this step [7,9,16]. UNC [9], proposes a modified ESIM that takes the concatenation of the retrieved sentences and claim along with ELMo embedding. Very recently, Dream [17] published the state of the art FEVER score using a graph reasoning module. It uses pre-trained XLNet [15] to just calculate contextual word embedding without fine-tuning. However, we show that using BERT as retrieval and verification components without any additional modules achieves comparable results.

3   Methods

FEVER provides N_D Wikipedia documents $D = \{d_i\}_{i=1}^{N_D}$. The document d_i consists of sentences $S^{d_i} = \{s_j^i\}_{j=1}^{N_{S^{d_i}}}$. The goal is to classify the claim c_l for $l = 1, \ldots, N_C = 145K$ as 'Supported', 'Refuted', or 'NEI'. For a prediction to be considered correct, a complete set of ground-truth evidence must be retrieved for the claim c_l. The 'NEI' labels do not have an evidence set. We explain the proposed system in the three-step pipeline: document retrieval, sentence retrieval, and claim verification. Figure 2 demonstrates the proposed BERT architectures for the second and third steps.

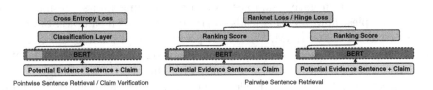

Fig. 2. Pointwise sentence retrieval and claim verification (left), Pairwise sentence retrieval (right). Orange boxes indicate the last hidden state of the [CLS] token. (Color figure online)

Following the UKP-Athene promising document retrieval component [7] (MediaWiki API), which results in more than 93% document recall, we use their method to collect a set of top documents $D_{top}^{c_l}$ for the claim c_l. We use all the retrieved documents as $D_{top}^{c_l}$.

The sentence retrieval step extracts the top five evidence sentences $S_{top}^{c_l}$. The training set consists of claims and the sentences from $D_{top}^{c_l}$ corresponding to c_l ($S_{all}^{c_l} = \{S^{d_i} | d_i \in D_{top}^{c_l}\}$).

BERT is a multi-layer transformer pre-trained on next sentence prediction and masked word prediction using extremely large datasets. BERT takes the input with a special classification embedding ([CLS]) followed by the tokens representations of the first and second sentences separated by another specific token ([SEP]). To use BERT for classification, a softmax is added on the last hidden state of the classification token ([CLS]) and trained together with the pre-trained layers.

We adopt the pre-trained BERT model and fine-tune using two different pointwise and pairwise approaches. By default, we use the BERT base (12 layers) in all the experiments. In order to compensate for the missed co-reference pronouns in the sentences [8], we add the page titles at the beginning of sentences. We use a batch size of 32, a learning rate of $2e-5$, and one epoch of training.

In the pointwise approach, we use cross entropy loss, and every single input is classified as evidence or non-evidence. At testing time, S_{top}^{cl} sentences are selected by their probability values. Alternatively, a threshold can also be used on the values to filter out uncertain results and trade-off the recall against the precision.

In the pairwise approach, a pair of positive and negative samples are compared against each other (Fig. 2 (right)) using the Ranknet loss function [3]. We do not force the positive and negative samples to be selected from the same claims because the number of sentences per claim is significantly different and this results in oversampling sentences from the claims with limited sentences. In addition, we experiment with the modified Hinge loss functions like [7]. At testing time, for both pairwise loss functions, the top five sentences S_{top}^{cl} are selected based on their output probability values.

The ratio of negative to positive sentences is high, thus it is not reasonable to train on all the negative samples. Random sampling limits the number of negative samples, however, this leads to training on trivial samples. Similar to [13], we focus on online HNM. We fix the positive samples batch size of 16 but heuristically increase negative batch from 16 to 64 and train on the positive samples and only the 16 negative samples with the highest loss values. In the case of pairwise retrieval, HNM selects the 32 hardest pairs out of 128 pairs. Loss values are computed in the no-gradient mode, and thus there is no need for more GPUs than normal training without HNM.

Table 1. Development set sentence retrieval performance. For * we calculated the scores using the official code, and for ** we used the F1 formula to calculate the score.

Model	Precision (%)	Recall@5 (%)	F1 (%)
UNC [9]	36.39	86.79	51.38
UCL [16]	22.74**	84.54	35.84
UKP-Athene [7]	23.67*	85.81*	37.11*
Pointwise	25.14	88.25	39.13
Pointwise+Threshold	**38.18**	88.00	**53.25**
Pointwise+HNM	25.13	88.29	39.13
Pairwise Ranknet	24.97	88.20	38.93
Pairwise Ranknet+HNM	24.97	**88.32**	38.93
Pairwise Hinge	24.94	88.07	38.88
Pairwise Hinge+HNM	25.01	88.28	38.98

In the claim verification step, each claim c_l is compared against $S_{top}^{c_l}$ and the final claim classification label is determined by aggregating the five individual decisions. Like [8], the default label is 'NEI' unless there is any supporting evidence to predict the claim label as 'Supported'. If there is at least one rejecting evidence and no supporting fact, the label is 'Refuted'. We train a pre-trained BERT as a three-class classifier (Fig. 2(left)). We train the model on 722K evidence-claim pairs provided by the first two steps. We use the batch size of 32, the learning rate of $2e-5$, and two epochs of training.

4   Results

Table 1 compares the development set performance of different retrieval variants. It indicates that both pointwise and pairwise BERT sentence retrieval improve the recall. Note that although precision and F1 are of value, as discussed by UNC [9], recall is the most important factor because the retrieval predictions are the samples that the verification system is trained on, and low recall leaves many claims with no probable evidence. Additionally, recall is weighted by the FEVER score requiring evidence for 'Supported' and 'Refuted' claims. A threshold can also regulate the recall and precision and shows that our method can achieve the best precision and F1 too. We opt to focus more on recall and train the claim verification model on the predictions with the maximum recall.

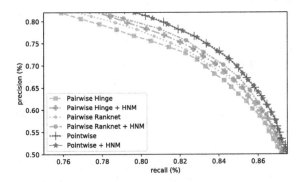

Fig. 3. Recall and precision results on the development set.

Although the pairwise Ranknet with HNM marginally has the best recall, we cannot conclude that it is necessarily better for this task. This is more clear by a trade-off between the precision and recall displayed in Fig. 3. The pointwise methods surpass the pairwise methods in terms of recall-precision performance. It also shows that HNM enhances both Ranknet and Hinge pairwise and preserves the pointwise performance.

Table 2 compares the development set results of the previous methods with the BERT models. The BERT claim verification even if it is trained on the

Table 2. Development set verification scores.

Model	FEVER score (%)	Label Acc. (%)
UNC [9]	66.14	69.60
UCL [16]	65.41	69.66
UKP-Athene [7]	64.74	–
BERT & UKP-Athene	69.79	71.70
BERT Large & UKP-Athene	70.64	72.72
BERT & BERT (Pointwise)	71.38	73.51
BERT & BERT (Pointwise+HNM)	71.33	73.54
BERT (Large) & BERT (Pointwise)	**72.42**	74.58
BERT (Large) & BERT (Pointwise+HNM)	**72.42**	**74.59**
BERT & BERT (Pairwise Ranknet)	71.02	73.22
BERT & BERT (Pairwise Ranknet+HNM)	70.99	73.02
BERT & BERT (Pairwise Hinge)	71.60	72.74
BERT & BERT (Pairwise Hinge+HNM)	70.70	72.76

Table 3. Results on the test set (October 2019).

Model	FEVER score (%)	Label Acc. (%)
DREAM [17]	**70.60**	**76.85**
BERT (Large) & BERT (Pointwise+HNM)	69.66	71.86
abcd_zh (unpublished)	69.40	72.81
BERT (Large) & BERT (Pointwise)	69.35	71.48
cunlp (unpublished)	68.80	72.47
BERT & BERT (Pointwise)	68.50	70.67
BERT (Large) & UKP-Athene	68.36	70.41
BERT & FEVER UKP-Athene	67.49	69.40
UNC [9]	64.21	68.21
UCL [16]	62.52	67.62
UKP-Athene [7]	61.58	65.46

UKP-Athene sentence retrieval predictions, the previous method with the highest recall, improves both label accuracy and FEVER score. Training based on the BERT retrieval predictions significantly enhances the verification because while it improves the FEVER score by providing more correct evidence, it provides a better training set for the verification system. It also shows that pointwise retrieval leads to more accurate claim verification. The large BERTs are only trained on the best retrieval systems, and as expected significantly improve the performance. Finally, we report the blind test set results in Table 3 using the FEVER leaderboard[3]. Our best model ranks at the second place that indicates the importance of using pre-trained language models for both sentence retrieval and claim verification.

---

[3] https://competitions.codalab.org/competitions/18814#results.

5   Conclusion

We demonstrated the BERT promising performance for the sentence retrieval and claim verification pipeline. In the retrieval step, we compared the pointwise and pairwise approaches and concluded that although the pairwise Ranknet approach achieved the highest recall, pairwise approaches are not necessarily superior to the pointwise approach particularly if precision is taken into account. Our results showed that training BERT on the pointwise retrieved sentences results in a better performance. We also examined HNM for training the retrieval systems and showed that it improves the retrieval and verification performance. We suspect that HNM can also make the training faster and leave the investigation for future work. Furthermore, using BERT as an end-to-end framework for the entire evidence-based claim verification pipeline can be investigated in the future.

References

1. Alberti, C., Lee, K., Collins, M.: A BERT baseline for the natural questions. arXiv preprint arXiv:1901.08634 (2019)
2. Bajaj, P., et al.: MS MARCO: a human generated machine reading comprehension dataset. arXiv preprint arXiv:1611.09268 (2016)
3. Burges, C., et al.: Learning to rank using gradient descent. In: Proceedings of the 22nd International Conference on Machine learning, ICML 2005, pp. 89–96 (2005)
4. Chen, D., Fisch, A., Weston, J., Bordes, A.: Reading Wikipedia to answer open-domain questions. In: Proceedings of the 55th Annual Meeting of the Association for Computational Linguistics (Volume 1: Long Papers), pp. 1870–1879 (2017)
5. Chen, Q., Zhu, X., Ling, Z., Wei, S., Jiang, H.: Enhancing and combining sequential and tree LSTM for natural language inference. arXiv preprint arXiv:1609.06038 (2016)
6. Devlin, J., Chang, M.W., Lee, K., Toutanova, K.: BERT: pre-training of deep bidirectional transformers for language understanding. arXiv preprint arXiv:1810.04805 (2018)
7. Hanselowski, A., et al.: UKP-Athene: multi-sentence textual entailment for claim verification. In: Proceedings of the First Workshop on Fact Extraction and VERification (FEVER), pp. 103–108. Association for Computational Linguistics, Brussels, November 2018. https://doi.org/10.18653/v1/W18-5516
8. Malon, C.: Team Papelo: transformer networks at FEVER. In: Proceedings of the First Workshop on Fact Extraction and VERification (FEVER), pp. 109–113. Association for Computational Linguistics, Brussels, November 2018. https://doi.org/10.18653/v1/W18-5517
9. Nie, Y., Chen, H., Bansal, M.: Combining fact extraction and verification with neural semantic matching networks. In: Proceedings of the AAAI Conference on Artificial Intelligence, vol. 33, no. 01, pp. 6859–6866 (2019). https://doi.org/10.1609/aaai.v33i01.33016859
10. Parikh, A.P., Täckström, O., Das, D., Uszkoreit, J.: A decomposable attention model for natural language inference. arXiv preprint arXiv:1606.01933 (2016)
11. Radford, A., Narasimhan, K., Salimans, T., Sutskever, I.: Improving language understanding by generative pre-training. Technical report, OpenAI (2018)

12. Sakata, W., Shibata, T., Tanaka, R., Kurohashi, S.: FAQ retrieval using query-question similarity and BERT-based query-answer relevance. In: SIGIR 2019: 42nd International ACM SIGIR Conference on Research and Development in Information Retrieval, pp. 1113–1116 (2019)
13. Schroff, F., Kalenichenko, D., Philbin, J.: FaceNet: a unified embedding for face recognition and clustering. In: Proceedings of the IEEE Conference on Computer Vision and Pattern Recognition, pp. 815–823 (2015)
14. Thorne, J., Vlachos, A., Christodoulopoulos, C., Mittal, A.: FEVER: a large-scale dataset for fact extraction and verification. arXiv preprint arXiv:1803.05355 (2018)
15. Yang, Z., Dai, Z., Yang, Y., Carbonell, J., Salakhutdinov, R., Le, Q.V.: XLNet: generalized autoregressive pretraining for language understanding. arXiv preprint arXiv:1906.08237 (2019)
16. Yoneda, T., Mitchell, J., Welbl, J., Stenetorp, P., Riedel, S.: UCL machine reading group: four factor framework for fact finding (HexaF). In: Proceedings of the First Workshop on Fact Extraction and VERification (FEVER), pp. 97–102. Association for Computational Linguistics, Brussels, November 2018. https://doi.org/10.18653/v1/W18-5515
17. Zhong, W., et al.: Reasoning over semantic-level graph for fact checking. arXiv preprint arXiv:1909.03745 (2019)

BiOnt: Deep Learning Using Multiple Biomedical Ontologies for Relation Extraction

Diana Sousa[(✉)] and Francisco M. Couto

LASIGE, Departamento de Informática, Faculdade de Ciências,
Universidade de Lisboa, 1749–016 Lisbon, Portugal
dfsousa@lasige.di.fc.ul.pt

Abstract. Successful biomedical relation extraction can provide evidence to researchers and clinicians about possible unknown associations between biomedical entities, advancing the current knowledge we have about those entities and their inherent mechanisms. Most biomedical relation extraction systems do not resort to external sources of knowledge, such as domain-specific ontologies. However, using deep learning methods, along with biomedical ontologies, has been recently shown to effectively advance the biomedical relation extraction field. To perform relation extraction, our deep learning system, BiOnt, employs four types of biomedical ontologies, namely, the Gene Ontology, the Human Phenotype Ontology, the Human Disease Ontology, and the Chemical Entities of Biological Interest, regarding gene-products, phenotypes, diseases, and chemical compounds, respectively. We tested our system with three data sets that represent three different types of relations of biomedical entities. BiOnt achieved, in F-score, an improvement of 4.93% points for drug-drug interactions (DDI corpus), 4.99% points for phenotype-gene relations (PGR corpus), and 2.21% points for chemical-induced disease relations (BC5CDR corpus), relatively to the state-of-the-art. The code supporting this system is available at https://github.com/lasigeBioTM/BiONT.

Keywords: Relation extraction · Biomedical ontologies · Deep learning · Text mining

1 Introduction

The description of the mechanisms that are responsible for the behavior of biological systems is non-trivial, and each step towards the understanding of those mechanisms often constitutes a scientific achievement [2,26]. Typical examples describe diseases that are associated with mechanisms that originate phenotypic

This work was supported by FCT through project DeST: Deep Semantic Tagger, ref. PTDC/CCI-BIO/28685/2017, LASIGE Research Unit, ref. UIDP/00408/2020, and PhD Scholarship, ref. SFRH/BD/145221/2019.

© Springer Nature Switzerland AG 2020
J. M. Jose et al. (Eds.): ECIR 2020, LNCS 12036, pp. 367–374, 2020.
https://doi.org/10.1007/978-3-030-45442-5_46

abnormalities as a result of modified gene expression, as well as the action of drugs on those diseases [4], among others. One significant step to fully understand biological systems mechanisms is to extract and classify the relations that exist between the different biomedical entities, namely chemicals, diseases, genes, and phenotypes. In literature, authors classify this problem as a Relation Extraction (RE) task. Biomedical RE aims to extract and classify relations between biomedical entities in highly heterogeneous or unstructured scientific or clinical text.

Deep learning is widely used to solve problems such as speech recognition, visual object recognition, and object detection. Lately, deep learning based-systems have started to tackle RE problems. These systems are becoming increasingly more complex, namely the MIMLCNN [10], and PCNN + Att [17] systems, that mark recent turning points in the deep learning RE field. Both of these systems use Word2Vec [19] that aims to capture the syntactic and semantic information about the word [11]. However, deep learning methods that effectively extract and classify relations between biomedical entities in the text are still scarce [13, 15].

Ontologies play an important role in biomedical research through a variety of applications and are used primarily as a source of vocabulary for standardization and integration purposes [3]. Word embeddings can learn how to detect relations between entities but manifest difficulties in grasping the semantics of each entity and their specific domain. Domain-specific ontologies provide and formalize this knowledge. Thus, a structured representation of the semantics between entities and their relations, an ontology, allows us to use it as an added feature to a machine learning classifier. Some of the biomedical entities structured in publicly available ontologies are genes properties/attributes (Gene Ontology (**GO**)) (45003 terms) [1, 23], phenotypes (Human Phenotype Ontology (**HPO**)) (25810 terms) [12], diseases (Human Disease Ontology (**DO**)) (18114 terms) [21], and drugs/chemicals (Chemical Entities of Biological Interest (**ChEBI**)) (133104 terms) [8][1].

This work presents the **BiOnt** system, a biomedical RE system built using bidirectional Long Short-Term Memory (LSTM) networks. The BiOnt system incorporates the state-of-the-art Word2Vec word embeddings [19] and makes use of different combinations of input channels to maximize performance. Our system is based on the work of Lamurias et al. [13] and Xu et al. [25]. Both of these models make use of biomedical resources as embedding layers for their respective systems. Lamurias et al. [13] uses the Xu et al. [25] model has a baseline with an added ontological embedding layer (BO-LSTM model). However, the BO-LSTM model is limited to two types of relations, namely, drug-drug, and gene-phenotype relations.

External sources of knowledge, such as biomedical ontologies, can provide highly valuable information for the detection of relations between entities in the text, as described previously by Lamurias et al. [13]. These knowledge-bases provide not only relevant characteristics about the respective entities but also

---

[1] Term counts at *09/09/2019*.

about the underlying semantics of the relations they establish. This information is not expressed directly in the training data but usually reinforces a relation between two entities in the text. The novelty of our system is that expands the previous work done by Lamurias et al. [13] by using four types of domain-specific ontologies, and combine them to extract new types of relations, along with word embeddings [19] and WordNet hypernyms [5]. BiOnt successfully replicates the results of the BO-LSTM application, using different types of ontologies. Our system can extract new relations between four biomedical entities, namely, genes, phenotypes, diseases, and chemicals. Figure 1 shows how these four types of biomedical ontologies can be combined to aid the relation extraction of ten different combinations of biomedical entities. The BiOnt system also explores the use of entities that are not direct entries in an ontology (e.g., genes), linking each entity to their most informative annotation concept within a corresponding ontology (e.g., GO). Our method incorporates more ontologies than the previously mentioned systems and is evaluated using three state-of-the-art data sets. The BiOnt system can be used to effectively populate knowledge bases regarding gold standard relations. Ultimately, it can be used to explore new experimental hypotheses providing evidence to researchers and clinicians about possible unknown associations between biomedical entities.

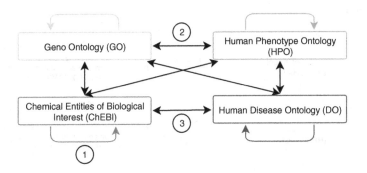

Fig. 1. The ten possible combinations between the four biomedical ontologies (the Gene Ontology (GO) [1], the Human Phenotype Ontology (HPO) [12], the Human Disease Ontology (DO) [21], and the Chemical Entities of Biological Interest (ChEBI) [8]). The **1** represents the DDI corpus, the **2** the PGR corpus, and the **3** the BC5CDR corpus (described in Sect. 3).

2   Methodology

This section describes the BiOnt model with an emphasis on the enhancements done to BO-LSTM [13] model to allow multi-ontology integration, expanding the number of different type candidate pairs from two to ten. The BiOnt model uses a combination of different language and entity related data representations, that feed individual channels creating a multichannel architecture. The input data is used to generate instances to be classified by the model. Each instance corresponds to a candidate pair of entities in a sentence. To each instance, the model assigns a positive or negative class. A positive class corresponds to an identified relation between two biomedical concepts, where the nature of this relation depends of the data set being used to perform the evaluation, and a negative class implies no relation between the different entities.

An instance should condense all relevant information to classify a candidate pair. To create an instance the BiOnt model relies on three primary data information layers. After sentence tokenization, these layers are: Shortest Dependency Path (SDP) [18,20], WordNet Classes [5], and **Ontology Embeddings**. The latter represents the relations between the ancestors for each ontology concept corresponding to an entity (Fig. 2). The model assumes that the input data already has the offsets of the relevant entities identified and their respective concept ID, the Named-Entity Recognition and Linking tasks. However, while most entities already corresponded to an ontology concept ID, some entities, such as genes do not have a direct entry in an ontology. BiOnt matches these entities to their most representative concept in the Gene Ontology [1]. To match the gene to their most representative GO term the priority was given to concepts inferred from experiments, for having a more sustained background and usually be more descriptive. For tie-breaking, if we have several GO terms inferred from experiments, the choice is the term that is the most specific (i.e., with the longer ancestry line).

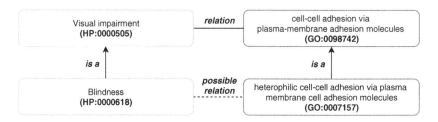

Fig. 2. BiOnt ontology embedding illustration based on the HPO and the GO ontologies, for the candidate relation between the human phenotype *blindness* and the gene *CRB1* (represented by the GO term *heterophilic cell-cell adhesion via plasma membrane cell adhesion molecules*).

As stated previously, our system expands the work done by Lamurias et al. [13] by using four types of domain-specific ontologies, and combine them to extract new types of relations. Therefore, to allow this diversity of relations, we adapted the BO-LSTM model common ancestors and the concatenation of ancestors channels. Since the common ancestors' channel could only be used for relations between the same type of biomedical entities, we only use the concatenation of ancestors channel for the relations between different biomedical entities.

3   Evaluation

To showcase our systems' performance, we used three different state-of-the-art data sets. These data sets represent three out of the ten possible combinations of the biomedical entities used in this work, drug-drug interactions, phenotype-gene relations, and chemical-induced disease relations. With these data sets, we intend to show the flexibility of our model to the different types of biomedical entities represented by biomedical ontologies. Figure 1 illustrates how the entities present in the three data sets (**1**, **2**, and **3**) are connected to the different biomedical entities.

Drug-Drug Interactions (1). The SemEval 2013: Task 9 DDI Extraction Corpus [9] is a corpus that describes drug-drug interactions (DDIs) focused on both pharmacokinetic (PK) and pharmacodynamic (PD) DDIs. The manually annotated corpus created by Herrero et al. [9] combines 5028 DDIs, from selected texts of the DrugBank database and Medline abstracts.

Phenotype-Gene Relations (2). The Phenotype-Gene Relations Corpus (PGR) [22] is a corpus that describes human phenotype-gene relations, created in a fully automated manner. Due to being a silver standard corpus is not expected to be as reliable as manually annotated corpora. Nonetheless, the authors show the system efficiency by training two state-of-the-art relation extraction deep learning systems. The PGR corpus combines 4283 human phenotype-gene relations.

Chemical-Induced Disease Relations (3). The BioCreative V CDR Corpus (BC5CDR) [16] is a corpus of chemical-induced disease (CID) relations. The BC5CDR corpus consists of 3116 chemical-disease interactions annotated from PubMed articles. To use the BC5CDR corpus, we had to preprocess the documents linking the annotations of the relations to their sentences. We assumed that if two entities share a relation in the document, they will continue to share that relation if present in the same sentence of that document.

4   Results and Discussion

Table 1 presents the relation extraction results of our system, BiOnt, for each data set. For all three data sets, our system performs better using the ontological embeddings layer (+ Ontologies), than just using the word embeddings and WordNet classes layers (State-of-the-art), by an average of 0.0404. The most relevant contribution for this metric was an increase in recall for the DDI and PGR corpus, and in precision for the BC5CDR corpus. The ontology embeddings contribute to the identification of more correct relations, with a small trade-off in precision, for the DDI corpus. For the other two data sets, the ontological embedding layer does not damage the precision, while more correct relations are identified.

Table 1. Relation extraction results with the BiOnt system, for each data set, expressing drug-drug interactions (DDI Corpus), phenotype-gene relations (PGR Corpus), and chemical-induced disease relations (BC5CDR Corpus).

Data Set	Configuration	Precision	Recall	F-score
DDI Corpus	State-of-the-art	0.7134	0.6410	0.6753
	+ Ontologies	0.6784	0.7775	**0.7246**
PGR Corpus	State-of-the-art	0.8421	0.6666	0.7442
	+ Ontologies	0.8438	0.7500	**0.7941**
BC5CDR Corpus	State-of-the-art	0.5371	0.7264	0.6175
	+ Ontologies	0.5770	0.7173	**0.6396**

For the DDI corpus, the BiOnt system, due to the inherent variability of the preprocessing phase (by randomizing the division between training and test sets), when comparing with the BO-LSTM system, performed slightly worse (0.7246 in F-score) than the previously reported results (0.7290 in F-score). The paper supporting the PGR corpus [22] reported some deep learning applications results, including with the BERT [7] based BioBERT [14] pre-trained biomedical language representation model (0.6716 in F-score). Our system outperformed those results with an F-score of 0.7941. Regarding the BC5CDR corpus, our system outperformed the best system (0.5703 in F-score) in the challenge task chemical-induced disease (CID) relation extraction of BioCreative V, by 0.0693 [24], with 0.6396 in F-score. The differences in F-score, for the distinct data sets, are mostly due to how they were built, and the completeness and complexity of the respective ontologies. For instance, the PGR corpus is a silver standard corpus, therefore, could have entities that were poorly identified, not identified at all, or not linked to the right identifier. The BC5CDR corpus was annotated for documents, not regarding the offsets of the entities that shared a relation in each document, which is also a possible limitation.

5   Conclusions and Future Work

This work showed that the knowledge encoded in biomedical ontologies plays a vital part in the development of learning systems, providing semantic and ancestry information for entities, such as genes, phenotypes, chemicals, and diseases. We evaluated BiOnt using three state-of-the-art data sets (DDI, PGR, and BC5CDR corpus), obtaining improvements in F-score (4.93, 4.99, and 2.21% points, respectively), by using an ontological information layer. Our system successfully enhances the results of Lamurias et al. [13] to other entities and ontologies. BiOnt shows that integrating biomedical ontologies instead of relying solely on the training data for creating classification models will allow us not only to find relevant information for a particular problem quicker but possibly also to find unknown associations between biomedical entities.

Regarding future work, it is possible to integrate more ontological information, and in different ways. For instance, one could consider only the relations between the ancestors with the highest information content (more relevant for the candidate pair they characterize). The information content could be inferred from the probability of each term in each ontology or resorting to an external data set. Also, a semantic similarity measurement could account for non-transitive relations (within the same ontology). Relatively to biomedical concepts that do not constitute ontology entries, we could explore quantitative evidence values, choose more than one representative term, and we could also employ semantic similarity measures [6].

References

1. Ashburner, M., et al.: Gene ontology: tool for the unification of biology. The gene ontology consortium. Nature Genet. **25**(1), 25–29 (2000)
2. Bechtel, W.: Biological mechanisms: organized to maintain autonomy. In: Boogerd, F.C., Bruggeman, F.J., Hofmeyr, J.H.S., Westerhoff, H.V. (eds.) Systems Biology, pp. 269–302. Elsevier, Amsterdam (2007)
3. Bodenreider, O.: Biomedical ontologies in action: role in knowledge management, data integration and decision support. In: IMIA Yearbook Medical Informatics, pp. 67–79 (2008)
4. Campaner, R.: Understanding mechanisms in the health sciences. Theor. Med. Bioeth. **32**(1), 5–17 (2011). https://doi.org/10.1007/s11017-010-9166-5
5. Ciaramita, M., Altun, Y.: Broad-coverage sense disambiguation and information extraction with a supersense sequence tagger. In: Proceedings of the 2006 Conference on Empirical Methods in Natural Language Processing, EMNLP 2006, pp. 594–602. Association for Computational Linguistics, Stroudsburg (2006)
6. Couto, F.M., Lamurias, A.: Semantic similarity definition. In: Ranganathan, S., Gribskov, M., Nakai, K., Schönbach, C. (eds.) Encyclopedia of Bioinformatics and Computational Biology, pp. 870–876. Academic Press, Oxford (2019)
7. Devlin, J., Chang, M.W., Lee, K., Toutanova, K.: BERT: pre-training of deep bidirectional transformers for language understanding. In: Proceedings of the 2019 Conference of the North American Chapter of the Association for Computational Linguistics: Human Language Technologies, Volume 1 (Long and Short Papers), pp. 4171–4186. Association for Computational Linguistics, Minneapolis (2019)
8. Hastings, J., et al.: ChEBI in 2016: improved services and an expanding collection of metabolites. Nucleic Acids Res. **44**(D1), D1214–D1219 (2015)
9. Herrero-Zazo, M., Segura-Bedmar, I., Martínez, P., Declerck, T.: The DDI corpus: an annotated corpus with pharmacological substances and drug-drug interactions. J. Biomed. Inform. **46**(5), 914–920 (2013)
10. Jiang, X., Wang, Q., Li, P., Wang, B.: Relation extraction with multi-instance multi-label convolutional neural networks. In: Proceedings of COLING 2016, the 26th International Conference on Computational Linguistics: Technical Papers, pp. 1471–1480. The COLING 2016 Organizing Committee, Osaka (2016)
11. Kumar, S.: A survey of deep learning methods for relation extraction. CoRR abs/1705.03645 (2017)
12. Köhler, S., Vasilevsky, N., Engelstad, M., Foster, E., et al.: The human phenotype ontology. Nucleic Acids Res. **45**, D865–D876 (2017)

13. Lamurias, A., Sousa, D., Clarke, L.A., Couto, F.M.: BO-LSTM: classifying relations via long short-term memory networks along biomedical ontologies. BMC Bioinform. **20**(1), 10 (2019)
14. Lee, J., et al.: BioBERT: a pre-trained biomedical language representation model for biomedical text mining. arXiv e-prints preprint arXiv:1901.08746 (2019)
15. Li, F., Zhang, M., Fu, G., Ji, D.H.: A neural joint model for entity and relation extraction from biomedical text. BMC Bioinform. **18**, 198 (2017)
16. Li, J., et al.: BioCreative V CDR task corpus: a resource for chemical disease relation extraction. Database **2016**, 1–10 (2016)
17. Lin, Y., Shen, S., Liu, Z., Luan, H., Sun, M.: Neural relation extraction with selective attention over instances. In: ACL (2016)
18. Mikolov, T., Sutskever, I., Chen, K., Corrado, G., Dean, J.: Distributed representations of words and phrases and their compositionality. In: Proceedings of the 26th International Conference on Neural Information Processing Systems - Volume 2, NIPS 2013, pp. 3111–3119. Curran Associates Inc., USA (2013)
19. Mikolov, T., Sutskever, I., Chen, K., Corrado, G.S., Dean, J.: Distributed representations of words and phrases and their compositionality. In: Burges, C.J.C., Bottou, L., Welling, M., Ghahramani, Z., Weinberger, K.Q. (eds.) Advances in Neural Information Processing Systems 26, pp. 3111–3119. Curran Associates Inc., New York (2013)
20. Pyysalo, S., Ginter, F., Moen, H., Salakoski, T., Ananiadou, S.: Distributional semantics resources for biomedical text processing. In: Proceedings of LBM 2013, pp. 39–44 (2013)
21. Schriml, L.M., et al.: Human Disease Ontology 2018 update: classification, content and workflow expansion. Nucleic Acids Res. **47**(D1), D955–D962 (2018)
22. Sousa, D., Lamurias, A., Couto, F.M.: A silver standard corpus of human phenotype-gene relations. In: Proceedings of the 2019 Conference of the North American Chapter of the Association for Computational Linguistics: Human Language Technologies, Volume 1 (Long and Short Papers), pp. 1487–1492. Association for Computational Linguistics, Minneapolis (2019)
23. The Gene Ontology Consortium: The Gene Ontology Resource: 20 years and still GOing strong. Nucleic Acids Res. **47**(D1), D330–D338 (2018)
24. Wei, C.H., et al.: Overview of the BioCreative V chemical disease relation (CDR) task. In: Proceedings of the Fifth BioCreative Challenge Evaluation Workshop, vol. 14 (2015)
25. Xu, B., Shi, X., Zhao, Z., Zheng, W.: Leveraging biomedical resources in Bi-LSTM for drug-drug interaction extraction. IEEE Access **6**, 33432–33439 (2018)
26. Yu, A.C.: Methods in biomedical ontology. J. Biomed. Inform. **39**(3), 252–266 (2006)

On the Temporality of Priors
in Entity Linking

Renato Stoffalette João[(✉)]

L3S Research Center, Appelstraße 9A, 30167 Hannover, Germany
`joao@L3S.de`

Abstract. Entity linking is a fundamental task in natural language processing which deals with the lexical ambiguity in texts. An important component in entity linking approaches is the mention-to-entity *prior* probability. Even though there is a large number of works in entity linking, the existing approaches do not explicitly consider the time aspect, specifically the *temporality* of an entity's prior probability. We posit that this prior probability is temporal in nature and affects the performance of entity linking systems. In this paper we systematically study the effect of the prior on the entity linking performance over the temporal validity of both texts and KBs.

Keywords: Entity linking · Entity disambiguation · Knowledge base

1 Introduction and Motivation

Entity linking is a well studied problem in natural language processing which involves the process of identifying ambiguous entity mentions (i.e persons, locations and organisations) in texts and linking them to their corresponding unique entries in a reference knowledge base. There has been numerous approaches and eventually systems proposing solutions to the task at hand. To mention a few, AIDA [9], Babelfy [14], WAT [15] and AGDISTS [18] for example, rely on graph based algorithms and the most recent approaches rely on techniques such as deep neural networks [8] and semantic embeddings [10,19].

An important component in most approaches is the probability that a mention links to one entity in the knowledge base. The prior probability, as suggested by Fader et al. [4], is a strong indicator to select the correct entity for a given mention, and consequently adopted as a baseline. Computation of this prior is typically done over knowledge sources such as Wikipedia. Wikipedia in fact provides useful features and has grounded several works on entity linking [1,2,9,12,13].

An entity's popularity is temporally sensitive and may change due to short term events. Fang and Chang [6] noticed the probability of entities mentioned in texts often change across time and location in micro blogs, and in their work they modeled spatio-temporal signals for solving ambiguity of entities. We, on

© Springer Nature Switzerland AG 2020
J. M. Jose et al. (Eds.): ECIR 2020, LNCS 12036, pp. 375–382, 2020.
https://doi.org/10.1007/978-3-030-45442-5_47

the other hand, take a macroscopic account of time, where perceivably a larger fraction of mention to entity bindings might not be observable in the short time duration but are only evident over a longer period of time, i.e., over a year. These changes might be then reflected in a reference knowledge base and disambiguation methods can produce different results for a given mention at different times.

When using a 2006 Wikipedia edition as a reference knowledge base for example, the mention *Amazon* shows different candidates as linking destinations, but the most popular one is the entity page referring to *Amazon River*, whilst when using a 2016 Wikipedia edition, the same term leads to the page about the e-commerce company *Amazon.com* as the most popular entity to link to.

In this paper, we systematically study the effect of temporal priors on the disambiguation performance by considering priors computed over snapshots of Wikipedia at different points in time. We also consider benchmarks that contain documents created and annotated at different points in time to better understand the potential change in performance with respect to the temporal priors.

We firstly show that the priors change over time and the overall disambiguation performance using temporal priors show high variability. This strongly indicates that temporal effects should be not only taken into account in (a) building entity linking approaches, but have major implications in (b) evaluation design, when baselines that are trained on temporally distant knowledge sources are compared.

2   Problem Definition

In this section we briefly define the entity linking task as well as describe our methodology.

Consider a document d from a set of documents $D = \{d_1, d_2, \ldots, d_n\}$, and a set of mentions $M = \{m_1, m_2, \ldots, m_n\}$ extracted from d. The goal of the entity linking is to find a unique identity represented by an entity e from a set of entities $E = \{e_1, e_2, \ldots, e_n\}$, with relation to each mention m. The set of entities E is usually extracted from a reference knowledge base KB.

A typical entity linking system generally performs the following steps: (1) mention detection which extracts terms or phrases that may refer to real world entities, and (2) entity disambiguation which selects the corresponding knowledge-base KB entries for each ambiguous mention.

Since we take into account the time effect on the disambiguation task, we now pose entity linking at a specific time t as follows. Given a document $d^t \in D^t$ and a set of mentions $M = \{m_1, m_2, \ldots, m_n\}$ from document d^t, the goal of the entity linking at time t is to find the correct mapping entity $e^t \in E^t$ with relation to the mention m. The difference now is that the set of entities E^t is extracted from the reference knowledge base KB at different time periods.

2.1   Candidate Entities Generation and Ranking

As suggested by Fader et al. [4], the entity's prior probability is a strong indicator to select the correct entity for a given mention. In our case the entity's prior probability is directly obtained from the Wikipedia corpus. To calculate entities' probability, we parsed all the articles from a Wikipedia corpus and collected all terms that were inside double square brackets in the Wikipedia articles. [[*Andy Kirk (footballer)* | *Kirk*]] for instance, represents a pair of mention and entity where *Kirk* is the mention term displayed in the Wikipedia article and *Andy Kirk (footballer)* is the title of the Wikipedia article corresponding to the real world entity. In this way we created a list of mentions and possible candidate entities according to each Wikipedia snapshot used in this paper.

Table 1. Information about the Wikipedia editions used for mining mention and entities. #Pages refers only to the number of entities' pages, excluding special pages.

Year	Date	#Pages
2006	30/11/2006	~1.4 M
2008	03/01/2008	~1.9 M
2010	15/03/2010	~2.8 M
2012	02/09/2012	~3.5 M
2014	06/11/2014	~4.1 M
2016	01/07/2016	~5.1 M

The probability of a certain entity e^t given a mention m was only calculated if the entity had a corresponding article inside Wikipedia at time t. Thus, the probability $P(e^t|m)$ that a mention m links to a certain entity e^t is given by the number of times the mention m links to the entity e^t over the number of times that m occurs in the whole corpus at time t.

We created dictionaries of mentions and their referring entities ranked by popularity of occurrence for every Wikipedia edition as seen on Table 1. As an example of mention and its ranked candidate entities, in the KB created from the 2016 Wikipedia edition, the mention *Obama* refers in 86.15% of the cases to the president *Barack Obama*, 6.47% to the city *Obama, Fukui* in Japan, 1.79% to the genus of planarian species *Obama (genus)*, and so on and so forth.

We filtered out mentions that occurred less than 100 times for simplicity matters in the whole corpus and for every mention we checked whether the referring candidate entities pointed to existing pages inside the Wikipedia corpus at a given time, and only after these steps we calculated the prior probability values the entities.

Our framework supports multiple selection of mention-entity dictionaries created from different *KB*s based on Wikipedia snapshots from different years.

3 Experiments and Results

3.1 Datasets

In order to evaluate our experiments we employed some data sets that are widely used benchmark datasets for entity linking tasks. *ACE04* is a news corpus introduced by Ratinov et al. [16] and it is a subset from the original ACE co-reference data set [3]. *AIDA/CONLL* is proposed by Hoffart et al. [9] and it is based on the data set from the CONLL 2003 shared task [17]. *AQUAINT50* was created in the work proposed by Milne and Witten [13], and is a subset from the original AQUAINT newswire corpus [7]. *IITB* is a dataset extracted from popular web pages about sports, entertainment, science and technology, and health[1][2], and it was created in the work proposed by Kulkarni et al. [11]. *MSNBC* was introduced by Cucerzan [2] and contains news documents from 10 MSNBC news categories. Table 2 shows more details about these datasets including the number of documents, documents' publication time, number of annotations as well as the reference knowledge base time.

3.2 Prior Probability Changes

In many entity linking systems, the entity mentions that should be linked are given as the input, hence the number of mentions generated by the systems equals the number of entity mentions that should be linked. For this reason most researchers use accuracy to evaluate their method's performance. Accuracy is a straightforward measure calculated as the number of correctly linked mentions divided by the total number of mentions.

Since we take into account the time variation, we only calculated accuracy over the total number of annotations that persisted across time, i.e. the entities from the ground truth that were also present in every Wikipedia edition used in this paper. Table 3 shows the accuracy calculated on the ground truth datasets using the prior probability model from different time periods. We can observe

Table 2. *#Docs* is the number of documents. *Docs Year* is the documents' publication time. *#Annotations* is the number of annotations (Number of *non-NIL* annotations). *Annot. Year* is the reference KB time period where the annotations were taken from.

Dataset	#Docs	Docs Year	#Annotations	Annot. Year
ACE04 [16]	57	2000	257	2010
AIDA/CONLL [9]	231	1996	4.485	2010
AQUAINT50 [13]	50	1998–2000	727	2007
IITB [11]	107	2008	12.099	2008
MSNBC [2]	20	2007	747	2006

---

[1] http://news.google.com/.

[2] http://www.espnstar.com/.

an accuracy change from 77.19% to 82.63% on *ACE04* using models created from Wikipedia 2006 and 2010 editions respectively, from 64.80% to 69.07% for *AQUAINT50* using models from 2006 and 2012 editions, from 64.13% to 68.16% for *AIDA/CONLL* using models from 2008 and 2014, from 46.60% to 49.76% on *IITB* using models from 2014 and 2006, and for *MSNBC* a change from 63.82% to 65.86% using models from Wikipedia 2012 and 2008 editions respectively.

Even though it is out of the scope of this work to spot a temporal trend on the entities changes when using knowledge bases from different time periods, we can clearly see there is some temporal variability which is easily observed by the influence on the accuracy calculated over the ground truth datasets. We observe that a simplistic popularity only based method that takes into account reference *KB*s from different time periods can produce an improvement of 5.4% points in the best case for the *ACE04* dataset and 2.0% points in the worst case for *MSNBC* dataset.

3.3   Comparing Ranked Entities

We detected distinct changes when it comes to entity linking using Wikipedia as a knowledge base. The first case occurs when the entity page title changes but still refers to the same entity in the real world. For example in the 2006 Wikipedia edition the mention *Hillary Clinton* showed higher probability of linking to the referring entity page titled *Hillary Rodham Clinton* and in the 2016 Wikipedia edition, the same mention was most likely to be linked to the entity page titled *Hillary Clinton*. In this case only the entity page title changed but they both refer to the same entity in the real world.

The second case happens when an entity's popularity actually changes over time. For example in the 2006 Wikipedia edition, the mention *Kirk* was most likely to be linked to the entity page titled *James T. Kirk* whereas in the 2016 Wikipedia edition the same mention showed a higher probability of linking to the entity page titled *Andy Kirk (footballer)*.

Another observation is the case when an entity mention that was considered unambiguous in the past and became ambiguous in a newer Wikipedia edition due to the addition of new information to Wikipedia. For example in the 2006 Wikipedia edition the mention *Al Capone* showed a single candidate entity, the north american gangster and businessman *Al Capone*, while in the newer 2016 Wikipedia edition, the same mention showed more candidate entities, including the former one plus a movie, a song, and other figures with the same name.

Top Ranked Entity Changes. Initially we were only concerned with the top ranked candidate entity for each mention. We made comparisons between the dictionaries of mentions from Wikipedia editions 2006 and 2016 and despite the fact of observing 33,531 mentions in the 2006 version and 161,264 mentions in the

2016 version, only 31,123 mentions appeared in both editions. Moreover, when we take into consideration both the ambiguous and unambiguous mentions, in 9.44% of the cases the mentions change their top ranked candidate entities, whilst when removing the unambiguous mentions this number increases to 15.36%. This is mainly due to the fact that most of the unambiguous mentions keep the same entity bindings, even though we spotted cases of mentions that were unambiguous and became ambiguous in a more recent knowledge base.

Table 3. Accuracy of the models on different data sets across different time periods.

Dataset	2006	2008	2010	2012	2014	2016
ACE2004	77.19	81.17	**82.63**	80.96	80.54	79.49
AIDA/CONLL_testb	61.86	64.13	66.47	67.78	**68.16**	68.14
AQUAINT50	64.80	68.18	68.92	**69.07**	67.30	66.86
IITB	**49.76**	49.43	49.50	47.78	46.60	47.60
MSNBC	65.30	**65.86**	65.67	63.82	64.56	65.67

Top 5 Entities Changes. In another experiment we wanted to calculate the entities rank correlation. One way to calculate rank correlation for lists that do not have all the element in common, is to ignore the non conjoint elements, but unfortunately this approach is not satisfactory since it throws away information. Hence, a more satisfactory approach, as proposed by Fagin et al. [5], is to treat an element i which appears ranked in list L_1 and does not appear in list L_2, at position $k+1$ or beyond, considering L_2's depth is k. This measure was used to assess the changes in the top 5 candidate entities rank positions.

We calculated the rank correlation for 18,727 mentions, since this is the number of mentions that are ambiguous and appears both in the 2006 and 2016 Wikipedia corpus. We normalized our results so the values would lie between [0,1]. Any value close to 0 means total agreement while any value close to 1 means total disagreement. Thus we observed an average value of 0.59 with a variance of 0.05 and a standard deviation of 0.21. We noticed that in 71.98% of the cases the rank correlation values are greater than 0.5. That tells us there is some significant number of changes in the candidate entities' rank's positions. Table 4 shows the mention *Watson* and its top 5 candidate entities together with their respective prior probabilities extracted from two different Wikipedia editions, one from 2006 and one from 2016.

Table 4. A mention example and its top 5 ranked candidate entities captured from two Wikipedia editions.

Mention	Entity	$P(e^t)$	Year
Watson	Doctor Watson	0.146	2006
	James D. Watson	0.130	
	Watson, Australian Capital Territory	0.115	
	Division of Watson	0.076	
	Watson	0.061	
Watson	Watson (computer)	0.068	2016
	Ben Watson (footballer, born July 1985)	0.054	
	Je-Vaughn Watson	0.050	
	Jamie Watson (soccer)	0.047	
	Arthur Watson (footballer, born 1870)	0.043	

4 Conclusions and Future Work

In this work we conducted experiments with different Wikipedia editions and also created an entity linking model that uses the entity's prior probability calculated over different Wikipedia snapshots. One limitation of previous works is the fact that the systems are trained on a fixed time Wikipedia edition. An entity's prior probability is temporal in nature, and we have observed in our experiments that mention to entity bindings change over time. We could clearly see some temporal variability which should be taken into account for entity linking system's evaluations. As future work we plan to extend this paper's experimental setup and build a ground truth for temporal entity linking as well as try to create an adaptive entity linking system.

References

1. Bunescu, R.C., Pasca, M.: Using encyclopedic knowledge for named entity disambiguation. In: EACL, Trento, Italy, 3–7 April 2006 (2006)
2. Cucerzan, S.: Large-scale named entity disambiguation based on Wikipedia data. In: EMNLP-CoNLL, Prague, Czech Republic, 28–30 June 2007, pp. 708–716 (2007)
3. Doddington, G.R., Mitchell, A., Przybocki, M.A., Ramshaw, L.A., Strassel, S.M., Weischedel, R.M.: The automatic content extraction (ACE) program-tasks, data, and evaluation. In: LREC, Lisbon, Portugal, vol. 2, p. 1 (2004)
4. Fader, A., Soderland, S., Etzioni, O.: Scaling Wikipedia-based named entity disambiguation to arbitrary web text. In: WIKIAI (2009)
5. Fagin, R., Kumar, R., Sivakumar, D.: Comparing top k lists. In: ACM-SIAM, Baltimore, Maryland, USA, 12–14 January 2003, pp. 28–36 (2003)
6. Fang, Y., Chang, M.: Entity linking on microblogs with spatial and temporal signals. TACL **2**, 259–272 (2014)

7. Graff, D.: The AQUAINT Corpus of English News Text: [Content Copyright] Portions 1998–2000 New York Times Inc., 1998–2000. Associated Press Inc., 1996–2000 Xinhua News Service. Linguistic Data Consortium (2002)
8. He, Z., Liu, S., Li, M., Zhou, M., Zhang, L., Wang, H.: Learning entity representation for entity disambiguation. In: ACL 2013, Sofia, Bulgaria, 4–9 August 2013, Volume 2: Short Papers, pp. 3–34 (2013)
9. Hoffart, J., et al.: Robust disambiguation of named entities in text. In: EMNLP 2011, Edinburgh, UK, 27–31 July 2011, pp. 782–792 (2011)
10. Huang, H., Heck, L., Ji, H.: Leveraging deep neural networks and knowledge graphs for entity disambiguation. CoRR abs/1504.07678 (2015)
11. Kulkarni, S., Singh, A., Ramakrishnan, G., Chakrabarti, S.: Collective annotation of Wikipedia entities in web text. In: SIGKDD, Paris, France, 28 June–1 July 2009, pp. 457–466 (2009)
12. Mihalcea, R., Csomai, A.: Wikify!: linking documents to encyclopedic knowledge. In: CIKM 2007, Lisbon, Portugal, 6–10 November 2007, pp. 233–242 (2007)
13. Milne, D.N., Witten, I.H.: Learning to link with Wikipedia. In: CIKM 2008, Napa Valley, California, USA, 26–30 October 2008, pp. 509–518 (2008)
14. Moro, A., Raganato, A., Navigli, R.: Entity linking meets word sense disambiguation: a unified approach. TACL **2**, 231–244 (2014)
15. Piccinno, F., Ferragina, P.: From TagME to WAT: a new entity annotator. In: ERD 2014, Gold Coast, Queensland, Australia, 11 July 2014, pp. 55–62 (2014)
16. Ratinov, L., Roth, D., Downey, D., Anderson, M.: Local and global algorithms for disambiguation to Wikipedia. In: The 49th Annual Meeting of the Association for Computational Linguistics: Human Language Technologies, Proceedings of the Conference, Portland, Oregon, USA, 19–24 June 2011, pp. 1375–1384 (2011)
17. Sang, E.F., De Meulder, F.: Introduction to the CoNLL-2003 shared task: language-independent named entity recognition. arXiv preprint cs/0306050 (2003)
18. Usbeck, R., et al.: AGDISTIS - graph-based disambiguation of named entities using linked data. In: Mika, P., et al. (eds.) ISWC 2014. LNCS, vol. 8796, pp. 457–471. Springer, Cham (2014). https://doi.org/10.1007/978-3-319-11964-9_29
19. Zwicklbauer, S., Seifert, C., Granitzer, M.: Robust and collective entity disambiguation through semantic embeddings. In: SIGIR 2016, Pisa, Italy, 17–21 July 2016, pp. 425–434 (2016)

Contextualized Embeddings in Named-Entity Recognition: An Empirical Study on Generalization

Bruno Taillé[1,2(✉)], Vincent Guigue[1], and Patrick Gallinari[1,3]

[1] Sorbonne Université, CNRS, Laboratoire d'Informatique de Paris 6, LIP6,
Paris, France
{bruno.taille,vincent.guigue,patrick.gallinari}@lip6.fr
[2] BNP Paribas, Paris, France
[3] Criteo AI Lab, Paris, France

Abstract. Contextualized embeddings use unsupervised language model pretraining to compute word representations depending on their context. This is intuitively useful for generalization, especially in Named-Entity Recognition where it is crucial to detect mentions never seen during training. However, standard English benchmarks overestimate the importance of lexical over contextual features because of an unrealistic lexical overlap between train and test mentions. In this paper, we perform an empirical analysis of the generalization capabilities of state-of-the-art contextualized embeddings by separating mentions by novelty and with out-of-domain evaluation. We show that they are particularly beneficial for unseen mentions detection, especially out-of-domain. For models trained on CoNLL03, language model contextualization leads to a +1.2% maximal relative micro-F1 score increase in-domain against +13% out-of-domain on the WNUT dataset (The code is available at https://github.com/btaille/contener).

Keywords: NER · Contextualized embeddings · Domain adaptation

1 Introduction

Named-Entity Recognition (NER) consists in detecting textual mentions of entities and classifying them into predefined types. It is modeled as sequence labeling, the standard neural architecture of which is BiLSTM-CRF [8]. Recent improvements mainly stem from using new types of representations: learned character-level word embeddings [9] and contextualized embeddings derived from a language model (LM) [1,6,14].

LM pretraining enables to obtain contextual word representations and reduce the dependency of neural networks on hand-labeled data specific to tasks or domains [7,16]. This contextualization ability can particularly benefit to NER domain adaptation which is often limited to training a network on source data and either feeding its predictions to a new classifier or finetuning it on target

© Springer Nature Switzerland AG 2020
J. M. Jose et al. (Eds.): ECIR 2020, LNCS 12036, pp. 383–391, 2020.
https://doi.org/10.1007/978-3-030-45442-5_48

data [10,17]. All the more as classical NER models have been shown to poorly generalize to unseen mentions or domains [2].

In this paper, we quantify the impact of ELMo [14], Flair [1] and BERT [6] representations on generalization to unseen mentions and new domains in NER. To better understand their effectiveness, we propose a set of experiments to distinguish the effect of unsupervised LM contextualization (C_{LM}) from task supervised contextualization (C_{NER}). We show that the former mainly benefits unseen mentions detection, all the more out-of-domain where it is even more beneficial than the latter.

2   Lexical Overlap

Neural NER models mainly rely on lexical features in the form of word embeddings, either learned at the character-level or not. Yet, standard NER benchmarks present a large lexical overlap between mentions in the train set and dev/test sets which leads to a poor evaluation of generalization to unseen mentions as shown by Augenstein et al. [2]. They separate seen from unseen mentions and evaluate out-of-domain to focus on generalization but only study models designed before 2011 and no longer in use.

We propose to use a similar setting to analyze the impact of state-of-the-art LM pretraining methods on generalization in NER. We introduce a slightly more fine-grained novelty partition by separating unseen mentions in *partial match* and *new* categories. A mention is an *exact match* (EM) if it appears in the exact same case-sensitive form in the train set, tagged with the same type. It is a *partial match* (PM) if at least one of its non stop words appears in a mention of same type. Every other mentions are *new*.

We study lexical overlap in CoNLL03 [18] and OntoNotes [20], the two main English NER datasets, as well as WNUT17 [5] which is smaller, specific to user generated content (tweets, comments) and was designed without exact overlap. For out-of-domain evaluation, we train on CoNLL03 (news articles) and test on the larger and more diverse OntoNotes (see Table 4 for genres) and the

Table 1. Per type lexical overlap of test mention occurrences with respective train set in-domain and with CoNLL03 train set in the out-of-domain scenario. (EM/PM = *exact/partial match*)

		CoNLL03					ON	OntoNotes*					WNUT	WNUT*			
		LOC	MISC	ORG	PER	ALL	ALL	LOC	MISC	ORG	PER	ALL	ALL	LOC	ORG	PER	ALL
Self	EM	82%	67%	54%	14%	52%	67%	87%	93%	54%	49%	69%	-	-	-	-	-
	PM	4%	11%	17%	43%	20%	24%	6%	2%	32%	36%	20%	12%	11%	5%	13%	12%
	New	14%	22%	29%	43%	28%	9%	7%	5%	14%	15%	11%	88%	89%	95%	87%	88%
CoNLL	EM	-	-	-	-	-	-	70%	78%	18%	16%	42%	-	26%	8%	1%	7%
	PM	-	-	-	-	-	-	7%	10%	45%	46%	28%	-	9%	15%	16%	14%
	New	-	-	-	-	-	-	23%	12%	38%	38%	30%	-	65%	77%	83%	78%

very specific WNUT. We remap OntoNotes and WNUT entity types to match CoNLL03's[1] and denote the obtained dataset with *.

As reported in Table 1, the two main benchmarks for English NER mainly evaluate performance on occurrences of mentions already seen during training, although they appear in different sentences. Such lexical overlap proportions are unrealistic in real-life where the model must process orders of magnitude more documents in the inference phase than it has been trained on, to amortize the annotation cost. On the contrary, WNUT proposes a particularly challenging low-resource setting with no exact overlap.

Furthermore, the overlap depends on the entity types: Location and Miscellaneous are the most overlapping types, even out-of-domain, whereas Person and Organization present a more varied vocabulary, also more subject to evolve with time and domain.

3 Word Representations

Word Embeddings map each word to a single vector which results in a lexical representation. We take **GloVe 840B** embeddings [13] trained on Common Crawl as the pretrained word embeddings baseline and fine-tune them as done in related work.

Character-Level Word Embeddings are learned by a word-level neural network from character embeddings to incorporate orthographic and morphological features. We reproduce the **Char-BiLSTM** from [9]. It is trained jointly with the NER model and its outputs are concatenated to GloVe embeddings. We also experiment with the Char-CNN layer from ELMo to isolate the effect of LM contextualization and denote it **ELMo[0]**.

Contextualized Word Embeddings take into account the context of a word in its representation, contrary to previous representations. A LM is pretrained and used to predict the representation of a word given its context. **ELMo** [14] uses a Char-CNN to obtain a context-independent word embedding and the concatenation of a forward and backward two-layer LSTM LM for contextualization. These representations are summed with weights learned for each task as the LM is frozen after pretraining. **BERT** [6] uses WordPiece subword embeddings [21] and learns a representation modeling both left and right contexts by training a Transformer encoder [19] for Masked LM and next sentence prediction. For a fairer comparison, we use the $BERT_{LARGE}$ feature-based approach where the LM is not fine-tuned and its last four hidden layers are concatenated. **Flair** [1] uses a character-level LM for contextualization. As in ELMo, they train two opposite LSTM LMs, freeze them and concatenate the predicted states of the first and last characters of each word. Flair and ELMo are pretrained on the 1 Billion Word Benchmark [3] while BERT uses Book Corpus [22] and English Wikipedia.

---

[1] We map LOC + GPE in OntoNotes to LOC in CoNLL03 and NORP + LANGUAGE to MISC. Other mappings are self-explanatory. We drop types that have no correspondence.

4 Experiments

In order to compare the different embeddings, we feed them as input to a classifier. We first use the state-of-the-art BiLSTM-CRF [8] with hidden size 100 in each direction and present in-domain results on all datasets in Table 2.

We then report out-of-domain performance in Table 3. To better capture the intrinsic effect of LM contextualization, we introduce the Map-CRF baseline from [1] where the BiLSTM is replaced by a simple linear projection of each word embedding. We only consider domain adaptation from CoNLL03 to OntoNotes* and WNUT* assuming that labeled data is scarcer, less varied and more generic than target data in real use cases.

We use the IOBES tagging scheme for NER and no preprocessing. We fix a batch size of 64, a learning rate of 0.001 and a 0.5 dropout rate at the embedding layer and after the BiLSTM or linear projection. The maximum number of epochs is set to 100 and we use early stopping with patience 5 on validation global micro-F1. For each configuration, we use the best performing optimization method between SGD and Adam with $\beta_1 = 0.9$ and $\beta_2 = 0.999$. We report the mean and standard deviation of five runs.

Table 2. In-domain micro-F1 scores of the BiLSTM-CRF. We split mentions by novelty: *exact match* (EM), *partial match* (PM) and *new*. Average of 5 runs, subscript denotes standard deviation.

Embedding	Dim	CoNLL03				OntoNotes*				WNUT*		
		EM	PM	New	All	EM	PM	New	All	PM	New	All
BERT	4096	95.7_1	88.8_3	82.2_3	90.5_1	96.9_2	88.6_3	81.1_5	$\mathbf{93.5}_2$	$77.0_{4.6}$	53.9_9	$\mathbf{57.0}_{1.0}$
ELMo	1024	95.9_1	89.2_5	85.8_7	$\mathbf{91.8}_3$	97.1_2	88.0_2	79.9_7	$\mathbf{93.4}_2$	$67.7_{3.2}$	49.5_9	$52.1_{1.0}$
Flair	4096	95.4_1	88.1_6	83.5_5	90.6_2	96.7_1	85.8_5	75.0_6	92.1_2	64.9_7	$48.2_{2.0}$	$50.4_{1.8}$
ELMo[0]	1024	95.8_1	87.2_2	83.5_4	90.7_1	96.9_1	85.9_3	75.5_6	92.4_1	$72.8_{1.3}$	$45.4_{2.8}$	$49.1_{2.3}$
GloVe + char	350	95.3_3	85.5_7	83.1_7	89.9_5	96.3_1	83.3_2	69.9_6	91.0_1	$63.2_{4.6}$	$33.4_{1.5}$	$38.0_{1.7}$
GloVe	300	95.1_4	85.3_5	81.1_5	89.3_4	96.2_2	82.9_2	63.8_5	90.4_2	$59.1_{2.9}$	$28.1_{1.5}$	$32.9_{1.2}$

4.1 General Observations

ELMo, BERT and Flair. Drawing conclusions from the comparison of ELMo, BERT and Flair is difficult because there is no clear hierarchy across datasets and they differ in dimensions, tokenization, contextualization levels and pretraining corpora. However, although BERT is particularly effective on the WNUT dataset in-domain, probably due to its subword tokenization, ELMo yields the most stable results in and out-of-domain.

Furthermore, Flair globally underperforms ELMo and BERT, particularly for unseen mentions and out-of-domain. This suggests that LM pretraining at a lexical level (word or subword) is more robust for generalization than at a character level. In fact, Flair only beats the non contextual ELMo[0] baseline with Map-CRF which indicates that character-level contextualization is less beneficial than word-level contextualization with character-level representations.

ELMo[0] Vs GloVe + char. Overall, ELMo[0] outperforms the GloVe + char baseline, particularly on unseen mentions, out-of-domain and on WNUT*. The main difference is the incorporation of morphological features: in ELMo[0] they are learned jointly with the LM on a huge dataset whereas the char-BiLSTM is only trained on the source NER training set. Yet, morphology is crucial to represent words never encountered during pretraining and in WNUT* around 20% of words in test mentions are out of GloVe vocabulary against 5% in CoNLL03 and 3% in OntoNotes*. This explains the poor performance of GloVe baselines on WNUT*, all the more out-of-domain, and why a model trained on CoNLL03 with ELMo outperforms one trained on WNUT* with GloVe + char. Thus, ELMo's improvement over previous state-of-the-art does not only stem from contextualization but also an effective non-contextual word representation.

Seen Mentions Bias. In every configuration, $F1_{exact} > F1_{partial} > F1_{new}$ with more than 10 points difference. This gap is wider out-of-domain where the context differs more from training data than in-domain. NER models thus poorly generalize to unseen mentions, and datasets with high lexical overlap only encourage this behavior. However, this generalization gap is reduced by two types of contextualization described hereafter.

Table 3. Micro-F1 scores of models trained on CoNLL03 and tested in-domain and out-of-domain on OntoNotes* and WNUT*. Average of 5 runs, subscript denotes standard deviation.

	Emb	CoNLL03				OntoNotes*				WNUT*			
		EM	PM	New	All	EM	PM	New	All	EM	PM	New	All
BiLSTM-CRF	BERT	95.7_1	88.8_3	82.2_3	90.5_1	95.1_1	82.9_5	73.5_4	$\mathbf{85.0_3}$	57.4_{10}	56.3_{12}	32.4_8	37.6_8
	ELMo	95.9_1	89.2_5	85.8_7	$\mathbf{91.8_3}$	94.3_1	79.2_2	72.4_4	83.4_2	55.8_{12}	52.7_{11}	36.5_{15}	$\mathbf{41.0_{12}}$
	Flair	95.4_1	88.1_6	83.5_5	90.6_2	94.0_3	76.1_{11}	62.1_5	79.0_5	56.2_{22}	49.4_{34}	29.1_{33}	34.9_{29}
	ELMo[0]	95.8_1	87.2_2	83.5_4	90.7_1	93.6_1	76.8_6	66.1_3	80.5_2	52.3_{12}	50.8_{15}	32.6_{22}	37.6_{18}
	G + char	95.3_3	85.5_7	83.1_7	89.9_5	93.9_2	73.9_{11}	60.4_7	77.9_5	55.9_8	46.8_{18}	19.6_{16}	27.2_{13}
	GloVe	95.1_4	85.3_5	81.1_5	89.3_4	93.7_2	73.0_{12}	57.4_{18}	76.9_9	53.9_{12}	46.3_{15}	13.3_{14}	27.1_{10}
Map-CRF	BERT	93.2_3	85.8_4	73.7_8	86.2_4	93.5_2	77.8_5	67.8_9	80.9_4	57.4_3	53.5_{26}	33.9_6	38.4_4
	ELMo	93.7_2	87.2_6	80.1_3	$\mathbf{88.7_2}$	93.6_1	79.1_5	69.5_4	$\mathbf{82.2_3}$	61.1_7	53.0_9	37.5_7	$\mathbf{42.4_6}$
	Flair	94.3_1	85.1_3	78.6_3	88.1_{03}	93.2_1	74.0_3	59.6_2	77.5_2	52.5_{12}	50.6_4	28.8_5	33.7_5
	ELMo[0]	92.2_3	80.5_1	68.6_4	83.4_4	91.6_4	69.6_{10}	56.8_{15}	75.0_{10}	51.9_{11}	42.6_9	32.4_3	35.8_4
	G + char	93.1_3	80.7_9	69.8_7	84.4_4	91.8_3	69.3_3	55.6_{11}	74.8_5	50.6_9	42.5_{14}	20.6_{28}	28.7_{25}
	GloVe	92.2_1	77.0_4	61.7_3	81.5_{05}	89.6_3	62.8_6	38.5_4	68.1_4	46.8_8	41.3_5	3.2_2	18.9_7

4.2 LM and NER Contextualizations

The ELMo[0] and Map-CRF baselines enable to strictly distinguish contextualization due to LM pretraining (C_{LM}: ELMo[0] to ELMo) from task supervised contextualization induced by the BiLSTM network (C_{NER}: Map to BiLSTM). In both cases, a BiLSTM incorporates syntactic information which improves generalization to unseen mentions for which context is decisive, as shown in Table 3.

Comparison. However, because C_{NER} is specific to the source dataset, it is more effective in-domain whereas C_{LM} is particularly helpful out-of-domain. In the latter setting, the benefits from C_{LM} even surpass those from C_{NER}, specifically on domains further from source data such as web text in OntoNotes* (see Table 4) or WNUT*. This is again explained by the difference in quantity and quality of the corpora on which these contextualizations are learned. The much larger and more generic unlabeled corpora on which LM are pretrained lead to contextual representations more robust to domain adaptation than C_{NER} learned on a small and specific NER corpus.

Similar behaviors can be observed when comparing BERT and Flair to the GloVe baselines, although we cannot separate the effects of representation and contextualization.

Complementarity. Both in-domain and out-of-domain on OntoNotes*, the two types of contextualization transfer complementary syntactic features leading to the best configuration. However, in the most difficult case of zero-shot domain adaptation from CoNLL03 to WNUT*, C_{NER} is detrimental with ELMo and BERT. This is probably due to the specificity of the target domain, excessively different from source data.

Table 4. Per-genre micro-F1 scores of the BiLSTM-CRF model trained on CoNLL03 and tested on OntoNotes* (broadcast conversation, broadcast news, news wire, magazine, telephone conversation and web text). C_{LM} mostly benefits genres furthest from the news source domain.

	bc	bn	nw	mz	tc	wb	All
BERT	87.2_5	88.4_4	84.7_2	82.4_{12}	84.5_{11}	79.5_{10}	$\mathbf{85.0}_3$
ELMo	85.0_6	88.6_3	82.9_3	78.1_7	84.0_8	79.9_5	83.4_2
Flair	78.0_{11}	86.5_4	80.4_6	71.1_4	73.5_{18}	72.1_8	79.0_5
ELMo[0]	82.6_5	88.0_3	79.6_5	73.4_6	79.2_{12}	75.1_3	80.5_2
GloVe + char	80.4_8	86.3_4	77.0_{10}	70.7_4	79.7_{18}	69.2_8	77.9_5

5   Related Work

Augenstein et al. [2] perform a quantitative study of two CRF-based models and a CNN with classical word embeddings [4] over seven NER datasets including CoNLL03 and OntoNotes. They separate performance on seen (*exact match*) and unseen mentions and show a drop in F1 on unseen mentions and out-of-domain. Although comprehensive in experiments, this analysis is limited to models dating back from 2005 to 2011. We use a similar experimental setting to draw new insights on state-of-the art architectures and word representations. We limit to the two main English NER benchmarks as well as WNUT which was specifically designed to tackle this generalization problem in the Twitter domain. These three datasets cover all the domains studied in [2].

Moosavi and Strube raise a similar lexical overlap issue in Coreference Resolution on the CoNLL2012 dataset. They first show that for out-of-domain evaluation the performance gap between Deep Learning models and a rule-based system fades away [11]. They then add linguistic features (such as gender, NER, POS...) to improve out-of-domain generalization [12]. Nevertheless, such features are obtained using models in turn based on lexical features and at least for NER the same lexical overlap issue arises.

Finally, Pires et al. [15] concurrently evaluate the cross-lingual generalization capability of Multilingual BERT for NER and POS tagging. Our work on mono-lingual generalization to unseen mentions and domains naturally complements this study.

6   Conclusion

NER benchmarks are biased towards seen mentions, at the opposite of real-life applications. Hence the necessity to disentangle performance on seen and unseen mentions and test out-of-domain. In such setting, we show that contextualization from LM pretraining is particularly beneficial for generalization to unseen mentions, all the more out-of-domain where it surpasses supervised contextualization.

Despite this improvement, unseen mentions detection remains challenging and further work could explore attention or regularization mechanisms to better incorporate context and improve generalization. Furthermore, we can investigate how to best incorporate target data to improve this LM pretraining zero-shot domain adaptation baseline.

References

1. Akbik, A., Blythe, D., Vollgraf, R.: Contextual string embeddings for sequence labeling. In: Proceedings of the 27th International Conference on Computational Linguistics, pp. 1638–1649 (2018). http://www.aclweb.org/anthology/C18-1139
2. Augenstein, I., Derczynski, L., Bontcheva, K.: Generalisation in named entity recognition: a quantitative analysis. Comput. Speech Lang. **44**, 61–83 (2017). https://linkinghub.elsevier.com/retrieve/pii/S088523081630002X

3. Chelba, C., et al.: One Billion Word Benchmark for Measuring Progress in Statistical Language Modeling. arXiv preprint arXiv:1312.3005 (2013). https://arxiv.org/pdf/1312.3005.pdf
4. Collobert, R., Weston, J.: Natural language processing (almost) from scratch. J. Mach. Learn. Res. **12**, 2493–2537 (2011). http://www.jmlr.org/papers/volume12/collobert11a/collobert11a.pdf
5. Derczynski, L., Nichols, E., Van Erp, M., Limsopatham, N.: Results of the WNUT2017 shared task on novel and emerging entity recognition. In: 3rd Workshop on Noisy User-generated Text, pp. 140–147 (2017). https://www.aclweb.org/anthology/W17-4418
6. Devlin, J., Chang, M.W., Lee, K., Toutanova, K.: BERT: pre-training of deep bidirectional transformers for language understanding. In: Proceedings of the 2019 Conference of the North American Chapter of the Association for Computational Linguistics: Human Language Technologies, vol. 1 (Long and Short Papers), pp. 4171–4186 (2019). https://www.aclweb.org/anthology/N19-1423
7. Howard, J., Ruder, S.: Universal language model fine-tuning for text classification. In: Proceedings of the 56th Annual Meeting of the Association for Computational Linguistics, pp. 328–339 (2018). http://aclweb.org/anthology/P18-1031
8. Huang, Z., Xu, W., Yu, K.: Bidirectional LSTM-CRF Models for Sequence Tagging. arXiv preprint arXiv:1508.01991 (2015). https://arxiv.org/pdf/1508.01991.pdf
9. Lample, G., Ballesteros, M., Subramanian, S., Kawakami, K., Dyer, C.: Neural architectures for named entity recognition. In: Proceedings of NAACL-HLT 2016, pp. 260–270 (2016). https://www.aclweb.org/anthology/N16-1030
10. Lee, J.Y., Dernoncourt, F., Szolovits, P.: Transfer learning for named-entity recognition with neural networks. In: Proceedings of the Eleventh International Conference on Language Resources and Evaluation (LREC-2018), pp. 4470–4473 (2018). http://www.lrec-conf.org/proceedings/lrec2018/pdf/878.pdf
11. Moosavi, N.S., Strube, M.: Lexical features in coreference resolution: to be used with caution. In: Proceedings of the 55th Annual Meeting of the Association for Computational Linguistics, pp. 14–19 (2017). https://www.aclweb.org/anthology/P17-2003
12. Moosavi, N.S., Strube, M.: Using linguistic features to improve the generalization capability of neural coreference resolvers. In: Proceedings of the 2018 Conference on Empirical Methods in Natural Language Processing, pp. 193–203 (2018). http://aclweb.org/anthology/D18-1018
13. Pennington, J., Socher, R., Manning, C.D.: GloVe: global vectors for word representation. In: EMNLP 2014 (2014). https://nlp.stanford.edu/pubs/glove.pdf
14. Peters, M.E., et al.: Deep contextualized word representations. In: Proceedings of the 2018 Conference of the North American Chapter of the Association for Computational Linguistics: Human Language Technologies, vol. 1 (Long Papers), pp. 2227–2237 (2018). http://www.aclweb.org/anthology/N18-1202
15. Pires, T., Schlinger, E., Garrette, D.: How multilingual is Multilingual BERT? In: Proceedings of the 57th Annual Meeting of the Association for Computational Linguistics, pp. 4996–5001 (2019). https://www.aclweb.org/anthology/P19-1493
16. Radford, A., Salimans, T., Narasimhan, K., Salimans, T., Sutskever, I.: Improving Language Understanding by Generative Pre-Training, p. 12 (2018). https://s3-us-west-2.amazonaws.com/openai-assets/research-covers/language-unsupervised/language_understanding_paper.pdf
17. Rodriguez, J.D., Caldwell, A., Liu, A.: Transfer learning for entity recognition of novel classes. In: Proceedings of the 27th International Conference on Computational Linguistics, pp. 1974–1985 (2018). http://aclweb.org/anthology/C18-1168

18. Sang, E.F.T.K., De Meulder, F.: Introduction to the CoNLL-2003 shared task. In: Proceedings of the Seventh Conference on Natural Language Learning at NAACL-HLT 2003, vol. 4, pp. 142–147 (2003). https://www.aclweb.org/anthology/W03-0419

19. Vaswani, A., et al.: Attention is all you need. In: Advances in Neural Information Processing Systems. pp. 5998–6008 (2017), https://papers.nips.cc/paper/7181-attention-is-all-you-need.pdf

20. Weischedel, R., et al.: OntoNotes Release 5.0 LDC2013T19. In: Linguistic Data Consortium, Philadelphia, PA (2013). https://catalog.ldc.upenn.edu/docs/LDC2013T19/OntoNotes-Release-5.0.pdf

21. Wu, Y., et al.: Google's Neural Machine Translation System: Bridging the Gap between Human and Machine Translation. arXiv preprint arXiv:1609.08144 (2016). https://arxiv.org/pdf/1609.08144.pdf

22. Zhu, Y., et al.: Aligning books and movies: towards story-like visual explanations by watching movies and reading books. In: Proceedings of the IEEE International Conference on Computer Vision, pp. 19–27 (2015). https://www.cv-foundation.org/openaccess/content_iccv_2015/papers/Zhu_Aligning_Books_and_ICCV_2015_paper.pdf

DAKE: Document-Level Attention for Keyphrase Extraction

Tokala Yaswanth Sri Sai Santosh[1], Debarshi Kumar Sanyal[2(✉)],
Plaban Kumar Bhowmick[3], and Partha Pratim Das[1]

[1] Department of Computer Science and Engineering, IIT Kharagpur,
Kharagpur 721302, India
santoshtyss@gmail.com, ppd@cse.iitkgp.ac.in
[2] National Digital Library of India, IIT Kharagpur, Kharagpur 721302, India
debarshisanyal@gmail.com
[3] Centre for Educational Technology, IIT Kharagpur, Kharagpur 721302, India
plaban@cet.iitkgp.ac.in

Abstract. Keyphrases provide a concise representation of the topical content of a document and they are helpful in various downstream tasks. Previous approaches for keyphrase extraction model it as a sequence labelling task and use local contextual information to understand the semantics of the input text but they fail when the local context is ambiguous or unclear. We present a new framework to improve keyphrase extraction by utilizing additional supporting contextual information. We retrieve this additional information from other sentences within the same document. To this end, we propose Document-level Attention for Keyphrase Extraction (DAKE), which comprises Bidirectional Long Short-Term Memory networks that capture hidden semantics in text, a document-level attention mechanism to incorporate document level contextual information, gating mechanisms which help to determine the influence of additional contextual information on the fusion with local contextual information, and Conditional Random Fields which capture output label dependencies. Our experimental results on a dataset of research papers show that the proposed model outperforms previous state-of-the-art approaches for keyphrase extraction.

Keywords: Keyphrase extraction · Sequence labelling · LSTM · Document-level attention

1 Introduction

Keyphrase extraction is the task of automatically extracting words or phrases from a text, which concisely represent the essence of the text. Because of the succinct expression, keyphrases are widely used in many tasks like document retrieval [13,25], document categorization [9,12], opinion mining [3] and summarization [24,31]. Figure 1 shows an example of a title and the abstract of a research paper along with the author-specified keyphrases highlighted in bold.

© Springer Nature Switzerland AG 2020
J. M. Jose et al. (Eds.): ECIR 2020, LNCS 12036, pp. 392–401, 2020.
https://doi.org/10.1007/978-3-030-45442-5_49

Present methods for keyphrase extraction follow a two-step procedure where they select important phrases from the document as potential keyphrase candidates by heuristic rules [18,28,29] and then the extracted candidate phrases are ranked either by unsupervised approaches [17,21,27] or supervised approaches [18,22,29]. Unsupervised approaches score those candidate phrases based on individual words comprising the candidate phrases. They utilize various scoring measures based on the informativeness of the word with respect to the whole document [10]. Other paradigms utilize graph-based ranking algorithms wherein each word in the document is mapped to a node in the graph and the connecting edges in the graph represent the association patterns among the words in the document. Then, the scores of the individual words are estimated using various graph centrality measures [6,21,27]. On the other hand, supervised approaches [4,14] use binary classification to label the extracted candidate phrases as keyphrases or non-keyphrases, based on various features such as, tf-idf, part-of-speech (POS) tags, and the position of phrases in the document. The major limitation of these supervised approaches is that they classify the labels of each candidate phrase independently without taking into account the dependencies that could potentially exist between neighbouring labels and they also ignore the semantic meaning of the text. To overcome the above stated limitation, [8] formulated keyphrase extraction as a sequence labeling task and used linear-chain Conditional Random Fields for this task. However, this approach does not explicitly take into account the long-term dependencies and semantics of the text. More recently, to capture both the semantics of the text as well as the dependencies among the labels of neighboring words [1] used a deep learning-based approach called BiLSTM-CRF which combines a bi-directional Long Short-Term Memory (BiLSTM) layer that models the sequential input text with a Conditional Random Field (CRF) layer that captures the dependencies in the output.

Title: **DCE-MRI** data analysis for cancer area classification.
Abstract: The paper aims at improving the support of medical researchers in the context of in-vivo cancer imaging. [..] The proposed approach is based on a three-step procedure: i) robust feature extraction from raw time-intensity curves, ii) voxel segmentation, and iii) voxel **classification** based on a learning-by-example approach. Finally, in the third step, a support vector machine (**SVM**) is trained to classify voxels according to the labels obtained by the clustering phase. [..]

Fig. 1. An example of keyphrase extraction with author-specified keyphrases highlighted in bold.

The above mentioned approaches treat keyphrase extraction as a sentence-level task where sentences in the same document are viewed as independent. When labeling a word, local contextual information from the surrounding words is crucial because the context gives insight to the semantic meaning of the word. However, there are many instances in which the local context is ambiguous or lacks sufficient information. If the model has access to supporting information that provides additional context, the model may use this additional supporting information to predict the label correctly. Such additional supporting information may be found from other sentences in the same document from which the query sentence is taken. To utilize this additional supporting information, we propose a document-level attention mechanism inspired from [20,30]; it dynamically weights the additional supporting information emphasizing the most relevant information from each supporting sentence with respect to the local context. But leveraging this additional supporting information has a downside of introducing noise into the representations. To alleviate this problem, we use a gating mechanism [20,30] that balances the influence of the local contextual representations and the additional supporting information from the document-level contextual representations.

To this end, in this paper, we propose Document-level Attention for Keyphrase Extraction (DAKE). It initially produces representations for each word that encode the local context from the query sentence using BiLSTM, then uses a document-level attention mechanism to incorporate the most relevant information from each supporting information with respect to the local context, and employs a gating mechanism to filter out the irrelevant information. Finally, it uses a CRF layer which captures output label dependencies to decode the gated local and the document-level contextual representations to predict the label. The main contributions of this paper are as follows:

- We propose DAKE, a BiLSTM-CRF model augmented with document-level attention and a gating mechanism for improved keyword extraction from research papers.
- Experimental results on a dataset of research papers show that DAKE outperforms previous state-of-the-art approaches.

2   Problem Formulation

We formally describe the keyphrase extraction task as follows: Given a sentence, $s = \{w_1, w_2, \ldots, w_n\}$ where n is the length of the sentence, predict the labels sequence $y = \{y_1, y_2, \ldots, y_n\}$ where y_i is the label corresponding to word w_i and it can KP (keyphrase word) or Not-KP (not a keyphrase word). Every longest sequence of KP words in a sentence is a keyphrase.

3   Proposed Method

The main components in our proposed architecture, DAKE, are: Word Embedding Layer, Sentence Encoding Layer, Document-level Attention mechanism, Gating mechanism, Context Augmenting Layer and Label Sequence Prediction Layer. The first layer produces word embeddings of the sentence from which the second layer generates word representations that encode the local context from the query sentence. Then the document-level attention mechanism extracts supporting information from other sentences in the document to enrich the current word representation. Subsequently, we utilize a gating mechanism to filter out the irrelevant information from each word representation. The next layer fuses the local and the global contexts into each word representation. Finally, we feed these word representations into the CRF layer which acts as a decoder to predict the label, KP or Not-KP, associated with each word. The model is trained in an end-to-end fashion.

3.1   Word Embedding Layer

Given a document $D = \{s_1, s_2, \ldots, s_m\}$ of m sentences, where a sentence $s_i = \{w_{i1}, w_{i2}, \ldots, w_{in}\}$ is a sequence of n words, we transform each word w_{ij} in the sentence s_i into a vector $\mathbf{x}_{ij}$ using pre-trained word embeddings.

3.2   Sentence Encoding Layer

We use a BiLSTM [11] to obtain the hidden representation H_i of the sentence s_i. A BiLSTM comprises a forward-LSTM which reads the input sequence in the original direction and a backward-LSTM which reads it in the opposite direction. We apply forward-LSTM on the sentence $s_i = (\mathbf{x}_{i1}, \mathbf{x}_{i2}, \ldots, \mathbf{x}_{in})$ to obtain $\overrightarrow{H_i} = (\overrightarrow{\mathbf{h}_{i1}}, \overrightarrow{\mathbf{h}_{i2}}, \ldots, \overrightarrow{\mathbf{h}_{in}})$. The backward-LSTM on s_i produces $\overleftarrow{H_i} = (\overleftarrow{\mathbf{h}_{i1}}, \overleftarrow{\mathbf{h}_{i2}}, \ldots, \overleftarrow{\mathbf{h}_{in}})$. We concatenate the outputs of the forward and the backward LSTMs to obtain the local contextual representation $H_i = \{\mathbf{h}_{i1}, \mathbf{h}_{i2}, \ldots, \mathbf{h}_{in}\}$ where $\mathbf{h}_{ij} = [\overrightarrow{\mathbf{h}_{ij}} : \overleftarrow{\mathbf{h}_{ij}}]$; here, : denotes concatenation operation. Succinctly, $\mathbf{h}_{ij} = \text{BiLSTM}(\mathbf{x}_{ij})$

3.3   Document-Level Attention

Many keyphrase mentions are tagged incorrectly in current approaches including the BiLSTM-CRF model [1] due to ambiguous contexts present in the input sentence. In cases where a sentence is short or highly ambiguous, the model may either fail to identify keyphrases due to insufficient information or make wrong predictions by using noisy context. We hypothesize that this limitation can be alleviated using additional supporting information from other sentences within the same document. To extract this global context, we need vector representations of other sentences in the same document D. We utilize BERT [5] as a sentence encoder to obtain representations for the sentences in D. Given an input sentence s_l in D, we extract the final hidden state of the [CLS] token as the representation $\mathbf{h}_l'$ of the sentence, where [CLS] is the special classification embedding in BERT. Then, for each word, w_{ij} in the input sentence s_i, we apply an attention mechanism to weight the supporting sentences in D as follows

$$e_{ij}^l = \mathbf{v}^\top \tanh(W_1 \mathbf{h}_{ij} + W_2 \mathbf{h}_l' + \mathbf{b}_1) \tag{1}$$

$$\alpha_{ij}^l = \frac{\exp(e_{ij}^l)}{\sum_{p=1}^m \exp(e_{ij}^p)} \tag{2}$$

where W_1, W_2 are trainable weight matrices and $\mathbf{b}_1$ is a trainable bias vector. We compute the final representation of supporting information as $\tilde{\mathbf{h}}_{ij} = \sum_{l=1}^{m} \alpha_{ij}^l \mathbf{h}_l'$. For each word w_{ij}, $\tilde{\mathbf{h}}_{ij}$ captures the document-level supporting evidence with regard to w_{ij}.

3.4   Gating Mechanism

Though the above supporting information from the entire document is valuable to the prediction, we must mitigate the influence of the distant supporting information as the prediction should be made primarily based on the local context. Therefore, we apply a gating mechanism to constrain this influence and enable the model to decide the amount of the supporting information that should be incorporated in the model, which is given as follows:

$$\mathbf{r}_{ij} = \sigma(W_3\tilde{\mathbf{h}}_{ij} + W_4\mathbf{h}_{ij} + \mathbf{b}_2) \tag{3}$$

$$\mathbf{z}_{ij} = \sigma(W_5\tilde{\mathbf{h}}_{ij} + W_6\mathbf{h}_{ij} + \mathbf{b}_3) \tag{4}$$

$$\mathbf{g}_{ij} = \tanh(W_7\tilde{\mathbf{h}}_{ij} + \mathbf{z}_{ij} \odot (W_8\tilde{\mathbf{h}}_{ij} + \mathbf{b}_4)) \tag{5}$$

$$\mathbf{d}_{ij} = \mathbf{r}_{ij} \odot \mathbf{h}_{ij} + (1 - \mathbf{r}_{ij}) \odot \mathbf{g}_{ij} \tag{6}$$

where $\odot$ denotes Hadamard product and $W_3, W_4, W_5, W_6, W_7, W_8$ are trainable weight matrices and $\mathbf{b}_2, \mathbf{b}_3, \mathbf{b}_4$ are trainable bias vectors. $\mathbf{d}_{ij}$ is the representation for the gated supporting evidence for w_{ij}.

3.5   Context Augmenting Layer

For each word w_{ij} of sentence s_i, we concatenate its local contextual representation $\mathbf{h}_{ij}$ and gated document-level supporting contextual representation $\mathbf{d}_{ij}$ to obtain its final representation $\mathbf{a}_{ij} = [\mathbf{h}_{ij} : \mathbf{d}_{ij}]$, where : denotes concatenation operation. These final representations $A_i = \{\mathbf{a}_{i1}, \mathbf{a}_{i2}, \ldots, \mathbf{a}_{in}\}$ of sentence s_i are fed to another BiLSTM to further encode the local contextual features along with supporting contextual information into unified representations $C_i = \{\mathbf{c}_{i1}, \mathbf{c}_{i2}, \ldots, \mathbf{c}_{in}\}$ where $\mathbf{c}_{ij} = \text{BiLSTM}(\mathbf{a}_{ij})$. The output of this encoding captures the interaction among the context words conditioned on the supporting information. This is different from the initial encoding layer, which captures the interaction among words of the sentence independent of the supporting information.

3.6   Label Sequence Prediction Layer

The obtained contextual representations C_i of query sentence s_i are given as input sequence to a CRF layer [16] that produces a probability distribution over the output label sequence using the dependencies among the labels of the entire input sequence. In order to efficiently find the best sequence of labels for an input sentence, the Viterbi algorithm [7] is used.

4 Experiments

4.1 Dataset

We use the dataset from [19] which comprises metadata of papers from several online digital libraries. The dataset contains metadata for 567,830 papers with a clear split as train, validation, and test sets provided by the authors, as follows: 527,830 were used for model training, 20,000 were used for validation and the rest 20,000 were used for testing. We refer to these sets as kp527k, kp20k-v and kp20k respectively. The metadata of each paper consists of title, abstract, and author-assigned keyphrases. The title and abstract of each paper are used to extract keyphrases, whereas the author-input keyphrases are used as gold-standard for evaluation.

4.2 Baselines and Evaluation Metrics

We compare our approach, DAKE with the following baselines: Bi-LSTM-CRF [1], CRF [8], Bi-LSTM [1], copy-RNN [19], KEA [29], Tf-Idf, TextRank [21] and SingleRank [27]. We also carry out an ablation test to understand the effectiveness of document-level attention and gating mechanism components by removing them. Similar to previous works, we evaluate the predictions of each method against the author-specified keyphrases that can be located in the corresponding paper abstracts in the dataset ("gold standard"). We present results for all our experiments using the precision, recall, and F1-score measures. For comparison of the methods, we choose the F1-score, which is the harmonic mean of precision and recall.

4.3 Implementation Details

We use pre-trained word embedding vectors obtained using GloVe [23]. We use SciBERT [2], a BERT model trained on scientific text for the sentence encoder. For word representations, we use 300-dimensional pre-trained word embeddings and for sentence encoder, we use 768 dimensional representation obtained using SciBERT. The hidden state of the LSTM is set to 300 dimensions. The model is trained end-to-end using the Adam optimization method [15]. The learning rate is initially set as 0.001 and decayed by 0.5 after each epoch. For regularization to avoid over-fitting, dropout [26] is applied to each layer. We select the model with the best F1-score on the validation set, kp20k-v.

5 Results and Discussion

Table 1a shows the results of our approach in comparison to various baselines. Our approach, DAKE outperforms all baselines in terms of the F1-score. Tf-Idf, TextRank and SingleRank are unsupervised extractive approaches while KEA, Bi-LSTM-CRF, CRF, Bi-LSTM follow supervised extractive approach.

copyRNN is a recently proposed generative model based on sequence-to-sequence learning along with a copying mechanism. For the unsupervised models and the sequence-to-sequence learning model, we report the performance at top-5 predicted keyphrases since top-5 showed highest performance in the previous works for these models. From Table 1a, we observe that the deep learning-based approaches perform better than the traditional feature-based approaches. This indicates the importance of understanding the semantics of the text for keyphrase extraction. BiLSTM-CRF yields better results in terms of the F1-score over CRF (improvement of F1-score by 18.17% from 17.46% to 35.63%) and BiLSTM (improvement of F1-score by 18.88% from 16.75% to 35.63%) models alone. This result indicates that the combination of BiLSTM, which is powerful in capturing the semantics of the textual content, with CRF, which captures the dependencies among the output labels, helped boost the performance in identifying keyphrases. Our proposed method, DAKE outperforms the BiLSTM-CRF (improvement of F1-score by 6.67% from 35.63% to 42.30%) approach, which indicates that the incorporation of additional contextual information from other sentences in the document into the BiLSTM-CRF model helps to further boost the performance.

Table 1. Performance analysis of DAKE

(a) Performance of different keyphrase extraction algorithms.

Method	Precision	Recall	F1-score
Tf-Idf	8.97	13.49	10.77
TextRank	15.29	23.01	18.37
SingleRank	8.42	12.70	10.14
KEA	15.14	22.78	18.19
copyRNN	27.71	41.79	33.29
CRF	66.67	10.04	17.46
BiLSTM	9.41	**76.24**	16.75
BiLSTM-CRF	64.19	24.66	35.63
DAKE	**68.21**	30.66	**42.30**

(b) Ablation Study: BiLSTM-CRF used as baseline.

Method	Precision	Recall	F1-score
DAKE without document-level attention	64.19	24.66	35.63
DAKE without gating mechanism	65.26	25.31	36.47
DAKE without context augmenting layer	66.74	26.45	38.09
DAKE without CRF layer	61.38	28.81	39.21
DAKE	**68.21**	**30.66**	**42.30**

Table 1b shows the ablation study. We observe that document-level attention increases the F1-score of the baseline BiLSTM-CRF by 0.84% (from 35.63% to 36.47%). This validates our hypothesis that additional supporting information boosts the performance for keyphrase extraction. But leveraging this additional supporting information has a downside of introducing noise into the representations, and to alleviate this, we used a gating mechanism which boosted the F1-score by 1.62% (from 36.47% to 38.09%). Document-level attention did not show great improvement when it has only one layer of BiSLTM because the final tagging predictions mainly depend on the local context of each word while additional context only supplements extra information. Therefore, our model needs another layer of BiLSTM to encode the sequential intermediate vectors containing additional context and local context, as evidenced from our F1-score improvement by 4.21% (from 38.09% to 42.30%). When CRF is removed from DAKE, the F1-score falls by 3.09%, showing that CRF successfully captures the output label dependencies.

6 Conclusion and Future Work

We proposed an architecture, DAKE, for keyword extraction from documents. It uses a BiLSTM-CRF network enhanced with a document-level attention mechanism to incorporate contextual information from the entire document, and gating mechanisms to balance between the global and the local contexts. It outperforms existing keyphrase extraction methods on a dataset of research papers. In future, we would like to integrate the relationships between documents such as those available from a citation network by enhancing our approach with contexts in which the document is referenced within a citation network.

Acknowledgements. This work is supported by *National Digital Library of India* Project sponsored by Ministry of Human Resource Development, Government of India at IIT Kharagpur.

References

1. Alzaidy, R., Caragea, C., Giles, C.L.: Bi-LSTM-CRF sequence labeling for keyphrase extraction from scholarly documents. In: Proceedings of The World Wide Web Conference, pp. 2551–2557. ACM (2019)
2. Beltagy, I., Cohan, A., Lo, K.: Scibert: pretrained contextualized embeddings for scientific text. arXiv preprint arXiv:1903.10676 (2019)
3. Berend, G.: Opinion expression mining by exploiting keyphrase extraction. In: Proceedings of the 5th International Joint Conference on Natural Language Processing. Asian Federation of Natural Language Processing (2011)
4. Caragea, C., Bulgarov, F.A., Godea, A., Gollapalli, S.D.: Citation-enhanced keyphrase extraction from research papers: a supervised approach. In: Proceedings of the 2014 Conference on Empirical Methods in Natural Language Processing, pp. 1435–1446 (2014)
5. Devlin, J., Chang, M.W., Lee, K., Toutanova, K.: Bert: pre-training of deep bidirectional transformers for language understanding. arXiv preprint arXiv:1810.04805 (2018)
6. Florescu, C., Caragea, C.: PositionRank: an unsupervised approach to keyphrase extraction from scholarly documents. In: Proceedings of the 55th Annual Meeting of the Association for Computational Linguistics, vol. 1: Long Papers, pp. 1105–1115 (2017)
7. Forney, G.D.: The Viterbi algorithm. Proc. IEEE **61**(3), 268–278 (1973)
8. Gollapalli, S.D., Li, X.L., Yang, P.: Incorporating expert knowledge into keyphrase extraction. In: Proceedings of the 31st AAAI Conference on Artificial Intelligence (2017)

9. Hammouda, K.M., Matute, D.N., Kamel, M.S.: CorePhrase: keyphrase extraction for document clustering. In: Perner, P., Imiya, A. (eds.) MLDM 2005. LNCS (LNAI), vol. 3587, pp. 265–274. Springer, Heidelberg (2005). https://doi.org/10.1007/11510888_26

10. Hasan, K.S., Ng, V.: Conundrums in unsupervised keyphrase extraction: making sense of the state-of-the-art. In: Proceedings of the 23rd International Conference on Computational Linguistics: Posters, pp. 365–373. Association for Computational Linguistics (2010)

11. Hochreiter, S., Schmidhuber, J.: Long short-term memory. Neural Comput. **9**(8), 1735–1780 (1997)

12. Hulth, A., Megyesi, B.B.: A study on automatically extracted keywords in text categorization. In: Proceedings of the 21st International Conference on Computational Linguistics and the 44th Annual Meeting of the Association for Computational Linguistics, pp. 537–544. Association for Computational Linguistics (2006)

13. Jones, S., Staveley, M.S.: Phrasier: a system for interactive document retrieval using keyphrases. In: Proceedings of the 22nd Annual International ACM SIGIR Conference on Research and Development in Information Retrieval, pp. 160–167. ACM (1999)

14. Kim, S.N., Medelyan, O., Kan, M.Y., Baldwin, T.: Automatic keyphrase extraction from scientific articles. Lang. Res. Eval. **47**(3), 723–742 (2013)

15. Kingma, D.P., Ba, J.: Adam: a method for stochastic optimization. arXiv preprint arXiv:1412.6980 (2014)

16. Lafferty, J., McCallum, A., Pereira, F.C.: Conditional random fields: probabilistic models for segmenting and labeling sequence data. In: Proceedings of the 18th International Conference on Machine Learning, pp. 282–289 (2001)

17. Le, T.T.N., Nguyen, M.L., Shimazu, A.: Unsupervised keyphrase extraction: introducing new kinds of words to keyphrases. In: Kang, B.H., Bai, Q. (eds.) AI 2016. LNCS (LNAI), vol. 9992, pp. 665–671. Springer, Cham (2016). https://doi.org/10.1007/978-3-319-50127-7_58

18. Medelyan, O., Frank, E., Witten, I.H.: Human-competitive tagging using automatic keyphrase extraction. In: Proceedings of the 2009 Conference on Empirical Methods in Natural Language Processing, vol. 3, pp. 1318–1327. Association for Computational Linguistics (2009)

19. Meng, R., Zhao, S., Han, S., He, D., Brusilovsky, P., Chi, Y.: Deep keyphrase generation. arXiv preprint arXiv:1704.06879 (2017)

20. Miculicich, L., Ram, D., Pappas, N., Henderson, J.: Document-level neural machine translation with hierarchical attention networks. arXiv preprint arXiv:1809.01576 (2018)

21. Mihalcea, R., Tarau, P.: Textrank: bringing order into text. In: Proceedings of the 2004 Conference on Empirical Methods in Natural Language Processing, pp. 404–411 (2004)

22. Nguyen, T.D., Kan, M.-Y.: Keyphrase extraction in scientific publications. In: Goh, D.H.-L., Cao, T.H., Sølvberg, I.T., Rasmussen, E. (eds.) ICADL 2007. LNCS, vol. 4822, pp. 317–326. Springer, Heidelberg (2007). https://doi.org/10.1007/978-3-540-77094-7_41

23. Pennington, J., Socher, R., Manning, C.: GloVe: global vectors for word representation. In: Proceedings of the 2014 Conference on Empirical Methods in Natural Language Processing, pp. 1532–1543 (2014)

24. Qazvinian, V., Radev, D.R., Ozgur, A.: Citation summarization through keyphrase extraction. In: Proceedings of the 23rd International Conference on Computational Linguistics (COLING 2010), pp. 895–903 (2010)

25. Sanyal, D.K., Bhowmick, P.K., Das, P.P., Chattopadhyay, S., Santosh, T.Y.S.S.: Enhancing access to scholarly publications with surrogate resources. Scientometrics **121**(2), 1129–1164 (2019). https://doi.org/10.1007/s11192-019-03227-4

26. Srivastava, N., Hinton, G., Krizhevsky, A., Sutskever, I., Salakhutdinov, R.: Dropout: a simple way to prevent neural networks from overfitting. J. Mach. Learn. Res. **15**(1), 1929–1958 (2014)

27. Wan, X., Xiao, J.: Single document keyphrase extraction using neighborhood knowledge. In: AAAI, vol. 8, pp. 855–860 (2008)

28. Wang, M., Zhao, B., Huang, Y.: PTR: phrase-based topical ranking for automatic keyphrase extraction in scientific publications. In: Hirose, A., Ozawa, S., Doya, K., Ikeda, K., Lee, M., Liu, D. (eds.) ICONIP 2016. LNCS, vol. 9950, pp. 120–128. Springer, Cham (2016). https://doi.org/10.1007/978-3-319-46681-1_15

29. Witten, I.H., Paynter, G.W., Frank, E., Gutwin, C., Nevill-Manning, C.G.: KEA: practical automated keyphrase extraction. In: Design and Usability of Digital Libraries: Case Studies in the Asia Pacific, pp. 129–152. IGI Global (2005)

30. Zhang, B., Whitehead, S., Huang, L., Ji, H.: Global attention for name tagging. In: Proceedings of the 22nd Conference on Computational Natural Language Learning, pp. 86–96 (2018)

31. Zhang, Y., Zincir-Heywood, N., Milios, E.: World wide web site summarization. Web Intell. Agent Syst. Int. J. **2**(1), 39–53 (2004)

Understanding Depression
from Psycholinguistic Patterns
in Social Media Texts

Alina Trifan$^{(\boxtimes)}$ ⓘ, Rui Antunes ⓘ, Sérgio Matos ⓘ, and Jose Luís Oliveira ⓘ

IEETA/DETI, University of Aveiro, Aveiro, Portugal
`alina.trifan@ua.pt`

Abstract. The World Health Organization reports that half of all mental illnesses begin by the age of 14. Most of these cases go undetected and untreated. The expanding use of social media has the potential to leverage the early identification of mental health diseases. As data gathered via social media are already digital, they have the ability to power up faster automatic analysis. In this article we evaluate the impact that psycholinguistic patterns can have on a standard machine learning approach for classifying depressed users based on their writings in an online public forum. We combine psycholinguistic features in a rule-based estimator and we evaluate their impact on this classification problem, along with three other standard classifiers. Our results on the Reddit Self-reported Depression Diagnosis dataset outperform some previously reported works on the same dataset. They stand for the importance of extracting psychologically motivated features when processing social media texts with the purpose of studying mental health.

Keywords: Mental health · Depression · Social media · Machine learning · Psycholinguistic features

1 Introduction

Suicide ideation, anxiety and depression are some of the most spread mental health diseases among adolescents and young adults, with a little under 800 000 people dying by suicide each year [1,2]. Fortunately, in the recent years there has been an increasing acknowledgement of this reality and a better understanding of the importance of enabling young people to improve their mental resilience, from early stages on.

Communication is at the core of society and currently the written digital communication is one of the most popular forms of expressing ourselves. We use social networks to detail our activities or routines, to describe our feelings, mental states, hopes and desires [6]. Young adults suffering of mental illness are more likely to express themselves online, either through blogging, social networks or specific public forums [15]. As we write digitally more and more, these large

© Springer Nature Switzerland AG 2020
J. M. Jose et al. (Eds.): ECIR 2020, LNCS 12036, pp. 402–409, 2020.
https://doi.org/10.1007/978-3-030-45442-5_50

volumes of data can be processed automatically with the purpose of inferring relevant information about one's well-being, such as mental health status.

Prevention and early identification of mental health diseases by means that are complimentary to traditional medical approaches have the ability to mitigate the under-supply of mental health facilities by advancing different types of counseling or support for the ones in need, such as connecting a depressed person to resources or peer support when they most need it [10]. Using social data has yet another advantage with respect to the stigma associated to mental health screening, as such approaches can provide new opportunities for early detection and intervention and have the potential to open new insights on research of the causes and mechanisms of mental health [4,5].

In this work we address the challenge of identifying depressed users of the Reddit social platform. We present encouraging results that demonstrate that social data has the potential for complementing standard clinical procedures. We base our methodology on a combination of a tf-idf weighting scheme for bag of words features and a rule-based estimator, that takes into account several psycholinguistic features that characterize depressed users. Our goal is to assess the extent up to which standard classifiers take into consideration psycholinguistic patterns and if by specifically contemplating them in a classification pipeline we can obtain better results. The dataset in use was the one proposed by Yates et al. [25].

This paper is structured in 4 more sections. We present a background on the subject in Sect. 2, followed by the methods we have employed in Sect. 3. Detailed results are presented and discussed in Sect. 4. Finally, Sect. 5 concludes the paper.

2   Background

A large volume of written data in the form of accessible common language is available through social media. This attracted the attention of natural language processing researchers. Among them, those who study the language of individuals in relation to their mental health conditions. Social media data has been identified as an emerging opportunity for revolutionizing in-the-moment measures of a broad range of people's thoughts and feelings [16]. Several studies focusing on mental health understanding through social network data have been conducted using Twitter[1] texts. Coppersmith et al. [9] presented a method for gathering data for a range of mental illnesses along with proof-of-concept results that focus on the analysis of four mental disorders: post-traumatic stress disorder, depression, bipolar disorder, and seasonal affective disorder. Their ultimate goal was to enable the ethical discussion regarding the balance between the utility of such data and the privacy of mental health related information. Later on, Coppersmith et al. [10] released a Twitter dataset of users who have attempted suicide, matched by neurotypical control users. Language modeling techniques were employed to classify these users, along with open government

---

[1] https://twitter.com/.

data to identify quantifiable signals that can relate them to psychometrically validated concepts associated to suicide. Nadeem et al. [18] used the same dataset to predict Major Depressive Disorder among online personas based on a Bag of Words approach and several statistical classifiers. More recently, Vioulès et al. [24] combined natural language processing features with a martingale framework to detect Twitter posts containing suicide-related content. The results were comparable to traditional machine learning classifiers.

While the previous examples proved that even short texts from Twitter can provide some insight into the relation between language and mental health conditions, longer-form content is nowadays explored for further insight into this matter. Yates et al. [25] introduced a large Reddit[2] dataset of self-reported depressed users, matched by similar control users. The release of the dataset was coupled with some preliminary results on their classification. The same research group included exact temporal spans that relate to the date of diagnosis, in an attempt to show that this type of diagnosis is not static [17]. Another Reddit dataset recently released by Cohan et al. [8], based on which authors investigated extended self-diagnoses matching patterns derived from mental health-related synonyms with focus on nine different mental health conditions. De Choudhury et al. [13] applied a logistic regression classifier that led to high accuracy results in order to predict suicidal ideation in Reddit users. One of the first demonstration of suicidice risk assesment through Reddit posts, matched with clinical knowledge was very recently reported by Shing et al. [23].

Initiatives such as CLEF Early Risk[3] or CLPsych[4], just to name a few, emerged over the last years as a proof of the importance of this research interest. These projects fostered collaborative work on the topic of mental health and social data and push forward new discoveries and insights. As a practical outcome that these initiative encouraged, triaging content in online social networks or public forums enabled the identification of content that requires the attention of moderators to ensure that urgent content can be responded to more quickly and consistently. Over the last years, the focus of these shared tasks was the early identification of people with suicidal inclinations or people susceptible to depression.

3   Methods

The dataset used in this work is the one proposed by Yates et al. [25], publicly available based on a signed user agreement with emphasis on data protection and proper acknowledgements. In order to get an understanding on the impact of the previously described patterns of depressed users we experimented standard feature extraction methods, which we have complemented with the design of a rule-based estimator that solely relies on these psycholinguistic features. We have experimented several classifiers that have been identified in the literature

---

[2] https://www.reddit.com/.

[3] http://early.irlab.org/.

[4] http://clpsych.org/.

as appropriate for this classification task. All experiments were managed using the scikit-learn machine learning framework (release version 0.21) [20]. The texts in the dataset were curated for any direct link to mental health. We considered this curation relevant as it relates to the possibility of identifying people that are unaware of their mental health status through heterogeneous texts.

Dataset Description. The dataset consists of all Reddit users who made a post between January and October 2016, matching high-precision patterns of self-reported diagnosis (e.g. "I was diagnosed with depression"). The depressed users were matched by control users, who have never posted in a subreddit related to mental health and never used a term related to it. In order to avoid a straight-forward separation of the two groups, all posts of diagnosed users related to depression or mental health were removed. In the end, 9210 diagnosed users were matched by 107 274 control users. Each user in the dataset has an average of 969 posts (median 646) and the mean post length is 148 tokens (median 74).

Data Preprocessing. The preprocessing of the Reddit posts follows standard approaches in text classification. The posts are lowercased and tokenized, after removing all non-alphabetic characters. Stopwords are filtered, by using an altered version of the stopwords list of the Natural Language Toolkit[5]. The alteration consists of removing from the original stopwords list self-related words and words that belong to the list of absolutist words, as described next. We are not interested in discarding these words as they may convey valuable psycholinguistic content, as detailed in the following subsections.

3.1   Experiments

The dataset was originally split into similar size chunks of training, validation and test samples, each of them containing roughly posts of 39 000 users. Because of the large size of each of these chunks (7 GB), we explored both incremental and online training with the following three classifiers: Multinomial Naive Bayes (MNB), linear Support Vector Machine with Stochastic Gradient Descent (SGD) and Passive Aggressive (PA). For the out of core classification, we trained the classifiers with batches of 500 users data. The batch size is not expected to have an impact on the performance of the classifiers[6]. For each of these classifiers, we have performed a grid search over the validation dataset in order to identify the best parameters that characterize them.

The first approach followed a standard processing stream for text classification. We considered Bag of Words (BoW) features for the three classifiers and we applied counts and tf-idf based feature weighting. A further study into psycholinguistic literature revealed possible patterns in the language of depressed users, that we modelled as features of a rule-based estimator:

---

[5] https://www.nltk.org/.

[6] https://scikit-learn.org/stable/modules/computing.html.

Absolutist Words. A recent study on absolutist thinking, which is considered a cognitive distortion by most cognitive therapies for anxiety and depression, showed that anxiety, depression, and suicidal ideation forums contained more absolutist words than control forums [3]. The study, conducted as a text analysis of 63 Internet forums with over 6400 members resulted in a validation of an absolutist words dictionary, presented in Table 1. Their usage frequency was considered for the rule-based estimator.

Table 1. Absolutist words validated by Al-Mosaiwi et al. [3].

Absolutely	Constant	Every	Never
All	Constantly	Everyone	Nothing
Always	Definitely	Everything	Totally
Complete	Entire	Full	Whole
Completely	Ever	Must	

Analysis of Lexical Categories. Depressed users frequently use negative emotion words and anger words on social networks [12,19]. Empath [14] is a text categorization open-source software that analyzes text across 200 built-in, pre-validated categories that were generated from common topics in a web dataset. Empath's categories have been human validated and are highly correlated ($r = 0.906$) with similar categories in the Linguistic Inquiry and Word Count [21]. As depressed users tend to have an overall more negative connotation of their texts, we used Empath's lexical category (version release 0.41) of a user's overall set of posts for the rule based estimator.

Self-related Speech. Depressed users tend to use them more often self-related words (such as: I, me, myself, mine) [7,22].

Posts Length. Depressed and suicidal people tend to write more words than control users [11]. We consider this information relevant for an heuristic that takes into consideration the number of tokens of a user for the rule-based estimator.

4 Results

The statistical analysis of the training dataset revealed that on average, depressed users have 770 mentions of absolutist words in their writings, while the average mentions for a control user are 210. Posts belonging to depressed users contain 2888 self-related words, while posts of control users contain on average 716 of them. The average number of tokens for a control user is at 20 551 tokens, while for a depressed one reaches 69 000. The categorization of the posts by means of Empath has little impact on the final results, given that most posts express negative emotions. The most relevant results are summarized in

Table 2. Comparative results on detecting depressed Reddit users based on multiple approaches. The following abbreviations are considered: MNB = Multinomial Naive Bayes, PA = Passive Aggressive Classifier, SGD = Support Vector Machine with Stochastic Gradient Descent, RE = Rule-based Estimator.

Method	Prec.	Rec.	F1	Acc
Tf-idf MNB (alpha=1)	0.61	0.47	0.53	0.94
Tf-idf PA (loss=$sqrt\_hinge$, tol=e^{-3}) batch	**0.82**	0.64	**0.72**	**0.96**
Tf-idf PA (loss=$sqrt\_hinge$, tol=e^{-3}) online	0.64	0.64	0.64	0.94
Tf-idf SGD (l1=0.95, loss=$hinge$) batch	0.76	0.62	0.68	0.95
Tf-idf SGD (l1=0.95, loss=$hinge$) online	0.70	0.65	0.68	0.95
RE PA (loss=$sqrt\_hinge$, tol=e^{-3})	0.63	0.13	0.22	0.95
Feature Union PA (loss=$sqrt\_hinge$, tol=e^{-3})	0.68	**0.72**	0.70	0.95
[25] CNN	**0.75**	0.57	0.65	N/A
[25] FastText	0.37	**0.70**	0.49	N/A

Table 2. Apart from assessing each model separately, we considered a feature union of equal weights for tf-idf and the output of the rule-based estimator, combined with a Passive Aggressive classifier.

We achieve the best results in terms of precision when using tf-idf weighting with a Passive Aggressive classifier. One explanation for this result may reside in the fact that the Passive Aggressive classifier is an online classifier that learns sequentially. Batch training led to better results for SGD and PA. We believe this happens because the models are being updated along the way and are less prone to overfit. When taking into account the psycholinguistic features we manage to improve the results in terms of recall, while not losing too much precision.

5   Conclusions

Analysis of social media texts has the potential to provide methods for understanding a user's mental health status and for the early detection of possible related diseases. We have presented in this paper preliminary results on the use of hand-crafted psycholinguistic features as possible improvements to standard classification approaches of depressed online personas.

As future work, we are interested in extending these psycholinguistic features with others that can be further revealed by clinical publications. Moreover, we want to analyze statistical information about the weights that each of these feature has in the final classification results. We also plan to investigate the lexical variability of posts of depressed users so as to infer possible new insights on the matter. If an automated process can predict or detect depressive users, they can be targeted for further medical assessment or they could be provided with alternative means of support and treatment. A great outcome of automatic processing of social networking data is the detection of users that are unaware of their condition.

Acknowledgements. This work was supported by the Integrated Programme of SR&TD SOCA (Ref. CENTRO-01-0145-FEDER-000010), co-funded by Centro 2020 program, Portugal 2020, European Union, through the European Regional Development Fund. Rui Antunes is supported by the Fundação para a Ciência e a Tecnologia (PhD Grant SFRH/BD/137000/2018).

References

1. World Health Organization Mental Health. http://www.who.int/mental_health/en/. Accessed 10 Oct 2019
2. Mental health atlas (2017). (Geneva: World Health Organization, 2018)
3. Al-Mosaiwi, M., Johnstone, T.: In an absolute state: elevated use of absolutist words is a marker specific to anxiety, depression, and suicidal ideation. Clin. Psychol. Sci., 2167702617747074 (2018)
4. Arseniev-Koehler, A., Mozgai, S., Scherer, S.: What type of happiness are you looking for? - A closer look at detecting mental health from language. In: Proceedings of the Fifth Workshop on Computational Linguistics and Clinical Psychology: From Keyboard to Clinic, pp. 1–12 (2018)
5. Bruffaerts, R., et al.: Mental health problems in college freshmen: prevalence and academic functioning. J. Affect. Disord. **225**, 97–103 (2018)
6. Calvo, R.A., Milne, D.N., Hussain, M.S., Christensen, H.: Natural language processing in mental health applications using non-clinical texts. Nat. Lang. Eng. **23**(5), 649–685 (2017)
7. Chung, C., Pennebaker, J.W.: The psychological functions of function words. Soc. Commun. **1**, 343–359 (2007)
8. Cohan, A., Desmet, B., Yates, A., Soldaini, L., MacAvaney, S., Goharian, N.: SMHD: a large-scale resource for exploring online language usage for multiple mental health conditions. In: The 27th International Conference on Computational Linguistics (COLING 2018), pp. 1485–1497. ACL (2018)
9. Coppersmith, G., Dredze, M., Harman, C.: Quantifying mental health signals in Twitter. In: Proceedings of the Workshop on Computational Linguistics and Clinical Psychology: From Linguistic Signal to Clinical Reality, pp. 51–60 (2014)
10. Coppersmith, G., Leary, R., Whyne, E., Wood, T.: Quantifying suicidal ideation via language usage on social media. In: Joint Statistics Meetings Proceedings, Statistical Computing Section, JSM (2015)
11. Coppersmith, G., Ngo, K., Leary, R., Wood, A.: Exploratory analysis of social media prior to a suicide attempt. In: Proceedings of the Third Workshop on Computational Linguisitics and Clinical Psychology, pp. 106–117 (2016)
12. De Choudhury, M., Gamon, M., Counts, S., Horvitz, E.: Predicting depression via social media. ICWSM **13**, 1–10 (2013)
13. De Choudhury, M., Kiciman, E., Dredze, M., Coppersmith, G., Kumar, M.: Discovering shifts to suicidal ideation from mental health content in social media. In: Proceedings of the 2016 CHI Conference on Human Factors in Computing Systems, pp. 2098–2110. ACM (2016)
14. Fast, E., Chen, B., Bernstein, M.S.: Empath: understanding topic signals in large-scale text. In: Proceedings of the 2016 CHI Conference on Human Factors in Computing Systems, pp. 4647–4657. ACM (2016)
15. Gowen, K., Deschaine, M., Gruttadara, D., Markey, D.: Young adults with mental health conditions and social networking websites: seeking tools to build community. Psych. Rehabil. J. **35**(3), 245 (2012)

16. Guntuku, S.C., Yaden, D.B., Kern, M.L., Ungar, L.H., Eichstaedt, J.C.: Detecting depression and mental illness on social media: an integrative review. Curr. Opin. Behav. Sci. **18**, 43–49 (2017)
17. MacAvaney, S., et al.: RSDD-Time: temporal annotation of self-reported mental health diagnoses. In: Proceedings of the Fifth Workshop on Computational Linguistics and Clinical Psychology: From Keyboard to Clinic, pp. 168–173 (2018)
18. Nadeem, M.: Identifying depression on Twitter. CoRR **abs/1607.07384** (2016)
19. Park, M., Cha, C., Cha, M.: Depressive moods of users portrayed in Twitter. In: Proceedings of the ACM SIGKDD Workshop on Healthcare Informatics (HI-KDD), vol. 2012, pp. 1–8. ACM, New York (2012)
20. Pedregosa, F., et al.: Scikit-learn: machine learning in python. J. Mach. Learn. Res. **12**, 2825–2830 (2011)
21. Pennebaker, J.W., Francis, M.E., Booth, R.J.: Linguistic inquiry and word count: LIWC 2001. Mahway: Lawrence Erlbaum Assoc. **71**(2001), 2001 (2001)
22. Rude, S., Gortner, E.M., Pennebaker, J.: Language use of depressed and depression-vulnerable college students. Cogn. Emot. **18**(8), 1121–1133 (2004)
23. Shing, H.C., Nair, S., Zirikly, A., Friedenberg, M., Daumé III, H., Resnik, P.: Expert, crowdsourced, and machine assessment of suicide risk via online postings. In: Proceedings of the Fifth Workshop on Computational Linguistics and Clinical Psychology: From Keyboard to Clinic, pp. 25–36 (2018)
24. Vioulès, M.J., Moulahi, B., Azé, J., Bringay, S.: Detection of suicide-related posts in Twitter data streams. IBM J. Res. Dev. **62**(1), 1–7 (2018)
25. Yates, A., Cohan, A., Goharian, N.: Depression and self-harm risk assessment in online forums. In: Proceedings of the 2017 Conference on Empirical Methods in Natural Language Processing, pp. 2968–2978. Association for Computational Linguistics (2017)

Predicting the Size of Candidate Document Set for Implicit Web Search Result Diversification

Yasar Baris Ulu and Ismail Sengor Altingovde$^{(\boxtimes)}$ (iD)

Middle East Technical University, Ankara, Turkey
{yasar.ulu,altingovde}@ceng.metu.edu.tr

Abstract. Implicit result diversification methods exploit the content of the documents in the *candidate set*, i.e., the initial retrieval results of a query, to obtain a relevant and diverse ranking. As our first contribution, we explore whether recently introduced word embeddings can be exploited for representing documents to improve diversification, and show a positive result. As a second improvement, we propose to automatically predict the size of candidate set on *per query* basis. Experimental evaluations using our BM25 runs as well as the best-performing ad hoc runs submitted to TREC (2009–2012) show that our approach improves the performance of implicit diversification up to 5.4% wrt. initial ranking.

1 Introduction

Diversification of web search results is a well-known approach to handle queries that are ambiguous, underspecified, or including multiple aspects [23]. Diversification methods in the literature are broadly categorized as *implicit* or *explicit*. The implicit methods essentially make use the documents in the *candidate set*, i.e., the initial retrieval results for the query. In contrary, explicit methods exploit the knowledge of query aspects, which is usually inferred from a topic taxonomy [1] or query log [20]. The exhaustive experiments in the literature confirm that the latter type of additional information is very useful, as explicit methods consistently outperform the implicit ones [10,23]. This finding does not render the implicit diversification less valuable, as in many scenarios the query aspects are not readily available or not easy to infer (e.g., for the rare queries in Web search) [15], but rather calls for approaches to improve their performance.

Our contributions in this paper are two-fold: First, we re-visit the implicit diversification using recently introduced word embeddings, and show that using the latter to represent documents is superior to traditional vector space model (with tf-idf weights). However, we observe that using either type of representations, implicit diversification can hardly beat even the initial –non-diversified– ranking (confirming the observations of [10]). These findings, obtained using the best-performing trade-off parameter λ (i.e., used to tune the weight of relevance vs. diversity in a ranking, as explained in Sect. 2) and a candidate set size $N = 100$ documents (an ad hoc yet intuitive choice made in several earlier works

© Springer Nature Switzerland AG 2020
J. M. Jose et al. (Eds.): ECIR 2020, LNCS 12036, pp. 410–417, 2020.
https://doi.org/10.1007/978-3-030-45442-5_51

[10,15,21]), imply that a more customized tuning of parameters is required for implicit diversification. An earlier work also recognized such a need for selective diversification, and proposed to predict the trade-off parameter λ on a per query basis [21]. However, the second parameter that is equally important, the size of the candidate set (N), on which diversification is applied, is left unexplored. We believe that for the implicit methods, where the evidence used for diversification is based solely on the content of the documents, tuning the candidate set size is crucial: a small set with documents relevant to only the main query might not cover any alternative intents (aspects) of the query, while a too large set is likely to include several noisy documents and hence, mislead the implicit methods.

In the light of above discussion, as our second contribution, we propose to predict the candidate set size, N, on a *per query* basis, to achieve a more customized diversification of query results. To this end, we employ a rich set of features that capture the retrieval effectiveness (i.e., query performance predictors [6,14,21,24]) and pairwise similarity of documents (using alternative document representations). All features are computed over the candidate set at several rank-cutoffs (actually, from 10 to 100 with a step size of 10). Before the diversification for a query, we predict N (as well as λ, as in [21]), based on these features.

In our evaluations, we employ MMR [5] as a representative implicit method, as it is widely employed in the literature, has fewer parameters to tune and fast. Our findings over the homemade runs (based on the BM25 function) as well as the representative runs from the previous TREC campaigns (2009 to 2012) are promising. By predicting the parameters on a per query basis and employing word embeddings, implicit diversification can outperform the non-diversified baselines (with relative gains up to 5.4%), as well as the diversification baseline with parameters based on majority voting (with even larger gains).

Related Work. Word embeddings are employed for various tasks related to diversification of search results (such as expanding the queries in tweet search [16], generating diversified query expansions [13], inferring query aspects [25]). However, as far as we know, the impact of employing word embeddings to represent the documents for implicit diversification has not been explored.

Earlier works proposed several implicit diversification methods [23]. While most of these works employ a fixed N, such as 50 or 100 (e.g., [7,10,15,16,18]), a few works (such as [12]) identified N (and/or λ) over a training set, (i.e., as our Majority Voting baseline presented in Sect. 4). Santos et al. [21] suggested a selective diversification approach, where only λ is predicted for each query, using kNN approach. None of these works predict the candidate set size on a per query basis for result diversification. Finally, in an *unpublished* thesis work [2], preliminary experiments for candidate set size prediction are presented for explicit diversification. In contrary, our work addresses implicit diversification, which requires features that capture inter-document similarity and are not used in the latter work. Furthermore, we predict both parameters N and λ (consecutively) using the same set of features, which is different than the setup in [2].

2 Document Representation for Implicit Diversification

There are several implicit methods in the literature [23], and in this work, we use MMR [5] as a representative method due to two reasons. First, being a simple, intuitive and efficient method, MMR is employed as a baseline and/or representative approach in a large number of works (e.g., [10,21,26,27]). Secondly, we conducted preliminary experiments with some other candidates (namely, MSD [8], MMC and GNE [26]) and did not observe meaningful performance differences wrt. MMR. Actually, only GNE produced slightly better results, however as it is based on a greedy local search, its execution time is considerably longer than MMR. Therefore, we proceed with MMR as a representative method. In what follows, we first briefly review MMR and then discuss how word embeddings are employed to represent documents in this context.

Maximal Marginal Relevance (MMR) [5]. This is a greedy best-first search approach that aims to choose the document that maximizes the following scoring function in each iteration.

$$MMR(d, q, S) = \lambda * rel(d, q) - (1 - \lambda) * \max_{d_j \in S} sim(d, d_j) \qquad (1)$$

Given a query q and a candidate result set D of size N, MMR constructs a diversified ranking S of size s (typically, $s < N$) as follows. At first, the document with the highest relevance score is inserted into S. Then, in each iteration, the document that maximizes Eq. 1 is added to S. While computing the score of a document $d \in D - S$, its relevance to q, denoted as $rel(q, d)$, is discounted by the d's maximum similarity the previously selected documents into S. In Eq. 1, $sim(d, d_j)$ is typically computed by the Cosine distance of documents that are represented as tf-idf weighted vectors. Finally, λ is a trade-off parameter to balance the relevance and diversity in the final result set S.

Word Embeddings for Document Representation. In this preliminary work, we take a simplistic approach and represent each document d based on the embedding vectors of their terms $t \in d$. In the literature, different approaches are proposed for this purpose, such as computing the minimum, maximum or average for each dimension of the embedding vectors over all terms in the document [4]. The aggregation operation can also be weighted, e.g., by IDF values of the terms. In this work, based on our preliminary experiments, we represent each document as a concatenation of minimum and maximum vectors, as in [4]. Thus, $sim(d, d_j)$ in Eq. 1 is computed as the Cosine distance between the latter type of vectors.

3 Predicting the Candidate Set Size

Implicit diversification methods do not exploit any external information (in contrary to their explicit competitors) and their diversification decision is essentially based on the content of the documents. While the size of the candidate set, N, is an important parameter for all diversification approaches, it is more crucial for the implicit methods: In particular, a very large candidate set is likely to have more irrelevant documents towards the tail of the ranking, yet such documents -yielding smaller similarity to the relevant ones that are ranked higher- are more likely to be scored high by Eq. 1, and hence, would decrease the relevance of the final ranking. In contrary, setting N too small will risk to have any diverse document in the final ranking, and hence, reduce the diversity. This implies that the value of N should be determined on a *per query* basis.

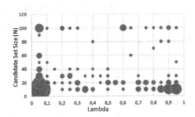

Fig. 1. Best-performing (N, λ) pair over BM25 runs for 198 TREC topics.

Table 1. List of features computed over each ranking.

Feature	Description	Count
scoreRatio [17]	Ratio of top to last document's score	10
scoreMean [14]	Mean of scores in document set	10
scoreMeanDecrease	Decrease in mean of scores in document set	9
scoreMedian	Median of scores in document set	10
scoreStandardDev [14,24]	Standard deviation of scores	10
scoreVariance	Variance of scores	10
coefficientOfVariation	Coefficient of variation	10
NQC [24]	Scores of Normalized Query Commitment	10
PairwiseTfIdfSimilarity	Pairwise (td-idf vector) similarity (min, max, avg)	30
PairwiseWESimilarity	Pairwise (WE vector) similarity (min, max, avg)	30
PairwiseEntitySimilarity	Pairwise (Entity list) similarity (min, max, avg)	30

As a further motivation, consider Fig. 1, a bubble chart that presents the best-performing (λ, N) pairs (for diversification with MMR) for 198 queries used in the TREC Diversity track (2009-2012). The initial runs are obtained using BM25 and the size of the bubble denotes the frequency of a pair. Clearly, there is no single λ or N that optimizes all queries, and indeed, values are quite scattered.

In this work, we propose to predict N. Since our approach requires determining an optimal cut-off point in the candidate ranking, we compute each of the following features over a ranking of top-n documents, where $n \in \{10, 20, \ldots N\}$. Our features (shown in Table 1) can be grouped into two categories. The first group are based on well-known query performance predictors [6,14,24], and as in [21], they are intended to capture the quality of the ranking (i.e., in terms of relevance). The second group of features is intended to reflect the diversity of a ranking. To this end, we propose to compute the pairwise similarity of the documents, and aggregate these scores using minimum, maximum and average functions. While computing such similarities, we employ both tf-idf and word embedding based document representations (as discussed in Sect. 2). Finally, as entities are found helpful in earlier works [21], as a third option, we represent each document based on the named entities it contains (see Sect. 4 for details).

A training instance for a query includes a vector of these features computed for each top-n ranking ($n \in \{10, 20, \ldots 100\}$), i.e., including 10 variants for each feature. For each query, we apply parameter sweeping over $N \in \{10, 20, \ldots, 100\}$ and $\lambda \in \{0.05, 1.0, \ldots, 0.95\}$, and determine the best performing values (for diversification with MMR), to serve as the ground truth (categorical) class labels. For a test query, we first predict N as the class label. Next, we predict λ (as in [21]) by using the aforementioned features and the *predicted* N value, as an additional feature (i.e., as in the *classifier chain* approach in [19]). Our preliminary experiments with Weka [9] revealed that best results are obtained using a lazy learning algorithm, kNN (as in [21]). Thus, for a test query, N (and then, λ) are predicted using majority voting among the class labels of its k neighbors.

Table 2. Diversification performance (α-nDCG@10) of MMR using TF-IDF vectors (MMR$_{\text{TfIdf}}$) vs. word embedding vectors (MMR$_{\text{WordEmb}}$) for BM25 and TREC runs.

	BM25 runs			TREC runs		
Topic set	NonDiv	MMR$_{\text{TfIdf}}$	MMR$_{\text{WordEmb}}$	NonDiv	MMR$_{\text{TfIdf}}$	MMR$_{\text{WordEmb}}$
2009	0,2520	0,2360	**0,2531**	0,2530	0,2533	**0,2544**
2010	0,2427	0,2461	**0,2573**	0,3716	0,3634	**0,3718**
2011	0,4680	0,4581	**0,4693**	0,5312	0,5312	**0,5315**
2012	**0,3218**	0,2911	0,3215	0,4942	0,4926	**0,4962**

4   Evaluation Setup and Results

Dataset and Runs. We employ topic sets that are introduced in "Diversity Task" of TREC Web Track between 2009 and 2012. Each topic set includes 50 queries (except 2010, which has 48), their official aspects and the relevance judgments at the aspect level. We have two types of runs created as follows. First, we used our own retrieval system to index ClueWeb09 collection Part-B (with 50M documents) and then, for each topic set, we generated an an initial ranking of top-100 documents per query, using the well-known BM25 function. These are referred to as BM25 runs. Secondly, we selected the best-performing run (again, on ClueWeb-B) submitted to ad hoc retrieval track of TREC (2009-2012). As in [3,10], as these runs are not diversified, they can safely serve as initial retrieval results (with various ranking methods beyond BM25), and best-performing run is the one that yields the highest α-nDCG@10 score. These are referred to as TREC runs. The ids of the selected runs for each year are as follows: Ucdsift (2009), Uogtr (2010), Srchvs11b (2011) and Qutparabline (2012).

Experimental Parameters. To represent documents with embeddings (Sect. 2), we employ the pre-trained Glove word vectors (with 100 dimensions) for 400K words. Tf-idf vectors are based on the document and collection statistics, as usual. To extract the named entities in documents (to compute some features in Table 1), we used an entity list (of people, locations, etc.) from DBpedia together with the dictionary-based entity recognition approach of [22].

For prediction of N and λ, kNN algorithm is applied with 5-fold cross validation over each run. We employ the best performing k. We observed that, especially for the TREC runs, setting k to 1 is adequate in several cases. The size of the final ranking S is 10, and we report α-nDCG@10 scores.

Results for Document Representation Experiments. As our first research question, we focus on the impact of using word embeddings for representing documents during diversification. In this experiment, we set $N = 100$ as typical (e.g., [10,21]), and report results for the best-performing λ for each run.

Table 2 shows that the performance of MMR$_{\text{Tfidf}}$ is inferior to the non-diversified ranking for the majority of the cases, i.e., its application yields even less diverse rankings. This finding confirms [10], where MMR is rarely found to provide any significant gains. In contrary, MMR$_{\text{WordEmb}}$ outperforms the MMR$_{\text{Tfidf}}$ in all cases (underlined cases in Table 2 are statistically significant using paired t-test at 0.05 confidence level). Furthermore, MMR$_{\text{WordEmb}}$ is also superior to the non-diversified baseline for seven (out of eight) runs, but with a small difference in most cases. These findings indicate that word embeddings are useful for MMR, but not adequate for impressive diversification performance.

Table 3. Diversification performance (α-nDCG@10) of MMR$_{\text{WordEmb}}$ with parameters obtained via Orcl$_{100,\lambda}$ (best λ [21]), Orcl$_{N,\lambda}$ (best N and λ), Majority Voting and kNN.

	BM25 runs					TREC runs				
TSet	NonDiv	Orcl$_{100,\lambda}$	Orcl$_{N,\lambda}$	MV	kNN	NonDiv	Orcl$_{100,\lambda}$	Orcl$_{N,\lambda}$	MV	kNN
2009	0,2520	0,2917	0,3044	0,2469	**0,2612**	0,2530	0,2853	0,3075	0,2562	**0,2589**
2010	0,2427	0,2876	0,3137	0,2443	**0,2554**	0,3716	0,4020	0,4032	0,3595	**0,3768**
2011	0,4680	0,4888	0,5194	0,4279	**0,4750**	0,5312	0,5468	0,5507	0,5284	**0,5379**
2012	0,3218	0,3270	0,4452	0,3257	**0,3392**	**0,4942**	0,5066	0,5102	0,4772	0,4890

Results for Predicting Parameters. We evaluate the performance of predicting the parameters N and λ only for MMR$_{\text{WordEmb}}$ (due to the aforementioned findings). Table 3 reports the results both for BM25 runs and TREC best runs. We provide α-nDCG@10 scores for non-diversified (NonDiv) ranking, as well as two oracle approaches (discussed later). The traditional baseline Majority Voting (MV) sets N and λ to the most frequent value in training folds, respectively.

We make several observations from Table 3. First, the MV baseline cannot beat the initial non-diversified ranking (NonDiv) for several cases. When kNN is applied to predict the parameters N and λ, the diversification performance is superior to MV baseline (in all cases), and outperforms the NonDiv ranking in all runs but one (i.e., *Qutparabline* from 2012). For BM25 runs, kNN based diversification provides relative gains wrt. the non-diversified ranking ranging from 1.5% to 5.4%. For more competitive TREC runs, the relative gains are in the range 1.2% to 2.3% (except the 2012 run, where there is a relative degradation of 1%). Given that the latter runs are employing sophisticated approaches far beyond BM25, our findings are promising. Note that, in some cases (underlined in Table 3) improvements wrt. MV are statistically significant (using paired t-test at 0.05 confidence level), while there is no significant degradation wrt. MV or NonDiv. The latter is a contrary and encouraging finding in comparison to [10], where MMR is observed to yield only significant degradation in most cases.

Table 3 also reports two oracle approaches: In $Oracle_{N,\lambda}$, the best-performing N and λ is used for each query. In $Oracle_{100,\lambda}$, we fixed N as 100, and only employed the best-performing λ. The latter oracle aims to provide an upper-bound for predicting only λ as in [21], while the former one presents the upper-bound for our approach, predicting both parameters. We see that our approach can yield higher performance (for all runs), and in certain cases, the possible gain is considerably larger than that of predicting only λ. As a further observation, a comparison of kNN performance to $Oracle_{N,\lambda}$ indicates that there is still room for improvement, i.e., if better prediction of parameters can be achieved.

Conclusion. We showed that implicit diversification benefits from word embeddings based document representation, but it still yields rather small gains in diversification effectiveness wrt. the initial ranking. As a remedy, we proposed to predict N, the candidate set size, using a rich set of features. By predicting N (together with λ, as in [21]) and employing word embeddings, we achieved better diversification. In our future work, we plan to use document embeddings (e.g., Doc2Vec [11]) for document representation. We will also exploit additional (e.g., click-based) features for better prediction of the diversification parameters.

Acknowledgements. This work is partially funded by The Scientific and Technological Research Council of Turkey (TÜBİTAK) grant 117E861 & TÜBA GEBIP (2016) award.

References

1. Agrawal, R., Gollapudi, S., Halverson, A., Ieong, S.: Diversifying search results. In: Proceedings of WSDM, pp. 5–14. ACM (2009)
2. Akcay, M.: Analyzing and boosting the performance of explicit result diversification methods for web search. Master's thesis, Middle East Technical University (METU) (2016)
3. Akcay, M., Altingovde, I.S., Macdonald, C., Ounis, I.: On the additivity and weak baselines for search result diversification research. In: Proceedings of ICTIR, pp. 109–116 (2017)
4. Boom, C.D., Canneyt, S.V., Demeester, T., Dhoedt, B.: Representation learning for very short texts using weighted word embedding aggregation. Pattern Recogn. Lett. **80**, 150–156 (2016)

5. Carbonell, J.G., Goldstein, J.: The use of MMR, diversity-based reranking for reordering documents and producing summaries. In: Proceedings of SIGIR, pp. 335–336 (1998)
6. Carmel, D., Kurland, O.: Query performance prediction for IR. In: Proceedings of SIGIR, pp. 1196–1197 (2012)
7. Dang, V., Croft, W.B.: Diversity by proportionality: an election-based approach to search result diversification. In: Proceedings of SIGIR, pp. 65–74 (2012)
8. Gollapudi, S., Sharma, A.: An axiomatic approach for result diversification. In: Proceedings of WWW, pp. 381–390 (2009)
9. Hall, M., Frank, E., Holmes, G., Pfahringer, B., Reutemann, P., Witten, I.H.: The WEKA data mining software: an update. SIGKDD Explor. 11(1), 10–18 (2009)
10. Kharazmi, S., Scholer, F., Vallet, D., Sanderson, M.: Examining additivity and weak baselines. Trans. Inf. Syst. 34(4), 23 (2016)
11. Le, Q.V., Mikolov, T.: Distributed representations of sentences and documents. In: Proceedings of ICML, pp. 1188–1196 (2014)
12. Limsopatham, N., McCreadie, R., Albakour, M., Macdonald, C., Santos, R.L.T., Ounis, I.: University of glasgow at TREC 2012: experiments with terrier in medical records, microblog, and web tracks. In: Proceedings of TREC (2012)
13. Liu, X., Bouchoucha, A., Sordoni, A., Nie, J.: Compact aspect embedding for diversified query expansions. In: Proceedings of AAAI, pp. 115–121 (2014)
14. Markovits, G., Shtok, A., Kurland, O., Carmel, D.: Predicting query performance for fusion-based retrieval. In: Proceedings of CIKM, pp. 813–822 (2012)
15. Naini, K.D., Altingovde, I.S., Siberski, W.: Scalable and efficient web search result diversification. ACM Trans. Web 10(3), 15:1–15:30 (2016)
16. Onal, K.D., Altingovde, I.S., Karagoz, P.: Utilizing word embeddings for result diversification in tweet search. In: Proceedings of AIRS, pp. 366–378 (2015)
17. Ozdemiray, A.M., Altingovde, I.S.: Query performance prediction for aspect weighting in search result diversification. In: Proceedings of CIKM, pp. 1871–1874 (2014)
18. Ozdemiray, A.M., Altingovde, I.S.: Explicit search result diversification using score and rank aggregation methods. JASIST 66(6), 1212–1228 (2015)
19. Read, J., Pfahringer, B., Holmes, G., Frank, E.: Classifier chains for multi-label classification. Mach. Learn. 85(3), 333–359 (2011)
20. Santos, R.L.T., Macdonald, C., Ounis, I.: Exploiting query reformulations for web search result diversification. In: Proceedings of WWW, pp. 881–890 (2010)
21. Santos, R.L.T., Macdonald, C., Ounis, I.: Selectively diversifying web search results. In: Proceedings of CIKM, pp. 1179–1188 (2010)
22. Santos, R.L.T., Macdonald, C., Ounis, I.: Voting for related entities. In: Proceedings of RIAO, pp. 1–8 (2010)
23. Santos, R.L.T., Macdonald, C., Ounis, I.: Search result diversification. Found. Trends Inf. Retrieval 9(1), 1–90 (2015)
24. Shtok, A., Kurland, O., Carmel, D., Raiber, F., Markovits, G.: Predicting query performance by query-drift estimation. ACM Trans. Inf. Syst. 30(2), 11 (2012)
25. Ullah, M.Z., Shajalal, M., Chy, A.N., Aono, M.: Query subtopic mining exploiting word embedding for search result diversification. In: Proceedings of AIRS, pp. 308–314 (2016)
26. Vieira, M.R., et al.: On query result diversification. In: Proceedings of ICDE, pp. 1163–1174 (2011)
27. Zuccon, G., Azzopardi, L., Zhang, D., Wang, J.: Top-k retrieval using facility location analysis. In: Proceedings of ECIR, pp. 305–316 (2012)

Aspect-Based Academic Search
Using Domain-Specific KB

Prajna Upadhyay[1(✉)], Srikanta Bedathur[1], Tanmoy Chakraborty[2],
and Maya Ramanath[1]

[1] IIT Delhi, Hauz Khas, New Delhi 110016, India
{prajna.upadhyay,srikanta,ramanath}@cse.iitd.ac.in
[2] IIIT-Delhi, Okhla Industrial Estate, Phase III, New Delhi 110020, India
tanmoy@iiitd.ac.in

Abstract. Academic search engines allow scientists to explore related work relevant to a given query. Often, the user is also aware of the *aspect* to retrieve a relevant document. In such cases, existing search engines can be used by expanding the query with terms describing that aspect. However, this approach does not guarantee good results since plain keyword matches do not always imply relevance. To address this issue, we define and solve a novel academic search task, called *aspect-based retrieval*, which allows the user to specify the aspect along with the query to retrieve a ranked list of relevant documents. The primary idea is to estimate a language model for the aspect as well as the query using a domain-specific knowledge base and use a mixture of the two to determine the relevance of the article. Our evaluation of the results over the Open Research Corpus dataset shows that our method outperforms keyword-based expansion of query with aspect with and without relevance feedback.

Keywords: Academic retrieval · Aspect · Technical knowledge base

1 Introduction

Academic search engines such as Google Scholar, PubMed, and Semantic Scholar play a central role in the lives of researchers dealing with the ever growing flood of related work. To further improve the academic search experience, there have been proposals to either re-rank the results using user's interests [12] and the set of papers assessed relevant [2], or to recommend new articles based on a query article [4].

In this paper, we define and solve a novel academic search task, called *aspect-based retrieval*, which is targeted towards enabling the academic search user to specify the *aspect* along with the query to retrieve a ranked list of scientific articles that are (i) relevant to the query, and (ii) the relevance relation [3] between the query and retrieved documents is semantically close to the specified aspect. We illustrate this expected behavior with a concrete example: consider the query autoencoder and an aspect of interest, say, application, then we aim to rank high the articles which are related to the concept of autoencoder *and* are about the *applications* of autoencoders. If there were two papers, titled (a) "Complex-valued Autoencoders" and (b) "Exploring autoencoders for unsupervised feature selection", our system should rank the paper (b) higher than paper (a) since it specifically deals with applications of the specified query rather than its variants.

J. M. Jose et al. (Eds.): ECIR 2020, LNCS 12036, pp. 418–424, 2020.
https://doi.org/10.1007/978-3-030-45442-5_52

Note that both papers are indeed relevant to the central concept being queried. Aspect-based retrieval of scientific documents is not straightforward at all since the relation semantics do not manifest as simple keyword matches. Simply expanding the query with terms that define the aspect fails to retrieve relevant articles. In the above example, the paper (b) does not contain the application in the title as well as the abstract. On the other hand, a document, titled "Evaluating the Performance of Dynamic Database Applications" is not related to the query RDBMS along the *application* aspect although its title contains the term applications. The way in which the specified semantic relationship manifests between the query and a document is highly dependent on the domain we operate in. For example, given the query autoencoder, for a document to be related along *application* aspect, the presence of terms like feature or selection with terms similar to application would be highly suggestive because feature_selection is known to be an application of autoencoder. But this can not be easily determined by just analysing the documents without prior knowledge of the domain.

We address the challenge of aspect-oriented retrieval for scientific documents by minimizing the risk of returning a document whose language model diverges from the model estimated for domain-specific query and aspect specified by the user. The query and aspect models are derived using domain-specific knowledge bases (KBs), which is challenging in itself due to the inherent sparsity of relations in these KBs. By using a domain-specific KB of computer science, like TeKnowbase [14], we show how to overcome the sparsity issue in KB via inference using meta-paths derived from the KB. Our results over the Open Research Corpus [1] dataset containing more than 39 million published research papers show that our proposed approach outperforms variants of query likelihood language models with/without relevance feedback.

2   Aspect Based Retrieval Model

2.1   System Overview

Our system takes a query and an aspect as input, and returns a ranked list of relevant documents. Users express their information need as strings of words called **queries**. In our case, a query is a technical entity. An **aspect** is the relevance relation specified by the user between the query and the relevant document. Given a query, the documents are ranked using retrieval models [9]. A **retrieval model** transforms the document space into an intermediate representation and returns a ranked list of documents according to some scoring function.

2.2   Retrieval Model and Estimation

Language modelling techniques [10] model the relevance of a document as the probability of generating the query from the document. Expanding the query with aspect terms and doing relevance feedback [7] will only retrieve documents containing those terms. Given a query q and an aspect a, a relevant document d consists of terms determined by both q and a.

Aspect Dependent Prior Probability. $P(w|a)$ is the prior, which is the probability of a term w appearing in d given an aspect a, independent of the query.

Query and Aspect Dependent Probability. The probability of a term appearing in d given a query q and aspect a is denoted by $P(w|q, a)$.

Mixture of the Two Probability Distributions. The relevance of d will be determined by a mixture model of both the probability distributions. Equation (1) describes our probability distribution. It is denoted as MM.

$$MM(w) = \lambda P(w|a) + (1 - \lambda)P(w|q, a) \qquad (1)$$

Scoring of Documents. The language model M_d of a candidate document is expressed by Eq. (2). Dirichlet smoothing is used for M_d.

$$M_d(w) = \frac{tf(w, d) + \mu P(w|C)}{length(d) + \mu} \qquad (2)$$

where $tf(w, d)$ is the frequency of w in d, and $P(w|C)$ is the probability of w appearing in the entire collection. The risk associated with using MM to approximate M_d is expressed by KL-divergence between MM and M_d and the documents are returned in an order of increasing KL-divergence.

$$KL(MM||M_d) = \sum_w P(w|MM) \log \frac{P(w|MM)}{P(w|M_d)} \qquad (3)$$

Estimation of Query-Independent Component. $P(w|a)$ is estimated using a narrow set of documents from our dataset acquired as follows. We chose 10 queries and retrieved the top 10 documents for them using the standard query likelihood model. Additionally, we fired queries of the form "query+aspect" to retrieve the top-10 documents using the same model. We recruited evaluators to annotate about 1500 documents with the aspect labels (described in details in Sect. 3.2). In order to increase the size of our document set, we used heuristics to collect more documents given an aspect. We formulated a query containing only the aspect as a keyword and retrieved documents for it using the standard likelihood model. But, we retained only those documents which contained the name of the aspect in the title on the intuition that such documents are highly likely (though not guaranteed) to be about those aspects. Having a set of ground-truth documents D, $P(w|a)$ is estimated according to Eq. (4).

$$P(w|a) = \frac{1}{|D|} \sum_{d \in D} \frac{tf(w, d)}{\sum_{w' \in d} tf(w', d)} \qquad (4)$$

Estimation of Query-Dependent Component. We used relationships in TeKnowbase (TKB) [14] to represent aspects. TKB consists of entities such as hidden_markov_model or speech_recognition and other domain-specific relationships like application, implementation or algorithm. The triple ⟨speech_recognition, application, hidden_markov_model⟩ conveys information that speech_recognition is an application of hidden_markov_model. The entities connected via application relation in TKB have a higher probability of appearing in documents addressing *application* aspect. However, TKB is sparse. To automatically infer the entities participating in a particular relationship type, we used meta-paths. A meta-path is a sequence of edges with labels connecting two nodes which have been used previously for KB completion tasks [6], link prediction [8] as well as to find similarity between two nodes [11, 13]. Figure 1 shows how meta-paths can be used to infer relationships between entities.

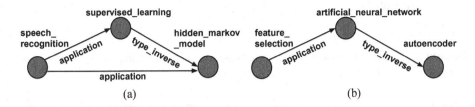

Fig. 1. Examples of meta-paths for **application** relation. (a) The meta-path ⟨application, type_inverse⟩ exists between **speech_recognition** and **hidden_markov_model**. (b) **feature_selection** and **autoencoder** are not related by **application** relation but still it can be inferred because of the existence of the same meta-path ⟨application, type_inverse⟩ between them.

(1) Direct inference using meta-paths. To automatically determine entities that participate in a given relationship type with entity e_i, we used the path-constrained random walk algorithm (PRA) proposed in [6]. Given a set E of entities in TKB, a source node e_i and a meta-path P, a path-constrained random walk defines a probability distribution $h_{e_i,P}(e_j)$ to all entities in E which is the probability of reaching e_j from e_i by doing a random walk along P. The key idea is to acquire the set of meta-paths representing the given relationship type and use PRA for inferencing. To do so, we retrieved the set of all meta-paths (MP) connecting the given relationship type in TKB and scored them according to their frequency. Given a set of meta-paths $MP = P_1, P_2, ..., P_n$, the score for each node reachable from source e_i is given by: $score_{e_i}(e_j) = \alpha_1 h_{e_i,P_1}(e_j) + \alpha_2 h_{e_i,P_2}(e_j) + ... + \alpha_n h_{e_i,P_n}(e_j)$, where $\alpha_l (l = 1 ... n)$ is the frequency of meta-path P_l. $score_{e_i}(e_j)$ is converted to a probability distribution using *softmax* and denoted by Eq. (5).

(2) Indirect inference using meta-paths. PRA assigns zero probability to nodes that are not reachable via any meta-paths in MP. To address this issue, we used MetaPath2Vec [5]. It takes a meta-path as input and constructs embeddings of entities such that entities that are likely to be connected via the meta-path (and not necessarily having a meta-path between them) are assigned vector representations closer to each other. We used the top-k meta-paths in MP as input to metapath2vec and obtained vector representations V_e for entity e. We used the softmax function to convert cosine similarities between entities into a probability distribution as described in Eq. (6).

$$DI_{e_i}(e_j) = \frac{e^{score_{e_i}(e_j)}}{\sum_{e_k \in |E|} e^{score_{e_i}(e_k)}} \tag{5}$$

$$h'_{e_i}(e_j) = \frac{e^{\text{sim}(V(e_i),V(e_j))}}{\sum_{e_k \in |E|} e^{\text{sim}(V(e_i),V(e_k))}}. \tag{6}$$

The probability distribution for inferencing is a mixture of $DI_{e_i}(e_j)$ and $h'_{e_i}(e_j)$ using β as given in Eq. (7). Since the documents are represented as bag of words, we defined the distribution over terms instead of entities using Eq. (8). $terms(e)$ is the set of words present in the entity e, and e_q is the entity that q represents. The final probability is a mixture given by Eq. (1).

$$P_a(e_i|e_j) = \beta * DI_{e_i}(e_j) + (1 - \beta) * h'_{e_i}(e_j) \tag{7}$$

$$P(w|q, a) = \sum_e P_a(e|e_q), \text{s.t. } w \in terms(e) \tag{8}$$

3   Experiments

3.1   Setup

Dataset. We used the Open Research Corpus dataset and indexed it using Galago. The baseline models (described below) are already implemented in Galago.

Aspects. We experimented with 3 different aspects – *application, algorithm* and *implementation*. We set λ and β to 0.5 in Eqs. (1) and (7). We restricted ourselves to meta-paths of size at most 3. We set $k = 5$ for choosing the top-k meta-paths for generating embeddings using MetaPath2Vec (described in Sect. 2.2).

Benchmarks. Benchmark queries were taken from a set of 100 queries released by [15] out of which 43 existed as whole entities in TKB, shown in Fig. 2.

---

artificial intelligence, augmented reality, autoencoder, big data, category theory, clojure, cnn, computer vision, cryptography, data mining, data science, deep learning, differential evolution, dirichlet process, duality, genetic algorithm, graph drawing, graph theory, hashing, information geometry, information retrieval, information theory, knowledge graph, machine learning, memory hierarchy, mobile payment, natural language, neural network, ontology, personality trait, prolog, question answering, recommender system, reinforcement learning, sap, semantic web, sentiment analysis, smart thermostat, social media, speech recognition, supervised learning, variable neighborhood search, word embedding

---

Fig. 2. Benchmark queries

Baselines. We explicitly added the keyword representing the aspect to the query and used standard retrieval models with/without relevance feedback techniques as baselines described below:

(1) **Query likelihood model with query only (QL+query).** Query likelihood [10] estimates a language model for each document in the collection and ranks them by the likelihood of seeing the query terms as a random sample given that document model.

(2) **Query likelihood model with query + aspect name (QL + query + aspect).** We used the same model as above but added the terms application, algorithm or implementation to the query based on the aspect and retrieved the results.

(3) **Query expansion with pseudo relevance feedback on QL + query + aspect (QL + query + aspect + QE).** We chose top-100 terms to expand the query for the query used in the previous baseline using relevance feedback model [7]. Top-1000 documents were used as feedback documents. The weight of the original query was set as 0.75.

(4) **Mixture Model (MM).** This is our retrieval model described in Sect. 2.

3.2   Evaluation Scheme and Metrics

Evaluation Scheme. In the absence of an extensive ground-truth dataset, we conducted a crowd-sourced user-evaluation exercise (involving Computer Science students and researchers, not related to the project) to measure the performance of our model. We formulated domain-specific questions, and depending on the answers marked by the evaluators, the documents were assigned a particular score for a query and aspect pair.

Evaluation Metrics. Each query and abstract pair was graded by at least 2 evaluators. We converted the response from each of them into a graded relevance scale and averaged the relevance values marked by them for each query-abstract pair. We used **Discounted Cumulative Gain (DCG)** and **Precision** to evaluate top-5 documents.

3.3   Results and Discussions

Table 1 shows the results for *algorithm, application* and *implementation* aspects. We observe that our model outperforms the rest of the baselines in terms of precision@5 and DCG for all of the 3 aspects. **QL + query + aspect + QE** comes second after our retrieval model. By modelling the aspect and query dependant probability explicitly, we were able to address the problems of simple keyword-based match for aspects described in Sects. 1 and 2. For example, the top-2 papers retrieved for **genetic_algorithm** for application aspect by our model were *Genetic Ant Algorithm for Continuous Function Optimization and Its MATLAB Implementation* and *Solve Zero-One Knapsack Problem by Greedy Genetic Algorithm*. The top-2 papers retrieved by **QL + query + aspect + QE** for *application* aspect do not describe any application of **genetic_algorithm** but contained a few terms like "a wide application prospect" in the abstract due to which it was retrieved in the top positions. Adding relevance feedback terms also did not work because the list of pseudo relevant documents did not contain relevant documents in the first place due to plain keyword-based retrieval. Both the papers retrieved by our method address application aspect for **genetic_algorithm** even if "application" is not mentioned in the title.

Table 1. Results for algorithm, application and implementation aspect.

Approach	Algorithm			Application			Implementation		
	DCG@5	P@5	P@1	DCG@5	P@5	P@1	DCG@5	P@5	P@1
MM	**6.27**	**0.70**	**0.75**	**2.64**	**0.45**	**0.47**	**2.33**	**0.44**	0.40
QL+query	2.69	0.3	0.33	1.42	0.25	0.22	1.05	0.16	0.23
QL+query+aspect	5.03	0.56	0.59	2.38	0.41	0.35	1.92	0.30	*0.43*
QL+query+aspect+QE	*5.12*	*0.58*	*0.61*	*2.5*	*0.43*	*0.41*	2.29	*0.37*	**0.49**

4   Conclusion

In this paper, we built an aspect-based retrieval model for scientific literature using TeKnowbase. Given a query and an aspect, this model returns a ranked list of documents that address that aspect for the query. We tested our model for 43 queries and 3 aspects with satisfactory results. We could beat the results obtained by adding aspect name explicitly to the query and doing pseudo-relevance feedback on those documents.

Acknowledgements. This work was partially supported by IIT Delhi-IBM Research AI Horizons Network collaborative grant; Ramanujan Fellowship, DST (ECR/2017/001691) and the Infosys Centre for AI, IIIT-Delhi, India. T. Chakraborty would like to thank the support of Google India Faculty Award.

References

1. Ammar, W., et al.: Construction of the literature graph in semantic scholar. In: NAACL (2018)
2. Raamkumar, A.S, Foo, S., Pang, N.: Can I have more ofthese please?: assisting researchers in finding similar research papers from a seed basket of papers. The Electronic Library (2018)
3. Bean, C., Green, R.: Relevance Relationships. Information Science and Knowledge Management (2001)
4. Chakraborty, T., Krishna, A., Singh, M., Ganguly, N., Goyal, P., Mukherjee, A.: FeRoSA: a faceted recommendation system for scientific articles. In: PAKDD (2016)
5. Dong, Y., Chawla, N.V., Swami, A.: Metapath2Vec: scalable representation learning for heterogeneous networks. In: KDD (2017)
6. Lao, N., Mitchell, T., Cohen, W.W.: Random walk inference and learning in a large scale knowledge base. In: EMNLP (2011)
7. Lavrenko, V., Croft, W.B.: Relevance based language models. In: SIGIR (2001)
8. Ley, M.: DBLP - some lessons learned. In: PVLDB (2009)
9. Manning, C.D., Raghavan, P., Schütze, H.: Introduction to Information Retrieval. Cambridge University Press, Cambridge (2008)
10. Ponte, J.M., Croft, W.B.: A language modeling approach to information retrieval. In: SIGIR (1998)
11. Shi, C., Kong, X., Yu, P.S., Xie, S., Wu, B.: Relevance search in heterogeneous networks. In: EDBT (2012)
12. Sugiyama, K., Kan, M.-Y.: A comprehensive evaluation of scholarly paper recommendation using potential citation papers. Int. J. Digit. Libr. **16**(2), 91–109 (2014). https://doi.org/10. 1007/s00799-014-0122-2
13. Sun, Y., Barber, R., Gupta, M., Aggarwal, C.C., Han, J.: Co-author relationship prediction in heterogeneous bibliographic networks. In: ASONAM (2011)
14. Upadhyay, P., Bindal, A., Kumar, M., Ramanath, M.: Construction and applications of teknowbase: a knowledge base of computer science concepts. In: Companion Proceedings of the The Web Conference (2018)
15. Xiong, C., Power, R., Callan, J.: Explicit semantic ranking for academic search via knowledge graph embedding. In: WWW (2017)

Dynamic Heterogeneous Graph Embedding Using Hierarchical Attentions

Luwei Yang$^{(\boxtimes)}$, Zhibo Xiao, Wen Jiang, Yi Wei, Yi Hu, and Hao Wang

Alibaba Group, Hangzhou, China
{luwei.ylw,xiaozhibo.xzb,wen.jiangw}@alibaba-inc.com

Abstract. Graph embedding has attracted many research interests. Existing works mainly focus on static homogeneous/heterogeneous networks or dynamic homogeneous networks. However, dynamic heterogeneous networks are more ubiquitous in reality, e.g. social network, e-commerce network, citation network, etc. There is still a lack of research on dynamic heterogeneous graph embedding. In this paper, we propose a novel dynamic heterogeneous graph embedding method using hierarchical attentions (DyHAN) that learns node embeddings leveraging both structural heterogeneity and temporal evolution. We evaluate our method on three real-world datasets. The results show that DyHAN outperforms various state-of-the-art baselines in terms of link prediction task.

Keywords: Graph embedding · Heterogeneous network · Dynamic graph embedding

1 Introduction

Graph (Network) embedding has attracted tremendous research interests. It learns the projection of nodes in a network into a low-dimensional space by encoding network structures or/and node properties. This technique has been successfully applied to various domains, such as recommendation [11,18], node classification [8], link prediction [1] and biology [7].

In real-world, graphs often not only evolve over time but also contain multiple types of nodes and edges. For instance, e-commerce network has two types of nodes, user and item, and multiple types of edges, click, buy, add-to-preference and add-to-cart. The nodes and edges may change over time. In social network, users may develop their multiple-type connections (follow, reply, retweet, etc) with others over time. The dynamics of a network and the structural heterogeneity provide abundant information for encoding nodes.

Recent research mainly focuses on static graph embedding which has a fixed set of nodes and edges. DeepWalk [9] and node2vec [6] leverage a random walk/ biased random walk and skip-gram model. LINE [12] preserves both first-order and second-order proximities. GCN [8] uses convolutional operations on node's neighborhood. GraphSAGE [7] or PinSAGE [18] proposes an inductive method

© Springer Nature Switzerland AG 2020
J. M. Jose et al. (Eds.): ECIR 2020, LNCS 12036, pp. 425–432, 2020.
https://doi.org/10.1007/978-3-030-45442-5_53

to aggregate structural information with node features. Further works consider heterogeneity. metapath2vec [2] takes meta-path into account when generating random walks. GATNE [1] aggregates node embedding by separating network into different views according to edge types. HAN [16] uses two-level attentions to learn the importance of neighbor nodes and meta-paths.

Dynamic graph embedding is an emerging area [17]. DynamicTriad [19] uses triadic closure to improve node embeddings. DySAT [10] extends the original GAT [15] to temporal graph snapshots. MetaDynaMix [4] proposes a metapath-based technique for dynamic heterogeneous information network embedding. More works may refer to [3,5,13].

Nonetheless, there is still a lack of research taking into account both temporal evolution and structural heterogeneity. Inspired by works on [16] and [10], we propose a novel dynamic heterogeneous graph embedding approach using hierarchical attention layers (DyHAN), which is able to capture the importance in different level aggregations. To be specific, for an arbitrary node, node-level attention intends to learn the importance of its neighbor for a specific edge type. Edge-level attention aims to learn the importance of every edge-type for this node. Temporal-level attention is able to fuse the final embedding by figuring out the importance of each time step graph snapshot. We evaluate our method on three real-world dynamic heterogeneous network datasets, EComm, Twitter and Aliaba.com. The results show that DyHAN outperforms several state-of-the-art baselines in link prediction task.

2   Problem Definition

In this section, we provide necessary information throughout this paper. We consider a dynamic heterogeneous network is defined as a series of snapshots, $\{G^1, G^2, ..., G^T\}$. A snapshot at time t is defined as $G^t = (\mathcal{V}^t, \mathcal{E}^t, \mathcal{W}^t)$, where $\mathcal{V}^t$ is the node set with node type $o \in \mathcal{O}$. $\mathcal{E}^t$ is the edge set with edge type $r \in \mathcal{R}$. $\mathcal{O}$ and $\mathcal{R}$ are node type set and edge type set respectively, and $|\mathcal{O}| + |\mathcal{R}| > 2$. We assume for each time snapshot the nodes and links both can be changed.

Dynamic heterogeneous graph embedding aims to learn a mapping function $f : \mathcal{V} \rightarrow \mathbb{R}^d$, such that it preserves the structural similarity among nodes and their temporal tendencies in developing link relationships.

3   Proposed Method

In this section, we introduce our proposed approach DyHAN employing hierarchical attentions on dynamic heterogeneous graph embedding which combines the basic ideas proposed in [10,16]. It has three main components, node-level attention, edge-level attention and temporal-level attention. All of these three components aggregate different layer of information using different attention layers. The overall architecture of DyHAN is represented by Fig. 1.

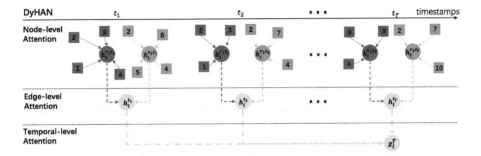

Fig. 1. Architecture of DyHAN.

Node-Level Attention. For each time step snapshot, we separate it into different subgraphs according to edge types. A self-attention is employed to aggregate node embedding for each subgraph. The importance of node pair (i, j) for edge type r and time step t can be expressed by,

$$\alpha_{ij}^{rt} = \frac{\exp(\sigma(\mathbf{a}_r^\top[\mathbf{W}_{nl}^r\mathbf{x}_i||\mathbf{W}_{nl}^r\mathbf{x}_j]))}{\sum_{k \in N_i^{rt}} \exp(\sigma(\mathbf{a}_r^\top[\mathbf{W}_{nl}^r\mathbf{x}_i||\mathbf{W}_{nl}^r\mathbf{x}_k]))}, \tag{1}$$

where σ is an activation function, $\mathbf{x}_i$ is the input representation of node i, $\mathbf{W}_{nl}^r$ is a linear transformation matrix, $||$ denotes the concatenation. N_i^{rt} denotes the sampled neighbor nodes for node i for edge type r and time step t. Different from [15] which uses all immediate neighbors, instead, for the sake of induction, we follow the framework described in [7] to use the sampled neighbors. $\mathbf{a}_r$ is a weight vector that parameterizes the attention function for edge type r. Then the embedding of node i for edge type r and time step t is obtained as,

$$\mathbf{h}_i^{rt} = \sigma \left(\sum_{j \in N_i^{rt}} \alpha_{ij}^{rt} \cdot \mathbf{W}_{nl}^r\mathbf{x}_j \right). \tag{2}$$

Note that the parameters are shared among different time step snapshots.

Edge-Level Attention. We assume the edge-specific node embedding expresses one semantic type of information in a heterogeneous graph. To aggregate these information more efficiently and robustly, we employ an attention layer to learn the importance of different edge types automatically. The importance of each edge type is calculated by an one-layer MLP.

$$\beta_i^{rt} = \frac{\exp(\mathbf{q}^\top \cdot \sigma(\mathbf{W}_{el}\mathbf{h}_i^{rt} + \mathbf{b}_{el}))}{\sum_{l=1}^{R} \exp(\mathbf{q}^\top \cdot \sigma(\mathbf{W}_{el}\mathbf{h}_i^{lt} + \mathbf{b}_{el}))} \tag{3}$$

where σ is an activation function, $\mathbf{q}^\top$ is the edge-level attention vector. $\mathbf{W}_{el}$ and $\mathbf{b}_{el}$ are the one-layer MLP's parameters. All parameters are shared across different time steps and different edge types. Then the fused embedding of node i is,

$$\mathbf{h}_i^t = \sum_{r=1}^{R} \beta_i^{rt} \cdot \mathbf{h}_i^{rt}. \tag{4}$$

Temporal-Level Attention. Once obtained the node embeddings for each time step snapshot, the next is to aggregate these node embeddings across a series of time snapshots. To compute the final node embedding, we use $\mathbf{h}_i^T$ to attend over all its historically-temporal representations, $\{\mathbf{h}_i^1, \mathbf{h}_i^2, ..., \mathbf{h}_i^{T-1}\}$. The Scaled Dot-Product Attention [14] is used by assuming that it is able to capture temporal evolution characteristics. We pack the representation of node i across time as $\mathbf{H}_i \in \mathbb{R}^{T \times D}$. Then, $\mathbf{H}_i$ is transformed into queries $\mathbf{Q} = \mathbf{H}_i \mathbf{W}_q$, keys $\mathbf{K} = \mathbf{H}_i \mathbf{W}_k$ and values $\mathbf{V} = \mathbf{H}_i \mathbf{W}_v$, where $\mathbf{W}_q \in \mathbb{R}^{D \times D'}$, $\mathbf{W}_k \in \mathbb{R}^{D \times D'}$ and $\mathbf{W}_v \in \mathbb{R}^{D \times D'}$. The temporal attention is defined as,

$$\mathbf{Z}_i = \text{softmax}(\frac{\mathbf{Q}\mathbf{K}^\top}{\sqrt{D'}} + \mathbf{M}) \cdot \mathbf{V}, \tag{5}$$

where $\mathbf{M} \in \mathbb{R}^{T \times T}$ is a mask matrix so that $\mathbf{h}_i$ only attends over time steps $\leq t$.

$$M_{ij} = \begin{cases} 0 & \text{if } i \leq j, \\ -\infty, & \text{otherwise} \end{cases} \tag{6}$$

We will use the $\mathbf{z}_i^T$ as the final node embedding. Note that multi-head attention could be applied to node-level and temporal-level attentions.

Optimization. In order to train the model capturing both structural and temporal information, we encourage nearby nodes at the last time step to have similar representations. A cross entropy loss is employed,

$$L(\mathbf{z}_u^T) = -\log(\sigma(<\mathbf{z}_u^T, \mathbf{z}_v^T>)) - Q \cdot \mathbb{E}_{v_n \sim P_n(v)} \log(\sigma(<-\mathbf{z}_u^T, \mathbf{z}_{v_n}^T>)) \tag{7}$$

where σ is the sigmoid function and $<,>$ denotes the inner product. v is the node that co-occurs near u on fixed-length random walk in the last time step. P_n is a negative sampling distribution, here we use the node's degree in the last time step. Q defines the number of negative samples.

4   Experiments

Datasets. We use three real-world datasets for evaluation. The statistics of them are summarized in Table 1.

$EComm^1$ dataset is sampled from the dataset of CIKM 2019 EComm AI contest from a category. There are two types of nodes, user and item. It has four types of edges including click, collect, add-to-cart and buy.

---

[1] https://tianchi.aliyun.com/competition/entrance/231719/introduction.

Twitter[2] dataset is sampled from the user behavior logs in Twitter about the discovery of elusive Higgs boson between 1st and 7th July 2012. There are three types of edges: retweet, reply and mention. Note that there is only one type of node.

Alibaba.com dataset is sampled from the user behavior logs in the alibaba.com e-commerce platform. A network from customer electronics category between 11th July and 21st July 2019 is sampled. It consists of interactions between users and items. There are three types of interactions, click, enquiry and contact.

Table 1. Statistics of datasets.

Dataset	# nodes	# edges	# node types	# edge types	# time steps
EComm	37724	91033	2	4	11
Twitter	100000	63410	1	3	7
Alibaba.com	16620	93956	2	3	11

Experimental Setup. We learn node embeddings based on graph snapshots $\{G^1, G^2, ..., G^t\}$, then a link prediction experiment is conducted on the last graph snapshot G^{t+1}.

A link prediction task aims to predict whether there is an existing link between any two nodes. We follow the evaluation framework for link prediction as stated in [10,19]. We create a Logistic Regression classifier for dynamic link predictions. We sample 20% of edges from the last time step snapshot as the held-out validation set for hyper-parameter tuning. The rest of edges of the last time step snapshot are used for link prediction task. In specific, we choose randomly 25% of links and the remaining 75% of links as training and test set respectively. An equal number of randomly sampled pairs of nodes without link as negative examples for each training and test set respectively. We use the inner product of the node embeddings of the node-pair as the representation feature of the link. Then Area Under the ROC Curve (AUC) [9] score and accuracy are used to report the performance.

Baselines. Considering availability of code and the effort of reimplementation, we compare our proposed DyHAN with following state-of-the-art static/dynamic and homogeneous/heterogeneous graph embedding algorithms. *DeepWalk* [9], we use the implementation provided by [7]. *Metapath2Vec* [2], the original implementation provided by the authors are dedicated to specific dataset. As a result, it is not convenient to directly generalize to other datasets. We reimplemented it in python. *GAT* [15], the original implementation provided by the authors is designed for node classification. We reimplemented it in the GraphSAGE framework. Note that the nodes to be attended over are sampled from immediate neighbors. *GraphSAGE* [7], we use the implementation provided by the

---

[2] http://snap.stanford.edu/data/higgs-twitter.html.

authors and use the default settings. Four variants with different node aggregation techniques are tested, namely, mean, mean-pooling, max-pooling and LSTM. *DynamicTriads* [19] and *DySAT* [10], we use the implementation provided by the authors. A method named *DyGAT* which ignores the structural heterogeneity was also implemented for comparison of incorporating heterogeneity. For random-walk based methods, we set the number of walks for each node as 50 and the length of each walk is set to 5. All training epoch is set to 1. All node embedding dimension is set to 32.

Results. The experimental results are shown by Table 2. DyHAN achieves the highest AUC score and accuracy among competitors. To be more specific, DyHAN obtains gains of 2.8%–4.9% on AUC and gains of 0.7%–7.8% on accuracy comparing the best baseline (exluding DyGAT). The gains of DyGAT over GAT show the efficacy of incorporating temporal information. Furthermore, the gains of DyHAN over DyGAT shows the efficacy of considering heterogeneity.

Table 2. Experimental results on three real-world datasets.

Method	EComm		Twitter		Alibaba.com	
	ROC-AUC	Accuracy	ROC-AUC	Accuracy	ROC-AUC	Accuracy
DeepWalk	0.573	0.554	0.571	0.661	0.558	0.538
Metapath2Vec	0.613	0.574	0.571[a]	0.661[a]	0.577	0.570
GraphSAGE-mean	0.640	0.602	0.557	0.640	0.549	0.533
GraphSAGE-meanpool	0.584	0.554	0.574	0.661	0.571	0.547
GraphSAGE-maxpool	0.638	0.606	0.559	0.622	0.568	0.551
GraphSAGE-LSTM	0.579	0.551	0.564	0.635	0.563	0.542
GAT	0.656	0.601	0.580	0.634	0.557	0.533
DynamicTriad	0.595	0.567	0.641	0.661	0.571	0.524
DySAT	0.504	0.496	0.652	0.661	0.523	0.527
DyGAT	0.680	0.638	0.645	0.633	0.569	0.539
DyHAN	**0.688**	**0.653**	**0.659**	**0.672**	**0.601**	**0.574**

[a]Note that Metapath2Vec is same as DeepWalk when the number of node type is one.

5   Conclusions

In this paper, we have proposed a novel hierarchical attention neural networks named DyHAN to learn node embeddings in dynamic heterogeneous graphs. DyHAN is able to effectively capture both structural heterogeneity and temporal evolution. Experimental results on three real-world datasets show that DyHAN outperforms several state-of-the-art techniques. One interesting future direction is exploring more temporal aggregation techniques.

References

1. Cen, Y., Zou, X., Zhang, J., Yang, H., Zhou, J., Tang, J.: Representation learning for attributed multiplex heterogeneous network. In: Proceedings of the 25th ACM SIGKDD Conference on Knowledge Discovery and Data Mining (2019)
2. Dong, Y., Chawla, N.V., Swami, A.: metapath2vec: scalable representation learning for heterogeneous networks. In: Proceedings of the 23rd ACM SIGKDD International Conference on Knowledge Discovery and Data Mining, pp. 135–144 (2017)
3. Du, L., Wang, Y., Song, G., Lu, Z., Wang, J.: Dynamic network embedding: an extended approach for skip-gram based network embedding. In: Proceedings of the Twenty-Seventh International Joint Conference on Aritificial Intelligence (2018)
4. Milani Fard, A., Bagheri, E., Wang, K.: Relationship prediction in dynamic heterogeneous information networks. In: Azzopardi, L., Stein, B., Fuhr, N., Mayr, P., Hauff, C., Hiemstra, D. (eds.) ECIR 2019. LNCS, vol. 11437, pp. 19–34. Springer, Cham (2019). https://doi.org/10.1007/978-3-030-15712-8_2
5. Goyal, P., Kamra, N., He, X., Liu, Y.: DynGEM: Deep embedding method for dynamic graphs. arXiv preprint arXiv:1805.11273 (2018)
6. Grover, A., Leskovec, J.: node2vec: scalable feature learning for networks. In: Proceedings of the 22nd ACM SIGKDD International Conference on Knowledge Discovery and Data Mining (2016)
7. Hamilton, W.L., Ying, R., Leskovec, J.: Inductive representation learning on large graphs. In: Advances in Neural Information Processing Systems, pp. 1024–1034 (2017)
8. Kipf, T.N., Welling, M.: Semi-supervised classification with graph convolutional networks. In: Proceedings of the International Conference on Learning Representations (2017)
9. Perozzi, B., Al-Rfou, R., Skiena, S.: Deepwalk: online learning of social representations. In: Proceedings of the 20th ACM SIGKDD International Conference on Knowledge Discovery and Data Mining, pp. 701–710 (2014)
10. Sankar, A., Wu, Y., Gou, L., Zhang, W., Yang, H.: Dynamic graph representation learning via self-attention networks. In: Proceedings of Workshop on Representation Learning on Graphs and Manifolds in the Seventh International Conference on Learning Representation (2019)
11. Shi, C., Hu, B., Zhao, W.X., Yu, P.S.: Heterogeneous information network embedding for recommendation. IEEE Trans. Knowl. Data Eng. **31**(2), 357–370 (2018)
12. Tang, J., Qu, M., Wang, M., Zhang, M., Yan, J., Mei, Q.: Line: large-scale information network embedding. In: Proceedings of the 24th International Conference on World Wide Web (2015)
13. Trivedi, R.S., Farajtabar, M., Biswal, P., Zha, H.: Learning dynamic graph representations. In: Proceedings of Workshop on Modeling and Decision-Making in the Spatiotemporal Domain in the 32nd Conference on Neural Information Processing Systems (2018)
14. Vaswani, A., et al.: Attention is all you need. In: Advances in Neural Information Processing Systems, pp. 5998–6008 (2017)
15. Veličković, P., Cucurull, G., Casanova, A., Romero, A., Liò, P., Bengio, Y.: Graph attention networks. In: Proceedings of the 6th International Conference on Learning Representations (2018)
16. Wang, X., et al.: Heterogeneous graph attention network. In: Proceedings of the Web Conference (2019)

17. Wu, Z., Pan, S., Chen, F., Long, G., Zhang, C., Yu, P.S.: A comprehensive survey on graph neural networks. arXiv preprint arXiv:1901.00596 (2019)
18. Ying, R., He, R., Chen, K., Eksombatchai, P., Hamilton, W.L., Leskovec, J.: Graph convolutional neural networks for web-scale recommender systems. In: Proceedings of the 24th ACM SIGKDD International Conference on Knowledge Discovery & Data Mining, pp. 974–983 (2018)
19. Zhou, L., Yang, Y., Ren, X., Wu, F., Zhuang, Y.: Dynamic network embedding by modeling triadic closure process. In: Proceedings of the Thirty-Second AAAI Conference on Artificial Intelligence (2018)

DSR: A Collection for the Evaluation of Graded Disease-Symptom Relations

Markus Zlabinger[1](✉), Sebastian Hofstätter[1], Navid Rekabsaz[2], and Allan Hanbury[1]

[1] TU Wien, Vienna, Austria
{markus.zlabinger,sebastian.hofstatter,allan.hanbury}@tuwien.ac.at
[2] Johannes Kepler University, Linz, Austria
navid.rekabsaz@jku.at

Abstract. The effective extraction of ranked disease-symptom relationships is a critical component in various medical tasks, including computer-assisted medical diagnosis or the discovery of unexpected associations between diseases. While existing disease-symptom relationship extraction methods are used as the foundation in the various medical tasks, no collection is available to systematically evaluate the performance of such methods. In this paper, we introduce the *Disease-Symptom Relation Collection* (DSR-collection), created by five physicians as expert annotators. We provide graded symptom judgments for diseases by differentiating between *relevant symptoms* and *primary symptoms*. Further, we provide several strong baselines, based on the methods used in previous studies. The first method is based on word embeddings, and the second on co-occurrences of MeSH-keywords of medical articles. For the co-occurrence method, we propose an adaption in which not only keywords are considered, but also the full text of medical articles. The evaluation on the DSR-collection shows the effectiveness of the proposed adaption in terms of nDCG, precision, and recall.

Keywords: Disease-symptom relationship · Medical expert data annotation · Disease-symptom information extraction

1 Introduction

Disease-symptom knowledge bases are the foundation for many medical tasks – including medical diagnosis [9] or the discovery of unexpected associations between diseases [12,14]. Most knowledge bases only capture a binary relationship between diseases and symptoms, neglecting the degree of the importance between a symptoms and a disease. For example, abdominal pain and nausea are both symptoms of an appendicitis, but while abdominal pain is a key differentiating factor, nausea does only little to distinguish appendicitis from other diseases of the digestive system. While several disease-symptom extraction methods have been proposed that retrieve a ranked list of symptoms for a disease [7,10,13,14],

© Springer Nature Switzerland AG 2020
J. M. Jose et al. (Eds.): ECIR 2020, LNCS 12036, pp. 433–440, 2020.
https://doi.org/10.1007/978-3-030-45442-5_54

no collection is available to systematically evaluate the performance of such methods [11]. While these method are extensively used in downstream tasks, e.g., to increase the accuracy of computer-assisted medical diagnosis [9], their effectiveness for disease-symptom extraction remains unclear.

In this paper, we introduce the **D**isease-**S**ymptom **R**elation Collection (DSR-collection) for the evaluation of graded disease-symptom relations. The collection is annotated by five physicians and contains 235 symptoms for 20 diseases. We label the symptoms using graded judgments [5], where we differentiate between: *relevant symptoms* (graded as 1) and *primary symptoms* (graded as 2). Primary symptoms—also called cardinal symptoms—are the leading symptoms that guide physicians in the process of disease diagnosis. The graded judgments allow us for the first time to measure the importance of different symptoms with grade-based metrics, such as nDCG [4].

As baselines, we implement two methods from previous studies to compute graded disease-symptom relations: In the first method [10], the relation is the cosine similarity between the word vectors of a disease and a symptom, taken from a word embedding model. In the second method [14], the relation between a disease and symptom is calculated based on their co-occurrence in the MeSH-keywords[1] of medical articles. We describe limitations of the keyword-based method [14] and propose an adaption in which we calculate the relations not only on keywords of medical articles, but also on the full text and the title.

We evaluate the baselines on the DSR-collection to compare their effectiveness in the extraction of graded disease-symptom relations. As evaluation metrics, we consider precision, recall, and nDCG. For all three metrics, our proposed adapted version of the keyword-based method outperforms the other methods, providing a strong baseline for the DSR-collection.

The contributions of this paper are the following:

- We introduce the DSR-collection for the evaluation of graded disease-symptom relations. We make the collection freely available to the research community.[2]
- We compare various baselines on the DSR-collection to give insights on their effectiveness in the extraction of disease-symptom relations.

2   Disease-Symptom Relation Collection

In this section, we describe the new **D**isease-**S**ymptom **R**elation Collection (DSR-collection) for the evaluation of disease-symptom relations. We create the collection in two steps: In the first step, relevant disease-symptom pairs (e.g. *appendicitis-nausea*) are collected by two physicians. They collect the pairs in a collaborative effort from high-quality sources, including medical textbooks and an online information service[3] that is curated by medical experts.

---

[1] MeSH-keywords are meta-data that indicates the core topics of an medical article.

[2] Contact this paper's first author to gain access.

[3] The website *netdoktor.at* which is certificated by the Health on the Net Foundation.

In the second step, the primary symptoms of the collected disease-symptom pairs are annotated. The annotation of primary symptoms is conducted to incorporate a graded relevance information into the collection. For the annotation procedure, we develop guidelines that briefly describe the task and an online annotation tool. Then, the annotation of primary symptoms is conducted by three physicians. The final label is obtained by a majority voting. Based on the labels obtained from the majority voting, we assign the relevance score 2 to *primary symptoms* and 1 to the other symptoms, which we call *relevant symptoms*.

In total, the DSR-collection contains relevant symptoms and primary symptoms for 20 diseases. We give an overview of the collection in Table 1. For the 20 diseases, the collection contains a total of 235 symptoms, of which 55 are labeled as primary symptom (about 25%). The top-3 most occurring symptoms are: *fatigue* which appears for 15 of the 20 diseases, *fever* which appears for 10, and *coughing* which appears for 7. Notice that the diseases are selected from different medical disciplines: mental (e.g. *Depression*), dental (e.g. *Periodontitis*), digestive (e.g. *Appendicitis*), and respiration (e.g. *Asthma*).

Table 1. Overview of the DSR-collection. For each disease, we display the number of relevant symptoms (#S), the number of primary symptoms (#P), and the Fleiss' inter-annotator agreement (κ).

Disease	#S	#P	κ	Disease	#S	#P	κ
Anorexia Nervosa	7	2	1.00	Influenza	11	2	0.57
Appendicitis	7	2	1.00	Measles	9	4	0.38
Asthma	9	4	0.76	Mental Depression	13	3	0.21
Bronchitis	9	1	0.71	Migraine Disorders	12	4	0.37
Cholecystitis	12	1	0.55	Myocardial Infarction	11	4	0.44
COPD	7	3	0.83	Periodontitis	3	4	0.46
Diabetes Mellitus	11	3	0.72	Pulmonary Embolism	13	2	0.83
Epididymitis	8	2	0.67	Sleep Apnea Syndromes	13	2	0.31
Erysipelas	7	3	0.69	Tonsillitis	7	4	0.63
GERD	8	2	0.76	Trigeminal Neuralgia	3	3	0.28

We calculate the inter-annotator agreement using Fleiss' kappa [2], a statistical measure to compute the agreement for three or more annotators. For the annotation of the primary symptoms, we measure a kappa value of $\kappa = 0.61$, which indicates a substantial agreement between the three annotators [6]. Individual κ-values per disease are reported in Table 1. By analyzing the disagreements, we found that the annotators labeled primary symptoms with varying frequencies: The first annotator annotated on average 2.1 primary symptoms per disease, the second 2.8, and the third 3.8.

Vocabulary Compatibility: We map each disease and symptom of the collection to the Unified Medical Language System (UMLS) vocabulary. The UMLS is a compendium of over 100 vocabularies (e.g. ICD-10, MeSH, SNOMED-CT) that are cross-linked with each other. This makes the collection compatible with the UMLS vocabulary and also with each of the over 100 cross-linked vocabularies.

Although the different vocabularies are compatible with the collection, a fair comparison of methods is only possible when the methods utilize the same vocabulary since the vocabulary impacts the evaluation outcome. For instance, the symptom *loss of appetite* is categorized as a symptom in MeSH; whereas, in the cross-linked UMLS vocabulary, it is categorized as a disease. Therefore, the symptom *loss of appetite* can be identified when using the MeSH vocabulary, but it cannot be identified when using the UMLS vocabulary.

Evaluation: We consider following evaluation metrics for the collection: Recall@k, Precision@k, and nDCG@k at the cutoff $k = 5$ and $k = 10$. *Recall* measures how many of the relevant symptoms are retrieved, *Precision* measures how many of the retrieved symptoms are relevant, and finally, *nDCG* is a standard metric to evaluate graded relevance [5].

3 Disease-Symptom Extraction Methods

3.1 Related Methods

In this section, we discuss disease-symptom extraction methods used in previous studies. A commonly used resource for the extraction of disease-symptom relations are the articles of the PubMed database. PubMed contains more than 30 million biomedical articles, including the abstract, title, and various metadata. Previous work [3, 7] uses the abstracts of the PubMed articles together with rule-based approaches. In particular, Hassan et al. [3] derive patterns of disease-symptom relations from dependency graphs, followed by the automatic selection of the best patterns based on proposed selection criteria. Martin et al. [7] generate extraction rules automatically, which are then inspected for their viability by medical experts. Xia et al. [13] design special queries that include the name and synonyms of each disease and symptom. They use these queries to return the relevant articles, and use the number of retrieved results to perform a ranking via Pointwise Mutual Information (PMI).

The mentioned studies use resources that are not publicly available, i.e., rules in [3,7] and special queries in [13]. To enable reproducibility in future studies, we define our baselines based on the methods that only utilize publicly available resources, described in the next section.

3.2 Baseline Methods

Here, we first describe two recently proposed methods [10,14] for the extraction of disease-symptom relations as our baselines. Afterwards, we describe limitations of the method described in [14] and propose an adapted version in which the limitations are addressed. We apply the methods on the open-access subset of the PubMed Central (PMC) database, containing 1,542,847 medical articles. To have a common representation for diseases/symptoms across methods (including an unique name and identifier), we consider the 382 symptoms and 4,787 diseases from the Medical Subject Headings (MeSH) vocabulary [14]. Given the set of diseases (X) and symptoms (S), each method aims to compute a relation scoring function $\lambda(x, s) \in \mathbb{R}$ between a disease $x \in X$ and a symptom $s \in S$. In the following, we explain each method in detail.

EMBEDDING: Proposed by Shah et al. [10], the method is based on the cosine similarity of the vector representations of a disease and a symptom. We first apply MetaMap [1], a tool for the identification of medical concepts within a given text, to the full text of all PMC articles to substitute the identified diseases/symptoms by their unique names. Then, we train a word2vec model [8] with 300 dimensions and a window size of 15, following the parameter setting in [10]. Using the word embedding, the disease-symptom relation is defined as $\lambda(x, s) = \cos(\mathbf{e}_x, \mathbf{e}_s)$, where $\mathbf{e}$ refers to the vector representation of a word.

COOCCUR: This method, proposed by Zhou et al. [14], calculates the relation of a disease and a symptom, by measuring the degree of their co-occurrences in the MeSH-keywords of medical articles. The raw co-occurrence of the disease x and symptom s, is denoted by $co(x, s)$. The raw co-occurrence does not consider the overall appearance of each symptom across diseases. For instance, symptoms like *pain* or *obesity* tend to co-occur with many diseases, and are therefore less informative. Hence, the raw co-occurrence is normalized by an *Inverse Symptom Frequency (ISF)* measure, defined as $\text{ISF}(s) = \frac{|X|}{n_s}$, where $|X|$ is the total number of diseases and n_s is the number of diseases that co-occur with s at least in one of the articles. Finally, the disease-symptom relation is defined as $\lambda(x, s) = co(x, s) \times \text{ISF}(s)$. We compute three variants of the COOCCUR method:

- KWD: The disease-symptom relations are computed using the MeSH-keywords of the ≈1.5 million PMC articles.
- KWDLARGE: While KWD uses the 1.5 million PMC articles, Zhou et al. [14] apply the exact same method on the ≈30 million articles of the PubMed database. While they did not evaluate the effectiveness of their disease-symptom relation extraction method, they published their relation scores which we will evaluate in this paper.
- FULLTEXT: Applying the COOCCUR method only on MeSH-keywords has two disadvantages: First, keywords are not available for all articles (e.g. only 30% of the ≈1.5 million PMC articles have keywords) and second, usually only the core topics of an article occur as keywords. We address these limitations by proposing an adaption of the COOCCUR method, in which we use the full text, the title, and the keywords of the ≈1.5 million PMC articles. Specifically, we adapt the computation of the co-occurrence $co(x, s)$, as follows: We first retrieve a set of relevant articles to a disease x, where an article is relevant if the disease exists in either the keyword, or the title section of the article. Given these relevant articles and a symptom s, we compute the adapted co-occurrence $co(x, s)$, which is the number of relevant articles in that the

symptom occurs in the full text. The identification of the diseases in the title and symptoms in the full text is done using the MetaMap tool [1].

4   Evaluation Results and Discussion

We now compare the disease-symptom extraction baselines on the proposed DSR-collection. The results for various evaluation metrics are shown in Table 2. The FULLTEXT-variant of the COOCCUR method outperforms the other baselines on all evaluation metrics. This demonstrates the high effectiveness of our proposed adaption to the COOCCUR method.

Further, we see a clear advantage of the COOCCUR-method with MeSH-keywords from ≈30 million PubMed articles as the resource (KWDLARGE) – in comparison to the same method with keywords from approximately 1.5 million PMC articles (KWD). This highlights the importance of the number of input samples to the method.

Table 2. Comparison of the disease-symptom extraction methods using our proposed DSR-collection. We show significant improvements with: a refers to EMBEDDING, b to KWD, and c to KWDLARGE (two-sided, paired t-test: $p < 0.01$).

Method	nDCG@5	P@5	R@5	nDCG@10	P@10	R@10
EMBEDDING	0.20	0.18	0.08	0.19	0.15	0.13
COOCCUR-KWD	0.27	0.22	0.10	0.22	0.14	0.12
COOCCUR-KWDLARGE	0.32^a	0.27	0.12	0.28^{ab}	0.19	0.17
COOCCUR-FULLTEXT	$\mathbf{0.41}^{abc}$	$\mathbf{0.39}^{abc}$	$\mathbf{0.17}^{abc}$	$\mathbf{0.36}^{abc}$	$\mathbf{0.28}^{abc}$	$\mathbf{0.25}^{ab}$

Error Analysis: A common error source is a result of the fine granularity of the symptoms in the medical vocabularies. For example, the utilized MeSH vocabulary contains the symptoms *abdominal pain* and *abdomen, acute*[4]. Both symptoms can be found in the top ranks of the evaluated methods for the disease appendicitis (see Table 3). However, since the corpus is not labeled on such a fine-grained level, the symptom *abdomen, acute* is counted as a false positive.

Table 3. Top-4 extracted symptoms of each method for the disease *appendicitis*. The retrieved relevant symptoms and primary symptoms are highlighted.

EMBEDDING	KWD	KWDLARGE	FULLTEXT
Abdomen, Acute	Abdominal Pain	Abdomen, Acute	Abdomen, Acute
Abdominal Pain	Abdomen, Acute	Abdominal Pain	Abdominal Pain
Fever of Unknown Origin	Obesity	Pelvic Pain	Vomiting
Renal Colic	Thinness	Pain, Postoperative	Nausea

---

[4] Symptom for acute abdominal pain.

Another error source is a result of the bias in medical articles towards specific disease-symptom relationships. For instance, between the symptom *obesity* and *periodontitis*[5] a special relationship exists, which is the topic of various publications. Despite obesity not being a characteristic symptom of a periodontitis, all methods return the symptom in the top-3 ranks. A promising research direction is the selective extraction of symptoms from biomedical literature by also considering the context (e.g. in a sentence) in that a disease/symptom appears.

5   Conclusion

We introduced the *Disease-Symptom Relation Collection* (DSR-collection) for the evaluation of graded disease-symptom relations. We provided baseline results for two recent methods, one based on word embeddings and the second on the co-occurrence of MeSH-keywords of medical articles. We proposed an adaption to the co-occurrence method to make it applicable to the full text of medical articles and showed significant improvement of effectiveness over the other methods.

References

1. Aronson, A.R.: Effective mapping of biomedical text to the UMLS Metathesaurus: the MetaMap program. In: Proceedings of the American Medical Informatics Association Symposium, pp. 17–21 (2001)
2. Fleiss, J.L.: Measuring nominal scale agreement among many raters. Psychol. Bull. **76**(5), 378 (1971)
3. Hassan, M., Makkaoui, O., Coulet, A., Toussain, Y.: Extracting disease-symptom relationships by learning syntactic patterns from dependency graphs. In: Proceedings of BioNLP 2015, pp. 71–80. Association for Computational Linguistics, Beijing (2015)
4. Järvelin, K., Kekäläinen, J.: Cumulated gain-based evaluation of IR techniques. ACM Trans. Inf. Syst. (TOIS) **20**(4), 422–446 (2002)
5. Kekäläinen, J.: Binary and graded relevance in IR evaluations - comparison of the effects on ranking of IR systems. Inf. Process. Manag. **41**(5), 1019–1033 (2005)
6. Landis, J.R., Koch, G.G.: The measurement of observer agreement for categorical data. Biometrics **33**, 159–174 (1977)
7. Martin, L., Battistelli, D., Charnois, T.: Symptom extraction issue. In: Proceedings of BioNLP 2014, pp. 107–111 (2014). Association for Computational Linguistics, Baltimore (2014)
8. Mikolov, T., Sutskever, I., Chen, K., Corrado, G.S., Dean, J.: Distributed representations of words and phrases and their compositionality. In: Advances in Neural Information Processing Systems, pp. 3111–3119 (2013)
9. Ni, J., Fei, H., Fan, W., Zhang, X.: Automated medical diagnosis by ranking clusters across the symptom-disease network. In: 2017 IEEE International Conference on Data Mining (ICDM), pp. 1009–1014, November 2017
10. Shah, S., Luo, X., Kanakasabai, S., Tuason, R., Klopper, G.: Neural networks for mining the associations between diseases and symptoms in clinical notes. Health Inf. Sci. Syst. **7**(1), 1–9 (2018). https://doi.org/10.1007/s13755-018-0062-0

---

[5] A dental disease where the gum that surrounds the teeth retreats.

11. Shen, Y., Li, Y., Zheng, H.T., Tang, B., Yang, M.: Enhancing ontology-driven diagnostic reasoning with a symptom-dependency-aware Naïve Bayes classifier. BMC Bioinform. **20**(1), 330 (2019)
12. del Valle, E.P.G., García, G.L., Santamaría, L.P., Zanin, M., Ruiz, E.M., González, A.R.: Evaluating Wikipedia as a source of information for disease understanding. In: 2018 IEEE 31st International Symposium on Computer-Based Medical Systems (CBMS), pp. 399–404, June 2018
13. Xia, E., Sun, W., Mei, J., Xu, E., Wang, K., Qin, Y.: Mining disease-symptom relation from massive biomedical literature and its application in severe disease diagnosis. In: 45 - AMIA 2018 Annual Symposium, pp. 1118–1126 (2018)
14. Zhou, X., Menche, J., Barabási, A.L., Sharma, A.: Human symptoms-disease network. Nat. Commun. **5**(1), 1–10 (2014)

Demonstration Papers

A Web-Based Platform for Mining and Ranking Association Rules

Addi Ait-Mlouk$^{(\boxtimes)}$ and Lili Jiang

Department of Computing Science, Umeå University, Umeå, Sweden
{addi.ait-mlouk,lili.jiang}@cs.umu.se

Abstract. In this demo, we introduce an interactive system, which effectively applies multiple criteria analysis to rank association rules. We first use association rules techniques to explore the correlations between variables in given data (i.e., database and linked data (LD)), and secondly apply multiple criteria analysis (MCA) to select the most relevant rules according to user preferences. The developed system is flexible and allows intuitive creation and execution of different algorithms for an extensive range of advanced data analysis topics. Furthermore, we demonstrate a case study of association rule mining and ranking on road accident data.

Keywords: Data mining · Association rules · Multiple criteria analysis

1 Introduction

Association rules mining is a powerful technique in data mining [3] for discovering correlation and relationships between objects. This technique like other data mining techniques has been integrated into a bunch of open-source tools, such as WEKA [11], TANAGRA [10], fpm [4] and DM-MCDA [1]. Most public tools are desktop-based and don't provide intuitive use for the data mining community, hence reduce the chance of re-usability and/or upgradability by developers.

In this demo, we (1) implement a baseline approach to extract association rules from given data, (2) propose an approach based on multiple criteria analysis to rank the extracted rules according to the decision maker's preferences, (3) demonstrate the idea of association rules mining through road accident data as a case study, where the dataset containing information on location, drivers, and the accident characteristics, the vehicles involved and victims [2], and (4) present the preliminary results of the proposed system. Our implementation will be open-source and a live demo can be found at https://youtu.be/QILaVUghlsM.

2 Mining and Ranking Association Rules

Figure 1 shows our framework architecture which includes three main modules: *data preprocessing*, *association rules mining*, and *multiple criteria analysis*. *Data preprocessing* allows users to process and prepare data to be used by *data mining*

© Springer Nature Switzerland AG 2020
J. M. Jose et al. (Eds.): ECIR 2020, LNCS 12036, pp. 443–448, 2020.
https://doi.org/10.1007/978-3-030-45442-5_55

algorithms to extract association rules based on some input thresholds (minimum support and minimum confidence). Afterward, we use *multiple criteria analysis* to rank the relevant rules based on user preferences and assign the result to different relevant categories. The system was implemented on R [7] and r shiny [9], providing interactive and user-friendly interfaces for different modules. The system can be extended with new data sets on demand, rendering information retrieval and query linked data using SPARQL queries (data linkage).

2.1   Quality Measurement of Association Rules

In order to select interesting rules from the large set of extracted rules, constraints on various measures of significance and interest are used [5]. The best-known constraints are minimum support, minimum confidence and lift. Due to page limitations, we refer readers to [5] for further references.

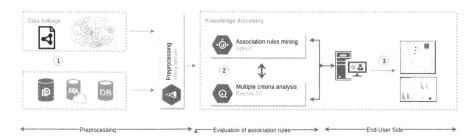

Fig. 1. Overview of system architecture: (1) Data preprocessing and data linkage, (2) Association rules mining with multiple criteria analysis, and (3) Visualization

2.2   Multiple Criteria Analysis (MCA)

MCA is a sub-field of operational research, and management science, dedicated to the development of decision support tools in order to solve complex decision problems involving multiple criteria objectives. When modeling a real decision problem using multiple criteria analysis, several issues [8] can be considered: choice, sorting and ranking. In our context of a large number of extracted rules, we apply the multiple criteria ranking precisely, in a method called ELECTRE (ELimination and Choice Expressing REality) TRI [5,6]. We conduct the following three computations to present the process of ranking:

1. The partial concordance indices $c_j(a, b_h)$: (it represents the degree of concordance with the hypothesis of outranking of a over b where a is rule and b is profile):

$$C_j(a, b_h) = \begin{cases} 0 \ if \ g_j(b_h) - g_j(a) \succeq p_j(b_h) \\ 1 \ if \ g_j(b_h) - g_j(a) \preceq q_j(b_h) \\ \quad if \ not \ \frac{p_j(b_h) + g_j(a)}{p_j(b_h) - q_j(b_h)} \end{cases} \quad (1)$$

2. The discordance indices $d_j(a, b_h)$: (it represents the degree of discordance with the hypothesis that a outranks b on elementary criterion g_j).

$$d_j(a, b_h) = \begin{cases} 0 \ if \ g_j(a_h) \preceq g_j(b_h) + p_j(b_h) \\ 1 \ if \ g_j(a_h) \succ g_j(b_h) + v_j(b_h) \\ \qquad if \ not \ \in [0, 1] \end{cases} \tag{2}$$

3. The credibility indices $\sigma(a, b_h)$: (it represents the credibility of the outranking of a over b on criterion j):

$$\sigma(a, b_h) = C(a, b_h) \prod_{j \in \overline{F}} \frac{1 - d_j(a, b_h)}{1 - C(a, b_h)} \tag{3}$$

where, K_j is the weight of criteria j, $C_j(a, b_h)$ is the partial concordance index of criteria j, and $F = \{j \in F : d_j(a, b_h) > C(a, b_h)\}$

The pseudo-code of the proposed algorithm takes as input a list of extracted rules $(a_1, a_2,...,a_n)$, a list of quality measurement (support, confidence,lift), a list of profiles (b_1, b_2) and the preference (p), indifference (q), veto (v), cutting level λ thresholds. It computes the concordance, discordance and credibility indices for each rule under a given quality measurement. After iterating all extracted rules, the algorithm compares the credibility indices $\sigma(a, b_h)$ and the default value of λ (0.7) then generates a list of relevant rules [5].

3   Demonstration

Firstly, users are allowed to upload a new CSV dataset or use SPARQL query to retrieve data from linked data (Dbpedia, wikidata, etc.). As shown in Fig. 2, the left panel displays data file input for a dataset (support CSV file) and parameter settings for some thresholds (attributes, minsup, minconf, etc). Secondly, users extract association rules and the main panel displays seven different tabs to characterize association rules across data sources and their summaries information. *Summary* tab shows the summary of rules mining algorithm, *ScatterPlot* tab shows the scatter-plot of extracted rules, *FrequentItemsets* tab shows frequent itemsets from data, and the rest of tabs shows interactive techniques to visualize association rules.

Figure 3 shows the association rules evaluation interface. The parameters configuration panel is displayed on the left side including method choice and the list of association rules with quality measurement. The main panel presents different steps for ELECTRE TRI (Eqs *(1)–(3)*), afterward, user can move to the bottom panels and clicks 'Decision matrix' tab to see the performance table of association rules associated with predefined thresholds (Table 2) and quality

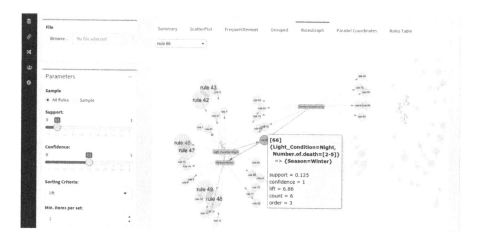

Fig. 2. Association rules mining and visualization interface

measurement. Furthermore, user can move to other tabs for different computations and assign association rules to different categories (The set of categories to which the rules must be assigned to is completely ordered from the best to the worst), the rules in the first category will be the most relevant ones to user preferences (Table 1).

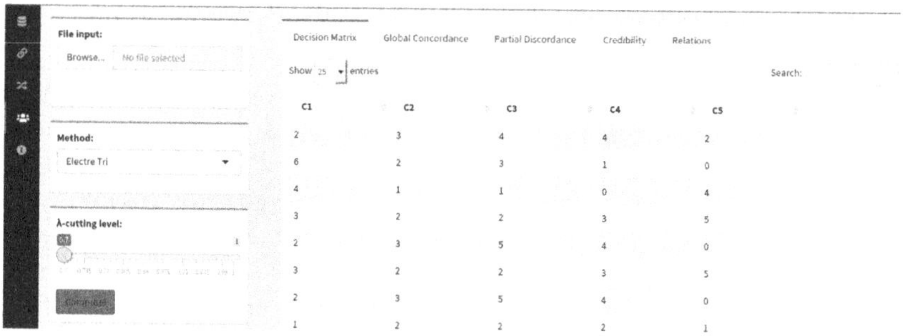

Fig. 3. Select and rank extracted association rules through MCA

Table 1. Profiles defining the category limits

Profile	Support	Confidence	Lift
b_1	0.3	0.6	1.0
b_1	0.4	0.7	0.9

Table 2. Thresholds for ELECTRE TRI method

Thresholds	Support	Confidence	Lift
$weight(k_j)$	0.4	0.7	0.9
$p_j(b_1)$	0.3	0.6	1.0
$q_j(b_1)$	0.4	0.5	0.9
$v_j(b_1)$	0.4	0.8	0.8
$p_j(b_2)$	0.2	0.7	1.0
$q_j(b_2)$	0.5	0.6	0.9
$v_j(b_2)$	0.4	0.5	1.0

Association rules mining algorithms produce a large number of rules. It is, therefore, necessary to choose relevant ones according to the decision-makers' preferences and predefined thresholds p_j, q_j, v_j. Based on this study, the integration of multiple criteria analysis within the association rules process performs well and produces relevant rules. After eliminating the rules users are not interested in, eleven significant rules were obtained in this case study (R1, R2, R5, R10, R15, R20, R30, R31, R15, R22, R13).

4 Conclusion

This demo paper presents an open-source software for information retrieval by combining data mining especially association rules analysis and multiple criteria analysis. The system starts with data processing by uploading a new CSV dataset or uses a SPARQL query to retrieve data from linked data. Afterward, users obtain association rules extracted by the Apriori algorithm and the relevant rules are chosen using multiple criteria analysis approach. In the future, we aim to extend the system with more data mining algorithms in the context of big data.

Acknowledgement. This research is funded by Umeå University in Sweden on federated database research.

References

1. Ait-Mlouk, A., Agouti, T.: DM-MCDA: a web-based platform for data mining and multiple criteria decision analysis: a case study on road accident. SoftwareX **10**, 100323 (2019). https://doi.org/10.1016/j.softx.2019.100323
2. Ait-Mlouk, A., Gharnati, F., Agouti, T.: An improved approach for association rule mining using a multi-criteria decision support system: a case study in road safety. Eur. Transp. Res. Rev. **9**(3), 40 (2017)
3. Fayyad, U.M., Piatetsky-Shapiro, G., Smyth, P.: Advances in Knowledge Discovery and Data Mining, pp. 1–34 (1996)
4. FPM (2019). http://www.borgelt.net/software.html

5. Le Bras, Y., Meyer, P., Lenca, P., Lallich, S.: A robustness measure of association rules. In: Balcázar, J.L., Bonchi, F., Gionis, A., Sebag, M. (eds.) ECML PKDD 2010. LNCS (LNAI), vol. 6322, pp. 227–242. Springer, Heidelberg (2010). https://doi.org/10.1007/978-3-642-15883-4_15

6. Mousseau, V., Figueira, J., Naux, J.P.: Using assignment examples to infer weights for ELECTRE TRI method: some experimental results. Eur. J. Oper. Res. **130**(2), 263–275 (2001)

7. R (2019). https://www.r-project.org/

8. Roy, B., Vincke, P.: Multicriteria analysis: survey and new directions. Eur. J. Oper. Res. **8**(3), 207–218 (1981)

9. Rstudio: R shiny (2019). https://shiny.rstudio.com/

10. TANAGRA (2019). http://eric.univ-lyon2.fr/~ricco/tanagra/fr/tanagra.html

11. WEKA (2019). https://www.cs.waikato.ac.nz/~ml/weka/index.html

Army ANT: A Workbench for Innovation in Entity-Oriented Search

José Devezas$^{(\boxtimes)}$ and Sérgio Nunes

INESC TEC and Faculty of Engineering, University of Porto,
Rua Dr. Roberto Frias, s/n, 4200-465 Porto, Portugal
{jld,ssn}@fe.up.pt

Abstract. As entity-oriented search takes the lead in modern search, the need for increasingly flexible tools, capable of motivating innovation in information retrieval research, also becomes more evident. Army ANT is an open source framework that takes a step forward in generalizing information retrieval research, so that modern approaches can be easily integrated in a shared evaluation environment. We present an overview on the system architecture of Army ANT, which has four main abstractions: (*i*) readers, to iterate over text collections, potentially containing associated entities and triples; (*ii*) engines, that implement indexing and searching approaches, supporting different retrieval tasks and ranking functions; (*iii*) databases, to store additional document metadata; and (*iv*) evaluators, to assess retrieval performance for specific tasks and test collections. We also introduce the command line interface and the web interface, presenting a learn mode as a way to explore, analyze and understand representation and retrieval models, through tracing, score component visualization and documentation.

Keywords: Evaluation framework · Entity-oriented search · Representation modeling · Retrieval modeling

1 Introduction

Army ANT is an experimental workbench, built as a centralized codebase for research work in entity-oriented search. Over the years, there have been several experimental frameworks in information retrieval. Some of the most notable include the Lemur Project [1], Terrier [10] and, more recently, Nordlys [8], which is also focused on entity-oriented search. Army ANT was created as a structured framework for testing novel retrieval approaches in a comprehensive manner, even when potentially deviating from traditional paradigms. This required a flexible structure, that we developed by iteratively satisfying the requirements of multiple engine implementations for representing and retrieving combined data [4, Definition 2.3]. An important step in research, that we also motivate and support through our framework, is the continuous documentation of models and collections, which is fundamental for reproducibility, but also useful to advance research, by exploring, learning and building on previous approaches.

J. M. Jose et al. (Eds.): ECIR 2020, LNCS 12036, pp. 449–453, 2020.
https://doi.org/10.1007/978-3-030-45442-5_56

2   System Architecture

The basic unit of Army ANT is the engine, which must implement the representation model for indexing and the retrieval model for searching. The indexing method has access to one of multiple collection readers and can optionally consider external features. The search method is based on a keyword query, pagination parameters and, optionally, a task identifier, a ranking function and its parameters, and a debug flag. For searching and evaluating over the web interface, each engine is required to have a unique identifier, which frequently describes the representation model and indexed collection (e.g., *lucene-wapo* for a Lucene index over the TREC Washington Post Corpus (WaPo)[1]). Each engine has an entry in the YAML configuration file (*config.yaml*), so that it is visible to the web interface. Supported ranking functions, their parameter names and specific values can also be defined in the configuration file. Combinations of selected parameter values can then be used by the evaluation module to launch individual runs, known as evaluation tasks. When completed, each task will provide a performance overview, based on efficiency and effectiveness metrics for each parameter configuration, as well as complementary visualizations and a ZIP archive with intermediate results. Intermediate results include elements like the average precisions for each topic, used in the calculation of the mean average precision, or the results for each individual topic, along with the relevance per retrieved item, according to a ground truth (e.g., qrels from TREC or INEX). This means that, even if Army ANT evolves and no backward compatibility is maintained, the archive can still be downloaded and independently used to compute other metrics, such as statistical tests, or to correct any wrong calculations. Additionally, an overall table, comparing the performance among different runs, is also available for download as a CSV or LaTeX file.

Out-of-the-box, Army ANT[2] provides reader implementations for INEX 2009 Wikipedia Collection [12], TREC Washington Post Corpus, and Living Labs API [3] documents. It also provides a Lucene baseline engine, supporting TF-IDF, BM25 and divergence from randomness, a Lucene features helper, to index and combine external features using the sigmoid approach by Craswell et al. [5], a TensorFlow Ranking [11] engine, which uses Lucene to compute features, and other experimental engines, such as graph-of-entity [6] and hypergraph-of-entity [7]. The latter model supports several tasks, including ad hoc document retrieval (with entities) [2, Ch. 8], ad hoc entity retrieval [2, §3.1], related entity finding [2, §4.4.3] and entity list completion [2, p. 91], that are not easily explored through conventional evaluation frameworks with the concept of retrieval task. Finally, evaluators are available for the INEX Ad Hoc track and the INEX XER track, as well as for the TREC Common Core track and for the Living Labs API team-draft interleaving online evaluation. On a smaller scale, Army ANT also provides several utility functions, covering DBpedia and Wikidata access, as well as statistics for the measurement of rank concordance and

---

[1] https://trec.nist.gov/data/wapost/.
[2] https://github.com/feup-infolab/army-ant-install/tree/ecir-2020.

correlation. Several index inspection, debugging tools and documentation strategies are also integrated into Army ANT's workflow. The workbench is written in Python, providing integrated implementations for engines written in Java and C++, which we use as examples of cross-language interoperability.

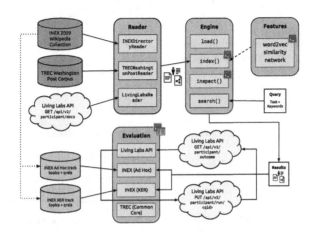

Fig. 1. Army ANT system architecture. Solid arrows represent information flow, while dashed arrows represent optional interactions. Dotted arrows are simply used to indicate subcomponents of test collections (i.e., topics and relevance judgments).

2.1 Overview

We divided the system into what we consider the atomic components of information retrieval research:

1. Iterate over the units of information in a collection (*reader*);
2. Index and search for those units of information (*engine*),
3. Eventually decorate them with additional metadata (*database*);
4. Assess the effectiveness and efficiency of the retrieval (*evaluator*);
5. Obtain as much additional information as possible about the system, in order to reiterate and improve (*web interface* ⇒ *learn mode*).

Figure 1 provides an overview of the components in Army ANT, illustrating how they interact with test collections or APIs, as well as with each other. It shows some of the supported implementations, namely readers and evaluators, for both disk-based and REST-based data, and it illustrates feature providers, such as word2vec similarities, that can also be integrated into an index (e.g., providing contextual similarity links to the hypergraph-of-entity). Finally, we can see that a query is defined as a task and a sequence of keywords, and that results can be based on documents, entities, and their relations. Each component may have a command line icon, as well as a web interface icon, showing how it is available to the user.

Fig. 2. Evaluation task submission, showing ranking function parameter selection.

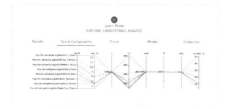

Fig. 3. Learn mode: parallel coordinates visualization of the score components for a query to graph-of-word.

Fig. 4. Exporting evaluation results.

Listing 1.1. CLI indexing example.

```
./army-ant.py index \
    --source-path "data/inex-2009" \
    --source-reader "inex_dir" \
    --index-location "index/lucene" \
    --index-type "lucene" \
    --db-name "aa_inex"
```

2.2   Interface

The command line interface can be used for instance for indexing a collection, as seen in Listing 1.1, where the command **index** was issued along with arguments for the source collection, target index and an optional database. A web interface is also available, with modules for accessing search and learn modes, and managing evaluation tasks. Figure 2 illustrates a run for the topics and qrels of the INEX Ad Hoc track, based on the hypergraph-of-entity and the random walk score, configuring values for four parameters. Figure 4 shows the preview dialog for exporting a selection of effectiveness metrics, for all runs. Figure 3 illustrates the score component visualization, a part of the learn mode, which is based on the parallel coordinates system [9].

3   Conclusion

We have presented Army ANT, a flexible workbench for innovation in entity-oriented search and a general platform to support information retrieval research. It promotes reusability by separating collection reading from indexing and by structuring the process of implementing new representation and retrieval models with minimal constraints. One of the biggest strengths of Army ANT is its web interface, where researchers can demo their search engine, as well as explore, understand and analyze several of its facets, either tracing the ranking process for particular queries or visualizing the score components for those same queries.

At the same time, we also provide a way for researchers to document their models and collections, using the learn mode to transfer knowledge to other researchers or even to students in a classroom.

Acknowledgements. This work was financed by the Portuguese funding agency, FCT – Fundação para a Ciência e a Tecnologia, through national funds, and co-funded by the FEDER, where applicable. José Devezas is supported by research grant PD/BD/128160/2016, provided by FCT, within the scope of POCH, supported by the European Social Fund and by national funds from MCTES.

References

1. The Lemur Toolkit for Language Modeling and Information Retrieval | Center for Intelligent Information Retrieval | UMass Amherst. https://ciir.cs.umass.edu/lemur. Accessed 20 Dec 2019
2. Balog, K.: Entity-Oriented Search. Springer, Cham (2018). https://doi.org/10.1007/978-3-319-93935-3
3. Balog, K., Kelly, L., Schuth, A.: Head first: living labs for ad-hoc search evaluation. In: Proceedings of the 23rd ACM International Conference on Conference on Information and Knowledge Management, CIKM 2014, pp. 1815–1818 (2014)
4. Bast, H., Buchhold, B., Haussmann, E., et al.: Semantic search on text and knowledge bases. Found. Trends® Inf. Retrieval **10**(2–3), 119–271 (2016)
5. Craswell, N., Robertson, S.E., Zaragoza, H., Taylor, M.J.: Relevance weighting for query independent evidence. In: Proceedings of the 28th Annual International ACM SIGIR Conference on Research and Development in Information Retrieval, SIGIR 2005, Salvador, Brazil, pp. 416–423, August 2005
6. Devezas, J., Lopes, C., Nunes, S.: Graph-of-entity: a model for combined data representation and retrieval. In: 8th Symposium on Languages, Applications and Technologies, SLATE 2019, Coimbra, Portugal, June 2019
7. Devezas, J., Nunes, S.: Hypergraph-of-entity: a unified representation model for the retrieval of text and knowledge. Open Comput. Sci. J. **9**(1), 103–127 (2019)
8. Hasibi, F., Balog, K., Garigliotti, D., Zhang, S.: Nordlys: a toolkit for entity-oriented and semantic search. In: Proceedings of the 40th International ACM SIGIR Conference on Research and Development in Information Retrieval, SIGIR 2017, New York, NY, USA, pp. 1289–1292 (2017)
9. Inselberg, A.: Parallel coordinates: visual multidimensional geometry and its applications. In: Proceedings of the International Conference on Knowledge Discovery and Information Retrieval, KDIR 2012, Barcelona, Spain, October 2012
10. Ounis, I., Amati, G., Plachouras, V., He, B., Macdonald, C., Lioma, C.: Terrier: a high performance and scalable information retrieval platform. In: Proceedings of ACM SIGIR 2006 Workshop on Open Source Information Retrieval, OSIR 2006, Seattle, Washington, USA, August 2006
11. Pasumarthi, R.K., et al.: TF-ranking: scalable TensorFlow library for learning-to-rank. In: Proceedings of the 25th ACM SIGKDD International Conference on Knowledge Discovery & Data Mining, KDD 2019, Anchorage, AK, USA, pp. 2970–2978, August 2019. https://doi.org/10.1145/3292500.3330677
12. Schenkel, R., Suchanek, F., Kasneci, G.: YAWN: A semantically annotated Wikipedia XML corpus. In: Datenbanksysteme in Business, Technologie und Web (BTW 2007) – 12. Fachtagung des GI-Fachbereichs "Datenbanken und Informationssysteme" (DBIS), Gesellschaft für Informatik e.V., Bonn, pp. 277–291 (2007)

A Search Engine for Police Press Releases to Double-Check the News

Maik Fröbe[1]([✉]), Nina Schwanke[1], Matthias Hagen[1], and Martin Potthast[2]

[1] Martin-Luther-Universität Halle-Wittenberg, Halle, Germany
maik.froebe@informatik.uni-halle.de
[2] Leipzig University, Leipzig, Germany

Abstract. Many people have doubts about the factual accuracy of online news, while still trusting the press releases of police departments. To enable an easy corroboration of online news about police-related events, we build a search engine for press releases of police departments. Addressing the German "market", the search engine takes the URL of a German piece of online news as input and retrieves relevant press releases of the German police. Comparing different query-by-document strategies in a TREC-style evaluation on 105 topics, we show that our system is able to accurately identify relevant press releases if there are any.

1 Introduction

We introduce Police PR Search[1], a search system that allows for easily double-checking a piece of online news about police-related events (e.g., serious crimes but also accidents or demonstrations) against the content of relevant police press releases. Our current prototype indexes and retrieves press releases of virtually the entire German police force when queried with the URL of some (German) online news article. Readers of online news about police-related events can use the system to retrieve official background information.

An illustrating example is given in Fig. 1. On June 23, 2016, a hostage situation occurred in a cinema in Viernheim, a small town south of Frankfurt and Darmstadt, Germany. The yellow press German newspaper BILD quickly picked up the incident (cf. Fig. 1 (left)), reporting about a rampage and the involvement of explosives. Soon after, the BBC tweeted about 20 casualties in the cinema (cf. Fig. 1 (right)). Indeed, the police did shoot and kill the hostage-taker, but luckily no one else was injured, and no explosives were involved, as explained in the official press release of the police, which was published later after the incident. The sensationally exaggerated and erroneous facts were then removed from the online articles.

Not every piece of (online) news is wrong or sensationally exaggerated. Still, wrong news articles are published frequently [8], and sometimes even intentionally [7]. Readers might thus want to double-check news articles against official

---

[1] Demo: https://demo.webis.de/police-pr.

© Springer Nature Switzerland AG 2020
J. M. Jose et al. (Eds.): ECIR 2020, LNCS 12036, pp. 454–458, 2020.
https://doi.org/10.1007/978-3-030-45442-5_57

Fig. 1. Excerpts from the coverage of a hostage taking in a cinema in Viernheim on June 23, 2016, in the German newspaper BILD (left) and a tweet from the BBC (right).

statements, which can be a rather time-consuming process of searching for other trustworthy sources. Manually selecting text fragments as queries, for instance, may still yield the same wrong information in the search results.

To offer some (semi-) automatic support in such situations, we have developed a search engine that can be queried directly with the URL of an online news article. As results to be retrieved, we index official press releases from police and fire departments. They offer information on a lot of local events—topics that many readers are interested in anyway [6]—and the police is a trusted source of information for many [3]. Currently addressing the German market with the prototype, we have crawled and indexed press releases from the German press portal Blaulicht. In our evaluation, we compare different strategies of formulating search queries given a URL. In a TREC-style setup [5] on 105 topics covering 7 classes of police-related events, we show that even the most simple querying strategy (searching the title of the news article in the titles and bodies of the press releases) substantially outperforms the search facility offered by the press portal itself. It turns out that the best (and more involved) automatic querying strategy implemented in our system achieves precision@1 and nDCG@5 scores of about 0.9—a performance clearly indicating practical applicability.

2   Search System and Query-by-Document Strategies

We extract the title, body, date, and police department location from all press releases as fields for retrieval with Elasticsearch's BM25F implementation. As querying strategies against this index, we basically follow a query-by-document approach (the news article as the "query"). Somewhat following previous works that try to identify the most important keyphrases from an input query document to find similar content [2,9], we compare three query formulation strategies.

Our three "query-by-document" strategies extract information from an input news article and combine them as follows: (1) Only the title of the article, (2) title and main content of the article, and, (3) title, main content, and publication date and locations mentioned in the article (if any). Since publication dates and locations can not be extracted accurately for all potential input articles (third strategy), we resort to only title and body information in such cases.

The queries against the Elasticsearch index are formulated as follows: The title of a given news article is queried against the title and body of the police press releases. When a news article has a body, it is queried against that of the indexed press releases. When a publication date for the news article can be extracted, it is queried against the body of the press releases and used as a filter to remove press releases that were published more than two weeks before, and more than eight weeks after that date. The potential locations extracted from a news article are used as queries against the police department name field as well as against the title and body of the police press releases.

3   Evaluation

To test our system, we follow the TREC evaluation paradigm [5]: 1,172,703 press releases form the document collection[2], covering virtually all German police departments. Topics are formed by news articles about police-related incidents, and relevance judgments are obtained in a depth-5 pooling of different search system rankings.

To create topics representative of the "importance" of police-related incidents, we use the German crime statistics of 2018 [1]. The seven categories we select to cover are murder, theft, migration, related to sports events, thunderstorms, traffic accidents, and general capital offenses. As per the results of a G*Power t-test [4], based on a small pilot experiment, we create 15 topics (105 in total).

The individual topics in the form of news articles were compiled as follows: Given a random police press release from one of the aforementioned categories, an expert tried to identify a related news article using various online news search engines. The expert also was instructed to rate a topic's difficulty during the creation. A difficulty of "Level 1" indicates that the news article and the press release use very similar vocabulary, "Level 2" that the titles greatly differ but there are similarities in the bodies, and "Level 3" for larger differences in the titles and the bodies. If no news article was found for some press release after 5 min, the expert continued with another random press release.

The qrels for the 105 topics (i.e., news articles) have been created as follows: The initial press release used to create the topic is judged as highly relevant (score of 2). Then a depth-5 pool of the rankings returned by different querying strategies and retrieval systems is completely judged: the Blaulicht portal's original search facility using the title or the title and the body as query, and our

---

[2] They are publicly available under www.presseportal.de/blaulicht/.

Table 1. Evaluation of the query-strategies: title (T), title and body (TB), and a combination of title, body, place, and date (TBPD) for various difficulty levels. We compare our search-engine (PPR) with the search engine of the German police press portal (ORI), reporting nDCG@5 and precision@1 (P@1).

Method	Time	All levels		Level 1		Level 2		Level 3	
		nDCG	P@1	nDCG	P@1	nDCG	P@1	nDCG	P@1
T@ORI	0.7 s	0.13	0.14	0.28	0.31	0.01	0.01	0.00	0.00
TB@ORI	0.9 s	0.04	0.04	0.10	0.10	0.02	0.02	0.00	0.00
T@PPR	0.6 s	0.21	0.21	0.44	0.48	0.14	0.11	0.04	0.07
TB@PPR	9.2 s	0.75	0.76	0.81	0.86	0.78	0.79	0.47	0.47
TBPD@PPR	9.1 s	0.92	0.88	0.93	0.90	0.94	0.98	0.59	0.60

three strategies detailed above. The top-5 results of each ranking are judged on a graded scale from 0 (irrelevant) to 2 (highly relevant). A press release is judged as "highly relevant" (score 2) if it directly deals with the event described in the news article, and as "relevant" (score 1) if the news article's event refers to the police press release. Most topics only have one highly relevant press release.

Table 1 shows the aggregated effectiveness of the different systems/strategies in terms of precision@1 and nDCG@5. The performance of the Blaulicht portal's search facility is rather low: even for easy topics (Level 1), hardly any relevant documents are found using a news article's title as the query. A reason might be that some exact match retrieval is used since for many topics no result is returned. This trend gets even worse if additional information in the form of the bodies of the news articles is incorporated into the query.

Our Elasticsearch-based system outperforms the portal's search facility by far on every category (differences significant) even when only the title is used as the query. Adding more information to the query (body, location, date) results in slower search as the news articles' body texts produce very long queries. Still, the effectiveness greatly improves by adding body as well as location and date information to the query with precision@1 and nDCG@5 reaching 0.9 on average. However, this gain in effectiveness comes at the cost of an increased average response time of more than 9 s. Testing and implementing strategies selecting the most informative keywords and phrases from the body to reduce query length thus form an interesting direction for future efficiency improvements.

4   Conclusion and Future Work

Our prototype shows that using a news article as the query against an index of police press releases can often very accurately deliver background information about police-related incidents. Facts can directly be double-checked against official statements, a source of trust for many. For future work, we envision improvements in a number of directions. The efficiency for long queries involving a news article's body text can possibly be improved by keyphrase extraction methods, reducing the queries to maybe several tens of words only. It would also be interesting to more closely analyze cases where no press release can be identified; in our user study, this often was the case when the vocabulary greatly differs and broadening this analysis might help to avoid showing only irrelevant or no results at all.

Acknowledgements. We thank Ahmad Dawar Hakimi, Anita Susheva, Brennan Nicholson, Christopher Pfeiffer, Christoph Traser, and David Sturm for their work on the first prototype and for crawling the press releases. Our special thanks go to the Blaulicht press portal for supplying us with their collection of police press releases for research.

References

1. Polizeiliche Kriminalstatistik, Bundeskriminalamt (2018). https://www.bka.de/ SharedDocs/Downloads/DE/Publikationen/PolizeilicheKriminalstatistik/2018/pks 2018Jahrbuch1Faelle.pdf?__blob=publicationFile&v=6. Accessed 23 Sept 2019
2. Dasdan, A., D'Alberto, P., Kolay, S., Drome, C.: Automatic retrieval of similar content using search engine query interface. In: Proceedings of the 18th ACM Conference on Information and Knowledge Management, CIKM 2009, Hong Kong, China, 2–6 November 2009, pp. 701–710 (2009)
3. European Commission: Wie sehr vertrauen Sie der Polizei? (2019). https:// de.statista.com/statistik/daten/studie/377233/umfrage/umfrage-in-deutschland-zum-vertrauen-in-die-polizei/. Accessed 19 Jan 2020
4. Faul, F., Erdfelder, E., Buchner, A., Lang, A.G.: Statistical power analyses using G*Power 3.1: tests for correlation and regression analyses. Behav. Res. Methods **41**, 1149–1160 (2009)
5. Harman, D.: TREC-style evaluations. In: Agosti, M., Ferro, N., Forner, P., Müller, H., Santucci, G. (eds.) PROMISE 2012. LNCS, vol. 7757, pp. 97–115. Springer, Heidelberg (2013). https://doi.org/10.1007/978-3-642-36415-0_7
6. Schrøder, K.: What do news readers really want to read about? How relevance works for news audiences. Digital News Publications (2019). http://www. digitalnewsreport.org/publications/2019/news-readers-really-want-read-relevance-works-news-audiences/. Accessed 19 Jan 2020
7. Shu, K., Sliva, A., Wang, S., Tang, J., Liu, H.: Fake news detection on social media: a data mining perspective. SIGKDD Explor. Newsl. **19**(1), 22–36 (2017)
8. Tandoc Jr., E., Lim, Z., Ling, R.: Defining "fake news". Digit. Journal. **6**(2), 137–153 (2018)
9. Yang, Y., Bansal, N., Dakka, W., Ipeirotis, P.G., Koudas, N., Papadias, D.: Query by document. In: Proceedings of the Second International Conference on Web Search and Web Data Mining, WSDM 2009, Barcelona, Spain, 9–11 February 2009, pp. 34–43 (2009)

Neural-IR-Explorer: A Content-Focused Tool to Explore Neural Re-ranking Results

Sebastian Hofstätter[✉], Markus Zlabinger, and Allan Hanbury

TU Wien, Vienna, Austria
{sebastian.hofstaetter,markus.zlabinger,allan.hanbury}@tuwien.ac.at

Abstract. In this paper we look beyond metrics-based evaluation of Information Retrieval systems, to explore the reasons behind ranking results. We present the content-focused Neural-IR-Explorer, which empowers users to browse through retrieval results and inspect the inner workings and fine-grained results of neural re-ranking models. The explorer includes a categorized overview of the available queries, as well as an individual query result view with various options to highlight semantic connections between query-document pairs.
The Neural-IR-Explorer is available at:
https://neural-ir-explorer.ec.tuwien.ac.at/.

1 Introduction

The prevalent evaluation of Information Retrieval systems, based on metrics that are averaged across a set of queries, distills a large variety of information into a single number. This approach makes it possible to compare models and configurations, however it also decouples the explanation from the evaluation. With the adoption of neural re-ranking models, where the scoring process is arguably more complex than traditional retrieval methods, the divide between result score and the reasoning behind it becomes even stronger. Because neural models learn based on data, they are more likely to evade our intuition about how their components should behave. Having a thorough understanding of neural re-ranking models is important for anybody who wants to analyze or deploy these models [6,7].

In this paper we present the Neural-IR-Explorer: a system to explore the output of neural re-ranking models. The explorer complements metrics based evaluation, by focusing on the content of queries and documents, and how the neural models relate them to each other. We enable users to efficiently browse the output of a batched retrieval run. We start with an overview page showing all evaluated queries. We cluster the queries using their term representations taken from the neural model. Users can explore each query result in more detail: We show the internal partial scores and content of the returned documents with different highlighting modes to surface the inner workings of a neural re-ranking

© Springer Nature Switzerland AG 2020
J. M. Jose et al. (Eds.): ECIR 2020, LNCS 12036, pp. 459–464, 2020.
https://doi.org/10.1007/978-3-030-45442-5_58

model. Here, users can also select different query terms to individually highlight their connections to the terms in each document.

In our demo we focus on the kernel-pooling models KNRM [14] and TK [8] evaluated on the MSMARCO-Passage [2] collection. The kernel-pooling makes it easy to analyze temporary scoring results. Finally, we discuss some of the insights we gained about the KNRM model using the Neural-IR-Explorer. The Neural-IR-Explorer is available at: https://neural-ir-explorer.ec.tuwien.ac.at/.

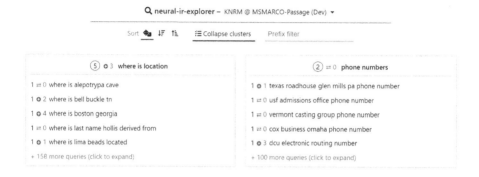

Fig. 1. Screenshot of the query cluster view with the controls at the top and two clusters

2   Related Work

Our work sits at the intersection of visual IR evaluation and the interpretability of neural networks with semantic word representations. The IR community mainly focused on tools to visualize result metrics over different configurations: *CLAIRE* allows users to select and evaluate a broad range of different settings [1]; *AVIATOR* integrates basic metric visualization directly in the experimentation process [4]; and the *RETRIEVAL* tool provides a data-management platform for multimedia retrieval including differently scoped metric views [9]. Lipani et al. [11] created a tool to inspect different pooling strategies, including an overview of the relevant result positions of retrieval runs.

From a visualization point of view term-by-term similarities are similar to attention, as both map a single value to a token. Lee et al. [10] created a visualization system for attention in a translation task. Transformer-based models provide ample opportunity to visualize different aspects of the many attention layers used [3,13]. Visualizing simpler word embeddings is possible via a neighborhood of terms [5].

3 Neural-IR-Explorer

Now we showcase the capabilities of the Neural-IR-Explorer (Sect. 3.1) and how we already used it to gain novel insights (Sect. 3.2). The explorer displays data created by a batched evaluation run of a neural re-ranking model. The back-end is written in Python and uses Flask as web server; the front-end uses Vue.js. The source code is available at: github.com/sebastian-hofstaetter/neural-ir-explorer.

3.1 Application

When users first visit our website they are greeted with a short introduction to neural re-ranking and the selected neural model. We provide short explanations throughout the application, so that that new users can effectively use our tool. We expect this tool's audience to be not only neural re-ranking experts, but anyone who is interested in IR.

The central hub of the Neural-IR-Explorer is the query overview (Fig. 1). We organize the queries by clustering them in visually separated cards. We collapse the cards to only show a couple of queries per default. This is especially useful for collections with a large number of queries, such as the MSMARCO collection we use in this demo (the DEV set contains over 6.000 queries). In the cluster header we display a manually assigned summary title, the median result of the queries, and median difference to the initial BM25 ranking, as this is the basis for the re-ranking. Each query is displayed with the rank of the first relevant document, the difference to BM25, and the query text. The controls at the top allow to sort the queries and clusters – including a random option to discover new queries. Users can expand all clusters or apply a term-prefix filter to search for specific words in the queries.

Once a user clicks on a query, they are redirected to the query result view (Fig. 2). Here, we offer an information rich view of the top documents returned by the neural re-ranking model. Each document is displayed in full with its rank, overall and kernel-specific scores. The header controls allow to highlight the connections between the query and document terms in two different ways. First, users can choose a minimum cosine similarity that a term pair must exceed to be colored, which is a simple way of exploring the semantic similarity of the word representations. Secondly, for kernel-pooling models that we support, we offer a highlight mode much closer to how the neural model sees the document: based on the association of a term to a kernel. Users can select one or more kernels and terms are highlighted based on their value after the kernel transformation.

Additionally, we enable users to select two documents and compare them side-by-side (Fig. 3). Users can highlight query-document connections as in the list view. Additionally, we display the different kernel-scores in the middle, so that users can effectively investigate which kernels of the neural model have the deciding influence of the different scores for the two documents.

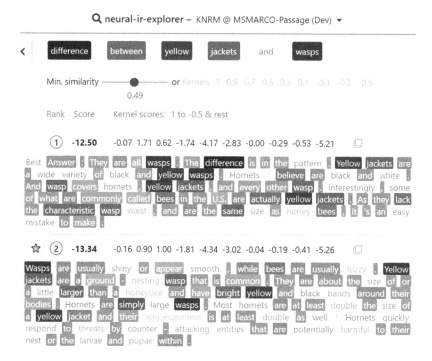

Fig. 2. Screenshot of the detailed query result view (with min. similarity highlighting)

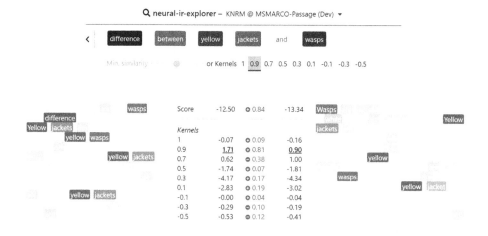

Fig. 3. Screenshot of the detailed side-by-side view of two documents

3.2 Neural Re-ranking Model Analysis

We already found the Neural-IR-Explorer to be a useful tool to analyze the KNRM neural model and understand its behaviors better. The KNRM model includes a kernel for exact matches (cosine similarity of exactly 1), however judging from the displayed kernel scores this kernel is not a deciding factor. Most of the time the kernels for 0.9 & 0.7 (meaning quite close cosine similarities) are in fact the deciding factor for the overall score of the model. We assume this is due to the fact, that every candidate document (retrieved via exact matched BM25) contains exact matches and therefore it is not a differentiating factor anymore – a specific property of the re-ranking task.

Additionally, the Neural-IR-Explorer also illuminates the pool bias [12] of the MSMARCO ranking collection: The small number of judged documents per query makes the evaluation fragile. Users can see how relevant unjudged documents are actually ranked higher than the relevant judged documents, wrongly decreasing the model's score.

4 Conclusion

We presented the content-focused Neural-IR-Explorer to complement metric based evaluation of retrieval models. The key contribution of the Neural-IR-Explorer is to empower users to efficiently explore retrieval results in different depths. The explorer is a first step to open the black-boxes of neural re-ranking models, as it investigates neural network internals in the retrieval task setting. The seamless and instantly updated visualizations of the Neural-IR-Explorer offer a great foundation for future work inspirations, both for neural ranking models as well as how we evaluate them.

Acknowledgements. This work has received funding from the European Union's Horizon 2020 research and innovation program under grant agreement No. 822670.

References

1. Angelini, M., Fazzini, V., Ferro, N., Santucci, G., Silvello, G.: CLAIRE: a combinatorial visual analytics system for information retrieval evaluation. Inform. Process. Manage. **54**, 1077–1100 (2018)
2. Bajaj, P., et al.: MS MARCO: a human generated MAchine Reading COmprehension dataset. In: Proceedings of NIPS (2016)
3. Coenen, A., et al.: Visualizing and measuring the geometry of BERT. arXiv:1906.02715 (2019)
4. Giachelle, F., Silvello, G.: A progressive visual analytics tool for incremental experimental evaluation. arXiv preprint arXiv:1904.08754 (2019)
5. Heimerl, F., Gleicher, M.: Interactive analysis of word vector embeddings. In: Computer Graphics Forum, vol. 37. Wiley Online Library (2018)
6. Hofstätter, S., Hanbury, A.: Let's measure run time! Extending the IR replicability infrastructure to include performance aspects. In: Proceedings of OSIRRC (2019)

7. Hofstätter, S., Rekabsaz, N., Eickhoff, C., Hanbury, A.: On the effect of low-frequency terms on neural-IR models. In: Proceedings of SIGIR (2019)
8. Hofstätter, S., Zlabinger, M., Hanbury, A.: Interpretable & time-budget-constrained contextualization for re-ranking. In: Proceedings of ECAI (2020)
9. Ioannakis, G., Koutsoudis, A., Pratikakis, I., Chamzas, C.: RETRIEVAL: an online performance evaluation tool for information retrieval methods. IEEE Trans. Multimed. **20**, 119–127 (2017)
10. Lee, J., Shin, J.-H., Kim, J.-S.: Interactive visualization and manipulation of attention-based neural machine translation. In: Proceedings of EMNLP (2017)
11. Lipani, A., Lupu, M., Hanbury, A.: Visual pool: a tool to visualize and interact with the pooling method. In: Proceedings of SIGIR (2017)
12. Lipani, A., Zuccon, G., Lupu, M., Koopman, B., Hanbury, A.: The impact of fixed-cost pooling strategies on test collection bias. In: Proceedings of ICTIR (2016)
13. Vig, J.: A multiscale visualization of attention in the transformer model. arXiv:1906.05714 (2019)
14. Xiong, C., Dai, Z., Callan, J., Liu, Z., Power, R.: End-to-end neural ad-hoc ranking with kernel pooling. In: Proceedings of SIGIR (2017)

Revisionista.PT: Uncovering the News Cycle Using Web Archives

Flávio Martins[✉][ID] and André Mourão[ID]

NOVA LINCS, School of Science and Technology, Universidade NOVA de Lisboa,
Lisbon, Portugal
flavio.martins@fct.unl.pt, a.mourao@campus.fct.unl.pt

Abstract. In this demo, we present a meta-journalistic tool that reveals post-publication changes in articles of Portuguese online news media. *Revisionista.PT* can uncover the news cycle of online media, offering a glimpse into an otherwise unknown dynamic edit history. We leverage on article snapshots periodically collected by Web archives to reconstruct an approximate timeline of the changes: additions, edits, and corrections. *Revisionista.PT* is currently tracking changes in about 140,000 articles published by 12 selected news sources and has a user-friendly interface that will be familiar to users of version control systems. In addition, an *open source* browser extension can be installed by users so that they can be alerted of changes to articles they may be reading. Initial work on this demo was started as an entry submitted into *Arquivo.PT*'s 2019 Prize, where it received an award for second place.

1 Introduction

Nowadays, online media plays a critical role in the dissemination of the news. In the age of *online first* and the *24 h news cycle*, news publishing has been deeply transformed to allow more agility and flexibility, and (online) news articles are now updated seamlessly as new information becomes available. However, the desire to be first to publish and achieve high click-through rates has led to an increase in *clickbait*, both pre- and post-publication editorialization of titles and headlines that may misrepresent the actual content of the articles [1]. These trends may have contributed to the increased media mistrust that is discussed today under the controversial term *fake news*, described as "a poorly-defined and misleading term that conflates a variety of false information, from genuine error through to foreign interference in democratic processes" [2].

In some instances, undisclosed post-publication changes have swung the overall message and tone of news articles and, when revealed, have angered readers [4]. How can news readers be sure that the articles they are reading were not altered secretly after their original publication? Can we trust publishers to guarantee that a link you share with a friend will contain the same content you have read?

© Springer Nature Switzerland AG 2020
J. M. Jose et al. (Eds.): ECIR 2020, LNCS 12036, pp. 465–469, 2020.
https://doi.org/10.1007/978-3-030-45442-5_59

Revisionista.PT[1] allows readers to see all kinds of post-publication changes, and allows scholars to research this phenomenon and reason about how these changes affect the readers perception of news in the age of social media.

2   Proposed Solution

In this demo paper, we propose a meta-journalistic tool that reveals post-publication changes in articles published by Portuguese online news media. The goal of *Revisionista.PT* is to discover significant post-publication changes in news articles and to present them in a transparent auditable timeline. The most important raw materials to achieve this objective are historical collections of web page snapshots from news outlets that are preserved by the Portuguese Web Archive, *Arquivo.PT*[2].

To create this project, we had to build solutions to solve a number of technical challenges: *Given a web page, how to extract the textual content of the article?*, *What are the significant differences between article revisions?*, and, most importantly, *What dimensions can be used to classify the changes found?*.

In Fig. 1 we presents a simplified overview of the pipeline that we use to process a collection of news article URLs and generate the summaries of the changes that will be shown in the user interface. The pipeline can be described in three main steps: (1) fetching of HTML snapshots, (2) text extraction from HTML, and (3) changes extraction from text.

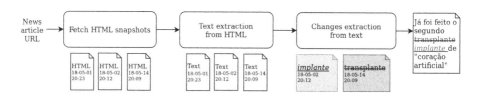

Fig. 1. From a news article URL to a set of edits

News Article URLs. To build a list of news article URLs our strategy was to develop URL *slug* templates that are matched when fetching news articles of a given news source (e. g., `https://observador.pt/<YYYY>/<MM>/<DD>/<title>`. We used a semi-automatic interactive process to create the *slug* templates that we then used to whitelist URLs that will be fetched from each news source. The Portuguese Web Archive (*Arquivo.PT*) is currently the main data source. Using an established Web archive allowed us to focus on tasks that are orthogonal to the preservation of the web and leverage on their preserved web page snapshots.

---

[1] https://revisionista.pt.

[2] https://arquivo.pt.

Fetch HTML Snapshots. Given a list of news article URLs, our crawler downloads multiple snapshots for each of the URLs in the list. We adopted a more conservative crawling strategy that can be described as a bisection method over the largest time span (furthest pair of snapshots), with downloads proceeding only on the branches where changes are found.

Text Extraction from HTML. At this stage, the goal is to extract the textual content and other interesting metadata such as title, author names and publication date, that we want to consider for version control. News web pages contain a myriad of extra content, such as sidebars with breaking news, and comment sections that are undesirable for our purposes and not to be included.

To guarantee a high-fidelity at this stage, we opted to build site-specific specialised content extractors for each target news source. The extractors essentially select the main HTML element containing the article (e. g., <div class="article">) and extracts all the text inside all the inner HTML elements except from known site-specific undesired elements identified semi-automatically.

In addition, we normalised relative time expressions such as, *Hoje, 20:21* (Today, 20:21) into a standard format (27-01-2018, 20:21:00).

Changes Extraction from Text. Diffs for prose are different from code diffs. Character-based diffs are not as useful on prose. Diffs on word boundaries or even concept boundaries are more useful. After extracting diff at multiple levels (char, word and line), we decided that word based level offered the best balance between conciseness and usability. For *Revisionista.PT*, we opted to customise an algorithm that is used in Google Docs to Track Changes Google [3] but in word-based mode. Word-based level diffs are important to allow interpreting the differences between versions (e. g., entities changes, typos corrected, etc.), improving readability and enabling simple browsing and search on diff text.

Changes Categorisation. Finally, once all the changes are computed we can do a broad categorisation of revisions found using an automatic rule-based approach:

- **Republished:** The publishing date of the article changed.
- **Undisclosed correction:** Text changes not explicitly disclosed.
- **Disclosed correction:** Text changes are accompanied by a note.
- **Disclosed update:** Updates the article with a note and new text only.
- **Live:** Coverage of live events or ongoing developments.

In Fig. 2, we can see a representation of the changes categorised as *Live.* "AO MINUTO" can be roughly translated to "minute by minute".

AO MINUTO: Atentado terrorista em Barcelona faz 13 mortos - Mundo - Correio da Manhã

Text added appears highlighted in green.
Text removed appears highlighted in red.

Notícia

AO MINUTO: Atentado terrorista em Barcelona faz 13 mortosDois homens armados barricados mortosUm suspeito detido, outro abatido. Governo catalão confirma 80 feridos, 15 em restaurante junto às Ramblas. Há dezenas de feridos.
estado grave.

Pelo menos 13 pessoas morreram e dezenas e mais de 50 ficaram

Fig. 2. Inline differences in *Revisionista.PT*

Fig. 3. Searching and browsing in *Revisionista.PT*

3   Revisionista.PT

Revisionista.PT allows search and browsing of online news articles that were found to have post-publication changes. We examined 139,561 articles, of which 6,793 were edited after publication (4.87%), for a total number of edits of 7,405 (some articles were edited more than once). These articles are a part of 12 online Portuguese news platforms, from 2015 to 2018. In Fig. 3, we show the search interface, where relevance-based ranking can be overridden (edit date, number of edits, edit size) and results can be filtered by publication and edit type. In addition to showing added and removed text inline with the revisions, a special article view presents the changes contextualised within the article (see Fig. 2).

Revisionista.PT *Companion Extension*[3] is *open source* and can check on *Arquivo.PT* for different snapshots of the current page and show the changes. Our goal is to enable users to find undisclosed post-publication edits in the wild.

We intend to continue building new features into *Revisionista.PT* to provide more assurances and help journalists do their job with confidence. Our goal is to build automatic classifiers to help identify articles that need correction and significant changes that are too important to not include an explicit disclosure. Following this direction, we will create an interface to allow expert journalists to annotate changes found by *Revisionista.PT* according to different parameters to build a useful dataset. In addition, we will create a new tool that will allow journalists to subscribe to alerts helping them to keep their articles up-to-date with fresh information arriving from other news sources and published articles.

References

1. Andrew, B.C.: Media-generated shortcuts: do newspaper headlines present another roadblock for low-information rationality? Harv. Int. J. Press/Polit. **12**(2), 24–43 (2007). https://doi.org/10.1177/1081180X07299795
2. Digital, Culture, Media and Sport Committee: Disinformation and "fake news": Final report, vol. HC 1791. British House of Commons (2018)
3. Google: google/diff-match-patch: diff match patch is a high-performance library in multiple languages that manipulates plain text (2019). https://github.com/google/diff-match-patch
4. VillageVoice: Why did the New York times change their Brooklyn bridge arrests story? (2011). https://web.archive.org/web/20111003160551/blogs.villagevoice.com/runn inscared/2011/10/why_did_the_new_1.php

---

[3] https://github.com/revisionista.

MathSeer: A Math-Aware Search Interface with Intuitive Formula Editing, Reuse, and Lookup

Gavin Nishizawa[✉], Jennifer Liu, Yancarlos Diaz, Abishai Dmello,
Wei Zhong, and Richard Zanibbi

Rochester Institute of Technology, Rochester, NY 14623, USA
{ghn6069,jwt7689,yxd3549,ad7527,wxz8033,rxzvcs}@rit.edu
https://www.cs.rit.edu/~dprl/mathseer

Abstract. There has been growing interest in math-aware search engines that support retrieval using both formulas and keywords. An important unresolved issue is the design of search interfaces: for wide adoption, they must be engaging and easy-to-use, particularly for non-experts. The MathSeer interface addresses this with straightforward formula creation, editing, and lookup. Formulas are stored in 'chips' created using handwriting, LaTeX, and images. MathSeer sessions are also stored at automatically generated URLs that save all chips and their editing history. To avoid re-entering formulas, chips can be reused, edited, or used in creating other formulas. As users enter formulas, our novel auto-completion facility returns entity cards searchable by formula or entity name, making formulas easy to (re)locate, and descriptions of symbols and notation available before queries are issued.

Keywords: Mathematical information retrieval · User interface design · Multimodal input

1 Introduction

Math-aware search engines supporting keyword and formula search have been around since at least 2003, when the Digital Library of Mathematical Functions[1] supported LaTeX in queries [13]. The new information that sophisticated math retrieval would provide, such as more easily locating definitions of symbols and other notations, finding usage, proofs and mathematical properties across disciplines, and compiling information on applications (e.g., variations of the log loss for machine learning) has stimulated work in math-aware search, alongside parallel developments in math question answering within the Natural Language Processing community [2]. To realize their full potential, math-aware search interfaces must be engaging and easy-to-use for different levels of expertise, and particularly for non-experts (e.g., students in middle school).

---

[1] https://dlmf.nist.gov.

Supported by National Science Foundation (USA) and Alfred P. Sloan Foundation.

J. M. Jose et al. (Eds.): ECIR 2020, LNCS 12036, pp. 470–475, 2020.
https://doi.org/10.1007/978-3-030-45442-5_60

2   Interface Design Elements

Formula Entry. Let's first consider the problem of creating formulas. While formulas such as '$y = x^2 + 1$' can be easily written in LaTeX, others such as:

$$\nabla \times F = \left(\frac{\partial F_z}{\partial y} - \frac{\partial F_y}{\partial z} \right) \mathbf{i} + \left(\frac{\partial F_x}{\partial z} - \frac{\partial F_z}{\partial x} \right) \mathbf{j} + \left(\frac{\partial F_y}{\partial x} - \frac{\partial F_x}{\partial y} \right) \mathbf{k} \qquad (1)$$

are large, complex, and contain symbols that many non-experts cannot name let alone express in a query. Despite this, most math-aware search engines are restricted to two forms of input: (1) LaTeX (or MathML) entry in text boxes, and (2) visual template editors similar to the Microsoft Equation Editor [11,13]. Many users find template editors confining, and so the text box approach is the most common, often in combination with a palette used to insert symbols and structures in the entry box. Text input is used by most online math-aware systems, including DLMF [8], WebMIAS [10], Math WebSearch [3], Wolfram Alpha, SymboLab, SearchOnMath, and the (now-defunct) Springer LaTeX Search.

Two challenges for text-based input are (1) most users are unfamiliar with LaTeX (even fewer know MathML), and (2) rendered formulas are shown separately from input, leading to users having difficulty locating entry errors [14]. Appealing solutions to these issues are handwritten formula input, formula image upload, and supporting the analogy of physically moving symbols around on a page [15]. These are key design elements in the MathSeer interface. In one study, a majority of the undergraduate participants reported preferring drawing over typing formulas given a choice between the two [12]. They also expressed formulas with handwriting that they could not using a keyboard (e.g., $\binom{4}{2}$).

To address these issues, our MathSeer search interface (see Fig. 1) allows formula input using a combination of typing LaTeX, uploading formula images, and drawing formulas by hand.[2] In MathSeer handwritten symbols are recognized each time a user stops drawing for a short time. Pressing a button recognizes formula structure, and copies the LaTeX result into the panel at the bottom-left of the interface. The LaTeX can then be edited, with a rendering of the formula updated in real-time (e.g., to quickly change 'p' to 'P'). At bottom-right, palettes containing symbols and structures may be used to insert corresponding LaTeX at the cursor position in the LaTeX panel.

Images may be dragged-and-dropped on the canvas or uploaded using a button that presents a file navigation pop-up window. This produces a formula 'chip' on the canvas, which can be used directly in a query, edited, or used in constructing other formulas. A line-of-sight graph-based parsing technique is used to recognize formula images and handwritten formulas [4].

Users can freely alternate between drawing and manipulating symbols on the canvas, uploading images, and editing the LaTeX panel contents. Robust undo/redo operations are provided to easily reverse operations. Formulas in the query bar can be chosen for editing by clicking on them, allowing for quick

---

[2] Video Demonstration: https://cs.rit.edu/~dprl/mathseer/demos.

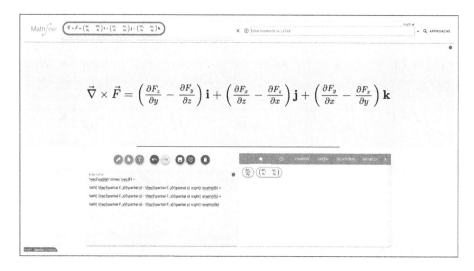

Fig. 1. MathSeer interface. Query formulas and keywords are 'chips' at top left; keywords are entered using the box at top right. Formulas are created by manipulating symbols on the canvas, uploading formula images, and editing LaTeX in the panel at bottom left. At bottom-right is a panel for 'favorite' formulas (two are shown here), the formula history, and palettes for symbols and structures to insert in the LaTeX panel.

switching between formulas. Mansouri et al. found that users search for math with keywords or in the context of a question [6]. In order to help the user add additional information for their query, MathSeer also supports keywords in their search queries (see Fig. 1).

Formula Containers and Reuse ('Chips'). Handwritten formula entry is convenient for small expressions, but for large expressions such as Eq. 1 handwriting is slow [12], and accurate recognition is challenging [5]. Users may also want to avoid re-entering formulas, and to share formulas with others [12].

MathSeer introduces a new model for formula reuse, flexible containers that we call formula 'chips'. Figure 1 shows a chip in the query bar, and there are two 'favorite' chips in a list at bottom-right. Chips can be created and used in a number of ways. In addition to the formula creation operations described above, chips can be created by selecting symbols on the canvas and 'popping' them up into a chip. All formula chips have their creation history automatically recorded, and are stored in a 'history' menu in the symbol palette panel. On the canvas, chips may be easily moved, resized, and 'pushed' onto the canvas (i.e., the symbols on the chip are added to the canvas, and the chip disappears).

Chips have two possible states: 'recognized' chips containing a LaTeX string, and 'template' chips representing only symbols on a canvas. Chips that are 'favorites' are shown using an orange border, and are either a recognized or template chip. As an example use for **template (grey) chips**, '$\int dx$' with a large space in the middle can be used as a template to quickly create other

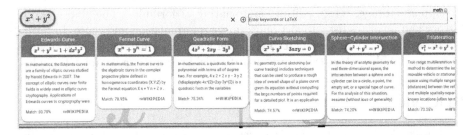

Fig. 2. Expanded auto-complete results displaying entity cards with similar formulas.

formulas with an integral, by dragging and dropping the chip from the favorites or history tab in the palette panel to the canvas. **Recognized (blue) chips** in the history and favorites tabs in the palette panel can also be used like palette buttons - clicking on them inserts their interpretation in the LaTeX panel, making it easy to re-use and insert large formulas. Chips may also be exported as images with metadata containing all chip data, allowing chip images to be later reused in MathSeer (e.g., using drag-and-drop) or shared with others (e.g., over email). Using chips for formula containers was inspired by the Approach0 interface.[3]

MathSeer records the entire editing session, including all formula chips using an automatically generated URL that users can revisit later. The idea to use a URL to record editing state came from discussions with the creators of 2dsearch [9].

Math Entity Cards. To support formula autocompletion using online data (e.g., from Wikidata), we use a new type of entity card that provides concept names and descriptions for formulas. We use these to provide names and descriptions for individual symbols and formulas in real-time as they are entered [1]. Formula search over the card collection is done using Tangent-CFT embedding vectors [7]. In addition, we will soon allow formulas to be quickly found by searching concept names on cards (e.g., typing 'Pyt' brings up the card and formula for the Pythagorean Theorem). Further, we plan to allow users to create their own entity cards for formula chips. An illustration of math entity cards is shown in Fig. 2. This view is expanded to show the full cards; in the unexpanded view only the formulas and titles are visible.

3   Conclusion and Future Work

The MathSeer interface addresses limitations of the standard text box + symbol palette formula entry technique common in math-aware search interfaces. MathSeer's interface supports multimodal formula editing through handwritten, LaTeX, and image input. We have introduced formula chips, a new container

---

[3] https://www.approach0.xyz/search.

to support storage, reuse, editing, and sharing of formulas. The chip creation history and favorites list support quick query reformulation and reuse. In future work, we are considering manual editing operations to define spatial relationships between symbols and/or sub-expressions to avoid recognizing complex formulas.

References

1. Dmello, A.: Representing mathematical concepts associated with formulas using math entity cards. Master's thesis, Rochester Institute of Technology (2019). https://scholarworks.rit.edu/theses/10238
2. Hopkins, M., Bras, R.L., Petrescu-Prahova, C., Stanovsky, G., Hajishirzi, H., Koncel-Kedziorski, R.: SemEval-2019 Task 10: Math question answering. In: Proceedings of SemEval@NAACL-HLT 2019, Minneapolis, MN, USA, pp. 893–899 (2019)
3. Kohlhase, M., Matican, B., Prodescu, C.: MathWebSearch 0.5: Scaling an open formula search engine. In: Proceedings of CICM, Bremen, Germany, pp. 342–357 (2012)
4. Mahdavi, M., Condon, M., Davila, K., Zanibbi, R.: LPGA: Line-of-sight parsing with graph-based attention for math formula recognition. In: Proceedings of ICDAR, Sydney, Australia, pp. 647–654 (2019)
5. Mahdavi, M., Zanibbi, R., Mouchère, H., Viard-Gaudin, C., Garain, U.: ICDAR 2019 CROHME + TFD: Competition on Recognition of Handwritten Mathematical Expressions and Typeset Formula Detection. In: Proceedings of ICDAR, Sydney, Australia, pp. 1533–1538 (2019)
6. Mansouri, B., Zanibbi, R., Oard, D.W.: Characterizing searches for mathematical concepts. In: 2019 ACM/IEEE Joint Conference on Digital Libraries (JCDL), pp. 57–66 (2019)
7. Mansouri, B., Rohatgi, S., Oard, D.W., Wu, J., Giles, C.L., Zanibbi, R.: Tangent-CFT: An embedding model for mathematical formulas. In: Proceedings of ICTIR 2019, pp. 11–18 (2019)
8. Miller, B.R., Youssef, A.: Technical aspects of the Digital Library of Mathematical Functions. Ann. Math. Artif. Intell. **38**(1–3), 121–136 (2003)
9. Russell-Rose, T., Chamberlain, J., Kruschwitz, U.: Rethinking 'advanced search': A new approach to complex query formulation. In: Azzopardi, L., Stein, B., Fuhr, N., Mayr, P., Hauff, C., Hiemstra, D. (eds.) ECIR 2019. LNCS, vol. 11438, pp. 236–240. Springer, Cham (2019). https://doi.org/10.1007/978-3-030-15719-7_31
10. Sojka, P., Ruzicka, M., Novotný, V.: MiaS: math-aware retrieval in digital mathematical libraries. In: Proceedings of CIKM, Torino, Italy, pp. 1923–1926 (2018)
11. Wangari, K.: Discovering real-world usage scenarios for a multimodal math search interface. Master's thesis, Rochester Institute of Technology, December (2013)
12. Wangari, K., Zanibbi, R., Agarwal, A.: Discovering real-world use cases for a multimodal math search interface. In: Proceedings of SIGIR, Gold Coast, Australia, pp. 947–950 (2014)
13. Zanibbi, R., Blostein, D.: Recognition and retrieval of mathematical expressions. IJDAR **15**(4), 331–357 (2012). https://doi.org/10.1007/s10032-011-0174-4

14. Zanibbi, R., Novins, K., Arvo, J., Zanibbi, K.: Aiding manipulation of handwritten mathematical expressions through style-preserving morphs. In: Proceedings of Graphics Interface, Ottawa, Canada, pp. 127–134 (2001)

15. Zanibbi, R., Orakwue, A.: Math search for the masses: multimodal search interfaces and appearance-based retrieval. In: Kerber, M., Carette, J., Kaliszyk, C., Rabe, F., Sorge, V. (eds.) CICM 2015. LNCS (LNAI), vol. 9150, pp. 18–36. Springer, Cham (2015). https://doi.org/10.1007/978-3-319-20615-8_2

NLPEXPLORER: Exploring the Universe of NLP Papers

Monarch Parmar, Naman Jain, Pranjali Jain, P. Jayakrishna Sahit,
Soham Pachpande, Shruti Singh, and Mayank Singh[(✉)]

Indian Institute of Technology Gandhinagar, Gandhinagar, Gujarat, India
`singh.mayank@iitgn.ac.in`

Abstract. Understanding the current research trends, problems, and their innovative solutions remains a bottleneck due to the ever-increasing volume of scientific articles. In this paper, we propose NLPEXPLORER, a completely automatic portal for indexing, searching, and visualizing Natural Language Processing (NLP) research volume. NLPEXPLORER presents interesting insights from papers, authors, venues, and topics. In contrast to previous topic modelling based approaches, we manually curate five course-grained non-exclusive topical categories namely *Linguistic Target* (Syntax, Discourse, etc.), *Tasks* (Tagging, Summarization, etc.), *Approaches* (unsupervised, supervised, etc.), *Languages* (English, Chinese, etc.) and *Dataset types* (news, clinical notes, etc.). Some of the novel features include a list of young popular authors, popular URLs and datasets, list of topically diverse papers and recent popular papers. Also, it provides temporal statistics such as yearwise popularity of topics, datasets, and seminal papers. To facilitate future research and system development, we make all the processed dataset accessible through API calls. The current system is available at http://nlpexplorer.org.

Keywords: Natural language processing · Search · Research papers

1 Introduction

Effective scientific literature understanding plays a critical role towards the research community's common goal of *"March for Science"*. However, the yearly generated research volume shows an upward trend with an estimated increase of 8–9% per year. This results in information overload, impacting the literature review process and often, leads to *'reinventing the wheel'* syndrome. This negatively affects the efficiency of scientific progress on the knowledge frontier.

In the past, significant efforts have been made to curate peer-reviewed open access scientific information. In particular, the field of Natural Language Processing (*NLP*) witnessed the development of *ACL Anthology* since the year 2001, which curates papers (in PDF format) from more than 70 NLP venues

---

M. Parmar, N. Jain, P. Jain, P. Jayakrishna Sahit and S. Pachpande—Alphabetically ordered with equal contribution

© Springer Nature Switzerland AG 2020
J. M. Jose et al. (Eds.): ECIR 2020, LNCS 12036, pp. 476–480, 2020.
https://doi.org/10.1007/978-3-030-45442-5_61

including popular conferences like ACL, NAACL, EMNLP, etc. In the year 2008, Bird et al. [2] released the *ACL Anthology Reference Corpus (ACL ARC)*, consisting of OCRed extracted text and metadata of PDF articles. In the year 2009, Radev et al. [8] developed *ACL Anthology Network (AAN)* by manually constructing paper citation, author citation and author collaboration network, along with other interesting statistics and citation summaries of the articles. The recently updated AAN system [9] indexes articles published till 2014. Another initiative, ACL Anthology SearchBench [10], provides bibliographic metadata filtering and full-text structured semantic search in ACL Anthology. CL Scholar [12] periodically crawls ACL Anthology and constructs a computational linguistic knowledge graph. It supports natural language queries and entity specific queries about the authors, venue, and paper. Almost all systems described above are either in dormant condition [2,8–10] or not available [12].

In this paper, we discuss the development of NLPExplorer. NLPExplorer provides paper, author, venue and topic-specific temporal as well as aggregated statistics. For the first time, we also showcase, the usage of URLs over the years, popular top-level and sub-domains, survey papers, new authors, etc. The system also presents a visualization of the timeline for the first occurrences of sub-topics.

Our main contributions are as follows: (i) We periodically download, preprocess, and index the ACL Anthology dataset consisting of more than 55 thousand full-text PDF articles. (ii) We invest extensive effort in automatic curation, structured information extraction, cleaning, indexing, and other related preprocessing steps. (iii) We classify papers into a first-of-its-kind detailed list of NLP topics and subtopics. (iv) The proposed system presents content as well as bibliographic statistics along with basic keyword-based search facilities. (v) We deploy the current system in Google Cloud with an API-based retrieval facility.

2 Dataset

We leverage the ACL Anthology dataset [1] that hosts articles dedicated to Computational Linguistics and Natural Language Processing. Each article is present as a PDF file along with the metadata namely author list, venue name, year of publication and a unique eight-character identifier. Overall, the dataset has 55,565 papers, 39,555 unique authors, 78 venues and 723,976 citations.

3 Architecture

The architecture of NLPExplorer is composed of three interdependent modules, (i) data acquisition, (ii) storage and query processing, and (iii) user interface module. Figure 1 presents the detailed description of the architecture. The *data acquisition* module periodically curates newly added semi-structured information such as metadata and the corresponding PDF articles from ACL Anthology. It leverages OCR++ [11] tool to extract structural and bibliographical information from each of the PDF articles, and then passes the metadata, structural and bibliographical information to the storage and query processing module.

The *storage and query processing* module indexes the periodical updates into MongoDB [5] database and Elasticsearch [3] engine. MongoDB database handles the basic storage and retrieval, whereas Elasticsearch supports full-text based query search and ranking. The third module, *the user interface*, fetches processed search results from the storage and query processing module. The user interface is designed using Python's Flask library, and JavaScript libraries Plotly [7] and Timeline [13] are used to render statistics and graphical components of the interface.

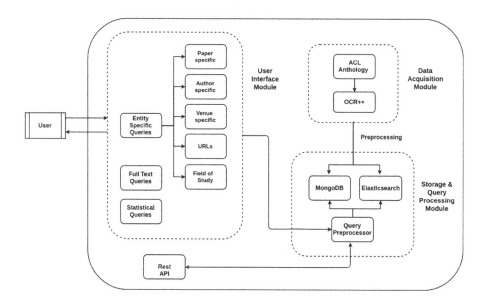

Fig. 1. Architecture of NLPEXPLORER. Arrows represent data flow.

The current system is deployed at Google cloud [4]. The infrastructure consists of a 4 CPU - 7.5 GB RAM, Linux VM instance that can be extended on demand. The system supports REST API requests served using MongoDB database.

4   Current Features

We categorize the wide range of functionalities of NLPEXPLORER as follows:

– **Entity-specific & Full-Text search:** NLPEXPLORER supports basic keyword based search leveraging the metadata and the full-text of articles. The search can be chosen to output results to any of the following domains - authors, papers, venues, URLs, and field of study. We also extend full-text search of articles in order to visualise n-gram trends over the period of time.

- **Paper statistics:** We provide standard paper related statistics such as the publication year and venue, author information, citation distribution over the years and the link to the corresponding PDF article. Additionally, we provide interesting insights like similar papers, topical distribution and mentioned URLs. We also provide statistics such as the list of popular recent papers, popular survey papers, seminal papers, papers with diverse topics, and publication count of last five years.
- **Author statistics:** We provide author statistics such as publication and citation distribution over the years, topical distribution of papers and venue preference. Additionally, we provide the list of popular recent authors, popular authors in the lifetime of ACL Anthology, authors with top publication counts, recent authors with high publication counts, and highly diverse authors.
- **Field of study statistics:** We curate a comprehensive list of topics categorized into five broad categories, (i) *Linguistic Targets* (Syntax, Discourse, Pragmatics, etc.), (ii) *Tasks* (Tagging, Summarization, Chunking, etc.), (iii) *Approaches* (unsupervised, supervised, etc.), (iv) *Languages* (English, Chinese, Hindi, etc.), and (v) *Dataset Types* (news, clinical notes, etc.). NLP-EXPLORER provides interesting insights such as the temporal distribution of papers and authors in each topic and subtopics, distribution of papers in each conference, and a timeline to visualize the introduction of new topics. We make the list of topics and the processed data of the system available at the systems' webpage [6].
- **Venue statistics:** Venue-related statistics include temporal distribution of publications and citations, topical distribution, and the list of papers in a year. We also include insights such as the top NLP venues citing and cited by the candidate venue, popular authors publishing in the candidate venue, and the shift in topical distribution over the years.
- **URL statistics:** We analyse URLs reported in the research papers. The URL-related statistics include top URLs in different categories such as universities, digital libraries, datasets and research groups, alongwith the analysis of top-level domains (TLDs) and corresponding sub-domains. Additionally, NLPEXPLORER provides year-wise usage distribution, total usage, and the list of top-most subdomains and associated papers.

5   Future Extensions and Conclusion

The current system provides a basic functionality for knowledge exploration in NLP domain. In future, we plan to incorporate advanced set of functionalities leveraging natural language understanding of research papers. Some of the main proposals include natural language query retrieval, intelligent ranking by leveraging citation sentiments and discourse-level citation information, visualization of topical flow from cited papers to the main text of citing papers, automatic generation of leaderboard for NLP tasks, visualization of author collaboration networks, paper citation networks, venue interaction networks, etc.

In this paper, we present an end-to-end automated system that periodically mines the ACL Anthology and serves as a tool to aid researchers in knowledge exploration and discovery. The goal of NLPEXPLORER is to serve as a retrieval engine for research papers, as well as a tool to assist researchers in knowledge discovery by helping them to better understand the problem domain, the top researchers in the field of study, and the latest research in the domain. Even though, the current system supports NLP research domain, we claim that similar systems can be built for any domain given the availability of full text articles and basic metadata.

References

1. ACL Anthology Homepage. https://www.aclweb.org/anthology/. Accessed 13 Oct 2019
2. Bird, S., et al.: The ACL anthology reference corpus: a reference dataset for bibliographic research in computational linguistics. In: Language Resources and Evaluation Conference (2008)
3. Elasticsearch Homepage. https://www.elastic.co. Accessed 13 Oct 2019
4. Google Cloud Homepage. https://cloud.google.com/. Accessed 13 Oct 2019
5. MongoDB Homepage. https://www.mongodb.com. Accessed 13 Oct 2019
6. NLPExplorer Homepage. http://nlpexplorer.org/. Accessed 13 Oct 2019
7. Plotly JavaScript Homepage. https://plot.ly/javascript. Accessed 13 Oct 2019
8. Radev, D.R., Muthukrishnan, P., Qazvinian, V.: The ACL anthology network corpus. In: ACL-IJCNLP p. 54 (2009)
9. Radev, D.R., Muthukrishnan, P., Qazvinian, V., Abu-Jbara, A.: The ACL anthology network corpus. Lang. Resour. Eval. **47**(4), 919–944 (2013). https://doi.org/10.1007/s10579-012-9211-2
10. Schäfer, U., Kiefer, B., Spurk, C., Steffen, J., Wang, R.: The ACL anthology searchbench. In: Proceedings of the 49th Annual Meeting of the Association for Computational Linguistics: Human Language Technologies: Systems Demonstrations, pp. 7–13. Association for Computational Linguistics (2011)
11. Singh, M., et al.: OCR++: a robust framework for information extraction from scholarly articles. In: Proceedings of COLING 2016, the 26th International Conference on Computational Linguistics: Technical Papers, pp. 3390–3400. The COLING 2016 Organizing Committee (2016)
12. Singh, M., et al.: CL scholar: the ACL anthology knowledge graph miner. In: Proceedings of the NAACL (2018)
13. TimelineJS Homepage. https://cdnjs.com/libraries/timelinejs. Accessed 13 Oct 2019

Personal Research Assistant for Online Exploration of Historical News

Lidia Pivovarova[1(✉)], Axel Jean-Caurant[2], Jari Avikainen[1], Khalid Alnajjar[1], Mark Granroth-Wilding[1], Leo Leppänen[1], Elaine Zosa[1], and Hannu Toivonen[1]

[1] University of Helsinki, Helsinki, Finland
{lidia.pivovarova,jari.avikainen,khalid.alnajjar,mark.granroth-wilding,
leo.leppanen,elaine.zosa,hannu.toivonen}@helsinki.fi
[2] University of La Rochelle, La Rochelle, France
axel.jean-caurant@univ-lr.fr

Abstract. We present a novel environment for exploratory search in large collections of historical newspapers developed as a part of the NewsEye project. In this paper we focus on the intelligent Personal Research Assistant (PRA) component in the environment and the web interface. The PRA is an interactive exploratory engine that combines results of various text analysis tools in an unsupervised fashion to conduct autonomous investigations on the data according to users' needs. The PRA is freely available online together with some datasets of European historical newspapers. The methods used by the assistant are of potential benefit to other exploratory search applications.

Keywords: Exploratory search · Intelligent personal assistant

1 Introduction

We present the NewsEye Personal Research Assistant (PRA)[1], able to analyse large collections of historical news using an extensible inventory of text-processing tools. These include query-based document search, finding related documents, named entity recognition, stance detection and describing the topics in a collection. The core component – the *Investigator* – performs exploratory corpus analysis on behalf of the user to discover potentially interesting phenomena in the data. The Investigator acts within the modern exploratory search paradigm [2,10], though it uses a broad inventory of text processing tools that can be applied to various document sets depending on the query.

Intelligent personal assistants have been employed in various applications, due to their ability to provide context-based support to users efficiently, saving time and allowing them to focus on important tasks: e.g. navigation [5], time management [4], e-mail organization [7] or patient healthcare [6].

---

[1] This work has been supported by the European Union's Horizon 2020 research and innovation programme under grant 770299 (NewsEye).

J. M. Jose et al. (Eds.): ECIR 2020, LNCS 12036, pp. 481–485, 2020.
https://doi.org/10.1007/978-3-030-45442-5_62

It has been noted that scholars have special information needs and require support for corpus management [8]. Historians are typically interested in analyzing historical data on a level of abstraction that computational models cannot fully learn on their own. Applying potentially informative computational analyses on multiple sub-collections is not only tedious and time-consuming, but sometimes ruled out by the lack of easy-to-use tools and specialist skills (e.g. programming). As a result, a tool is required that is capable of automatically analyzing historical data while giving historians the freedom to dynamically adjust the parameters and context of the analysis.

The Personal Research Assistant is implemented as a part of the NewsEye Project, which aims to develop novel methods facilitating access to digitized historical newspapers for a broad range of users, including professional historians as well as the general public. Computer scientists, historians and librarians are involved in the project, which allows developing and testing computational solutions that meet the needs of digital humanities research studying historical newspapers[2].

A platform has been built for the NewsEye project that incorporates a broad range of features such as text recognition [3], semantic annotation [9], advanced textual analytics [11] and an intelligent personal assistant. It includes a web interface that permits users to find relevant documents based on queries[3].

Users interact with the PRA through a web-interface, where the PRA returns requested information and analysis, as well as the results of the Investigator's autonomous search, along with automatically generated natural language reports, when applicable. Though the NewsEye Investigator is developed specifically for historical research, we believe the same design principles are applicable in other humanities disciplines, where objectivity is a crucial issue.

Though it is still under development, the PRA already performs independent analysis and produces meaningful results.

2   NewsEye Data Analysis Platform

The NewsEye platform provides access to a number of Austrian, French and Finnish newspapers from 19th and early 20th centuries and provides a number of analytical tools to facilitate historical research. These come in various levels of complexity, from straightforward word counts to more sophisticated probabilistic models. The data set and the tool inventory are easily extensible.

The general information flow within the infrastructure is presented in Fig. 1. Images of scanned newspapers are provided by National Libraries of Austria, France and Finland. The images are processed to extract text and separate pages into articles. Articles are then semantically annotated by a number of NLP methods including named entity recognition, sentiment analysis, and novelty and event detection. All these operations are performed offline and the results are

---

[2] For additional information on the project, its datasets, tools and publications visit https://www.newseye.eu/.

[3] Free accessible through https://platform.newseye.eu/.

stored and made accessible through a Solr index. Dynamic text analysis is run on demand and performs query-specific analysis of sets of documents, document linking and comparative analysis of multiple document sets.

Thus the PRA deals with a hetero-geneous data and a variety of analyt-ical tools. The goal of the PRA is to make effective use of theese tools to find peculiarities of potential interest for historical research. The PRA pro-duces a set of natural language reports detailing its findings. These are pro-duced by an automatic natural lan-guage generation system [1] and can be generated in English, French, Finnish or German.

The user interface allows users to query data on various levels. First, it is possible to directly query the database index for simple data col-lection. The search outputs can be saved and combined to build users' own sub-corpora. Then the Investi-gator starts autonomous exploratory analysis based on a sub-corpus. The requirement of autonomy comes from

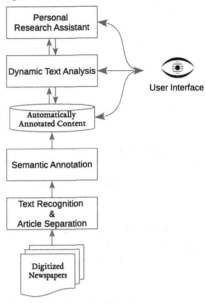

Fig. 1. Information flow within the News-Eye infrastructure.

the needs of humanities studies, where the option to approach history without predefined questions is seen as a key advantage of modern data-driven methods. In adition, the user can directly call a specific analysis tool on the sub-corpus.

This functionality is exemplified in Fig. 2. Figure 2(a) presents the search interface that allows the user to browse the collection and create sub-corpora. Figure 2(b) shows analysis output, organized as a set of analysis tasks. Using icons to the right of each task the user can request a natural-language report, raw results or task parameters. A report for one task is shown.

3   Current Status and Further Work

Main parts of the data processing pipeline are implemented, at least at a proto-type level. Future work will include development and integration of more sophis-ticated methods for text analysis. We also plan to make more newspapers avail-able through the NewsEye platform. Thus, the PRA data and tool inventory will be expanded. This expansion does not theoretically require any changes in the interface, since most of the user forms in the interface are produced automatically based on the tool specification provided by the PRA API.

Nevertheless, some data analysis instruments can be more efficiently exploited via a more specific interface. For example, the NewsEye interface has a

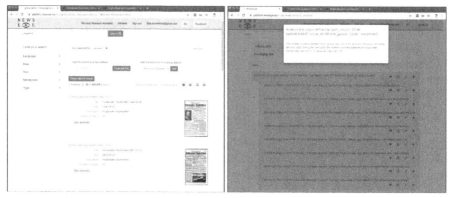

(a) Search interface   (b) Analysis interface

Fig. 2. Example screenshots from the NewsEye user interface.

special section to represent *topic models*, where the user may request word clouds for each topic (Fig. 3). This and other analysis tools, e.g. time series analysis, require specialized visual support.

Fig. 3. Topic modelilng representation in the NewsEye user inteface.

The core PRA component, the autonomous Investigator, is due to change. The current investigator uses patterns – predefined sequences of tools that are run in parallel. In the future, it should be able to adjust its exploration plan on the fly. In principle, the output of its work could be presented in the simple list of tasks, as in Fig. 2(b), but we plan to develop more friendly interface for the investigator.

In this paper we presented the NewsEye exploratory platform, which facilitates historical newspapers studies. The platform provides access to a number of search and text analysis tools. The current interface allows users to access to a large collection of newspapers from the 19th-20th centuries and to analyse them using the autonomous Investigator, which processes data using a variety of analysis tools. The data collection and the tool inventory will be expanded in the near future.

References

1. Leppänen, L., Munezero, M., Granroth-Wilding, M., Toivonen, H.: Data-driven news generation for automated journalism. In: Proceedings of the 10th International Conference on Natural Language Generation, pp. 188–197 (2017)
2. Marchionini, G., White, R.: Find what you need, understand what you find. Int. J. Hum.-Comput. Interact. **23**(3), 205–237 (2007)
3. Michael, J., Labahn, R., Gruning, T., Zollner, J.: Evaluating sequence-to-sequence models for handwritten text recognition. In: International Conference on Document Analysis and Recognition (ICDAR) (2019)
4. Myers, K., et al.: An intelligent personal assistant for task and time management. AI Mag. **28**(2), 47 (2007)
5. Page, L.C., Gehlbach, H.: How an artificially intelligent virtual assistant helps students navigate the road to college. AERA Open **3**(4), 2332858417749220 (2017)
6. Santos, J., Rodrigues, J.J., Silva, B.M., Casal, J., Saleem, K., Denisov, V.: An IoT-based mobile gateway for intelligent personal assistants on mobile health environments. J. Netw. Comput. Appl. **71**, 194–204 (2016)
7. Segal, R.B., Kephart, J.O.: SwiftFile: an intelligent assistant for organizing e-mail. In: In AAAI 2000 Spring Symposium on Adaptive User Interfaces, Stanford, CA (2000)
8. Singh, J., Nejdl, W., Anand, A.: Expedition: a time-aware exploratory search system designed for scholars. In: Proceedings of the 39th International ACM SIGIR Conference on Research and Development in Information Retrieval, pp. 1105–1108. ACM (2016)
9. Sumikawa, Y., Jatowt, A., Doucet, A., Moreux, J.P.: Large scale analysis of semantic and temporal aspects in cultural heritage collection's search. In: 2019 ACM/IEEE Joint Conference on Digital Libraries (JCDL), pp. 77–86. IEEE (2019)
10. White, R.W., Roth, R.A.: Exploratory search: Beyond the query-response paradigm. Synth. Lect. Inform. Concepts Retrieval Serv. **1**(1), 1–98 (2009)
11. Zosa, E., Granroth-Wilding, M.: Multilingual dynamic topic model. In: Recent Advances in Natural Language Processing (RANLP) (2019)

QISS: An Open Source Image Similarity Search Engine

Maxime Portaz[✉], Adrien Nivaggioli, Hicham Randrianarivo[✉], Ilyes Kacher, and Sylvain Peyronnet

Qwant Research, Paris, France
{m.portaz,h.randrianarivo}@qwant.com

Abstract. Qwant Image Similarity Search (QISS) is a multi-lingual image similarity search engine based on a dual path neural networks that embed texts and images into a common feature space where they are easily comparable. Our demonstrator, available at http://research. qwant.com/images, allows real-time searches in a database of approximately 100 million images.

Keywords: Neural networks · Image retrieval

1 Introduction

Qwant Image Similarity Search (QISS) is a multi-lingual image search engine. It allows users to make queries both textually or by using images. QISS relies on similarity search. This means that it will compare the content of a query with the data in its index and returns the elements it considers most similar visually or semantically. In our case, we consider the index elements whose Euclidean distance from the query is closest to zero to be the most similar.

If an image and its describing text are close to one another in the representing space, it is possible to query either one with the other. QISS aims to allow the user to use text or image to query a set of images. While other search engines are based on text surrounding images or tags, QISS evaluates the semantic similarity between the query and each element of the database.

In order to process a query, QISS projects it with a Deep Neural Network. QISS is using a dual path Neural Network, that embed different languages and images into on semantic space [5]. It relies on Nvidia TensorRT server[1] for inference. The indexation of roughly 100 millions images, all available through QISS, is done using the Facebook AI Similarity Search (FAISS) library [3]. The QISS project is open source: the code for neural network training is available at https://github.com/QwantResearch/text-image-similarity while the servers that compose QISS are available at https://hub.docker.com. All dockers are accessible at the address https://hub.docker.com/r/<docker_name> and

---

[1] https://docs.nvidia.com/deeplearning/sdk/tensorrt-inference-server-guide/docs/index.html

© Springer Nature Switzerland AG 2020
J. M. Jose et al. (Eds.): ECIR 2020, LNCS 12036, pp. 486–490, 2020.
https://doi.org/10.1007/978-3-030-45442-5_63

can be obtained using the command `docker pull docker_name`. Docker names for this project are: `chicham/text_server`, `chicham/image_server`, `chicham/language_server`, `chicham/index_server` and `chicham/lmdb_server`. We also use nvcr.io/nvidia/tensorrtserver:19.06-py3 as the model server.

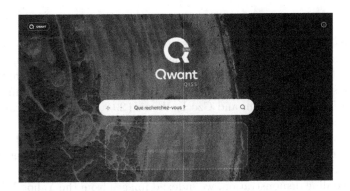

Fig. 1. QISS homepage, where the user can query the index by uploading an image or typing a sentence.

2   System Description

QISS can be used to query the images index by either using texts or images as queries. Also, the representation of the texts is multi-lingual. This means that words from different languages but with similar meanings will have close representations in the semantic space.

2.1   Multi-lingual Text Representation

One of QISS's constraints is to be available in several languages. Instead of using a translation of textual image descriptions, we propose to use multi-lingual word embeddings to cope with multiple languages. Word embeddings are used to project words into a semantic space, where distance and semantic similarity are related. Multi-lingual embeddings, such as Multilingual Unsupervised or Supervised word Embeddings (MUSE) [1], allow for the representation of different languages into one common space. Thanks to this alignment, a neural network can extract information from the embedded words in all learned languages. This allows QISS to have only one index that contains every image, and that can answer to queries expressed into several languages. This is a strong difference with classic search engines that have one index for each language.

2.2   Model for Image and Text Representation

To project both images and texts into the same space, we use two networks trained simultaneously. The image branch of the network uses a Convolutional

Neural Network (CNN) followed by a fully connected layer that embed images. The second branch is a multi-layer Recurrent Neural Network (RNN) that compose a multi-lingual word embeddings list into the same space. This list corresponds to a given sentence.

2.3 Data

We use two datasets to train the models used by QISS. Each dataset is composed of images and their corresponding captions. The first dataset is Common Objects in COntext (COCO) [4]. It contains 123 287 images with 5 English captions per image. The second dataset is called Multi30K [2]. It contains 31 014 images with captions in French, German, and Czech. We use 29 000 images for training and 1014 for validation and 1000 for testing.

MUSE allows for a common representation for 110 languages. Once we trained our model in English using COCO, we used MUSE to transfer the computed embeddings to any language supported by MUSE, at no cost.

For the online demonstration, we indexed images from the Yahoo Flickr Creative Commons (YFCC) [6] image dataset. This dataset contains roughly 100 million images under Creative Commons license.

2.4 Overview

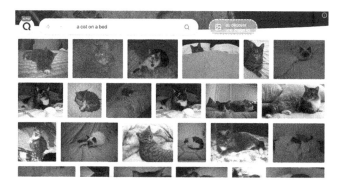

Fig. 2. Result page when the user search for "a cat on a bed", showing the images closest to this text.

As said above, QISS is a full image search engine based on similarity search. Figure 1 shows the interface, where it is possible to search using a text query or by uploading an image. The results are shown in Fig. 2. The images that our method evaluates as the most similar to the query (either text or image) are returned.

The Fig. 3 shows the overview of the system. In our general system, images are taken from a Web Crawler. However, in the context of the online demonstrator research.qwant.com/images, we are using only images from the YFCC dataset. These images go through TensorRT features Extractor, to be then indexed with FAISS.

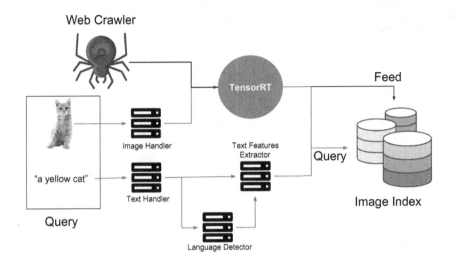

Fig. 3. Overview of the QISS system.

At the query time, the user can:

- Upload an Image. It is sent to the Image Handler and the inference is realized with NVidia TensorRT.
- Search a text. The text goes to the language detector and the Text Features Extractor.

References

1. Conneau, A., Lample, G., Ranzato, M., Denoyer, L., Jégou, H.: Word translation without parallel data. In: International Conference on Learning Representations (2018). https://doi.org/10.1111/j.1540-4560.2007.00543.x. http://arxiv.org/abs/1710.04087
2. Elliott, D., Frank, S., Sima'an, K., Specia, L.: Multi30K: multilingual English-German image descriptions. In: Proceedings of the 5th Workshop on Vision and Language. Association for Computational Linguistics, Stroudsburg (2016). https://doi.org/10.18653/v1/W16-3210. http://arxiv.org/abs/1605.00459. http://aclweb.org/anthology/W16-3210
3. Johnson, J., Douze, M., Jégou, H.: Billion-scale similarity search with GPUs. arXiv preprint arXiv:1702.08734 (2017)

4. Lin, T.-Y., et al.: Microsoft COCO: common objects in context. In: Fleet, D., Pajdla, T., Schiele, B., Tuytelaars, T. (eds.) ECCV 2014. LNCS, vol. 8693, pp. 740–755. Springer, Cham (2014). https://doi.org/10.1007/978-3-319-10602-1_48. http://arxiv.org/abs/1405.0312
5. Portaz, M., Randrianarivo, H., Nivaggioli, A., Maudet, E., Servan, C., Peyronnet, S.: Image search using multilingual texts: a cross-modal learning approach between image and text. Ph.D. thesis, Qwant Research (2019)
6. Shamma, D.A.: One hundred million creative commons Flickr images for research, 24 June (2014)

EveSense: What Can You Sense from Twitter?

Zafar Saeed[1], Rabeeh Ayaz Abbasi[1], and Imran Razzak[2(✉)]

[1] Department of Computer Science, Quaid-i-Azam University, Islamabad, Pakistan
zsaeed@cs.qau.edu.pk, rabbasi@qau.edu.pk
[2] Deakin University, Geelong, VIC, Australia
imran.razzak@deakin.edu.au

Abstract. Social media has become a useful source for detecting real-life events. This paper presents an event detection application *EveSense*. It detects real-life events and related trending topics from the Twitter stream and allows users to find interesting events that have recently occurred. It uses a novel Dynamic Heartbeat Graph (DHG) approach, which efficiently extracts distinguishing features and performs better than the existing event detection methods. We tested and evaluated the application on three case studies, including a sports event (FA cup Final) and two political events (Super Tuesday and US Election).

1 Introduction and Motivation

Social media is the fastest breaking news reporting media and provides plenty of ways to share information [6]. Twitter is one of the most popular social media. Sometimes, it can break the news before newswire—a well-known electronic service that transmits the latest news stories via the Internet. Research conducted at Universities of Edinburgh and Glasgow showed that mainstream media ignored a large number of minor news events [4]. Detecting events from Twitter provides new insight into the searchable information related to real-life. The area of event detection has sparsely studied and is not a new phenomenon. However, the characteristics of Twitter data make it a non-trivial task. Existing systems [1–3,5] typically focus on the burstiness of the data. It is naive to employ burstiness as a key feature to detect the occurrence of events. A rise in tweets frequency related to a long term event often dominates other small but newsworthy events [7,8]. People report major events more often for an extended period. In results, the existing system does not correspond to other events that might be interesting to several users. In this work, we developed an application based on a novel Dynamic Heartbeat Graph (DHG) approach [9] that exploits the dynamic nature of Twitter data for event detection. It detects newsworthy events from the Twitter stream by capturing the change and cohesiveness among the event-related topics. In a nutshell, Fig. 1 presents the work-flow of the event detection approach with the help of a toy example. The next section collectively describes the framework and system design in detail.

ⓒ Springer Nature Switzerland AG 2020
J. M. Jose et al. (Eds.): ECIR 2020, LNCS 12036, pp. 491–495, 2020.
https://doi.org/10.1007/978-3-030-45442-5_64

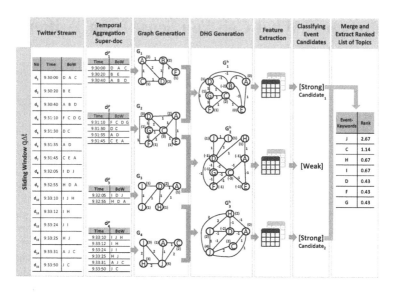

Fig. 1. A toy example describing the work-flow of event detection approach.

2   System Design and Evaluation

The goal of this paper is to describe how the EveSense processes data and produces event descriptions from the Twitter stream. Event detection is performed using an unsupervised graph-based approach devised in our previous studies[1].

The system architecture of EveSense application, as shown in Fig. 2 consists of two major modules, i.e., (1) background processing unit and (2) interactive unit.

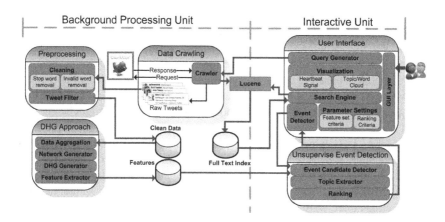

Fig. 2. System architecture of EveSense

---

[1] Event detection approach, and performance comparison with existing approaches are available in [7–9].

Background Processing Unit: consists of a crawler, pre-processor, and DHG formulation. The crawler takes seed words as input to collect the tweets from the Twitter stream. Based on the live or retrospective orientation of the event, the crawler gathers tweets and creates a full-text index using Lucene API[2]. The raw tweets are forwarded to the pre-processing module. Filtration is applied to the data based on heuristics to remove specific tweets, i.e., duplicate, re-tweets, containing URLs, having less than three words, and tweets that do not contain any words other than hashtag(s) and mention(s). The classic IR approach is then used for tokenization, stop/common word removal, and stemming. The clean data is passed on to the DHG approach module that performs four significant tasks, which are the backbone of the EventSense. DHG approach module transforms the data stream into a series of difference graph called the DHG series and extracts three unique features that are later used to detect emerging events[3].

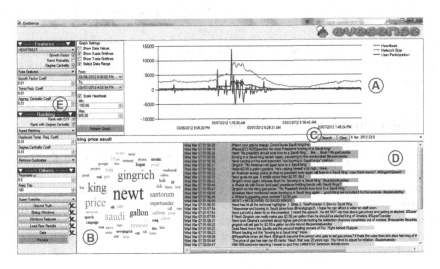

Fig. 3. GUI for event detection, observation and representation

Interactive Unit: consists of Event Detector and User Interface (UI) modules. The event detector module uses a binary classifier to label the event candidate graphs. Topic extractor combines top trending topics from the candidate graphs, and then a ranked list is generated. All the results are presented and visualized on the user interface. UI is one of the major modules controlling the services of all other modules and provides support for customizing different parameter settings corresponding to crawler, and pre-processor. Some of the parameters (Fig. 3E) associated with the DHG approach that allows various modes of building the graph's structure in which temporal aggregation (batch) of tweets and relationships between words are the most important among the others. Events concerning the type, user participation, and region, varies in

---

[2] https://lucenenet.apache.org/.
[3] The detail about feature design can be found in our previous studies [7,8].

popularity and life span hence needed to adjust some of the tuning parameters. The UI can customize the usage and fusion of feature set to observe the optimum results.

Visualizer: The visualization functionality of EveSense produces three temporal signals based (Fig. 3A) on heartbeat score, network size, and user participation. It improves the information seeking and observation process. The UI allows users to analyze different time-slots to observe the event(s) in that particular time interval by generating an interactive word cloud (Fig. 3B) of ranked topics.

Searching Micro-documents: Multiple words from the cloud can be selected to generate a query to retrieve the actual tweets from the corpus matching with the search term(s) (Fig. 3C). The system uses .Net version of a well-known Lucence Library V2.3 to generate a full-text index and facilitate search engine operation within the context of system design and ranking the retrieved tweets (Fig. 3D) with ten unique color codes. Each color covers 10% of the matched tweets facilitating the process of user's information needs.

Performance: Our study shows that the DHG approach, which is the foundation of the event detection method in the system design, is superior in terms of both execution time and accuracy. The detail performance comparison is discussed in [8].

3   Contributions and Conclusion

In this paper, we presented EveSense that detects real-life events from the Twitter stream. It uses a novel approach that repeatedly senses the change-patterns in the Twitter stream and captures newsworthy events efficiently. We evaluated the application on three benchmark datasets FA Cup final, Super Tuesday, and US Election [1]. In addition to the convincing results, EveSense also detected small but newsworthy events that are ignored by the mainstream media. A few of the significant examples of such cases are given in Table 1.

Table 1. Event related trending topics ignored in mainstream media

Case study	Time	Top 10 detected keywords	Relevant Tweets from the Corpus
Super Tuesday	01:50–02:00	newt, gas, president, price, gingrich, king, saudi, speech, make, bow	Gingrich announces 49 step plan to stop Americans from bowing to Saudi King I want to have an energy policy in America so no president will ever again bow to a Saudi king
FA Cup	17:48–17:50	drogba, ankle, torr, groin, physio, hurt, time, wast, ball, injury, stop	He's hurt his ankle, but he's just having a quick wank while he waits Drogba has started the fake injury time wasting

The EveSense also visualize the topics to depict the theme of different events. Generally, the system is useful for individuals who are interested in discovering interesting events from the Twitter stream. It can be helpful for News agencies trying to shape the news story around significant real-life events. Additionally, it can effectively contribute to help state institutions for efficient decision-making and policy-making after analyzing recent local events of interest such as traffic jams, security threats, and epidemics in a specific region. It is an open-source application[4] developed in .Net framework and is fairly easy to use.

References

1. Aiello, L.M., et al.: Sensing trending topics in Twitter. IEEE Trans. Multimedia **15**(6), 1268–1282 (2013)
2. Choi, H.-J., Park, C.H.: Emerging topic detection in twitter stream based on high utility pattern mining. Expert Syst. Appl. **115**, 27–36 (2019)
3. Marcus, A., Bernstein, M.S., Badar, O., Karger, D.R., Madden, S., Miller, R.C.: TwitInfo: aggregating and visualizing microblogs for event exploration. In: Proceedings of the SIGCHI Conference on Human Factors in Computing Systems, CHI 2011, pp. 227–236. ACM, New York (2011)
4. Petrovic, S., Osborne, M., McCreadie, R., Macdonald, C., Ounis, I., Shrimpton, L.: Can Twitter replace newswire for breaking news? In: Proceedings of the Seventh International AAAI Conference on Weblogs and Social Media, USA, pp. 713–716. AAAI Press, July 2013
5. Rill, S., Reinel, D., Scheidt, J., Zicari, R.V.: PoliTwi: early detection of emerging political topics on Twitter and the impact on concept-level sentiment analysis. Knowl.-Based Syst. **69**, 24–33 (2014)
6. Saeed, Z., et al.: What's happening around the world? A survey and framework on event detection techniques on Twitter. J. Grid Comput. **17**(2), 279–312 (2019)
7. Saeed, Z., Abbasi, R.A., Razzak, I., Maqbool, O., Sadaf, A., Xu, G.: Enhanced heartbeat graph for emerging event detection on Twitter using time series networks. Expert Syst. Appl. **136**, 115–132 (2019)
8. Saeed, Z., Abbasi, R.A., Razzak, M.I., Xu, G.: Event detection in Twitter stream using weighted dynamic heartbeat graph approach. IEEE Comput. Intell. Mag. **14**(3), 29–38 (2019)
9. Saeed, Z., Abbasi, R.A., Sadaf, A., Razzak, M.I., Xu, G.: Text stream to temporal network - a dynamic heartbeat graph to detect emerging events on Twitter. In: Phung, D., Tseng, V.S., Webb, G.I., Ho, B., Ganji, M., Rashidi, L. (eds.) PAKDD 2018. LNCS (LNAI), vol. 10938, pp. 534–545. Springer, Cham (2018). https://doi.org/10.1007/978-3-319-93037-4_42

---

[4] https://github.com/Zafar-Saeed/EveSenseApplication.

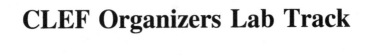

CLEF Organizers Lab Track

CheckThat! at CLEF 2020: Enabling the Automatic Identification and Verification of Claims in Social Media

Alberto Barrón-Cedeño[1]([⊠]), Tamer Elsayed[2], Preslav Nakov[3],
Giovanni Da San Martino[3], Maram Hasanain[2], Reem Suwaileh[2],
and Fatima Haouari[2]

[1] DIT–Università di Bologna, Forlì, Italy
a.barron@unibo.it
[2] Qatar University, Doha, Qatar
{telsayed,maram.hasanain,rs081123,200159617}@qu.edu.qa
[3] Qatar Computing Research Institute, HBKU, Doha, Qatar
{pnakov,gmartino}@hbku.edu.qa

Abstract. We describe the third edition of the CheckThat! Lab, which is part of the 2020 Cross-Language Evaluation Forum (CLEF). CheckThat! proposes four complementary tasks and a related task from previous lab editions, offered in English, Arabic, and Spanish. Task 1 asks to predict which tweets in a Twitter stream are worth fact-checking. Task 2 asks to determine whether a claim posted in a tweet can be verified using a set of previously fact-checked claims. Task 3 asks to retrieve text snippets from a given set of Web pages that would be useful for verifying a target tweet's claim. Task 4 asks to predict the veracity of a target tweet's claim using a set of potentially-relevant Web pages. Finally, the lab offers a fifth task that asks to predict the check-worthiness of the claims made in English political debates and speeches. CheckThat! features a full evaluation framework. The evaluation is carried out using mean average precision or precision at rank k for ranking tasks, and F_1 for classification tasks.

1 Introduction

The mission of the CheckThat! lab is to foster the development of technology that would enable the automatic verification of claims. Automated systems for claim identification and verification can be very useful as supportive technology for investigative journalism, as they could provide help and guidance, thus saving time [14,22,24,33]. A system could automatically identify check-worthy claims, make sure they have not been fact-checked already by a reputable fact-checking organization, and then present them to a journalist for further analysis in a ranked list. Additionally, the system could identify documents that are potentially *useful* for humans to perform manual fact-checking of a claim, and it could also estimate a *veracity score* supported by evidence to increase the journalist's understanding and the trust in the system's decision.

© Springer Nature Switzerland AG 2020
J. M. Jose et al. (Eds.): ECIR 2020, LNCS 12036, pp. 499–507, 2020.
https://doi.org/10.1007/978-3-030-45442-5_65

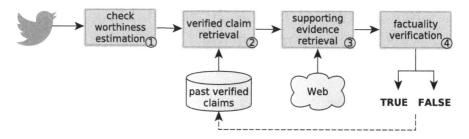

Fig. 1. Information verification pipeline. Our tasks cover all four steps. (Box 1 maps to task 1 whereas boxes 3–4 map to task 2 of the 2018 and 2019 editions [10, 29].)

CheckThat! at CLEF 2020 is the third edition of the lab.[1] The 2018 edition [29] of CheckThat! focused on the identification and verification of claims in political debates.[2] Whereas the 2019 edition [9, 10] also focused on political debates, isolated claims were considered as well, in conjunction with a closed set of Web documents to retrieve evidence from.[3]

In 2020, CheckThat! turns its attention to social media—in particular to *Twitter*—as information posted on that platform is not checked by an authoritative entity before publication and such information tends to disseminate very quickly. Moreover, social media posts lack context due to their short length and conversational nature; thus, identifying a claim's context is sometimes key for enabling effective fact-checking [7].

2   Description of the Tasks

The lab is mainly organized around four tasks, which correspond to the four main blocks in the verification pipeline, as illustrated in Fig. 1. Tasks 1, 3, and 4 can be seen as reformulations of corresponding tasks in 2019, which enables re-use of training data and systems from previous editions of the lab (cf. Sect. 3). Task 2 runs for the first time. While Tasks 1–4 are focused on Twitter, Task 5 (not in Fig. 1) focuses on political debates as in the previous two editions of the lab. All tasks are run in English. Additionally, Tasks 1, 3, and 4 are also offered in Arabic and/or Spanish.

2.1   Task 1: Check-Worthiness on Tweets

Task 1 is formulated as follows: *Given a topic and a stream of potentially-related tweets, rank the tweets according to their check-worthiness for the topic.*

Previous work on check-worthiness focused primarily on political debates and speeches, but here we focus on tweets instead.

---

[1] https://sites.google.com/view/clef2020-checkthat/.
[2] http://alt.qcri.org/clef2018-factcheck/.
[3] https://sites.google.com/view/clef2019-checkthat.

Dataset. We include "topics" this year, as we want to have a scenario that is close to that from 2019; a topic gives a context just like a debate did. We construct the dataset by tracking a set of manually-created topics in Twitter. A sample of tweets from the tracked stream (per topic) is shared with the participating systems as input for Task 1. The systems are asked to submit a ranked list of the tweets for each topic. Finally, using pooling, a set of tweets is selected and then judged by in-house annotators.

Evaluation. We treat Task 1 as a ranking problem. Systems are evaluated using ranking evaluation measures, namely Mean Average Precision (MAP) and precision at rank k (P@k). The official measure is P@30.

2.2 Task 2: Verified Claim Retrieval

Task 2 is defined as follows: *Given a check-worthy claim and a dataset of verified claims, rank the verified claims, so that those that verify the input claim (or a sub-claim in it) are ranked on top.*

Given an input claim c and a set $V_c = \{v_i\}$ of verified claims, we consider each pair (c, v_i) as *Relevant* if v_i would save the process of verifying c from scratch, and as *Irrelevant* otherwise. Note that there might be more than one *relevant* verified claim per input claim, e.g., because the input claim might be composed of multiple claims. The task is similar to paraphrasing and textual similarity tasks, as well as to textual entailment [8,12,30].

Dataset. Verified claims are retrieved from fact-checking websites such as *Snopes* and *PolitiFact*.

Evaluation. Mean Average Precision on the first 5 retrieved claims (MAP@5) is used to assess the quality of the rankings submitted by the participants. A perfect ranking will have on top all v_i such that (c, v_i) is *Relevant*, in any order, followed by all *Irrelevant* claims. In addition to MAP@5, we also report MRR, MAP@k ($k = 3, 10, 20, all$) and Recall@k for $k = 3, 5, 10, 20$ in order to provide participants with more information about their systems.

2.3 Task 3: Evidence Retrieval

Task 3 is defined as follows: *Given a check-worthy claim on a specific topic and a set of text snippets extracted from potentially-relevant webpages, return a ranked list of all evidence snippets for the claim. Evidence snippets are those snippets that are useful in verifying the given claim.*

Dataset. While tracking on-topic tweets, we search the Web to retrieve top-m Web pages using topic-related queries. This would ensure the freshness of the retrieved pages and enable reusability of the dataset for real-time verification tasks. Once we acquire annotations for Task 1, we share with participants the Web pages and text snippets from them solely for the check-worthy claims, which would enable the start of the evaluation cycle for Task 3. In-house annotators will label each snippet as evidence or not for a target claim.

Evaluation. Tasks 3 is a ranking problem. We evaluate the ranked list per topic using MAP and P@k. The official measure is P@10.

2.4   Task 4: Claim Verification

Task 4 is defined as follows: *Given a check-worthy claim on a specific topic and a set of potentially-relevant Web pages, predict the veracity of the claim.* This task closes the verification pipeline.

Dataset. The dataset for this task is the same as for Task 3. The only difference is that the in-house annotators judge each claim as true or false.

Evaluation. Task 4 is a binary classification problem. Therefore, it is evaluated using standard classification evaluation measures: Precision, Recall, F_1, and Accuracy. The official measure is macro-averaged F_1.

2.5   Task 5: Check-Worthiness on Debates

Task 5 is defined as follows: *Given a debate segmented into sentences, together with speaker information, prioritize sentences for fact-checking.* This is a ranking task and each sentence should be associated with a score.

Dataset. This is the third iteration of this task. We believe it is important to keep it alive as we have a large body of annotated data already and new material arrives with the coming 2020 US Presidential elections.

Evaluation. Task 5 is yet another ranking problem. We use MAP as the official evaluation measure. We further report P@k for $k \in \{5, 10, 20, 50\}$.

3   Previously on CheckThat!

Two editions of CheckThat! have been held so far. While the datasets come from different genres, some of the tasks in the 2020 edition are reformulated. Hence, considering some of the most successful approaches applied in the past represents a good starting point to address the current challenges.

3.1   CheckThat! 2019

The 2019 edition featured two tasks [10]:

Task 1$_{2019}$. Given a political debate, interview, or speech, transcribed and segmented into sentences, rank the sentences by the priority with which they should be fact-checked.

The most successful approaches used neural networks for the individual classification of the instances. For example, Hansen et al. [19] learned domain-specific word embeddings and syntactic dependencies and applied an LSTM classifier.

Using some external knowledge paid off—they pre-trained the network with previous Trump and Clinton debates, supervised weakly with the ClaimBuster system. Some efforts were carried out in order to consider context. Favano et al. [11] trained a feed-forward neural network, including the two previous sentences as context. Whereas many approaches opted for embedding representations, feature engineering was also popular [13].

Task 2$_{2019}$. Given a claim and a set of Web pages potentially relevant with respect to the claim, identify which of the pages (and passages thereof) are useful for assisting a human in fact-checking the claim. Finally, determine the factuality of the claim.

The systems for evidence passage identification followed two approaches. BERT was trained and used to predict whether an input passage is useful to fact-check a claim [11]. Other participating systems used classifiers (e.g., SVM) with a variety of features including similarity between the claim and a passage, bag of words, and named entities [20]. As for predicting claim veracity, the most effective approach used a textual entailment model. The input was represented using word embeddings and external data was also used in training [15].

In the 2020 edition, Task 1$_{2019}$ becomes Task 5, and Task 1 is a reformulation based on tweets (cf. Sect. 2.1). See [2] for further details. Task 2$_{2019}$ becomes Tasks 3 and 4 (cf. Sects. 2.3 and 2.4). See [21] for further details.

3.2   CheckThat! 2018

The 2018 edition featured two tasks [29]:

Task 1$_{2018}$ was identical to Task 1$_{2019}$.

The most successful approaches used either a multilayer perceptron or an SVM. Zuo et al. [36] enriched the dataset by producing *pseudo-speeches* as a concatenation of all interventions by a debater. They used averaged word embeddings and bag-of-words as representations. Hansen et al. [18] represented the entries with embeddings, part of speech tags, and syntactic dependencies. They used a GRU neural network with attention. See [1] for further details.

Task 2$_{2018}$. Given a check-worthy claim in the form of a (transcribed) sentence, determine whether the claim is likely to be true, half-true, or false.

The best way to address this task was to retrieve relevant information from the Web, followed by a comparison to the claim in order to assess its factuality.[4] After retrieving such *evidence*, it is fed into the supervised model, together with the claim in order to assess its veracity. In the case of [18], they fed the claim and the most similar Web-retrieved text to convolutional neural networks and SVMs. Meanwhile, Ghanem et al. [16] computed features, such as the similarity between the claim and the Web text, and the Alexa rank for the website. See [4] for further details.

4   Related Work

There has been work on checking the factuality/credibility of a claim, of a news article, or of an information source [3,25,26,28,31,35]. Claims can come from different sources, but special attention has been given to those from social media [17,27,32,34]. Check worthiness estimation is still a fairly-new problem especially in the context of social media [14,22–24].

CheckThat! further shares some aspects with other initiatives that have been run with high success in the past, e.g., stance detection (Fake News[5]), semantic textual similarity (STS at SemEval[6]), and community question answering (cQA at SemEval[7]).

5   Conclusion

We have presented the 2020 edition of the CheckThat! Lab, which features tasks that span the full verification pipeline: from spotting check-worthy claims to checking whether they have been fact-checked elsewhere already, to retrieving useful passages within relevant pages, to finally making a prediction about the factuality of a claim. To the best of our knowledge, this is the first shared task that addresses all steps of the fact-checking process. Moreover, unlike previous editions of the CheckThat! Lab, our main focus here is on social media, which are the center of "fake news" and disinformation. We further feature a more realistic information retrieval scenario with pooling for evaluation, as done at IR venues such as TREC. Last but not least, in-line with the general mission of CLEF, we promote multi-linguality by offering our tasks in different languages.

We hope that these tasks and the associated datasets will serve the mission of the CheckThat! initiative, which is to foster the development of datasets, tools and technology that would enable the automatic verification of claims and will support human fact-checkers in their fight against "fake news" and disinformation.

---

[4] While this year a similar procedure had to be carried out, we decompose it into three tasks (cf. Sect. 2).

[5] Official Challenge website: http://www.fakenewschallenge.org/.

[6] STS task at the SemEval 2017 edition: http://alt.qcri.org/semeval2017/task1/.

[7] cQA task at the SemEval 2017 edition: http://alt.qcri.org/semeval2017/task3/.

Acknowledgments. The work of Tamer Elsayed and Maram Hasanain was made possible by NPRP grant# NPRP 11S-1204-170060 from the Qatar National Research Fund (a member of Qatar Foundation). The work of Reem Suwaileh was supported by GSRA grant# GSRA5-1-0527-18082 from the Qatar National Research Fund and the work of Fatima Haouari was supported by GSRA grant# GSRA6-1-0611-19074 from the Qatar National Research Fund. The statements made herein are solely the responsibility of the authors. This research is also part of the Tanbih project, developed by the Qatar Computing Research Institute, HBKU and MIT-CSAIL, which aims to limit the effect of "fake news", propaganda, and media bias.

References

1. Atanasova, P., et al.: Overview of the CLEF-2018 CheckThat! Lab on automatic identification and verification of political claims. Task 1: check-worthiness. In: Cappellato et al. [6]
2. Atanasova, P., Nakov, P., Karadzhov, G., Mohtarami, M., Da San Martino, G.: Overview of the CLEF-2019 CheckThat! Lab on automatic identification and verification of claims. Task 1: check-worthiness. In: Cappellato et al. [5]
3. Ba, M.L., Berti-Equille, L., Shah, K., Hammady, H.M.: VERA: a platform for veracity estimation over web data. In: Proceedings of the 25th International Conference Companion on World Wide Web, WWW 2016 Companion, pp. 159–162 (2016)
4. Barrón-Cedeño, A., et al.: Overview of the CLEF-2018 CheckThat! Lab on automatic identification and verification of political claims. Task 2: factuality. In: Cappellato et al. [6]
5. Cappellato, L., Ferro, N., Losada, D., Müller, H. (eds.): Working Notes of CLEF 2019 Conference and Labs of the Evaluation Forum. CEUR Workshop Proceedings. CEUR-WS.org (2019)
6. Cappellato, L., Ferro, N., Nie, J.Y., Soulier, L. (eds.): Working Notes of CLEF 2018-Conference and Labs of the Evaluation Forum. CEUR Workshop Proceedings. CEUR-WS.org (2018)
7. Cazalens, S., Lamarre, P., Leblay, J., Manolescu, I., Tannier, X.: A content management perspective on fact-checking. In: Proceedings of The Web Conference 2018, WWW 2018, pp. 565–574 (2018)
8. Cer, D., Diab, M., Agirre, E., Lopez-Gazpio, I., Specia, L.: SemEval-2017 Task 1: semantic textual similarity multilingual and crosslingual focused evaluation. In: Proceedings of the 11th International Workshop on Semantic Evaluation, SemEval 2017, pp. 1–14 (2017)
9. Elsayed, T., et al.: CheckThat! at CLEF 2019: automatic identification and verification of claims. In: Azzopardi, L., Stein, B., Fuhr, N., Mayr, P., Hauff, C., Hiemstra, D. (eds.) ECIR 2019. LNCS, vol. 11438, pp. 309–315. Springer, Cham (2019). https://doi.org/10.1007/978-3-030-15719-7_41
10. Elsayed, T., et al.: Overview of the CLEF-2019 CheckThat! Lab: automatic identification and verification of claims. In: Crestani, F., et al. (eds.) CLEF 2019. LNCS, vol. 11696, pp. 301–321. Springer, Cham (2019). https://doi.org/10.1007/978-3-030-28577-7_25
11. Favano, L., Carman, M., Lanzi, P.: TheEarthIsFlat's submission to CLEF'19 CheckThat! Challenge. In: Cappellato et al. [5]
12. Filice, S., Da San Martino, G., Moschitti, A.: Structural representations for learning relations between pairs of texts, pp. 1003–1013 (2015)

13. Gasior, J., Przybyła, P.: The IPIPAN team participation in the check-worthiness task of the CLEF2019 CheckThat! Lab. In: Cappellato et al. [5]
14. Gencheva, P., Nakov, P., Màrquez, L., Barrón-Cedeño, A., Koychev, I.: A context-aware approach for detecting worth-checking claims in political debates. In: Proceedings of the International Conference Recent Advances in Natural Language Processing, RANLP 2017, pp. 267–276 (2017)
15. Ghanem, B., Glavaš, G., Giachanou, A., Ponzetto, S., Rosso, P., Rangel, F.: UPV-UMA at CheckThat! Lab: verifying Arabic claims using cross lingual approach. In: Cappellato et al. [5]
16. Ghanem, B., Montes-y Gómez, M., Rangel, F., Rosso, P.: UPV-INAOE-Autoritas - Check That: preliminary approach for checking worthiness of claims. In: Cappellato et al. [6]
17. Gupta, A., Kumaraguru, P., Castillo, C., Meier, P.: TweetCred: real-time credibility assessment of content on Twitter. In: Aiello, L.M., McFarland, D. (eds.) SocInfo 2014. LNCS, vol. 8851, pp. 228–243. Springer, Cham (2014). https://doi.org/10.1007/978-3-319-13734-6_16
18. Hansen, C., Hansen, C., Simonsen, J., Lioma, C.: The Copenhagen team participation in the check-worthiness task of the competition of automatic identification and verification of claims in political debates of the CLEF-2018 fact checking lab. In: Cappellato et al. [6]
19. Hansen, C., Hansen, C., Simonsen, J., Lioma, C.: Neural weakly supervised fact check-worthiness detection with contrastive sampling-based ranking loss. In: Cappellato et al. [5]
20. Haouari, F., Ali, Z., Elsayed, T.: bigIR at CLEF 2019: automatic verification of Arabic claims over the web. In: Cappellato et al. [5]
21. Hasanain, M., Suwaileh, R., Elsayed, T., Barrón-Cedeño, A., Nakov, P.: Overview of the CLEF-2019 CheckThat! Lab on automatic identification and verification of claims. Task 2: evidence and factuality. In: Cappellato et al. [5]
22. Hassan, N., Li, C., Tremayne, M.: Detecting check-worthy factual claims in presidential debates. In: Proceedings of the 24th ACM International on Conference on Information and Knowledge Management, CIKM 2015, pp. 1835–1838 (2015)
23. Hassan, N., Tremayne, M., Arslan, F., Li, C.: Comparing automated factual claim detection against judgments of journalism organizations. In: Computation+Journalism Symposium (2016)
24. Hassan, N., et al.: ClaimBuster: the first-ever end-to-end fact-checking system. Proc. VLDB Endow. **10**(12), 1945–1948 (2017)
25. Karadzhov, G., Nakov, P., Màrquez, L., Barrón-Cedeño, A., Koychev, I.: Fully automated fact checking using external sources. In: Proceedings of the International Conference Recent Advances in Natural Language Processing, RANLP 2017, pp. 344–353 (2017)
26. Ma, J., et al.: Detecting rumors from microblogs with recurrent neural networks. In: Proceedings of the International Joint Conference on Artificial Intelligence, IJCAI 2016, pp. 3818–3824 (2016)
27. Mitra, T., Gilbert, E.: CREDBANK: a large-scale social media corpus with associated credibility annotations. In: Proceedings of the Ninth International AAAI Conference on Web and Social Media, ICWSM 2015 (2015)
28. Mukherjee, S., Weikum, G.: Leveraging joint interactions for credibility analysis in news communities. In: Proceedings of the 24th ACM International on Conference on Information and Knowledge Management, CIKM 2015, pp. 353–362 (2015)

29. Nakov, P., et al.: Overview of the CLEF-2018 lab on automatic identification and verification of claims in political debates. In: Working Notes of CLEF 2018 – Conference and Labs of the Evaluation Forum, Avignon, France, CLEF 2018 (2018)
30. Nakov, P., et al.: SemEval-2016 Task 3: community question answering. In: Proceedings of the 10th International Workshop on Semantic Evaluation, SemEval 2015, pp. 525–545 (2016)
31. Popat, K., Mukherjee, S., Strötgen, J., Weikum, G.: Credibility assessment of textual claims on the web. In: Proceedings of the 25th ACM International Conference on Information and Knowledge Management, CIKM 2016, pp. 2173–2178 (2016)
32. Shu, K., Sliva, A., Wang, S., Tang, J., Liu, H.: Fake news detection on social media: a data mining perspective. SIGKDD Explor. Newsl. **19**(1), 22–36 (2017)
33. Vasileva, S., Atanasova, P., Màrquez, L., Barrón-Cedeño, A., Nakov, P.: It takes nine to smell a rat: neural multi-task learning for check-worthiness prediction. In: Proceedings of the International Conference on Recent Advances in Natural Language Processing, RANLP 2019 (2019)
34. Zhao, Z., Resnick, P., Mei, Q.: Enquiring minds: early detection of rumors in social media from enquiry posts. In: Proceedings of the 24th International Conference on World Wide Web, WWW 2015, pp. 1395–1405 (2015)
35. Zubiaga, A., Liakata, M., Procter, R., Hoi, G.W.S., Tolmie, P.: Analysing how people orient to and spread rumours in social media by looking at conversational threads. PLoS ONE **11**(3), 1–29 (2016)
36. Zuo, C., Karakas, A., Banerjee, R.: A hybrid recognition system for check-worthy claims using heuristics and supervised learning. In: Cappellato et al. [6]

Shared Tasks on Authorship Analysis
at PAN 2020

Janek Bevendorff[1], Bilal Ghanem[2], Anastasia Giachanou[2], Mike Kestemont[3],
Enrique Manjavacas[3], Martin Potthast[4(✉)], Francisco Rangel[5], Paolo Rosso[2],
Günther Specht[6], Efstathios Stamatatos[7], Benno Stein[1], Matti Wiegmann[1],
and Eva Zangerle[6]

[1] Bauhaus-Universität Weimar, Weimar, Germany
[2] Universitat Politècnica de València, Valencia, Spain
[3] University of Antwerp, Antwerp, Belgium
[4] Leipzig University, Leipzig, Germany
pan@webis.de, martin.potthast@uni-leipzig.de
[5] Symanto Research, Nuremberg, Germany
[6] University of Innsbruck, Innsbruck, Austria
[7] University of the Aegean, Samos, Greece
http://pan.webis.de

Abstract. The paper gives a brief overview of the four shared tasks that are to be organized at the PAN 2020 lab on digital text forensics and stylometry, hosted at CLEF conference. The tasks include author profiling, celebrity profiling, cross-domain author verification, and style change detection, seeking to advance the state of the art and to evaluate it on new benchmark datasets.

1 Introduction

PAN is a series of scientific events and shared tasks on digital text forensics and stylometry, bringing together scientists, industry professionals, and public institutions from information retrieval and NLP to work on challenges in authorship analysis, originality, and computational ethics. Since its inception in 2007, PAN has hosted 22 shared tasks at 21 different events with continually increasing reception within the community. The latest installment of PAN at CLEF 2019 had a strong focus on authorship analysis, featuring tasks on author profiling, celebrity profiling, authorship attribution, and style change detection. Continuing in 2020, PAN will again organize four shared tasks in these domains. The first task, profiling fake news spreaders on Twitter, addresses the critical societal problem of fake news from the perspective of author profiling, by studying stylistic deviations of users inclined to spread them. The second task, cross-domain authorship verification, studies the stylistic association between authors and their works in a setting without the interference of domain-specific vocabulary. The third task, celebrity profiling, analyzes the presumed influence that celebrities have on their followers to study whether celebrities can be profiled based on their followership. The fourth task, style change detection, continues the research on multi-author documents by attempting to separate segments of a document based on authorship.

© Springer Nature Switzerland AG 2020
J. M. Jose et al. (Eds.): ECIR 2020, LNCS 12036, pp. 508–516, 2020.
https://doi.org/10.1007/978-3-030-45442-5_66

A milestone in PAN's development has been the development of the TIRA platform, switching from the traditional submission of answers to *software* submissions. The guaranteed availability of all submitted software greatly enhances the reproducibility of methods and PAN is committed to continue this endeavor.

2   Author Profiling

Author profiling distinguishes between classes of authors by studying how language is shared by people. This helps in identifying profiling aspects such as age, gender, and language variety, among others. In the years 2013–2018, we addressed several aspects in the shared tasks we organized at PAN.[1] In 2013, the aim was to identify gender and age in social media texts for English and Spanish [22]. The corpus included chat lines of potential pedophiles with the purpose of investigating the robustness of the best-performing systems also from this perspective (i.e., identifying the age of the pedophiles). Age classes included a gap in between: 10s (13–17), 20s (23–27), 30s (33–48). Results in both languages and in both subtasks were below 70% accuracy.

In 2014, the aims of the shared task were twofold: to address age identification from a continuous perspective (without gaps between the age classes), and to include other genres such as blogs, Twitter and reviews (in Trip Advisor), both in English and Spanish. The best results were obtained on Twitter, where users showed a more spontaneous way to communicate [20]. In 2015, apart from age and gender identification, we addressed also personality recognition in Twitter in English, Spanish, Dutch and Italian. The best results (above 80% accuracy) were obtained on English data [24]. In 2016, we addressed the problem of cross-genre gender and age identification (training on Twitter data and testing on blogs and social media data), in English, Spanish, and Dutch. The best results were obtained on blogs for English with an accuracy above 75% for gender and below 60% for age identification [25]. In 2017, we addressed gender and language variety identification in Twitter, in English, Spanish, Portuguese and Arabic. The lowest results were obtained for Arabic with an accuracy of 80% for gender and 83% for language variety identification [23]. In 2018, our aim was to investigate if approaching gender identification in Twitter from a multimodal perspective (e.g., considering also images of the links in tweets) could improve results. The corpus was composed of English, Spanish, and Arabic tweets. Only for Arabic it was possible to improve accuracy (albeit less than 2%) [21].

Last, in 2019, in the shared task on bots and gender profiling, we aimed at investigating how difficult it is to discriminate bots from humans on the basis only of textual data, and what were the most difficult types of bots. We used Twitter data both in English and Spanish and the best-performing systems showed that it is possible to profile bots with an accuracy above 90%. Advanced bots that generated human-like language, also with metaphors, were the most difficult to be profiled. It is interesting to mention that when bots were profiled as humans, they were mostly confused with males [19]. The number of the participants in the several editions of the author profiling task can be seen in Fig. 1.

---

[1] To generate the corpora, we followed a methodology that complies with the EU General Data Protection Regulation [18].

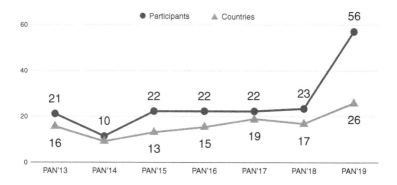

Fig. 1. Evolution of the number of participants and countries in the author profiling task.

Profiling Fake News Spreaders on Twitter at PAN'20

Fake news can be very harmful since it is usually created with the aim to manipulate public opinions and beliefs. Recently fake news detection has gained a lot of attention from the research community. Indeed, their early detection can prevent further dissemination of false claims and rumors, but it's a hard and time-consuming task, since they involve manual annotation. Recent approaches that have been proposed [6,7,16] are effective in detecting false claims that already have been disseminated, but not the newly emerging ones. In addition, these models do not take into account the role of users that unintentionally or intentionally share the false claims and who play a critical role in their propagation. To this end, in this task, we aim at identifying and profiling fake news spreaders on social media as a first step towards preventing fake news from being propagated among online users.

We propose a new task that focuses on fake and real news spreaders detection. The detection of accounts that are possible spreaders of fake news is very important for the field of misinformation detection. These accounts could be operated by laymen [2], "professional" trolls [4], and even bots [14]. The fake news spreaders might be identified from several possible perspectives: textual, semantic, sentiment, social variables, etc. Previous work [5] showed that word embeddings and style features are important to profile such accounts, whereas other information, such as hashtags, is not useful.

Given a users with her corresponding tweet stream, the task is to identify the users as a faker (fake news spreader), or a legitimate users (real news spreader). For the evaluation setup, we create a collection of Twitter accounts, each with a sample of tweets from her timeline. The collection has been created in English and Spanish, and it is balanced. Thus, we are going to use accuracy to evaluate the performance of the systems.

3   Celebrity Profiling

Celebrity profiling is author profiling applied to celebrities. Celebrities can contribute much to author profiling research: they are prolific social media users, often supplying extensive writing samples as well as personal details. Celebrities build a consistent public persona either themselves or with the help of public relations agents. In addition, celebrities are in a unique position within their communities: they are highly influential on their followers, frequently considered trustworthy and reliable, and they act as hubs for like-minded people on social media. Celebrity Profiling [30] is the newest addition to PAN's shared tasks. In 2019 [31], the goal was to determine the demographics age, gender, occupation, and fame from the timelines of celebrities on Twitter. Eight participants submitted solutions, which, given sufficient training data, performed well on demographics with a coherent separability by topic or domain. Poor performance was achieved in cases where certain demographics are rare (e.g., non-binary genders), or where they are underrepresented (e.g., age groups for very low and high ages). Also domain-invariant demographics, like the scientific creative occupations, posed problems. The results of the first shared task on celebrity profiling are coherent with most of the related work in author profiling, authorship analysis, and computational stylometry in general: the domain-specific vocabulary is the primary discriminator and demographic differences are often reflected by topics.

Celebrity Profiling at PAN'20

The unique contributions of celebrities on social media towards author profiling research is their domain-variant claim-to-fame and the varying degree of influence they exert on their followers. The formation of closely connected communities around celebrities, who are also under their influence, allows us to investigate the role of author characteristics, domain, and demographic on language use. For the upcoming edition of celebrity profiling, we focus on separating a celebrity author's textual characteristics from domain-specific language use, using the demographics as an indicator. Instead of predicting the authors demographics from his text alone, we use the texts of highly influenced individuals, while the prediction targets remain largely the same as last year (age, gender, occupation). The results of this shared task will help us to determine for the first time, whether and to what extent an influencer's demographics and characteristics can be predicted from his or her followers. Tangible applications, besides academic interest, include methods to profile users with few own text samples, and to judge influence exerted between users in a community.

4   Author Identification

Authentication is a major concern in today's global information society and in this sense it does not come as a surprise that author identification has been a long-running task at PAN. Author identification still poses a challenging empirical problem in fields related to information and computer science, but the underlying methods are nowadays also increasingly used as an auxiliary technology in more applied domains, such as literary studies or forensic linguistics. These communities crucially rely on trustworthy, transparent benchmark initiatives that reliably establish the state of the art in the field [17]. Author identification is concerned with the automated identification of the individual(s) who authored an anonymous document on the basis of text-internal properties related to language and writing style [9,12,27]. At different editions of PAN (since 2007),

author identification has been studied in multiple incarnations: AUTHORSHIP ATTRI-
BUTION: given a document and a set of candidate authors, determine which of them
wrote the document (2011–2012, 2016–2020); AUTHORSHIP VERIFICATION: given
a pair of documents, determine whether they are written by the same author (2013–
2015); AUTHORSHIP OBFUSCATION: given a document and a set of documents from
the same author, paraphrase the former so that its author cannot be identified anymore
(2016–2018); OBFUSCATION EVALUATION: devise and implement performance measures that quantify safeness, soundness, and/or sensibleness of an obfuscation software
(2016–2018).

For the next edition, we shall continue working with 'fanfiction' [10,11]. This term
refers to the global phenomenon of non-professional authors taking up the production
of fiction in the tradition of well-known cultural domains, called 'fandoms', such as
J.K. Rowling's Harry Potter or Sherlock Holmes [8]. The abundance of data is a major
advantage, as fanfiction is nowadays estimated to form the fastest growing form of
online writing [3]. Fan writers actively aim to increase their readership and on most platforms (e.g., archiveofourown.org or fanfiction.net), the bulk of writings can be openly
accessed, although the intellectual rights are not unproblematic [29]. The multilingual-
ism of the phenomenon is another asset, extending far beyond the Indo-European languages that are the traditional focus of shared tasks. Finally, fanfiction is characterized
by a relative wealth of author-provided metadata, relating to the textual domain (the
fandom), period of production, and intended audience.

Cross-domain Authorship Verification at PAN'20

In 2020, we shall visit the task of authorship verification again: as opposed to authorship
attribution, which requires a carefully balanced classification setup, authorship verification is a more fundamental task. Authorship verification can be formalized as the task
of approximating the target function $\phi : (D_k, d_u) \rightarrow \{T, F\}$, where D_k is a set of
documents of known authorship by the same author and d_u is a document of questioned
authorship. If $\phi(D_k, d_u) = T$, then the author of D_k is also the author of d_u and if
$\phi(D_k, d_u) = F$, then the author of D_k is not the same with the author of d_u. In cross-
domain settings, D_k and d_u do not share topic, genre or even language (in our case the
fandom is different). A simple form of the verification task is to only consider the case
where D_k is singleton, thus only pairs of documents are examined. Given a training
set of such text pairs, verification systems can be trained and calibrated to analyze the
authorship of unseen pairs. Such verifiers produce a score in the form of a bounded
scalar between 0 and 1, indicating the probability of the test item being a same-author
pair (rather than a binary choice).

The nature of the relationship between the training set and test set and their exact
composition is crucial to the difficulty of the task. For PAN'20, we shall vary these
along a number of dimensions. (I) The ratio of same-author pairs (SA) over the number
of different-author (DA) pairs: while this ratio is extremely low in real-world settings,
computational systems benefit from under-sampling DAs to achieve a better balance.
(II) Systems are known to be very sensitive to changes in domain and topic: whether
or not train and test pairs are extracted from the same fandom(s) will strongly affect
performance [1]. Including multiple fandoms into training and/or test pairs is another

valuable aspect for experimentation. (III) Overfitting on specific authors is a real danger during training: allowing authors to contribute more than one text during the construction of training pairs might affect performance. Likewise, one explicitly can vary the number of test authors (if any) that have not been encountered in training. (IV) Text length is another challenge [13]: short documents are more difficult to analyze and text pairs that significantly differ in length also present an important obstacle.

We shall extract a number of datasets exploring these aspects from a recent large-scale crawl from an established fan platform fanfiction.net, that contains over 5.8M stories, in 44 languages, distributed over about 10,300 fandoms. We intend to apply various techniques to estimate the degree of topical divergence between individual fandoms. These estimates will be useful to construct datasets of varying complexity. The large size of these datasets will be a novel contribution to the state of the art: whereas a larger number of different authors typically degrades the performance of authorship attributors [15], the same is not necessarily true for verification systems, that are intrinsically better suited to learn from a variety of authorial styles [13]. Finally, our aim is to also release these datasets outside of the strict TIRA environment, in order to further lower the barrier for experimentation and stimulate the data's wider adoption in the community.

5   Style Change Detection

The goal of the style change detection task is to identify the text positions within a given multi-author document at which the author switches, based on an intrinsic style analysis. Detecting these positions is a crucial part of the authorship identification process, and for multi-author document analysis in general—documents which have not been studied a lot to date.

This task has been part of PAN since 2016, with varying task definitions, datasets and evaluation procedures. In 2016, participants were asked to identify and group fragments of a given document that correspond to individual authors [26]. In 2017, we asked participants to detect whether a given document is multi-authored and if this is indeed the case, to determine the positions at which authorship changes [28]. However, this task was deemed as highly complex and hence, was relaxed in 2018, asking participants to predict whether a given document is single- or multi-authored [11]. Given the promising results achieved, in 2019, participants were asked to firstly detect whether a document was single- or multi-authored and, if it was indeed written by multiple authors, to predict the number of authors [32].

Style Change Detection at PAN'20

Given the key role of this task and the progress made in previous years, at PAN'20, we will continue to advance research in this direction. We aim to steer the task back to its original goal: detecting the exact position of authorship changes. Therefore, the task for PAN'20 is to find the positions of style changes at the paragraph-level. For each pair of consecutive paragraphs of a document, we ask participants to estimate whether there is indeed a style change between those two paragraphs. This binary classification task will be performed on a dataset curated based on a publicly available dump of a Q&A platform to cover different types of documents at different lengths and topics. We will distill two different datasets: one featuring a rather narrow set of topics being discussed, and a second dataset containing a broad variety of topics. This setup allows for analyzing the performance of the developed approaches in dimensions of text length, topics, and the number of contributing authors.

Acknowledgments. The work of Anastasia Giachanou is supported by the SNSF Early Post-doc Mobility grant P2TIP2_181441 under the project Early Fake News Detection on Social Media, Switzerland. The work of Paolo Rosso was partially funded by the Spanish MICINN under the research project MISMIS-FAKEnHATE on Misinformation and Miscommunication in social media: FAKE news and HATE speech (PGC2018-096212-B-C31).

References

1. Bevendorff, J., Hagen, M., Stein, B., Potthast, M.: Bias analysis and mitigation in the evaluation of authorship verification. In: Proceedings of the 57th Annual Meeting of the Association for Computational Linguistics, Florence, Italy, pp. 6301–6306. Association for Computational Linguistics, July 2019. https://doi.org/10.18653/v1/P19-1634. https://www.aclweb.org/anthology/P19-1634

2. Darwish, K., Alexandrov, D., Nakov, P., Mejova, Y.: Seminar users in the Arabic Twitter sphere. In: Ciampaglia, G.L., Mashhadi, A., Yasseri, T. (eds.) SocInfo 2017. LNCS, vol. 10539, pp. 91–108. Springer, Cham (2017). https://doi.org/10.1007/978-3-319-67217-5_7

3. Fathallah, J.: Fanfiction and the Author. How FanFic Changes Popular Cultural Texts. Amsterdam University Press, Amsterdam (2017)

4. Ghanem, B., Buscaldi, D., Rosso, P.: TexTrolls: identifying Russian trolls on Twitter from a textual perspective. arXiv preprint arXiv:1910.01340 (2019)

5. Ghanem, B., Paolo Ponzetto, S., Rosso, P.: FacTweet: profiling fake news twitter accounts. arXiv preprint arXiv:1910.06592 (2019)

6. Ghanem, B., Rosso, P., Rangel, F.: An emotional analysis of false information in social media and news articles. arXiv preprint arXiv:1908.09951 (2019)

7. Giachanou, A., Rosso, P., Crestani, F.: Leveraging emotional signals for credibility detection. In: Proceedings of the 42nd International ACM SIGIR Conference on Research and Development in Information Retrieval, pp. 877–880. ACM (2019)

8. Hellekson, K., Busse, K. (eds.): The Fan Fiction Studies Reader. University of Iowa Press, Iowa (2014)

9. Juola, P.: Authorship attribution. Found. Trends Inf. Retr. **1**(3), 233–334 (2006)

10. Kestemont, M., Stamatatos, E., Manjavacas, E., Daelemans, W., Potthast, M., Stein, B.: Overview of the cross-domain authorship attribution task at PAN 2019. In: Cappellato, L., Ferro, N., Losada, D., Müller, H. (eds.) CLEF 2019 Labs and Workshops, Notebook Papers. CEUR-WS.org, September 2019. http://ceur-ws.org/Vol-2380/

11. Kestemont, M., et al.: Overview of the author identification task at PAN-2018: cross-domain authorship attribution and style change detection. In: Cappellato, L., et al. (eds.) Working Notes Papers of the CLEF 2018 Evaluation Labs, Avignon, France, 10–14 September 2018, pp. 1–25 (2018)

12. Koppel, M., Schler, J., Argamon, S.: Computational methods in authorship attribution. J. Am. Soc. Inform. Sci. Technol. **60**(1), 9–26 (2009)

13. Koppel, M., Winter, Y.: Determining if two documents are written by the same author. J. Assoc. Inf. Sci. Technol. **65**(1), 178–187 (2014)

14. Lazer, D.M., et al.: The science of fake news. Science **359**(6380), 1094–1096 (2018)
15. Luyckx, K., Daelemans, W.: The effect of author set size and data size in authorship attribution. Digit. Scholarsh. Humanit. **26**(1), 35–55 (2010). https://doi.org/10.1093/llc/fqq013
16. Popat, K., Mukherjee, S., Yates, A., Weikum, G.: DeClarE: debunking fake news and false claims using evidence-aware deep learning. arXiv preprint arXiv:1809.06416 (2018)
17. Potthast, M., et al.: Who wrote the web? Revisiting influential author identification research applicable to information retrieval. In: Ferro, N., et al. (eds.) ECIR 2016. LNCS, vol. 9626, pp. 393–407. Springer, Cham (2016). https://doi.org/10.1007/978-3-319-30671-1_29
18. Rangel, F., Rosso, P.: On the implications of the general data protection regulation on the organisation of evaluation tasks. Language and Law=Linguagem e Direito **5**(2), 95–117 (2019)
19. Rangel, F., Rosso, P.: Overview of the 7th author profiling task at PAN 2019: bots and gender profiling. In: Cappellato, L., Ferro, N., Müller, H., Losada, D. (eds.) CLEF 2019 Labs and Workshops, Notebook Papers. CEUR Workshop Proceedings. CEUR-WS.org (2019)
20. Rangel, F., et al.: Overview of the 2nd author profiling task at PAN 2014. In: Cappellato, L., Ferro, N., Halvey, M., Kraaij, W. (eds.) CLEF 2014 Labs and Workshops, Notebook Papers, vol. 1180. CEUR-WS.org (2014)
21. Rangel, F., Rosso, P., Montes-y-Gómez, M., Potthast, M., Stein, B.: Overview of the 6th author profiling task at PAN 2018: multimodal gender identification in Twitter. In: Cappellato, L., Ferro, N., Nie, J.Y., Soulier, L. (eds.) Working Notes Papers of the CLEF 2018 Evaluation Labs. CEUR Workshop Proceedings. CLEF and CEUR-WS.org, September 2018
22. Rangel, F., Rosso, P., Moshe Koppel, M., Stamatatos, E., Inches, G.: Overview of the author profiling task at PAN 2013. In: Forner, P., Navigli, R., Tufis, D. (eds.) CLEF 2013 Labs and Workshops, Notebook Papers, vol. 1179. CEUR-WS.org (2013)
23. Rangel, F., Rosso, P., Potthast, M., Stein, B.: Overview of the 5th author profiling task at PAN 2017: gender and language variety identification in Twitter. In: Working Notes Papers of the CLEF (2017)
24. Rangel, F., Rosso, P., Potthast, M., Stein, B., Daelemans, W.: Overview of the 3rd author profiling task at PAN 2015. In: Cappellato, L., Ferro, N., Jones, G., San Juan, E. (eds.) CLEF 2015 Labs and Workshops, Notebook Papers. CEUR Workshop Proceedings, vol. 1391. CEUR-WS.org (2015)
25. Rangel, F., Rosso, P., Verhoeven, B., Daelemans, W., Potthast, M., Stein, B.: Overview of the 4th author profiling task at PAN 2016: cross-genre evaluations. In: Working Notes Papers of the CLEF 2016 Evaluation Labs. CEUR Workshop Proceedings. CLEF and CEUR-WS.org, September 2016
26. Rosso, P., Rangel, F., Potthast, M., Stamatatos, E., Tschuggnall, M., Stein, B.: Overview of PAN'16: new challenges for authorship analysis: cross-genre profiling, clustering, diarization, and obfuscation. In: Fuhr, N., et al. (eds.) CLEF 2016. LNCS, vol. 9822, pp. 332–350. Springer, Cham (2016). https://doi.org/10.1007/978-3-319-44564-9_28
27. Stamatatos, E.: A survey of modern authorship attribution methods. JASIST **60**(3), 538–556 (2009). https://doi.org/10.1002/asi.21001
28. Tschuggnall, M., et al.: Overview of the author identification task at PAN-2017: style breach detection and author clustering. In: Cappellato, L., et al. (eds.) Working Notes Papers of the CLEF 2017 Evaluation Labs, pp. 1–22 (2017)
29. Tushnet, R.: Legal fictions: copyright, fan fiction, and a new common law. Loyola Los Angeles Entertain. Law Rev. **17**(3), 651 (1997)
30. Wiegmann, M., Stein, B., Potthast, M.: Celebrity profiling. In: 57th Annual Meeting of the Association for Computational Linguistics (ACL 2019). Association for Computational Linguistics, July 2019

31. Wiegmann, M., Stein, B., Potthast, M.: Overview of the celebrity profiling task at PAN 2019. In: Cappellato, L., Ferro, N., Losada, D., Müller, H. (eds.) CLEF 2019 Labs and Workshops, Notebook Papers. CEUR-WS.org, September 2019
32. Zangerle, E., Tschuggnall, M., Specht, G., Stein, B., Potthast, M.: Overview of the style change detection task at PAN 2019. In: Working Notes of CLEF 2019 - Conference and Labs of the Evaluation Forum, Lugano, Switzerland, 9–12 September 2019. CEUR Workshop Proceedings, vol. 2380. CEUR-WS.org (2019). http://ceur-ws.org/Vol-2380

Touché: First Shared Task on Argument Retrieval

Alexander Bondarenko[1]([✉]), Matthias Hagen[1], Martin Potthast[2],
Henning Wachsmuth[3], Meriem Beloucif[4], Chris Biemann[4],
Alexander Panchenko[5], and Benno Stein[6]

[1] Martin-Luther-Universität Halle-Wittenberg, Halle, Germany
touche@webis.de
[2] Leipzig University, Leipzig, Germany
[3] Paderborn University, Paderborn, Germany
[4] Universität Hamburg, Hamburg, Germany
[5] Skolkovo Institute of Science and Technology, Moscow, Russia
[6] Bauhaus-Universität Weimar, Weimar, Germany

Abstract. Technologies for argument mining and argumentation processing are maturing continuously, giving rise to the idea of retrieving arguments in search scenarios. We introduce *Touché*, the first lab on *Argument Retrieval* featuring two subtasks: (1) the retrieval of arguments from a focused debate collection to support argumentative conversations, and (2) the retrieval of arguments from a generic web crawl to answer comparative questions with argumentative results. The goal of this lab is to perform an evaluation of various strategies to retrieve argumentative information from the web content. In this paper, we describe the setting of each subtask: the motivation, the data, and the evaluation methodology.

1 Introduction

Decision making processes, be it at the societal or at the personal level, eventually come to a point where one side will challenge the other with a *why*-question, i.e., a prompt to justify one's stance. In its most basic form, the answer to a why-question is a plain fact, but, more commonly, it requires a formulation of an argument, which is a justified claim.

The web is rife with documents comprising arguments, from news articles, blog posts, and discussion threads to advertisements and reviews of products and services. While the leading web search engines support the retrieval of plain facts often fairly well, hardly any support is currently provided for the retrieval of argumentative text, let alone the retrieval and ranking of individual arguments. This is particularly unfortunate in today's political and corresponding societal discourse, where well-reasoned argumentation on all kinds of controversial topics is of utmost importance. Yet, especially search results on such topics are often riddled with populism, conspiracy theories, and one-sidedness. This is not to say that these are not valid argumentative techniques, but rather that they arguably

© Springer Nature Switzerland AG 2020
J. M. Jose et al. (Eds.): ECIR 2020, LNCS 12036, pp. 517–523, 2020.
https://doi.org/10.1007/978-3-030-45442-5_67

do not lead to the kind of information and insights one wants to support. Also at the personal level, users are very interested in (and often search for) the most relevant arguments that speak for or against a possible decision.

We propose a CLEF lab that aims for a better support of argument retrieval, making it possible to find "strong" arguments for decisions at the societal level (e.g., "Is climate change real and what to do?") and at the personal level (e.g., "Should I study abroad if I've never left my country before?"). In particular, the lab includes two subtasks, both of which are explained in detail below:

1. Argument retrieval from a focused debate collection to support argumentative conversations by providing justifications for the claims.
2. Argument retrieval from a generic web crawl to answer comparative questions with argumentative results and to support decision making.

With the proposed lab and its subtasks, our goal is to establish an understanding of how to evaluate argument retrieval processes as well as what kind of approaches effectively retrieve arguments that are beneficial for a variety of information needs and that are well-conceived in selected scenarios. An important component of the retrieval pipeline will be developing computational methods for the argument quality assessment (whether a given argument is a "strong" one), which is discussed in Sect. 4. This will not only allow for a better support of the argumentative information needs by search engines, but also, in the long run, may become an enabling technology for automatic open-domain agents that convincingly discuss and interact with humans.

2   Task Definition

The proposed lab will consist of the two subtasks to cover various information needs found in two different kinds of argumentative scenarios. The lab follows the classical TREC-style[1] evaluation methodology, where a dataset and a set of topics are provided to the participants. Each topic contains a search query and a detailed search scenario description. The task is to retrieve relevant documents satisfying conditions provided in the topic. Participants then submit their ranked retrieval results for each topic to be judged. Participating teams will be able to submit up to three runs with different approaches to be evaluated by expert assessors. We will make all the runs from this year's task available to the community—to have a basis for training models in light of a potential second edition of the lab as well as to enable reproducibility and independent research.

Task 1: Conversational Argument Retrieval
The first subtask is motivated by the support of users who search for arguments directly, e.g., by supporting their stance, or by aiding them in building a stance on topics of a general societal interest. Examples of such (usually controversial) topics are the abandonment of plastic bottles, animal experiments, immigration, and abortion. Multiple online portals are centered around exactly such topics,

---

[1] https://trec.nist.gov/tracks.html.

such as Yahoo! Answers or Quora. Surprisingly, however, search engines nowadays do not provide any effective way to retrieve reliable arguments from these platforms, even though the search engines provide snippets for direct answers to factoid questions, among others. Presumably, the main reason is that search engines often do not grasp the argumentative nature of the underlying information need.

This subtask targets argumentative conversations. We will provide a focused crawl with content from online debate portals (idebate.org, debatepedia.org, debatewise.org) and from Reddit's ChangeMyView.[2] As a baseline retrieval model, we resort to the search engine args.me [10]. The lab participants will have to retrieve "strong" arguments (refer to Sect. 4 for a more detailed description of "strong") from the provided dataset for the 50 given topics, covering a wide range of controversial issues collected from the debate portals and ChangeMyView.

Task 2: Comparative Argument Retrieval

The second task is motivated by the support of users in personal decisions from everyday life where choices need to be made. In particular, the goal of the task is to find relevant arguments when comparing several objects with different options (e.g., "Is X better than Y for Z?"). Evidently, comparative information needs seem to be important: Question answering platforms such as Quora are filled with topics such as "How Python compares to PHP for web development?[3]" or "Is Germany better to live in compared to US?[4]". Still, in their current form, search engines such as Google and DuckDuckGo[5] do not provide much support for such comparative queries besides the "ten blue links". Retrieval of comparative information is thus eminent in web search and, according to [1], appears in about 10% of all search sessions. The retrieval topics for this task are based on real-world comparative questions that were submitted to commercial search engines and posted to question answering platforms.

The participants of this task will retrieve documents from a general web crawl (namely, ClueWeb12[6]) that help users to answer their comparative question. The task will be to identify documents that, ideally, comprise convincing argumentation for or against one or the other option underlying the comparative question. Two BM25F-based Elasticsearch systems, ChatNoir [2] and TARGER [5], will be available as baseline retrieval systems to participants that face problems or just want to avoid indexing the whole dataset on their side. Furthermore, TARGER's API can also be used to identify argumentative units in free text input. Additionally, we provide a baseline argument retrieval approach proposed by [3], which

---

[2] https://www.reddit.com/r/changemyview.
[3] https://www.quora.com/How-does-Python-compare-to-PHP-for-server-side-web-development.
[4] https://www.quora.com/Is-Germany-better-to-live-in-compared-to-the-US.
[5] https://duckduckgo.com.
[6] https://lemurproject.org/clueweb12/.

integrates the TARGER's API to capture "argumentativeness" in text documents. The basic idea applied in the approach is to axiomatically re-rank the top-50 results of BM25F for those topics that seem to be argumentative.

3   Data Description

For both subtasks, the topics have already been defined, ensuring that respective information is available in the focused crawl of debate portals and Reddit, and in the ClueWeb12, respectively. In total, we prepared 100 topics, 50 for each subtask. Each topic consists of a *title* representing either a search query or a choice question, a *description* providing a detailed definition of the search task, and a *narrative* accurately defining relevant documents.

Example topic for Task 1:

```
<title>
    climate change real
</title>
<description>
    You read an opinion piece on how climate change is a hoax
    and disagree. Now you are looking for arguments supporting
    the claim that climate change is in fact real.
</description>
<narrative>
    Relevant arguments will support the given stance that
    climate change is real or attack a hoax side's argument.
</narrative>
```

Example topic for Task 2:

```
<title>
    What are advantages and disadvantages of PHP over Python
    and vice versa?
</title>
<description>
    The user is looking for differences and similarities of PHP
    and Python and wants to know about scenarios that favor one
    over the other.
</description>
<narrative>
    Relevant documents may contain an overview of more than
    these two programming languages but must include both of
    them with an explicit comparison of these two.
</narrative>
```

The topics for the first subtask were chosen from the online debate portals having the largest number of the user-generated comments, and thus represent the matters of the highest societal interest. As for the second subtask, two types of web sources have been used to create the topics: Questions posted to Yahoo! Answers, as well as question queries submitted to either of two search engines, Yandex or Google. We thoroughly selected question queries which correspond to a choice problem, and where the answers should contain a sufficient number of pro and con arguments. We also ensured that relevant documents can be found in the provided search collection.

4 Evaluation

For the first subtask, the evaluation is based on the pooled top-k results of the participants' runs. For these, human assessors will label argumentative text passages or documents manually, both for their general topical relevance, and for argument quality dimensions, which have been found to be important for the evaluation of arguments [9]: Whether an argumentative text is logically cogent, whether it is rhetorically well-written, and whether it contributes to the users' stance-building process (i.e., somewhat similar to the concept of "utility").

For the second subtask, the human assessors will judge, in addition to document relevance, whether a sufficient argumentative support is provided as defined by [4] and will evaluate the trustworthiness and credibility of the web documents as in [8]. Thus, a "strong" argument is defined to be the one fulfilling certain argument quality criteria such as logical cogency, rhetorical well-writtenness, contribution to a stance-building, level of support, and credibility.

For both subtasks, the performance and ordering of the submitted runs will be measured in traditional ranking-based ways with respect to relevance (e.g., graded relevance judgments nDCG [6], but ignoring repeated and near-duplicate entries). Moreover, we will include in the judgment the qualitative aspects of arguments. We plan to do assessment of the submitted runs via crowdsourcing. The study conducted by [7] shows that argument assessment is feasible via crowdsourcing; however, especially the argument quality-oriented aspects will be further developed over the course of prospective future editions of the lab: Our goal is to establish a widely accepted way of evaluating argument retrieval and to support targeted improvements of the retrieval technology developed by the lab participants.

5 Participants and Lab Organization

In essence, the first task is a general retrieval task with a focused collection. Participants of respective ad-hoc setups from previous years at TREC, CLEF, and NTCIR can participate with ease, either with their favorite system, or enriched with a pipeline taking argumentativeness into account. The second task will be centered around information needs expressed as questions, which is attractive to participants from previous question-oriented tracks/labs. Both subtasks ultimately aim to support the building of (retrieval) systems that can discuss and argue with human users. Hence, also the rapidly growing community around conversational search, with successful workshops at the recent SIGIR, WWW, IJCAI, and EMNLP (CAIR and SCAI), might be interested.

Argument mining has grown rapidly in the NLP community over the last couple of years, with flourishing publications and workshops at the premier conferences (e.g., the ArgMining workshop at ACL and EMNLP as well as the COMMA series of focused conferences). Using the baseline retrieval systems that we will provide for the lab, groups from the NLP community not familiar with setting up a search system can contribute runs (e.g., by re-ranking the baseline's initial result set regarding their ideas of what forms a good argument).

The proposed CLEF 2020 lab requires human assessors to annotate different aspects of the retrieved argumentative passages or documents. For both subtasks, the pooling depths of the participants' runs can be adjusted to fit the available assessment resources.

With regard to organization, we plan to release data and the baseline solutions early, which will allow for a quick start for participants. The received solutions and notebooks will be reviewed for quality by the organizing committee. Notebooks that do not meet rigorous quality standards (e.g., quality of method presentation, novelty factor, soundness of approach) may be limited in size, or entirely rejected. All participants will be encouraged to present a poster at the lab session. A selection of the contributions will be invited to present their systems orally.

Overall, the lab session will be divided into two major parts. The first part will be comprised of an introduction to all the participants' solutions, an invited keynote by a speaker from the area, and the best solution talks. The second part of the lab will include a second invited keynote, a poster session, and a plenary discussion to wrap-up the lab results and to discuss possible objectives for improvements and future directions. The expected length of the lab session at the conference is one day (≈ 8 h) with an estimated amount of 20–25 participants, including 5 talks and 10 poster presentations plus the aforementioned keynotes and overview presentations.

6   Conclusion

We propose Touché, the first lab on Argument Retrieval at CLEF 2020 to support a growing interest in argumentation technologies. In particular, by providing data, baseline retrieval systems, and the evaluation of the participants' approaches, we encourage researchers to contribute ideas to be shared with the community. We suggest to integrate a notion of a "strong" argumentation, or argument quality, into a retrieval pipeline by exploiting definite metrics such as level of support, credibility, cogency, style, and contribution to a stance. Our lab is aimed at inducing a discussion and developing computational approaches to the underlying aspects of the argument retrieval from an application-driven perspective.

Acknowledgments. This work has been partially supported by the DFG through the project "ACQuA: Answering Comparative Questions with Arguments" (grants

BI 1544/7-1 and HA 5851/2-1) as part of the priority program "RATIO: Robust Argumentation Machines" (SPP 1999).

References

1. Bailey, P., et al.: User task understanding: a web search engine perspective. In: NII Shonan Meeting on Whole-Session Evaluation of Interactive Information Retrieval Systems (2012)
2. Bevendorff, J., Stein, B., Hagen, M., Potthast, M.: Elastic ChatNoir: search engine for the ClueWeb and the common crawl. In: Pasi, G., Piwowarski, B., Azzopardi, L., Hanbury, A. (eds.) ECIR 2018. LNCS, vol. 10772, pp. 820–824. Springer, Cham (2018). https://doi.org/10.1007/978-3-319-76941-7_83
3. Bondarenko, A., Völske, M., Panchenko, A., Biemann, C., Stein, B., Hagen, M.: Webis at TREC 2018: common core track. In: Voorhees, E., Ellis, A. (eds.) Proceedings of the 27th International Text Retrieval Conference (TREC) (2018)
4. Braunstain, L., Kurland, O., Carmel, D., Szpektor, I., Shtok, A.: Supporting human answers for advice-seeking questions in CQA sites. In: Ferro, N., et al. (eds.) ECIR 2016. LNCS, vol. 9626, pp. 129–141. Springer, Cham (2016). https://doi.org/10.1007/978-3-319-30671-1_10
5. Chernodub, A., et al.: TARGER: neural argument mining at your fingertips. In: Proceedings of the 57th Annual Meeting of the Association for Computational Linguistics (ACL), pp. 195–200 (2019)
6. Kanoulas, E., Aslam, J.A.: Empirical justification of the gain and discount function for nDCG. In: Proceedings of the 18th ACM Conference on Information and Knowledge Management (CIKM), pp. 611–620 (2009)
7. Potthast, M., et al.: Argument search: assessing argument relevance. In: Proceedings of the 42nd International ACM Conference on Research and Development in Information Retrieval (SIGIR), pp. 1117–1120 (2019)
8. Rafalak, M., Abramczuk, K., Wierzbicki, A.: Incredible: is (almost) all web content trustworthy? Analysis of psychological factors related to website credibility evaluation. In: Proceedings of the 23rd International World Wide Web Conference (WWW), Companion Volume, pp. 1117–1122 (2014)
9. Wachsmuth, H., et al.: Computational argumentation quality assessment in natural language. In: Proceedings of the 15th Conference of the European Chapter of the Association for Computational Linguistics (EACL), pp. 176–187 (2017)
10. Wachsmuth, H., et al.: Building an argument search engine for the web. In: Proceedings of the Fourth Workshop on Argument Mining (ArgMining), pp. 49–59 (2017)

Introducing the CLEF 2020 HIPE Shared Task: Named Entity Recognition and Linking on Historical Newspapers

Maud Ehrmann[1(✉)] ⓘ, Matteo Romanello[1] ⓘ, Stefan Bircher[2], and Simon Clematide[2] ⓘ

[1] Digital Humanities Laboratory, EPFL, Lausanne, Switzerland
maud.ehrmann@epfl.ch
[2] Institute of Computational Linguistics, University of Zurich, Zurich, Switzerland

Abstract. Since its introduction some twenty years ago, named entity (NE) processing has become an essential component of virtually any text mining application and has undergone major changes. Recently, two main trends characterise its developments: the adoption of deep learning architectures and the consideration of textual material originating from historical and cultural heritage collections. While the former opens up new opportunities, the latter introduces new challenges with heterogeneous, historical and noisy inputs. If NE processing tools are increasingly being used in the context of historical documents, performance values are below the ones on contemporary data and are hardly comparable. In this context, this paper introduces the CLEF 2020 Evaluation Lab HIPE (Identifying Historical People, Places and other Entities) on named entity recognition and linking on diachronic historical newspaper material in French, German and English. Our objective is threefold: strengthening the robustness of existing approaches on non-standard inputs, enabling performance comparison of NE processing on historical texts, and, in the long run, fostering efficient semantic indexing of historical documents in order to support scholarship on digital cultural heritage collections.

Keywords: Named entity processing · Text understanding · Information extraction · Historical newspapers · Digital Humanities

1 Introduction

Recognition and identification of real-world entities is at the core of virtually any text mining application. As a matter of fact, referential units such as names of persons, locations and organizations underlie the semantics of texts and guide their interpretation. Around since the seminal Message Understanding Conference (MUC) evaluation cycle in the 1990s [11], named entity-related tasks have undergone major evolutions until now, from entity recognition and classification to entity disambiguation and linking [21,25]. Besides the general domain of well-written newswire data, named entity (NE) processing is also applied to

© Springer Nature Switzerland AG 2020
J. M. Jose et al. (Eds.): ECIR 2020, LNCS 12036, pp. 524–532, 2020.
https://doi.org/10.1007/978-3-030-45442-5_68

specific domains, particularly bio-medical [10,14], and on more noisy inputs such as speech transcriptions [9] and tweets [26].

Recently, two main trends characterise developments in NE processing. First, at the technical level, the adoption of deep learning architectures and the usage of embedded language representations greatly reshapes the field and opens up new research directions [1,16,17]. Second, with respect to application domain and language spectrum, NE processing has been called upon to contribute to the field of Digital Humanities (DH), where massive digitization of historical documents is producing huge amounts of texts [30]. Thanks to large-scale digitization projects driven by cultural institutions, millions of images are being acquired and, when it comes to text, their content is transcribed, either manually via dedicated interfaces, or automatically via Optical Character Recognition (OCR). Beyond this great achievement in terms of document preservation and accessibility, the next crucial step is to adapt and develop appropriate language technologies to search and retrieve the contents of this 'Big Data from the Past' [13]. In this regard, information extraction techniques, and particularly NE recognition and linking, can certainly be regarded as among the first steps.

This paper introduces the CLEF 2020 Evaluation Lab[1] HIPE (Identifying Historical People, Places and other Entities)[2]. With the aim of supporting the development and progress of NE systems on historical documents (Sect. 2), this lab proposes two tasks, namely named entity recognition and linking, on historical newspapers in French, German and English (Sect. 3). We additionally report first results on French historical newspapers (Sect. 4), which comfort the idea of various benefits of such lab for both NLP and DH communities.

2 Motivation and Objectives

NE processing tools are increasingly being used in the context of historical documents. Research activities in this domain target texts of different nature (e.g. museum records, state-related documents, genealogical data, historical newspapers) and different tasks (NE recognition and classification, entity linking, or both). Experiments involve different time periods , focus on different domains, and use different typologies. This great diversity demonstrates how many and varied the needs—and the challenges—are, but also makes performance comparison difficult, if not impossible.

Furthermore, as per language technologies in general [29], it appears that the application of NE processing on historical texts poses new challenges [7,23]. First, inputs can be extremely noisy, with errors which do not resemble tweet misspellings or speech transcription hesitations, for which adapted approaches have already been devised [5,27]. Second, the language under study is mostly of earlier stage(s), which renders usual external and internal evidences less effective (e.g., the usage of different naming conventions and presence of historical spelling variations) [2,3]. Further, beside historical VIPs, texts from the past contain

---

[1] https://clef2020.clef-initiative.eu/.
[2] https://impresso.github.io/CLEF-HIPE-2020.

rare entities which have undergone significant changes (esp. locations) or do no longer exist, and for which adequate linguistic resources and knowledge bases are missing [12]. Finally, archives and texts from the past are not as anglophone as in today's information society, making multilingual resources and processing capacities even more essential [22].

Overall, and as demonstrated by Vilain et al. [31], the transfer of NE tools from one domain to another is not straightforward, and the performance of NE tools initially developed for homogeneous texts of the immediate past are affected when applied on historical material. This echoes the proposition of Plank [24], according to whom what is considered as standard data (i.e. contemporary news genre) is more a historical coincidence than a reality: in NLP non-canonical, heterogeneous, biased and noisy data is rather the norm than the exception.

Even though many evaluation campaigns on NE were organized over the last decades[3], only one considered French historical texts [8]. To the best of our knowledge, no NE evaluation campaign ever addressed multilingual, diachronic historical material. In the context of new needs and materials emerging from the humanities, we believe that an evaluation campaign on historical documents is timely and will be beneficial. In addition to the release of a multilingual, historical NE-annotated corpus, the objective of this shared task is threefold: strengthening the robustness of existing approaches on non-standard inputs; enabling performance comparison of NE processing on historical texts; and fostering efficient semantic indexing of historical documents.

3   Overview of the Evaluation Lab

3.1   Task Description

The HIPE shared task puts forward 2 NE processing tasks with sub-tasks of increasing level of difficulty. Participants can submit up to 3 runs per sub-task.

Task 1: Named Entity Recognition and Classification (NERC)

Subtask 1.1 - NERC Coarse-Grained: this task includes the recognition and classification of entity mentions according to high-level entity types (Person, Location, Organisation, Product and Date).

Subtask 1.2 - NERC Fine-Grained: this task includes the classification of mentions according to finer-grained entity types, nested entities (up to one level of depth) and the detection of entity mention components (e.g. function, title, name).

Task 2: Named Entity Linking (EL). This task requires the linking of named entity mentions to a unique referent in a knowledge base (a frozen dump of Wikidata) or to a NIL node if the mention does not have a referent.

---

[3] MUC, ACE, CONNL, KBP, ESTER, HAREM, QUAERO, GERMEVAL, etc.

3.2   Data Sets

Corpus. The HIPE corpus is composed of items from the digitized archives of several Swiss, Luxembourgish and American newspapers on a diachronic basis.[4] For each language, articles of 4 different newspapers were sampled on a decade time-bucket basis, according to the time span of the newspaper (longest duration spans ca. 200 years). More precisely, articles were first randomly sampled from each year of the considered decades, with the constraints of having a title and more than 100 characters. Subsequently to this sampling, a manual triage was applied in order to keep journalistic content only and to remove undesirable items such as feuilleton, cross-words, weather tables, time-schedules, obituaries, and what a human could not even read because of OCR noise.

Alongside each article, metadata (journal, date, title, page number, image region coordinates), the corresponding scan(s) and an OCR quality assessment score is provided. Different OCR versions of same texts are not provided, and the OCR quality of the corpus therefore corresponds to real-life setting, with variations according to digitization time and preservation state of original documents.

For each task and language—with the exception of English—the corpus is divided into training, dev and test data sets, released in IOB format with hierarchical information. For English, only dev and test sets will be released.

Annotation. HIPE *annotation guidelines* [6] are derived from the Quaero annotation guide[5]. Originally designed for the annotation of "extended" named entities (i.e. more than the 3 or 4 traditional entity classes) in French speech transcriptions, Quaero guidelines have furthermore been used on historic press corpora [28]. HIPE slightly recasts and simplifies them, considering only a subset of entity types and components, as well as of linguistic units eligible as named entities. HIPE guidelines were iteratively consolidated via the annotation of a "mini-reference" corpus, where annotation decisions were tested and difficult cases discussed. Despite these adaptations, HIPE annotated corpora will mostly remain compatible with Quaero guidelines.

The *annotation campaign* is carried out by the task organizers with the support of trilingual collaborators. We use INCEpTION as an annotation tool [15], with the visualisation of image segments alongside OCR transcriptions.[6] Before starting annotating, each annotator is first trained on a mini-reference corpus, where the inter-annotator agreement (IAA) with the gold reference is computed. For each language, a sub-sample of the corpus is annotated by 2 annotators and IAA is computed, before and after an adjudication. Randomly selected articles will also be controlled by the adjudicator. Finally, HIPE will provide *complementary resources* in the form of in-domain word-level and character-level embeddings

---

[4] From the Swiss National Library, the Luxembourgish National Library, and the Library of Congress, respectively.

[5] See the original Quaero guidelines: http://www.quaero.org/media/files/bibliographie/quaero-guide-annotation-2011.pdf.

[6] HIPE is one of the official INCEpTION project use cases, see https://inception-project.github.io/use-case-gallery/impresso/.

acquired from historical newspaper corpora. In the same vein, participants will be encouraged to share any external resource they might use. HIPE corpus and resources will be released under a CC-BY-SA-NC 4.0 license.

Table 1. Results and examples from exploratory experiments on NER for French.

Type	Freq.	CRF			Neural		
		P	R	F_1	P	R	F_1
pers	3,638	75.6	71.0	73.2	84.6	84.7	84.6
func	3,133	69.7	43.7	53.7	76.5	66.8	71.3
loc	2,369	67.8	63.6	65.6	73.1	77.2	75.1
amount	1,716	63.4	53.8	58.2	75.0	71.0	72.9
time	1,675	70.4	59.3	64.4	73.4	71.3	72.3
org	1,412	63.9	44.4	52.4	67.6	52.4	59.1
prod	612	54.1	30.2	38.8	60.7	51.6	55.8
all	14,555	69.4	56.2	62.1	76.2	72.0	74.0

Token		IOB		
OCR	correct	Gold	CRF	Neural
Tńi*louse	Toulouse	B-loc	O	B-loc
Caa.Qrs	Cahors	B-loc	O	B-loc
o˚an	Jean	B-pers	O	B-pers
Chêne	Chêne	I-pers	B-pers	I-pers
Ïcmp0	Temps	B-prod	O	B-prod
f&tvxps	Temps	B-prod	O	B-prod
f&tïtipB	Temps	B-prod	O	B-prod

3.3   Evaluation

Named Entity Recognition and Classification (Task 1) will be evaluated in terms of macro and micro Precision, Recall, F-measure, and Slot Error Rate [20]. Two evaluation scenarios will be considered: strict (exact boundary matching) and relaxed (fuzzy boundary matching). Entity linking (Task 2) will be evaluated in terms of Precision, Recall, F-measure taking into account literal and metonymic senses.

4   Exploratory Experiments on NER for Historical French

We made an exploratory study in order to assess whether the massive improvements in neural NER [1,17] on modern texts carry over to historical material with OCR noise. The data of our experiments is the *Quaero Old Press* (QOP) corpus, 295 OCRed[7] newspaper documents dating from December 1890 annotated according to the *Quaero* guidelines [28], split by us into train (1.45 m tokens) and dev/test (each 0.2 m). We only consider the outermost entity level (no nested entities or components) and train on the fine-grained subcategories (e.g., *loc.adm.town*) of the 7 main classes.

Modeling NER as a sequence labeling problem and applying Bi-LSTM networks is state of the art [1,4,17,19]. Our experiments follow [1] in using character-based contextual string embeddings as input word representations, allowing to "better handle rare and misspelled words as well as model subword structures such as prefixes and endings". These contextualized word embeddings rely on neural forward and backward character-level language models that have been

---

[7] [28] reports a character error rate of 5.09% and a word error rate of 36.59%.

trained by us on a large collection (500 m tokens) of late 19th and early 20th centuries Swiss-French newspapers. In accordance to the literature, a Bi-LSTM NER model with an on-top CRF layer (Bi-LSTM-CRF) works best for our data.

As a baseline system, which will also be provided for the shared task, we train a traditional CRF sequence classifier [18] using basic spelling features such as a token's character prefix and suffix, the casing of the initial character and the presence of punctuation marks and digits. The baseline classifier shows fairly modest overall performance scores of 69.4% recall, 56.2% precision and 62.1 F_1 (see Table 1).

Trained and evaluated on the QOP data, the neural model relying on contextual string embeddings clearly outperforms the baseline classifier. As shown in Table 1, the Bi-LSTM-CRF model achieves better F_1 for all of the 7 entity types and surpasses the feature-based classifier by nearly 12 points F_1. Examples in Table 1 evidence that the CRF model frequently struggles with entities containing miss-recognized special characters and/or punctuation marks. In many such cases, the Bi-LSTM-CRF classifier is capable of assigning the correct label. These results indicate that the new neural methods are ready to enable substantial progress in NER on noisy historical texts.

5   Conclusion

From the perspective of natural language processing (NLP), the HIPE evaluation lab provides the opportunity to test the robustness of existing NERC and EL approaches against challenging historical material and to gain new insights with respect to domain and language adaptation. From the perspective of digital humanities, the lab's outcomes help DH practitioners in mapping state-of-the-art solutions for NE processing on historical texts, and in getting a better understanding of what is already possible as opposed to what is still challenging. Most importantly, digital scholars are in need of support to explore the large quantities of digitized text they currently have at hand, and NE processing is high on the agenda. Such processing can support research questions in various domains (e.g. history, political science, literature, historical linguistics) and knowing about their performance is crucial in order to make an informed use of the processed data. Overall, HIPE will contribute to advance the state of the art in semantic indexing of historical material, within the specific domain of historical newspaper processing, as in e.g. the "*impresso* - Media Monitoring of the Past" project[8] and, more generally, within the domain of text understanding of historical material, as in the Time Machine Europe project[9] which ambitions the application of AI technologies on cultural heritage data.

Acknowledgements. This CLEF evaluation lab is part of the research activities of the project "*impresso* – Media Monitoring of the Past", for which authors gratefully acknowledge the financial support of the Swiss National Science Foundation under

---

[8] https://impresso-project.ch.
[9] https://www.timemachine.eu.

grant number CR-SII5_173719. We would also like to thank C. Watter, G. Schneider and A. Flückiger for their invaluable help with the construction of the data sets, as well as R. Eckart de Castillo, C. Neudecker, S. Rosset and D. Smith for their support and guidance as part of the lab's advisory board.

References

1. Akbik, A., Blythe, D., Vollgraf, R.: Contextual string embeddings for sequence labeling. In: Proceedings of the 27th International Conference on Computational Linguistics (COLING 2018), Santa Fe, New Mexico, USA, pp. 1638–1649. Association for Computational Linguistics (2018)
2. Bollmann, M.: A large-scale comparison of historical text normalization systems. In: Proceedings of the 2019 Conference of the North American Chapter of the Association for Computational Linguistics: Human Language Technologies, Volume 1 (Long and Short Papers), Minneapolis, Minnesota, pp. 3885–3898. Association for Computational Linguistics, June 2019
3. Borin, L., Kokkinakis, D., Olsson, L.: Naming the past: named entity and animacy recognition in 19th century Swedish literature. In: Proceedings of the Workshop on Language Technology for Cultural Heritage Data (LaTeCH 2007), pp. 1–8 (2007)
4. Chiu, J.P.C., Nichols, E.: Named entity recognition with bidirectional LSTM-CNNs. Trans. Assoc. Comput. Linguist. (TACL) **4**, 357–370 (2016)
5. Dinarelli, M., Rosset, S.: Tree-structured named entity recognition on OCR data: analysis, processing and results. In: 2012, editor, Proceedings of the Eighth International Conference on Language Resources and Evaluation, Istanbul, Turkey, May 2012. European Language Resources Association (ELRA) (2012). ISBN 978-2-9517408-7-7
6. Ehrmann, M., Watter, C., Romanello, M., Clematide, S., Flückiger: Impresso Named Entity Annotation Guidelines, January 2020. https://doi.org/10.5281/zenodo.3604227
7. Ehrmann, M., Colavizza, G., Rochat, Y., Kaplan, F.: Diachronic evaluation of NER systems on old newspapers. In: Proceedings of the 13th Conference on Natural Language Processing (KONVENS 2016), pp. 97–107. Bochumer Linguistische Arbeitsberichte (2016). https://infoscience.epfl.ch/record/221391?ln=en
8. Galibert, O., Rosset, S., Grouin, C., Zweigenbaum, P., Quintard, L.: Extended named entity annotation on OCRed documents : from corpus constitution to evaluation campaign. In: Proceedings of the Eighth Conference on International Language Resources and Evaluation, Istanbul, Turkey, pp. 3126–3131 (2012)
9. Galibert, O., Leixa, J., Adda, G., Choukri, K., Gravier, G.: The ETAPE speech processing evaluation. In: LREC, pp. 3995–3999. Citeseer (2014)
10. Goulart, R.R.V., de Lima, V.S., Xavier, C.C.: A systematic review of named entity recognition in biomedical texts. J. Braz. Comput. Soc. **17**(2), 103–116 (2011). https://doi.org/10.1007/s13173-011-0031-9. ISSN 1678–4804
11. Grishman, R., Sundheim, B.: Design of the MUC-6 evaluation. In: Sixth Message Understanding Conference (MUC-6): Proceedings of a Conference Held in Columbia, Maryland (1995)

12. Van Hooland, S., De Wilde, M., Verborgh, R., Steiner, T., Van de Walle, R.: Exploring entity recognition and disambiguation for cultural heritage collections. Digit. Scholarsh. Humanit. **30**(2), 262–279 (2015). https://doi.org/10.1093/llc/fqt067. ISSN 2055-7671
13. Kaplan, F., di Lenardo, I.: Big data of the past. Front. Digit. Humanit. **4** (2017). https://doi.org/10.3389/fdigh.2017.00012. https://www.frontiersin.org/articles/10.3389/fdigh.2017.00012/full. ISSN 2297-2668
14. Kim, J.-D., Ohta, T., Tateisi, Y., Tsujii, J.: Genia corpus-a semantically annotated corpus for bio-textmining. Bioinformatics **19**(suppl 1), i180–i182 (2003)
15. Klie, J.-C., Bugert, M., Boullosa, B., de Castilho, R.E., Gurevych, I.: The inception platform: machine-assisted and knowledge-oriented interactive annotation. In: Proceedings of the 27th International Conference on Computational Linguistics: System Demonstrations, pp. 5–9 (2018)
16. Labusch, K., Neudecker, C., Zellhöfer, D.: BERT for named entity recognition in contemporary and historic German. In: Preliminary proceedings of the 15th Conference on Natural Language Processing (KONVENS 2019): Long Papers, Erlangen, Germany, pp. 1–9. German Society for Computational Linguistics & Language Technology (2019)
17. Lample, G., Ballesteros, M., Subramanian, S., Kawakami, K., Dyer, C.: Neural architectures for named entity recognition. In: Proceedings of the 2016 Conference of the North American Chapter of the Association for Computational Linguistics: Human Language Technologies, pp. 260–270. Association for Computational Linguistics, San Diego, June 2016
18. Lavergne, T., Cappé, O., Yvon, F.: Practical very large scale CRFs. In: Hajič, J., Carberry, S., Clark, S., Nivre, J. (eds.) Proceedings the 48th Annual Meeting of the Association for Computational Linguistics (ACL), Uppsala, Sweden, pp. 504–513. Association for Computational Linguistics, July 2010
19. Ma, X., Hovy, E.: End-to-end sequence labeling via bi-directional LSTM-CNNs-CRF. In: Proceedings of the 54th Annual Meeting of the Association for Computational Linguistics (Volume 1: Long Papers), pp. 1064–1074. Association for Computational Linguistics, Berlin, August 2016
20. Makhoul, J., Kubala, F., Schwartz, R., Weischedel, R.: Performance measures for information extraction. In: Proceedings of DARPA Broadcast News Workshop, pp. 249–252 (1999)
21. Nadeau, D., Sekine, S.: A survey of named entity recognition and classification. Lingvisticae Investigationes **30**(1), 3–26 (2007)
22. Neudecker, C., Antonacopoulos, A.: Making Europe's historical newspapers searchable. In: 2016 12th IAPR Workshop on Document Analysis Systems (DAS), pp. 405–410. IEEE (2016)
23. Piotrowski, M.: Natural language processing for historical texts. Synth. Lect. Hum. Lang. Technol. **5**(2), 1–157 (2012)
24. Plank, B.: What to do about non-standard (or non-canonical) language in NLP. In: Proceedings of the 13th Conference on Natural Language Processing (KONVENS 2016). Bochumer Linguistische Arbeitsberichte (2016)
25. Rao, D., McNamee, P., Dredze, M.: Entity linking: finding extracted entities in a knowledge base. In: Poibeau, T., Saggion, H., Piskorski, J., Yangarber, R. (eds.) Multi-source, Multilingual Information Extraction and Summarization. Theory and Applications of Natural Language Processing. Springer, Heidelberg (2013). https://doi.org/10.1007/978-3-642-28569-1_5

26. Ritter, A., Clark, S., Etzioni, O., et al.: Named entity recognition in Tweets: an experimental study. In: Proceedings of the Conference on Empirical Methods in Natural Language Processing, pp. 1524–1534 (2011)

27. Rodriquez, K.J., Bryant, M., Blanke, T., Luszczynska, M.: Comparison of named entity recognition tools for raw OCR text. In: Jancsary, J. (ed.) Proceedings of KONVENS 2012, pp. 410–414. ÖGAI, September 20. http://www.oegai.at/konvens2012/proceedings/60_rodriquez12w/

28. Rosset, S., Grouin, C., Fort, K., Galibert, O., Kahn, J., Zweigenbaum, P.: Structured named entities in two distinct press corpora: contemporary broadcast news and old newspapers. In: Proceedings of the Sixth Linguistic Annotation Workshop The LAW VI, pp. 40–48. Association for Computational Linguistics, Jeju, July 2012

29. Sporleder, C.: Natural language processing for cultural heritage domains. Lang. Linguist. Compass **4**(9), 750–768 (2010). https://doi.org/10.1111/j.1749-818X.2010.00230.x. ISSN 1749–818X

30. Terras, M.M.: The rise of digitization. In: Rikowski, R. (ed.) Digitisation Perspectives, pp. 3–20. SensePublishers, Rotterdam (2011). https://doi.org/10.1007/978-94-6091-299-3_1. www.emeraldinsight.com.ezproxy.lancs.ac.uk/doi/full/10.1108/OIR-06-2015-0193. ISBN 978-94-6091-299-3

31. Vilain, M., Su, J., Lubar, S.: Entity extraction is a boring solved problem: or is it? In: Human Language Technologies 2007: The Conference of the North American Chapter of the Association for Computational Linguistics; Companion Volume, Short Papers, NAACL-Short 2007, Rochester, New York, pp. 181–184. Association for Computational Linguistics (2007). http://dl.acm.org/citation.cfm?id=1614108.1614154

ImageCLEF 2020: Multimedia Retrieval in Lifelogging, Medical, Nature, and Internet Applications

Bogdan Ionescu[1]([✉]), Henning Müller[2], Renaud Péteri[3],
Duc-Tien Dang-Nguyen[4], Liting Zhou[5], Luca Piras[6], Michael Riegler[7],
Pål Halvorsen[7], Minh-Triet Tran[8], Mathias Lux[9], Cathal Gurrin[5],
Jon Chamberlain[10], Adrian Clark[10], Antonio Campello[11],
Alba G. Seco de Herrera[10], Asma Ben Abacha[12], Vivek Datla[13],
Sadid A. Hasan[14], Joey Liu[13], Dina Demner-Fushman[12], Obioma Pelka[15],
Christoph M. Friedrich[15], Yashin Dicente Cid[16], Serge Kozlovski[17],
Vitali Liauchuk[17], Vassili Kovalev[17], Raul Berari[18], Paul Brie[18],
Dimitri Fichou[18], Mihai Dogariu[1], Liviu Daniel Stefan[1],
and Mihai Gabriel Constantin[1]

[1] University Politehnica of Bucharest, Bucharest, Romania
bogdan.ionescu@upb.ro
[2] University of Applied Sciences Western Switzerland (HES-SO),
Delémont, Switzerland
[3] University of La Rochelle, La Rochelle, France
[4] University of Bergen, Bergen, Norway
[5] Dublin City University, Dublin, Ireland
[6] Pluribus One & University of Cagliari, Cagliari, Italy
[7] SimulaMet, Oslo, Norway
[8] University of Science, Ho Chi Minh City, Vietnam
[9] Klagenfurt University, Klagenfurt, Austria
[10] University of Essex, Colchester, UK
[11] Wellcome Trust, London, UK
[12] National Library of Medicine, Bethesda, USA
[13] Philips Research Cambridge, Cambridge, USA
[14] CVS Health, Monroeville, USA
[15] University of Applied Sciences and Arts Dortmund, Dortmund, Germany
[16] University of Warwick, Coventry, UK
[17] United Institute of Informatics Problems, Minsk, Belarus
[18] teleportHQ, Cluj-Napoca, Romania

Abstract. This paper presents an overview of the 2020 ImageCLEF lab that will be organized as part of the Conference and Labs of the Evaluation Forum—CLEF Labs 2020 in Thessaloniki, Greece. ImageCLEF is an ongoing evaluation initiative (run since 2003) that promotes the evaluation of technologies for annotation, indexing and retrieval of visual data with the aim of providing information access to large collections of images in various usage scenarios and domains. In 2020, the 18th edition of ImageCLEF will organize four main tasks: (i) a *Lifelog* task (videos, images and other sources) about daily activity understanding, retrieval

© Springer Nature Switzerland AG 2020
J. M. Jose et al. (Eds.): ECIR 2020, LNCS 12036, pp. 533–541, 2020.
https://doi.org/10.1007/978-3-030-45442-5_69

and summarization, (ii) a *Medical* task that groups three previous tasks (caption analysis, tuberculosis prediction, and medical visual question answering) with new data and adapted tasks, (iii) a *Coral* task about segmenting and labeling collections of coral images for 3D modeling, and a new (iv) *Web user interface* task addressing the problems of detecting and recognizing hand drawn website UIs (User Interfaces) for generating automatic code. The strong participation, with over 235 research groups registering and 63 submitting over 359 runs for the tasks in 2019 shows an important interest in this benchmarking campaign. We expect the new tasks to attract at least as many researchers for 2020.

Keywords: Lifelogging retrieval and summarization · Medical image classification · Coral image segmentation and classification · Recognition of hand drawn website UIs · ImageCLEF benchmarking · Annotated data

1   Introduction

The ImageCLEF evaluation campaign was started as part of the CLEF (Cross Language Evaluation Forum) in 2003 [7,8]. It has been held every year since then and delivered many results in the analysis and retrieval of images [20,21]. Medical tasks started in 2004 and have in some years been the majority of the tasks in ImageCLEF [18,19].

The objectives of ImageCLEF have always been the multilingual or language-independent analysis of visual content. A focus has often been on multimodal data sets, so combining images with structured information, free text or other information that helps in the decision making.

Since 2018, ImageCLEF uses the crowdAI (now migrated to AIcrowd[1] starting with 2020) platform to distribute the data and receive the submitted results. The system allows having an online leader board and gives the possibility to keep data sets accessible beyond competition, including a continuous submission of runs and addition to the leader board.

Over the years, ImageCLEF and also CLEF have shown a strong scholarly impact that was captured in [27,28]. This underlines the importance of evaluation campaigns for disseminating best scientific practices.

In the following, we introduce the four tasks that are going to run in the 2020 edition[2], namely: ImageCLEFlifelog, ImageCLEFmedical, ImageCLEFcoral, and the new ImageCLEFdrawnUI. Figure 1 captures with a few images the specificity of the tasks.

---

[1] https://www.aicrowd.com/.
[2] https://www.imageclef.org/2020.

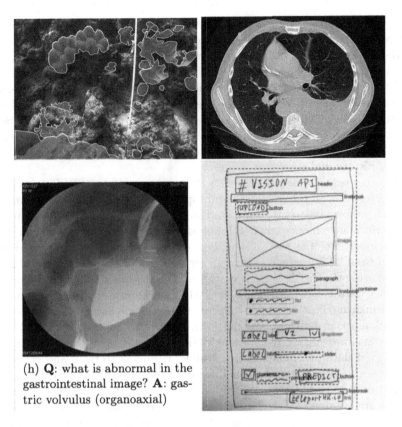

(h) **Q**: what is abnormal in the gastrointestinal image? **A**: gastric volvulus (organoaxial)

Fig. 1. Sample images from (left to right, top to bottom): ImageCLEFcoral with coral image segmentation and labeling, ImageCLEFmedical, e.g., tuberculosis prediction and Visual Question Answering, and ImageCLEFdrawnUI with recognition of hand drawn website UIs.

2   ImageCLEFlifelog

The main goal of the Lifelog task since its first edition [9] has been to advance the state-of-the-art research in lifelogging as an application of information retrieval. Different personal devices, such as smartphones, video cameras as well as wearable devices that allow collection of different data about our daily life are available. Large amounts of data are created by these devices containing videos, images, audio and sensor data. To be able to organize such a vast amount of data, there is a clear need of systems that can do this automatically.

As in the previous three editions, the task focuses mainly on images. The 2020 task will again be split into two subtasks: the lifelog moment retrieval and a new sports performance lifelog task. The first subtask includes new and enriched data, focusing on daily living activities and the chronological order of the moments. The second subtask provides a completely new dataset for assessing sports performance.

For the *Lifelog Core Task: Lifelog Moment Retrieval* the participants are required to retrieve several predefined activities in a lifelogger's life. For example, they are asked to return the relevant moments for the query "Find the moment(s) when the lifelogger was having a beer on the beach with his/her friends". Particular attention will be paid to the diversification of the selected moments with respect to the target scenario. To make the task possible and interesting a rich multimodal dataset will be used. The data are completely new and contain about 4.5 months of data from three lifeloggers, 1,500–2,500 images per day, visual concepts, semantic content, biometric information, music listening history and computer usage.

The other task, *Lifelog Task: Sports Performance Lifelog* (SPLL, 1st edition), is completely new in terms of data and topic. Teams are required to predict the expected performance (e.g., estimated finishing time, average heart rate and other performance measurements) for a non-professional athlete who trained for a sports event. For the task, a new dataset is provided containing information collected from 20–24 people that train for a 5 km run. Objective sensor data is collected using the FitBit Versa 2 sport watch[3]; subjective wellness, training load and injury data is collected using the PMSYS system[4]; and information about meals, drinks, medication, etc. is collected using Google Forms. The data contain information about daily sleeping patterns, daily heart rate, sport activities, logs of food consumed during the training period (from at least 2 participants) and self reported data like mode, stress, fatigue, readiness to train and other measurements also used for professional soccer teams. The data are collected over a period of four to five month. The copyright and ethical approval to release the data are obtained by the task organizers. For the sports task data, we have the data approved by the Norwegian Center for Research Data[5]. For assessing the performance of the approaches, classic metrics will be used, e.g., precision, cluster recall (to account for the diversification), etc. For this sports task, we will also utilize metrics such as Mean Absolute Error (MAE) and Root mean squared error (RMSE).

3   ImageCLEFmedical

The ImageCLEF medical task has been running every year since 2004 [22]. In 2020, it will follow a similar format as in the previous edition [18] containing the same three subtasks with some modifications. The three tasks will be: *tuberculosis analysis* [10–12], *figure caption analysis* [13,17,23], and *Visual Question Answering* [2,16].

The *tuberculosis task* will use, as in previous editions, Computed Tomography (CT) scans of patients with tuberculosis and more clinical data. In this edition, the task will concentrate only on generating an automatic report based on the CT and not assessing a TB severity score. The new report will be more detailed

---

[3] https://www.fitbit.com/versa.
[4] https://forzasys.com/pmsys.html.
[5] https://nsd.no.

than in the previous edition, containing more specific information, such as in which region each TB-related finding is located.

The *caption analysis task* will include more data compared to 2019. In 2020, an extension of the Radiology Objects in Context (ROCO) [24] data set is used and manually curated to reduce the data variability. The collection includes images from the medical literature including caption information, concepts and 7 sub-classes denoting the image radiology modality. The task concentrates on extracting Unified Medical Language System (UMLS®) Concept Unique Identifiers (CUIs) and can also be used as a first step towards the *Medical Visual Question Answering (VQA-Med)* task.

The *medical Visual Question Answering (VQA-Med) task* poses a challenging problem that involves both natural language processing and computer vision. In continuation of the two previous editions, the task consists of answering a natural language question from the visual content of an associated radiology image. VQA-Med 2020 will focus further on questions about abnormalities and will include a new subtask on visual question generation from radiology images.

4   ImageCLEFcoral

Coral reefs are important ecosystems because they are the most biodiverse parts of the oceans. However, corals thrive in narrow temperature ranges and ocean warming trends, among other factors, indicate that many of them will be lost within the next 30 years [3]. This would be a catastrophe, not only because of the extinction of many of the marine species they host but also because they provide an income and an essential food source to the people who live nearby [4,25]. Monitoring changes in reef composition and structural complexity on a large-scale is crucial to understanding and prioritizing conservation efforts.

Key to conservation work is knowledge of the state of reefs. Autonomous underwater vehicles are able to collect large amounts of data, more than can be annotated by a human. Although there have been promising attempts at automatically annotating imagery of reefs for complexity and benthic composition [15,26], it is fair to say that the problem is far from being solved. The aim of this competition is to encourage researchers to improve techniques for automatically identifying areas of interest and label them in a way that helps marine biologists and ecologists.

Following the success of the first edition of the ImageCLEFcoral task [5], in 2020, participants are required to devise and implement algorithms for automatically annotating regions in a collection of images with types of benthic substrate, such as hard coral or sponge. The dataset comprises 440 human-annotated training images and a further 200 unseen test images of a region of coral reef in Indonesia. The images were captured in high-quality JPEG format using an innovative underwater image capture system developed at the Marine Technology Research Unit at the University of Essex, UK.

The ground truth annotations of the training and test sets were made largely by marine biology MSc students at Essex and checked by an experienced coral reef researcher. The annotations were performed using a web-based tool developed in a collaborative project with London-based company Filament Ltd which allowed many people to work concurrently and which was carefully designed to be simple and quick to use; this proved so effective that we are exploring whether the tool can be made publicly available for other tasks in the future.

As in the first edition, algorithmic performance will be evaluated on the unseen test data using the intersection over union metric popularized in the PASCAL VOC[6] exercise. This computes the area of intersection of the output of an algorithm and the corresponding ground truth, normalizing that by the area of their union to ensure its maximum value remains bounded.

5   ImageCLEFdrawnUI

The user interface (UI) is the space where interactions between humans and computers occur. The increasing dependence of web and mobile applications have led many enterprises to increase the priority of developing user interfaces, in an effort to improve the overall user experience. Currently, the performance of any modern digital product is strongly correlated to the quality and usability of its user interface. However, building a user interface for digital applications is a complex process involving the interaction between multiple specialists, each with its own specific domain knowledge.

Generally, a business owner is setting a business goal. Then, a project owner or product owner is refining the requirements and builds a prototype of the application using wireframes. Once validated, the designer transforms the wireframes into designs which are transformed into code by the developer. This process is time consuming and expensive. Moreover, as more people get involved, the process is increasingly error prone. In addition, user interface experts are in limited supply. Globally, there are about 22 million developers[7], among which only 10 million are estimated to be also JavaScript UI developers[8].

Recently the use of machine learning to facilitate this process has been demonstrated as a viable solution. In 2018, pix2code, a machine learning based approach to generate low fidelity domain specific languages from screenshots, was published and open sourced [1]. Also, in 2018 Chen Chunyang et al. created their own dataset from android apps with 185,277 pairs of UI images and GUI skeletons. The dataset and code were also open-sourced [6].

In this context, in the 2020 ImageCLEFdrawnUI task, given a set of images of hand drawn UIs, participants are required to develop machine learning techniques that are able to predict the exact position and type of UI elements. The provided dataset consists of 3,000 hand drawn images inspired from mobile application screenshots and actual web pages containing 1,000 different templates. Each image was manually labelled with the positions of the bounding

---

[6] http://host.robots.ox.ac.uk/pascal/VOC/.

[7] http://evansdata.com.

[8] http://appdevelopermagazine.com.

boxes corresponding to each UI element and its type. To avoid any ambiguity, a predefined shape dictionary with 21 classes is used (e.g., paragraph, label, header). The performance of the algorithms will be evaluated using the standard mean Average Precision over IoU .5, commonly used in object detection [14].

6   Conclusions

In this paper, we present an overview of the upcoming ImageCLEF 2020 campaign. ImageCLEF has organized many tasks in a variety of domains over the past 18 years, from general stock photography, medical and biodiversity data to multimodal lifelogging. The focus has always been on language independent or multi-lingual approaches and most often on multimodal data analysis. 2020 has a set of interesting tasks that are expected to again draw a large number of participants. As in 2019, the focus for 2020 has been on the diversity of applications and on creating clean data sets to provide a solid basis for the evaluations.

Acknowledgement. Mihai Dogariu and Liviu Daniel Stefan work has been funded by the Operational Programme Human Capital of the Ministry of European Funds through the Financial Agreement 51675/09.07.2019, SMIS code 125125.

References

1. Beltramelli, T.: pix2code: generating code from a graphical user interface screenshot. In: Proceedings of the ACM SIGCHI Symposium on Engineering Interactive Computing Systems, pp. 1–9 (2018)
2. Ben Abacha, A., Hasan, S.A., Datla, V.V., Liu, J., Demner-Fushman, D., Müller, H.: VQA-Med: overview of the medical visual question answering task at Image-CLEF 2019. In: Working Notes of CLEF 2019 - Conference and Labs of the Evaluation Forum, Lugano, Switzerland, 9–12 September 2019 (2019). http://ceur-ws.org/Vol-2380/paper_272.pdf
3. Birkeland, C.: Global status of coral reefs: in combination, disturbances and stressors become ratchets. In: World Seas: An Environmental Evaluation, pp. 35–56. Elsevier, Amsterdam (2019)
4. Brander, L.M., Rehdanz, K., Tol, R.S., Van Beukering, P.J.: The economic impact of ocean acidification on coral reefs. Clim. Chang. Econ. **3**(01), 1250002 (2012)
5. Chamberlain, J., Campello, A., Wright, J.P., Clift, L.G., Clark, A., García Seco de Herrera, A.: Overview of ImageCLEFcoral 2019 task. In: CLEF2019 Working Notes. CEUR Workshop Proceedings, vol. 2380. CEUR-WS.org (2019)
6. Chen, C., Su, T., Meng, G., Xing, Z., Liu, Y.: From UI design image to GUISkeleton: a neural machine translator to bootstrap mobile GUI implementation. In: International Conference on Software Engineering, vol. 6 (2018)
7. Clough, P., Müller, H., Sanderson, M.: The CLEF 2004 cross-language image retrieval track. In: Peters, C., Clough, P., Gonzalo, J., Jones, G.J.F., Kluck, M., Magnini, B. (eds.) CLEF 2004. LNCS, vol. 3491, pp. 597–613. Springer, Heidelberg (2005). https://doi.org/10.1007/11519645_59
8. Clough, P., Sanderson, M.: The CLEF 2003 cross language image retrieval task. In: Proceedings of the Cross Language Evaluation Forum (CLEF 2003) (2004)

9. Dang-Nguyen, D.T., Piras, L., Riegler, M., Boato, G., Zhou, L., Gurrin, C.: Overview of ImageCLEFlifelog 2017: lifelog retrieval and summarization. In: Cappellato, L., Ferro, N., Goeuriot, L., Mandl, T. (eds.) CLEF 2017 Working Notes. CEUR Workshop Proceedings (CEUR-WS.org) http://ceur-ws.org/Vol-1866/ (2017). ISSN 1613–0073

10. Dicente Cid, Y., Kalinovsky, A., Liauchuk, V., Kovalev, V., Müller, H.: Overview of ImageCLEFtuberculosis 2017 - predicting tuberculosis type and drug resistances. In: CLEF2017 Working Notes. CEUR Workshop Proceedings, Dublin, Ireland, 11–14 September 2017. CEUR-WS.org http://ceur-ws.org

11. Dicente Cid, Y., Liauchuk, V., Klimuk, D., Tarasau, A., Kovalev, V., Müller, H.: Overview of ImageCLEF tuberculosis 2019 - automatic CT-based report generation and tuberculosis severity assessment. In: CLEF2019 Working Notes. CEUR Workshop Proceedings, Lugano, Switzerland, 9–12 September 2019. CEUR-WS.org http://ceur-ws.org

12. Dicente Cid, Y., Liauchuk, V., Kovalev, V., Müller, H.: Overview of ImageCLEF-tuberculosis 2018 - detecting multi-drug resistance, classifying tuberculosis type, and assessing severity score. In: CLEF2018 Working Notes. CEUR Workshop Proceedings, Avignon, France, 10–14 September 2018. CEUR-WS.org http://ceur-ws.org

13. Eickhoff, C., Schwall, I., García Seco de Herrera, A., Müller, H.: Overview of Image-CLEFcaption 2017 - the image caption prediction and concept extraction tasks to understand biomedical images. In: CLEF2017 Working Notes. CEUR Workshop Proceedings, Dublin, Ireland, 11–14 September 2017. CEUR-WS.org http://ceur-ws.org

14. Everingham, M., Gool, L.V., Williams, C.K.I., Winn, J., Zisserman, A.: The PASCAL Visual Object Classes (VOC) challenge. Int. J. Comput. Vis. **88**, 303–338 (2010). https://doi.org/10.1007/s11263-009-0275-4

15. Gonzalez-Rivero, M., et al.: The Catlin Seaview Survey-kilometre-scale seascape assessment, and monitoring of coral reef ecosystems. Aquat. Conserv. Mar. Freshwater Ecosyst. **24**, 184–198 (2014)

16. Hasan, S.A., Ling, Y., Farri, O., Liu, J., Lungren, M., Müller, H.: Overview of the ImageCLEF 2018 medical domain visual question answering task. In: CLEF 2018 Working Notes. CEUR Workshop Proceedings, Avignon, France, 10–14 September 2018. CEUR-WS.org http://ceur-ws.org

17. Seco de Herrera, A.G., Eickhoff, C., Andrearczyk, V., Müller, H.: Overview of the ImageCLEF 2018 caption prediction tasks. In: CLEF 2018 Working Notes. CEUR Workshop Proceedings, Avignon, France, 10–14 September 2018. CEUR-WS.org http://ceur-ws.org

18. Ionescu, B., et al.: ImageCLEF 2019: multimedia retrieval in medicine, lifelogging, security and nature. In: Crestani, F., et al. (eds.) CLEF 2019. LNCS, vol. 11696, pp. 358–386. Springer, Cham (2019). https://doi.org/10.1007/978-3-030-28577-7_28

19. Ionescu, B., et al.: Overview of ImageCLEF 2018: challenges, datasets and evaluation. In: Bellot, P., et al. (eds.) CLEF 2018. LNCS, vol. 11018, pp. 309–334. Springer, Cham (2018). https://doi.org/10.1007/978-3-319-98932-7_28

20. Kalpathy-Cramer, J., de Herrera, A.G.S., Demner-Fushman, D., Antani, S., Bedrick, S., Müller, H.: Evaluating performance of biomedical image retrieval systems: overview of the medical image retrieval task at ImageCLEF 2004–2014. Comput. Med. Imaging Graph. **39**(0), 55–61 (2015)

21. Müller, H., Clough, P., Deselaers, T., Caputo, B. (eds.): ImageCLEF - Experimental Evaluation in Visual Information Retrieval. INRED, vol. 32. Springer, Heidelberg (2010). https://doi.org/10.1007/978-3-642-15181-1

22. Müller, H., Kalpathy-Cramer, J., García Seco de Herrera, A.: Experiences from the ImageCLEF medical retrieval and annotation tasks. Information Retrieval Evaluation in a Changing World. TIRS, vol. 41, pp. 231–250. Springer, Cham (2019). https://doi.org/10.1007/978-3-030-22948-1_10

23. Pelka, O., Friedrich, C.M., García Seco de Herrera, A., Müller, H.: Overview of the ImageCLEFmed 2019 concept prediction task. In: CLEF2019 Working Notes. CEUR Workshop Proceedings, Lugano, Switzerland, 09–12 September 2019, vol. 2380. CEUR-WS.org http://ceur-ws.org

24. Pelka, O., Koitka, S., Rückert, J., Nensa, F., Friedrich, C.M.: Radiology objects in context (ROCO): a multimodal image dataset. In: Stoyanov, D., et al. (eds.) LABELS/CVII/STENT -2018. LNCS, vol. 11043, pp. 180–189. Springer, Cham (2018). https://doi.org/10.1007/978-3-030-01364-6_20

25. Speers, A.E., Besedin, E.Y., Palardy, J.E., Moore, C.: Impacts of climate change and ocean acidification on coral reef fisheries: an integrated ecological-economic model. Ecol. Econ. **128**, 33–43 (2016)

26. Stokes, M., Deane, G.: Automated processing of coral reef benthic images. Limnol. Oceanogr. Methods **7**, 157–168 (2009)

27. Tsikrika, T., de Herrera, A.G.S., Müller, H.: Assessing the scholarly impact of ImageCLEF. In: Forner, P., Gonzalo, J., Kekäläinen, J., Lalmas, M., de Rijke, M. (eds.) CLEF 2011. LNCS, vol. 6941, pp. 95–106. Springer, Heidelberg (2011). https://doi.org/10.1007/978-3-642-23708-9_12

28. Tsikrika, T., Larsen, B., Müller, H., Endrullis, S., Rahm, E.: The scholarly impact of CLEF (2000–2009). In: Forner, P., Müller, H., Paredes, R., Rosso, P., Stein, B. (eds.) CLEF 2013. LNCS, vol. 8138, pp. 1–12. Springer, Heidelberg (2013). https://doi.org/10.1007/978-3-642-40802-1_1

LifeCLEF 2020 Teaser: Biodiversity Identification and Prediction Challenges

Alexis Joly[1]([✉])(iD), Hervé Goëau[2], Stefan Kahl[7], Christophe Botella[1,3](iD),
Rafael Ruiz De Castaneda[9](iD), Hervé Glotin[4](iD), Elijah Cole[10](iD),
Julien Champ[1], Benjamin Deneu[1](iD), Maximillien Servajean[8](iD),
Titouan Lorieul[1](iD), Willem-Pier Vellinga[5], Fabian-Robert Stöter[1](iD),
Andrew Durso[9](iD), Pierre Bonnet[2](iD), and Henning Müller[6](iD)

[1] Inria, LIRMM, Montpellier, France
alexis.joly@inria.fr
[2] CIRAD, UMR AMAP, Montpellier, France
[3] INRA, UMR AMAP, Montpellier, France
[4] Aix Marseille Univ, Université de Toulon, CNRS, LIS, DYNI, Marseille, France
[5] Xeno-canto Foundation, Amsterdam, The Netherlands
[6] HES-SO, Sierre, Switzerland
[7] Chemnitz University of Technology, Chemnitz, Germany
[8] LIRMM, Université Paul Valéry, University of Montpellier, CNRS,
Montpellier, France
[9] University of Geneva, Geneva, Switzerland
[10] Caltech, Pasadena, USA

Abstract. Building accurate knowledge of the identity, the geographic
distribution and the evolution of species is essential for the sustainable
development of humanity, as well as for biodiversity conservation. However, the difficulty of identifying plants and animals in the field is hindering the aggregation of new data and knowledge. Identifying and naming
living plants or animals is almost impossible for the general public and
is often difficult even for professionals and naturalists. Bridging this gap
is a key step towards enabling effective biodiversity monitoring systems.
The LifeCLEF campaign, presented in this paper, has been promoting
and evaluating advances in this domain since 2011. The 2020 edition
proposes four data-oriented challenges related to the identification and
prediction of biodiversity: (i) PlantCLEF: cross-domain plant identification based on herbarium sheets, (ii) BirdCLEF: bird species recognition
in audio soundscapes, (iii) GeoLifeCLEF: location-based prediction of
species based on environmental and occurrence data, and (iv) Snake-
CLEF: image-based snake identification.

Keywords: Biodiversity · Machine learning · IA · Species
identification · Species prediction · Plant identification · Bird
identification · Snake identification · Species distribution model

© Springer Nature Switzerland AG 2020
J. M. Jose et al. (Eds.): ECIR 2020, LNCS 12036, pp. 542–549, 2020.
https://doi.org/10.1007/978-3-030-45442-5_70

1   Introduction

Accurately identifying organisms observed in the wild is an essential step in ecological studies. Unfortunately, observing and identifying living organisms requires high levels of expertise. For instance, plants alone account for more than 400,000 different species and the distinctions between them can be quite subtle. Since the Rio Conference of 1992, this *taxonomic gap* has been recognized as one of the major obstacles to the global implementation of the Convention on Biological Diversity [6]. In 2004, Gaston and O'Neill [27] discussed the potential of automated approaches for species identification. They suggested that, if the scientific community were able to (i) produce large training datasets, (ii) precisely evaluate error rates, (iii) scale up automated approaches, and (iv) detect novel species, then it would be possible to develop a generic automated species identification system that would open up new vistas for research in biology and related fields.

Since the publication of [27], automated species identification has been studied in many contexts [26, 29, 30, 38, 41, 43, 44, 48]. This area continues to expand rapidly, particularly due to recent advances in deep learning [25, 28, 31, 42, 45–47]. In order to measure progress in a sustainable and repeatable way, the LifeCLEF [15] research platform was created in 2014 as a continuation and extension of the plant identification task [37] that had been run within the ImageCLEF lab [12] since 2011 [32–34]. LifeCLEF expanded the challenge by considering animals in addition to plants, and including audio and video content in addition to images. LifeCLEF 2020 consists of four challenges (PlantCLEF, BirdCLEF, GeoLifeCLEF, and SnakeCLEF), which we will now describe in turn.

2   PlantCLEF 2020 Challenge: Identifying Plant Pictures from Herbarium Sheets

Motivation: For several centuries, botanists have collected, catalogued and systematically stored plant specimens in herbaria. These physical specimens are used to study the variability of species, their phylogenetic relationship, their evolution, or phenological trends. One of the key step in the workflow of botanists and taxonomists is to find the herbarium sheets that correspond to a new specimen observed in the field. This task requires a high level of expertise and can be very tedious. Developing automated tools to facilitate this work is thus of crucial importance. More generally, this will help to convert these invaluable centuries-old materials into FAIR [23] data.

Data Collection: The task will rely on a large collection of more than 60,000 herbarium sheets that were collected in French Guyana (The "Herbier IRD de Guyane", CAY [14]) and digitized in the context of the e-ReColNat project [8]. iDigBio [11] hosts millions of images of herbarium specimens. Several tens of thousands of these images, illustrating the French Guyana flora, will be used for the PlantCLEF task this year. A valuable asset of this collection is that several herbarium sheets are accompanied by a few pictures of the same specimen in the field. For the test set, we will use in-the-field pictures coming different sources including Pl@ntNet [20] and Encyclopedia of Life [9].

Task Description: The challenge will be evaluated as a cross-domain classification task. The training set will consist of herbarium sheets whereas the test set will be composed of field pictures. To enable learning a mapping between the herbarium sheets domain and the field pictures domain, we will provide both herbarium sheets and field pictures for a subset of species. The metrics used for the evaluation of the task will be the classification accuracy and the mean reciprocal rank.

3   BirdCLEF 2020 Challenge: Bird Species Recognition in Audio Soundscapes

Motivation: Monitoring birds by sound is important for many environmental and scientific purposes. Birds are difficult to photograph and sound offers better possibilities for inventory coverage. A number of participatory science projects have focused on recording a very large number of bird sounds, making it possible to recognize most species by their sound and to train deep learning models to automate this process. It was shown in previous editions of BirdCLEF [35,36] that systems for identifying birds from mono-directional recordings are now performing very well and several mobile applications implementing this are emerging today. However, there is also interest in identifying birds from *omnidirectional* or *binaural* recordings. This would enable more passive monitoring scenarios like networks of static recorders that continuously capture the surrounding sound environment. The advantage of this type of approach is that it introduces less sampling bias than the opportunistic observations of citizen scientists. However, recognizing birds in such content is much more difficult due to the high vocal activity with signal overlap (e.g. during the dawn chorus) and high levels of ambient noise.

Data Collection: The training set used for the challenge will be a version of the 2019 training set [40] enriched by new contributions from the Xeno-canto network and a geographic extension. It will contain approximately 80K recordings covering between 1500 and 2000 species from North, Central and South America, as well as Europe. This will be the largest bioacoustic dataset used in the literature. For the test set, three sources of soundscapes will be used: (i) 100+ h of manually annotated soundscapes recorded using 30 field recorders between January and June of 2017 in Ithaca, NY, USA by the Cornell Lab of Ornithology [7], (ii) 10+ h of fully annotated dawn chorus soundscapes recorded using solar-powered field recorders between January and June 2018 near Frankfurt (a.M.), Germany by OekoFor [19], (iii) 2 h acquired at high sampling rate (250 kHz) by binaural antenna in Côte d'Azur, France.

Task Description: In 2020, two scenarios will be evaluated: (i) the recognition of all specimens singing in a long sequence (up to one hour) of raw soundscapes that can contain tens of birds singing simultaneously, and (ii) chorus source separation in complex soundscapes that were recorded in stereo at very high sampling rate (250 kHz SR). For the first scenario, participants will be asked to provide time intervals of recognized singing birds. Participants will be allowed to use any of the provided metadata complementary to the audio content (.wav format, 44.1 kHz, 48 kHz, or 96 kHz sampling rate). The task is focused on developing real-world applicable solutions and therefore requires participants to submit single models trained on none other than the mono-species recordings provided as training data. For the second task on stereophonic recordings, the goal will be to determine the species singing in chorus simultaneously during a time interval. In contrast to task one, the challengers are invited to run automatic source separation before or jointly to the bird species classification, taking advantage of the multi-channel recordings. Participants will be allowed to use any other data than the provided recordings, but will have to provide the scripts to check that their solution is fully automatic. For both tasks, the evaluation measure will be the classification mean average precision (c-mAP, [39]).

4   GeoLifeCLEF 2020 Challenge: Location-Based Prediction of Species Based on Environmental and Occurrence Data

Motivation: Automatic prediction of the list of species most likely to be observed at a given location is useful for many scenarios related to biodiversity management and conservation. First, it could improve species identification tools (whether automatic, semi-automatic or based on traditional field guides) by reducing the list of candidate species observable at a given site. More generally, this could facilitate biodiversity inventories through the development of location-based recommendation services (*e.g.* on mobile phones), encourage the involvement of citizen scientist observers, and accelerate the annotation and validation of species observations to produce large, high-quality data sets. Last but not least, this could be used for educational purposes through biodiversity discovery applications with features such as contextualized educational pathways.

Data Collection: The challenge will rely on a collection of millions of occurrences of plants and animals in the US and France (primarily from GBIF [10], iNaturalist [13], Pl@ntNet [20] and a few expert collections). In addition to geocoordinates and species name, each occurrence will be matched with a set of geographic images characterizing the local landscape and environment around the occurrence. In more detail, this will include: (i) high resolution (about $1 \, m^2$/pixel) remotely sensed imagery (from NAIP [18] for the US and from IGN [2] for France), (ii) bio-climatic rasters ($1 \, km^2$/pixel, from WorldClim [24]) and (iii), land cover rasters ($30 \, m^2$/pixel from NLCD [17] for the US, $10 \, m^2$/pixel from CESBIO [3] for France).

Task Description: The occurrence dataset will be split in a training set with known species name labels and a test set used for the evaluation. For each occurrence (with geographic images) in the test set, the goal of the task will be to return a candidate set of species with associated confidence scores. The evaluation metrics will be the top-K accuracy (for different values of K) and a set-based prediction metric which will be specified later.

5   SnakeCLEF 2020 Challenge: Image-Based Snake Identification

Motivation: Developing a robust system for identifying species of snakes from photographs is an important goal in biodiversity and global health. With over half a million victims of death and disability from venomous snakebite annually, understanding the global distribution of the $>3,700$ species of snakes and differentiating species from images (particularly images of low quality) will significantly improve epidemiology data and treatment outcomes. The goals and usage of image-based snake identification are complementary with those of other challenges: classifying snake species in images, predicting the list of species that are the most likely to be observed at a given location, and eventually developing automated tools that can facilitate integration of changing taxonomies and new discoveries.

Data Collection: Images of about 100 snake species from all around the world (between 300 and 150,000 images per species) will be aggregated from different data sources (including iNaturalist [13]). This will extend the dataset used in a previous challenge [22] hosted on the AICrowd platform. The distribution of the number of images between the classes is highly imbalanced.

Task Description: Given the set of images and corresponding geographic location information, the goal of the task will be to return for each image a ranked list of species sorted according to the likelihood that they are in the image and might have been observed at that location.

6   Timeline and Registration Instructions

All information about the timeline and participation in the challenges is provided on the LifeCLEF 2020 web pages [16]. The system used to run the challenges (registration, submission, leaderboard, etc.) is the AIcrowd platform [1].

7   Discussion and Conclusion

The long-term societal impact of boosting research on biodiversity informatics is difficult to overstate. To fully reach its objective, an evaluation campaign such as LifeCLEF requires a long-term research effort so as to (i) encourage non-incremental contributions, (ii) measure consistent performance gaps, (iii) progressively scale-up the problem and (iv), enable the emergence of a strong community. The 2020 edition of the lab will support this vision and will include the following innovations:

- SnakeCLEF, a challenging new task related to the identification of snake species, will be introduced. It will be organized by the University of Geneva, Switzerland.
- The PlantCLEF task will focus on a brave new challenge: the identification of plant pictures based on a training set of digitized herbarium sheets (cross-domain image classification).
- The BirdCLEF challenge will be enriched with new annotated soundscape data, and with challenging new high sampling rate stereophonic recordings (from SEAMED, SMILES ANR, and SABIOD [21] projects).
- The GeoLifeCLEF challenge will be enriched with new plant and animal occurrences from two continents and high resolution remotely sensed imagery.

The results of this challenge will be published in the proceedings of the CLEF 2020 conference [5] and in the CEUR-WS workshop proceedings [4].

Acknowledgements. This work is supported in part by the SEAMED PACA project, the SMILES project (ANR-18-CE40-0014), and an NSF Graduate Research Fellowship (DGE-1745301).

References

1. AIcrowd. https://www.aicrowd.com/
2. BD ORTHO. http://professionnels.ign.fr/bdortho
3. Carte d'ccupation des sols 2016. http://osr-cesbio.ups-tlse.fr/~oso/posts/2017-03-30-carte-s2-2016/
4. CEUR-WS. http://ceur-ws.org/
5. CLEF 2020. https://clef2020.clef-initiative.eu/
6. Convention on Biodiversity. https://www.cbd.int/
7. Cornell Lab of Ornithology. https://www.birds.cornell.edu/home/
8. e-ReColNat. https://www.recolnat.org/
9. Encyclopedia of Life. https://eol.org/
10. Global Biodiversity Information Facility. https://www.gbif.org/
11. iDigBio. https://www.idigbio.org/
12. ImageCLEF. http://www.imageclef.org/
13. iNaturalist. https://www.inaturalist.org/
14. L'Herbier IRD de Guyane. http://publish.plantnet-project.org/project/caypub
15. LifeCLEF. http://www.lifeclef.org/
16. LifeCLEF 2020. https://www.imageclef.org/lifeclef2020
17. Multi-Resolution Land Characteristics Consortium. https://www.mrlc.gov/
18. National Agricultural Imagery Program. https://www.fsa.usda.gov/programs-and-services/aerial-photography/imagery-programs/naip-imagery/
19. OekoFor. http://oekofor.de
20. Pl@ntNet. https://plantnet.org/
21. Scaled Acoustic BIODiversity (SABIOD) Platform. http://sabiod.telemeta.org/
22. Snake Species Identification Challenge. https://www.aicrowd.com/challenges/snake-species-identification-challenge
23. The FAIR Data Principles. https://www.force11.org/group/fairgroup/fairprinciples

24. WorldClim Bioclimatic Variables. https://www.worldclim.org/bioclim
25. Bonnet, P., et al.: Plant identification: experts vs. machines in the era of deep learning. In: Joly, A., Vrochidis, S., Karatzas, K., Karppinen, A., Bonnet, P. (eds.) Multimedia Tools and Applications for Environmental & Biodiversity Informatics. MSA, pp. 131–149. Springer, Cham (2018). https://doi.org/10.1007/978-3-319-76445-0_8
26. Cai, J., Ee, D., Pham, B., Roe, P., Zhang, J.: Sensor network for the monitoring of ecosystem: bird species recognition. In: 2007 3rd International Conference on Intelligent Sensors, Sensor Networks and Information, ISSNIP 2007 (2007). https://doi.org/10.1109/ISSNIP.2007.4496859
27. Gaston, K.J., O'Neill, M.A.: Automated species identification: why not? Philos. Trans. R. Soc. London B Biol. Sci. **359**(1444), 655–667 (2004)
28. Ghazi, M.M., Yanikoglu, B., Aptoula, E.: Plant identification using deep neural networks via optimization of transfer learning parameters. Neurocomputing **235**, 228–235 (2017)
29. Glotin, H., Clark, C., LeCun, Y., Dugan, P., Halkias, X., Sueur, J.: Proceedings of the 1st workshop on Machine Learning for Bioacoustics - ICML4B. ICML, Atlanta USA (2013)
30. Glotin, H., LeCun, Y., Artiéres, T., Mallat, S., Tchernichovski, O., Halkias, X.: Proceedings of the Neural Information Processing Scaled for Bioacoustics, from Neurons to Big Data. NIPS International Conference, Tahoe, USA (2013). http://sabiod.org/nips4b
31. Goeau, H., Bonnet, P., Joly, A.: Plant identification based on noisy web data: the amazing performance of deep learning (LifeCLEF 2017). In: CLEF 2017-Conference and Labs of the Evaluation Forum, pp. 1–13 (2017)
32. Goëau, H., et al.: The ImageCLEF 2013 plant identification task. In: CLEF, Valencia, Spain (2013)
33. Goëau, H., et al.: The ImageCLEF 2011 plant images classification task. In: CLEF 2011 (2011)
34. Goëau, H., et al.: ImageCLEF2012 plant images identification task. In: CLEF 2012, Rome (2012)
35. Goëau, H., Glotin, H., Planqué, R., Vellinga, W.P., Stefan, K., Joly, A.: Overview of BirdCLEF 2018: monophone vs. soundscape bird identification. In: CLEF Working Notes 2018 (2018)
36. Goëau, H., Glotin, H., Vellinga, W., Planqué, B., Joly, A.: LifeCLEF bird identification task 2017. In: Working Notes of CLEF 2017 - Conference and Labs of the Evaluation Forum, Dublin, Ireland, 11–14 September 2017 (2017)
37. Goëau, H., et al.: The ImageCLEF plant identification task 2013. In: Proceedings of the 2nd ACM International Workshop on Multimedia Analysis for Ecological Data, pp. 23–28. ACM (2013)
38. Joly, A., et al.: Interactive plant identification based on social image data. Ecol. Inform. **23**, 22–34 (2014)
39. Joly, A., et al.: Overview of LifeCLEF 2018: a large-scale evaluation of species identification and recommendation algorithms in the era of AI. In: Bellot, P., et al. (eds.) CLEF 2018. LNCS, vol. 11018, pp. 247–266. Springer, Cham (2018). https://doi.org/10.1007/978-3-319-98932-7_24
40. Kahl, S., Stöter, F.R., Glotin, H., Planqué, R., Vellinga, W.P., Joly, A.: Overview of BirdCLEF 2019: large-scale bird recognition in soundscapes. In: CLEF (Working Notes) (2019)

41. Lee, D.J., Schoenberger, R.B., Shiozawa, D., Xu, X., Zhan, P.: Contour matching for a fish recognition and migration-monitoring system. In: Optics East, pp. 37–48. International Society for Optics and Photonics (2004)
42. Lee, S.H., Chan, C.S., Remagnino, P.: Multi-organ plant classification based on convolutional and recurrent neural networks. IEEE Trans. Image Process. **27**(9), 4287–4301 (2018)
43. Towsey, M., Planitz, B., Nantes, A., Wimmer, J., Roe, P.: A toolbox for animal call recognition. Bioacoustics **21**(2), 107–125 (2012)
44. Trifa, V.M., Kirschel, A.N., Taylor, C.E., Vallejo, E.E.: Automated species recognition of antbirds in a Mexican rainforest using hidden Markov models. J. Acoust. Soc. Am. **123**, 2424 (2008)
45. Van Horn, G., et al.: The inaturalist species classification and detection dataset. In: CVPR (2018)
46. Wäldchen, J., Mäder, P.: Machine learning for image based species identification. Methods in Ecol. Evol. **9**(11), 2216–2225 (2018)
47. Wäldchen, J., Rzanny, M., Seeland, M., Mäder, P.: Automated plant species identification–trends and future directions. PLoS Comput. Biol. **14**(4), e1005993 (2018)
48. Yu, X., Wang, J., Kays, R., Jansen, P.A., Wang, T., Huang, T.: Automated identification of animal species in camera trap images. EURASIP J. Image Video Process. **2013**(1), 1–10 (2013). https://doi.org/10.1186/1687-5281-2013-52

BioASQ at CLEF2020: Large-Scale Biomedical Semantic Indexing and Question Answering

Martin Krallinger[1], Anastasia Krithara[2], Anastasios Nentidis[2,3(✉)],
Georgios Paliouras[2], and Marta Villegas[1]

[1] Barcelona Supercomputing Center, Barcelona, Spain
{martin.krallinger,marta.villegas}@bsc.es
[2] National Center for Scientific Research "Demokritos", Athens, Greece
{akrithara,tasosnent,paliourg}@iit.demokritos.gr
[3] Aristotle University of Thessaloniki, Thessaloniki, Greece
nentidis@csd.auth.gr

Abstract. This paper describes the eighth edition of the BioASQ Challenge, which will run as an evaluation Lab in the context of CLEF2020. The aim of BioASQ is the promotion of systems and methods for highly precise biomedical information access. This is done through the organization of a series of challenges (shared tasks) on large-scale biomedical semantic indexing and question answering, where different teams develop systems that compete on the same demanding benchmark datasets that represent the real information needs of biomedical experts. In order to facilitate this information finding process, the BioASQ challenge introduced two complementary tasks: (a) the automated indexing of large volumes of unlabelled data, primarily scientific articles, with biomedical concepts, (b) the processing of biomedical questions and the generation of comprehensible answers. Rewarding the most competitive systems that outperform the state of the art, BioASQ manages to push the research frontier towards ensuring that the biomedical experts will have direct access to valuable knowledge.

Keywords: Biomedical information · Semantic indexing · Question answering

1 Introduction

The availability of biomedical data increases rapidly with scientific literature being a major data resource for biomedical knowledge. MEDLINE/PubMed currently comprises more than 20 million articles with more than 2 new articles published in biomedical journals every minute[1]. This wealth of new knowledge is essential for scientific progress in biomedicine and can have a high impact on public health. However, ensuring that this knowledge is used in a timely manner by the biomedical experts is necessary for maximizing the benefit of the society.

---

[1] https://www.nlm.nih.gov/bsd/medline_pubmed_production_stats.html.

© Springer Nature Switzerland AG 2020
J. M. Jose et al. (Eds.): ECIR 2020, LNCS 12036, pp. 550–556, 2020.
https://doi.org/10.1007/978-3-030-45442-5_71

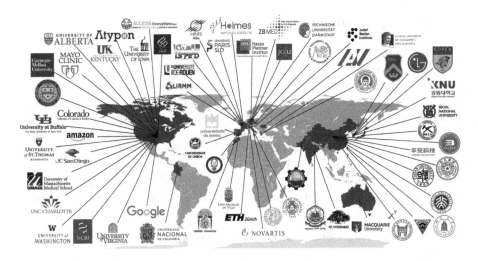

Fig. 1. Distribution of participating teams for the seven years of the BioASQ challenges

BioASQ[2] is a series of international challenges (shared tasks) and workshops focusing on biomedical semantic indexing and question answering. The BioASQ challenges [7] are structured into complementary tasks and sub-tasks so that participating teams can focus on tasks relevant to their area of expertise, including hierarchical text classification, machine learning, information retrieval and multi-document summarization amongst many other areas. BioASQ is also unique in requiring that the participating systems answer biomedical natural language questions by searching in both structured data (e.g. ontologies, databases) and unstructured data (e.g., biomedical articles).

As BioASQ consistently rewards highly precise biomedical information access systems developed by teams around the world, ensures that the biomedical experts eventually have more and more direct access to valuable knowledge that will help them avoid costly (sometimes even fatal) mistakes and provide high quality health services. The worldwide distribution of more than 60 teams from 20 counties participating in the challenges through the seven years of BioASQ is shown in Fig. 1. The BioASQ challenge has been running on an annual basis since 2012. The workshop has been taking place in the CLEF conference till 2015. In 2016 and 2017 it took place in ACL, in conjunction with BioNLP. In 2018 and 2019, it took place as an independent workshop in EMNLP and ECML-PKDD conferences respectively. In 2020, an independent BioASQ CLEF lab will run and the results will be presented in the context of CLEF 2020 conference.

---

[2] http://www.bioasq.org.

2 BioASQ Evaluation Lab 2020

The BioASQ challenge assesses the performance of information systems in supporting the following tasks that are central in the biomedical question answering process: (a) the indexing of large volumes of unlabeled data, primarily scientific articles, with biomedical concepts (in English and Spanish), (b) the processing of biomedical questions and the generation of answers and supporting material. Both these tasks have been running since the first year of BioASQ, but this is the first year for BioASQ to extend beyond the English language, by challenging the community to semantically index biomedical content in Spanish. Therefore, after the introduction of the new BioASQ task MESINESP in early 2020, the eighth BioASQ challenge will consist of the three tasks described in this section.

2.1 Task 8a: Large-Scale Biomedical Semantic Indexing

BioASQ task 8a requires systems to automatically assign MeSH terms to biomedical articles added to the MEDLINE database, thus assisting the indexing of biomedical literature. In effect, this is a classification task that requires documents to be automatically classified into a hierarchy of classes. A training dataset of about 14,9 million annotated articles is already available for task 8a and testsets with newly published articles will be released weekly, before the NLM curators annotate them. The systems will assign MeSH labels to them, which will be compared against the labels assigned by the curators.

Evaluation in Task 8a: As the manual annotations become gradually available, the scores of the systems are updated. In this manner, the evaluation of the systems participating in task 8a is fully automated on the side of BioASQ and thus can run on a weekly basis throughout the year. The performance of the systems taking part in task 8a is assessed with a range of different measures. Some of them are variants of standard information retrieval measures for multi-label classification problems (e.g. precision, recall, f-measure accuracy). Additionally, measures that use the MeSH hierarchy to provide a more refined estimate of the systems' performance are used. The official measures for identifying the winners of the task are micro-averaged F-measure (MiF) and the Lowest Common Ancestor F-measure (LCA-F) [2].

2.2 Task 8b: Biomedical Question Answering

BioASQ task 8b takes place in two phases. In the first phase, the participants are given English questions formulated by biomedical experts. For each question, the participating systems have to retrieve relevant MEDLINE documents, relevant snippets (passages) of the documents, relevant concepts (from five designated ontologies), and relevant RDF triples (from the Linked Life Data platform). This is also a classification task that requires questions to be classified into classes from multiple hierarchies. Subsequently, in the second phase of task

8b, the participants are given some relevant documents and snippets that the experts themselves have identified (using tools developed in BioASQ [6]), and they are required to return 'exact' answers (e.g., names of particular diseases or genes) and 'ideal' answers (a paragraph-sized summary of the most important information of the first phase per question). A training dataset of 3,243 biomedical questions is already available for participants of task 8b to train their systems and about 500 new biomedical questions, with corresponding golden annotations and answers, will be developed for the five testsets of task 8b.

Evaluation in Task 8b: The responses of the systems are evaluated both automatically and manually by the experts employing a variety of evaluation measures [3]. In phase A, on the retrieval of relevant material, both ordered and unordered measures are calculated but the official evaluation is based on the Mean Average Precision (MAP). For the exact answers in phase B, different evaluation measures are used depending on the type of the question. For yes/no questions the official evaluation measure is the macro-averaged F-Measure on questions with answers *yes* and *no*. For factoid questions, where the participants are allowed to return up to five answers, the Mean Reciprocal Rank (MRR) is used. For List questions, the official measure is the mean F-Measure. Finally, for ideal answers, the official evaluation is based on manual scores assigned by experts estimating the readability, recall, precision and repetition of each response provided by the participating systems.

2.3 Task MESINESP8: Medical Semantic Indexing in Spanish

Currently, most of the Biomedical NLP and IR research is being done on English documents, and only a few tasks have been carried out on non-English texts. Nonetheless, it is important to note that there is also a considerable amount of medically relevant content published in other languages than English and particularly clinical texts that are entirely written in the native language of each country, with a few exceptions. For example, there is a large subset of medical content published in Spanish each year. Resources like PubMed do only contain a fraction of the biomedical and medical literature originally published in Spanish, which is also stored in other resources such as IBECS[3], SCIELO[4] or LILACS[5].

In this task, the participants are asked to classify new IBECS and LILACS documents in Spanish. The classes come from the DeCS vocabulary[6] which was developed from the MeSH hierarchy. A dataset of more than 300,000 articles in Spanish with DeCS annotations retrieved from the Virtual Health Library (VHL) Portal[7] will be available for the participants to train their systems.

---

[3] http://ibecs.isciii.es/.
[4] https://scielo.org/en/.
[5] http://lilacs.bvsalud.org/en/.
[6] http://decs.bvs.br/I/decsweb2019.htm.
[7] https://bvsalud.org/en/.

Evaluation in Task MESINESP8: A set of about 1,500 articles in Spanish will be annotated by two human expert annotators with golden DeCS labels. Some of them will be provided to the participants as development dataset and some of them will be used as a testset to evaluate the performance of the systems. The responses of the systems in this task will be evaluated with the same variety of flat evaluation measures used for task 8a [2], with the micro-averaged F-measure (MiF) as the official one.

2.4   BioASQ Datasets and Tools

A major contribution of BioASQ is the development and maintenance of benchmark datasets for biomedical semantic indexing and question answering. The dataset for the semantic indexing task includes more than 14 millions articles from PubMed. This year, given the new MESINESP task, a new dataset of more than 300 thousands Spanish semantically indexed articles has been created. Furthermore, a set of 3,243 realistic questions and answers have been generated, constituting a unique resource for the development of question answering systems.

In addition, BioASQ has created a lively ecosystem, supported by tools and systems that facilitate research, such as the BioASQ Annotation Tool [6] for dataset development on question answering and a range of evaluation measures for automated assessment of system performance in all tasks. All software and data that are produced are open to the public[8]. It is worth mentioning, that this year we plan to create a repository, through which, several participating systems will also be available. This will allow new participants and teams, to build on existing models.

3   The Impact of BioASQ Results

BioASQ has reportedly had a very large impact, both in research and in industry; it has vastly helped advance the field of text mining in bioinformatics and has enabled researchers and practitioners to create novel computational models for life and health sciences. By bringing people together who work on the same benchmark data, BioASQ significantly facilitates the exchange and fusion of ideas and eventually accelerates progress in the field.

For example, the Medical Text Indexer (MTI) [5], which is developed by the NLM to assist in the indexing of biomedical literature, has improved its performance by almost 10% in the last 7 years (Fig. 2). NLM has announced that improvement in MTI is largely due to the adoption of ideas from the systems that compete in the BioASQ challenge [4]. Recently, MTI has reached a performance level that allows it to be used in the fully automated indexing of articles of specific types [1]. In general, a variety of biomedical semantic indexing and question answering systems has been developed based on the BioASQ datasets which are continuously maintained and extended. In addition, BioASQ keeps evolving considering the inclusion of new tasks and the development of new datasets.

---

[8] https://github.com/bioasq.

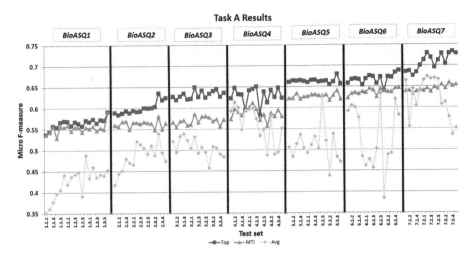

Fig. 2. Performance of the participating systems in task 8a, on semantic indexing. Each year, the participating systems push the state-of-the-art to higher levels

Acknowledgments. Google was a proud sponsor of the BioASQ Challenge in 2019. The eighth edition of BioASQ is also sponsored by the Atypon Systems inc. BioASQ is grateful to NLM for providing the baselines for task 8a and to the CMU team for providing the baselines for task 8b. The MESINESP task is sponsored by the Spanish Plan for advancement of Language Technologies (Plan TL) and the Secretaría de Estado para el Avance Digital (SEAD). BioASQ is also grateful to LILACS, SCIELO and Biblioteca virtual en salud and Instituto de salud Carlos III for providing data for the BioASQ MESINESP task.

References

1. Incorporating values for indexing method in medline/pubmed xml. https://www.nlm.nih.gov/pubs/techbull/ja18/ja18_indexing_method.html. Accessed 1 Oct 2019
2. Kosmopoulos, A., Partalas, I., Gaussier, E., Paliouras, G., Androutsopoulos, I.: Evaluation measures for hierarchical classification: a unified view and novel approaches. Data Mining Knowl. Discov. **29**(3), 820–865 (2014). https://doi.org/10.1007/s10618-014-0382-x
3. Malakasiotis, P., Pavlopoulos, I., Androutsopoulos, I., Nentidis, A.: Evaluation measures for task b. Technical report, Technical report BioASQ (2018). http://participants-area.bioasq.org/Tasks/b/eval_meas_2018
4. Mork, J., Aronson, A., Demner-Fushman, D.: 12 years on–is the nlm medical text indexer still useful and relevant? J. Biomed. Semant. **8**(1), 8 (2017)
5. Mork, J., Jimeno-Yepes, A., Aronson, A.: The NLM medical text indexer system for indexing biomedical literature (2013)

6. Ngomo, A.C.N., Heino, N., Speck, R., Ermilov, T., Tsatsaronis, G.: Annotation tool. Project deliverable D3.3 (2013). http://www.bioasq.org/sites/default/files/PublicDocuments/2013-D3.3-AnnotationTool.pdf
7. Tsatsaronis, G., et al.: An overview of the BioASQ large-scale biomedical semantic indexing and question answering competition. BMC Bioinformatics **16**, 138 (2015). https://doi.org/10.1186/s12859-015-0564-6

eRisk 2020: Self-harm and Depression Challenges

David E. Losada[1]([✉]) [ID], Fabio Crestani[2] [ID], and Javier Parapar[3] [ID]

[1] Centro Singular de Investigación en Tecnoloxías Intelixentes (CiTIUS),
Universidade de Santiago de Compostela, Santiago de Compostela, Spain
`david.losada@usc.es`
[2] Faculty of Informatics, Universitá della Svizzera italiana (USI),
Lugano, Switzerland
`fabio.crestani@usi.ch`
[3] Information Retrieval Lab, Centro de Investigación en Tecnoloxías da Información
e as Comunicacións (CITIC), Universidade da Coruña, A Coruña, Spain
`javierparapar@udc.es`

Abstract. This paper describes eRisk, the CLEF lab on early risk prediction on the Internet. eRisk started in 2017 as an attempt to set the experimental foundations of early risk detection. Over the last three editions of eRisk (2017, 2018 and 2019), the lab organized a number of early risk detection challenges oriented to the problems of detecting depression, anorexia and self-harm. We review in this paper the main lessons learned from the past and we discuss our future plans for the 2020 edition.

1 Introduction

eRisk is a CLEF lab whose main goal is to explore issues of evaluation methodology, performance metrics and other challenges related to building testbeds for early risk detection [4–6]. The predictive tools developed under eRisk's shared tasks could be potentially useful in different areas, particularly those related to health and safety. For example, warning alerts can be sent when an individual starts broadcasting suicidal thoughts on Social Media. eRisk tries to instigate interdisciplinary research (e.g. related to information retrieval, machine learning, psychology, and computational linguistics) and the advances developed under this challenge would be potentially applicable to support a number of socially important problems.

The lab casts early risk prediction as a process of *sequential accumulation of evidence*. In other words, given a stream of data (e.g. real-time Social Media entries), alerts should be fired when there is enough evidence about a certain type of risk. The participants have access to a stream of social media entries and they have to balance between making *early* alerts (e.g., based on few entries or posts) or *not-so-early* (late) alerts (e.g., if participants opt to see a wider range of entries and only emit alerts after thoroughly analyzing the available pieces of evidence). The testset building methodology and the evaluation strategies proposed

© Springer Nature Switzerland AG 2020
J. M. Jose et al. (Eds.): ECIR 2020, LNCS 12036, pp. 557–563, 2020.
https://doi.org/10.1007/978-3-030-45442-5_72

under eRisk are general and, thus, potentially applicable to multiple application domains (for example, health, security, or cybergrooming). However, all previous eRisks have focused on tasks and data related to psychological disorders.

2   Previous Editions of eRisk

eRisk 2017 included an exploratory task on early detection of signs of depression. This shared task was defined using the test collection and evaluation metrics proposed in [3]. The interactions between depression and natural language use is an intriguing problem and the eRisk participants approached the challenge from multiple perspectives.

The 2017 task was demanding both for the participants and the organisers because it had ten releases of data, and, after each release, the teams had one week to submit their predictions. Furthermore, eRisk was new to all participants and the research groups were not familiar with this novel evaluation metrics. As a result, only eight teams (out of 30 registered participants) were able to follow the tight schedule, submitting thirty different system variants (or runs).

In 2018, two shared tasks were organized: task 1, on early detection of signs of depression (which was a continuation of the 2017's pilot task), and task 2 on early detection of signs of anorexia. The two tasks had the same overall organization and evaluation method of the previous year. eRisk 2018 had 11 active participants (out of 41 registered teams) who submitted 45 and 35 system variants (for Task 1 and Task 2, respectively).

In 2019, three shared tasks were organized. Two of them were oriented to the same early detection technologies (task 1, early detection of signs of anorexia; task 2, early detection of signs of self-harm), while a new task (task 3) was introduced oriented to automatically filling a depression questionnaire based on user interactions in social media. eRisk 2019 had 14 active participants (out of 62 registered teams) who submitted 54, 33 and 33 system variants (respectively, for the 3 tasks). The increasing numbers of participants and runs submitted suggest that eRisk is slowly becoming an experimental reference for early risk research.

2.1   Early Risk Prediction Tasks

Previous eRisk's early prediction tasks consisted of sequentially processing writings –posts or comments– published by social media users and learn to detect signs of risk as soon as possible. The participating systems had to process the writings in chronological order (oldest writings are given first to the participants). In this way, algorithms that effectively perform this shared task could be applied to monitor interactions in blogs, social networks, or other types of Social Media. Table 1 reports the main statistics of the collections utilized in the early prediction tasks of eRisk 2017–2019.

Reddit was the source of data for these shared tasks. It is a social media platform where users (*redditors*) post and vote submissions which are organized by communities of interests (*subreddits*). Reddit has a large set of users and many of them have a large thread of submissions (covering several years). Reddit has active subreddits about psychological disorders, such as depression or eating disorders. Reddit's terms and conditions permit to use its contents for research purposes[1].

The test collections used in the eRisk early detection tasks have the same format as the collection described in [3]. It is a collection of writings (posts or comments) published by redditors. For each task, there are two classes of redditors: the positive class (e.g., depression or anorexia) and the negative class (control group). The positive class was obtained following the extraction method proposed by Coppersmith and colleagues [2] (an automatic approach to identify people diagnosed with depression in Twitter). We adapted this extraction approach to Reddit as follows. Self-expressions related to medical diagnoses (e.g. "Today, I was diagnosed with depression") can be obtained by running specific phrases against the platform search tool. Next, we manually reviewed the retrieved results to verify that they were really genuine. Our confidence on the reliability of these labels is high. There are many subreddits oriented to people suffering from psychological disorders and, usually, many redditors are active on these subreddits. These users tend to be very explicit about their problems and medical condition. This extraction approach is semi-automatic (requires manual revision of the retrieved posts), but it is an effective way to extract a group of people that are diagnosed with a given disorder. The manual reviews were thorough and strict. Expressions such as "I have anorexia", "I think I have anorexia", or "I am anorexic" were not considered as explicit expressions of a diagnosis. We only included a user into the positive set when there was a mention of a diagnosis that was clear and explicit (e.g., "Last month, I was diagnosed with anorexia nervosa", "After struggling with anorexia for a long time, last week I was diagnosed"). For each redditor, the test collection contains his sequence of writings (in chronological order) and each task was organized into a **Training stage**, where the participants had access to training data (we released the full history of writings published by a set of training redditors), and a **Test stage**. In eRisk 2017 and eRisk 2018, the test stage was organized in a 10-week format as follows. The sequence of writings published by each user was split into 10 chunks (the first chunk has the oldest 10% of the user's writings, the second chunk has the second oldest 10%, and so forth). The test stage had 10 releases of data (one release per week). The first week we gave the first chunk of data to the participants, the second week we gave the second chunk of data, and so forth. After each release, the participants had to process the data and, before the next release, each participant had to choose between: (a) emitting a decision on the redditor (positive or negative), or (b) making no decision (i.e. waiting to see more chunks). This choice had to be made for each redditor in the test set.

---

[1] Reddit privacy policy states explicitly that the submitted posts and comments are not private and will still be accessible after the user's account is deleted. Reddit does not permit unauthorized commercial use of its contents or redistribution, except as permitted by the doctrine of fair use. These research activities are an example of fair use.

If the participant emitted a decision then the decision was considered as final. The systems were evaluated based on the accuracy of the decisions and the number of chunks required to take the decisions (see below). In 2019, we moved from this "chunk-based" release of test data to a "item-by-item" release of test data. We set up a REST server that iteratively provided the user's writings to the participants[2]. In this way, each participant could stop and make an alert at any point of the user's chronology (the server waited for the responses of the participants and only gave new user data after receiving the participants' input).

Evaluation Metrics for Early Risk Detection. The evaluation of these tasks considered standard classification measures, such as F1, Precision and Recall, computed with respect to the positive group. These standard classification measures evaluate the participants' estimations with respect to golden truth labels. eRisk included them in the evaluation reports because these measures are well-known and interpretable. However, these three measures are time-unaware and, thus, do not penalize late alerts. In order to reward early detection algorithms, we introduced in [3] a new measure called ERDE (Early Risk Detection Error). ERDE takes into account the correctness of the (binary) decision and the delay, which is measured by counting the number (k) of writings seen before making the decision.

In eRisk 2019 the set of evaluation metrics was extended. We complemented the evaluation report with additional decision-based metrics that try to capture additional aspects of the problem. We adopted $F_{latency}$, an alternative evaluation metric for early risk prediction that was proposed by Sadeque and colleagues [7]. Another novelty introduced in 2019 was that user's data was processed by the participants in a post by post basis (as opposed to the old chunk-based approach). Besides decision-based evaluation metrics, eRisk 2019 incorporated a ranking-based approach to evaluate the participants. This form of evaluation was based on rankings of users by decreasing estimated risk. These rankings were produced after each round of writings and were evaluated with standard information retrieval measures, such as P@10 or NDCG. A full description of this ranking-based evaluation approach can be found in the eRisk 2019's overview report [6].

2.2   Depression Level Estimation Task

Introduced in 2019, the task consisted of estimating the level of depression from a thread of user submissions. For each user, the participants were given a full history of writings (in a single release of data) and the participants had to fill a standard depression questionnaire based on the evidence found in the history of postings. The questionnaire is derived from the Beck's Depression Inventory (BDI) [1], which assesses the presence of feelings like sadness, pessimism, loss of

---

[2] More information about the server and the modality of release of the date can be found at the eRisk's website on http://early.irlab.org/server.html.

Table 1. Main statistics of the train and test collections used in the early prediction tasks of eRisk 2017–2019.

	Training stage		Test stage	
	eRisk 2017 - Depression task			
	Depressed	*Control*	*Depressed*	*Control*
Num. subjects	83	403	52	349
Num. submissions (posts & comments)	30,851	264,172	18,706	217,665
Avg num. of submissions per subject	371.7	655.5	359.7	623.7
Avg num. of days from first to last submission	572.7	626.6	608.31	623.2
Avg num. words per submission	27.6	21.3	26.9	22.5
	eRisk 2018 - Depression task			
	Depressed	*Control*	*Depressed*	*Control*
Num. subjects	135	752	79	741
Num. submissions (posts & comments)	49,557	481,837	40,665	504,523
Avg num. of submissions per subject	367.1	640.7	514.7	680.9
Avg num. of days from first to last submission	586.43	625.0	786.9	702.5
Avg num. words per submission	27.4	21.8	27.6	23.7
	eRisk 2018 - Anorexia task			
	Anorexia	*Control*	*Anorexia*	*Control*
Num. subjects	20	132	41	279
Num. submissions (posts & comments)	7,452	77,514	17,422	151,364
Avg num. of submissions per subject	372.6	587.2	424.9	542.5
Avg num. of days from first to last submission	803.3	641.5	798.9	670.6
Avg num. words per submission	41.2	20.9	35.7	20.9
	eRisk 2019 - Anorexia task			
	Anorexia	*Control*	*Anorexia*	*Control*
Num. subjects	61	411	73	742
Num. submissions (posts & comments)	24,874	228,878	17,619	552,890
Avg num. of submissions per subject	407.8	556.9	241.4	745.1
Avg num. of days from first to last submission	≈800	≈650	≈510	≈930
Avg num. words per submission	37.3	20.9	37.2	21.7
	eRisk 2019 - Self-harm task			
	Self-harm	*Control*	*Self-harm*	*Control*
Num. subjects	–	–	41	299
Num. submissions (posts & comments)	–	–	6,927	163,506
Avg num. of submissions per subject	–	–	169.0	546.8
Avg num. of days from first to last submission	–	–	≈495	≈500
Avg num. words per submission	–	–	24.8	18.8

energy, etc. for the detection of depression. The questionnaire contains 21 questions and each question has a set of at least four possible responses, ranging in intensity. For example, the question on sadness has these four possible responses: (0) I do not feel sad, (1) I feel sad, (2) I am sad all the time and I can't snap out of it, and (3) I am so sad or unhappy that I can't stand it.

The task aimed at exploring the viability of automatically estimating the severity of the multiple symptoms associated with depression. Given the user's history of writings, the algorithms had to estimate the user's response to each individual question. We collected questionnaires filled by social media users together with their history of writings (we extracted each history of writings right after the user provided us with the filled questionnaire). The questionnaires filled by the users (ground truth) were used to assess the quality of the responses provided by the participants.

Four evaluation measures were introduced to evaluate the participants' estimations. The Average Hit Rate (AHR) computes the ratio of cases where the *automatic questionnaire* has exactly the same answer as the real questionnaire. The Average Closeness Rate (ACR) is a less stringent measure that considers the distance between each real answer and the answer submitted by the participating team. The two other measures, ADODL and DHCR, were oriented to compute how effective the systems are at estimating the overall depression level of the individual. These two measures compute the deviation between the total depression score (sum of all responses in the questionnaire) of the real questionnaire vs the questionnaire submitted by the participants.

2.3   Results

A full description and analysis of the results can be found in the lab overviews [4–6] and working note proceedings. For the early risk prediction tasks, most of the participating teams focused on classification aspects (i.e. how to learn effective classifiers from the training data) and no much attention was paid to the tradeoff between accuracy and delay. For the depression level estimation task, the results show that an automatic analysis of the user's writings is useful at extracting some signals or symptoms related to depression (e.g., some participant had a hit rate of 40%).

Although the effectiveness of the proposed solutions is still modest, the experiments performed under these shared tasks suggest that evidence extracted from social media is valuable. Automatic or semi-automatic screening tools are indeed promising to detect at-risk individuals. This result encouraged us to continue with the lab in 2020 and further explore the creation of new benchmarks for text-based screening of signs of such risks.

Another important outcome of the previous eRisk labs is related to the evaluation methodology. How to define appropriate metrics for early risk prediction is a challenge by itself and eRisk labs have already instigated the development of new early prediction metrics [7,8].

3   Conclusions and Future Work

eRisk will continue at CLEF 2020. Our plan is to organize two shared tasks. The first task will be a continuation of 2019's eRisk task on early detection of signs of self-harm. The second task will be a continuation of 2019's task on depression level estimation. In 2019, these two tasks were really challenging and the participants had no training data. In 2020, we will use the eRisk 2019 data as training data, and new test cases will be collected and included into the 2020 test split. By running these two tasks again we expect to further gain insight into the main factors and issues related to extracting signs of self-harm and depression from Social Media entries.

Acknowledgements. We thank the support obtained from the Swiss National Science Foundation (SNSF) under the project "Early risk prediction on the Internet: an evaluation corpus", 2015. We also thank the financial support obtained from the (i) "Ministerio de Ciencia, Innovación y Universidades" of the Government of Spain (research grants RTI2018-093336-B-C21 and RTI2018-093336-B-C22), (ii) "Consellería de Educación, Universidade e Formación Profesional", Xunta de Galicia (grants ED431C 2018/29, ED431G/08 and ED431G/01 – "Centro singular de investigación de Galicia" –). All grants were co-funded by the European Regional Development Fund (ERDF/FEDER program).

References

1. Beck, A.T., Ward, C.H., Mendelson, M., Mock, J., Erbaugh, J.: An inventory for measuring depression. JAMA Psychiatry **4**(6), 561–571 (1961)
2. Coppersmith, G., Dredze, M., Harman, C.: Quantifying mental health signals in Twitter. In: ACL Workshop on Computational Linguistics and Clinical Psychology (2014)
3. Losada, D.E., Crestani, F.: A test collection for research on depression and language use. In: Fuhr, N., et al. (eds.) CLEF 2016. LNCS, vol. 9822, pp. 28–39. Springer, Cham (2016). https://doi.org/10.1007/978-3-319-44564-9_3
4. Losada, D.E., Crestani, F., Parapar, J.: eRISK 2017: CLEF lab on early risk prediction on the internet: experimental foundations. In: Jones, G.J.F., et al. (eds.) CLEF 2017. LNCS, vol. 10456, pp. 346–360. Springer, Cham (2017). https://doi.org/10.1007/978-3-319-65813-1_30
5. Losada, D.E., Crestani, F., Parapar, J.: Overview of eRisk: early risk prediction on the internet. In: Bellot, P., et al. (eds.) CLEF 2018. LNCS, vol. 11018, pp. 343–361. Springer, Cham (2018). https://doi.org/10.1007/978-3-319-98932-7_30
6. Losada, D.E., Crestani, F., Parapar, J.: Overview of eRisk 2019 early risk prediction on the internet. In: Crestani, F., et al. (eds.) CLEF 2019. LNCS, vol. 11696, pp. 340–357. Springer, Cham (2019). https://doi.org/10.1007/978-3-030-28577-7_27
7. Sadeque, F., Xu, D., Bethard, S.: Measuring the latency of depression detection in social media. In: Proceedings of the Eleventh ACM International Conference on Web Search and Data Mining, WSDM 2018, pp. 495–503. ACM, New York (2018)
8. Trotzek, M., Koitka, S., Friedrich, C.: Utilizing neural networks and linguistic metadata for early detection of depression indications in text sequences. IEEE Trans. Knowl. Data Eng. **32**(3), 588–601 (2018)

Finding Old Answers to New Math Questions: The ARQMath Lab at CLEF 2020

Behrooz Mansouri[1(✉)], Anurag Agarwal[1], Douglas Oard[2], and Richard Zanibbi[1]

[1] Rochester Institute of Technology, Rochester, NY, USA
{bm3302,axasma,rxzvcs}@rit.edu
[2] University of Maryland, College Park, MD, USA
oard@umd.edu

Abstract. The ARQMath Lab at CLEF 2020 considers the problem of finding answers to *new* mathematical questions among posted answers on a community question answering site (Math Stack Exchange). Queries are question postings held out from the test collection, each containing both text and at least one formula. We expect this to be a challenging task, as both math and text may be needed to find relevant answer posts. While several models have been proposed for text question answering, math question answering is in an earlier stage of development. To advance math-aware search and mathematical question answering systems, we will create a standard test collection for researchers to use for benchmarking. ARQMath will also include a formula retrieval sub-task: individual formulas from question posts are used to locate formulas in earlier answer posts, with relevance determined by narrative fields created based on the original question. We will use these narrative fields to explore diverse information needs for formula search (e.g., alternative notation, applications in specific fields or definition).

Keywords: Community question answering · Formula retrieval · Mathematical Information Retrieval · Math-aware search

1 Introduction

In a recent study, Mansouri et al. found that 20% of mathematical queries in a general-purpose search engine were expressed as well-formed questions, a rate ten times higher than that for all queries submitted [7]. Results such as these and the presence of Community Question Answering sites such as Math Stack Exchange[1] (MSE) and Math Overflow [11] suggest that there is a great public interest in finding answers to mathematical questions posed in natural language, using *both* text and mathematical notation. Related to this, there has also been increasing

---

[1] https://math.stackexchange.com.

J. M. Jose et al. (Eds.): ECIR 2020, LNCS 12036, pp. 564–571, 2020.
https://doi.org/10.1007/978-3-030-45442-5_73

work on math retrieval and math question answering in both the Information Retrieval (IR) and Natural Language Processing (NLP) communities.

In light of this growing interest, we are organizing a new lab at the Conference and Labs of the Evaluation Forum (CLEF) on Answer Retrieval for Questions about Math (ARQMath).[2] Using the mathematics and free text in posts from Math Stack Exchange, participating systems will be given a question, and asked to return a ranked list of potential answers. Relevance will be determined by how well the returned posts answer the provided question. Through this task we will explore leveraging math notation together with text to improve the quality of retrieval results. This is one case of what we generically call math-aware information retrieval, in which the focus is on leveraging the ability to process mathematical notation to enhance, rather than to replace, other information retrieval techniques. We will also include a query-by-example task on formula retrieval in which relevance will be determined by the degree to which a retrieved formula is useful for the searcher's intended purpose.

Question answering (QA) was among the earliest target applications for Artificial Intelligence. Techniques for answering one specific type of mathematical question, automated theorem proving, date back to 1956 when Newell and Simon [9] introduced the *logic theorist* that proved theorems in symbolic logic. Three years later, they introduced the General Problem Solver [10], which attempted to mimic students' behavior in discovering proofs. For math QA, an important recent development was the work of Ling et al. [6], who solved algebraic word problems by generating answer rationales and human-readable mathematical expressions that derive a final answer along with a description of the method used to solve the problem. Kushman et al. [5] presented an approach for automatically learning to solve algebra word problems expressed in text *and* math by defining a joint log-linear distribution over full systems of equations and aligning their variables and numbers to the problem text.

More recently, machine learning has been applied to answering a broader range of questions. For example, the *Arosti* system of Clark et al. [3] achieved a score of over 90% on the (non-diagram) multiple choice portion of the New York Regents 8th Grade Science Exam, the first such system to pass this test. Natural Language Processing has, in recent years, focused on Reading Comprehension QA tasks requiring an answer to be located in a single document. One recent shared task that involved processing mathematical notation was Task 10 at SemEval 2019 [4], which provided a question set derived from the Math-SAT (Scholastic Achievement Test) practice exams that included 2,778 training questions and 1,082 test questions from three major categories: Closed Algebra, Open Algebra and Geometry. A majority of the questions were multiple choice, with only a minority having a numeric answer.

Within the IR community, much of the recent work on QA has focused on Community Question Answering (CQA), with the goal of augmenting human talent by finding earlier answers that people have already given that can serve as answers to newly asked questions. One important line of work enabled by CQA

---

[2] https://www.cs.rit.edu/~dprl/ARQMath.

is that it becomes possible not just to search directly for potential answers, but also to search for prior questions that could lead to potential answers. Because CQA systems include social media features such as voting for answer quality, some types of non-text features can also be leveraged.

Math-Aware Information Retrieval. The existing evaluation resources for math-aware information retrieval were initially developed over a five-year period at the National Institute of Informatics (NII) Testbeds and Community for Information Access Research (at NTCIR-10 [1], NTCIR-11 [2] and NTCIR-12 [12]). The NTCIR Mathematical Information Retrieval (MathIR) tasks developed evaluation methods and allowed participating teams to establish baselines for both "text + math" queries and isolated formula retrieval. NTCIR-12 ultimately made use of two collections, one a set of arXiv papers from physics that is split into paragraph-sized documents, and the other a set of articles from English Wikipedia. Interest in these tasks is global; at the NTCIR-12 MathIR task, for example, there were participating groups from around the world, including Europe (Czech Republic, Germany), Asia (China, India, Japan) and North America (Canada, USA). The NTCIR-12 isolated formula retrieval test collection was also later used by participants for the 2016 Competition on Recognition of Online Handwritten Mathematical Expressions (CROHME) [8] at the International Conference on Frontiers in Handwriting Recognition (ICFHR).

ARQMath Goals. The ARQMath lab will provide an opportunity to push mathematical question answering in a new direction, where informal language is frequently used, and where answers provided by a community are selected and

Table 1. Example queries and results for question answering and formula retrieval.

QUESTION ANSWERING	FORMULA RETRIEVAL
QUESTION	QUERY
I've spent the better part of this day trying to show from first principles that this sequence tends to 1. Could anyone give me an idea of how I can approach this problem? $$\lim_{n \to +\infty} n^{\frac{1}{n}}$$	$\lim_{n \to +\infty} n^{\frac{1}{n}}$
RELEVANT	RELEVANT
You can use AM $\geq$ GM. $$\frac{1 + 1 + \cdots + 1 + \sqrt{n} + \sqrt{n}}{n} \geq n^{1/n} \geq 1$$ $$1 - \frac{2}{n} + \frac{2}{\sqrt{n}} \geq n^{1/n} \geq 1$$	$\lim_{n \to \infty} \sqrt[n]{n}$
NON-RELEVANT	NON-RELEVANT
If you just want to show it converges, then the partial sums are increasing but the whole series is bounded above by $$1 + \int_1^\infty \frac{1}{x^2} dx = 2$$	$\sum_{k=1}^\infty \frac{1}{k^2} = \frac{\pi^2}{6}$

ranked rather than generated. One goal is to produce test collections; a second is to drive innovation in evaluation methods; and a third is to drive innovation in the development of math-aware information retrieval systems.

2   Overview of Tasks

There will be two tasks in the first year: (1) a question answering task (QA), where systems are provided a question post from MSE and then return a ranked list of answer posts and (2) an isolated formula retrieval task. Table 1 illustrates these two tasks, and we provide details about each task below.

Finding Answers to Math Questions (Main Task). For the QA task, at least 50 questions from MSE will be sampled, with the requirement that each question contains both text and at least one formula. Participants will have the option to run queries using only the text or math portions in each question, or to use both math and text. We will ask participants to label each run with which of these conditions they chose. One challenge inherent in this design is that the expressive power of text and formulas are sometimes complementary; so although all topics will include both text and formula(s), some may be better suited to text-based or math-based retrieval. We plan to accommodate this by reporting results for all participants that are averaged over three topic sets: (1) all topics, (2) topics for which the assessor believes the text alone to be an adequate characterization of the topic, and (3) topics for which the assessor believes the formula(s) alone to be an adequate characterization of the topic.

Formula Search (Secondary Task). In this task individual formulas are used as queries, and systems return a ranked list of similar and/or related formulas. As with the NTCIR-12 Wikipedia Formula Browsing Task, this task has the goal of fostering development of component technology for computing math similarity. We envision two improvements over what was done at NTCIR: further developing the concept of "formula relevance" and creating a collection with a larger number of formula queries (NTCIR-12 has only 20 formula queries + 20 modified versions of the same formulas with wildcards added).

Each formula query will be a single formula extracted from a question used in the main task. For each query, annotators will write a short human-readable narrative field – not available to participating systems – that reflects their understanding of the type(s) of similarity the person who asked the original question would have found useful. This may include alternative notation, simplification, specialization, or applications in specific fields, and we expect to extend those categories further based on suggestions from participating teams. Because participating systems won't have access to this narrative field in for their "standard condition" run, we expect this task to support research on diversity ranking for formula retrieval. We are also aiming to have at least 50 formula queries in the first year, with the intent to expand both query sets in subsequent years.

3   The Math Stack Exchange Collection

Our collection will be comprised of question and answer postings from Math Stack Exchange (MSE). These postings are freely available as data dumps from the Internet Archive. At the time of this writing, there are 1.1 million questions.

Figure 1 (left) shows the distribution of the number of formulas per question post. For query development, only the 45% of questions containing at least one formula,[3] will be considered. As Fig. 1 (right) shows, question production on Math Stack Exchange has been fairly steady in recent years. We plan to release an annotated version of the complete Math Stack Exchange data dump containing questions and answers produced through December 2018 as the test collection, holding out questions submitted in 2019 for query development.

We plan to stratify the questions by predicted difficulty (in the opinion of the collection developers) so as to avoid, for example, having too many common questions that are nearly identical to questions from 2018 or earlier. Each question will include a unique ID, the asker-entered title for their question, the asker-entered body of the question, three formats for formulas found in the body of the question (LaTeX, Presentation MathML, and Content Math ML; see below), a list containing any edits to the question that were subsequently done by the asker, one or more asker-entered type tags (e.g., "calculus"), and comments on the question entered by other users, the average score assigned by other MSE users to the question (which one might interpret as a measure of how "good" a question it is), and the asker's reputation score (which is estimated from scores given to their prior questions). Only the first four of these (ID, title, body, formulas) will be used for the standard condition, but the additional features will be available for use in contrastive runs.

We will use open source tools such as LaTeXML,[4] to label and convert LaTeX formulas from posts and convert them to XML markup, including both Presentation (appearance-based) and Content (semantic) MathML. We will perform this extraction centrally and distribute the extracted formulas as standoff annotations with references to the location of the formula in each XML question or answer post. Converting LaTeX to Presentation MathML is a straightforward transformation between formula appearance representations (i.e., symbols on writing lines). Producing Content MathML from LaTeX requires inference, and is thus potentially errorful. However, Content MathML supports a higher level of abstraction by representing operator structure explicitly. Centralizing this conversion will remove one possible source of variation, but conversion scripts will also be made available to participants who wish to experiment with extended conversion capabilities.

---

[3] In Math Stack Exchange formulas almost invariably appear between two '$' signs in LaTeX notation (e.g., $a+b=c$).

[4] https://dlmf.nist.gov/LaTeXML/.

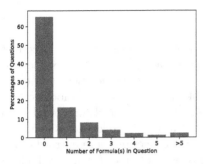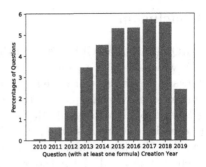

Fig. 1. Formulas in math stack exchange question postings. **Left:** formula counts for questions. **Right:** creation years for questions containing at least one formula.

4   Relevance Judgments

For both the QA and formula retrieval tasks, manual and automatic runs will be allowed. For each topic, the top-N (e.g., top-20) results from each participant run, along with additional manual runs conducted by the organizers, will be pooled. We will trade off pool depth and number of topics based on the available annotation resources.

Because specialized mathematical knowledge may be needed for assessment, the pooled documents will be assessed for relevance by volunteers from participating teams, augmented by assessors hired by the organizers. Evaluation will be performed using a web-based system (e.g., Sepia[5]). Assessors for the main task will be asked to identify relevant answers using pools from the main task. Assessors for the formula retrieval task will work with merged pools from both the formula retrieval task and (where appropriate for the question) from the main task to identify similar formulas. Most pools will be judged by a single assessor, but some will be dual-assessed to observe annotator agreement. For the formula retrieval task, queries will be selected for dual annotation using stratified random sampling so as to cover a broad range of similarity types. We also plan for some limited experimentation with alternative annotation strategies (e.g., additionally annotating the most useful parts of a relevant answer, or annotating the preference order between relevant answers) with the goal of informing evaluation design in future years.

We will use trec_eval to compute ranked document retrieval measures for each run for both tasks, with inferred Average Precision (infAP) as the standard measure for comparing systems. This choice of infAP is intended to provide results that can support future experimentation with the test collection by future systems that did not contribute to the judgment pools, but we will provide the full range of trec_eval measures to participating systems for use when different evaluation measures could provide additional insights.

---

[5] https://code.google.com/archive/p/sepia/.

5   Conclusion

The ARQMath lab at CLEF 2020 is the first in a three-year sequence of labs through which we aim to push the state of the art in evaluation design for math-aware IR, and in which we seek to support the development and ultimate deployment of new techniques for that task. We have chosen to focus on CQA, using Math Stack Exchange, both because that task models an actual employment scenario, and because the scale of the available collection is sufficient to also support the development of a second test collection more narrowly focused on formula retrieval. Math is, of course, only one example of structured notation, and we might reasonably hope to one day leverage similar ideas in other domains that also frequently use specialized notation, such as biology or chemistry.

Acknowledgements. This material is based upon work supported by the Alfred P. Sloan Foundation under Grant No. G-2017-9827 and the National Science Foundation (USA) under Grant No. IIS-1717997.

References

1. Aizawa, A., Kohlhase, M., Ounis, I.: NTCIR-10 math pilot task overview. In: NTCIR (2013)
2. Aizawa, A., Kohlhase, M., Ounis, I., Schubotz, M.: NTCIR-11 Math-2 task overview. In: NTCIR (2014)
3. Clark, P., et al.: From 'F' to 'A' on the NY regents science exams: An overview of the Aristo. arXiv preprint arXiv:1909.01958 (2019)
4. Hopkins, M., Le Bras, R., Petrescu-Prahova, C., Stanovsky, G., Hajishirzi, H., Koncel-Kedziorski, R.: SemEval-2019 Task 10: Math question answering. In: Proceedings of the 13th International Workshop on Semantic Evaluation (2019)
5. Kushman, N., Artzi, Y., Zettlemoyer, L., Barzilay, R.: Learning to automatically solve algebra word problems. In: Proceedings of the 52nd Annual Meeting of the Association for Computational Linguistics (2014)
6. Ling, W., Yogatama, D., Dyer, C., Blunsom, P.: Program induction by rationale generation: Learning to solve and explain algebraic word problems. In: Proceedings of the 55th Annual Meeting of the Association for Computational Linguistics (2017)
7. Mansouri, B., Zanibbi, R., Oard, D.W.: Characterizing searches for mathematical concepts. In: Proceedings of JCDL (2019)
8. Mouchère, H., Viard-Gaudin, C., Zanibbi, R., Garain, U.: ICFHR 2016 CROHME: competition on recognition of online handwritten mathematical expressions. In: Proceedings of ICFHR (2016)
9. Newell, A., Simon, H.: The logic theory machine-A complex information processing system. IRE Trans. Inf. Theor. **2**, 61–79 (1956)
10. Newell, A., Shaw, J.C., Simon, H.A.: Report on a general problem solving program. In: IFIP Congress (1959)

11. Tausczik, Y.R., Kittur, A., Kraut, R.E.: Collaborative problem solving: a study of MathOverflow. In: CSCW 2014, pp. 355–367 (2014)
12. Zanibbi, R., Aizawa, A., Kohlhase, M., Ounis, I., Topic, G., Davila, K.: NTCIR-12 MathIR task overview. In: NTCIR (2016)

ChEMU: Named Entity Recognition and Event Extraction of Chemical Reactions from Patents

Dat Quoc Nguyen[1,3], Zenan Zhai[1], Hiyori Yoshikawa[1,4], Biaoyan Fang[1],
Christian Druckenbrodt[2], Camilo Thorne[2], Ralph Hoessel[2],
Saber A. Akhondi[2], Trevor Cohn[1], Timothy Baldwin[1], and Karin Verspoor[1(✉)]

[1] The University of Melbourne, Melbourne, Australia
{zenan.zhai,hiyori.yoshikawa,biaoyanf,trevor.cohn,tbaldwin,
karin.verspoor}@unimelb.edu.au
[2] Elsevier, Amsterdam, The Netherlands
{c.druckenbrodt,c.thorne.1,r.hoessel,s.akhondi}@elsevier.com
[3] VinAI Research, Hanoi, Vietnam
v.datnq9@vinai.io
[4] Fujitsu Laboratories Ltd., Kanagawa, Japan

Abstract. We introduce a new evaluation lab named ChEMU (Chem-informatics Elsevier Melbourne University), part of the 11th Conference and Labs of the Evaluation Forum (CLEF-2020). ChEMU involves two key information extraction tasks over chemical reactions from patents. Task 1—Named entity recognition—involves identifying chemical compounds as well as their types in context, i.e., to assign the label of a chemical compound according to the role which the compound plays within a chemical reaction. Task 2—Event extraction over chemical reactions—involves event trigger detection and argument recognition. We briefly present the motivations and goals of the ChEMU tasks, as well as resources and evaluation methodology.

Keywords: Named entity recognition · Event extraction · Chemical reactions · Patents

1 Introduction

The chemical industry undoubtedly depends on the discovery of new chemical compounds. However, new chemical compounds are often initially disclosed in patent documents, and only a small fraction of these compounds are published in journals, usually taking an additional 1–3 years after the patent [13]. Therefore, most chemical compounds are only available through patent documents [3]. In addition, chemical patent documents contain unique information, such as reactions, experimental conditions, mode of action, which is essential for the understanding of compound prior art, providing a means for novelty checking and validation as well as pointers for chemical research in both academia and

© Springer Nature Switzerland AG 2020
J. M. Jose et al. (Eds.): ECIR 2020, LNCS 12036, pp. 572–579, 2020.
https://doi.org/10.1007/978-3-030-45442-5_74

industry [1,2]. As the number of new chemical patent applications has been drastically increasing [11], it is becoming crucial to develop natural language processing (NLP) approaches that enable automatic extraction of key information from the chemical patents [2].

In this paper, we propose a new evaluation lab (called ChEMU) focusing on information extraction over chemical reactions from patents. In particular, we will focus on two key information extraction tasks of *chemical named entity recognition* (NER) and *chemical reaction event extraction*. While previous related shared tasks focusing on chemicals or drugs such as CHEMDNER [7] have also included chemical named entity recognition as a task, those have primarily focused on PubMed abstracts. The CHEMDNER patents task [8] was limited to entity mentions and chemical entity passage detection, and only considered titles and abstracts of patents. For our ChEMU lab, we extend the existing corpora in several directions: first, we go beyond chemical NER to require labeling of the role of a chemical with respect to a reaction, and to consider complete chemical reactions in addition to entities. The ChEMU website is available at: http:// chemu.eng.unimelb.edu.au.

2   Goals and Importance

What are the Goals of This Evaluation Lab? Our goals are: (1) To develop tasks that impact chemical research in both academia and industry, (2) To provide the community with a new dataset of chemical entities, enriched with relational links between chemical event triggers and arguments, and (3) To advance the state-of-the-art in information extraction over chemical patents.

Why is This Lab Needed? For evaluating information extraction developments in the scientific literature domain, there have been a large number of labs/shared tasks offered within previous i2b2/n2c2, SemEval, BioNLP, BioCreative, TREC and CLEF workshops. However, less attention has been paid to the chemical patent domain. In particular, there has previously been only one shared task on this domain, which is the CHEMDNER patents task at the BioCreative V workshop, involving detection of mentions of chemical compounds and genes/proteins in patent text [8].

Information extraction approaches developed for the scientific literature domain may not apply directly to the chemical patent domain. This is because as legal documents, patents are written very differently as compared to scientific literature. When writing scientific papers, authors strive to make their words as clear and straightforward as possible, whereas patent authors often seek to protect their knowledge from being fully disclosed [15]. In tension with this is the need to claim broad scope for intellectual property reasons, and hence patents typically contain more details and are more exhaustive than scientific papers [9].

There are also a number of characteristics of patent texts that create challenges for NLP in this context. Long sentences listing names of compounds are frequently used in chemical patents. The structure of sentences in patent claims

Table 1. Brief definitions of ChEMU chemical entity types, organised into chemical entity types, a reaction label introduced in the text, and reaction properties.

Entity type	Definition
REACTION_PRODUCT	A product is a substance that is formed during a chemical reaction
STARTING_MATERIAL	A substance that is consumed in the course of a chemical reaction providing atoms to products is considered as starting material
REAGENT_CATALYST	A reagent is a compound added to a system to cause or help with a chemical reaction. Compounds like catalysts, bases to remove protons or acids to add protons must be also annotated with this tag
SOLVENT	A solvent is a chemical entity that dissolves a solute resulting in a solution
OTHER_COMPOUND	Other chemical compounds that are not the products, starting materials, reagents, catalysts and solvents
EXAMPLE_LABEL	A label associated with a reaction specification
TEMPERATURE	The temperature at which the reaction was carried out must be annotated with this tag
TIME	The reaction time of the reaction
YIELD_PERCENT	Yield given in percent values
YIELD_OTHER	Yields provided in other units than %

is usually complex, and syntactic parsing in patents can be difficult [4]. A quantitative analysis from [16] showed that the average sentence length in a patent corpus is much longer than in general language use. That work also showed that the lexicon used in patents usually includes domain-specific and novel terms that are difficult to understand.

How Will the Community Benefit from the Lab? The ChEMU lab will provide a new challenging set of tasks, in an area of significant pharmacological importance. The lab will focus attention on more complex analysis of chemical patents, provide strong baselines as well as providing a useful resource for future research.

What are Usage Scenarios? Automatically identifying compounds which serve as the starting material or are a product of a chemical reaction would allow more targeted extraction of chemical information from patents and can improve the usefulness of patent resources. Automatic extraction of chemical reaction events supports the construction of cheminformatics databases, capturing key information about chemicals and how they are produced, from the patent resources.

3   Tasks

The ChEMU lab at CLEF-2020[1] offers the two information extraction tasks of Named entity recognition (**Task 1**) and Event extraction (**Task 2**) over chemical reactions from patent documents. Teams may participate in one or both tasks.

Table 2. An example of a chemical reaction snippet and BRAT annotations in a standoff format [14] w.r.t. Task 1.

(step 1) Synthesis of 3-chloro-6-(trifluoromethyl)pyridazine

To 3-(trifluoromethyl)-1H-pyridazin-6-one (1.1 g, 6.7 mmol) was added phosphorus oxychloride (10 mL) and the mixture was stirred at 100°C for 2.5 hr, and concentrated under reduced pressure. To the obtained residue were added dichloromethane and water, and the mixture was stirred at room temperature for 5 min. After stirring, the mixture was alkalified with potassium carbonate and partitioned. The organic layer was washed with saturated brine, dried over sodium sulfate and the desiccant was filtered off. The solvent was evaporated and the obtained residue was purified by silica gel column chromatography (petroleum ether / ethyl acetate) to give the title compound (0.77 g , 4.2 mmol , 63%).

ID	Entity type	Offsets		Text span
T1	EXAMPLE_LABEL	6	7	1
T2	REACTION_PRODUCT	22	60	3-chloro-6-(trifluoromethyl)pyridazine
T3	STARTING_MATERIAL	64	102	3-(trifluoromethyl)-1H-pyridazin-6-one
T4	REAGENT_CATALYST	131	153	phosphorus oxychloride
T5	TEMPERATURE	193	203	100°C
T6	TIME	208	214	2.5 hr
T7	SOLVENT	292	307	dichloromethane
T8	SOLVENT	312	317	water
T9	TEMPERATURE	350	366	room temperature
T10	TIME	371	376	5 min
T11	OTHER_COMPOUND	426	445	potassium carbonate
T12	OTHER_COMPOUND	507	512	brine
T13	OTHER_COMPOUND	525	539	sodium sulfate
T14	OTHER_COMPOUND	678	693	petroleum ether
T15	OTHER_COMPOUND	694	707	ethyl acetate
T16	REACTION_PRODUCT	721	735	title compound
T17	YIELD_OTHER	737	743	0.77 g
T18	YIELD_OTHER	745	753	4.2 mmol
T19	YIELD_PERCENT	755	758	63%

---

[1] https://clef2020.clef-initiative.eu.

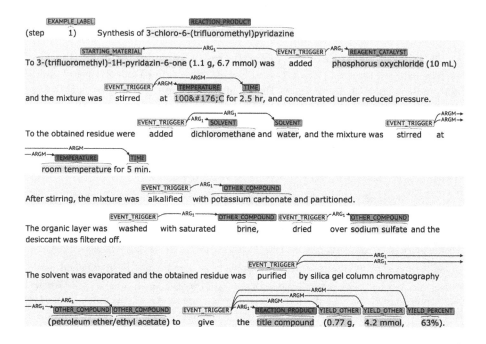

Fig. 1. BRAT visualization of chemical reaction events.

3.1   Task 1: Named Entity Recognition

In general, a chemical reaction is a process leading to the transformation of one set of chemical substances to another [10]. Task 1 involves identifying chemical compounds and their specific types, i.e. to assign the label of a chemical compound according to the role which it plays within a chemical reaction. In addition to chemical compounds, this task also requires identification of the temperatures and reaction times at which the chemical reaction is carried out, as well as yields obtained for the final chemical product and the label of the reaction.

This task involves both entity boundary prediction and entity label classification. We define 10 different entity type labels as shown in Table 1. See examples of those entity types in Table 2.

3.2   Task 2: Event Extraction

As illustrated in Figs. 1 and 2, a chemical reaction leading to an end product often consists of a sequence of individual event steps. Task 2 is to identify those steps which involve chemical entities recognized from Task 1. Unlike a conventional event extraction problem [6] which involves event trigger word detection, event typing and argument prediction, our Task 2 requires identification of event trigger words (e.g. "added" and "stirred") which all have the same type of

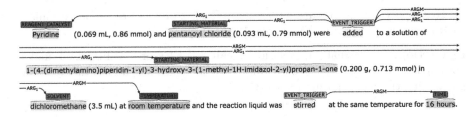

Fig. 2. BRAT visualization of a more complex event with the trigger word "added" involving five arguments.

"EVENT_TRIGGER", and then determination of the chemical entity arguments of these events.[2]

When predicting event arguments, we adapt semantic argument role labels **Arg1** and **ArgM** from the Proposition Bank [12] to label the relations between the trigger words and the chemical entities: **Arg1** is used to label the relation between an event trigger word and a chemical compound. Here, **Arg1** represents argument roles of being causally affected by another participant in the event [5]. **ArgM** represents adjunct roles with respect to an event, used to label the relation between a trigger word and a temperature, time or yield entity.

An end-to-end process incorporating both Task 1 and Task 2 can be equivalently viewed as a relation extraction task which identifies 11 entity types including 10 types defined in Table 1 plus "EVENT_TRIGGER", and extracts relations between the "EVENT_TRIGGER" entities and the remaining entities.

4   Data and Evaluation

Data: For system development and evaluation, a new corpus of 1500 chemical reaction snippets will be provided for both tasks (an example of a chemical reaction snippet is shown in Table 2). These snippets are sampled from 170 English document patents from the European Patent Office and the United States Patent and Trademark Office. We will mark up every chemical compound or event trigger with both text spans and IDs, and highlight relations and event arguments, as illustrated in Figs. 1 and 2. We have begun preparing the corpus and will make available strong baselines for the tasks. Initial publications related to the data and Task 1 appear at the 2019 ALTA and BioNLP workshops, respectively [18,19].

The corpus will be split into 70%/10%/20% training/development/test. Gold annotations for the training and development sets will be provided to task participants in the BRAT standoff format [14] during the development phase. The raw test set will be provided for final test phase.

---

[2] Note that those individual event steps are sequentially ordered, thus we do not consider cases where an event is an argument of another event, i.e. we do not label the relationship between two event triggers.

To support teams who are interested in Task 2 only, a pre-trained chemical NER tagger is provided as a resource [19].

Evaluation: For evaluation, precision, recall and F1 scores will be used, under both strict and relaxed span matching conditions. F1 will be the main metric for ranking the participating teams [17].[3]

5   Conclusion

In this paper, we have presented a brief description of the upcoming ChEMU lab at CLEF-2020. ChEMU will focus on two new tasks of named entity recognition and event extraction over chemical reactions from patents. We expect participants from both academia and industry. We will advertise our ChEMU lab via social media as well as NLP-related mailing lists.

Acknowledgments. This work is supported by an Australian Research Council Linkage Project, LP160101469, and Elsevier. We would like to thank Estrid He, Zubair Afzal and Mark Sheehan for supporting this work, as well as the anonymous reviewers for their feedback.

References

1. Akhondi, S.A., et al.: Annotated chemical patent corpus: a gold standard for text mining. PLoS ONE **9**, 1–8 (2014)
2. Akhondi, S.A., et al.: Automatic identification of relevant chemical compounds from patents. Database **2019**, baz001 (2019)
3. Bregonje, M.: Patents: a unique source for scientific technical information in chemistry related industry? World Pat. Inf. **27**(4), 309–315 (2005)
4. Hu, M., Cinciruk, D., Walsh, J.M.: Improving automated patent claim parsing: dataset, system, and experiments. CoRR abs/1605.01744 (2016)
5. Jurafsky, D., Martin, J.H.: Semantic Role Labeling and Argument Structure. In: Speech and Language Processing, 3rd edn. (2019)
6. Kim, J.D., Ohta, T., Pyysalo, S., Kano, Y., Tsujii, J.: Overview of BioNLP'09 shared task on event extraction. In: Proceedings of the BioNLP 2009 Workshop Companion Volume for Shared Task, pp. 1–9 (2009)
7. Krallinger, M., Leitner, F., Rabal, O., Vazquez, M., Oyarzabal, J., Valencia, A.: CHEMDNER: the drugs and chemical names extraction challenge. J. Cheminform. **7**(1), S1 (2015)
8. Krallinger, M., et al.: Overview of the CHEMDNER patents task. In: Proceedings of the Fifth BioCreative Challenge Evaluation Workshop, pp. 63–75 (2015)
9. Lupu, M., Mayer, K., Tait, J., Trippe, A.J.: Current Challenges in Patent Information Retrieval, 1st edn. Springer, Heidelberg (2011). https://doi.org/10.1007/978-3-642-19231-9
10. Muller, P.: Glossary of terms used in physical organic chemistry (IUPAC Recommendations 1994). Pure Appl. Chem. **66**(5), 1077–1184 (2009)

---

[3] https://bitbucket.org/nicta_biomed/brateval/src/master/.

11. Muresan, S., et al.: Making every SAR point count: the development of chemistry connect for the large-scale integration of structure and bioactivity data. Drug Discovery Today **16**(23), 1019–1030 (2011)
12. Palmer, M., Gildea, D., Kingsbury, P.: The proposition bank: an annotated corpus of semantic roles. Comput. Linguist. **31**(1), 71–106 (2005)
13. Senger, S., Bartek, L., Papadatos, G., Gaulton, A.: Managing expectations: assessment of chemistry databases generated by automated extraction of chemical structures from patents. J. Cheminformatics **7**, 49:1–49:12 (2015)
14. Stenetorp, P., Pyysalo, S., Topić, G., Ohta, T., Ananiadou, S., Tsujii, J.: brat: a web-based tool for NLP-assisted text annotation. In: Proceedings of the Demonstrations Session at EACL 2012 (2012)
15. Valentinuzzi, M.E.: Patents and scientific papers: quite different concepts: the reward is found in giving, not in keeping [Retrospectroscope]. IEEE Pulse **8**(1), 49–53 (2017)
16. Verberne, S., D'hondt, E., Oostdijk, N., Koster, C.: Quantifying the challenges in parsing patent claims. In: Proceedings of the 1st International Workshop on Advances in Patent Information Retrieval at ECIR 2010, pp. 14–21 (2010)
17. Verspoor, K., et al.: Annotating the biomedical literature for the human variome. Database **2013**, bat019 (2013)
18. Yoshikawa, H., et al.: Detecting chemical reactions in patents. In: Proceedings of the 17th Annual Workshop of the Australasian Language Technology Association, pp. 100–110 (2019)
19. Zhai, Z., et al.: Improving chemical named entity recognition in patents with contextualized word embeddings. In: Proceedings of the 18th BioNLP Workshop, pp. 328–338 (2019)

Living Labs for Academic Search
at CLEF 2020

Philipp Schaer[1]([✉])[iD], Johann Schaible[2][iD], and Bernd Müller[3]

[1] TH Köln - University of Applied Sciences, Cologne, Germany
philipp.schaer@th-koeln.de
[2] GESIS - Leibniz Institute for the Social Sciences, Cologne, Germany
johann.schaible@gesis.org
[3] ZB MED - Information Centre for Life Sciences, Cologne, Germany
muellerb@zbmed.de

Abstract. The need for innovation in the field of academic search and
IR, in general, is shown by the stagnating system performance in controlled evaluation campaigns, as demonstrated in TREC and CLEF
meta-evaluation studies, as well as user studies in real systems of scientific information and digital libraries. The question of what constitutes
relevance in academic search is multi-layered and a topic that drives
research communities for years. The Living Labs for Academic Search
(LiLAS) workshop has the goal to inspire the discussion on research and
evaluation of academic search systems by strengthening the concept of
living labs to the domain of academic search. We want to bring together
IR researchers interested in online evaluations of academic search systems
and foster knowledge on improving the search for academic resources like
literature, research data, and the interlinking between these resources.
The employed online evaluation approach based on a living lab infrastructure allows the direct connection to real-world academic search systems from the life sciences and the social sciences.

Keywords: Evaluation · Living labs · Academic search · CLEF

1 Introduction

The search for scientific information is a challenge that the field of Information
Retrieval has been dealing with for many years in the unique environment of
digital libraries or academic search systems. Its roots go back to the fundamental papers of IR research, such as the early Cranfield studies where Cleverdon
et al. conducted their experiments in the context of scientific document retrieval.
Even today, the search for academic material is a broad field of research with
international conferences such as the ACM/IEEE Joint Conference on Digital
Libraries (JCDL) or the Theories and Practices in Digital Libraries (TPDL),
which always have a strong IR-specific background.

The need for innovation in this field of academic search and IR, in general,
is shown by the stagnating system performance in controlled evaluation campaigns, as demonstrated in TREC and CLEF meta-evaluation studies [1,11], as

© Springer Nature Switzerland AG 2020
J. M. Jose et al. (Eds.): ECIR 2020, LNCS 12036, pp. 580–586, 2020.
https://doi.org/10.1007/978-3-030-45442-5_75

well as user studies in real systems of scientific information and digital libraries. Even though large amounts of data are available in highly specialized subject databases (such as ArXiV or PubMed), digital libraries (like the ACM Digital Library), or web search engines (Google Scholar or SemanticScholar), many user needs and requirements remain unsatisfied. The central concern is to find both relevant and high-quality documents - if possible, directly on the first result page - for the actual user of an academic search system. The question of what constitutes relevance in academic search is multi-layered [4] and a topic that drives research communities like the Bibliometrics-enhanced Information Retrieval (BIR) workshops [10].

The Living Labs for Academic Search (LiLAS) workshop would like to inspire the discussion, research, and evaluation of academic search systems by strengthening the concept of living labs to the underrepresented domain of academic search. To do so, the CLEF 2020 workshop lab is about to bring together IR researchers interested in the online evaluation of academic search systems. The goal is to foster knowledge on improving the search for academic resources like literature (ranging from short bibliographic records to full-text papers), research data, and the interlinking between these resources. The employed online evaluation approach based on a living lab infrastructure allows the direct connection to real-world academic search systems from the Life Sciences and the Social Sciences.

The unique feature of the proposed living lab [6] is that the evaluations can be carried out in a productive online system. Contrary to the usual TREC tasks, no expert judgments are used for the evaluation of the information systems and the underlying retrieval systems. The lab employs A/B tests or more complex interleaving procedures to present the experimental systems to the users. However, this requires that the developers of the experimental systems also have access to the production systems, which is rarely the case. Living labs make this possible by bringing platform operators and researchers together. At the same time, they provide a methodological and technical framework for online experiments.

From a broader perspective, the motivation behind this lab is to (a) bring together interested researchers, (b) advertise the online evaluation campaign idea, and (c) develop ideas, best practices, and guidelines for a full online evaluation campaign at CLEF 2021.

2   Background of the LiLAS Workshop

We describe two aspects that form the basis of the LiLAS workshop at CLEF: (1) Evaluating IR systems based on the so-called living labs paradigm and (2) the focus on academic search.

2.1   Living Lab Evaluations at CLEF and TREC

The Living Labs for Information Retrieval (LL4IR) and Open Search initiatives were launched as part of the TREC and CLEF evaluation campaigns to promote cooperation between industrial and academic research in the field of living labs. As part of these initiatives, an initial evaluation infrastructure was created in the form of an API[1] that allows academic researchers to access the search systems of other platforms. A hard requirement is that the platforms actively participate in the initiative and have implemented the corresponding API for their systems. The API developed under a free license is publicly available and represents the greatest success of the campaign to date. Within CLEF, LL4IR was first organized as a lab in the form of a challenge at the CLEF 2015 conference series and continued with TREC OpenSearch 2016 and 2017 [2].

Another living lab campaign that ran at CLEF was NewsREEL that provides the possibility to predict user interactions in an offline dataset as well as to recommend news articles in real-time. Researchers could evaluate their models either with a test collection (offline) or by delivering the recommended content to real users of the news publishing platforms (online) via the Open Recommendation Platform (ORP)[2]. Since NewsREEL joined the MediaEval Benchmarking Initiative for Multimedia Evaluation[3] in 2018, the evaluation setup comprises offline evaluations only [8].

In a nutshell, the living lab evaluation paradigm represents a user-centric study methodology for researchers to evaluate the performance of retrieval systems within real-world applications. Thus, it offers a more realistic experiment and evaluation environment as offline test collections, and therefore should be further investigated to raise IR-evaluation to the next level.

2.2   Academic Search Evaluation

Since the early days of the Cranfield experiments, scientific retrieval has developed and improved relatively little in comparison and contrast to everyday web search, commercial search, or social network search. The common interest in academic search seems small compared to these scenarios. Many information systems are based on bibliographical metadata (e.g., WoS, Scopus, PubMed) and do not make use of full-text information, which in most cases is not available due to licensing hiccups of the academic publishing system. Even when full-text is searchable, the characteristics of scientific communication are not thoroughly operationalized and are even ignored. This leads to simple search systems that still heavenly rely on Boolean Retrieval techniques to get the most out of the available (meta-)data. Although during the last years, the BIR community added much traction for the overall topic of academic search, it is not focusing on the rigorous evaluation of proposed methods. The lack of an overall evaluation scenario leads to little understanding of the consequences for the different domains and fields of science.

---

[1] https://bitbucket.org/living-labs/ll-api, all online resources were checked for validity in January 2020.
[2] https://orp.plista.com/.
[3] http://www.multimediaeval.org/mediaeval2018/newsreelmm/.

Within TREC and CLEF, only very few tracks or labs focused on the evaluation of academic search systems, although some made use of scientific documents or use-cases to generate test collections. An example might be the CLEF Domain-specific track [7] (an offline evaluation lab). In this track, a corpus of bibliographic records and research project descriptions from the Social Sciences was created to test the special needs of scientific retrieval tasks. As stated by the organizers: "General-purpose news documents require very different search criteria than those used for reference retrieval in databases of scientific literature items, and also offer no possibility for comparable test runs with domain-specific terminology." The DS track ran for many years and developed a total of four versions of the GIRT test collection of scientific reference documents. More recently, the TREC Precision Medicine/Clinical Decision Support Track released a large test collection in 2016 based on 1.2 million open access full-text documents from PubMedCentral. Another test collection from the sciences is iSearch [9]. Both collections contain citation data.

3   Outline of the LiLAS Workshop

3.1   Roadmap for CLEF 2020 and Beyond

The timeline for LiLAS starts with a workshop lab at CLEF 2020 to kick-off the evaluation lab that is planned for CLEF 2021. The concrete details of the 2021 online experiments, like tasks, metrics, and technologies, are to be discussed at the workshop, but right now, we favor a Docker-based container infrastructure setup that is briefly described in Sect. 4.

The LiLAS workshop is designed to be a blend of the most successful parts of LL4IR, NewsREEL, and the DS track. The DS track had a strong focus on scientific search, thesauri, and multilingual search. NewsREEL had an active technological component, and LL4IR turned from product search to academic search but was not able to implement the scientific focus into the last iteration. There is much potential that is still not used in the question of how to online evaluate scientific search platforms.

In retrospective, the entrance barriers for LL4IR and NewsREEL for new participants were quite high. This was due to the unusual evaluation schema that raised many questions and might have put off some interested groups. To surpass these issues, we would like to offer different ways to participate in LiLAS: Simple non-technical submissions of pre-computed results and Docker-based containers that offer more evaluation possibilities but require the participants to develop a good-performing system. Our blueprint for this approach is TIRA (short for Testbed for Information Retrieval Algorithms), as it has proven itself to be a reliable infrastructure, although it does not support living lab functionalities.

As the lab specifications for 2021 are still open for discussion, we would like to use the workshop to address some technical and conceptional ideas for this. Questions are how to allow a low barrier entrance to this campaign (e.g., in the form of ready to use templates, reusing Docker containers from campaigns like OSIRRC, or offering tutorial sessions at ECIR or other venues). The unique

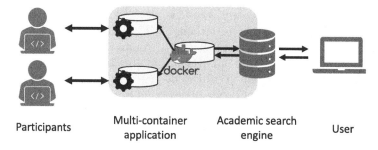

Fig. 1. **Infrastructure design for an online living lab evaluation with exper-imental academic search systems:** Participants package their systems with the help of Docker containers that are deployed in the backend of academic search engines. Users retrieve results from these systems and deliver feedback for later evaluation.

selling points of this lab should be advertised and communicated very clearly so that new participants see the potential both in the available data sets (literature *and* research data) and the possibility to have an online evaluation for their retrieval and recommendation approaches.

To prepare the workshop, we will release some sample data sets from the scientific search systems LIVIVO[4] and GESIS-wide Search [5][5] and some Docker templates to allow early adopters to implement first prototypes. At the workshop, we would like to have these early adopters who took part in this *open beta phase* to present their first-hand experiences to lay a foundation for a full-size campaign in 2021.

4   Evaluation Infrastructure

The evaluation infrastructure for this lab is called STELLA [3]. Figure 1 shows the general structure behind it. It would incorporate usage feedback like click-through rates and would allow participating research groups to package their systems with the help of Docker containers. These containers are deployed in the backend of academic search engines. Users search and interact with these systems and deliver feedback for later evaluation.

The infrastructure is based on the container virtualization Docker, which allows us to run and test different research algorithms on productive systems in a so-called multi-container environment. STELLA's main component is a central Living Lab API that connects data and content providers (sites) with research prototypes (participants) encapsulated in Docker containers. In addition to embedding in production systems and accessing their data, this solution also allows the backup of usage data (e.g., clickthrough or download data), which are necessary for an evaluation of the systems.

---

[4] https://livivo.de.
[5] https://search.gesis.org/.

Although STELLA and its predecessors, like the LL4IR API, have already been successfully evaluated, the software is far from being a fully-fledged production system. Through the discussion with future participants and other actors in the field, like the TIRA maintainers, we hope to achieve a more stable code base. Another practical benefit will result from the common requirements but different areas of application (online vs. offline evaluation). Future integration of the systems provides a unique opportunity to evaluate in two different stages of the systems: (1) Systems could measure themselves against static test collections, and later (2) prove themselves in online scenarios according to their basic performance. Finally, this would allow generating new components for future test collections through the recorded usage data.

This approach would surpass some of the shortcomings of the previous labs' implementations. The main advantages would be:

- No focus on head queries (the system's top-k most common queries)
- Retrieval and recommendation tasks within the same platform
- A technological platform that is built on open-source frameworks like Docker that allows easy distribution of implementation templates (like in the SIGIR 2019 OSIRRC workshop[6] that would lower the entrance barrier and foster reproducibility).

5   Outlook

We see academic search as a broader term for scientific and especially domain-specific retrieval tasks, which comprises document, dataset as well as bibliometric-enhanced retrieval. As web-search platforms like Google Scholar (or Google Dataset Search) or proprietary digital libraries like the ACM Digital Library are not open to public research and do not offer any domain-specific features, we focus on mid-size scientific search systems that offer domain-specific resources and use-cases. This focus allows for using many specific information types like bibliographic metadata, usage data, download rates, or citations in order to develop and evaluate innovative search applications. We see a clear connection to the Bibliographic-enhanced Information Retrieval (BIR) community that runs a series of successful workshops since 2013 at venues like ECIR or SIGIR. A proposed SMART lab on "Scientific mining and retrieval"[7] might be a good starting point to join forces. The common point of interest might be to create an integrated test collection for academic search that contains academic material and their citations, is useful for the scientific retrieval process, and is suitable for offline and online evaluation as well.

---

[6] https://github.com/osirrc.
[7] https://www.scientometrics-school.eu/images/esss_Programme2019.pdf.

References

1. Armstrong, T.G., Moffat, A., Webber, W., Zobel, J.: Improvements that don't add up: ad-hoc retrieval results since 1998. In: Proceeding of the 18th ACM Conference on Information and Knowledge Management, CIKM 2009, Hong Kong, China, pp. 601–610. ACM (2009). https://doi.org/10.1145/1645953.1646031

2. Balog, K., Schuth, A., Dekker, P., Schaer, P., Tavakolpoursaleh, N., Chuang, P.Y.: Overview of the TREC 2016 open search track. In: Proceedings of the Twenty-Fifth Text REtrieval Conference (TREC 2016). NIST (2016)

3. Breuer, T., Schaer, P., Tavalkolpoursaleh, N., Schaible, J., Wolff, B., Müller, B.: STELLA: towards a framework for the reproducibility of online search experiments. In: Proceedings of The Open-Source IR Replicability Challenge (OSIRRC) @ SIGIR (2019)

4. Carevic, Z., Schaer, P.: On the connection between citation-based and topical relevance ranking: results of a pretest using iSearch. In: Proceedings of the First Workshop on Bibliometric-Enhanced Information Retrieval co-located with 36th European Conference on Information Retrieval (ECIR 2014), Amsterdam, The Netherlands, 13 April 2014. CEUR Workshop Proceedings, vol. 1143, pp. 37–44. CEUR-WS.org (2014). http://ceur-ws.org/Vol-1143/paper5.pdf

5. Hienert, D., Kern, D., Boland, K., Zapilko, B., Mutschke, P.: A digital library for research data and related information in the social sciences. In: 19th ACM/IEEE Joint Conference on Digital Libraries, JCDL 2019, Champaign, IL, USA, 2–6 June 2019, pp. 148–157. IEEE (2019). https://doi.org/10.1109/JCDL.2019.00030

6. Kelly, D., Dumais, S., Pedersen, J.O.: Evaluation challenges and directions for information-seeking support systems. Computer $42(3)$, 60–66 (2009)

7. Kluck, M., Gey, F.C.: The domain-specific task of CLEF - specific evaluation strategies in cross-language information retrieval. In: Peters, C. (ed.) CLEF 2000. LNCS, vol. 2069, pp. 48–56. Springer, Heidelberg (2001). https://doi.org/10.1007/3-540-44645-1_5

8. Lommatzsch, A., Kille, B., Hopfgartner, F., Ramming, L.: Newsreel multimedia at mediaeval 2018: news recommendation with image and text content. In: Working Notes Proceedings of the MediaEval 2018 Workshop. CEUR-WS (2018)

9. Lykke, M., Larsen, B., Lund, H., Ingwersen, P.: Developing a test collection for the evaluation of integrated search. In: Gurrin, C., et al. (eds.) ECIR 2010. LNCS, vol. 5993, pp. 627–630. Springer, Heidelberg (2010). https://doi.org/10.1007/978-3-642-12275-0_63

10. Mayr, P., Scharnhorst, A., Larsen, B., Schaer, P., Mutschke, P.: Bibliometric-enhanced information retrieval. In: de Rijke, M., et al. (eds.) ECIR 2014. LNCS, vol. 8416, pp. 798–801. Springer, Cham (2014). https://doi.org/10.1007/978-3-319-06028-6_99

11. Yang, W., Lu, K., Yang, P., Lin, J.: Critically examining the "Neural Hype": weak baselines and the additivity of effectiveness gains from neural ranking models. In: Proceedings of the 42nd International ACM SIGIR Conference on Research and Development in Information Retrieval - SIGIR 2019, Paris, France, pp. 1129–1132. ACM Press (2019). https://doi.org/10.1145/3331184.3331340

CLEF eHealth Evaluation Lab 2020

Hanna Suominen[1,2,3] , Liadh Kelly[4] , Lorraine Goeuriot[5(✉)] ,
and Martin Krallinger[6]

[1] The Australian National University, Acton, ACT 2601, Australia
`hanna.suominen@anu.edu.au`
[2] Data61/Commonwealth Scientific and Industrial Research Organisation,
Acton, ACT, Australia
[3] University of Turku, Turku, Finland
[4] Maynooth University, Co., Kildare, Ireland
`liadh.kelly@mu.ie`
[5] Univ. Grenoble Alpes, CNRS, Grenoble INP, LIG, 38000 Grenoble, France
`lorraine.Goeuriot@imag.fr`
[6] Barcelona Supercomputing Center (BSC-CNS), 08034 Barcelona, Spain
`martin.krallinger@bsc.es`
`https://researchers.anu.edu.au/researchers/suominen-h`

Abstract. Laypeople's increasing difficulties to retrieve and digest valid
and relevant information in their preferred language to make health-
centred decisions has motivated CLEF eHealth to organize yearly labs
since 2012. These 20 evaluation tasks on *Information Extraction* (IE),
management, and *Information Retrieval* (IR) in 2013–2019 have been
popular—as demonstrated by the large number of team registrations,
submissions, papers, their included authors, and citations (748, 177, 184,
741, and 1299, respectively, up to and including 2018)—and achieved
statistically significant improvements in the processing quality. In 2020,
CLEF eHealth is calling for participants to contribute to the following
two tasks: The 2020 Task 1 on IE focuses on term coding for clinical
textual data in Spanish. The terms considered are extracted from clinical case records and they are mapped onto the Spanish version of the
International Classification of Diseases, the 10th Revision, including also
textual evidence spans for the clinical codes. The 2020 Task 2 is a novel
extension of the most popular and established task in CLEF eHealth on
CHS. This IR task uses the representative web corpus used in the 2018
challenge, but now also spoken queries, as well as textual transcripts
of these queries, are offered to the participants. The task is structured
into a number of optional subtasks, covering ad-hoc search using the
spoken queries, textual transcripts of the spoken queries, or provided
automatic speech-to-text conversions of the spoken queries. In this paper
we describe the evolution of CLEF eHealth and this year's tasks. The

---

HS, LK & LG co-chair the CLEF eHealth lab and contributed equally to this paper. MK
leads the 2020 IE task supported by the Spanish Plan TL while HS & LG lead the 2020
IR task. We gratefully acknowledge the contribution of the people and organizations
involved in CLEF eHealth in 2012–2020. We thank the CLEF Initiative, Dr Benjamin
Lecouteux (Université Grenoble Alpes), Dr João Palotti (Qatar Computing Research
Institute), and Dr Guido Zuccon (University of Queensland).

J. M. Jose et al. (Eds.): ECIR 2020, LNCS 12036, pp. 587–594, 2020.
https://doi.org/10.1007/978-3-030-45442-5_76

substantial community interest in the tasks and their resources has led to CLEF eHealth maturing as a primary venue for all interdisciplinary actors of the ecosystem for producing, processing, and consuming electronic health information.

Keywords: eHealth · Medical informatics · Information extraction · Information storage and retrieval · Speech recognition

1   Introduction

Improving the legibility of *Electronic Health Record* (EHR) can contribute to patients' right to be informed about their health and health care. The requirement to ensure that patients can understand their own privacy-sensitive, official health information in their EHR are stipulated by policies and laws. For example, the *Declaration on the Promotion of Patients' Rights in Europe* by *World Health Organization* (WHO) from 1994 obligates health care workers to communicate in a way appropriate to each patient's capacity for understanding and give each patient a legible written summary of these care guidelines. This patient education must capture the patient's health status, condition, diagnosis, and prognosis, together with the proposed and alternative treatment/non-treatment with risks, benefits, and progress. Patients' better abilities to understand their own EHR empowers them to take part in the related health/care judgment, leading to their increased independence from health care providers, better health/care decisions, and decreased health care costs [11]. Improving patients' ability to digest this content could mean enriching the EHR-text with hyperlinks to term definitions, paraphrasing, care guidelines, and further supportive information on patient-friendly and reliable websites, and the enabling methods for such reading aids can also release health care workers' time from EHR-writing to, for example, longer patient-education discussions [14].

Information access conferences have organized evaluation labs on related *Electronic Health* (eHealth) *Information Extraction* (IE), *Information Management* (IM), and *Information Retrieval* (IR) tasks for almost 20 years. Yet, with rare exception, they have targeted the health care experts' information needs only [1,2,6]. Such exception, the *CLEF eHealth Evaluation-lab and Lab-workshop Series*[1] has been organized every year since 2012 as part of the *Conference and Labs of the Evaluation Forum* (CLEF) [4,5,8–10,13,16,17]. In 2012, the inaugural scientific CLEF workshop took place, and from 2013–2019 this annual workshop has been supplemented with a lead-up evaluation lab, consisting of, on average, three shared tasks each year (Fig. 1). Although the tasks have been centered around the patients and their families' needs in accessing and understanding eHealth information, also *Automatic Speech Recognition* (ASR) and IE to aid clinicians in IM were considered in 2015–2016 and in 2017–2019, tasks on technology assisted reviews to support health scientists and health care policymakers' information access were organized.

---

[1] http://clef-ehealth.org/.

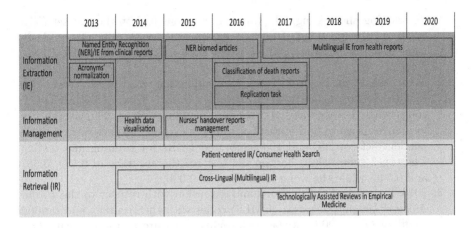

Fig. 1. Timeline of the CLEF eHealth tasks in 2013–2020

This paper presents first an overview of CLEF eHealth lab series from 2012 to 2019 and introduces its 2020 evaluation tasks. Then, it concludes by presenting our vision for CLEF eHealth beyond 2020.

2   CLEF eHealth Contributions and Growth in 2012–2019

CLEF eHealth tasks offered yearly from 2013 have brought together researchers working on related information access topics, provided them with resources to work with and validate their outcomes, and accelerated pathways from scientific ideas to societal impact. In 2013, 2014, 2015, 2016, 2017, 2018, and 2019 as many as 170, 220, 100, 116, 67, 70, and 67 teams have registered their expression of interest in the CLEF eHealth tasks, respectively, and the number of teams proceeding to the task submission stage has been 53, 24, 20, 20, 32, 28, and 9, respectively [4,5,8–10,16,17].[2]

According to our analysis of the impact of CLEF eHealth labs up to 2017 [15], the submitting teams have achieved statistically significant improvements in the processing quality in at least 1 out of the top-3 methods submitted to the following eight tasks:[3]

1. 2013 Task 1a on English disorder identification with $F1$ and random shuffling ($P = .009$) as the performance measure and statistical significance test, respectively, on independent sets of 200 and 100 annotated EHRs

---

[2] "Expressing an interest" for a CLEF task consists of filling in a form on the CLEF conference website with contact information, and tick boxes corresponding to the labs of interest. This is usually done several months before run submission, which explains the drop in the numbers.

[3] Some tasks have not presented a method ranking and/or statistical significance evaluation of this kind in the lab/task overviews. In other words, different kinds of improvements have been obtained in other tasks as well.

for training and testing. The top-3 submissions had on the test set $F1$ of 0.750, 0.737, and 0.707, whilst to illustrate the task difficulty, typically using a simple baseline method by the task organizers, the worst $F1$ was 0.428.

2. 2013 Task 1b on English disorder normalization with respect to the *Systematized Nomenclature of Medicine—Clinical Terms* (SNOMED-CT) codes with the accuracy and random shuffling ($P = .009$) as the performance measure and statistical significance test, respectively, on independent sets of 200 and 100 annotated EHRs for training and testing. The top-3 submissions had on the test set the accuracy of 0.589, 0.587, and 0.546, while the worst accuracy was 0.006.

3. 2013 Task 2 on English shorthand extension with respect to the *Unified Medical Language System* (UMLS) codes with the accuracy and random shuffling ($P = .009$) as the performance measure and statistical significance test, respectively, on independent sets of 200 and 100 annotated EHRs for training and testing. The top-3 submissions had on the test set the accuracy of 0.719, 0.683, and 0.664, while the worst accuracy was 0.426.

4. 2013 Task 3 on English IR with the *Precision at 10* (P@10) and Wilcoxon test ($P = .04$) as the performance measure and statistical significance test, respectively, on 50 test queries and the matching result set. The top-3 submissions had P@10 of 0.518, 0.504, and 0.484, while the worst P@10 was 0.006.

5. 2015 Task 1 on English nursing handover ASR with the error and Wilcoxon test ($P = .04$) as the performance measure and statistical significance test, respectively, on independent sets of 100 and 100 annotated EHRs for training and testing. The top-3 submissions had on the test set the error of 0.385, 0.523, and 0.528, while the worst error was 0.954.

6. 2016 Task 1 on English nursing handover IE with $F1$ and Wilcoxon test ($P = .04$) as the performance measure and statistical significance test, respectively, on independent sets of 200 and 100 annotated EHRs for training and testing. The top-3 submissions had on the test set $F1$ of 0.382, 0.374, and 0.345, while the worst $F1$ was 0.000.

7 & 8. 2016 Task 2 on French IE, with entity recognition and cause of death subtasks. Both subtasks used $F1$ and t-test ($P \leq .001$) as the performance measure and statistical significance test, respectively, on 1,668 titles of scientific articles and 6 full text drug monographs for training and testing. The corpus was split evenly between training data supplied to the participants at the beginning of the lab, and an unseen test set used to evaluate participants' systems. In the entity recognition subtask, the top-3 submissions had on the test set $F1$ of 0.749, 0.702, and 0.699, while the worst $F1$ was 0.126. In the cause of death subtasks, the top-3 submissions had on the test set $F1$ of 0.848, 0.844, and 0.752, while the worst $F1$ was 0.554.

The 2012–2017 contributions have been reported by October 2018 in 184 papers for the 741 included authors from 33 countries across the world, and the papers have attracted nearly 1, 300 citations, creating h-index and i10-index of 18 and 35, respectively, on Google Scholar [14]. CLEF eHealth 2012 lab workshop has resulted in 16 papers and each year CLEF eHealth 2013–2017 evaluation labs have increased this number from 31 to 35. In accordance with the CLEF eHealth mission to foster teamwork, the number of co-authors per paper has been from 1 to 15 (the mean and standard deviation of 4 and 3, respectively). In about a quarter of the papers, this co-authoring collaboration has been international, and sometimes even intercontinental.

This substantial community interest in the CLEF eHealth tasks and their resources has led to the evaluation campaign maturing and establishing its presence over the years. In 2020, CLEF eHealth is one of the primary venues for all interdisciplinary actors of the ecosystem for producing, processing, and consuming eHealth information [1,2,6]. Its niche is addressing health information needs of laypeople—and not health care experts only—in retrieving and digesting valid and relevant eHealth information to make health-centered decisions.

3   CLEF eHealth 2020 Information Extraction and Retrieval Tasks

The 2020 CLEF eHealth Task 1 on IE, called *CodiEsp* supported by the *Spanish National Plan for the Advancement of Language Technology* (Plan TL), builds upon the five previous editions of the task in 2015–2019 [4,5,8,10,16] that have already addressed the analysis of biomedical text in English, French, Hungarian, Italian, and German. This year, the CodiEsp task, will focus on the *International Classification of Diseases, the 10th Revision* (ICD10) coding for clinical case data in Spanish using the *Spanish version of ICD10* (CIE10).

The CodiEsp task will explore the automatic assignment of CIE10 codes to clinical case documents in Spanish, namely of two categories: procedure and diagnosis (known as 'Procedimiento' and 'Diagnostico' in Spanish). The following three subtasks will be posed: (1) *CodiEsp Diagnosis Coding* will consist of automatically assigning diagnosis codes to clinical cases in Spanish. (2) *CodiEsp Procedure Coding* will focus on assigning procedure codes to clinical cases in Spanish. (3) *CodiEsp Explainable Artificial Intelligence* (AI) will evaluate the explainability/interpretability of the proposed systems, as well as their performance by requesting to return the text spans supporting the assignment of CIE10 codes.

The CodiEsp corpus used for this task consists of a total of 1, 000 clinical cases that were manually annotated by clinical coding professionals with clinical procedure and diagnosis codes from the Spanish version of ICD10 together with the actual minimal text spans supporting the clinical codes. The CodiEsp corpus has around 18, 000 sentences, and contains about 411, 000 words and 19, 000 clinical codes. Code annotations will be released in a separate file together with the respective document code and the span of text that leads to the codification (the evidence). Additional data resources including medical literature abstracts in Spanish indexed with ICD10 codes, linguistic resources, gazetteers, and a background set of medical texts in Spanish will also be released to complement the CodiEsp corpus, together with annotation guidelines and details.

For the CodiEsp Diagnosis and Procedure Coding subtasks, participants will submit their coding predictions returning ranked results. For every document, a list of possible codes will be submitted, ordered by confidence or relevance. Since these subtasks are designed to be ranking competitions, they will be evaluated on a standard ranking metric: Mean Average Precision. For the CodiEsp Explainable AI subtask, explainability of the systems will be considered, in addition to their performance on the test set. Systems have to provide textual evidence from the clinical case documents that supports the code assignment and thus can be interpreted by humans. This automatically returned evidence will be evaluated against manually annotated text spans. True positive evidence texts are those that consist in a sub-match of the manual annotations. $F1$ will be used as the primary evaluation metric.

The 2020 CLEF eHealth Task 2 on IR builds on the tasks that have run at CLEF eHealth since its inception in 2012. This *Consumer Health Search* (CHS) task follows a standard IR shared challenge paradigm from the perspective that it provides participants with a test collection consisting of a set of documents and a set of topics to develop IR techniques for. Runs submitted by participants are pooled, and manual relevance assessments conducted. Performance measures are then returned to participants.

In the 2017 CLEF eHealth CHS task, similarly to 2016, we used the ClueWeb 12 B13[4] document collection [12,18]. This consisted of a collection of 52.3 million medically related web pages. Given the scale of this document collection participants reported that it was difficult to store and manipulate the document collection. In response, the 2018 CHS task introduced a new document collection, named *clefehealth2018*. This collection consists of over 5 million medical webpages from selected domains acquired from the CommonCrawl [7]. Given the positive feedback received for this document collection, it will be used again in the 2020 CHS task.

Historically the CLEF eHealth IR task has released text queries representative of layperson information needs in various scenarios. In recent years, query variations issued by multiple laypeople for the same information need have been offered. In this year's task we extend this to spoken queries. These spoken queries are generated by 6 individuals using the information needs derived for the 2018 challenge [7]. We also provide textual transcripts of these spoken queries and ASR translations.

Given the query variants for an information need, participants are challenged in the 2020 task with retrieving the relevant documents from the provided document collection. This is divided into a number of subtasks which can be completed using the spoken queries or their textual transcripts by hand or ASR. Similar to the 2018 CHS tasks, subtasks explored this year are: ad-hoc/personalized search, query variations, and search intent with Binary Preference, Mean Reciprocal Rank, Normalized Discounted Cumulative Gain@1–10,

---

[4] http://lemurproject.org/clueweb12/index.php.

and (Understandability-biased) Rank-biased Precision as subtask-dependent evaluation measures. Participants can submit multiple runs for each subtask.

4   A Vision for CLEF eHealth Beyond 2020

The general purpose of CLEF eHealth throughout the years, as its 2020 IE and IR tasks demonstrate, has been to assist laypeople in finding and understanding health information in order to make enlightened decisions. Breaking language barriers has been our priority over the years, and this will continue in our multilingual tasks. Text has been our major media of interest, but speech has been, and continues to be, included in tasks as a major new way of interacting with systems. Each year of the labs has enabled the identification of difficulties and challenges in IE, IM, and IR which have shaped our tasks. For example, popular IR tasks have considered multilingual, contextualized, and/or spoken queries and query variants. However, further exploration of query construction, aiming at a better understanding of CHS are still needed. The task into the future will also further explore relevance dimensions, and work toward a better assessment of readability and reliability, as well as methods to take these dimensions into consideration. As lab organizers, our purpose is to increase the impact and the value of the resources, methods and the community built by CLEF eHealth. Examining the quality and stability of the lab contributions will help the CLEF eHealth series to better understand where it should be improved and how. As future work, we intend continuing our analyses of the influence of the CLEF eHealth evaluation series from the perspectives of publications and data/software releases [3,14,15].

References

1. Demner-Fushman, D., Elhadad, N.: Aspiring to unintended consequences of natural language processing: a review of recent developments in clinical and consumer-generated text processing. Yearb. Med. Inform. **1**, 224–233 (2016)
2. Filannino, M., Uzuner, Ö.: Advancing the state of the art in clinical natural language processing through shared tasks. Yearb. Med. Inform. **27**(01), 184–192 (2018)
3. Goeuriot, L., et al.: An analysis of evaluation campaigns in ad-hoc medical information retrieval: CLEF eHealth 2013 and 2014. Inf. Retrieval J. **21**(6), 507–540 (2018). https://doi.org/10.1007/s10791-018-9331-4
4. Goeuriot, L., et al.: Overview of the CLEF eHealth evaluation lab 2015. In: Mothe, J., et al. (eds.) CLEF 2015. LNCS, vol. 9283, pp. 429–443. Springer, Cham (2015). https://doi.org/10.1007/978-3-319-24027-5_44
5. Goeuriot, L., et al.: CLEF 2017 eHealth evaluation lab overview. In: Jones, G.J.F., et al. (eds.) CLEF 2017. LNCS, vol. 10456, pp. 291–303. Springer, Cham (2017). https://doi.org/10.1007/978-3-319-65813-1_26
6. Huang, C.C., Lu, Z.: Community challenges in biomedical text mining over 10 years: success, failure and the future. Brief. Bioinf. **17**(1), 132–144 (2016)

7. Jimmy, Zuccon, G., Palotti, J.: Overview of the CLEF 2018 consumer health search task. In: Working Notes of Conference and Labs of the Evaluation (CLEF) Forum. CEUR Workshop Proceedings (2018)
8. Kelly, L., Goeuriot, L., Suominen, H., Névéol, A., Palotti, J., Zuccon, G.: Overview of the CLEF eHealth evaluation lab 2016. In: Fuhr, N., et al. (eds.) CLEF 2016. LNCS, vol. 9822, pp. 255–266. Springer, Cham (2016). https://doi.org/10.1007/978-3-319-44564-9_24
9. Kelly, L., et al.: Overview of the ShARe/CLEF eHealth evaluation lab 2014. In: Kanoulas, E., et al. (eds.) CLEF 2014. LNCS, vol. 8685, pp. 172–191. Springer, Cham (2014). https://doi.org/10.1007/978-3-319-11382-1_17
10. Kelly, L., et al.: Overview of the CLEF eHealth evaluation lab 2019. In: Crestani, F., et al. (eds.) CLEF 2019. LNCS, vol. 11696, pp. 322–339. Springer, Cham (2019). https://doi.org/10.1007/978-3-030-28577-7_26
11. McAllister, M., Dunn, G., Payne, K., Davies, L., Todd, C.: Patient empowerment: the need to consider it as a measurable patient-reported outcome for chronic conditions. BMC Health Serv. Res. **12**, 157 (2012)
12. Palotti, J., et al.: CLEF 2017 task overview: the IR task at the eHealth evaluation lab. In: Working Notes of Conference and Labs of the Evaluation (CLEF) Forum. CEUR Workshop Proceedings (2017)
13. Suominen, H.: CLEFeHealth2012 – the CLEF 2012 workshop on cross-language evaluation of methods, applications, and resources for ehealth document analysis. In: Forner, P., Karlgren, J., Womser-Hacker, C., Ferro, N. (eds.) CLEF 2012 Working Notes. vol. 1178. CEUR Workshop Proceedings (CEUR-WS.org) (2012)
14. Suominen, H., Kelly, L., Goeuriot, L.: Scholarly influence of the conference and labs of the evaluation forum eHealth initiative: review and bibliometric study of the 2012 to 2017 outcomes. JMIR Res. Protoc. **7**(7), e10961 (2018)
15. Suominen, H., Kelly, L., Goeuriot, L.: The scholarly impact and strategic intent of CLEF eHealth labs from 2012 to 2017. In: Ferro, N., Peters, C. (eds.) Information Retrieval Evaluation in a Changing World: Lessons Learned from 20 Years of CLEF, pp. 333–363. Springer, Cham (2019). https://doi.org/10.1007/978-3-030-22948-1_14
16. Suominen, H., et al.: Overview of the CLEF eHealth evaluation lab 2018. In: Bellot, P., et al. (eds.) Experimental IR Meets Multilinguality, Multimodality, and Interaction, pp. 286–301. Springer, Cham (2018). https://doi.org/10.1007/978-3-319-98932-7_26
17. Suominen, H., et al.: Overview of the ShARe/CLEF eHealth evaluation lab 2013. In: Forner, P., Müller, H., Paredes, R., Rosso, P., Stein, B. (eds.) CLEF 2013. LNCS, vol. 8138, pp. 212–231. Springer, Heidelberg (2013). https://doi.org/10.1007/978-3-642-40802-1_24
18. Zuccon, G., et al.: The IR task at the CLEF eHealth evaluation lab 2016: user-centred health information retrieval. In: CLEF 2016 Evaluation Labs and Workshop: Online Working Notes. CEUR-WS, September 2016

Doctoral Consortium Papers

Reproducible Online Search Experiments

Timo Breuer[1,2(✉)] [iD]

[1] TH Köln (University of Applied Sciences), Cologne, Germany
timo.breuer@th-koeln.de
[2] Universität Duisburg-Essen, Duisburg, Germany

Abstract. In the empirical sciences, the evidence is commonly manifested by experimental results. However, very often, these findings are not reproducible, hindering scientific progress. Innovations in the field of information retrieval (IR) are mainly driven by experimental results as well. While there are several attempts to assure the reproducibility of offline experiments with standardized test collections, reproducible outcomes of online experiments remain an open issue. This research project will be concerned with the reproducibility of online experiments, including real-world user feedback. In contrast to previous living lab attempts by the IR community, this project has a stronger focus on making IR systems and corresponding results reproducible. The project aims to provide insights concerning key components that affect reproducibility in online search experiments. Outcomes help to improve the design of reproducible IR online experiments in the future.

Keywords: Reproducibility · Online evaluation · Living lab

1 Motivation

Reproducible findings are fundamental for scientific progress and validity. In 2016, a Nature survey [2] revealed that lack of reproducibility nearly affects all scientific disciplines and can be considered as a general concern. Non-reproducible results limit the trustworthiness of publications and hinder progress. Besides investigating various reasons for non-reproducibility, the study showed that scientists mostly agree upon the importance of the problem that became known as *reproducibility crisis* during the last years. Especially in the field of information retrieval (IR), new findings are manifested by empirical studies and experiments. Innovations are assumed to be valid if their results are superior compared to those of previous findings. Despite this intuitive but rather naive assumption, achieving reproducibility in the field of IR is a many-faceted problem. For instance, the meta-evaluation by Armstrong et al. [1] reveal the illusory progress of ad-hoc retrieval performance over an entire decade, caused by comparisons to weak baselines. Ten years later, Yang et al. [16] report similar results as part of their meta-evaluation. The lacking upwards trend in retrieval

© Springer Nature Switzerland AG 2020
J. M. Jose et al. (Eds.): ECIR 2020, LNCS 12036, pp. 597–601, 2020.
https://doi.org/10.1007/978-3-030-45442-5_77

performance can be traced back to non-reproducible findings. If baselines of previous results are not or only laboriously reproducible, the community does not use them adequately.

We see a gap between reproducibility efforts for offline evaluations on the one side and online retrieval experiments trying to include real-world user interactions on the other side. While several initiatives are trying to establish reproducible IR research for offline evaluations on standard test collections, there is little research effort concerning the reproducibility of online experiments. This dissertation project will be concerned with the reproducibility of online experiments in the field of information retrieval.

2   Related Work

Progress in information retrieval revolves around the evaluation of experimental results. This research project will focus specifically on two aspects of evaluation in IR - reproducible experiments and the living lab paradigm. This section gives a brief overview of the two evaluation branches.

As mentioned in the previous section, meta-evaluations of IR systems revealed limited progress over the years [1,16]. During the last years, the IR community tried to tackle this problem with several attempts concerned with reproducibility. These can be broadly categorized into attempts on a conceptual level and initiatives in the form of workshops, infrastructures, and frameworks. Conceptually, Ferro and Kelly elaborate an implementation for the field of information retrieval [10] of the ACM Artifact and Review Badging[1]. The PRIMAD model [8] offers orientation which components of an IR experiment may affect reproducibility or have to be considered when trying to reproduce the corresponding experiment. The Evaluation-as-a-Service (EaaS) paradigm [13] reverses the conventional evaluation approach of a shared task like it is applied at the TREC conference. Instead of letting participants submit the results (runs) only, the complete retrieval system is submitted in a form such that it can be rerun independently by others to produce the results. Workshops deal with the reproducibility either re- or proactively. For example, the CENTRE workshop [9] challenges participants to reconstruct IR systems and their results, whereas The Open-Source IR Replicability Challenge (OSIRRC) [7] motivated participants to package their retrieval systems and corresponding software dependencies in advance to prepare them for appropriate reuse.

Compared to offline ad-hoc retrieval, online search experiments are affected by non-deterministic variables including user behavior, updated data collections, modifications of web interfaces, or traffic dependencies [11]. Balog et al. introduced the first living lab campaign in 2014 [3]. The infrastructure found application in several workshops and intiviates at the CLEF and TREC conferences from 2015 to 2017 [14]. Despite these elegant solutions for implementing living lab infrastructures, the aspect of reproducibility remained neglected, e.g., there was no specification of how the experiments could be archived for later use [12].

---

[1] https://www.acm.org/publications/policies/artifact-review-badging.

On the other hand, research efforts towards reproducible IR experiments have a strong focus on ad-hoc retrieval experiments and do not include any insights beyond offline environments at the time of writing.

3   Preliminary Work and Research Proposal

Preliminary Work. We participated in the CENTRE@CLEF2019 workshop dedicated to the replicability, reproducibility, and generalizability of ad-hoc retrieval experiments [5]. The workshop's organizers challenge the participants to reconstruct results of previous submissions to the CLEF, NTCIR, and TREC conferences. CENTRE defines replicability and reproducibility by using the same or another test collection of the original setup, respectively. The results of our experimental setups showed that we can replicate the outcomes fairly well, whereas reproduced outcomes are significantly lower. Having the reimplementation of an ad-hoc retrieval system at hand, we decided to contribute it to the OSIRRC@SIGIR2019 workshop [7]. All contributions resulted in an image library of Docker images to which we contributed the IRC-CENTRE2019 image [4]. Additionally, we introduced STELLA - a new interpretation of the living lab paradigm - at the OSIRRC workshop [6]. We propose to transfer the idea of encapsulating retrieval systems with Docker containers to the online search scenario. In order to underline the feasibility and benefits of this proposal, we aligned components of the STELLA framework to the PRIMAD model.

Based on this preliminary work, we investigate the reproducibility of retrieval systems with the main focus on online environments. In the following, we present the research questions of this project.

RQ1 - How is the ACM terminology of repeatability, replicability, and reproducibility applied to online search experiments? While the ACM definitions can be implemented for offline ad-hoc experiments, an analogy for the online case is less obvious. Did previous online search experiments consider reproducibility? If so, is it possible to go a step further and align them with the PRIMAD model?

RQ2 - How can simulations of search sessions, based on user logs, help to identify key components of reproducible retrieval performance? To what extent affect the identified components of an online experiment the reproducibility? Compared to offline experiments, the user of the search engine is a key component in the online case. User logs comprise implicit feedback such that they can be used to model the user component of an experiment. What influence do the user and other session-related components have on the reproducibility?

RQ3 - What requirements must a living lab infrastructure meet in order to guarantee reproducible online search experiments? By identifying key components that affect the reproducibility of online search experiments, we gain insights about the requirements for reproducible online search experiments. What kind of practical steps have to be considered when implementing a framework for reproducible retrieval experiments in production environments?

4   Methodology and Experiments

Addressing *RQ1*, we want to conduct a literature survey and evaluate how previous living lab approaches and online experiments paid attention to the topic of reproducibility. Since there exist different terminologies, we use the ACM definitions of repeatability, replicability, and reproducibility as a starting point. As a result, we do not only want to give an overview of how existing literature paid attention to these concepts, but also provide an ontology that is inspired by the PRIMAD model [8]. While the ACM terminology is defined by the two experimental components of the research team and setup, the PRIMAD model conceptualizes the experiment on a more granular level. More specifically, it pays attention to the platform, research goal, implementation, method, actor, and data. This point of view is mainly data-focused and applies well to the offline ad-hoc experiment. However, it could be extended such that it also considers the actual user of a retrieval system.

Regarding *RQ2*, we primarily focus on vertical search experiments. As a result, we want to provide insights concerning key components that affect the reproducibility of online search experiments. Beforehand, the reusability of user logs is particularly interesting, since reusable test collections are fundamental to offline retrieval experiments. Tan et al. [15] examine the reusability of user judgments that contributed to a relevance pool by performing a *leave-one-out* analysis. As a starting point, we propose to repeat this study with the user logs of another search engine. Assuming we have retrieved a fair amount of interaction logs that deliver relevance feedback in the form of clicks and other interactions [11], we systematically assess the influence of specific components. For instance, we can simulate sessions with different durations, tasks, or users. By comparing a diverse set of different session constellations and corresponding outcomes, we identify significant influences. Are specific components more important than others or even crucial for successful reproduction? Furthermore, it is of interest to relate to previous offline reproducibility efforts. Consider two rankers A and B, that are compared by the conventional offline ad-hoc experiment. The retrieval effectiveness of A outperforms that of B, which is denoted as $A \succ B$ and is confirmed to be reproducible. Under which circumstances and to which extent can $A \succ B$ be reproduced in an online environment? Which components affect the reproducibility?

Having identified major influences and key components, we address *RQ3* by deriving requirements that have to be met by an adequate living lab infrastructure. On a functional level, technical components of the infrastructure have to be included. Quality requirements play an essential role, as well. Since experimental systems will be deployed in production environments, a certain degree of quality has to be guaranteed. Subpar retrieval performance and latencies caused by long query processing may affect user behavior, and at the worst, damage the reputation of the sites. Furthermore, we have to consider general conditions like the ethical and juridical aspects of data logging. On an organizational level, it has to be specified, which prerequisites an embedded search engine provider has to fulfill.

References

1. Armstrong, T.G., Moffat, A., Webber, W., Zobel, J.: Improvements that don't add up: ad-hoc retrieval results since 1998. In: Proceedings of CIKM, pp. 601–610 (2009)
2. Baker, M.: 1,500 scientists lift the lid on reproducibility. Nature **533**, 452–454 (2016)
3. Balog, K., Kelly, L., Schuth, A.: Head first: living labs for ad-hoc search evaluation. In: Proceedings of CIKM, pp. 1815–1818 (2014)
4. Breuer, T., Schaer, P.: Dockerizing automatic routing runs for the open-source IR replicability challenge (osirrc 2019). In: Proceedings of the Open-Source IR Replicability Challenge (OSIRRC) @ SIGIR (2019)
5. Breuer, T., Schaer, P.: Replicability and reproducibility of automatic routing runs. In: Working Notes of CLEF. CEUR Workshop Proceedings (2019)
6. Breuer, T., Schaer, P., Tavalkolpoursaleh, N., Schaible, J., Wolff, B., Müller, B.: STELLA: towards a framework for the reproducibility of online search experiments. In: Proceedings of the Open-Source IR Replicability Challenge (OSIRRC) @ SIGIR (2019)
7. Clancy, R., Ferro, N., Hauff, C., Lin, J., Sakai, T., Wu, Z.Z.: The SIGIR 2019 open-source ir replicability challenge (OSIRRC 2019). In: Proceedings of SIGIR, pp. 1432–1434 (2019)
8. Ferro, N., Fuhr, N., Järvelin, K., Kando, N., Lippold, M., Zobel, J.: Increasing reproducibility in IR: findings from the dagstuhl seminar on "Reproducibility of Data-Oriented Experiments in e-Science". SIGIR Forum **50**, 68–82 (2016)
9. Ferro, N., Fuhr, N., Maistro, M., Sakai, T., Soboroff, I.: CENTRE@CLEF2019: overview of the replicability and reproducibility tasks. In: Working Notes of CLEF (2019)
10. Ferro, N., Kelly, D.: SIGIR initiative to implement ACM artifact review and badging. SIGIR Forum **52**, 4–10 (2018)
11. Hofmann, K., Li, L., Radlinski, F.: Online evaluation for information retrieval. Found. Trends Inf. Retrieval **10**, 1–117 (2016)
12. Hopfgartner, F., et al.: Continuous evaluation of large-scale information access systems: a case for living labs. In: Information Retrieval Evaluation in a Changing World - Lessons Learned from 20 Years of CLEF, pp. 511–543 (2019)
13. Hopfgartner, F., et al.: Evaluation-as-a-service for the computational sciences: overview and outlook. J. Data Inf. Qual. **10**, 15:1–15:32 (2018)
14. Jagerman, R., Balog, K., de Rijke, M.: OpenSearch: lessons learned from an online evaluation campaign. J. Data Inf. Qual. **1**, 13:1–13:15 (2018)
15. Tan, L., Baruah, G., Lin, J.: On the reusability of "Living Labs" test collections: a case study of real-time summarization. In: Proceedings of SIGIR, pp. 793–796 (2017)
16. Yang, W., Lu, K., Yang, P., Lin, J.: Critically examining the "Neural Hype": weak baselines and the additivity of effectiveness gains from neural ranking models. In: Proceedings of SIGIR, pp. 1129–1132 (2019)

Graph-Based Entity-Oriented Search: A Unified Framework in Information Retrieval

José Devezas[(⊠)] [iD]

INESC TEC and Faculty of Engineering, University of Porto,
Rua Dr. Roberto Frias, s/n, 4200-465 Porto, Portugal
jld@fe.up.pt

Abstract. Modern search engines have evolved beyond document retrieval. Nowadays, the information needs of the users can be directly satisfied through entity-oriented search, by taking into account the entities that better relate to the query, as opposed to relying exclusively on the best matching terms. Evolving from keyword-based to entity-oriented search poses several challenges, not only regarding the understanding of natural language queries, which are more familiar to the end-user, but also regarding the integration of unstructured documents and structured information sources such as knowledge bases. One opportunity that remains open is the research of unified frameworks for the representation and retrieval of heterogeneous information sources. The doctoral work we present here proposes graph-based models to promote the cooperation between different units of information, in order to maximize the amount of available leads that help the user satisfy an information need.

Keywords: Entity-oriented search · Graph-based models · Representation models · Retrieval models

1 Research Statement

We propose a graph-based unified framework to cover two fronts in entity-oriented search: representation and retrieval. On one side, we propose the joint representation of terms, entities and their relations, as a collection-based model of corpora and knowledge bases. Our idea is to provide a novel way to index documents and entities such that, from design, the aim is to seamlessly integrate units of information of different types, providing a higher-level of flexibility and expressiveness than the inverted index or the triplestore alone. On the other side, we propose a universal ranking function that should be issued over the representation model that we design, in order to rank the different units of information, based on other input units of information. The goal is to obtain a unique ranking function that, depending on the query and the target result, will be able to solve several tasks from entity-oriented search, be it document and entity retrieval or the recommendation-alike tasks of related entity finding and entity list completion.

© Springer Nature Switzerland AG 2020
J. M. Jose et al. (Eds.): ECIR 2020, LNCS 12036, pp. 602–607, 2020.
https://doi.org/10.1007/978-3-030-45442-5_78

2   Motivation

Classical information retrieval has focused on the representation of documents and their features, as well as the retrieval and ranking of those documents based on a representation model—usually the inverted index. Entity-oriented search, however, relies simultaneously on corpora and knowledge bases, which have mismatching representation models—usually the inverted index and the quad indexes in a triplestore. Moreover, if we look at the definitions of 'information retrieval' and 'entity-oriented search', which are ten years apart, we will find a collision between two paradigms that are seemingly different.

"Information retrieval (IR) is finding material (usually documents) of an unstructured nature (usually text) that satisfies an information need from within large collections (usually stored on computers)."

– Manning et al. [8, Ch. 1], 2008

"Entity-oriented search is the search paradigm of organizing and accessing information centered around entities, and their attributes and relationships."

– Balog [2, §1.3, Def. 1.5], 2018

While the first is focused on unstructured data (documents), the second largely relies on structured data (entities and relations). Moreover, entity-oriented search also incorporates ad hoc document retrieval, as long as it leverages entities [2, Ch. 8]. This collision of definitions, along with the consideration for both documents and entities, means that an integration of representation and/or retrieval models must happen at one of the stages of indexing and/or search. One way to tackle the problem is through separated subsystems based on an inverted index and a triplestore, which either independently respond to the query or contribute with different signals to the final scores. Another way to approach the problem is by considering data as a joint representation of corpora and knowledge bases—also known as combined data, according to Bast et al. [4, Def. 2.3].

Being able to jointly index and search over the documents and entities in such a collection of combined data, through their relations, should provide a way to harness all available information, both from unstructured and structured data, as well as from the cross-referencing of both. Bast and Buchhold [3] have justified the need for such a unified framework based on the example of a friendship relation that could only be found in the text, but should influence retrieved triples by establishing new connections. One way to tackle this problem and to build a seamless model of text and knowledge is to use graphs, since they have already been used for the retrieval of documents [5,11] and entities [1,14]. While models like the graph-of-word [11] were designed to answer keyword-based queries with a list of ranked documents, entities in a knowledge graph like DBpedia [1] are still more frequently retrieved through a structured query language like SPARQL [9]. Our goal is to propose a unified model that is able to support the ranking of documents and entities, for a given keyword query, entity, or set of entities, without the need for a structured query language. At the same time, the model should retain the complex relations between entities, exploiting them for all tasks.

3   Description

This doctoral work focuses on improving entity-oriented search performance by researching and proposing novel graph-based approaches to better harness the power of all available information sources. The goal is to satisfy the user's need to answer increasingly complex verbose queries by automatically cross-referencing and weighting related information from unstructured and structured data, at a common stage. It is particularly relevant to research and develop representation models that are able to support the integration of existing tasks through a general ranking function. This unified model should be able to provide a clear framework for entity-oriented search, similar to what the inverted index and TF-IDF are for ad hoc document retrieval, unlocking a larger universe of possibilities for the combination of heterogeneous and multimodal information sources.

The steps required to build such a model are the following:

1. Proposing a graph-based joint representation model of terms, entities and their relations, for indexing corpora and knowledge bases.
2. Proposing a universal ranking function, to be issued over the graph, that can support:
   (a) Ad hoc document retrieval (leveraging entities);
   (b) Ad hoc entity retrieval (leveraging documents);
   (c) At least one of the following recommendation-alike tasks:
       i. Related entity finding (leveraging documents and entities);
       ii. Entity list completion (leveraging documents and entities).
3. Validating the viability of the unified model based on:
   (a) The effectiveness of individual tasks;
   (b) Different combined data test collections.

The main research questions are linked to each of the previous three steps:

1. Which nodes, edges and respective weighting functions (if any) should we use to jointly represent documents and entities for the best overall performance of the considered retrieval tasks?
   – Is it constructive to consider synonyms?
   – And contextual similarity?
   – What about syntactic dependencies?
   – Which entity relations should we consider?
   – And which term-entity relations?
2. How can we model a ranking function, over the proposed representation model, such that the same ranking function can be applied for ranking different units of information, either regarding relevance to a keyword query, or relatedness with one or multiple input entities?

3. How can we evaluate the graph-based unified model?
   (a) Are there any combined data test collections that cover multiple tasks?
   (b) Which variations of the representation model should we consider?
   (c) Which parameter values of the ranking function should we assess?

4 Research Methodology

The research methodology is based on the well-established approach for information retrieval experimentation [10], supported on the empirical cycle and the test collections provided by evaluation initiatives [13]. An example of a combined data test collection is the INEX 2009 Wikipedia collection [12], which consists of a Wikipedia snapshot, formatted as XML, containing inter-document links, as well as entity and link annotations using XML tags based on the WordNet thesaurus. While this collection is from 2009, its great advantage is that there are topics and relevance judgments that can be used to evaluate the tasks of ad hoc document retrieval, ad hoc entity retrieval and entity list completion.

One of the main experiments we already carried was based on the graph-of-entity [6], a model that we proposed, inspired by the graph-of-word [11]. We also experimented in-depth with hypergraph-of-entity [7]. This model was built as an attempt to tackle the complexity and scalability issues of the graph-of-entity, as well as to improve the expressiveness and flexibility of the model and thus its generality. We evaluated both models based on subsets of the INEX 2009 Wikipedia collection, measuring mean average precision, precision at a cutoff of 10 and normalized discounted cumulative gain for the top-p results. While we were able to propose a hypergraph-based model and a universal ranking function (the random walk score), we have only tested ad hoc document retrieval and ad hoc entity retrieval. We are working on an experiment based on entity list completion, which we hope will further support our idea of a unified framework for entity-oriented search. All three experiments were designed around the topics and relevance judgments from the INEX 2010 Ad Hoc track and the INEX 2009 XML Entity Ranking track. Evaluation depends on the existence of test collections based on combined data [4, Def. 2.3], with multiple associated topics and relevance judgments corresponding to the different tasks to be supported and generalized by the model. So far, we have identified the INEX 2009 Wikipedia collection, as well as the TREC Washington Post Corpus as two potential collections to be used in this context.

5 Research Issues

- Can we jointly represent unstructured and structured data as a graph?
- Will this unlock novel ranking strategies?
- Is it possible for these ranking strategies to support the generalization of entity-oriented search tasks?

– Will the incorporation of explicit and implicit information derived from the relations between text, found in corpora, and entities, found in knowledge bases, improve overall retrieval effectiveness? Or is there, instead, a trade-off between generalization and effectiveness?

Acknowledgements. José Devezas is supported by research grant PD/BD/ 128160/2016, provided by the Portuguese national funding agency for science, research and technology, Fundação para a Ciência e a Tecnologia (FCT), within the scope of Operational Program Human Capital (POCH), supported by the European Social Fund and by national funds from MCTES.

References

1. Auer, S., Bizer, C., Kobilarov, G., Lehmann, J., Cyganiak, R., Ives, Z.: DBpedia: a nucleus for a web of open data. In: Aberer, K., et al. (eds.) ASWC/ISWC -2007. LNCS, vol. 4825, pp. 722–735. Springer, Heidelberg (2007). https://doi.org/10. 1007/978-3-540-76298-0_52
2. Balog, K.: Entity-Oriented Search: TIRS, vol. 39. Springer, Cham (2018). https:// doi.org/10.1007/978-3-319-93935-3_9
3. Bast, H., Buchhold, B.: An index for efficient semantic full-text search. In: Proceedings of the 22nd ACM International Conference on Conference on Information and Knowledge Management, pp. 369–378 (2013). https://doi.org/10.1145/ 2505515.2505689
4. Bast, H., Buchhold, B., Haussmann, E., et al.: Semantic search on text and knowledge bases. Found. Trends® Inf. Retrieval **10**(2–3), 119–271 (2016)
5. Blanco, R., Lioma, C.: Graph-based term weighting for information retrieval. Inf. Retrieval **15**(1), 54–92 (2012). https://doi.org/10.1007/s10791-011-9172-x
6. Devezas, J., Lopes, C., Nunes, S.: Graph-of-entity: a model for combined data representation and retrieval. In: 8th Symposium on Languages, Applications and Technologies, SLATE 2019, 27–28 June 2019, Coimbra, Portugal (2019). https:// doi.org/10.4230/OASIcs.SLATE.2019.1
7. Devezas, J., Nunes, S.: Hypergraph-of-entity: A unified representation model for the retrieval of text and knowledge. Open Comput. Sci. J. **10**(1) (2019)
8. Manning, C.D., Raghavan, P., Schütze, H., et al.: Introduction to information retrieval, vol. 1. Cambridge University Press, Cambridge (2008)
9. Pérez, J., Arenas, M., Gutiérrez, C.: Semantics and complexity of SPARQL. ACM Trans. Database Syst. **34**(3), 16:1–16:45 (2009). https://doi.org/10.1145/1567274. 1567278
10. Robertson, S.E.: The methodology of information retrieval experiment. In: Information Retrieval Experiment, pp. 9–31. Butterworth-Heinemann (1981)
11. Rousseau, F., Vazirgiannis, M.: Graph-of-word and TW-IDF: new approach to ad hoc IR. In: Proceedings of the 22nd ACM International Conference on Information & Knowledge Management, pp. 59–68. ACM (2013)
12. Schenkel, R., Suchanek, F.M., Kasneci, G.: YAWN: a semantically annotated Wikipedia XML corpus. In: Datenbank systeme in Business, Technologie und Web (BTW 2007), 12. Fachtagung des GI-Fachbereichs "Datenbanken und Informationssysteme" (DBIS), Proceedings, 7–9 März 2007, Aachen, Germany, pp. 277–291 (2007). http://subs.emis.de/LNI/Proceedings/Proceedings103/article1404.html

13. Voorhees, E.M.: The philosophy of information retrieval evaluation. In: Peters, C., Braschler, M., Gonzalo, J., Kluck, M. (eds.) CLEF 2001. LNCS, vol. 2406, pp. 355–370. Springer, Heidelberg (2002). https://doi.org/10.1007/3-540-45691-0_34

14. Vrandečić, D., Krötzsch, M.: Wikidata: a free collaborative knowledge base. Commun. ACM **57**, 78–85 (2014). http://cacm.acm.org/magazines/2014/10/178785-wikidata/fulltext

Graph Databases for Information Retrieval

Chris Kamphuis[(✉)]

Radboud University, Nijmegen, The Netherlands
`chris@cs.ru.nl`

Abstract. Graph models have been deployed in the context of information retrieval for many years. Computations involving the graph structure are often separated from computations related to the base ranking. In recent years, graph data management has been a topic of interest in database research. We propose to deploy graph database management systems to implement existing and novel graph-based models for information retrieval. For this a unifying mapping from a graph query language to graph based retrieval models needs to be developed; extending standard graph database operations with functionality for keyword search. We also investigate how data structures and algorithms for ranking should change in presence of continuous database updates. We want to investigate how temporal decay can affect ranking when data is continuously updated. Finally, can databases be deployed for efficient two-stage retrieval approaches?

Keywords: Graph databases · Information retrieval · Query languages

1 Motivation

Many IR systems make use of graph-based models. For social media Clements et al. [5] show this by using random walks over typed social media. In the context of websearch, Metzler et al. [9] and Craswell et al. [6] show the usefulness of using anchor text. Vuurens et al. [12] presented a news tracker for ad-hoc information needs. They used a temporal graph model; implemented as a standalone application over streaming graphs. Entity oriented search has been another topic of interest in the IR community [1,2], where researchers make extensive use of knowledge bases to improve ranking. It would be a natural choice to use graph databases to represent these models, especially if the system allows for data to be updated continuously. The goal of this PhD research is to represent graph-based models in a graph database that also supports keyword search. This would allow graph based methods to be combined with traditional keyword search, without using different system components.

2 Background and Related Work

There has been a long history of IR research in combination with relational databases. Recently, Mühleisen et al. [10] implemented such a system, where

© Springer Nature Switzerland AG 2020
J. M. Jose et al. (Eds.): ECIR 2020, LNCS 12036, pp. 608–612, 2020.
https://doi.org/10.1007/978-3-030-45442-5_79

they achieved efficiency and effectiveness on par with custom made inverted index systems. They argue that using relational databases instead of custom-made inverted indexes has the following advantages for IR.

Firstly, viewing IR as a database application offers a formal framework for complex query operators. This forces IR researchers to be precise on how to deal with more complex query operators and their edge cases. By expressing the ranking logic in query languages, it is not possible to resort to heuristics or shortcuts. This helps explaining the resulting rankings of documents. Secondly, a relational database offers a clean system architecture. Storage management is taken care of by the database engine, separating low level components shared in any data application from IR components. Thirdly, advances in database research offer benefits on systems build using databases. All performance gains in database engines will directly trickle down to the IR system. Fourthly, databases offer additional tools for error analysis. E.g., join together relevance judgement, document representations and the result set to generate scatter plots. Finally, it offers opportunities for rapid prototyping. Many IR researchers might not be interested in data management and query evaluation. This can be taken care of by the database engine, researches can focus on methods for ranking.

In previous work [7] we argued how a graph query languages for IR also yield benefits for reproducible IR research.

Industry has also shown an interest in graph databases in combination with full text IR. Neo4J, the most popular database[1], supports graph analysis in combination with full text analysis using Lucene[2]. Lucene is however embedded in their graph query language, while we envision retrieval to be carried out by the same system that does the graph processing. Busch et al. [4] presented Earlybird, a system that allows continuous updates to be searched real-time. They show how updates can be processed in the context of inverted indexes. A similar approach could be taken in the context of graph databases. Böttcher et al. [3] presented an approach how updates for compressed databases can be implemented efficiently by updating the data without decompressing all data.

3   Proposed Research and Methodology

How can keyword search be integrated with a graph query language?
Many retrieval models make use of graph information, often graph-based models support traditional keyword search. When implementing graph-based models in a graph database, it should be able to integrate graph-based results with keyword search seamlessly.

As shown by Mühleisen et al. [10], keyword search can be expressed as SQL queries over relational databases and executed efficiently. The goal of the proposed research is to extend this idea, and express keyword search in graph databases. This could be approached in two ways. One way would be to try

---

[1] https://db-engines.com/en/ranking/graph+dbms, Last Accessed: November 2019.

[2] https://neo4j.com/docs/cypher-manual/3.5/schema/index/#schema-index-fulltext-search, Last Accessed: November, 2019.

to express the keyword ranking functions as structured queries in a graph query language. This could for example be achieved by representing documents and terms as properties in the graph e.g. vertices themselves, and translate classic keyword search algorithms to functions over the graph. Especially in the context of entity ranking, where entities are part of a knowledge base, this would make sense. Another approach would be to follow Mühleisen et al., and represent the graph database as an extension on top of the relational database. Using the relational database as the core for a graph database would be a natural fit, as graph traversal algorithms can be expressed using recursive join operations. As keyword search can already be expressed in the relational database, combining keyword search with the additional graph information will computationally be cheap.

What data structures and algorithms should be used when data is continuously updated? Often when IR systems are being developed, they are evaluated on static data. When deploying such a system to production, it might not be as effective or efficient when data is constantly changing. For example, index compression algorithms might assign document identifiers to documents according to their contents; delta gaps should be as small as possible. When data is constantly updated, many of these algorithms will not work. Graph data does tend to appear in the context when data constantly is changing. Ideally the data does not have to be re-indexed all the time, the underlying index can just be updated when changes to the data are being presented.

In order for a graph database system to be useful in an IR context, it needs to be efficient in both handling updates to data and when it is used for ranking. In our research we want to investigate which algorithms and data structures can be used to efficiently handle both database transactions and retrieval.

How should edges/vertices be ranked in the presence of continuous updates? When graph databases do support continuous updates, temporal graphs could be represented in the graph database. One could add timestamps to vertices to store useful temporal information of the graph (e.g. when the is vertex added). The question then arises how the temporal evolution of the graph should affect the ranking scores of the edges and/or vertices. Examples of use-cases where this would be especially interesting include tweets and comments on videos, where not only their content but also their recency are relevant. Using a temporal graph database would also be an ideal setup to establish a connection between temporal decay and ranking in models for temporal summarizing as described in the Real-Time Summarization track [8].

Can a database be used in order to unify different stages in the ranking process? IR systems often consist of multiple parts that carry out different stages in the retrieval process. An initial ranker calculates an initial ranking score, e.g. BM25, over inverted indexes. After the initial ranking stage, a more effective, but computationally more expensive method re-ranks the top-k retrieved documents. This second stage re-ranker is often completely detached from the initial ranker, e.g. neural approaches and learning-to-rank methods are often implemented in different programming languages than the initial ranker. Data from the inverted index and results from the initial retrieval step, need to be copied in order for the re-ranking algorithm to work. Often re-ranking algorithms work with different data structures than inverted indexes do, so it might be necessary to restructure the data also. These mismatches between retrieval stages can introduce latency in the full retrieval pipeline. In order to avoid this latency, ideally the initial ranking stage and the re-ranking stage can be expressed in the same programming language using the same data structures such that latency introduced by re-ranking is only an effect of the re-ranking algorithm itself.

Recently Raasveldt and Mühleisen [11] presented DuckDB, an analytical embeddable database system. It has been specifically designed for executing analytical queries fast in an embedded environment. This research proposes to use DuckDB as a database backend for fast top-k retrieval by implementing methods as described by Mühleisen et al. [10]. DuckDB has bindings for Python, the *de facto* language used for experiments with neural methods. DuckDB supports extracting the results of the issued queries directly to NumPy arrays, allowing for (neural) re-rankers to quickly start the re-ranking without first having to move the data. We want to investigate whether the results of Mühleisen et al. [10] also hold for DuckDB. If so, latency introduced by moving data might be minimized, and re-ranking systems might be able to re-rank the top-k faster. Allowing more re-ranking methods to be deployed in production.

4   Discussion

Are graphs to correct abstraction level in the context for IR? One could argue that graphs are not the correct abstraction level in the context of IR. Specifically expressing keyword queries as graph database queries might make things more complex. We would argue that although this might be seen as a disadvantage, it would allow for more complex (graph inherent) structures to be integrated with keyword search more easily.

How should graphs be constructed for text documents? It would be possible to construct graphs from text documents in different ways. What would be the right granularity of the graph? Edges could represent terms in a document and link terms together when they appear in the same document. Maybe it would make more sense to express sentences or even documents as edges in the graph. For example in the context of the web, a document level graph could make sense. It would also make sense to change the granularity depending on the search task. When processing the text documents, should preprocessing (stopping, stemming) be integrated in the graph query language or should the graph query language only be used for querying the database retrieval time?

Acknowledgements. This work is part of the research program Commit2Data with project number 628.011.001 (SQIREL-GRAPHS), which is (partly) financed by the Netherlands Organisation for Scientific Research (NWO).

References

1. Balog, K.: Entity-Oriented Search. Springer, Cham (2018). https://doi.org/10.1007/978-3-319-93935-3
2. Balog, K., Serdyukov, P., De Vries, A.P.: Overview of the trec 2010 entity track. Tech. rep. Norwegian University of Science and Technology Trondheim (2010)
3. Böttcher, S., Bültmann, A., Hartel, R., Schlüßler, J.: Implementing efficient updates in compressed big text databases. In: Decker, H., Lhotská, L., Link, S., Basl, J., Tjoa, A.M. (eds.) DEXA 2013. LNCS, vol. 8056, pp. 189–202. Springer, Heidelberg (2013). https://doi.org/10.1007/978-3-642-40173-2_17
4. Busch, M., Gade, K., Larson, B., Lok, P., Luckenbill, S., Lin, J.J.: Earlybird: real-time search at twitter. In: 2012 IEEE 28th International Conference on Data Engineering, pp. 1360–1369 (2012)
5. Clements, M., De Vries, A.P., Reinders, M.J.T.: The task-dependent effect of tags and ratings on social media access. ACM Trans. Inf. Syst. **28**(4), 1–42 (2010). https://doi.org/10.1145/1852102.1852107
6. Craswell, N., Hawking, D., Robertson, S.: Effective site finding using link anchor information. In: Proceedings of the 24th Annual International ACM SIGIR Conference on Research and Development in Information Retrieval, SIGIR 2001, pp. 250–257. ACM, New York (2001). https://doi.org/10.1145/383952.383999
7. Kamphuis, C., de Vries, A.P.: Reproducible IR needs an (IR) (graph) query language. In: Proceedings of the Open-Source IR Replicability Challenge Co-Located with 42nd International ACM SIGIR Conference on Research and Development in Information Retrieval, OSIRRC@SIGIR 2019, Paris, France, 25 July 2019, pp. 17–20 (2019). http://ceur-ws.org/Vol-2409/position03.pdf
8. Lin, J., et al.: Overview of the TREC 2017 real-time summarization track. In: TREC (2017)
9. Metzler, D., Novak, J., Cui, H., Reddy, S.: Building enriched document representations using aggregated anchor text. In: Proceedings of the 32nd International ACM SIGIR Conference on Research and Development in Information Retrieval, SIGIR 2009, pp. 219–226. ACM, New York (2009). https://doi.org/10.1145/1571941.1571981
10. Mühleisen, H., Samar, T., Lin, J., de Vries, A.: Old dogs are great at new tricks: column stores for IR prototyping. In: Proceedings of the 37th Annual International ACM SIGIR Conference on Research and Development in Information Retrieval, SIGIR 2014, Gold Coast, Australia, pp. 863–866 (2014)
11. Raasveldt, M., Mühleisen, H.: [demo] DuckDB: an embeddable analytical database. In: Proceedings of the ACM SIGMOD International Conference on Management of Data, pp. 1981–1984, June 2019. https://doi.org/10.1145/3299869.3320212
12. Vuurens, J.B., de Vries, A.P., Blanco, R., Mika, P.: Online news tracking for ad-hoc information needs. In: Proceedings of the 2015 International Conference on The Theory of Information Retrieval, ICTIR 2015, pp. 221–230. ACM, New York (2015). https://doi.org/10.1145/2808194.2809474

Towards a Better Contextualization of Web Contents via Entity-Level Analytics

Amit Kumar[✉]

Department of Computer Science, Université de Caen Normandie,
Campus Côte de Nacre, 14032 Caen Cedex, France
amit.kumar@unicaen.fr

Abstract. With the abundance of data and wide access to the internet, a user can be overwhelmed with information. For an average Web user, it is very difficult to identify which information is relevant or irrelevant. Hence, in the era of continuously enhancing Web, organization and interpretation of Web contents are very important in order to easily access the relevant information. Many recent advancements in the area of Web content management such as classification of Web contents, information diffusion, credibility of information, etc. have been explored based on text and semantic of the document. In this paper, we propose a purely semantic contextualization of Web contents. We hypothesize that named entities and their types present in a Web document convey substantial semantic information. By extraction of this information, we aim to study the reasoning and explanation behind the Web contents or patterns. Furthermore, we also plan to exploit LOD (Linked Open Data) to get a deeper insight of Web contents.

Keywords: Contextualization · Knowledge extraction · Entity-level analytics

1 Introduction and Motivation

Even in the 30th year of World Wide Web, we can still observe the enormous amount of growth in the Web contents being created and subsequently available to any internet user. Because of the inconsistency of the data being generated, it is very hard for an ordinary user to distinguish the Web contents according to their societal relevance. With the availability of NER techniques [5] and LOD [1,11,13], we have access to a lot of information about the named entities described in a Web content. This contextualization of the entities contained in a text can help us to deal with Web contents. We observe that, for a text describing an event, there are specific recurring patterns of entity types appearing together. For instance, in the case of 'natural disasters', entities like organizations, countries, presidents appear together whereas in the case of 'political events', entities like parties, leaders, business-persons appear together. The availability of tools

© Springer Nature Switzerland AG 2020
J. M. Jose et al. (Eds.): ECIR 2020, LNCS 12036, pp. 613–618, 2020.
https://doi.org/10.1007/978-3-030-45442-5_80

like AIDA [16] or DBPedia Spotlight [12], which can interlink text documents to LOD has provided us efficient means to capture the semantics of a plain text using the entity-level.

The purpose of this study is to analyze those large amounts of data and to help the user in getting a better semantic understanding of a Web document. To this end, we aim to study a Web document semantically using entity-level analytics. Ultimately, we plan to exploit and aggregate external knowledge using LOD for the proper contextualization of a Web content.

2   Background and Related Work

Knowledge bases (KBs) are an effective way to store Web documents semantically in a structured format. Because of easy accessibility, these KBs are fruitful resources for many tasks in information retrieval [4] and natural language processing [9]. Recently, researchers from different domains have developed different knowledge acquisition approaches for the creation of knowledge graphs. This results in an emanation of large publicly accessible KBs such as Freebase [1], DBPedia [11], YAGO [13], which accommodate spatial and temporal information in addition to structural knowledge. Many applications such as complex event detection [17], named entities disambiguation [14] or social media topic classification [3] from various domains have acquired the benefit by integrating knowledge from LOD.

Entity-level analytics aggregates semantic information by incorporating knowledge about an entity or its types. The problem of event diffusion prediction into foreign language communities [8] has shown encouraging results with the assimilation of knowledge about the entities contained in a document. Here the introduced framework ELEVATE only utilizes the information about the entities in the document and resources from YAGO [13]. In [7], the authors address the task of Web content fine-grained hierarchical classification. They hypothesize that a document is symbolized by the named entities it comprised. They propose the idea of the 'semantic fingerprinting' method that expresses the overall semantics of a Web document by a compact vector. Entity-level analytics is also effective in computational fact checking of information [2]. The authors claim that human fact checking can be achieved by finding the shortest path on a conceptually or semantically defined network such as knowledge graphs (KGs).

Entity-level analytics provide a depth insight into Web contents. KGs carry a lot of information about entities, but all the information is not equally important for a given text. The novelty of this thesis is to discriminate between interesting and uninteresting semantic information about entities w.r.t. the context of a text.

3   Current Work

3.1   CALVADOS : For Entity-Level Content Analysis

The idea of semantic fingerprinting [6] - as an approach towards Web analytics was well acknowledged by researchers in the semantic Web community. Thus, we presented the CALVADOS system [7] as an extension of semantic fingerprinting. At first, this system filters all the named entities present in a given text. By utilization of type information for all these entities from YAGO, it creates a representative vector (called semantic fingerprint) for the text. In last, it predicts the fine-grained type of the content using machine learning techniques. Moreover, it reports the semantic building block of the text. Figure 1 outlines the conceptual pipeline. The notable contributions of the mentioned scientific article are:

- employ `semantic fingerprint` to represent document's semantics
- exploration and visualization of dependencies among entities comprised
- data digestion supported by providing contextual KB data (e.g., types)

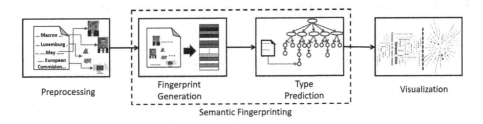

Fig. 1. Conceptual overview of the CALVADOS pipeline

3.2 Concise Entity-Type Extraction

Exploitation of named entities and their types are always valuable in getting a better contextualization of a Web document [7]. But, sometimes we can be easily drowned by too much information. For example, recent articles involving Donald Trump deal with his position of president. But Trump has 76 facets (considering Wordnet types) like communicator, president, business-person, etc., which are not equally relevant. For some entities, it is even more complicated. For instance, Arnold Schwarzenegger is famous to be an actor, a politician and a bodybuilder. When an article deals with him, the context of the article makes us understand which facet is relevant, e.g., 'actor' if the article deals with a film release. Hence, it arises two research questions:

- **RQ1:** What are the most relevant type(s) for an entity in general (i.e., without context)?
- **RQ2:** What are the most relevant type(s) for an entity in a given context?

Computational Model: Let $d_i \epsilon D$ represents a document. The named entities associated with a document d_i is given by $N(d_i)$. $T(n_j)$ represents all the k types associate with any entity n_j. Entity-level document type is represented by d_i^t as shown in Eq. 4. Our task is to select m number of types from $T(n_j)$, where

$m << k$. We define two types of models - First, for the calculation of T_{Gen} (*i.e.* **RQ1**, Eq. 5) and second, for the calculation of T_{Con} (*i.e.* **RQ2**, Eq. 6).

$$D = d_1, d_2.....d_p \tag{1}$$

$$N(d_i) = n_1^i, n_2^i, ...n_q^i \tag{2}$$

$$T(n_j) = t_1^j, t_2^j, ...t_k^j \tag{3}$$

$$\text{Entity-level typed document, } d_i^t = \varphi(n_1^i, n_2^i, ...n_q^i) \tag{4}$$

$$T_{Gen} = f_{gen}(T(n_j)) \tag{5}$$

$$T_{Con} = f_{con}(T(n_j), text) \tag{6}$$

Currently, we have focus on our first research question (**RQ1**). Our first challenge is to find or develop the appropriate data set for the aforementioned task.

Gold Standard Creation: It is not easy to find the most relevant type(s) for an entity in general. We create the gold standard based on the Wordnet hierarchy mentioned in YAGO (1981 types). We consider that the most relevant type(s) for any entity is mentioned in its Wikipedia page. Precisely, we extract the types that are mentioned in the first or second sentence of entity's Wikipedia page and map them to the mentioned hierarchy.

Example – Extracted Wikipedia labels for 'Arnold Schwarzenegger' are actor, filmmaker, businessman, author, bodybuilder and politician. After mapping of these labels to Schwarzenegger's Wordnet hierarchy in YAGO, the ground truths are *actor, film-maker, businessman, bodybuilder and politician.*

Experimental Pipeline: Our next challenge is to find the suitable mechanism for concise entity-type prediction in general. We rely only on the structural information, which we get by exploring knowledge graph of entity in YAGO. We implemented several techniques as baselines but none of these techniques show promising results. These models are:

- **Based on Leaf Node:** We had the intuition that the most specific or relevant type of an entity should be at the deepest in the YAGO Wordnet entity's hierarchy. So, we picked the type that is at the deepest in the hierarchy.
- **Based on Branching Factor:** While implementing the model based on leaf node, we observe that sometimes, we were selecting a too specific type, e.g., forward (child of football-player in the hierarchy) instead of football-player. So, we decided to pick the node that has the highest branching factor (number of direct children) and at the deepest in the hierarchy.
- **Based on ML Classifier:** We developed a model based on random forest. We used all the Wordnet types present in the entity's hierarchy as features.

We aim to develop a relevant types prediction model based on Graph Neural Network [15]. More specifically, we utilize the concept of Graph Convolutional Network (GCN) [10]. While implementation, we faced the following challenges:

- In GCN, Readout function [15] is used to get embedding for the graph based on an aggregation of node features from the final iteration. Finding the suitable Readout function for our task is one of the main challenges.
- Entity's graph is a sub-graph of YAGO Wordnet hierarchy. Only delivering the structure of the sub-graph is not sufficient. It needs label information along with the structure of sub-graph. Encoding node label is our next challenge. One hot encoding is one of the solutions for giving the label information.

4   Challenges and Next Steps

Based on our proposed research and current work progress, we find the following challenges to handle in the near future:

- In the early stage of our experiments, we realise that some of the types within a category are very hard to predict, e.g., in person category - there are entities with ground truth types 'intellectual' or 'military officer' where the model fails to predict it correctly. Our challenge is to find the common patterns among these sub-categories and to propose the solution for this failure.
- Our next challenge is to develop a gold standard for task 2 (**RQ2**).
- Our last challenge is to develop method for types prediction in a given context.

References

1. Bollacker, K., et al.: Freebase: a collaboratively created graph database for structuring human knowledge. In Proceedings of SIGMOD 2008, pp. 1247–1250. ACM (2008)
2. Ciampaglia, G.L., et al.: Computational fact checking from knowledge networks. PLoS ONE **10**(6), 1–13 (2015)
3. Cano, A.E., et al.: Harnessing linked knowledge sources for topic classification in social media. In: Proceedings of the 24th ACM Conference on Hypertext and Social Media, HT 2013, pp. 41–50. ACM (2013)
4. Dalton, J., Dietz, L., Allan, J.: Entity query feature expansion using knowledge base links. In: Proceedings of SIGIR 2014, pp. 365–374. ACM (2014)
5. Finkel, J. R., Grenager, T., Manning, C.: Incorporating non-local information into information extraction systems by Gibbs sampling. In: Proceedings of ACL 2005, pp. 363–370. ACL (2005)
6. Govind, Alec, C., Spaniol, M.: Semantic fingerprinting: a novel method for entity-level content classification. In: Mikkonen, T., Klamma, R., Hernández, J. (eds.) ICWE 2018. LNCS. Springer, Cham (2018). https://doi.org/10.1007/978-3-319-91662-0_21
7. Govind, Kumar, A., Alec, C., Spaniol, M.: CALVADOS: a tool for the semantic analysis and digestion of web contents. In: Hitzler, P., et al. (eds.) ESWC 2019. LNCS, pp. 1–6. Springer, Cham (2019). https://doi.org/10.1007/978-3-030-32327-1_17
8. Govind, Spaniol, M.: ELEVATE: a framework for entity-level event diffusion prediction into foreign language communities. In: Proceedings of WebSci 2017, pp. 111–120. ACM (2017)

9. Hao, Y., et al.: Pattern-revising enhanced simple question answering over knowledge bases. In: Proceedings of COLING 2018, pp. 3272–3282. ACL (2018)

10. Kipf, T.N., Welling, M.: Semi-supervised classification with graph convolutional networks. In: International Conference on Learning Representations, ICLR 2017 (2017)

11. Lehmann, J., et al.: DBpedia-a large-scale, multilingual knowledge base extracted from Wikipedia. Semant. Web J. **6**, 167–195 (2015)

12. Mendes, P.N., Jakob, M., García-Silva, A., Bizer, C.: Dbpedia spotlight: shedding light on the web of documents. In: Proceedings of I-Semantics 2011, pp. 1–8. ACM (2011)

13. Rebele, T., Suchanek, F., Hoffart, J., Biega, J., Kuzey, E., Weikum, G.: YAGO: a multilingual knowledge base from Wikipedia, Wordnet, and Geonames. In: Groth, P., et al. (eds.) ISWC 2016. LNCS, vol. 9982, pp. 177–185. Springer, Cham (2016). https://doi.org/10.1007/978-3-319-46547-0_19

14. Usbeck, R., et al.: AGDISTIS - graph-based disambiguation of named entities using linked data. In: Mika, P., et al. (eds.) ISWC 2014. LNCS, vol. 8796, pp. 457–471. Springer, Cham (2014). https://doi.org/10.1007/978-3-319-11964-9_29

15. Xu, K., Hu, W., Leskovec, J., Jegelka, S.: How powerful are graph neural networks? In: International Conference on Learning Representations, ICLR 2019 (2019)

16. Yosef, M.A., et al.: AIDA: an online tool for accurate disambiguation of named entities in text and tables. In: Proceedings of VLDB, vol. 2011, pp. 1450–1453 (2011)

17. Yan, Y., et al.: Event oriented dictionary learning for complex event detection. IEEE Trans. Image Process. **24**(6), 1867–1878 (2015)

Incremental Approach for Automatic Generation of Domain-Specific Sentiment Lexicon

Shamsuddeen Hassan Muhammad[1,2,3](✉) [ID], Pavel Brazdil[1] [ID],
and Alípio Jorge[1,2] [ID]

[1] LIAAD - INESC TEC, Porto, Portugal
{shamsuddeen.muhammad,pbrazdil,alipio.jorge}@inesctec.pt
[2] Faculty of Sciences - University of Porto, Porto, Portugal
[3] Bayero University, Kano, Nigeria

Abstract. Sentiment lexicon plays a vital role in lexicon-based sentiment analysis. The lexicon-based method is often preferred because it leads to more explainable answers in comparison with many machine learning-based methods. But, semantic orientation of a word depends on its domain. Hence, a general-purpose sentiment lexicon may gives sub-optimal performance compare with a domain-specific lexicon. However, it is challenging to manually generate a domain-specific sentiment lexicon for each domain. Still, it is impractical to generate complete sentiment lexicon for a domain from a single corpus. To this end, we propose an approach to automatically generate a domain-specific sentiment lexicon using a vector model enriched by weights. Importantly, we propose an incremental approach for updating an existing lexicon to either the same domain or different domain (domain-adaptation). Finally, we discuss how to incorporate sentiment lexicons information in neural models (word embedding) for better performance.

Keywords: Domain-specific · Sentiment analysis · Sentiment lexicon · Word embedding · Machine learning

1 Motivation

Sentiment lexicon is a dictionary of a lexical item with the corresponding semantic orientation. Recently, with the issue of growing concern about interpretable and explainable artificial intelligence, domains that require high explainability in sentiment analysis task (eg., health domain and financial domain), lexicon-based sentiment analysis approaches are often preferred over machine-learning-based approaches [12,13]. However, sentiment lexicons are domain-dependent, a word may convey two different connotations in a different domain. For example, the word *high* may have a positive connotation in economics (e.g., *he has a high salary*), and negative connotation in medicine (e.g., *he has a high blood pressure*). Therefore, general-purpose sentiment lexicon may not give the expected

© Springer Nature Switzerland AG 2020
J. M. Jose et al. (Eds.): ECIR 2020, LNCS 12036, pp. 619–623, 2020.
https://doi.org/10.1007/978-3-030-45442-5_81

predictive accuracy across different domains. Thus, a lexicon-based approach with domain-specific lexicons are used to achieve better performance [1,4].

Although research has been carried out on corpus-based approaches for automatic generation of a domain-specific lexicon [1,4,5,7,9,10,14], existing approaches focused on creation of a lexicon from a single corpus [4]. Afterwards, one cannot automatically update the lexicon with a new corpus. There are many reasons one would want to update an existing lexicon: (i) the existing lexicon may not contain sufficient number of sentiment-bearing words (i.e., it is limited) and it needs to be extended with a corpus from the same domain with a source corpus; (ii) the language may have evolved (new words and meaning changes) and it is necessary to update the existing lexicon with a new corpus. The new corpus may not be large to enable generation of a new lexicon from scratch. Thus, it is better to update the existing lexicon with the new corpus; and (iii) we need to update an existing lexicon to another domain (domain-adaptation) with a corpus from different domain with the source corpus. To this end, this work proposes an incremental approach for the automatic generation of a domain-specific sentiment lexicon.

2   Research Questions and Methodology

We aim to investigate an incremental technique for automatically generating domain-specific sentiment lexicon from a corpus. Specifically, we aim to answer the following three research questions:

RQ1: Can we automatically generate a sentiment lexicon from a corpus and improves the existing approaches?

RQ2: Can we automatically update an existing sentiment lexicon given a new corpus from the same domain (i.e., to extend an existing lexicon to have more entries) or from a different domain (i.e., to adapt the existing lexicon to a new domain - domain adaptation)?

RQ3: How can we enrich the existing sentiment lexicons using information obtained from neural models (word embedding)?

To the best of our knowledge, no one attempted to design an approach for automatic construction of a sentiment lexicon in an incremental fashion. But, incremental approaches are common in the area of data streaming [15]; thus, our work could fill this gap and represent a novel contribution. The research plan is structured as follows: Sect. 2.1 attempts to answer RQ1, Sect. 2.2 attempts to answer RQ2, and Sect. 2.3 attempts to answer RQ3.

2.1   Sentiment Lexicon Generation Using Weight Vector Model (non-Incremental)

Sattam et al. [4] introduced a novel domain agnostic sentiment lexicon-generation approach from a review corpus annotated with star-ratings. We propose an extended approach that includes the use of weight vector. Also, our approach includes verbs and nouns in the lexicon as studies show they contain sentiment [7,11]. The process includes the following four steps: (i) gathering data annotated with star-ratings; (ii) pre-processing the data; (iii) obtaining word-tag rating distribution, as in Fig. 1 from the corpus introduced in [16]; and (iv) generation of sentiment value for each word-tag pair using the equation: $SV_{w-T} = \sum_1^{10} FR_{w-T} * W$. Where FR_{w-T} represents the frequency of word-tag pair and W is a weight vector. If the result is positive, the word is categorize as positive, otherwise it is negative. This basic approach of sentiment lexicon generation forms the basis of the incremental approach proposes in Sect. 2.2.

2.2 Incremental Approach for Sentiment Lexicon Generation Using Sufficient Statistics

We propose an incremental approach for sentiment lexicon expansion to either the same domain or different domain (domain-adaptation). To illustrate the approaches, assume we have a sentiment lexicon L_i generated from a corpus C_i(using the approach described in Sect. 2.1). Then, we receive a new batch of corpus C_{i+1} (of the same or different domain with C_i). The incremental approach aims to generate an updated sentiment lexicon L_{i+1} that would improve the accuracy of the lexicon L_i.

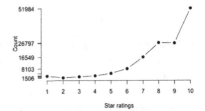

Fig. 1. Distribution of the word-tag pair(excellent, a)

Updating Lexicon Using a Corpus in the Same Domain: Assume we receive C_{i+1} and we want to update L_i. Assume we have the distributions of all the words in the previous corpus (C_i) saved. A naive approach would involve generating distributions of all the words in the new batch (C_{i+1}) without creating a new lexicon from it. Such a distribution represents the so-called "sufficient statistics" [15] and we can construct lexicon from each set of the distributions. To update L_i, the two sets of distributions (from C_i and C_{i+1}) are first merged and updated lexicon (L_{i+1}) is generated using the approach described in Sect. 2.1. However, this approach may be inefficient since we update all the words in the existing lexicon.

An enhanced and more efficient approach aims to update only subset of the words in L_i whose orientation may have changed. This approach use L_i to predict the user's sentiment rating scores on the new labelled corpus C_{i+1} sentences. If the predicted rating scores are the same with the user's sentiment ratings, we can skip those sentences and only consider those sentences where the predicted rating is significantly different from the user's sentiment rating scores. We extract the words from these sentences (reviews), elaborate the corresponding distribution of sentiment values, merge the distribution with the corresponding subset in the L_i and generate a new sentiment lexicon L_{i+1}.

Updating Lexicon Using a Corpus from Different Domain: Assume we receive C_{i+1} and we want to update L_i to a new domain. Firstly, we propose to detect if C_{i+1} and C_i are from different domain. To do this, we generate the distribution of C_{i+1} and compare it with the distribution of C_i. If the distributions of the two corpora differ significantly, it indicates a domain shift. Alternatively, we can use L_i to predict the user's sentiment rating scores on the new labelled corpus C_{i+1} sentences. If the prediction accuracy is below some predefined threshold, we can conclude there is a domain shift. After detecting the domain shift, we merge the distribution using a similar approach discussed (in updating using the same corpus) and generate the lexicon. However, in this case, we give different weight to the two distributions by taking into consideration not only their size, but also recency. More recent batches will be given more weight than the previous ones.

2.3 Word Embedding

The idea of word embedding have been widely used for generation of sentiment lexicon because of their advantage for giving semantic representation of words [9]. If two words appear in similar contexts, they will have similar embedding. We propose to use word embedding in the following way. Suppose we have seed words with their sentiment values, and we encounter some word, say Wx, for which we do not have a sentiment value (SVal) yet. But if we have its embedding, we can look for the most similar embedding in the embedding space and retrieve the corresponding word, Wy, retrieve its SVal and use it as a SVal of Wx. As reported in [11], neural models performance can increase by including lexicon information. We aim to further study litreture and find how to exploit combination of an existing sentiment lexicon (more explainable) and neural models performance.

2.4 Evaluation

We plan to evaluate our system and compare it with other five existing lexicons: SentiWords, SPLM, SO-CAL, Bing Liu's Opinion Lexicon, and SentiWordNet [14]. The evaluation task will be on three sentiment analysis tasks (movie review, polarity of tweets and hotel review). In these comparisons we will compare (1) the precision of the predictions of sentiment values and (2) runtime to carry out updates of the lexicon.

3   Research Issues for Discussion

We seek suggestions on how our proposal can be improved. More importantly, discussion on how to exploit combination of word embedding with sentiment lexicon. We also welcome comments.

Acknowledgement. This project was partially financed by the Portuguese funding agency, FCT - Fundação para a Ciência e a Tecnologia, through national funds, and co-funded by the FEDER.

References

1. Xing, F.Z., Pallucchini, F., Cambria, E.: Cognitive-inspired domain adaptation of sentiment lexicons. Inf. Process. Manag. (2019). https://doi.org/10.1016/j.ipm.2018.11.002
2. Liu, B.: Sentiment Lexicon Generation. In: Sentiment Analysis: Mining Opinions, Sentiments, and Emotions, pp. 189–201. Cambridge University Press, Cambridge (2015). https://doi.org/10.1017/CBO9781139084789.008
3. Alqasemi, F., Abdelwahab, A., Abdelkader, H.: Constructing automatic domain-specific sentiment lexicon using KNN search via terms discrimination vectors. Int. J. Comput. Appl. (2019). https://doi.org/10.1080/1206212X.2017.1409477
4. Almatarneh, S., Gamallo, P.: Automatic construction of domain-specific sentiment lexicons for polarity classification. In: Advances in Intelligent Systems and Computing. pp. 175–182 (2017). https://doi.org/10.1007/978-3-319-61578-3_17
5. Hamilton, W.L., Clark, K., Leskovec, J., Jurafsky, D.: Inducing domain-specific sentiment lexicons from unlabeled corpora. In: Proceedings of the 2016 Conference on Empirical Methods in Natural Language Processing (2016). https://doi.org/10.18653/v1/D16-1057
6. Forte, A.C., Brazdil, P.B.: Determining the level of clients' dissatisfaction from their commentaries. In: Proceedings of PROPOR-2015, vol. 9727, pp. 74–85 (2016). https://doi.org/10.1007/978-3-319-41552-9_7
7. Taboada, M., Brooke, J., Tofiloski, M., Voll, K., Stede, M.: Lexicon-based methods for sentiment analysis. Comput. Linguist. **37**, 267–307 (2011). https://doi.org/10.1162/COLI_a_00049
8. Sedinkina, M., Breitkopf, N., Schütze, H.: Automatic domain adaptation outperforms manual domain adaptation for predicting financial outcomes (2019). https://doi.org/10.18653/v1/p19-1034
9. Ano, E.C., Morisio, M.: Word embeddings for sentiment analysis: a comprehensive empirical survey. arXiv preprint arXiv:1902.00753 (2019)
10. Wang, L., Xia, R.: Sentiment lexicon construction with representation learning based on hierarchical sentiment supervision. In: Proceedings of EMNLP 2017 - Conference on Empirical Methods in Natural Language Processing (2017). https://doi.org/10.18653/v1/d17-1052
11. Barnes, J., Touileb, S., Øvrelid, L., Velldal, E.: Lexicon information in neural sentiment analysis: a multi-task learning approach. In: Proceedings of the 22nd Nordic Conference on Computational Linguistics, pp. 175–186 (2019)
12. Zucco, C., Liang, H., Fatta, G.D., Cannataro, M.: Explainable sentiment analysis with applications in medicine. In: Proceedings - 2018 IEEE International Conference on Bioinformatics and Biomedicine, BIBM 2018 (2019). https://doi.org/10.1109/BIBM.2018.8621359
13. Dosilovic, F.K., Brcic, M., Hlupic, N.: Explainable artificial intelligence: a survey. In: 2018 41st International Convention on Information and Communication Technology, Electronics and Microelectronics, MIPRO 2018 - Proceedings (2018). https://doi.org/10.23919/MIPRO.2018.8400040
14. Muhammad, S.H.: An overview of sentiment analysis approaches. In: MAP-i Seminar Proceedings, pp. 65–70 (2019)
15. Gama, J.: Knowledge Discovery From Data Streams. Chapman & Hall/CRC, New York (2010). https://doi.org/10.1201/EBK1439826119
16. Potts, C.: On the negativity of negation. Semant. Linguist. Theory. **20**, 636 (2015). https://doi.org/10.3765/salt.v0i20.2565

Time-Critical Geolocation for Social Good

Reem Suwaileh[✉]

Qatar University, Doha, Qatar
`rs081123@qu.edu.qa`

Abstract. Twitter has become an instrumental source of news in emergencies where efficient access, dissemination of information, and immediate reactions are critical. Nevertheless, due to several challenges, the current fully-automated processing methods are not yet mature enough for deployment in real scenarios. In this dissertation, I focus on tackling the lack of context problem by studying automatic geo-location techniques. I specifically aim to study the *Location Mention Prediction* problem in which the system has to extract location mentions in tweets and pin them on the map. To address this problem, I aim to exploit different techniques such as training neural models, enriching the tweet representation, and studying methods to mitigate the lack of labeled data. I anticipate many downstream applications for the Location Mention Prediction problem such as incident detection, real-time action management during emergencies, and fake news and rumor detection among others.

Keywords: Geolocation · Social good · Twitter

1 Introduction

Twitter, as a cost-effective and time-saving communication channel, plays an important role in emergencies. It becomes a substantial information source for response and recovery [28] and for grassroots management in relief activities [10,27]. Invaluable efforts have been made to utilize social media for preparedness, relief, and recovery during natural disasters and after they are over [11]. However, the current automatic solutions require human intervention in many stages due to many challenges. Among these challenges is the lack of geographical context that is important for response and relief. For example, during a disaster, responders need to locate the incidents (e.g., road closures, infrastructure damage, etc.) as they happen, tweets, and users discussing the disaster. However, people often tend to hide their geographical information due to privacy and safety concerns [16,22]. Anderson et al. [12] analysed tweet disaster datasets that span over a period of 6 years and showed only around 2% or less of the tweets are geo-referenced. Thus, developing automatic geolocation tools would enable real-time location-aware monitoring of the disaster which makes the decision-making process more reliable, effective, and efficient.

In my work, I am interested in tackling the ***Location Mention Prediction*** (LMP) problem during time-critical situations. The problem involves two tasks

© Springer Nature Switzerland AG 2020
J. M. Jose et al. (Eds.): ECIR 2020, LNCS 12036, pp. 624–629, 2020.
https://doi.org/10.1007/978-3-030-45442-5_82

that can be tackled separately or jointly: (1) *Location Recognition*: extracting the location mentions in tweets, and (2) *Location Disambiguation*: locating potential location mentions on the map. The *Disambiguation* task includes two identification sub-tasks: (2.1.) identifying the *intended location* from a set of location mentions sharing the same toponym. (2.2.) identifying the *locational focus* of a tweet containing different location mentions. Learning to predict the location mentions is a non-trivial task. The location taggers have to address many challenges including microblogging-specific challenges (e.g., tweet sparsity, noisiness, stream rapid-changing, hashtag riding, etc.) and the task-specific challenges (e.g., time-criticality of the solution, scarcity of labeled data, etc.). While tackling these challenges, I aim to address several research questions including: **RQ1.** Are deep learning approaches more effective compared to the state-of-the-art and traditional machine learning-based LMP approaches?, **RQ2.** Would context expansion (using user's tweets, on-topic tweets, etc) improve LMP?, **RQ3.** How can we reduce the effect of the scarcity of labeled data on the performance of the LMP system?, and **RQ4.** How can the LMP systems control the trade-off between effectiveness and efficiency during crisis scenarios?.

The remainder of this paper is organized as follows. The related literature is reviewed in Sect. 2. My proposed methodology is discussed in Sect. 3 followed by the evaluation setup in Sect. 4.

2   Related Work

In this section, I discuss the related work to LMP problem over tweets.

TwitterStand [23] is a tweet geo-tagging system for extracting breaking news and pinning them on the map. Lingad et al. [18] compared a few NER tools on disaster-related Twitter data and found their performance noticeably degraded over the Twitter stream. Li et al. [14], on the other hand, constructed their own noisy gazetteer using a crowdsourcing-like method to match extracted location mentions from tweets by the POI tagger. Malmasi et al. [19] extracted noun phrases (NPs) in tweets using a recursive rule-based tree parser and link potential locations with Geonames entries using fuzzy matching. Ghahremanlou et al. [7] explored combined techniques to identify the location mentions by both matching and StafordNER. The major weakness in gazetteer-based methods is the mismatch between the noisy Twitter stream and non-noisy gazetteer entries [17]. To address this issue, Li et al. [15], constructed their own noisy gazetteer using collected cross-posts on Twitter from Foursquare check-ins. Alternatively, Sultanik and Fink [25], used Information Retrieval (IR) based approach to identify the location mentions in tweets. Unlike Ghahremanlou et al. [7], Yin et al. [29] retrained StandfordNER using tweet dataset to effectively identify the location mentions in tweets. More interestingly, to achieve high coverage of recognized locations, a couple of studies [6,30] adopt an ensemble-based parser.

In 2014, the topic of the fifth Australasian Language Technology Association (ALTA) shared task was on identifying location mentions in tweets [21]. Participants explored several techniques such as feature engineering, ensemble classifiers, rule-based classification, knowledge infusion, CRFs sequence labelers, semi-supervision. Al-Olimat et al. [1] proposed identifying the location names by traversing a tree of the tweet's n-grams to extract valid locations that exist in their pre-build region-specific gazetteer. Moreover, Hoang and Mothe [8] combined syntactic and semantic features to train traditional ML-based models whereas Kumar and Singh [13] trained a Convolutional Neural Network (CNN) model that learns the continuous representation of tweet text and then identifies the location mentions.

The gab in existing solutions is two-fold. First, in relation to methods, a few studies investigated deep learning-based solutions, most of the proposed solutions are gazetteer-based, and most of them do not consider efficiency when developed. Second, there is not a unified evaluation framework in which a few small-scale datasets are available and different tools are compared. Additionally, the efficiency of the proposed methods is rarely evaluated.

3   Proposed Research

In this section, I describe the proposed solutions to address the research questions listed in Sect. 1.

Deep Location Prediction (RQ1): I perceive the location recognition task as a multi-label classification task. I opt to use the Neural Networks (NNs) algorithms due to their ability to learn features and model parameters simultaneously from incomplete or noisy training data [4]. I specifically plan to experiment with (1) the Bidirectional Long-Term Short-Term Memory [9,24], (2) Encoder-Decoder with attention [2,26], and (3) BERT with Fine-tuning [5]. For the disambiguation task, I plan to explore the effectiveness of Siamese Neural Networks [3] that was used recently in neural-based IR models, especially for short text matching [20]. I further plan to experiment with character n-grams to better capture the lexical information.

Context Expansion (RQ2): Due to the short length of tweets, systems lack the context that would enable them to detect the location mentions effectively. To enrich the context of tweet, I plan to explore four tweet expansion sources including (1) User's tweets: I hypothesize that tweets shared by the user within a time window, say 10 min, are most probably discussing the same topic, (2) On-topic tweets: I assume that tweets sharing the same trending related hashtag to the disaster to be topically relevant, (3) Linked webpages: I anticipate the URLs to be useful sources for enriching the tweet context, and (4) Knowledge-bases (KB): I hypothesize that entity recognition and analysis using external auxiliary data, e.g., knowledge-bases, can aid understanding the spatial focus of a tweet which in turn enables LMP. I plan to use general-purpose KBs and study the effectiveness of knowledge-base population and acceleration techniques to maintain an online up-to-date KBs during the disaster.

Handling Data Scarcity (RQ3): Deep learning algorithms are data-hungry which requires budget and expensive resources for acquiring labeled data. To address this challenge, I investigate possible ways to reduce its effect during disasters such as (1) Exploiting existing data: Using training data from past disasters of similar or a different disaster type to train prediction systems. I plan to leverage one-step domain adaptation techniques (e.g., divergence-based, etc.), (2) Acquiring cheaper data: I plan to explore the effectiveness of expanding the small labeled datasets of a current disaster using semi-supervision, weak supervision and active learning methods, and (3) Reusing pre-trained tools: I plan to study the effectiveness of already-trained tools (based on their availability) on old disasters for effective LMP on new disasters (e.g., transfer learning).

Effectiveness and Efficiency Trade-off (RQ4): As I plan to tackle the recognition task in the disaster domain, I aim to train my models in *real-time* while the disaster is happening. Thus, I plan to investigate the trade-off between effectiveness and efficiency of LMP systems. Possible paths to study this trade-off are: (1) Tuning system decision by, for example, prioritizing tweets for LMP instead of checking every tweet chronologically, (2) Analyzing time and space complexities, and (3) exploring possible ways to modify the effectiveness measures to account for efficiency.

4 Experimental Evaluation

To evaluate the LMP proposed approaches, precision, recall, and F1 scores will be computed. When a system manages to extract part of the location mention, it is penalized by counting the false positives and false negatives multiplied by the percentage of overlap between the system's output and the ground-truth. To conduct the initial evaluation, the publicly-available LMP English datasets, that are samples of disaster-specific streams [1,21,29], will be used. I anticipate the solutions to generalize to other data domains sharing the same properties with the Twitter stream. The Geonames[1] and OpenStreetMap[2] gazetteers will be utilized for the evaluation of the *Disambiguation task*. The approaches reviewed in Sect. 2, according to their availability and reproducibility, are the baselines against the proposed approaches for all tasks: (1) Twitter-based location mention detection and disambiguation tool (e.g., LNEx [1]), (2) Academic NER (e.g., StanfordNER[3], etc.), and (3) Commercial NER taggers (e.g., Google NL[4], etc.).

Acknowledgments. This work was made possible by GSRA grant# GSRA5-1-0527-18082 from the Qatar National Research Fund (a member of Qatar Foundation). The statements made herein are solely the responsibility of the authors.

References

1. Al-Olimat, H.S., Thirunarayan, K., Shalin, V., Sheth, A.: Location name extraction from targeted text streams using gazetteer-based statistical language models. In: COLING, pp. 1986–1997 (2018)

---

[1] http://www.geonames.org/.
[2] http://www.openstreetmap.org/.
[3] http://nlp.stanford.edu/software/CRF-NER.shtml.
[4] https://cloud.google.com/natural-language/.

2. Cho, K., et al.: Learning phrase representations using rnn encoder-decoder for statistical machine translation. arXiv preprint arXiv:1406.1078 (2014)
3. Chopra, S., Hadsell, R., LeCun, Y., et al.: Learning a similarity metric discriminatively, with application to face verification. In: CVPR, vol. 1, pp. 539–546 (2005)
4. Dernoncourt, F., Lee, J.Y., Szolovits, P.: NeuroNER: an easy-to-use program for named-entity recognition based on neural networks. In: EMNLP (2017)
5. Devlin, J., Chang, M.W., Lee, K., Toutanova, K.: Bert: pre-training of deep bidirectional transformers for language understanding. arXiv (2018)
6. Gelernter, J., Balaji, S.: An algorithm for local geoparsing of microtext. GeoInformatica 17(4), 635–667 (2013)
7. Ghahremanlou, L., Sherchan, W., Thom, J.A.: Geotagging twitter messages in crisis management. Comput. J. 58(9), 1937–1954 (2014)
8. Hoang, T.B.N., Mothe, J.: Location extraction from tweets. Inf. Process. Manag. 54(2), 129–144 (2018)
9. Hochreiter, S., Schmidhuber, J.: Long short-term memory. Neural Comput. 9(8), 1735–1780 (1997)
10. Hughes, A.L., Palen, L.: Twitter adoption and use in mass convergence and emergency events. IJEM 6, 248–260 (2009)
11. Imran, M., Castillo, C., Diaz, F., Vieweg, S.: Processing social media messages in mass emergency: a survey. ACM Comput. Surv. 47(4), 1–38 (2014)
12. Jurgens, D., Finethy, T., McCorriston, J., Xu, Y.T., Ruths, D.: Geolocation prediction in twitter using social networks: a critical analysis and review of current practice. In: ICWSM (2015)
13. Kumar, A., Singh, J.P.: Location reference identification from tweets during emergencies: a deep learning approach. IJDRR 33, 365–375 (2019)
14. Li, C., Sun, A.: Fine-grained location extraction from tweets with temporal awareness. In: SIGIR, pp. 43–52 (2014)
15. Li, C., Sun, A.: Extracting fine-grained location with temporal awareness in tweets: a two-stage approach. JASIST 68(7), 1652–1670 (2017)
16. Li, X., Caragea, D., Zhang, H., Imran, M.: Localizing and quantifying damage in social media images. In: IEEE/ACM ASONAM, pp. 194–201. IEEE (2018)
17. Lieberman, M.D., Samet, H., Sankaranarayanan, J.: Geotagging with local lexicons to build indexes for textually-specified spatial data. In: ICDE, pp. 201–212 (2010)
18. Lingad, J., Karimi, S., Yin, J.: Location extraction from disaster-related microblogs. In: Proceedings of the 22nd International Conference on World Wide Web, pp. 1017–1020. ACM (2013)
19. Malmasi, S., Dras, M.: Location mention detection in tweets and microblogs. In: PACLING, pp. 123–134 (2015)
20. Mitra, B., Craswell, N., et al.: An introduction to neural information retrieval. Found. Trends® Inf. Ret. 13(1), 1–126 (2018)
21. Molla, D., Karimi, S.: Overview of the 2014 alta shared task: identifying expressions of locations in tweets. In: ALTA Workshop, pp. 151–156 (2014)
22. Murdock, V.: Your mileage may vary: on the limits of social media. SIGSPATIAL Special 3(2), 62–66 (2011)
23. Sankaranarayanan, J., Samet, H., Teitler, B.E., Lieberman, M.D., Sperling, J.: Twitterstand: news in tweets. In: ACM SIGSPATIAL, pp. 42–51 (2009)
24. Schuster, M., Paliwal, K.K.: Bidirectional recurrent neural networks. IEEE Trans. Signal Process. 45(11), 2673–2681 (1997)
25. Sultanik, E.A., Fink, C.: Rapid geotagging and disambiguation of social media text via an indexed gazetteer. Proc. ISCRAM 12, 1–10 (2012)

26. Sutskever, I., Vinyals, O., Le, Q.V.: Sequence to sequence learning with neural networks. In: Advances in neural information processing systems, pp. 3104–3112 (2014)
27. Veil, S.R., Buehner, T., Palenchar, M.J.: A work-in-process literature review: Incorporating social media in risk and crisis communication. J. Contingencies Crisis Manag. **19**(2), 110–122 (2011)
28. Vieweg, S.E.: Situational awareness in mass emergency: a behavioral and linguistic analysis of microblogged communications. Ph.D. thesis (2012)
29. Yin, J., Karimi, S., Lingad, J.: Pinpointing locational focus in microblogs. In: ADCS, p. 66 (2014)
30. Zhang, W., Gelernter, J.: Geocoding location expressions in twitter messages: a preference learning method. JOSIS **9**, 37–70 (2014)

Bibliometric-Enhanced Legal Information Retrieval

Gineke Wiggers[1,2](✉) [ID]

[1] Leiden University, Leiden, The Netherlands
g.wiggers@law.leidenuniv.nl
[2] Legal Intelligence, Alphen aan den Rijn, The Netherlands

Abstract. This research project addresses user-focused ranking in legal information retrieval (IR). It studies the perception of relevance of search results for users of Dutch legal IR systems, the employment of usage and citation variables to improve the ranking of search results (bibliometric-enhanced information retrieval), and user-centred evaluation for ranking improvements. The goal of this project is improve the ranking in legal IR systems. Ultimately, this will help legal professionals find relevant information faster.

Keywords: Legal information retrieval · Ranking · User study

1 Motivation

In the legal domain, the amount of information available digitally is rapidly increasing. Legal scholars and professionals have to navigate this information to find the case law and articles relevant for them. Dutch legal information retrieval systems currently focus primarily on matching queries and documents for their ranking. In order to help users find the most relevant documents, these ranking algorithms should be expanded.

A possible solution is bibliometric-enhanced information retrieval, where citations are measured and used as a proxy for impact in the ranking algorithm. But the legal domain differs from other research domains due to the often strong interconnection between research and practice. In the Dutch legal domain this is demonstrated by the lack of distinction between legal scientific and professional publications. This is one of the reasons why bibliometrics, which in other fields measures impact of scientific publications, has not yet been established within the Dutch legal domain.

This research will cover both the theory (such as relevance factors in legal publications and the meaning of citations in legal publications) and experimentation of applying bibliometrics to legal documents. The result is a new bibliometrics-enhanced ranking algorithm for legal information retrieval systems.

With thanks to Legal Intelligence for providing the data for this research.

2 Research Questions

The main question in this research is: can we use citation and usage (click) metrics to improve ranking in legal IR? This question comprises five sub-questions:

1. What factors influence the perception of relevance of users of legal IR systems?
2. What does a citation signify in legal publications?
3. What is the relation between user interactions with documents (usage) and citations in legal publications?
4. What is the right balance between text-based relevance and user-based relevance in ranking algorithms for legal IR?
5. What is the appropriate user-focused rank evaluation metric for legal IR?

3 Related Work

The notion of using bibliometrics to enhance information retrieval is inspired by the groundwork of Garfield [5] on the theory of citations as a measure of impact, and more recent work of Beel and Gip [4]. The normalization of citations is based on the work of Waltman et al. [18]. Based on their work we have decided on time, field, and document type normalization. The context of Dutch legal publications and their citation culture is provided by Stolker [16] and Snel [15].

This research is inspired by work of Van Opijnen and Santos [10], who translate the theory of Saracevic [13] on the spheres of relevance to the legal domain. The work of Barry and Schamber [2,3] describes the different indicators of relevance as identified by users.

The choice for a user-focused evaluation metric is influenced by the work of Järvelin and Kekäläinen [6]. The decision to use click data as a form of implicit feedback is based on work of Joachims et al. [7,8].

4 Preliminary Work

4.1 What Factors Influence the Perception of Relevance of Usersof Legal IR Systems?[1]

[17] The goal of our study was to make explicit which document characteristics users consider as factors of relevance when assessing a search result (document representation; such as title and snippet) in legal IR. To achieve this, we conducted a user questionnaire. In this questionnaire we showed users of a legal IR system a query and two search results. The user has to choose which of the two results he/she would like to see ranked higher for the query and is asked to provide a reasoning for his/her choice. The questionnaire had eleven pairs of search results spread over two queries. A total of 43 legal professionals participated in our study.

---

[1] An earlier version of this paper was presented at the 17th Dutch-Belgian Information Retrieval Workshop.

The identified relevance factors were title relevance, document type, recency, level of depth, legal hierarchy, law area (topic), authority (credibility), bibliographical relevance, source authority, usability, whether the document is annotated, and the length of the document. The identified factors confirm previous research, such as the work of Barry and Schamber [2,3]. The identified factors also suggest that there are document characteristics (e.g. authority, legal hierarchy and whether the document is annotated) that are usually grouped under cognitive or situational relevance, and thereby considered to be personal, but that users in the legal domain agree on. The agreement within the field makes that these factors can be grouped under domain relevance as described by Van Opijnen and Santos [10].

4.2   What Does a Citation Signify in Legal Publications?[2]

[19] In our next paper we examined citations in legal information retrieval. Citation metrics can be a factor of relevance in the ranking algorithms of information retrieval systems. But the challenge in legal bibliometrics, and therefore legal bibliometric-enhanced IR, is that the legal domain differs from other research domains in two manners: (1) its strong national ties and (2) the often strong interconnection between research and practice [16].

First, we contrasted citations in the legal domain to citations in the hard sciences based on the literature on scholarly citations. Second, we applied quantitative analysis of legal documents and citations to test whether the theory described in literature (particularly the distinction between scholarly and practitioners oriented publications) is confirmed by the data.

An analysis of 52 cited (seed) documents and 3196 citing documents showed no strict separation in citations between documents aimed at scholars and documents aimed at practitioners. Our results suggest that citations in legal documents do not measure the impact on scholarly publications and scholars, but measure a broader scope of impact, or relevance, for the legal field.

5   Proposed Methods

5.1   What Is the Relation Between User Interactions with Documents (Usage) and Citations in Legal Publications?

Based on the outcome of the above research question, we wish to create a boost function to create the bibliometric-enhanced ranking algorithm. Because citations in legal documents measure a broader form of impact, it is likely that citations alone do not provide a complete overview of this broader impact, but that other factors, such as usage (through click data), create a more complete overview.

---

[2] An earlier version of this paper has been presented at the 8th International Bibliometric-enhanced Information Retrieval Workshop (BIR 2019) at the 41st European Conference on Information Retrieval.

However, the work of Perneger [12] suggests that there might be a correlation between usage and citations. Therefore we took all documents published and added to the Legal Intelligence system (the largest legal IR system in the Netherlands) in February 2017.[3] This led to a set of 43,218 documents.

For each document, based on their unique document ID, we collected all click data (usage) until 2019. We then computed the Spearman correlation between the usage and the citations. We chose Spearman correlation because the data, like all citation data, does not have a normal distribution. The results show a Spearman correlation of 0.57 ($p < 0.0001$). This means that there is a moderate positive correlation between citations and usage, which is highly significant.

For the ranking algorithm, this correlation means that two separate boost functions will boost some results too much and others too little. Therefore a single harmonized boost function appears to be a better choice.

5.2 What Is the Right Balance Between Text-Based Relevance and User-Based Relevance in Ranking Algorithms for Legal IR?

Next to balancing citations and usage, the boost function as a whole will also have to be balanced against the current text-based relevance score (TF-IDF, BM25 or similar). Furthermore, the usage (click) and citation data will not be reliable from the moment of publication of the document. These new documents will have to be given the benefit of the doubt, for example through a freshness score. The optimal balance of these variables will be tuned using the evaluation metric.

5.3 What Is the Appropriate User-Focused Rank Evaluation Metric for Legal IR?

One of the challenges for legal information retrieval is finding the correct evaluation metric. This has several causes:

1. The small user group when compared to web search. This makes A/B testing difficult.
2. Differences is results lists. Because of differences in journal subscriptions two users may not see the same results list when they do the same query.
3. The high tariffs of legal experts, which makes a golden answer set prohibitively expensive to create and maintain.

A metric based on an implicit feedback model is of particular interest, as this is affordable and user-focused. We have made a first attempt to create an implicit feedback DCG model, but there is data sparsity. This means it is not possible to find queries for which the entire results list has a relevance judgment. Possible solutions will be sought in the fields of patent search and e-discovery.

---

[3] 2017 was chosen to ensure that the documents have had time to gather citations. February was chosen as January often contains holidays and a lot of review articles over the past year.

References

1. Atzmüller, C., Steiner, P.M.: Experimental vignette studies in survey research. Methodology **6**(3), 128–138 (2010)
2. Barry, C.L.: User-defined relevance criteria: an exploratory study. J. Am. Soc. Inf. Sci. **45**(3), 149–159 (1994)
3. Barry, C.L., Schamber, L.: Users' criteria for relevance evaluation: a cross-situational comparison. Inf. Process. Manage. **34**(2/3), 219–236 (1998)
4. Beel, J., Gipp, B.: Google Scholar's ranking algorithm: the impact of citation counts (an empirical study). In: Third International Conference on Research Challenges in Information Science (RCIS), pp. 439–446. IEEE (2009)
5. Garfield, G.: Citation Indexing: Its Theory and Application in Science, Technology, and Humanities. Wiley, New York (1979)
6. Järvelin, K., Kekäläinen, J.: Cumulated gain-based evaluation of IR techniques. ACM Trans. Inf. Syst. **20**(4), 422–446 (2002)
7. Joachims, T.: Optimizing search engines using clickthrough data. In: Proceedings of the Eighth ACM SIGKDD International Conference on Knowledge Discovery and Data Mining, pp. 133–142 (2002)
8. Joachims, T., Granka, L.A., Pan, B., Hembrooke, H., Gay, G.: Accurately interpreting clickthrough data as implicit feedback. Sigir **5**, 154–161 (2005)
9. Joachims, T., Swaminathan, A., Schnabel, T.: Unbiased learning-to-rank with biased feedback. In: Proceedings of the Tenth ACM International Conference on Web Search and Data Mining (WSDM 2017), pp. 781–789 (2017)
10. van Opijnen, Marc, Santos, Cristiana: On the concept of relevance in legal information retrieval. Artif. Intell. Law **25**(1), 65–87 (2017). https://doi.org/10.1007/s10506-017-9195-8
11. Park, T.K.: The nature of relevance in information retrieval: an empirical study. Library Q. **63**(3), 318–351 (1993)
12. Perneger, T.V.: Relation between online "hit counts" and subsequent citations: prospective study of research papers in the BMJ. BMJ **2004**(329), 546 (2004)
13. Saracevic, T.: Relevance: a review of and framework for the thinking on the notion in information science. J. Am. Soc. Inf. Sci. **26**, 321–343 (1975)
14. Jones, E., Oliphant, E., Peterson, P., et al.: SciPy: Open Source Scientific Tools for Python. http://www.scipy.org/. Accessed 22 Oct 2019
15. Snel, M.V.R.: Hoera, een lijstje! Over bronvermelden. Ars Aequi **3**, 254–260 (2018)
16. Stolker, C.: Rethinking the Law School: Education, Research, Outreach and Governance. Cambridge University Press, Cambridge (2015)
17. Wiggers, G., Verberne, S., Zwenne, G.J.: Exploration of intrinsic relevance judgments by legal professionals in information retrieval systems. In: Proceedings of the 17th Dutch-Belgian Information Retrieval Workshop, pp. 5–8 (2018). https://arxiv.org/abs/1812.04265
18. Waltman, L., van Eck, N.J., van Leeuwen, T.N., Visser, M.S., van Raan, A.F.: Towards a new crown indicator: some theoretical considerations. J. Inform. **5**(1), 37–47 (2011)
19. Wiggers, G., Verberne, S.: Citation metrics for legal information retrieval systems. In: Proceedings of the 8th International Workshop on Bibliometric-Enhanced Information Retrieval (BIR) co-located with the 41st European Conference on Information Retrieval (ECIR 2019), pp. 39–50 (2019). http://ceur-ws.org/Vol-2345/

Workshops

International Workshop on Algorithmic Bias in Search and Recommendation (Bias 2020)

Ludovico Boratto[1]([✉])(iD), Mirko Marras[2](iD), Stefano Faralli[3](iD),
and Giovanni Stilo[4](iD)

[1] EURECAT, Barcelona, Spain
ludovico.boratto@acm.org
[2] University of Cagliari, Cagliari, Italy
mirko.marras@unica.it
[3] University of Rome Unitelma Sapienza, Rome, Italy
stefano.faralli@unitelma.it
[4] University of L'Aquila, L'Aquila, Italy
giovanni.stilo@univaq.it

Abstract. Both search and recommendation algorithms provide results based on their relevance for the current user. In order to do so, such a relevance is usually computed by models trained on historical data, which is biased in most cases. Hence, the results produced by these algorithms naturally propagate, and frequently reinforce, biases hidden in the data, consequently strengthening inequalities. Being able to measure, characterize, and mitigate these biases while keeping high effectiveness is a topic of central interest for the information retrieval community. In this workshop, we aim to collect novel contributions in this emerging field and to provide a common ground for interested researchers and practitioners.

Keywords: Bias · Algorithms · Search · Recommendation

1 Motivation

Search and *recommendation* are getting closer and closer as research areas. Though they require fundamentally different inputs, i.e., the user is asked to provide a query in search, while implicit and explicit feedback is leveraged in recommendation, existing search algorithms are being personalized based on users' profiles and recommender systems are optimizing their output on the ranking quality.

Both classes of algorithms aim to learn patterns from historical data that conveys biases in terms of *unbalances* and *inequalities*. These hidden biases are unfortunately captured in the learned patterns, and often emphasized in the results these algorithms provide to users [2]. When a bias affects a *sensitive attribute* of a user, such as their gender or religion, the inequalities that are

© Springer Nature Switzerland AG 2020
J. M. Jose et al. (Eds.): ECIR 2020, LNCS 12036, pp. 637–640, 2020.
https://doi.org/10.1007/978-3-030-45442-5_84

reinforced by search and recommendation algorithms even lead to *severe societal consequences*, like users' discrimination [4].

For this critical reason, being able to *detect, measure, characterize*, and *mitigate* these biases while keeping high effectiveness is a prominent and timely topic for the IR community. Mitigating the effects generated by popularity bias [1,5,6], ensuring results that are fair with respect to the users [3,7], and being able to interpret why a model provides a given recommendation or search result are examples of challenges that may be important in real-world applications. This workshop aims to collect new contributions in this emerging field and to provide a common ground for interested researchers and practitioners.

The workshop welcomes contributions in all topics related to algorithmic bias in search and recommendation, focused (but not limited) to:

- *Data Set Collection and Preparation*:
  - Managing imbalances and inequalities within data sets.
  - Devising collection pipelines that lead to fair and unbiased data sets.
  - Collecting data sets useful for studying potential biased and unfair situations.
  - Designing procedures for creating synthetic data sets for research on bias and fairness.
- *Countermeasure Design and Development*:
  - Conducting exploratory analysis that uncover biases.
  - Designing treatments that mitigate biases (e.g., popularity bias mitigation).
  - Devising interpretable search and recommendation models.
  - Providing treatment procedures whose outcomes are easily interpretable.
  - Balancing inequalities among different groups of users or stakeholders.
- *Evaluation Protocol and Metric Formulation*:
  - Conducting quantitative experimental studies on bias and unfairness.
  - Defining objective metrics that consider fairness and/or bias.
  - Formulating bias-aware protocols to evaluate existing algorithms.
  - Evaluating existing strategies in unexplored domains
- *Case Study Exploration*:
  - News channels.
  - E-commerce platforms.
  - Educational environments.
  - Entertainment websites.
  - Healthcare systems.
  - Social networks.

2   Objectives

The workshop has the following main objectives:

1. Raise awareness on the algorithmic bias problem within the IR community.
2. Identify social and human dimensions affected by algorithmic bias in IR.
3. Solicit contributions from researchers who are facing algorithmic bias in IR.
4. Get insights on existing approaches, recent advances, and open issues.
5. Familiarize the IR community with existing practices from the field.
6. Uncover gaps between academic research and real-world needs in the field.

3   Organizers Biography

Ludovico Boratto is senior research scientist in the Data Science and Big Data Analytics research group at Eurecat. His research interests focus on Data Mining and Machine Learning approaches, mostly applied to recommender systems and social media analysis. The results of his research have been published in top-tier conferences and journals. His research activity also brought him to give talks and tutorials at top-tier conferences (e.g., ACM RecSys 2016, IEEE ICDM 2017) and research centers (Yahoo! Research). He is editor of the book "Group Recommender Systems: An Introduction", published by Springer. He is editorial board member of the "Information Processing & Management" journal (Elsevier) and guest editor of several journal's special issues. He is regularly part of the program committee of the main Data Mining and Web conferences, such as RecSys, KDD, SIGIR, WSDM, ICWSM, and TheWebConf. In 2012, he got a Ph.D. at the University of Cagliari (Italy), where he was research assistant until May 2016. In 2010 and 2014 he spent 10 months at Yahoo! Research in Barcelona as a visiting researcher. He is member of the ACM and of the IEEE.

Mirko Marras is a PhD student in Computer Science at the Department of Mathematics and Computer Science of the University of Cagliari (Italy). He received the MSc Degree in Computer Science (summa cum laude) from the same University in 2016. His research interests focus on algorithmic bias in machine learning for educational platforms, specifically in the context of semantic-aware systems, recommender systems, biometric systems, and opinion mining systems. He has co-authored papers in top-tier international journals, such as Pattern Recognition Letters (Elsevier), Computers in Human Behavior (Elsevier), and IEEE Cloud Computing. He has given talks and demonstrations at several international conferences and workshops, such as The Web Conference 2018, ECIR 2019, ESWC2017, INTERSPEECH 2019. He is student member in several national and international associations, including CVPL, AIxIA, IEEE, and ACM.

Stefano Faralli is an assistant professor at University of Rome Unitelma Sapienza, Rome, Italy. His research interests include Ontology Learning, Distributional Semantics, Word Sense Disambiguation/Induction, Recommender Systems, Linked Opend Data. He co-organized the International Workshop: Taxonomy Extraction Evaluation (TexEval) Task 17 of Semantic Evaluation (SemEval-2015) and the International Workshop on Social Interaction-based Recommendation (SIR 2018).

Giovanni Stilo is an Assistant Professor in the Department of Information Engineering, Computer Science and Mathematics at the University of L'Aquila. He received his PhD. in Computer Science in 2013, and in 2014 he was a visiting researcher at Yahoo! Labs in Barcelona. Between 2015 and 2018, he was a researcher in the Computer Science Department at La Sapienza University, in Rome. His research interests are in the areas of machine learning and data mining, and specifically temporal mining, social network analysis, network medicine, semantics-aware recommender systems, and anomaly detection. He is a member of the steering committee of the Intelligent Information Mining research group (http://iim.disim.univaq.it/). He has organized several international workshops, held in conjunction with top-tier conferences (ICDM, CIKM, and ECIR), and he is involved as editor and reviewer of top-tier journals, such as TITS, TKDE, DMKD, AI, KAIS, and AIIM.

References

1. Abdollahpouri, H., Burke, R., Mobasher, B.: Controlling popularity bias in learning-to-rank recommendation. In: Proceedings of the Eleventh ACM Conference on Recommender Systems, pp. 42–46. ACM (2017)
2. Boratto, L., Fenu, G., Marras, M.: The effect of algorithmic bias on recommender systems for massive open online courses. In: Azzopardi, L., Stein, B., Fuhr, N., Mayr, P., Hauff, C., Hiemstra, D. (eds.) ECIR 2019. LNCS, vol. 11437, pp. 457–472. Springer, Cham (2019). https://doi.org/10.1007/978-3-030-15712-8_30
3. Burke, R., Sonboli, N., Ordonez-Gauger, A.: Balanced neighborhoods for multi-sided fairness in recommendation. In: Conference on Fairness, Accountability and Transparency, pp. 202–214 (2018)
4. Hajian, S., Bonchi, F., Castillo, C.: Algorithmic bias: from discrimination discovery to fairness-aware data mining. In: Krishnapuram, B., Shah, M., Smola, A.J., Aggarwal, C.C., Shen, D., Rastogi, R. (eds.) Proceedings of the 22nd ACM SIGKDD International Conference on Knowledge Discovery and Data Mining, San Francisco, CA, USA, 13–17 August 2016, pp. 2125–2126. ACM (2016). DOI: https://doi.org/10.1145/2939672.2945386
5. Jannach, D., Lerche, L., Kamehkhosh, I., Jugovac, M.: What recommenders recommend: an analysis of recommendation biases and possible countermeasures. User Modeling and User-Adapted Interaction 25(5), 427–491 (2015). https://doi.org/10.1007/s11257-015-9165-3
6. Kamishima, T., Akaho, S., Asoh, H., Sakuma, J.: Correcting popularity bias by enhancing recommendation neutrality. In: RecSys Posters (2014)
7. Zheng, Y., Dave, T., Mishra, N., Kumar, H.: Fairness in reciprocal recommendations: a speed-dating study. In: Adjunct Publication of the 26th Conference on User Modeling, Adaptation and Personalization, pp. 29–34. ACM (2018)

Bibliometric-Enhanced Information Retrieval 10th Anniversary Workshop Edition

Guillaume Cabanac[1]([⊠]), Ingo Frommholz[2]([⊠]), and Philipp Mayr[3]([⊠])

[1] Computer Science Department, University of Toulouse,
IRIT UMR 5505, Toulouse, France
`guillaume.cabanac@univ-tlse3.fr`
[2] Institute for Research in Applicable Computing, University of Bedfordshire,
Luton, UK
`ifrommholz@acm.org`
[3] GESIS – Leibniz-Institute for the Social Sciences, Cologne, Germany
`philipp.mayr@gesis.org`

Abstract. The Bibliometric-enhanced Information Retrieval workshop series (BIR) was launched at ECIR in 2014 [19] and it was held at ECIR each year since then. This year we organize the 10th iteration of BIR. The workshop series at ECIR and JCDL/SIGIR tackles issues related to academic search, at the crossroads between Information Retrieval, Natural Language Processing and Bibliometrics. In this overview paper, we summarize the past workshops, present the workshop topics for 2020 and reflect on some future steps for this workshop series.

Keywords: Academic search · Information retrieval · Digital libraries · Bibliometrics · Scientometrics · Multidisciplinary

1 Motivation and Relevance to ECIR

Searching for scientific information is a long-lived user need. In the early 1960s, Salton was already striving to enhance information retrieval by including clues inferred from bibliographic citations [23]. The development of citation indexes pioneered by Garfield [11] proved determinant for such a research endeavour at the crossroads between the nascent fields of Bibliometrics[1] and Information Retrieval (IR) — BIR. The pioneers who established these fields in Information Science — such as Salton and Garfield — were followed by scientists who specialised in one of these [30], leading to the two loosely connected fields we know of today.

The purpose of the BIR workshop series founded in 2014 is to tighten up the link between IR and Bibliometrics [20]. We strive to get the 'retrievalists'

---

[1] Bibliometrics refers to the statistical analysis of the academic literature [21] and plays a key role in scientometrics: the quantitative analysis of science and innovation [16].

© Springer Nature Switzerland AG 2020
J. M. Jose et al. (Eds.): ECIR 2020, LNCS 12036, pp. 641–647, 2020.
https://doi.org/10.1007/978-3-030-45442-5_85

Table 1. Overview of the BIR workshop series

Year	Conference	Venue	Papers	Proceedings
2014	ECIR	Amsterdam, NL	6	Vol-1143
2015	ECIR	Vienna, AT	6	Vol-1344
2016	ECIR	Padua, IT	8	Vol-1567
2016	JCDL	Newark, US	$10 + 10^a$	Vol-1610
2017	ECIR	Aberdeen, UK	12	Vol-1823
2017	SIGIR	Tokyo, JP	11	Vol-1888
2018	ECIR	Grenoble, FR	9	Vol-2080
2019	ECIR	Cologne, DE	14	Vol-2345
2019	SIGIR	Paris, FR	$16 + 10^b$	Vol-2414
2020	ECIR	Lisbon, PT	TBA	TBA

[a] with CL-SciSumm 2016 Shared Task; [b] with CL-SciSumm 2019 Shared Task

and 'citationists' [30] active in both academia and the industry together, who are developing search engines and recommender systems such as ArnetMiner, Dimensions, Google Scholar, Microsoft Academic Search, and Semantic Scholar, just to name a few.

These bibliometric-enhanced IR systems must deal with the multifaceted nature of scientific information by searching for or recommending academic papers, patents, venues (i.e., conferences or journals), authors, experts (e.g., peer reviewers), references (to be cited to support an argument), and datasets. The underlying models harness relevance signals from keywords provided by authors, topics extracted from the full-texts, co-authorship networks, citation networks, and various classifications schemes of science.

BIR is a hot topic with growing recognition in the community in recent years: see for instance the Initiative for Open Citations [25], the Google Dataset Search [5], the Indian JNU initiative for indexing the world's literature in full-text [22], the increasing number of retractions [4], and massive studies of self-citations [13,27]. We believe that BIR@ECIR is a much needed scientific event for the 'retrievalists', 'citationists' and others to meet and join forces pushing the knowledge boundaries of IR applied to literature search and recommendation.

2   Summarzing the Past BIR Workshops

The BIR workshop series was launched at ECIR in 2014 [19] and it was held at ECIR each year since then. As our workshop lies at the crossroads between IR and NLP, we also ran BIR as a joint workshop called BIRNDL (Bibliometric-enhanced IR and NLP for Digital Libraries) at the JCDL [7] and SIGIR [9] conferences. All past workshops had a large number of participants (between ~30 and ~60), demonstrating the relevance of the workshop's topics.

In the following, we present an overview of the past BIR workshops and keynotes at BIR (Tables 1 and 2). All pointers to the workshops and proceedings are hosted at sites.google.com/view/bir-ws. Many of the presented workshop papers appeared in extended form in one of our four BIR-related special issues (2015 [20], 2018 [8,18], 2019 [2]).

3 Workshop Topics

The call for papers for the 2020 workshop (the 10th BIR edition) addressed current research issues regarding 3 aspects of the search/recommendation process:

1. User needs and behaviour regarding scientific information, such as:
   - Finding relevant papers/authors for a literature review.
   - Measuring the degree of plagiarism in a paper.
   - Identifying expert reviewers for a given submission.
   - Flagging predatory conferences and journals.
   - Information seeking behaviour and HCI in academic search.
2. Mining the scientific literature, such as:
   - Information extraction, text mining and parsing of scholarly literature.
   - Natural language processing (e.g., citation contexts).
   - Discourse modelling and argument mining.
3. Academic search/recommendation systems:
   - Modelling the multifaceted nature of scientific information.
   - Building test collections for reproducible BIR.
   - System support for literature search and recommendation.

4 Target Audience

The target audience of the BIR workshops are researchers and practitioners, junior and senior, from Scientometrics as well as Information Retrieval and Natural Language Processing. These could be IR/NLP researchers interested in potential new application areas for their work as well as researchers and practitioners working with, for instance, bibliometric data and interested in how IR/NLP methods can make use of such data.

5 Peer Review Process and Workshop Format

Our peer review process is supported by Easychair. Each submission is assigned to 2 to 3 reviewers, preferably at least one expert in IR and one expert in Bibliometrics or NLP. The accepted papers are either long papers (15-min talks) or short papers (5-min talks). Two interactive sessions close the morning and afternoon sessions with posters and demos, allowing attendees to discuss the latest developments in the field and opportunities (e.g., shared tasks such as the CL-SciSumm [12] at the BIRNDL joint workshop, see Sect. 2). These interactive sessions serve as ice-breakers, sparking interesting discussions that usually continue during lunch and the cocktail party. The sessions are also an opportunity for our speakers to further discuss their work.

Table 2. Keynotes at BIR

Year	Area[a]	Title of the keynote presentation	Presenter
2015	SCIM	In Praise of Interdisciplinary Research through Scientometrics	Cabanac [6]
2016	IR	Bibliometrics in Online Book Discussions: Lessons for Complex Search Tasks	Koolen [14]
2016	SCIM	Bibliometrics, Information Retrieval and Natural Language Processing: Natural Synergies to Support Digital Library Research	Wolfram [31]
2017	IR	Real-World Recommender Systems for Academia: The Pain and Gain in Building, Operating, and Researching them	Beel [3]
2017	NLP	Do "Future Work" sections have a purpose? Citation links and entailment for global scientometric questions	Teufel [26]
2018	NLP	Trends in Gaming Indicators: On Failed Attempts at Deception and their Computerised Detection	Labbé [15]
2018	IR	Integrating and Exploiting Public Metadata Sources in a Bibliographic Information System	Schenkel [24]
2019	NLP	Beyond Metadata: the New Challenges in Mining Scientific Papers	Atanassova [1]
2019	IR	Personalized Feed/Query-formulation, Predictive Impact, and Ranking	Wade [28]
2019	NLP	Discourse Processing for Text Analysis: Recent Successes, Current Challenges	Webber [29]
2020	SCIM	Metrics and trends in assessing the scientific impact	Tsatsaronis

[a]SCIM: Scientometrics; NLP: Natural Language Processing; IR: Information Retrieval

6   Next Steps

Research on scholarly document processing has for many years been scattered across multiple venues like ACL, SIGIR, JCDL, CIKM, LREC, NAACL, KDD, and others. Our next strategic step is the First Workshop on Scholarly Document Processing (SDP)[2] will be held in November 2020 in conjunction with the 2020 Conference on Empirical Methods in Natural Language Processing. This workshop and initiative will be organized by a diverse group of researchers (organizers from BIR, BIRNDL, Workshop on Mining Scientific Publications/WOSP

---

[2] https://ornlcda.github.io/SDProc/.

and Big Scholar) which have expertise in NLP, ML, Text Summarization/Mining, Computational Linguistics, Discourse Processing, IR, and others.

Acknowledgement. We organizers wish to thank all those who contributed to this workshop series: The researchers who contributed papers, the many reviewers who generously offered their time and expertise, and the participants of the BIR and BIRNDL workshops. Since 2016, we maintain the Bibliometric-enhanced-IR Bibliography that collects scientific papers which appear in collaboration with the BIR/BIRNDL organizers.

References

1. Atanassova, I.: Beyond metadata: the new challenges in mining scientific papers. In: Cabanac, G., Frommholz, I., Mayr, P. (eds.) Proceedings of the 8th International Workshop on Bibliometric-Enhanced Information Retrieval (BIR 2019) Co-located with the 41st European Conference on Information Retrieval (ECIR 2019), Cologne, Germany, 14 April 2019. CEUR Workshop Proceedings, vol. 2345, pp. 8–13. CEUR-WS.org (2019). http://ceur-ws.org/Vol-2345/paper1.pdf
2. Atanassova, I., Bertin, M., Mayr, P.: Editorial: mining scientific papers: NLP-enhanced bibliometrics. Front. Res. Metrics Anal. (2019). https://doi.org/10.3389/frma.2019.00002
3. Beel, J., Dinesh, S.: Real-world recommender systems for academia: the pain and gain in building, operating, and researching them. In: Mayr, P., Frommholz, I., Cabanac, G. (eds.) Proceedings of the Fifth Workshop on Bibliometric-enhanced Information Retrieval (BIR) Co-located with the 39th European Conference on Information Retrieval (ECIR 2017), Aberdeen, UK, 9th April 2017. CEUR Workshop Proceedings, vol. 1823, pp. 6–17. CEUR-WS.org (2017). http://ceur-ws.org/Vol-1823/paper1.pdf
4. Brainard, J., You, J.: What a massive database of retracted papers reveals about science publishing's "death penalty". Science (2018). https://doi.org/10.1126/science.aav8384
5. Brickley, D., Burgess, M., Noy, N.: Google dataset search: building a search engine for datasets in an open Web ecosystem. In: The World Wide Web Conference on - WWW 2019, pp. 1365–1375. ACM Press (2019). https://doi.org/10.1145/3308558.3313685
6. Cabanac, G.: In praise of interdisciplinary research through scientometrics. In: Mayr, P., Frommholz, I., Mutschke, P. (eds.) Proceedings of the Second Workshop on Bibliometric-Enhanced Information Retrieval Co-located with the 37th European Conference on Information Retrieval (ECIR 2015), Vienna, Austria, 29th March 2015. CEUR Workshop Proceedings, vol. 1344, pp. 5–13. CEUR-WS.org (2015). http://ceur-ws.org/Vol-1344/paper1.pdf
7. Cabanac, G., et al. (eds.): BIRNDL 2016: Proceedings of the Joint Workshop on Bibliometric-enhanced Information Retrieval and Natural Language Processing for Digital Libraries co-located with the Joint Conference on Digital Libraries, vol. 1610. CEUR-WS, Aachen (2016)
8. Cabanac, G., Frommholz, I., Mayr, P.: Bibliometric-enhanced information retrieval: preface. Scientometrics **116**(2), 1225–1227 (2018). https://doi.org/10.1007/s11192-018-2861-0

9. Chandrasekaran, M.K., Mayr, P. (eds.): BIRNDL 2019: Proceedings of the 4th Joint Workshop on Bibliometric-enhanced Information Retrieval and Natural Language Processing for Digital Libraries co-located with the Joint Conference on Digital Libraries, vol. 2414. CEUR-WS, Aachen (2019)

10. Chandrasekaran, M.K., Mayr, P. (eds.): Proceedings of the 4th Joint Workshop on Bibliometric-enhanced Information Retrieval and Natural Language Processing for Digital Libraries (BIRNDL 2019) co-located with the 42nd International ACM SIGIR Conference on Research and Development in Information Retrieval (SIGIR 2019), Paris, France, 25 July 2019, CEUR Workshop Proceedings, vol. 2414. CEUR-WS.org (2019). http://ceur-ws.org/Vol-2414

11. Garfield, E.: Citation indexes for science: a new dimension in documentation through association of ideas. Science **122**(3159), 108–111 (1955). https://doi.org/10.1126/science.122.3159.108

12. Jaidka, K., Chandrasekaran, M.K., Rustagi, S., Kan, M.Y.: Insights from CL-SciSumm 2016: the faceted scientific document summarization shared task. Int. J. Dig. Lib. **19**(2–3), 163–171 (2018). https://doi.org/10.1007/s00799-017-0221-y

13. Kacem, A., Flatt, J.W., Mayr, P.: Tracking self-citations in academic publishing. Scientometrics (2020). https://doi.org/10.1007/s11192-020-03413-9

14. Koolen, M.: Bibliometrics in online book discussions: lessons for complex search tasks. In: Mayr, P., Frommholz, I., Cabanac, G. (eds.) Proceedings of the Third Workshop on Bibliometric-enhanced Information Retrieval co-located with the 38th European Conference on Information Retrieval (ECIR 2016), Padova, Italy, 20 March 2016. CEUR Workshop Proceedings, vol. 1567, pp. 5–13. CEUR-WS.org (2016). http://ceur-ws.org/Vol-1567/paper1.pdf

15. Labbé, C.: Trends in gaming indicators: on failed attempts at deception and their computerised detection. In: Mayr, P., et al. [17], pp. 6–15. http://ceur-ws.org/Vol-2080/paper1.pdf

16. Leydesdorff, L., Milojević, S.: Scientometrics. In: Wright, J.D. (ed.) International Encyclopedia of the Social & Behavioral Sciences, vol. 21, 2nd edn, pp. 322–327. Elsevier, Amsterdam (2015)

17. Mayr, P., Frommholz, I., Cabanac, G. (eds.): Proceedings of the 7th International Workshop on Bibliometric-enhanced Information Retrieval (BIR 2018) Co-located with the 40th European Conference on Information Retrieval (ECIR 2018), Grenoble, France, 26 March 2018, CEUR Workshop Proceedings, vol. 2080. CEUR-WS.org (2018). http://ceur-ws.org/Vol-2080

18. Mayr, P., et al.: Special issue on bibliometric-enhanced information retrieval and natural language processing for digital libraries. Int. J. Dig. Lib. **19**(2–3), 107–111 (2018). https://doi.org/10.1007/s00799-017-0230-x

19. Mayr, P., Schaer, P., Scharnhorst, A., Larsen, B., Mutschke, P. (eds.): BIR 2016 Proceedings of the 1st Workshop on Bibliometric-enhanced Information Retrieval co-located with the 36th European Conference on Information Retrieval, vol. 1143. CEUR-WS, Aachen (2014)

20. Mayr, P., Scharnhorst, A.: Scientometrics and information retrieval: weak-links revitalized. Scientometrics **102**(3), 2193–2199 (2014). https://doi.org/10.1007/s11192-014-1484-3

21. Pritchard, A.: Statistical bibliography or bibliometrics? [Documentation notes]. J. Document. **25**(4), 348–349 (1969). https://doi.org/10.1108/eb026482

22. Pulla, P.: The plan to mine the world's research papers. Nature **571**, 316–318 (2019). https://doi.org/10.1038/d41586-019-02142-1

23. Salton, G.: Associative document retrieval techniques using bibliographic information. J. ACM **10**(4), 440–457 (1963). https://doi.org/10.1145/321186.321188

24. Schenkel, R.: Integrating and exploiting public metadata sources in a bibliographic information system. In: Mayr, P., et al. [17], pp. 16–21. http://ceur-ws.org/Vol-2080/paper2.pdf

25. Shotton, D.: Funders should mandate open citations. Nature **553**(7687), 129 (2018). https://doi.org/10.1038/d41586-018-00104-7

26. Teufel, S.: Do "future work" sections have a purpose? citation links and entailment for global scientometric questions. In: Mayr, P., Chandrasekaran, M.K., Jaidka, K. (eds.) Proceedings of the 2nd Joint Workshop on Bibliometric-enhanced Information Retrieval and Natural Language Processing for Digital Libraries (BIRNDL 2017) Co-located with the 40th International ACM SIGIR Conference on Research and Development in Information Retrieval (SIGIR 2017), Tokyo, Japan, 11 August 2017. CEUR Workshop Proceedings, vol. 1888, pp. 7–13. CEUR-WS.org (2017). http://ceur-ws.org/Vol-1888/paper1.pdf

27. Van Noorden, R., Singh Chawla, D.: Hundreds of extreme self-citing scientists revealed in new database. Nature **572**(7771), 578–579 (2019). https://doi.org/10.1038/d41586-019-02479-7

28. Wade, A.D., Williams, I.: Personalized feed/query-formulation, predictive impact, and ranking. In: Chandrasekaran, M.K., Mayr, P., [10], pp. 6–7. http://ceur-ws.org/Vol-2414/paper1.pdf

29. Webber, B.: Discourse processing for text analysis: recent successes, current challenges. In: Chandrasekaran, M.K., Mayr, P., [10], pp. 8–14. http://ceur-ws.org/Vol-2414/paper2.pdf

30. White, H.D., McCain, K.W.: Visualizing a discipline: an author co-citation analysis of Information science, 1972–1995. J. Am. Soc. Inf. Sci. **49**(4), 327–355 (1998). b57vc7

31. Wolfram, D.: Bibliometrics, information retrieval and natural language processing: natural synergies to support digital library research. In: Cabanac, G., et al. (eds.) Proceedings of the Joint Workshop on Bibliometric-enhanced Information Retrieval and Natural Language Processing for Digital Libraries (BIRNDL) co-located with the Joint Conference on Digital Libraries 2016 (JCDL 2016), Newark, NJ, USA, 23 June 2016. CEUR Workshop Proceedings, vol. 1610, pp. 6–13. CEUR-WS.org (2016). http://ceur-ws.org/Vol-1610/paper1.pdf

The 3rd International Workshop on Narrative Extraction from Texts: Text2Story 2020

Ricardo Campos[1,2](✉) ⓘ, Alípio Jorge[1,3] ⓘ, Adam Jatowt[4] ⓘ,
and Sumit Bhatia[5] ⓘ

[1] INESC TEC, Porto, Portugal
[2] Ci2 - Smart Cities Research Center - Polytechnic Institute of Tomar,
Tomar, Portugal
ricardo.campos@ipt.pt
[3] University of Porto, Porto, Portugal
amjorge@fc.up.pt
[4] Kyoto University, Kyoto, Japan
adam@dl.kuis.kyoto-u.ac.jp
[5] IBM Research AI, New Delhi, India
sumitbhatia@in.ibm.com

Abstract. The Third International Workshop on Narrative Extraction from Texts (Text2Story'20) [text2story20.inesctec.pt] held in conjunction with the 42nd European Conference on Information Retrieval (ECIR 2020) gives researchers of IR, NLP and other fields, the opportunity to share their recent advances in extraction and formal representation of narratives. This workshop also presents a forum to consolidate the multi-disciplinary efforts and foster discussions around the narrative extraction task, a hot topic in recent years.

Keywords: Information extraction · Narrative extraction

1 Background and Motivation to ECIR

Searching for relevant information is a permanent need for those who want to stay informed about a given event, news or story. While having access to information is now easier than ever with the proliferation of devices and different means of accessing data, keeping up-to-date with all the developments and various aspects of the topic being followed is a difficult task. In many situations, it is hard for readers to connect the dots of a given story [21]. This is due not only to the widespread presence of the media outlets in the digital space [14], but also to the increasing participation of citizens, who produce and promote an unprecedented number of comments and discussions on social media (some of them lasting over days, weeks or months). Automatic narrative extraction from texts offers a compelling approach to this problem by automatically identifying the sub-set of

J. M. Jose et al. (Eds.): ECIR 2020, LNCS 12036, pp. 648–653, 2020.
https://doi.org/10.1007/978-3-030-45442-5_86

interconnected raw documents, extracting the critical narrative/story elements, and representing them in a more adequate manner that conveys the key points of the story in an easy to understand format to the readers. This could be done through text summarization [13], timelines [15,19], word clouds [3,20], visual textual analytics [7,18] or in an intermediate structured formalism (e.g., wiki-like page structures [1]) that can feed further steps (e.g., gamification [6,16] or story generation [12] (such as automatically generating finance [5,11] and sport reports [22])).

Although information extraction and natural language processing have made significant progress towards automatic interpretation of texts, the problem of fully identifying and relating the different elements of a narrative in a document (set) still presents significant unsolved challenges [17].

The purpose of the Text2Story workshop series is to shorten the distance between IR and people working on automatic narrative extraction and construction from texts, a vibrant line of research that has been conducted over the last few years by many research groups.

2 Past-Related Activities

The Text2Story workshop series had its first edition at ECIR'18 [8], followed by a second edition on ECIR'19 [9]. In these two first editions, we had an approximate number of 70 participants, 16 research papers presented, plus demo and poster sessions, and vibrant talks from our four invited keynotes: Udo Kruschwitz (University of Essex), Eric Gaussier (University Grenoble Alps), Iryna Gurevych (Technische Universität Darmstadt), and Miguel Martinez-Alvarez (Signal AI). In addition to this, we also edited the Text2Story Special Issue on IPM Journal [10] which had more than 30 submissions and 8 papers accepted, demonstrating the growing activity of this specific research area. The organizers of the workshop have also been actively involved in this research area with the proposal and the contribution of new methods and solutions, most notably the YAKE! keyword extraction algorithm [2–4] - best short paper of ECIR'18 - and the Tell me Stories temporal summarization tool [15] - best demo presentation at ECIR'19.

In the third edition of this workshop series, we aim to raise awareness to the problem of creating text-to-narrative-structures and its related tasks. We focus on researchers and practitioners working on identifying, extracting and producing narrative stories, but also on people from industry, particularly journalists and stakeholders working in traditional and social media.

3 Topic Outline

The call for papers aimed to cover original research at the intersection of IR and NLP on all aspects of storyline identification and generation from texts including but not limited to narrative and content generation, formal representation, and visualization of narratives. The topics of the workshop are in line with the previous editions of the Text2Story workshop series. In particular, we featured the following topics:

- Event Identification
- Narrative Representation Language
- Information Retrieval Models based on Story Evolution
- Narrative-focused Search in Text Collections
- Temporal Information Retrieval and Narrative Extraction
- Sentiment and Opinion Detection
- Argumentation Mining
- Narrative Summarization
- Multi-modal Summarization
- Storyline Visualization
- Temporal Aspects of Storylines
- Story Evolution and Shift Detection
- Causal Relation Extraction and Arrangement
- Evaluation Methodologies for Narrative Extraction
- Big data applied to Narrative Extraction
- Resources and Dataset showcase
- Personalization and Recommendation
- User Profiling and User Behavior Modeling
- Credibility
- Models for detection and removal of bias in generated stories
- Ethical and fair narrative generation
- Fact Checking
- Bots Influence
- Bias in Text Documents
- Automatic Timeline Generation

4   Program Committee and Support Chairs

All papers were refereed through a double-blind peer-review process by at least three reviewers and are planned to be published by CEUR. The program committee members consist of researchers from industry and academia. The following members formed the program committee of the Text2Story'20 workshop:

- Álvaro Figueira (INESC TEC & University of Porto)
- Andreas Spitz (École polytechnique fédérale de Lausanne)
- António Horta Branco (University of Lisbon)
- Arian Pasquali (Signal AI)
- Bruno Martins (IST and INESC-ID - Instituto Superior Técnico, University of Lisbon)
- Daniel Gomes (FCT/Arquivo.pt)
- Daniel Loureiro (University of Porto)
- Denilson Barbosa (University of Alberta)
- Dhruv Gupta (Max Planck Institute for Informatics)

- Dwaipayan Roy (ISI Kolkata, India)
- Dyaa Albakour (Signal AI)
- Gaël Dias (Normandie University)
- Henrique Lopes Cardoso (University of Porto)
- Ismail Sengor Altingovde (Middle East Technical University)
- Jeffery Ansah (BHP)
- Jeremy Pickens (OpenText)
- João Magalhães (Universidade Nova de Lisboa)
- Kiran Kumar Bandeli (Walmart Inc.)
- Ludovic Moncla (INSA Lyon)
- Marc Spaniol (Université de Caen Normandie)
- Mark Finlayson (Florida International University)
- Mengdie Zhuang (The University of Sheffield)
- Nina Tahmasebi (University of Gothenburg)
- Nuno Moniz (LIAAD/INESC TEC)
- Pablo Gamallo (University of Santiago de Compostela)
- Paulo Quaresma (Universidade de Évora)
- Preslav Nakov (Qatar Computing Research Institute (QCRI))
- Ross Purves (University of Zurich)
- Satya Almasian (Heidelberg University)
- Sebastiao Miranda (Priberam)
- Sérgio Nunes (INESC TEC & University of Porto)
- Udo Kruschwitz (University of Essex)
- Vítor Mangaravite (UFMG)
- Yihong Zhang (Kyoto University)

Proceedings Chair

- João Paulo Cordeiro (INESC TEC; Universidade da Beira do Interior, Covilhã, Portugal)
- Conceição Rocha (INESC TEC)

Web and Dissemination Chair

- Arian Pasquali (Signal AI)
- Behrooz Mansouri (Rochester Institute of Technology)

Acknowledgements. The first two authors of this paper are financed by the ERDF – European Regional Development Fund through the North Portugal Regional Operational Programme (NORTE 2020), under the PORTUGAL 2020 and by National Funds through the Portuguese funding agency, FCT - Fundação para a Ciência e a Tecnologia within project PTDC/CCI-COM/31857/2017 (NORTE-01-0145-FEDER-03185).

References

1. Alonso, O., Kandylas, V., Tremblay, S.-E.: How it happened: discovering and archiving the evolution of a story using social signals. In: Proceedings of JCDL 2018, pp. 193–202. ACM (2018). https://doi.org/10.1145/3197026.3197034

2. Campos, R., Mangaravite, V., Pasquali, A., Jorge, A.M., Nunes, C., Jatowt, A.: A text feature based automatic keyword extraction method for single documents. In: Pasi, G., Piwowarski, B., Azzopardi, L., Hanbury, A. (eds.) ECIR 2018. LNCS, vol. 10772, pp. 684–691. Springer, Cham (2018). https://doi.org/10.1007/978-3-319-76941-7_63

3. Campos, R., Mangaravite, V., Pasquali, A., Jorge, A.M., Nunes, C., Jatowt, A.: YAKE! collection-independent automatic keyword extractor. In: Pasi, G., Piwowarski, B., Azzopardi, L., Hanbury, A. (eds.) ECIR 2018. LNCS, vol. 10772, pp. 806–810. Springer, Cham (2018). https://doi.org/10.1007/978-3-319-76941-7_80

4. Campos, R., Mangaravite, V., Pasquali, A., Jorge, A., Nunes, C., Jatowt, A.: YAKE! keyword extraction from single documents using multiple local features. Inf. Sci. **509**, 257–289 (2020)

5. El-Haj, M., Rayson, P., Moore, A.: The first financial narrative processing workshop (FNP 2018). In: Proceedings of the LREC 2018 Workshop (2018)

6. Grobelny, J., Smierzchalska, J., Krzysztof, K.: Narrative gamification as a method of increasing sales performance: a field experimental study. Int. J. Acad. Res. Bus. Soc. Sci. **8**(3), 430–447 (2018). https://doi.org/10.6007/IJARBSS/v8-i3/3940

7. Hoque, E., Carenini, G.: MultiConVis: a visual text analytics system for exploring a collection of online conversations. In: Proceedings of IUI 2016, pp. 96–107. ACM (2016). https://doi.org/10.1145/2856767.2856782

8. Jorge, A., Campos, R., Jatowt, A., Nunes, S.: First international workshop on narrative extraction from texts (Text2Story 2018). In: Pasi, G., Piwowarski, B., Azzopardi, L., Hanbury, A. (eds.) ECIR 2018. LNCS, vol. 10772, pp. 833–834. Springer, Heidelberg (2018). https://doi.org/10.1007/978-3-319-76941-7

9. Jorge, A., Campos, R., Jatowt, A., Bhatia, S.: Second international workshop on narrative extraction from texts (Text2Story 2019). In: Azzopardi, L., Stein, B., Fuhr, N., Mayr, P., Hauff, C., Hiemstra, D. (eds.) ECIR 2019. LNCS, vol. 11438, pp. 389–393. Springer, Heidelberg (2019). https://doi.org/10.1007/978-3-030-15719-7_54

10. Jorge, A., Campos, R., Jatowt, A., Nunes, S.: Special issue on narrative extraction from texts (Text2Story): preface. Inf. Process. Manage. Int. J. **56**(5), 1771–1774 (2019). https://doi.org/10.1016/j.ipm.2019.05.004

11. Lewis, C., Young, S.: Fad or future? Automated analysis of financial text and its implications for corporate reporting. Account. Bus. Res. **49**(5), 587–615 (2019). https://doi.org/10.1080/00014788.2019.1611730

12. Li, B., Lee-Urban, S., Johnston, G., Riedl, M.: Story generation with crowdsourced plot graphs. In: Proceedings of the AAAI 2013, pp. 598–604. AAAI Press (2013). https://doi.org/10.5555/2891460.2891543

13. Liu, S., et al.: TIARA: interactive, topic-based visual text summarization and analysis. ACM Trans. Intell. Syst. Technol. **3**(2), 1–28 (2012). Article no. 25. https://doi.org/10.1145/2089094.2089101

14. Martinez-Alvarez, M., et al.: First international workshop on recent trends in news information retrieval (NewsIR 2016). In: Ferro, N., et al. (eds.) ECIR 2016. LNCS, vol. 9626, pp. 878–882. Springer, Heidelberg (2016). https://doi.org/10.1007/978-3-319-30671-1_85

15. Pasquali, A., Mangaravite, V., Campos, R., Jorge, A., Jatowt, A.: Interactive system for automatically generating temporal narratives. In: Azzopardi, L., Stein, B., Fuhr, N., Mayr, P., Hauff, C., Hiemstra, D. (eds.) ECIR 2019. LNCS, vol. 11438, pp. 251–255. Springer, Heidelberg (2019). https://doi.org/10.1007/978-3-030-15719-7_34

16. Pujolà, J-T., Argüello, A.: Stories or scenarios: implementing narratives in gamified language teaching. In: Arnedo-Moreno, J., González, C.S., Mora, A. (eds.) GamiLearn 2019 (2019)
17. Riedl, O.: Computational narrative intelligence: a human-centered goal for artificial intelligence. arxiv.org (2016)
18. Seifert, C., Sabol, V., Kienreich, W., Lex, E., Granitzer, M.: Visual analysis and knowledge discovery for text. In: Gkoulalas-Divanis, A., Labbi, A. (eds.) Large-Scale Data Analytics, pp. 189–218. Springer, New York (2014). https://doi.org/10.1007/978-1-4614-9242-9_7
19. Tran, G., Alrifai, M., Herder, E.: Timeline summarization from relevant headlines. In: Hanbury, A., Kazai, G., Rauber, A., Fuhr, N. (eds.) ECIR 2015. LNCS, vol. 9022, pp. 245–256. Springer, Cham (2015). https://doi.org/10.1007/978-3-319-16354-3_26
20. Viégas, F., Wattenberg, M.: Tag clouds and the case for vernacular visualization. Interactions 15(4), 49–52 (2008). https://doi.org/10.1145/1374489.1374501
21. Vossen, P., Caselli, T., Kontzopoulou, Y.: Storylines for structuring massive streams of news. In: Proceedings of the First Workshop on Computing News Storylines (CNewsStory 2015@ACL-IJCNLP 2015), pp. 40–49. ACL (2015). https://doi.org/10.18653/v1/W15-4507
22. Zhang, J., Yao, J.-G., Wan, X.: Towards constructing sports news from live text commentary. In: Proceedings of ACL 2016, pp. 1361–1371. ACL (2016). https://doi.org/10.18653/v1/P16-1129

Proposal of the First International Workshop on Semantic Indexing and Information Retrieval for Health from Heterogeneous Content Types and Languages (SIIRH)

Francisco M. Couto[1]([⊠])[iD] and Martin Krallinger[2][iD]

[1] LASIGE, Faculdade de Ciências, Universidade de Lisboa, Lisbon, Portugal
fcouto@di.fc.ul.pt
[2] Life Science Department, Barcelona Supercomputing Centre (BSC-CNS),
C/Jordi Girona 29-31, 08034 Barcelona, Spain
martin.krallinger@bsc.es

Abstract. The application of Information Retrieval (IR) and deep learning strategies to explore the vast amount of rapidly growing health-related content is of utmost importance, but is also particularly challenging, due to the very specialized domain language, and implicit differences in language characteristics depending on the content type.

This workshop aims at presenting and discussing current and future directions for IR and machine learning approaches devoted to the retrieval and classification of different types of health-related documents ranging from layman or patient generated texts to highly specialized medical literature or clinical records. It includes a session on the MESINESP shared task, supported by the Spanish National Language Technology plan (Plan TL), in order to address the importance and impact of community evaluation efforts, in particular BioASQ, BioCreative, eHealth CLEF, MEDIQA and TREC, as scenarios for exploring evaluation settings and generate data collections of key importance for promoting the development and comparison of IR resources. Additionally, an open session will address IR technologies for heterogeneous health-related content open to multiple languages with a particular interest in the exploitation of structured controlled vocabularies and entity linking, covering the following topics: multilingual and non-English health-related IR, concept indexing, text categorization, generation of evaluation resources biomedical document IR strategies; scalability, robustness and reproducibility of health IR and text mining resources; use of specialized machine translation and advanced deep learning approaches for improving health related search results; medical Question Answering search tools; retrieval of multilingual health related web-content; and other related topics.

Supported by FCT through funding of the DeST: Deep Semantic Tagger project, ref. PTDC/CCI-BIO/28685/2017, and LaSIGE Research Unit, ref. UIDB/00408/2020.

J. M. Jose et al. (Eds.): ECIR 2020, LNCS 12036, pp. 654–659, 2020.
https://doi.org/10.1007/978-3-030-45442-5_87

Keywords: Semantic indexing · Ontologies · Controlled vocabularies · Information Retrieval · Text mining · Natural language processing · Biomedical informatics

1 Introduction

There is an increasing interest in exploiting the vast amount of rapidly growing content related to health [7] by means of Information Retrieval [12] (IR) and deep learning strategies [14,18]. Health-related content is particularly challenging, due to the highly specialized domain language and implicit differences in language characteristics depending on the content type (patient-generated content like discussion forum [15], blogs [8], social media [17] and other Internet sources, healthcare documentation and clinical records [6], professional or scientific publications [9], clinical practice guidelines, clinical trials documentation, medical questionnaires, medical informed consent documents, etc.). Moreover, it is also critical to provide search solutions for non-English content as well as cross-language or multilingual IR solutions [4,10,16].

Efficient retrieval of biomedical documents is key for evidence-based medicine, preparing systematic reviews or retrieval of particular clinical case studies. Due to particular search conditions of caregivers and healthcare professionals (limited amount of time spent per patient), they are also in need of more sophisticated retrieval approaches applied to electronic health records [11], a type of content highly challenging due to its telegraphic and domain specific language and the presence of negations and abbreviations. There is also interest in processing patient-generated content like social media and patient fora, a key resource for rare disease research, clinical trials patient selection/stratification or for discovering new patient-reported symptoms and treatment-related adverse effects. In the health-domain, indexing strategies relying on structured controlled vocabularies, like MeSH/DeCS or SNOMED CT, represent a critical component for efficient biomedical search engines, enabling query expansion and refinement [2] and the improvement of recommender systems [3].

1.1 BioASQ MESINESP Session

Currently, most of the Biomedical NLP and IR research is being done on English documents [13], and only few tasks have been carried out on non-English texts [5]. Many structured controlled vocabularies are also available only in English [19]. Nonetheless, it is important to note that there is also a considerable amount of medically relevant content published in languages other than English and particularly clinical texts are entirely written in the native language of each country, with a few exceptions. The critical importance of semantic indexing with medical vocabularies motivated several-shared tasks in the past, in particular the BioASQ tracks[1], with a considerable number of participants and impact in the field. Following the outline of previous medical indexing efforts,

---

[1] http://bioasq.org/.

in particular the success of the BioASQ tracks centered on PubMed, the BioASQ MESINESP TASK[2], supported by the Spanish National Language Technology plan (Plan TL), proposes to carry out the first task on semantic indexing of Spanish medical texts.

This workshop will be a forum where the community can present and discuss current and future directions for the area based on the experience in participating at the MESINESP shared task or other medical IR, QA or text categorization evaluation campaigns, as well as the exploitation of evaluation settings and data collections generated through these kind of community evaluation efforts (both during and after the competition period).

1.2   Open Session

In addition to the MESINESP and shared task/evaluation campaign participation experience session, the workshop will include an Open Session covering IR technologies for heterogeneous health-related content open to multiple languages with a particular interest in the exploitation of structured controlled vocabularies and entity linking for document indexing and semantic search applications.

Among the proposed topics for the Open Session are: (1) multilingual and non-English health related IR, concept indexing and text categorization strategies, (2) generation of evaluation resources for biomedical document IR strategies, (3) scalability, robustness, reproducibility, utility and usability [1] of health IR and text mining resources, (4) use of specialized machine translation and advanced deep learning approaches for improving health related search results, (5) medical Question Answering search tools, (6) retrieval of multilingual health related web-content. Note that we will also consider other submissions related to innovative cutting-edge health and biomedical IR strategies, including evaluation and Gold Standard evaluation data set generation.

2   Planned Format and Structure

All the teams implementing systems for MESINESP will be invited to submit an article describing their participation strategy. The program committee will review the papers and select which of them will have a presentation slot at the workshop. For the Open Session we will invite researchers to submit novel IR approaches to process heterogeneous health-related content with particular interest in non-English content, novel content types as well as semantic indexing strategies exploiting structured controlled vocabularies and ontologies.

We expect that further investigation on the topics will continue after the workshop, based on new insights obtained through discussions during the event. As a venue to compile the results of the follow-up investigation, a journal special issue will be organized to be published a few months after the workshop.

---

[2] http://temu.bsc.es/mesinesp.

3   People Involved

3.1   Organizers

Martin Krallinger: head of the Text Mining unit at the Barcelona Supercomputing Center (BSC), Spain
Francisco M. Couto: LASIGE member and associate professor at the University of Lisbon, Portugal

3.2   Programme Committee

Alberto Lavelli: FBK, Trento, Italy
Alfonso Valencia: Barcelona Supercomputing Center, Spain
Analia Lourenco: Universidade de Vigo, Spain
Anastasios Nentidis: National Center for Scientific Research Demokritos, Greece
André Lamurias: LASIGE, Portugal
Anne:Lyse Minard - University of Orleans, France
Aron Henriksson: Stockholm University, Sweden
Bruno Martins: INESC-ID, Portugal
Carsten Eickhoff: Brown University, USA
Chih:Hsuan Wei - NCBI/NIH, National Library of Medicine, USA
Cyril Grouin: LIMSI, CNRS, Université Paris-Saclay, Orsay, France
Diana Sousa: LASIGE, Portugal
Dimitrios Kokkinakis: University of Gothenburg, Sweden
Eben Holderness: McLean Hosp., Harvard Med. School & Brandeis University, USA
Ellen Vorhees: National Institute of Standards and Technology (NIST), USA.
Fabio Rinaldi: IDSIA, University of Zurich, Switzerland & FBK, Trento, Italy
Fleur Mougin: University of Bordeaux, France
Georgeta Bordea: Université de Bordeaux, France
Georgios Paliouras: National Center for Scientific Research Demokritos, Greece
Goran Nenadic: University of Manchester, UK
Graciela Gonzalez: Hernandez - University of Pennsylvania, USA
Hanna Suominen: CSIRO, Australia
Henning Muller: University of Applied Sciences Western Switzerland, Switzerland
Hercules Dalianis: Stockholm University, Sweden
Hyeju Jang: University of British Columbia, Canada
James Pustejovsky: Brandeis University, USA
Jin:Dong Kim - Research Organization of Information and Systems, Japan
Jong C. Park: KAIST Computer Science, Korea
Kevin Bretonnel Cohen: University of Colorado School of Medicine, Colorado, USA
Maria Skeppstedt: Institute for Language and Folklore, Sweden

Marcia Barros: LASIGE, Portugal
Mariana Lara: Neves - German Federal Institute for Risk Assessment, Germany
Marta Villegas: BSC, Spain
Pedro Ruas: LASIGE, Portugal
Rafael Berlanga Llavori: Universitat Jaume I, Spain
Rezarta Islamaj: Dogan - NIH/NLM/NCBI, USA
Sérgio Matos: University of Aveiro, Portugal
Shyamasree Saha: Europe PubMed Central, EMBL-EBI, UK
Suzanne Tamang: Stanford University School of Medicine, USA
Thierry Hamon: LIMSI, CNRS, Université Paris-Saclay & Université Paris 13,
France
Thomas Brox Røst: Norwegian University of Science and Technology, Norway
Yifan Peng: NCBI/NIH, National Library of Medicine, USA
Yonghui Wu: University of Florida, USA
Yoshinobu Kano: Shizuoka University, Japan
Zhiyong Lu: NCBI/NIH, National Library of Medicine, USA
Zita Marinho: Priberam, Portugal

References

1. Arighi, C.N., et al.: BioCreative III interactive task: an overview. BMC Bioinformatics **12**(8), S4 (2011). https://doi.org/10.1186/1471-2105-12-S8-S4
2. Barros, M., Couto, F.M.: Knowledge representation and management: a linked data perspective. Yearb. Med. Inform. **25**(01), 178–183 (2016)
3. Barros, M., Moitinho, A., Couto, F.: Hybrid semantic recommender system for chemical compounds. In: European Conference on Information Retrieval. Springer (2020)
4. Bawden, R., et al.: Findings of the WMT 2019 biomedical translation shared task: evaluation for MEDLINE abstracts and biomedical terminologies. In: Proceedings of the Fourth Conference on Machine Translation (Volume 3: Shared Task Papers, Day 2), pp. 29–53 (2019)
5. Campos, L., Pedro, V., Couto, F.: Impact of translation on named-entity recognition in radiology texts. Database **2017** (2017)
6. Costumero, R., García-Pedrero, Á., Gonzalo-Martín, C., Menasalvas, E., Millan, S.: Text analysis and information extraction from Spanish written documents. In: Ślęzak, D., Tan, A.-H., Peters, J.F., Schwabe, L. (eds.) BIH 2014. LNCS (LNAI), vol. 8609, pp. 188–197. Springer, Cham (2014). https://doi.org/10.1007/978-3-319-09891-3_18
7. Couto, F.M.: Data and Text Processing for Health and Life Sciences. AEMB, vol. 1137. Springer, Cham (2019). https://doi.org/10.1007/978-3-030-13845-5
8. Denecke, K., Nejdl, W.: How valuable is medical social media data? Content analysis of the medical web. Inf. Sci. **179**(12), 1870–1880 (2009)
9. Intxaurrondo, A., et al.: Finding mentions of abbreviations and their definitions in Spanish clinical cases: the BARR2 shared task evaluation results. In: IberEval@ SEPLN, pp. 280–289 (2018)
10. Kelly, L., et al.: Overview of the CLEF eHealth evaluation lab 2019. In: Crestani, F., et al. (eds.) CLEF 2019. LNCS, vol. 11696, pp. 322–339. Springer, Cham (2019). https://doi.org/10.1007/978-3-030-28577-7_26

11. Koleck, T.A., Dreisbach, C., Bourne, P.E., Bakken, S.: Natural language processing of symptoms documented in free-text narratives of electronic health records: a systematic review. J. Am. Med. Inform. Assoc.J. Am. Med. Inform. Assoc. **26**(4), 364–379 (2019)

12. Krallinger, M., Rabal, O., Lourenco, A., Oyarzabal, J., Valencia, A.: Information retrieval and text mining technologies for chemistry. Chem. Rev. **117**(12), 7673–7761 (2017)

13. Lamurias, A., Couto, F.M.: Text mining for bioinformatics using biomedical literature. In: Encyclopedia of Bioinformatics and Computational Biology, vol. 1 (2019)

14. Lee, J., et al.: BioBERT: pre-trained biomedical language representation model for biomedical textmining. Bioinformatics **36**, 1234–1240 (2019)

15. Liu, X., Chen, H.: AZDrugMiner: an information extraction system for mining patient-reported adverse drug events in online patient forums. In: Zeng, D., et al. (eds.) ICSH 2013. LNCS, vol. 8040, pp. 134–150. Springer, Heidelberg (2013). https://doi.org/10.1007/978-3-642-39844-5_16

16. Marimon, M., et al.: Automatic de-identification of medical texts in Spanish: the meddocan track, corpus, guidelines, methods and evaluation of results. In: Proceedings of the Iberian Languages Evaluation Forum (IberLEF 2019), vol. TBA, p. TBA. CEUR Workshop Proceedings, Bilbao, Spain, September 2019, TBA. CEUR-WS. org (2019)

17. Segura-Bedmar, I., Revert, R., Martínez, P.: Detecting drugs and adverse events from Spanish social media streams. In: Proceedings of the 5th International Workshop on Health Text Mining and Information Analysis (LOUHI), pp. 106–115 (2014)

18. Sousa, D., Couto, F.: BiOnt: deep learning using multiple biomedical ontologies for relation extraction. In: European Conference on Information Retrieval. Springer (2020)

19. Villegas, M., Intxaurrondo, A., Gonzalez-Agirre, A., Marimon, M., Krallinger, M.: The MeSpEN resource for English-Spanish medical machine translation and terminologies: census of parallel corpora, glossaries and term translations. In: Proceedings of the LREC 2018 Workshop "MultilingualBIO: Multilingual Biomedical Text Processing", Paris, France. European Language Resources Association (ELRA) (2018)

Tutorials

Principle-to-Program: Neural Methods for Similar Question Retrieval in Online Communities

Muthusamy Chelliah[1]([✉]), Manish Shrivastava[2], and Jaidam Ram Tej[1]

[1] Flipkart, Bangalore, India
chelgeetha@yahoo.com
[2] IIIT Hyderabad, Hyderabad, India

Abstract. Similar question retrieval is a challenge due to lexical gap between query and candidates in archive and is very different from traditional IR methods for duplicate detection, paraphrase identification and semantic equivalence. This tutorial covers recent deep learning techniques which overcome feature engineering issues with existing approaches based on translation models and latent topics. Hands-on proposal thus will introduce each concept from end user (e.g., question-answer pairs) and technique (e.g., attention) perspectives, present state of the art methods and a walkthrough of programs executed on Jupyter notebook using real-world datasets demonstrating principles introduced.

Keywords: Question answering · Semantic similarity · Neural networks · Paraphrase identification · Duplicate detection

1 Introduction

Time lag between user posting a question and receiving its answer could be reduced by retrieving similar, historic questions from community question answering (cQA) archives. Two seemingly different questions may refer implicitly to a common problem with the same answer. Identifying semantic equivalence is thus critical for retrieving similar questions and to automate reusing answers available for such previous questions. This is a difficult task because different users may formulate the same question in a variety of ways, using different vocabulary (e.g., watersports vs. snorkeling) and structure. Similar questions hence vary in style, length and content quality (e.g., blood pressure vs. hypertension).

Lexical gap - different but related words between queried (e.g., knot) and existing (e.g., tangle) questions - rules out traditional IR models (e.g., BM25) as a solution. Text fragments in questions (e.g., disk full) could lead to correlated content (e.g., format) in answers. Similarity - different from relatedness (e.g., synonymy, antonymy) - has been addressed with strategies like machine translation, knowledge graphs and topic models though. Treating question-answer pairs

© Springer Nature Switzerland AG 2020
J. M. Jose et al. (Eds.): ECIR 2020, LNCS 12036, pp. 663–668, 2020.
https://doi.org/10.1007/978-3-030-45442-5_88

as parallel text, relationships can be established through translation probability (e.g., word-to-word) - asymmetry being a handicap. Latent topics aligned across question-answer pairs is another option - heterogeneity being an issue.

2   Background/Motivation

Title/body of a question is concatenated into problem definition and IR task is to decide if 2 such text fragments are similar. Detecting (almost) exact copies of the same document in corpora from Web crawling/search systems is not enough as equivalent questions may have very little (or no) overlap. Considering only surface form of a question without factoring in semantics makes it hard to identify duplicates. Keyword-based retrieval methods (e.g., language/vector-space models) hence predict questions with same meaning as different. Traditional similarity measures based on word overlap - even those on a graded scale from 0 to 5 - have thus proven to be inadequate for capturing semantic equivalence.

Paraphrase identification which helps determine if 2 sentences have the same meaning is not sufficient as well for similar question retrieval. Polysemy and word order are other challenges which similar question retrieval has to tackle like any other NLP tasks. Also, submitted questions have extraneous details - which obscures key information buried in the noise - in the body irrelevant to main question being asked. Title alone on the other hand lacks crucial detail present in question body. Building a large amount of training data with similar questions is expensive and careful feature engineering is time consuming. Deep learning techniques recently have been effective for sentence-level analysis of short texts in a variety of IR tasks. Size of a question in an online community however varies from a single sentence to detailed problem description with many sentences.

3   Convolutional Neural Networks (CNNs)/Question-Question Pairs

CNNs transform first words into embeddings with unlabeled data and then build distributed, vector representations for pairs of original and related questions. Questions are scored next with a metric (e.g., cosine similarity) and those pairs above a threshold based on a held-out set are considered equivalent. During training, CNN is induced to produce similar vector representations for equivalent questions. We discuss:

- [1] evaluates in-domain word embeddings vs. one trained with Wikipedia, estimates impact of training set size and evaluates aspects of domain adaptation,
- [2] combines bag of words (BoW) to retrieve equivalent questions while learning to rank them according to similarity with a loss function,

- [4] integrates sentence modeling and semantic matching into a single framework without syntactic analysis and prior knowledge (e.g., wordnet). Word tokens are converted into vectors by a lookup layer and useful information is captured with convolutional/pooling layers; finally, matching metric is learnt - better than traditional ones (e.g., inner-product, Euclidean distance) - between question/answer capturing their interaction with a tensor layer.

4   Recurrent Neural Networks (RNNs)/Similarity Features

Available annotations on similar questions however are noisy and fragmented. An encoder maps title/body combination of questions - treated as word sequences - into vector representation with a recurrent model. Complementary decoder is trained to reproduce title from noisy question body. We discuss:

- [6] incorporates adaptive gating in non-consecutive CNNs to focus temporal averaging on key pieces of questions. Training paradigm utilizes entire corpus of unannotated questions in a semi-supervised manner and fine tunes learning model discriminatively with limited annotations,
- [14,15] applies LSTM with attention to select entire sentences and subparts (word/chunk) from shallow syntactic trees towards question retrieval and tree kernels to filtered text representations exploiting implicit features of subtree space for learning question reranking.

5   Latent Space/Meta-data

User asking a question in cQA sites is required to choose a label from a predefined hierarchy of categories. This meta-data encodes attributes/properties of words from which similar words can be grouped according to categories. Language models represent words and question categories in a vector space and calculate question-question similarity with linear combinations of dot products of vectors - thus being heuristic on data or difficult to scale up. Each question is thus defined as a distribution which generates each word (embedding) independently and subsequently a kernel is used to assess question similarities. This design will require representation of words that belong to the same category to be close to each other thus benefiting embedding learning. We discuss:

- [3] learns variable-length word embeddings with category information and aggregates them into fixed-size vectors,
- [8] optimizes an objective which in turn applies a non-linear transformation considering only local-relatedness of words (i.e., category and small window in a question/associated answers),
- [12] outperforms text-based methods in misflagged duplicate detection with features like user authority, question quality and relational data between questions.

6   Representation Learning/Question-Answer Pairs

Due to relatively short text, question-question pairs have insignificant information to determine their relationship. To combat scarcity of similar question pairs for training, question-answer pairs from archives can be leveraged in a weakly supervised fashion without manual labeling. An added advantage of this approach is mapping simple terms used by novice askers (e.g., short sighted) to technical terms (e.g., myopia) and concepts (e.g., lasik/laser surgery, contact lens) used by expert answerers. We discuss:

- [5] learns shared parameters and similarity metric minimizing contrastive-loss energy function connecting twin networks,
- [7] preserves local neighborhood structure of and mirrors semantic similarity among question and answer spaces,
- [9] represents hierarchical structures of word and concept information with layer-by-layer composition and pooling leading to question embedding that captures semantics/syntax.

7   Attention/Constituent Matching

Essential constituents (e.g., destination) are those - name and value - important to meaning of the question (e.g., route). Units in a semantic parse can be leveraged to alleviate defining/labeling them in open domain. We discuss:

- [11] combines FrameNet with neural networks through ensemble and embedding approaches for question retrieval with constituent matching,
- [13] integrates shallow lexical mismatching information with initial rank by an external search engine to generate deep question representation with attention autoencoder,
- [10] leverages semantic information in paired answers while alleviating noise caused by adding answers with three heterogeneous attention mechanisms for modeling temporal interaction in a long sentence, capturing relevance between questions and relevance between answers and extracting knowledge from answers.

8   Conclusion

Distributed representations help tackle lexical gap in question retrieval as features based on word embeddings that enable similarity calculation through neural networks; gated convolutions map key question information from lengthy detail to semantic representations and LSTM with attention weights alleviates noise in syntactic structure selecting most significant parse tree fragments from question text.

Simultaneously embedding categories of questions into vector space helps model local relatedness of words in learning. Misflagging duplicate detection through user authority and question quality is more indicative of behavior problems (e.g., posting questions). Local linear embedding is leveraged to use collective corpus-level information for embedding historical question-answer pairs in a latent space without lexical correlation and separate topic/translation models. Attention encoders contain context information with focus on current word of input sequence thus avoiding bias towards sentence end.

Incorporating user ratings/reputation still remains unexplored. Semantic parsing techniques like abstract meaning representation is a future direction for essential constituent matching.

References

1. Bogdanova, D., dos Santos, C., Barbosa, L., Zadrozny, B.: Detecting semantically equivalent questions in online user forums. In: Proceedings of the Nineteenth Conference on Computational Natural Language Learning, pp. 123–131 (2015)
2. Dos Santos, C., Barbosa, L., Bogdanova, D., Zadrozny, B.: Learning hybrid representations to retrieve semantically equivalent questions. In: Proceedings of the 53rd Annual Meeting of the Association for Computational Linguistics and the 7th International Joint Conference on Natural Language Processing, pp. 694–699 (2015)
3. Zhou, G., He, T., Zhao, J., Hu, P.: Learning continuous word embedding with metadata for question retrieval in community question answering. In: Proceedings of the 53rd Annual Meeting of the Association for Computational Linguistics and the 7th International Joint Conference on Natural Language Processing (Volume 1: Long Papers), pp. 250–259 (2015)
4. Qiu, X., Huang, X.: Convolutional neural tensor network architecture for community-based question answering. In: Twenty-Fourth IJCAI (2015)
5. Das, A., Yenala, H., Chinnakotla, M., Shrivastava, M.: Together we stand: siamese networks for similar question retrieval. In: Proceedings of the 54th Annual Meeting of the Association for Computational Linguistics (Volume 1: Long Papers), pp. 378–387 (2016)
6. Lei, T., et al.: Semi-supervised question retrieval with gated convolutions. arXiv preprint arXiv:1512.05726 (2015)
7. Deepak, P., Garg, D., Shevade, S.: Latent space embedding for retrieval in question-answer archives. In: Proceedings of the 2017 Conference on Empirical Methods in Natural Language Processing, pp. 855–865 (2017)
8. Zhang, K., Wu, W., Wang, F., Zhou, M., Li, Z.: Learning distributed representations of data in community question answering for question retrieval. In: Proceedings of the Ninth ACM International Conference on Web Search and Data Mining, pp. 533–542. ACM (2016)
9. Wang, P., Zhang, Y., Ji, L., Yan, J., Jin, L.: Concept embedded convolutional semantic model for question retrieval. In: WSDM 2017 (2017)
10. Liang, D., et al.: Adaptive multi-attention network incorporating answer information for duplicate question detection (2019)
11. Zhang, X., Sun, X., Wang, H.: Duplicate question identification by integrating FrameNet with neural networks. In: 32nd AAAI Conference on Artificial Intelligence (2018)

12. Hoogeveen, D., Bennett, A., Li, Y., Verspoor, K.M., Baldwin, T.: Detecting mis-flagged duplicate questions in community question-answering archives. In: Twelfth International AAAI Conference on Web and Social Media (2018)

13. Zhang, M., Wu, Y.: An unsupervised model with attention autoencoders for question retrieval. In: Thirty-Second AAAI Conference on Artificial Intelligence (2018)

14. Romeo, S., et al.: Neural attention for learning to rank questions in community question answering. In: Proceedings of COLING 2016, the 26th International Conference on Computational Linguistics, pp. 1734–1745 (2016)

15. Barrórn-Cedeno, A., Da San Martino, G., Romeo, S., Moschitti, A.: Selecting sentences versus selecting tree constituents for automatic question ranking. In: Proceedings of COLING 2016, the 26th International Conference on Computational Linguistics, pp. 2515–2525 (2016)

Text Meets Space: Geographic Content Extraction, Resolution and Information Retrieval

Jochen L. Leidner[1,2,3(✉)], Bruno Martins[4], Katherine McDonough[5,6], and Ross S. Purves[7]

[1] Polygon Analytics Ltd., 19a Canning Street, Edinburgh EH3 8HE, Scotland, UK
[2] Refinitiv Labs, Refinitiv Ltd., 5 Canada Square, London E14 5AQ, UK
[3] Regents Court, University of Sheffield, 211 Portobello, Sheffield S1, UK
leidner@acm.org
[4] Computer Science and Engineering Department, IST, University of Lisbon, Lisbon, Portugal
[5] British Library, Alan Turing Institute, 96 Euston Rd, London NW1 2DB, UK
[6] School of History, Queen Mary University, London, UK
[7] Department of Geography, University of Zurich, Zurich, Switzerland

Abstract. In this half-day tutorial, we will review the basic concepts of, methods for, and applications of geographic information retrieval, also showing some possible applications in fields such as the digital humanities. The tutorial is organized in four parts. First we introduce some basic ideas about geography, and demonstrate why text is a powerful way of exploring relevant questions. We then introduce a basic end-to-end pipeline discussing geographic information in documents, spatial and multi-dimensional indexing [19], and spatial retrieval and spatial filtering. After showing a range of possible applications, we conclude with suggestions for future work in the area.

1 Introduction

The notion of geographic relevance and the role of geographic space in information access have been recognized for a long time [15]. For example, the PERSEUS digital library aimed to make humanities documents accessible spatially, while e.g. the SEQUOIA and SPIRIT projects [8], as well as the GeoCLEF shared task [1] aimed to study geographic information retrieval. More recently, the pervasiveness of mobile computing devices [11] and other developments associated to the Internet of Things (IoT) all necessitate reflection on the role of geographic space in making information collected and stored accessible, not just indexed using words and numbers but also spatially. However, to date, not ECIR nor other IR conferences have offered a tutorial for interested researchers and practitioners, making the body of research that make up the state of the art accessible.

© Springer Nature Switzerland AG 2020
J. M. Jose et al. (Eds.): ECIR 2020, LNCS 12036, pp. 669–673, 2020.
https://doi.org/10.1007/978-3-030-45442-5_89

To this end, we propose a half-day to address this gap. We will introduce or recap the core concepts from geography and its intersection with IR, and survey existing techniques to (a) construct spatial representations from textual documents and queries (typically exploiting geographic knowledge from gazetteers [7] in doing so), and to (b) utilize geographic knowledge (prior and extracted from data) to better access document collections in which geographic space place a substantial roles. We will also cover example applications [5], e.g. in fields such as the digital humanities [12], and discuss possible avenues for future work in the area.

2   Goals and Objectives

In this tutorial, we aim to give a survey of the concepts and methods used to make implicit spatial evidence contained in text collections accessible. We cover selected early and seminal attempts [3,8,10,13] and more recent Machine Learning (ML) methods [6,16–18], hoping to inspire students and fellow researchers to get interested in conducting their own research in this area. Bringing two seemingly disparate worlds like geographic space and text documents together is exciting!

By the end of the tutorial session, the attendees will have a clear sense of the key concepts in Geo-NLP and Geographical Information Retrieval (GIR), and they will understand some seminal methods as well as open problems.

3   Description and Structure of the Tutorial

This one day tutorial will be divided into five sessions:

- Geography and text: an introduction to the ways in which geographic concepts are reflected in natural language and in text;
- Toponym recognition and resolution [9]: key to most geographically inspired analysis are the use of place-names in text, their identification, disambiguation, and resolution to unique locations;
- Geographic relevance and ranking [4]: methods for incorporating geographic information in IR indexes and ranking algorithms. Discuss what is geographic relevance, and how it varies with context and application domain;
- Applications: Concrete examples for the application of the introduced methods, in fields ranging from Digital Humanities to Web search, together with a discussion on requirements and their implications on algorithmic and data choices;
- Future challenges: Where are the most likely applications of GIR in the future, and what are key societal and methodologically driven challenges;

The first four sessions will each present fundamental challenges, a selection of examples from the state of the art, and include interactive exercises (computer and/or paper based) to illustrate basic concepts to participants.

4 Prerequisites

In terms of prerequisites, some knowledge of basic IR and ML concepts will be helpful. However, the tutorial is designed for a broad audience, introducing key high level concepts, and providing participants with material to deepen knowledge subsequently.

5 Target Audience

The target audience for this tutorial includes the following three groups:

- students of computer science, especially in information retrieval, who want to learn about mobility-relevant spatial computation around search/IR (e.g. [2]);
- practicing IR engineers who would like to expand their areas of expertise so as to include geographic search;
- information retrieval researchers interested in and introduction and state-of-the-art review [14] on GIR and Geo-NLP;
- geographers or GIS experts who have not yet worked with text, and who would like to learn how the spatial knowledge implicit in text collections can be used to support geospatial analysis.

Beyond these directly targeted groups, the tutorial could be of interest to anyone who would like to understand better how the world of geographic space relates to the world of unstructured textual documents.

6 Presenters and Their Experience

Jochen L. Leidner is a computer scientist and research manager. He is Director of Research at Refinitiv Labs (formerly Thomson Reuters F&R) in London where he leads the Research & Development function and team. A computational linguist by training, he holds Master's degrees (Erlangen and Cambridge) and a Ph.D. (Edinburgh). His 2007 Ph.D. thesis "Toponym Resolution in Text" (published in book form in 2008) attracted over 200 hundred citations. He is a Fellow of the Royal Geographical Society and currently also the Academy of Engineering Visiting Professor of Data Analytics in the Department of Computer Science at the University of Sheffield to instill industry practice into engineering training.

Bruno Martins is an assistant professor at the Department of Computer Science and Engineering of Instituto Superior Técnico in the University of Lisbon and a researcher at INESC-ID, where he works on problems related to the general areas of information retrieval, text mining, and the geographical information sciences. He has been involved in several research projects related to geospatial aspects in information access and retrieval, and he has accumulated a significant expertise in addressing challenges at the intersection of information retrieval, machine learning, and the geographical information sciences.

Katherine McDonough is a Senior Research Associate at The Alan Turing Institute with the Living with Machines project and a Research Fellow at Queen Mary, University of London. She has formerly taught and worked on digital humanities projects at Stanford University, Western Sydney University, and Bates College. With a background in eighteenth-century French history, her early research focused on the politics of infrastructure. She has written on GIR challenges for humanities research and is a member of the GéoDisco project, which examines geographic discourse in historical French encyclopedias. Her current work explores new approaches to GIR informed by humanistic source criticism.

Ross Purves is a professor at the University of Zurich. His research focuses on the geographic analysis of text, exploring both methodological issues (e.g. gazetteer quality and representation of vernacular names) and analysis of text to better understand landscape. He collaborated on the SPIRIT project, which investigated a number of concepts fundamental to geographic information retrieval. Together with Chris Jones, he organises the workshop on Geographic Information Retrieval which has been hosted by CIKM, SIGIR and ACM SIGSPATIAL, and which has been an important incubator of many ideas related to GIR. He recently co-authored a comprehensive review of GIR [14].

7   Previous Events

This is a new tutorial, and therefore was never presented before. All of the presenters are experienced teachers and have given seminars at a range of international conferences on related material.

8   Summary and Conclusion

We have presented a tutorial proposal for geospatial content processing and retrieval. Geographic aspects in information access and retrieval have been increasing in relevance, given the interest in analysing huge volumes of unstructured data in fields such as the digital humanities or the computational social sciences, and given the pervasiveness of networked sensors, GPS-enabled mobile devices, and in-car navigation systems. Modern information systems need to spatially enable text to make it accessible to a variety of use cases that contain a notion of "geographic relevance". This suggests that our novel tutorial would be likely to be of interest to most attendees of ECIR 2020.

References

1. GeoCLEF (n.d.). http://www.clef-initiative.eu/track/geoclef. Accessed 2019
2. Al-Olimat, H.S., Shalin, V.L., Thirunarayan, K., Sain, J.P.: Towards geocoding spatial expressions. Technical report (unpublished). https://arxiv.org/pdf/1906. 04960.pdf. Accessed 2019

3. Amitay, E., Har'El, N., Sivan, R., Soffer, A.: Web-a-where: geotagging web content. In: SIGIR 2004: Proceedings of the 27th Annual International ACM SIGIR Conference on Research and Development in Information Retrieval, Sheffield, UK, 25–29 July 2004, pp. 273–280 (2004)
4. Andogah, G.: Geographically Constrained Information Retrieval. Ph.D. thesis, University of Groningen, Groningen, The Netherlands (2010)
5. Ding, J., Gravano, L., Shivakumar, N.: Computing geographical scopes of web resources. In: El Abbadi, A., et al. (eds.) VLDB 2000, Proceedings of 26th International Conference on Very Large Data Bases, 10–14 September 2000, Cairo, Egypt, pp. 545–556. Morgan Kaufmann (2000)
6. Gritta, M., Pilehvar, M.T., Collier, N.: Which Melbourne? Augmenting geocoding with maps. In: Proceedings of the 56th Annual Meeting of the Association for Computational Linguistics, vol. 1, pp. 1285–1296. Association for Computational Linguistics, Melbourne (2018)
7. Hill, L.: Georeferencing. MIT Press, Cambridge (2009)
8. Joho, H., Sanderson, M.: The SPIRIT collection: an overview of a large web collection (2004)
9. Leidner, J.L.: Toponym Resolution in Text. Universal Press, Irvine (2008)
10. Leidner, J.L., Sinclair, G., Webber, B.: Grounding spatial named entities for information extraction and question answering. In: Proceedings of the HLT-NAACL 2003 Workshop on Analysis of Geographic References held at HLT/NAACL 2003, pp. 31–38 (2003). https://www.aclweb.org/anthology/W03-0105
11. Mathew, W., Raposo, R., Martins, B.: Predicting future locations with hidden markov models. In: Proceedings of the 2012 ACM Conference on Ubiquitous Computing, UbiComp 2012, pp. 911–918. ACM, New York (2012)
12. McDonough, K., Moncla, L., van de Camp, M.: Named entity recognition goes to old regime France: geographic text analysis for early modern French corpora. Int. J. Geogr. Inf. Sci. 33(12), 2498–2522 (2019)
13. Overell, S.E., Rüger, S.M.: Using co-occurrence models for placename disambiguation. Int. J. Geogr. Inf. Sci. 22(3), 265–287 (2008)
14. Purves, R.S., Clough, P., Jones, C.B., Hall, M.H., Murdock, V.: Geographic information retrieval: progress and challenges in spatial search of text. Found. Trends Inf. Retrieval 12(2–3), 164–318 (2018)
15. Sanderson, M., Kohler, J.: Analyzing geographic queries. In: Proceedings of the Workshop on Geographical Information Retrieval held at SIGIR 2004. http://www.geounizh.ch/~rsp/gir/
16. Santos, R., Murrieta-Flores, P., Calado, P., Martins, B.: Toponym matching through deep neural networks. Int. J. Geogr. Inf. Sci. 32(2), 324–348 (2018)
17. Speriosu, M., Baldridge, J.: Text-driven toponym resolution using indirect supervision. In: Proceedings of the 51st Annual Meeting of the Association for Computational Linguistics, vol. 1, pp. 1466–1476. Association for Computational Linguistics, Sofia (2013)
18. Yan, B., Janowicz, K., Mai, G., Gao, S.: From ITDL to Place2Vec: reasoning about place type similarity and relatedness by learning embeddings from augmented spatial contexts. In: Proceedings of the 25th ACM SIGSPATIAL International Conference on Advances in Geographic Information Systems, SIGSPATIAL 2017, pp. 35:1–35:10. ACM, New York (2017)
19. Zhang, X., Du, Z.: Spatial indexing. In: Wilson, J.P. (ed.) The Geographic Information Science & Technology Body of Knowledge. UCGIS, 4th Quarter 2017 Edition (2017). https://gistbok.ucgis.org/bok-topics/spatial-indexing

The Role of Entity Repositories
in Information Retrieval

Marius Paşca[✉]

Google, 1600 Amphitheatre Parkway, Mountain View CA, 94043, USA
mars@google.com

Keywords: Knowledge repositories · Information retrieval

Tutorial Description

Web search queries may seek information on entities (*"gary oldman"*, *"fifth element movie"*) or lists of entities (*"french science fiction movies"*); or seek answers to fact-seeking questions (*"who plays leeloo in the fifth element"*, *"director of the fifth element"*) that may refer to or involve entities (*Fifth Element, Milla Jovovich, Luc Besson*). By finding relevant entities within queries and matching them to relevant text and entities within documents and other available textual data, search engines have been moving closer to returning results that are not just a set of links but more directly answer the queries. Such developments are made possible or supported by the availability of large-scale entity repositories, which capture salient attributes and properties of a variety of open-domain entities and explicitly connects them to other related entities. This tutorial reviews distinguishing characteristics and discusses applications of existing entity repositories in information retrieval.

© Springer Nature Switzerland AG 2020
J. M. Jose et al. (Eds.): ECIR 2020, LNCS 12036, p. 675, 2020.
https://doi.org/10.1007/978-3-030-45442-5

Author Index

Printed in the United States
By Bookmasters